SUPERSYMMETRY, SUPERGRAVITY, AND UNIFICATION

This unique book gives a modern account of particle physics and gravity based on supersymmetry and supergravity, two of the most significant developments in theoretical physics since general relativity. It begins with a brief overview of the history of unification and then goes into a detailed exposition of both fundamental and phenomenological topics. The topics in fundamental physics include Einstein gravity, Yang–Mills theory, anomalies, the standard model, supersymmetry, and supergravity and the construction of supergravity couplings with matter and gauge fields as well as computational techniques for $SO(10)$ couplings. The topics of phenomenological interest include implications of supergravity models at colliders, CP violation, and proton stability, as well as topics in cosmology such as inflation, leptogenesis, baryogenesis, and dark matter. The book is intended for graduate students and researchers seeking to master the techniques for building unified models.

PRAN NATH is Matthews University Distinguished Professor of Physics at Northeastern University in Boston. He has made pioneering contributions to the development of supergravity theory and its applications to particle physics including the mSUGRA model. He has also contributed to the development of effective Lagrangian methods, pion physics, current algebra, solving the U(1) problem, dark matter and proton stability in unified models. He is the founding chair of the Supersymmetry and Unification of Fundamental Interactions (SUSY) conference and the Particles, Strings and Cosmology (PASCOS) symposium. He is a Fellow of both the American Physical Society and of the American Association for the Advancement of Science, and a recipient of the Humboldt Prize in physics.

CAMBRIDGE MONOGRAPHS ON MATHEMATICAL PHYSICS

General Editors: P. V. Landshoff, D. R. Nelson, S. Weinberg

C. Itzykson and J. M. Drouffe *Statistical Field Theory Volume 1: From Brownian Motion to Renormalization and Lattice Gauge Theory*[†]

C. Itzykson and J. M. Drouffe *Statistical Field Theory Volume 2: Strong Coupling, Monte Carlo Methods, Conformal Field Theory and Random Systems*[†]

G. Jaroszkiewicz *Principles of Discrete Time Mechanics*

C. V. Johnson *D-Branes*[†]

P. S. Joshi *Gravitational Collapse and Spacetime Singularities*[†]

J. I. Kapusta and C. Gale *Finite-Temperature Field Theory: Principles and Applications, 2nd edition*[†]

V. E. Korepin, N. M. Bogoliubov and A. G. Izergin *Quantum Inverse Scattering Method and Correlation Functions*[†]

J. Kroon *Conformal Methods in General Relativity*

M. Le Bellac *Thermal Field Theory*[†]

Y. Makeenko *Methods of Contemporary Gauge Theory*[†]

N. Manton and P. Sutcliffe *Topological Solitons*[†]

N. H. March *Liquid Metals: Concepts and Theory*[†]

I. Montvay and G. Münster *Quantum Fields on a Lattice*[†]

P. Nath *Supersymmetry, Supergravity, and Unification*

L. O'Raifeartaigh *Group Structure of Gauge Theories*[†]

T. Ortín *Gravity and Strings, 2nd edition*

T. Ortín *Gravity and Strings*

A. M. Ozorio de Almeida *Hamiltonian Systems: Chaos and Quantization*[†]

L. Parker and D. Toms *Quantum Field Theory in Curved Spacetime: Quantized Fields and Gravity*

R. Penrose and W. Rindler *Spinors and Space-Time Volume 1: Two-Spinor Calculus and Relativistic Fields*[†]

R. Penrose and W. Rindler *Spinors and Space-Time Volume 2: Spinor and Twistor Methods in Space-Time Geometry*[†]

S. Pokorski *Gauge Field Theories, 2nd edition*[†]

J. Polchinski *String Theory Volume 1: An Introduction to the Bosonic String*[†]

J. Polchinski *String Theory Volume 2: Superstring Theory and Beyond*[†]

J. C. Polkinghorne *Models of High Energy Processes*[†]

V. N. Popov *Functional Integrals and Collective Excitations*[†]

L. V. Prokhorov and S. V. Shabanov *Hamiltonian Mechanics of Gauge Systems*

S. Raychaudhuri and K. Sridhar *Particle Physics of Brane Worlds and Extra Dimensions*

A. Recknagel and V. Schiomerus *Boundary Conformal Field Theory and the Worldsheet Approach to D-Branes*

R. J. Rivers *Path Integral Methods in Quantum Field Theory*[†]

R. G. Roberts *The Structure of the Proton: Deep Inelastic Scattering*[†]

C. Rovelli *Quantum Gravity*[†]

W. C. Saslaw *Gravitational Physics of Stellar and Galactic Systems*[†]

R. N. Sen *Causality, Measurement Theory and the Differentiable Structure of Space-Time*

M. Shifman and A. Yung *Supersymmetric Solitons*

H. Stephani, D. Kramer, M. MacCallum, C. Hoenselaers and E. Herlt *Exact Solutions of Einstein's Field Equations, 2nd edition*[†]

J. Stewart *Advanced General Relativity*[†]

J. C. Taylor *Gauge Theories of Weak Interactions*[†]

T. Thiemann *Modern Canonical Quantum General Relativity*[†]

D. J. Toms *The Schwinger Action Principle and Effective Action*[†]

A. Vilenkin and E. P. S. Shellard *Cosmic Strings and Other Topological Defects*[†]

R. S. Ward and R. O. Wells, Jr *Twistor Geometry and Field Theory*[†]

E. J. Weinberg *Classical Solutions in Quantum Field Theory: Solitons and Instantons in High Energy Physics*

J. R. Wilson and G. J. Mathews *Relativistic Numerical Hydrodynamics*[†]

[†] Available in paperback

Supersymmetry, Supergravity, and Unification

PRAN NATH

Northeastern University, Boston

CAMBRIDGE
UNIVERSITY PRESS

CAMBRIDGE
UNIVERSITY PRESS

University Printing House, Cambridge CB2 8BS, United Kingdom

One Liberty Plaza, 20th Floor, New York, ny 10006, USA

477 Williamstown Road, Port Melbourne, vic 3207, Australia

4843/24, 2nd Floor, Ansari Road, Daryaganj, Delhi 110002, India

79 Anson Road, #0604/06, Singapore 079906

Cambridge University Press is part of the University of Cambridge.

It furthers the University's mission by disseminating knowledge in the pursuit of education, learning and research at the highest international levels of excellence.

www.cambridge.org
Information on this title: www.cambridge.org/9780521197021

First published 2017

Printed in the United States of America by Sheridan Books, Inc.

A catalogue record for this publication is available from the British Library

Library of Congress Cataloguing in Publication data
Names: Pran Nath, 1939– author.
Title: Supersymmetry, supergravity, and unification / Pran Nath,
Northeastern University, Boston.
Other titles: Cambridge monographs on mathematical physics.
Description: Cambridge, United Kingdom;
New York, NY : Cambridge University Press, [2017] |
Series: Cambridge monographs on mathematical physics |
Includes bibliographical references and index.
Identifiers: LCCN 2016026402 |
ISBN 9780521197021 (hardback ; alk. paper) |
ISBN 0521197023 (hardback ; alk. paper)
Subjects: LCSH: Supersymmetry. | Supergravity. | Unified field theories.
Classification: LCC QC174.17.S9 P73 2017 | DDC 530.14/23–dc23
LC record available at https://lccn.loc.gov/2016026402

ISBN 978-0-521-19702-1 Hardback

To: Kedar Nath Sastri and Tara Devi
and Shashi, Rishi, Anjali

Contents

Preface

Over the past decades a remarkable evolution has occurred in the development of particle physics in relation to general relativity, astrophysics, and cosmology. Till the early 1960s, gravity was considered a force not strong enough to be a player along with the electroweak and the strong forces, and did not receive any significant attention in texts on particle theory till that time. However, these perceptions changed at a remarkably rapid pace in the second half of the last century, most directly due to the discovery of supersymmetry and supergravity and, later, string theory. Thus, supergravity theory and theories based on superstrings bring gravity into the mix in an intrinsic fashion. Models based on supergravity, specifically supergravity grand unification, indicate that electroweak phenomena are driven by supergravity-induced effects connecting in a remarkable way the Planck scale with the electroweak scale. Another way that the connection between particle physics and cosmology has deepened over the past decades is due to the experimental observations that the universe is composed mostly of dark matter and dark energy, and particle theory is the likely place to achieve an understanding of these phenomena. Thus, while dark energy could be sourced by a tiny cosmological term in the Einstein equations, models based on supersymmetry and supergravity provide dark matter candidates in the form of a neutral spin $1/2$ particle (the neutralino) and a spin $3/2$ particle (the gravitino). Further, the early history of the universe is driven by particle physics via phenomena such as baryogenesis and leptogenesis, which can provide the observed baryon excess in the universe. Thus, currently there exists an unprecedented degree of fusion of astrophysics, cosmology, and particle theory. Each of these disciplines has its own glorious history with texts that center on them. However, for a young researcher entering the field of particle theory, developing a workable knowledge of these diverse fields presents a challenge. This text is written with the hope that it may make that task a bit easier. The text is written with the assumption that the reader has a workable knowledge of field theory.

Since the underlying theme of the book is unification, a number of interconnected topics are discussed. These include Einstein and Yang–Mills theories, the standard model, supersymmetry and supergravity, and cosmology along with some specialized topics. The layout of the book is as follows. In Chapter 1 we give a brief history of unification and show that, purely on empirical grounds, one is guided towards the idea that the laws of nature are not random and disconnected but appear to flow from an underlying unified structure. The first part of the book discusses a variety of topics that are essential elements in understanding the main topics of the book – which are supersymmetry, supergravity, and the development of supergravity grand unification and its phenomenological applications. Thus, in Chapter 2 we discuss the Einstein theory of gravity because it is a successful theory of one of the fundamental interactions of physics and it is a gauge theory. In Chapter 3 we discuss the Yang–Mills gauge theory, which is a basic ingredient in the formulation of the standard model as well in the development of a grand unified theory (GUT). In Chapter 4, spontaneous breaking of global symmetries and the Goldstone phenomena are discussed. Also discussed is spontaneous breaking of local symmetries and the Higgs phenomenon. In Chapter 5 we utilize the frameworks of Chapter 3 and Chapter 4 to develop the standard model of electroweak interactions based on the gauge group $SU(2)_L \otimes U(1)_Y$ and then extend it to include the $SU(3)_C$ gauge group, which describes the strong interactions involving quarks and gluons. One of the stringent constraints on model building is the constraint of anomaly cancellation, and we discuss this constraint in Chapter 6. In Chapter 7 we discuss effective Lagrangians, as they form the foundation of any field theoretical model of particle theory. The effective Lagrangians are a powerful technique and follow from simple principles, which are discussed at some length in Chapter 7.

In Chapter 8 we discuss supersymmetry basics and the construction of supersymmetry-invariant Lagrangians. In Chapter 9 we discuss grand unification, which can unify the electroweak and the strong interactions within its framework. Here we consider the Pati–Salam group $SU(4)_C \times SU(2)_L \times SU(2)_R$ where we first encounter unification of quarks and leptons, with leptons treated as the fourth color. However, the Pati–Salam group is not fully unified since it is of the product form. The first fully unified group was proposed by Georgi and Glashow as the simple group $SU(5)$. Later in this chapter we discuss $SO(10)$, which is more unified in that it can accommodate a full generation of quarks and leptons within one irreducible representation. Calculational techniques for carrying out computations in $SO(10)$ are also discussed in this chapter. In Chapter 10 we discuss the Lagrangian for the minimal supersymmetric standard model.

In Chapter 11 we discuss pure $N = 1$ supergravity. However, in order to construct particle physics models we must couple $N = 1$ supergravity with $N = 1$ matter. This is done in Chapter 12, where $N = 1$ matter supermultiplets are coupled to $N = 1$ gauge supermultiplets, which are then coupled to $N = 1$ supergravity. The resulting Lagrangian is sometimes referred to as applied $N = 1$ supergravity. The development of supergravity grand unification is discussed in Chapter 13. Here we first focus on the spontaneous breaking of supersymmetry and then on the generation of soft terms via gravity mediation. Resolution of the GUT and Planck scale hierarchy problem is also discussed here. Phenomenology based on supergravity grand unification is discussed in Chapter 14.

The important topic of CP violation is discussed in Chapter 15. A hallmark of grand unified theories is that they lead to proton decay, and this topic is discussed in Chapter 16. Topics of astroparticle physics and cosmology and their overlap with particle physics are discussed in Chapter 17. In Chapter 18 we discuss supergravities in higher dimensions and supergravities arising from strings. In Chapter 19 we discuss specialized topics. These include models with extra dimensions and extensions of the standard model using the Stueckelberg mechanism. The Stueckelberg mechanism allows one to construct $U(1)$ extensions of the conventional models without use of the Higgs mechanism. In Chapter 20, some speculations regarding the future of unification are given. In Chapter 21 many appendices with supplementary material related to the discussion in the main text are given. Notations, conventions, and some useful formulae are given in Chapter 22. Numerical values of the physical constants that enter into the discussion in the text are given in Chapter 23. In Chapter 24, a list of books and review articles is given which the reader may find useful for cross-referencing the material discussed in this text and for further reading.

As mentioned at the beginning of the preface, this book is primarily geared towards young graduate students and researchers. The text grew out of several graduate-level courses that I taught over the years as well as summer school lectures that I have given at various institutions. I have benefitted greatly over the years from collaborations and discussions with a large number of colleagues. Chief among those are the late Richard Lewis Arnowitt and Ali Hani Chamseddine. I have also benefitted from many other colleagues, friends, and former students. These include K. S. Babu, Utpal Chattopadhyay, Daniel Feldman, WanZhe Feng, Haim Goldberg, Ilia Gogoladze, James Halverson, Tarek Ibrahim, Boris Kors, Zuowei Liu, Brent D Nelson, Jogesh Pati, Pavel Fileviez Perez, Raza Syed, Tomasz Taylor, and Michael Vaughn, among many others. I have also profited from discussions and communications with the late Norman Ramsay regarding the early history of CP violation discussed in Chapter 15. I would like to thank WanZhe Feng, James Halverson, Maksim Piskunov, and Brent D. Nelson for reading parts of the book and offering useful comments and suggestions. Help from Gregory Peim regarding figures in the book is acknowledged. Finally, I wish to acknowledge Dr Simon Capelin, Editorial Director (Physical Sciences) at Cambridge University Press, for his interest and patience while the book was being written.

Pran Nath

1

A Brief History of Unification

"Pure logical thinking cannot yield us any knowledge of the empirical world; all
knowledge of reality starts from experience and ends in it."

—Albert Einstein[*]

Unification is a common theme that runs through the development of particle
physics. Progress is made when one discovers an underlying unifying principle
that connects different particles, different forces, or different phenomena. One
of the early examples is the concept of iso-spin. Heisenberg [2] introduced it
to explain the very similar nuclear properties of the proton and of the neutron
despite the fact that one is charged and the other is neutral. Thus, here one
postulates an $SU(2)$ internal symmetry group with generators T_a ($a = 1, 2, 3$)
which satisfy the algebra

$$[T_a, \ T_b] = i\epsilon_{abc}T_c, \tag{1.1}$$

where ϵ_{abc} is $+1$ when a, b, and c are cyclic, -1 when they are acyclic, and
vanishes otherwise, and the proton (p) and the neutron (n) belong to the doublet
representation of this group so that

$$N \equiv \begin{pmatrix} p \\ n \end{pmatrix} \rightarrow \begin{pmatrix} +1/2 \\ -1/2 \end{pmatrix}, \tag{1.2}$$

where the column after the arrow gives T_3 quantum numbers of the components.
The column vector with p and n states is often called the nucleon, to emphasize
that the proton and the neutron are coming from a common multiplet. Similarly,
the three pseudo-scalar mesons $\pi^{\pm}$ and π^0 can be thought of as components of
a $T = 1$ multiplet ϕ_a where

$$\pi^{\pm} = (\phi_1 \mp i\phi_2)/\sqrt{2}, \ \pi^0 = \phi_3. \tag{1.3}$$

[*] From the Herbert Spencer Lecture 1933 as cited by A. Salam [1].

The pion–nucleon system then can be described by the interaction Lagrangian

$$\mathcal{L}_{\pi\pi N} = g_{\pi\pi N} \bar{N} \gamma_5 \frac{\tau_a}{2} N \phi_a, \tag{1.4}$$

where τ_a ($a = 1$, 2, 3) are the Pauli matrices. It is easily checked that this interaction is invariant under global iso-spin transformations. The interactions $\bar{p}p\pi^0$, $\bar{n}n\pi^0$, and $\bar{p}n\pi^+$ are now governed by a single coupling constant.

A more potent illustration is the classification of the pseudo-scalar mesons and baryons. Thus, the $J^P = 0^-$ mesons consisting of π^+, π^0, π^-, K^+, K^0, $\overline{K^0}$, K^-, and η^0 can be grouped into the octet representation [3–5] of an $SU(3)$ flavor group with assignment of iso-spin T and a new quantum number called hypercharge Y as follows:

0^- mesons	T	Y
π^+, π^0, π^-	1	0
K^+, K^0	$\dfrac{1}{2}$	$\dfrac{1}{2}$
$\overline{K^0}, K^-$	$\dfrac{1}{2}$	$-\dfrac{1}{2}$
η^0	0	0

$$\tag{1.5}$$

where the hypercharge is defined so that

$$Y = \frac{1}{2}(S + N_B). \tag{1.6}$$

Here, S is the strangeness quantum number and N_B is the baryon number, and the electric charge of a particle is given by

$$Q = T_3 + Y. \tag{1.7}$$

Similarly, the spin 1/2 baryons p, n, Σ^+, Σ^0, Σ^-, Ξ^0, and Ξ^- can also be classified in the octet representation of $SU(3)$ with the iso-spin and hypercharge assignments similar to the case for the pseudo-scalar bosons as shown below:

spin $\dfrac{1}{2}$ baryons	T	Y
$\Sigma^+, \Sigma^0, \Sigma^-$	1	0
p, n	$\dfrac{1}{2}$	$\dfrac{1}{2}$
Ξ^0, Ξ^-	$\dfrac{1}{2}$	$-\dfrac{1}{2}$
Λ^0	0	0

$$\tag{1.8}$$

Further, the interactions of the octet of baryons with the octet of pseudo-scalar mesons can be described by just two (F-type and D-type) coupling constants. This means that the couplings of interactions such as $\bar{N}\vec{\tau}N \cdot \vec{\pi}$ and $\bar{\Lambda}\Lambda\eta^0$, $\bar{N}\Lambda K$

become related, which is an enormous simplification over an otherwise huge number of possible couplings in the absence of a symmetry principle. Of course, the $SU(3)$ symmetry like the iso-spin symmetry is not exact, but simple assumptions on how the symmetry breaks allow one to derive sum rules known as Gell–Mann–Okubo mass relations [6] among the masses within the multiplets, and these sum rules are in reasonably good agreement with experimental data on the pseudo-scalar meson masses and on the baryon masses.

The $SU(3)$ symmetry also helps in the classification of the meson–baryon resonances. Since the meson–baryon resonances must decay into mesons and baryons which belong to octet representations, one expects that the meson–baryon resonances are likely to belong to one of the irreducible representations in the product of meson and baryon octets. Thus,

$$8 \otimes 8 = 1 \oplus 8 \oplus 8 \oplus 10 \oplus 10^* \oplus 27. \tag{1.9}$$

Experimentally, the meson–baryon resonance $N^*(1232)$ (where the number in the parentheses indicates the mass of the resonance in MeV) is a $J^P = 3/2^+$ and $T = 3/2$ object. The iso-spin assignment implies that the resonance should belong to either the 10-plet or the 27-plet, with the 10-plet being the simplest possibility. Indeed, observation of other $J^P = 3/2^+$ resonances allows one to fill in the full 10-plet as shown below. Historically, the existence of Ω^- as a missing piece in the 10-plet was a prediction which was subsequently verified.

$J^P = \dfrac{3}{2}^+$ Resonances	T	Y
N^*	$\dfrac{3}{2}$	$\dfrac{1}{2}$
Y^*	1	0
Ξ^*	$\dfrac{1}{2}$	$-\dfrac{1}{2}$
Ω^-	0	-1

$$(1.10)$$

In the above, the classification of multiplets as belonging to irreducible representations of $SU(3)$ allowed us to make further predictions regarding the nature of meson–baryon resonances. However, one may ask if the octets of mesons and baryons are truly fundamental. In 1964, Gell-Mann and Zweig [7–9] proposed that the mesons and baryons could themselves be composed of something more fundamental such as particles which belong to the 3-plet representation of $SU(3)$ (quarks) and their conjugates 3* (anti-quarks) with iso-spin and hypercharge assignments $(T)^Y$ as follows:

$$3 = \left(\frac{1}{2}\right)^{1/6} + (0)^{-1/3} \tag{1.11}$$

$$3^* = \left(\frac{1}{2}\right)^{-1/6} + (0)^{1/3}. \tag{1.12}$$

Using the above, one can generate pseudo-scalar meson states so that

$$3 \otimes 3^* = 1 \oplus 8. \tag{1.13}$$

We can label the quarks u, d, and s with the quantum numbers as follows:

Quark	Q	T	T_3	Y	
u	$\dfrac{2}{3}$	$\dfrac{1}{2}$	$\dfrac{1}{2}$	$\dfrac{1}{6}$	
d	$-\dfrac{1}{3}$	$\dfrac{1}{2}$	$-\dfrac{1}{2}$	$\dfrac{1}{6}$	
s	$-\dfrac{1}{3}$	0	0	$-\dfrac{1}{3}$	(1.14)

It is then easily seen that the iso-spin and hypercharge assignments for the pseudo-scalar meson octet follow directly from the iso-spin and hypercharge assignments of the 3 and 3*. Similarly, the baryons can be constructed from the product of three 3-plets so that

$$3 \otimes 3 \otimes 3 = 1 \oplus 8 \oplus 8 \oplus 10. \tag{1.15}$$

Again, the iso-spin and hypercharge assignment of the octet of baryons follows from the iso-spin and hypercharge assignment of the quarks.

The above picture, however, causes a problem. In this picture one has mesons as bound states of quarks and anti-quarks, i.e., $q\bar{q}$ states while the baryons are bound states of three quarks, i.e., qqq states. However, qq states and $qqqq$ states can also form. Thus, qq bound states would appear as fractionally charged particles but there is no experimental evidence for them. Further, the N^* which has $J = 3/2$ and $T = 3/2$, contains a doubly charged state N^{*++}, which can be thought of as made up of three u quarks, i.e., the state $u^\uparrow u^\uparrow u^\uparrow$, i.e., with all spins up. This makes the spin wave function totally symmetric. When coupled with the fact that N^{*++} is the ground state of three quarks, the wave function of the three quarks becomes totally symmetric – in contradiction with the Fermi–Dirac statistics. The solution to the problem is offered by the introduction of a hidden quantum number [10,11], generally known as color. Specifically, we can introduce a color group $SU(3)_C$ whose fundamental representation contains three states, i.e., states which we can label as red, green, and blue or simply 1, 2, and 3. In this case we can write $N^{*++} \sim \epsilon_{abc} u_a(x_1) u_b(x_2) u_c(x_3)$, where we have suppressed the spins. This makes the state consistent with Fermi–Dirac statistics.

Even more central to the development of particle physics are local symmetries, which appear to be far more fundamental. Local symmetry includes the principle of gauge invariance, and this principle has played a key role in the development of theories of particle physics. Examples of these are the Maxwell–Dirac theory, the Yang–Mills–Dirac theory, and the Einstein theory of gravitation However, a major weakness of gauge theories is that while the principle of gauge invariance determines the self-interactions of the gauge fields and also constraints the interactions of the gauge fields with matter fields, the gauge principle is not strong enough to determine the nature of matter fields themselves. Thus, consider the Dirac–Maxwell system described by the Lagrangian

$$\mathcal{L} = -\frac{1}{4}F_{\mu\nu}F^{\mu\nu} - \sum_i \bar{\psi}_i \left[\frac{1}{i}\gamma^\mu(\partial_\mu - igQ_iA_\mu) + m_i\right]\psi_i. \tag{1.16}$$

Here, $F_{\mu\nu} \equiv \partial_\mu A_\nu - \partial_\nu A_\mu$ is the field strength and ψ_i are spin 1/2 Dirac fields which carry the $U(1)$ gauge charges Q_i. The equation of motion for the gauge field in this case reads

$$\partial_\nu F^{\mu\nu} = gJ^\mu$$
$$J^\mu = \sum_i Q_i\bar{\psi}_i\gamma^\mu\psi_i, \tag{1.17}$$

The principle of gauge invariance determines the interaction of the gauge field A_μ with the spin half fields ψ_i but is not powerful enough to determine the nature and the number of fields or the charges Q_i. This turns out to be a major weakness in our understanding, i.e., we lack a principle that can determine the type and the number of fields that enter into Eq. (1.16). This problem is persistent and permeates essentially all sectors of particle theory. We consider two further examples: Yang–Mills theory [12–14] and the Einstein theory. For the coupled Yang–Mills–Dirac system for a gauge group G with generators T_a which satisfy the algebraic relation $[T_a, T_b] = iC_{abc}T_c$ where C_{abc} are the structure constants of the gauge group, and A_μ^a are the gauge fields belonging to the adjoint representation of the gauge group, Eq. (1.17) is replaced by

$$\partial_\nu F_a^{\mu\nu} + gC_{abc}A_{\nu b}F_c^{\mu\nu} = gJ_a^\mu. \tag{1.18}$$

Here $F_a^{\mu\nu}$ are the field strengths given by

$$F_{\mu\nu}^a = \partial_\mu A_\nu^a - \partial_\nu A_\mu^a + gC_{abc}A_\mu^b A_\nu^c, \tag{1.19}$$

while the source J_a^μ contains the matter fields, and is given by

$$J_a^\mu = \sum_i Q_i\bar{\psi}_i\gamma^\mu T_a\psi_i. \tag{1.20}$$

where ψ_i are assumed to be spin 1/2 fields which belong to the fundamental representation of the gauge group. Using Eq. (1.19) in Eq. (1.18), we find that the Yang–Mills gauge invariance fully determines the self-interactions of the gauge fields A_μ^a. However, the Yang–Mills gauge invariance does not fully determine the right-hand side of Eq. (1.18), which is given by Eq. (1.20) and depends on a number of fields ψ_i which can be chosen in an arbitrary fashion.

    The same problem reappears when we consider gravity described by the Einstein theory, which is another example of a gauge theory. Here, the analogue of Eq. (1.18) is the Einstein field equations, which read

$$R^{\alpha\beta} - \frac{1}{2}g^{\alpha\beta}R = -\kappa^2 T^{\alpha\beta}. \tag{1.21}$$

Here, $T^{\alpha\beta}$ is the stress tensor, and κ is defined so that

$$\kappa \equiv \sqrt{8\pi G_N}, \tag{1.22}$$

where G_N is Newton's constant. The Planck mass is defined as the inverse of κ so that

$$M_{Pl} \equiv \frac{\sqrt{\hbar c}}{\kappa} \simeq 2.43 \times 10^{18} \text{ GeV}/c^2. \tag{1.23}$$

Here, the Einstein gauge invariance along with the constraint that no more than two derivatives appear in the Lagrangian or in the equations of motion completely determines the left-hand side of Eq. (1.21). However, the right-hand side of Eq. (1.21) remains undetermined, where a variety of fields including spin 0, spin 1/2, and spin 1 enter. Thus, the stress tensor $T^{\alpha\beta}$ depends on the number of quark and lepton generations and the number of Higgs fields, and on the gauge fields as well as any other matter field that may enter a particle physics Lagrangian. Again, while the Einstein gauge invariance along with the number of derivatives constraint is powerful enough to determine the self-interactions of the gravitational field, it is not powerful enough to fix the number and type of matter and gauge fields that enter on the right-hand side of Eq. (1.21). This is a singular weakness of the Einstein theory, as was realized early on by Einstein himself, who described the gravitational equations as

Similar to a building, one wing of which is made of fine marble, but the other wing of which is built of low grade wood. [15]

String theory in principle resolves this problem, i.e., the problem of determining the right-hand side of Eq. (1.21). For a given string model and for a given vacuum structure, the matter and gauge content can be determined, and thus the right-hand side of Eq. (1.21) is determined. However, there remains the issue of too many possibilities from which one unique case which describes our world must be extracted (for a review of string theory see Green *et al.* [16, 17]).

Another issue that enters into unification is the problem of mass scales, and this idea leads us in a new direction. There is empirical evidence that several mass scales exist in nature. The discovery that the law of reflection symmetry or parity symmetry is violated in weak interactions [18] led to the emergence of a two component equation for the neutrino [19–21] and subsequently to the emergence of the V–A theory of weak interactions [22, 23]. Thus, the interactions of the V–A theory are given by

$$\mathcal{L}_W = \frac{G_F}{\sqrt{2}} J_\mu(x) J^{\mu\dagger}(x), \tag{1.24}$$

where

$$J_\mu = \ell_\mu + h_\mu. \tag{1.25}$$

Here ℓ_μ is the current containing leptons so that

$$\ell_\mu = \sum_{i=1}^{3} \bar{\nu}_i \gamma_\mu (1 - \gamma_5) e_i, \tag{1.26}$$

where the sum runs over the e, the μ and the τ generations and h_μ in Eq. (1.25) is the hadronic current, which is given by

$$h_\mu = \sum_{i=1}^{3} \bar{u}_i V_{ij} (1 - \gamma_5) d_j, \tag{1.27}$$

where V_{ij} is the Cabbibo–Kobayashi–Maskawa mixing matrix [24, 25], and G_F appearing in Eq. (1.24) is the Fermi constant and has the value

$$G_F^{-\frac{1}{2}} \simeq 292.8 \text{ GeV}. \tag{1.28}$$

The dimension-six weak interaction operator of Eq. (1.24) is non-renormalizable and points to a more fundamental theory at the scale $G_F^{-1/2}$. Such a theory is the $SU(2)_L \otimes U(1)_Y$ gauge theory of electroweak interactions [26–28]. The spectrum of this theory contains in it $SU(2)_L$ doublets and singlets of quarks and leptons, and gauge bosons of the gauge group $SU(2)_L$ and $U(1)_Y$, i.e.,

$$\begin{pmatrix} \nu_i \\ \ell_i \end{pmatrix}_L, \ e_{Li}^c \qquad \text{leptons} \tag{1.29}$$

$$\begin{pmatrix} u_i \\ d_i \end{pmatrix}_L, \ u_{Li}^c, \ d_{Li}^c \qquad \text{quarks} \tag{1.30}$$

$$\begin{pmatrix} H^+ \\ H^0 \end{pmatrix} \qquad \text{Higgs boson} \tag{1.31}$$

$$A_\alpha^\mu, \ B^\mu \qquad SU(2)_L, U(1)_Y \text{ gauge bosons}, \tag{1.32}$$

where $\alpha = 1$–3. The mystery of the scale $G_F^{-\frac{1}{2}}$ is solved as it becomes related to the vacuum expectation value of the Higgs field H^0 such that

$$G_F^{-\frac{1}{2}} = 2^{1/4} v$$
$$\langle H^0 \rangle = v/\sqrt{2}. \tag{1.33}$$

Using Eq. (1.28) one finds

$$v \simeq 246 \text{ GeV}. \tag{1.34}$$

The standard model of electroweak interactions predicts two massive gauge fields, one of which is charged ($W^\pm$) and the other is neutral (Z^0), whose masses are determined in terms of the vacuum expectation value v and the coupling constant g associated with $SU(2)_L$ and the coupling g' associated with $U(1)_Y$ so that

$$M_{W^\pm} = \frac{1}{2}gv,$$
$$M_Z = \frac{v}{2}\sqrt{g^2 + g'^2}. \tag{1.35}$$

The electroweak model has been shown to be renormalizable [29].

While the electroweak model solves the mystery of the Fermi constant, it creates a mystery of its own in terms of the ratio of the coupling constants g'/g or $\sin\theta_W$ (where θ_W is the weak angle) defined by

$$\sin\theta_W = \frac{g'}{\sqrt{g^2 + g'^2}}. \tag{1.36}$$

Experimentally,

$$\sin^2\theta_W \simeq 0.23. \tag{1.37}$$

Equation (1.37) has no explanation within the standard model. Indeed, a proper understanding of it requires us to consider an entirely new regime of physics involving a mass scale much larger than $G_F^{-1/2}$. The mystery of this ratio is connected as well with the value of the strong interaction coupling α_s, which we now discuss.

Free quarks are not seen as they occur only as bound states within nucleons. As already discussed, the quarks must carry color, and the simplest hypothesis is that quarks belong to the triplet representation of $SU(3)_C$. One may thus think of expanding the gauge group which describes quarks from $SU(2)_L \otimes U(1)_Y$ to $SU(3)_C \otimes SU(2)_L \otimes U(1)_Y$. An entirely unexpected new feature emerges regarding the group $SU(3)_C$ if one assumes that unlike $SU(2)_L \otimes U(1)_Y$ it is not spontaneously broken but is exactly preserved. In this case, one finds that there would be eight massless gauge bosons which belong to the adjoint representation of the $SU(3)_C$ gauge group, and they can mediate interactions between colored quarks. The interaction of the gluons with quarks in this case is given by

$$\mathcal{L}_{\text{int}} = g_3 \sum_{i,a} \bar{\psi}_i \gamma^\mu T_a A_\mu^a \psi_i, \tag{1.38}$$

where T_a ($a = 1-8$) are the generators of the group $SU(3)_C$. Quite remarkably, non-abelian gauge theories with the appropriate matter content exhibit the phenomenon of asymptotic freedom [30, 31] at large energy scales where the interactions among quarks get weaker and can explain the scaling property seen in high-energy collisions [32]. On the other hand, at low energy the interactions between quarks become stronger, which explains how quarks are confined in nucleons. Indeed, the standard model of electroweak and strong interactions based on the gauge group

$$SU(3)_C \otimes SU(2)_L \otimes U(1)_Y, \tag{1.39}$$

is one of the most successful models in all of particle physics.

We now turn to another puzzle which relates to the quantum numbers of the quarks and the leptons. The left-handed leptons and the quarks have the following set of $SU(3)_C$, $SU(2)_L$, and $U(1)_Y$ quantum numbers:

$$q\left(3, 2, \frac{1}{6}\right), \ u^c\left(\bar{3}, 1, -\frac{2}{3}\right), \ d^c\left(\bar{3}, 1, \frac{1}{3}\right),$$

$$L\left(1, 2, -\frac{1}{2}\right), \ e^c(1, 1, 1), \tag{1.40}$$

where q and L are the doublets of $SU(2)_L$ and u^c, d^c, and e^c are $SU(2)_L$ singlets where we have dropped the subscript L. The assignment of the hypercharges of the particles in the standard model appears strange on the surface. However, there is a interesting property that the hypercharges satisfy, which is

$$\sum_i m_i C_i Y_i = 0, \tag{1.41}$$

where m_i is the multiplicity which counts the number of states in a multiplet and C_i are the number of colors and the sum on i runs over the multiplets in the standard model. This is precisely one of the conditions needed for the cancellation of anomalies. It is known that a cancellation of anomalies (see Chapter 6) among multiplets which couple to gauge fields is needed for a theory to be an acceptable quantum field theory [33, 34]. The question then is, if beyond the standard model there is a larger framework in which an arrangement of colors, iso-spins, and hypercharges as given in Eq. (1.40) can arise in a natural way. Such a framework would need to be more unified, and one possibility is an enlarged gauge group [35, 36] which can accommodate the standard model gauge group of Eq. (1.39). A major step in this direction was taken by Pati and Salam [35], who proposed the group $SU(4)_C \times SU(2)_L \times SU(2)_R$ where leptons and quarks are unified with the leptons arising as the fourth color. $SU(4)_C \times SU(2)_L \times SU(2)_R$, however, is a product group, and it is interesting to look for a fully unified group which can accommodate the standard model gauge group. It turns out that $SU(5)$, which is rank four, is the minimal grand unified group [36] which can accommodate the gauge group of Eq. (1.39). In $SU(5)$ one needs a combination of two irreducible representations to accommodate one full generation of the standard model particles, i.e., one needs the irreducible representations $\bar{5} \oplus 10$. The particle content of these is as follows:

$$\bar{5}_L = \begin{pmatrix} d_1^c \\ d_2^c \\ d_3^c \\ e \\ -\nu_e \end{pmatrix}_L, \ 10_L = \begin{pmatrix} 0 & u_3^c & -u_2^c & -u_1 & -d_1 \\ -u_3^c & 0 & u_1^c & -u_2 & -d_2 \\ u_2^c & -u_1^c & 0 & -u_3 & -d_3 \\ u_1 & u_2 & u_3 & 0 & -e^c \\ d_1 & d_2 & d_3 & e^c & 0 \end{pmatrix}_L. \tag{1.42}$$

The combination of $\bar{5}$ and 10 is anomaly free, and accommodates one full generation of quarks and leptons. Also, since the electric charge is related to iso-spin and hypercharge by the relation Eq. (1.7), one finds that grand unification provides an explanation of the quantization of charge, i.e., that the quarks (u, d) have charges $(\frac{2}{3}, -\frac{1}{3})$ while (ν_e, e) have charges $(0, -1)$. This can be viewed as a major triumph of grand unification. One less attractive aspect of $SU(5)$ grand unification is that one generation of quarks and leptons requires two irreducible representations, i.e., $\bar{5} \oplus 10$. It is preferable to have just one irreducible representation where a full generation of quarks and leptons can be accommodated. To do so would require a higher-rank gauge group. The simplest such possibility is the gauge group $SO(10)$ [37, 38], which has a 16-dimensional spinor representation and which decomposes under $SU(5)$ so that

$$16 = 1 \oplus \bar{5} \oplus 10. \tag{1.43}$$

Here, $\bar{5}$ and 10 accommodate a full generation of quarks and leptons as given in Eq. (1.42), and in addition one has a singlet field which we may label as ν^c. Such a field plays a role in giving mass to the neutrino through the so-called see-saw mechanism [39]. One can then generate a neutrino mass matrix, which in the ν, ν^c basis has the form

$$\begin{pmatrix} 0 & m \\ m & M \end{pmatrix}, \tag{1.44}$$

where m is size the electroweak scale and M is size the grand unified theory (GUT) scale. Diagonalization of this mass matrix generates a light neutrino mass of size m^2/M. Assuming $m = O(M_Z)$ and $M = O(10^{15-16})$ GeV, one finds $m_\nu = O(10^{-2} - 10^{-3})$ eV, which is in the range of desired neutrino masses. Thus, $SO(10)$ is an attractive grand unification group although it has the drawback that currently there are a large number of models that one can build using different Higgs representations to break the grand unification group. In addition, attempts have been made to build models using groups $SU(N)$, $N > 5$, and $SO(N)$ for $N > 10$ as well as exceptional groups. Among the five exceptional groups G_2, F_4, E_6, E_7, and E_8, only E_6 has chiral representations, and, further, it contains $SU(5)$ and $SO(10)$ as subgroups so that

$$SU(5) \subset SO(10) \subset E_6. \tag{1.45}$$

One advantage of E_6 is that it allows inclusion of Higgs fields along with quarks and leptons in the same multiplet. But there are also some drawbacks relative to the $SO(10)$ models.

As seen above, an irreducible representation of the grand unification group contains both quarks and leptons in the same multiplet. This means that the quarks can change into leptons by exchange of the so-called lepto-quarks, which are the gauge bosons of the $SU(5)$, $SO(10)$, or E_6 gauge group carrying the quantum numbers of both quarks and leptons. These gauge bosons develop masses by spontaneous breaking of the grand unified symmetry to the symmetry of the standard model gauge group. Let us assume that the GUT symmetry breaks at the scale M_G. In this case, lepto-quarks will achieve masses of $\mathcal{O}(M_G)$. If we integrate out the lepto-quark fields, we see dimension-six operators of the form [40, 41]

$$\overline{u^C}_L \gamma^\mu Q_L \overline{e^C}_L \gamma_\mu Q_L, \quad \overline{u^C}_L \gamma^\mu Q_L \overline{d^C}_L \gamma_\mu L_L \tag{1.46}$$

$$\overline{d^C}_L \gamma^\mu Q_L \overline{u^C}_L \gamma_\mu L_L, \quad \overline{d^C}_L \gamma^\mu Q_L \overline{\nu^C}_L \gamma_\mu Q_L, \tag{1.47}$$

where we have suppressed the generation, color, and $SU(2)$ indices and a front factor $1/M_V^2$ where M_V is the lepto-quark vector boson mass. These operators lead to proton decay modes such as

$$p \to \pi^0 e^+, \quad p \to \pi^+ \bar{\nu}, \quad p \to K^0 e^+. \tag{1.48}$$

A rough estimate of the proton decay width gives

$$\Gamma(p \to e^+ \pi^0) \approx \alpha_G^2 \frac{m_p^5}{M_V^4}, \tag{1.49}$$

where $\alpha_G = g_G^2/4\pi$ and g_G is the coupling constant at the unification scale. If we use the current experimental limit [42] on the $p \to e^+ \pi^0$ lifetime of $\tau(p \to e^+ \pi^0) > 1.4 \times 10^{34}$ years and $\alpha_G = 1/25$, one finds

$$M_V > 3 \times 10^{15} \text{ GeV}. \tag{1.50}$$

Thus, we see that the unification of quarks and leptons brings in a new mass scale, which is 13 orders of magnitude larger than the Fermi scale. The $SU(5)$ theory also makes one more important prediction, which is that at the GUT scale one has

$$\sin^2 \theta_W = \frac{3}{8}. \tag{1.51}$$

Although the $SU(3)$, $SU(2)$, and $U(1)$ gauge couplings are all equal at the GUT scale, i.e., $\alpha_1 = (5/3)\alpha_Y = \alpha_2 = \alpha_3$, after spontaneous breaking to $SU(3) \otimes SU(2) \otimes U(1)$, the couplings will evolve differently below the GUT scale and can be compared with data. Alternatively, one can extrapolate the experimental values of the couplings to high scales and ask if they unify. Thus, experimentally at the Z-mass scale one has

$$\alpha_2 \simeq 0.033, \ \alpha_3 \simeq 0.12, \ \sin^2 \theta_W \simeq 0.23. \tag{1.52}$$

However, an analysis using the standard model particle spectrum shows that their extrapolation to high scales does not have a simultaneous convergence [43]. Rather, one finds pairwise intersection at scales between 10^{14} and 10^{16} GeV with an average of around $\sim 10^{15}$ GeV. This is very similar to the scale we saw in Eq. (1.50). Again the analysis emphasizes the existence of a new scale in physics of around 13 orders of magnitude larger than the Fermi scale. Thus, the particle theory now has the task of understanding the regime of physics between the electroweak scale and this new empirically derived scale which is much larger.

The existence of the larger scale gives further credence to the idea of unification because of the close encounter of the couplings to convergence.

A new direction in unification came about in an entirely different direction in the early 1970s through the emergence of supersymmetry [44–50]. Supersymmetry is a new kind of symmetry which connects bosons and fermions. Thus, supersymmetry contains Bose and Fermi particles in common multiplets. Supersymmetry can be realized algebraically by the introduction of fermionic generators Q_α in addition to the generators of the Poincaré algebra consisting of P_μ, which are the generators of translations, and $J_{\mu\nu}$, which are the generators of rotations in four-dimensional space-time. Thus, suppose Q_α is a left spinor which satisfies the algebraic relations

$$\{Q_\alpha, Q_\beta^\dagger\} = -2(P_L\gamma^\mu\gamma^0)_{\alpha\beta}P_\mu \qquad (1.53)$$

$$\{Q_\alpha, Q_\beta\} = 0 \qquad (1.54)$$

$$[Q_\alpha, P_\mu] = 0, \qquad (1.55)$$

where $P_L = (1 - \gamma_5)/2$. Significant progress occurred when four-dimensional supersymmetry was realized in a Lagrangian form [47, 48]. It was also realized that supersymmetry in part helps resolve the gauge hierarchy problem [51–54]. However, physics that we observe at low energies is certainly not supersymmetric since we do not see the supersymmetric partners of the electron or of the photon in experiments, which implies that if supersymmetry exists in nature it must be a broken symmetry. This is not surprising since even good symmetries in nature such as $SU(2)_L \otimes U(1)_Y$ are realized at low energies only as broken symmetries. However, breaking of supersymmetry turns out to be more difficult. To see why this is the case, one may look at the Hamiltonian. From Eq. (1.53) we find that

$$H \equiv P^0 = \frac{1}{4}(QQ^\dagger + Q^\dagger Q) \geq 0. \qquad (1.56)$$

The above implies that the Hamiltonian of a supersymmetric theory is a positive semi-definite operator. For a supersymmetric vacuum $Q_\alpha|0\rangle = 0 = Q_\alpha^\dagger|0\rangle$, which implies that $\langle 0|H|0\rangle = 0$ for a supersymmetric state. When supersymmetry is spontaneously broken, we have $Q_\alpha|0\rangle \neq 0$ and $Q_\alpha^\dagger|0\rangle \neq 0$, and in this case $\langle 0|H|0\rangle > 0$. This clearly implies that the vacuum with unbroken supersymmetry has a lower energy than the vacuum with broken supersymmetry. All of this indicates that the breaking of supersymmetry would not be easy.

An interesting formulation of supersymmetry exists where instead of using the four co-ordinates of space-time one uses in addition fermionic co-ordinates, and the resulting space is referred to as superspace [49, 50]. Supersymmetry formulated in superspace is elegant and is a very useful tool in supersymmetric analyses. This supersymmetry is of course a global symmetry, and, as discussed above, the successful theories of fundamental interactions thus far are gauge theories. Thus, it appears logical that one should try to gauge supersymmetry. It turns out that gauging of supersymmetry, i.e., making global supersymmetry into a local supersymmetry, brings in gravity [55,56]. This is not totally unexpected since, as seen from Eqs. (1.53)–(1.55), the algebra of the supersymmetry generators is connected with the Poincaré algebra. The first local supersymmetry was formulated in curved superspace [55,56], and subsequently a simpler formulation of local supersymmetry was found [57,58]. The simplest gauged supersymmetry known as supergravity has just the graviton and a Rarita–Schwinger spin 3/2 field [59]. An interesting feature of models with supergravity coupled with matter is that the scalar potential in the theory is no longer positive semi-definite. Because of this there is no obstruction in the breaking of supersymmetry in a reasonable way. Further, one finds that in supergravity-based models it is possible to construct grand unified theories which are phenomenologically viable [60]. In theories of this type, supersymmetry breaks in a hidden sector and is communicated to the visible sector by gravitational interactions. Supergravity models have many interesting features including the fact that they represent the first example of a class of models where gravity enters in a non-trivial way in particle physics [61].

Supersymmetric models require at least two Higgs doublets H_1, and H_2, where H_2 gives mass to the up quarks and H_1 gives mass to the down quarks and leptons. Thus, in addition to quarks and leptons and gauge bosons of the standard model, the low-energy particle spectrum of the minimal supersymmetric model before spontaneous breaking of the electroweak symmetry consists of

$$\begin{pmatrix} \tilde{\nu}_i \\ \tilde{\ell}_i \end{pmatrix}_L, \ \tilde{e}_{Ri}^* \qquad\qquad \text{sleptons} \qquad\qquad (1.57)$$

$$\begin{pmatrix} \tilde{u}_i \\ \tilde{d}_i \end{pmatrix}_L, \ \tilde{u}_{Ri}^*, \ \tilde{d}_{Ri}^* \qquad\qquad \text{squarks} \qquad\qquad (1.58)$$

$$\begin{pmatrix} \tilde{H}_1^0 \\ \tilde{H}_1^- \end{pmatrix}_L, \ \begin{pmatrix} \tilde{H}_2^+ \\ \tilde{H}_2^0 \end{pmatrix}_L \qquad\qquad \text{Higgsinos} \qquad\qquad (1.59)$$

$$\tilde{\lambda}_a, \ \tilde{\lambda}_\alpha, \ \tilde{\lambda}_B \qquad\qquad SU(3)_C, \ SU(2)_L, \ U(1)_Y \ \text{gauginos,}$$
$$a = 1 - 8, \ \alpha = 1 - 3. \qquad\qquad (1.60)$$

As mentioned earlier, in supergravity unified models one can obtain spontaneous breaking of the electroweak symmetry in a phenomenologically viable way. This breaking appears in the form of soft terms which do not affect the ultraviolet behavior of the theory. These include masses for the scalar particles in the theory, trilinear scalar terms, and masses for the gauginos. Under certain restrictive conditions the soft terms are universal and give rise to the minimal supergravity model (mSUGRA). But more generally they also give non-universal soft terms and non-universal supergravity models.

Supergravity grand unification also resolves another puzzle that appears in the standard model. This puzzle relates to the fact that in the standard model one needs a tachyonic mass term for the Higgs boson in order to break the electroweak symmetry. However, the origin of such a term is unexplained. In supergravity grand unified theories the tachyonic mass term arises quite naturally, by a renormalization group effect with a soft breaking and a heavy top. In fact, the requirement that the top be heavy, i..e, greater than a 100 GeV, to break $SU(2) \otimes U(1)$ was an important prediction of supergravity grand unification [62], made at a time when the top quark was still considered to be light, i.e., only in the tens of GeV.

After breaking of the $SU(2)_L \otimes U(1)_Y$ gauge symmetry and after mass growth for $W^{\pm}$ and the Z boson, one is left with the following set of residual Higgs fields:

$$h^0, \ H^0, \ A^0, \ H^{\pm}, \tag{1.61}$$

where h^0 and H^0 are CP even and A^0 is CP odd and where h^0 is the analog of the standard model Higgs, H^0 is the heavier CP Higgs boson, and $H^{\pm}$ is the charged Higgs field. In addition, the gauginos and the Higgsinos mix, producing two charged Dirac fields called charginos ($\tilde{\chi}_i^{\pm}$) and four neutral Majorana fields, called neutralinos ($\tilde{\chi}_j^0$), which are the mass eigenstates and are linear combinations of neutral gauginos and Higgsinos, i.e., one has

$$\tilde{\chi}_i^{\pm}, \ \tilde{\chi}_j^0, \ \ i = 1, \ 2; \ j = 1\text{--}4. \tag{1.62}$$

In addition, there would, in general, be mixing between the left chiral and the right chiral sfermion fields. These mixings are proportional to fermion masses, and are thus negligible for the first two generations, but could be very significant for the third-generation sfermions.

In supergravity grand unified models, the superpartners of the matter, Higgs, and gauge fields known as sparticles, are expected to have masses in the mass range of a few hundred GeV up to a few TeV. Quite remarkably, such a spectrum allows for the unification of the $SU(3)_C$, $SU(2)_L$, and $U(1)$ gauge coupling constants, extrapolating the gauge couplings from the low scales to high scales [63,64] with the unification occurring at a scale [43] of

$$M_G \sim 2 \times 10^{16} \ \text{GeV}. \tag{1.63}$$

There are other interesting implications of supergravity unified models. Supersymmetric grand unified models can allow for a very rapid decay of the proton via exchange of squarks unless they are eliminated by a constraint. Such a constraint is R-parity, defined by

$$R = (-1)^{3(B-L)+2s}, \tag{1.64}$$

where B and L are the baryon and lepton numbers and s is the spin of the particle. R-parity-conserving interactions arise automatically in certain $SO(10)$ grand unified theories if one assumes that the matter fields lie in the spinner representations and the Higgs fields lie in tensor representations of the $SO(10)$ gauge group. With the inclusion of the R-parity constraint, one finds that the least massive supersymmetric particle (LSP) must be absolutely stable. Further, it is found that in supergravity grand unified models over a large part of the parameter space of models the lightest sparticle is a neutralino. With R-parity conservation, such a particle then automatically becomes a candidate for dark matter. Other candidates for dark matter in supergravity are the gravitino and the sneutrino, which is a supersymmetric partner of the neutrino. The assumption of R-parity implies that we must have associated production of supersymmetric particles at colliders, i.e., one must always produce supersymmetric particles in pairs. Thus, for example, at the Large Hadron Collider, in pp collisions sparticles can be produced only in pairs such as

$$q\bar{q} \to \tilde{\chi}_i^+ \tilde{\chi}_j^-, \quad \tilde{\ell}^+ \tilde{\ell}^-; \quad gg \to \tilde{q}_i \tilde{q}_j^*, \quad \tilde{g}\tilde{g}. \tag{1.65}$$

The decay of the particle states will result in at least two neutral LSPs which will escape detection, and their disappearance will be recorded as missing energy. Thus, supersymmetry signals with R-parity conservation will be associated with a significant amount of missing energy. The presence of the extra missing energy for supersymmetric final states provides a distinguishing feature from the background generated by the standard model processes.

One of the particles of the standard model spectrum is the Higgs boson [65–68]. After a long search the Higgs boson has been discovered [69, 70] with a mass of $\sim$126 GeV in the search channel $h^0 \to \gamma\gamma$. The fact that the Higgs boson can decay into the two-photon final state indicates that it is not a spin 1 particle, since such a decay is forbidden by the Landau–Yang theorem [71, 72], but it could be spin 0 or spin 2. Experiments have ruled out spin 2, leaving spin 0 as the only option. While many further tests will come in the future, it is safe to assume that the discovered boson is indeed the one that participates in the spontaneous breaking of the electroweak symmetry. One prediction of such a boson is the relative strength with which it couples to quarks and leptons and to the dibosons. Thus the couplings of the Higgs boson at the tree level are given by

$$\mathcal{L}_{h^0} = -\frac{m_f}{v} h^0 \bar{f} f - \frac{2M_W^2}{v} h^0 W^{\mu+} W_\mu^- - \frac{M_Z^2}{v} h^0 Z^\mu Z_\mu + \cdots \tag{1.66}$$

The current data from the Large Hadron Collider give estimates of the couplings to heavy fermions and to the dibosons, and more sensitive tests are likely to come from experiments at the Large Hadron Collider at $\sqrt{s} = 13$ TeV and possibly from the proposed International Linear Collider, which is an e^+e^- machine [73]. The observation of a significant deviation from the standard model prediction would be a signal for new physics beyond the standard model.

Before the data from the Large Hadron Collider, the Higgs mass was assumed to lie between the experimental lower limit of 115 GeV from the Large Electron Positron (LEP) machine and an upper bound of around 800 GeV from unitarity. However, stronger bounds arise from considerations of vacuum stability. Including the next-to-next leading order corrections, one finds that in the standard model the vacuum can be stable only up to 10^9–10^{10} GeV with a Higgs boson mass of ~126 GeV [74]. The analysis implies that new physics is needed if one wants vacuum stability up to the Planck mass. Supersymmetry provides the possible new physics needed for the stability of the vacuum up to the Planck scale. In addition to the above, there is still the problem of large loop corrections to the Higgs mass in the standard model. Thus, in the standard model the exchange of $q\bar{q}$ in the loop gives a correction to the Higgs mass so that

$$m_h^2 = m_0^2 + \mathcal{O}(\Lambda^2), \tag{1.67}$$

where Λ is a cut-off. In a GUT such a cut-off would be the size of the grand unification scale, i.e., $\Lambda \sim 10^{16}$ GeV. So, a large fine tuning of size one part in 10^{28} is needed to obtain the physical Higgs boson mass. Supersymmetric theories stabilize the gauge hierarchy [51–54]. Thus, in supersymmetric theories the large quadratic corrections are cancelled by the corresponding squark loop, and the cut-off is replaced by the squark mass. Now in supersymmetry at the tree level the Higgs boson mass is predicted to be $\leq M_Z$. In view of the experimental result that the Higgs boson mass is ~126 GeV, one needs a significant loop correction to pull the tree value of the Higgs mass to the experimentally observed value.

In supersymmetry, a correction of this size requires that the average stop mass be a few TeV in size. If one assumes that all scalar masses start with a common universal mass at the grand unification scale, one finds that the average size of scalars would be high, which explains the non-observation of supersymmetry at colliders thus far. However, if one allows for non-universality of soft masses, then only the third-generation squarks which have the largest Yukawa couplings are strongly constrained by the Higgs boson mass, while other sparticles such as the sleptons and the gauginos are not constrained except by the LHC exclusion plots, and could be relatively light. All these sparticles are thus candidates for discovery at the Large Hadron Collider in experiments at $\sqrt{s} = 13$ TeV.

Over the past decades the fields of particle physics and cosmology have been drawing closer, and particle physics has come to play a central role in cosmology. Thus, for example, one of the central mysteries of cosmology relates to the fact that most of the universe, i.e., roughly 95%, is either dark matter or dark energy, the nature of which is yet undetermined. Particle physics offers candidates for dark matter, such as the neutralino or the gravitino in supergravity models. Another mystery relates to the origin of baryon asymmetry in the universe. According to Sakharov [75], such an asymmetry requires three basic ingredients: violation of baryon number, C and CP violation, and out-of-equilibrium decays. The standard model does not have sufficient CP violation to allow for the desired amount of baryon asymmetry in the universe. On the other hand, models based on supersymmetry and supergravity contain many new CP-violating phases above and beyond what the standard model has, which can provide sufficient CP violation for the generation of the desired baryon asymmetry in the universe. A related mystery concerns the ratio of dark matter to baryonic matter, which is roughly 5:1, which begs the question if these two have a common origin. Again an explanation of such phenomena requires one to look at physics beyond the standard model.

The ultimate unification must, of course, also be a quantum theory of gravity. String theory [15, 16] is currently the leading candidate for such a possibility. However, string theory is a framework much like field theory within which model building must still be carried out. Enormous progress has been made over the past three decades providing new directions in model building. If we are able to extract a unique model that fully describes our world, it would be an achievement of "logical thinking" tempered by constraints of the "empirical world" as Einstein envisioned in his Herbert Spencer Lecture in 1953.

References

[1] A. Salam, *Rev. Mod. Phys.* **52**, 511 (1980).

[2] W. Heisenberg, *Z. Phys.* **77**, 1 (1932).

[3] M. Gell-Mann, Caltech Report No. CTSL-20 (1961).

[4] Y. Ne'eman, *Nucl. Phys.* **26**, 222 (1961).

[5] M. Gell-Mann and Y. Ne'eman, eds., *The Eightfold Way*, Benjamin (1964).

[6] S. Okubo, *Progr. Theoret. Phys.*, **27**, 949 (1962).

[7] M. Gell-Mann, *Phys. Lett.* **8**, 214 (1964).

[8] G. Zweig, CERN Preprint CERN-TH-401 (1964).

[9] G. Zweig, in D. Lichtenberg and S. Rosen, eds., *Developments in the Quark Theory of Hadrons*, vol. 1, Hadronic Press (1980), p. 22.

[10] O. W. Greenberg, *Phys. Rev. Lett.* **13**, 598 (1964).

[11] M. Y. Han and Y. Nambu, *Phys. Rev.* **139**, B1006 (1965).

[12] C. N. Yang and R. L. Mills, *Phys. Rev.* **96**, 191 (1954).

[13] R. Shaw, "Invariance under general isotopic spin transformations," Ph.D. dissertation, Cambridge University (1955).

[14] R. Utiyama, *Phys. Rev.* **101**, 1597 (1956).

[15] A. Einstein, *J. Franklin Inst.* **221**, 313 (1936).

[16] M. B. Green, J. H. Schwarz, and E. Witten, *Superstring Theory*, Vol. 1: *Introduction*, Cambridge University Press (1987).

[17] M. B. Green, J. H. Schwarz, and E. Witten, *Superstring Theory*, Vol. 2: *Loop Amplitudes, Anomalies And Phenomenology*, Cambridge University Press (1987).

[18] T. D. Lee and C.-N. Yang, *Phys. Rev.* **104**, 254 (1956).

[19] A. Salam, *Nuovo Cim.* **5**, 299 (1957).

[20] L. D. Landau, *Nucl. Phys.* **3**, 127 (1957).

[21] T. D. Lee and C. -N. Yang, *Phys. Rev.* **105**, 1671 (1957).

[22] E. C. G. Sudarshan and R. E. Marshark, in *Proceedings of the Padua–Venice Conference on Mesona and Newly Discovered Particles* (1957, unpublished) and *Phys. Rev.* **109**, 1860 (1958).

[23] R. P. Feynman and M. Gell-Mann, *Phys. Rev.* **109**, 193 (1958).

[24] N. Cabibbo, *Phys. Rev. Lett.* **10**, 531 (1963).

[25] M. Kobayashi and T. Maskawa, *Prog. Theor. Phys.* **49**, 652 (1973).

[26] S. L. Glashow, *Nucl. Phys.* **22** (1961) 579.

[27] S. Weinberg, *Phys. Rev. Lett.* **19** (1967) 1264.

[28] A. Salam, in N. Svartholm, ed., *Elementary Particle Theory*, Almquist and Wiksells (1969) p. 367.

[29] G. 't Hooft, *Nucl. Phys. B* **35**, 167 (1971). See also G. 't Hooft, *Nucl. Phys. B* **33**, 173 (1971); G. 't Hooft and M. J. G. Veltman, *Nucl. Phys. B* **44**, 189 (1972).

[30] H. D. Politzer, *Phys. Rev. Lett.* **30** (1973) 1346.

[31] D. J. Gross and F. Wilczek, *Phys. Rev. Lett.* **30** (1973) 1343.

[32] J. D. Bjorken, *Phys. Rev.* **179**, 1547 (1969).

[33] J. S. Bell and R. Jackiw, *Nuovo Cim. A* **60**, 47 (1969).

[34] S. L. Adler, *Phys. Rev.* **177**, 2426 (1969).

[35] J. C. Pati and A. Salam, *Phys. Rev. D* **10**, 275 (1974) [erratum, *ibid.* D **11**, 703 (1975)].

[36] H. Georgi and S. L. Glashow, *Phys. Rev. Lett.* **32**, 438 (1974).

[37] H. Georgi, in C. E. Carlson, eds, Particles and Fields, American Institute of Physics (1975).

[38] H. Fritzsch and P. Minkowski, *Annals Phys.* **93** (1975) 193.

[39] For seesaw neutrino masses see P. Minkowski, *Phys. Lett. B* **67** (1977) 421; M. Gell-Mann, P. Ramond and R. Slansky, in P. van Nieuwenhuizen and D. Freedman, eds., *Supergravity*, North-Holland, (1979), p. 315; T. Yanagida, *KEK Report 79-18*, Tsukuba (1979); S.L. Glashow, in M. Lévy, J.-L. Basdevant, D. Speiser, J. Weyers, R. Gastmans and M. Jacobs, eds., *Quarks and Leptons*, Cargèse, Plenum Press (1980), p. 707; R. N. Mohapatra and G. Senjanović, *Phys. Rev. Lett.* **44** (1980) 912.

[40] S. Weinberg, *Phys. Rev. Lett.* **43**, 1566 (1979).

[41] F. Wilczek and A. Zee, *Phys. Rev. Lett.* **43**, 1571 (1979).

[42] K. A. Olive *et al.* (Particle Data Group), *Chin. Phys. C* 38, 090001 (2014).

[43] U. Amaldi, W. de Boer and H. Furstenau, *Phys. Lett. B* **260**, 447 (1991).

[44] P. Ramond, *Phys. Rev. D* **3**, 2415 (1971).

[45] Y. A. Golfand and E. P. Likhtman, *JETP Lett.* **13**, 323 (1971).

[46] D. V. Volkov and V. P. Akulov, *Phys. Lett. B* **46**, 109 (1973).

[47] J. Wess and B. Zumino, *Nucl. Phys. B* **78**, 1 (1974).

[48] J. Wess and B. Zumino, *Nucl. Phys. B* **70**, 39 (1974).

[49] A. Salam and J. A. Strathdee, *Nucl. Phys. B* **76**, 477 (1974).

[50] A. Salam and J. A. Strathdee, *Phys. Rev. D* **11**, 1521 (1975).

[51] E. Witten, *Nucl. Phys. B* **177**, 477 (1981); *Phys. Rev. B* **185**, 513 (1981).

[52] S. Dimopoulos and H. Georgi, *Nucl. Phys. B* **193**, 150 (1981).

[53] N. Sakai, *Z. Phys. C* **11**, 153 (1981).

[54] R. K. Kaul, *Phys. Lett. B* **109**, 19 (1982).

[55] P. Nath and R. L. Arnowitt, *Phys. Lett. B* **56**, 177 (1975).

[56] R. L. Arnowitt, P. Nath, and B. Zumino, *Phys. Lett. B* **56**, 81 (1975).

[57] D. Z. Freedman, P. van Nieuwenhuizen, and S. Ferrara, *Phys. Rev. D* **13**, 3214 (1976).

[58] S. Deser and B. Zumino, *Phys. Lett. B* **62**, 335 (1976).

[59] W. Rarita and J. Schwinger, *Phys. Rev.* **60**, 61 (1941).

[60] A. H. Chamseddine, R. L. Arnowitt, and P. Nath, *Phys. Rev. Lett.* **49**, 970 (1982).

[61] R. Arnowitt, A. H. Chamseddine, and P. Nath, *Int. J. Mod. Phys. A* **27**, 1230028 (2012).

[62] L. Alvarez-Gaume, J. Polchinski, and M. B. Wise, *Nucl. Phys. B* **221**, 495 (1983).

[63] H. Georgi, H. R. Quinn, and S. Weinberg, *Phys. Rev. Lett.* **33**, 451 (1974).

[64] S. Dimopoulos, S. Raby, and F. Wilczek, *Phys. Rev. D* **24**, 1681 (1981).

[65] F. Englert and R. Brout, *Phys. Rev. Lett.* **13** (1964) 321.

[66] P. W. Higgs, *Phys. Lett.* **12** (1964) 132.

[67] P. W. Higgs, *Phys. Rev. Lett.* **13** (1964) 508.

[68] G. Guralnik, C. Hagen, and T. Kibble, *Phys. Rev. Lett.* **13** (1964) 585.

[69] ATLAS Collaboration, *Phys. Lett. B* **716** (2012) 1.

[70] CMS Collaboration, *Phys. Lett. B* **716** (2012) 30.

[71] L. D. Landau, *Dokl. Akad. Nauk Ser. Fiz.* **60**, 207 (1948).

[72] C. N. Yang, *Phys. Rev.* **77**, 242 (1950).

[73] M. E. Peskin, arXiv:1207.2516 [hep-ph] (2013).

[74] G. Degrassi, S. Di Vita, J. Elias-Miro, J. R. Espinosa, G. F. Giudice, G. Isidori, and A. Strumia, *J. High Energy Phys.* **1208**, 098 (2012).

[75] A. D. Sakharov, *Pisma Zh. Eksp. Teor. Fiz.* **5**, 32 (1967).

2

Gravitation

2.1 The Equivalence Principle

The general theory of relativity proposed by Einstein in 1915 [1], is a theory that describes gravitational phenomena from subatomic scales to cosmic scales and represents one of the finest achievements of theoretical physics. This chapter is meant only as an introduction to Einstein's theory, which is an essential ingredient of supergravity to be discussed later. For a more comprehensive treatment of the subject the reader is directed to several excellent texts listed at the end of this chapter. Among these, the text that is specifically suited for particle theorists is *Gravitation and Cosmology* by Weinberg [2]. The general theory of relativity is based on the principle of equivalence. This principle postulates equality of the inertial and the gravitational mass. The inertial mass appears in the Newtonian equations of motion

$$m_i \frac{d^2\vec{r}}{dt^2} = \vec{F}, \tag{2.1}$$

where m_i is the inertial mass and $\vec{F}$ is an external force. Thus, a force, for example, could arise because the particle carries an electrical charge and is placed in an electric field. The force could also arise because the particle is placed in a gravitational field, in which case the force is given by

$$\vec{F} = m_g \vec{g} = -m_g \vec{\nabla}\phi_N, \tag{2.2}$$

where m_g is the gravitational mass, $\vec{g}$ is the acceleration due to gravity, and ϕ_N is the Newtonian potential. Insertion of Eq. (2.2) into Eq. (2.1) gives the result

$$m_i \vec{a} = m_g \vec{g}, \tag{2.3}$$

where $\vec{a} = d^2\vec{r}/dt^2$. More precisely, there are two types of gravitational masses: one passive such as the one that appears in Eqs. (2.2) and (2.3) and the other active (which we denote with a prime, m_g') which is responsible for producing a gravitational potential. Let us consider the gravitational interaction between two particles 1 and 2. In this case the force on particle 1 arises from the gravitational potential generated by particle 2, and this force is given by

$$\vec{F}_1 = m_g^{(1)} \left(-\vec{\nabla} \frac{m_g^{'(2)}}{|\vec{r}_1 - \vec{r}_2|} \right). \tag{2.4}$$

Using the equality of action and reaction, the force on particle 2 generated by the gravitational potential of particle 1 takes the form

$$\vec{F}_2 = m_g^{(2)} \left(\vec{\nabla} \frac{m_g^{'(1)}}{|\vec{r}_1 - \vec{r}_2|} \right). \tag{2.5}$$

Setting $\vec{F}_1 + \vec{F}_2 = 0$ gives the relation

$$m_g^{(1)} m_g^{'(2)} = m_g^{(2)} m_g^{'(1)}. \tag{2.6}$$

The above condition leads to the relation

$$\frac{m_g^{'(1)}}{m_g^{(1)}} = \frac{m_g^{'(2)}}{m_g^{(2)}} = C_{\text{univ}}, \tag{2.7}$$

where C_{univ} is a universal constant. Since m_g and m_g' are related by a universal constant we next focus on m_g versus m_i. Experiments starting with Eötvös [3] up to the most recent ones by the Eöt-Wash group [4,5] show that that m_g/m_i is a constant independent of the nature of the composition of the mass. A check on the constancy of m_g/m_i comes from measurement of the Eötvös parameter η defined by

$$\eta = \frac{m_g^{(1)}/m_i^{(1)} - m_g^{(2)}/m_i^{(2)}}{m_g^{(1)}/m_i^{(1)} + m_g^{(2)}/m_i^{(2)}}. \tag{2.8}$$

For beryllium and titanium test bodies at a variety of distances the Eötvös parameter η is constrained so that [5]

$$\eta(\text{Be}-\text{Ti}) = (0.3 \pm 1.8) \times 10^{-13}. \tag{2.9}$$

Thus, constancy of m_g/m_i is borne out to a very high precision. Further tests of the constancy of m_g/m_i have recently been proposed [6,7] and the limit of Eq. (2.9) is likely to increase by up to two orders of magnitude [7]. Let us choose units where the ratio m_g/m_i is 1. In these units the experimental value of C_{univ} is given by

$$G_N \equiv C_{\text{univ}} = 6.673 \times 10^{-11} \text{ Nm}^2/\text{kg}^2, \tag{2.10}$$

where G_N is the familiar Newton's constant. The results above are codified in the so-called weak principle of equivalence, which states that inertial and gravitational masses are equal for all varieties of inertial masses. The strong principle of equivalence states that one can always find local inertial frames where the gravitational forces are absent and Poincaré invariance holds. It is clear then that while gravitational forces are absent in inertial frames they will appear in non-inertial frames. Since the transition from an inertial to a non-inertial frame brings in gravity, it appears logical to investigate the dynamics in non-inertial frames to achieve a deeper understanding of the gravitational phenomena. Thus, we may begin with an inertial frame with co-ordinates labeled by $\bar{x}^\mu$ ($\mu = 0,\ 1,\ 2,\ 3$) where the co-ordinate transformations preserve the line element

$$d\tau^2 = -\eta_{\mu\nu}\, d\bar{x}^\mu\, d\bar{x}^\nu. \tag{2.11}$$

Let us now carry out a transformation to a non-inertial frame $\bar{x} = \bar{x}(x)$. It is then easily seen that the line element in a general non-inertial frame is given by

$$d\tau^2 = -g_{\alpha\beta}\, dx^\alpha\, dx^\beta, \tag{2.12}$$

where $g_{\alpha\beta}$ is the metric in the non-inertial frame and is given by

$$g_{\alpha\beta} = \frac{\partial \bar{x}^\mu}{\partial x^\alpha}\frac{\partial \bar{x}^\nu}{\partial x^\beta}\eta_{\mu\nu}. \tag{2.13}$$

We can introduce an inverse metric $g^{\beta\gamma}$ defined by

$$g_{\alpha\gamma}g^{\gamma\beta} = \delta^\beta_\alpha. \tag{2.14}$$

Let us carry out the differentiation of the metric given by Eq. (2.13). We find

$$g_{\alpha\beta,\gamma} = \frac{\partial^2 \bar{x}^\mu}{\partial x^\gamma \partial x^\alpha}\frac{\partial \bar{x}^\nu}{\partial x^\beta}\eta_{\mu\nu} + \frac{\partial \bar{x}^\mu}{\partial x^\alpha}\frac{\partial^2 \bar{x}^\nu}{\partial x^\gamma \partial x^\beta}\eta_{\mu\nu}. \tag{2.15}$$

Next let us introduce the notation

$$\Gamma^\delta_{\gamma\alpha} = \frac{\partial x^\delta}{\partial \bar{x}^\mu}\frac{\partial^2 \bar{x}^\mu}{\partial x^\gamma \partial x^\alpha}, \tag{2.16}$$

which allows us to write

$$\frac{\partial^2 \bar{x}^\mu}{\partial x^\gamma \partial x^\alpha} = \Gamma^\delta_{\gamma\alpha}\frac{\partial \bar{x}^\mu}{\partial x^\delta}. \tag{2.17}$$

Using relations of the type in Eq. (2.17), we have

$$g_{\alpha\beta,\gamma} = \Gamma^\delta_{\gamma\alpha}\frac{\partial \bar{x}^\mu}{\partial x^\delta}\frac{\partial \bar{x}^\nu}{\partial x^\beta}\eta_{\mu\nu} + \Gamma^\delta_{\gamma\beta}\frac{\partial \bar{x}^\mu}{\partial x^\alpha}\frac{\partial \bar{x}^\nu}{\partial x^\delta}\eta_{\mu\nu}$$
$$= \Gamma^\delta_{\gamma\alpha}g_{\delta\beta} + \Gamma^\delta_{\gamma\beta}g_{\alpha\delta}. \tag{2.18}$$

By the interchange of α and γ and then by the interchange of β and γ in Eq. (2.18) we get three equations which can be used to solve for $\Gamma^\gamma_{\alpha\beta}$ using Eq. (2.14). Thus, the interchange of α and γ and β and γ in Eq. (2.18) gives us the following two equations:

$$g_{\gamma\beta,\alpha} = \Gamma^\delta_{\alpha\gamma}g_{\delta\beta} + \Gamma^\delta_{\alpha\beta}g_{\gamma\delta} \tag{2.19}$$

$$g_{\alpha\gamma,\beta} = \Gamma^\delta_{\beta\alpha}g_{\delta\gamma} + \Gamma^\delta_{\beta\gamma}g_{\alpha\delta}. \tag{2.20}$$

Adding Eqs. (2.18) and (2.19) and subtracting Eq. (2.20) gives us the following result:

$$g_{\alpha\beta,\gamma} + g_{\gamma\beta,\alpha} - g_{\alpha\gamma,\beta} = \Gamma^{\delta}_{\gamma\alpha} g_{\delta\beta} + \Gamma^{\delta}_{\alpha\gamma} g_{\delta\beta}. \tag{2.21}$$

Using the symmetry property on the lower indices of $\Gamma^{\delta}_{\alpha\gamma}$ and the relation Eq. (2.14), we can solve for the affinity:

$$\Gamma^{\gamma}_{\alpha\beta} = \frac{1}{2} g^{\gamma\delta} \left(g_{\alpha\delta,\beta} + g_{\delta\beta,\alpha} - g_{\alpha\beta,\delta} \right). \tag{2.22}$$

The object $\Gamma^{\gamma}_{\alpha\beta}$ is often referred to as the Christoffel symbol.

Tensors are useful concepts for formulating laws of physics in inertial frames. They also prove to be useful in non-inertial frames. Thus, we define tensors under the general co-ordinate transformations as follows. A field $\phi(x)$ remains unchanged under general coordinate transformations as we go from a frame S to a frame S' so that $\phi'(x') = \phi(x)$, so the value of the field is numerically the same in frames S and S' at a given point, which is x in frame S and x' in frame S'. We define a contravariant vector $V^{\mu}(x)$ and a covariant vector $U_{\mu}(x)$ by the following transformation properties:

$$V'^{\mu}(x') = \frac{\partial x'^{\mu}}{\partial x^{\nu}} V^{\nu} \tag{2.23}$$

$$U'_{\mu}(x') = \frac{\partial x^{\nu}}{\partial x'^{\mu}} U_{\nu}. \tag{2.24}$$

More generally, one may have tensors with both upper and lower indices, and each upper and lower index will appropriately transform according to Eq. (2.23) and Eq. (2.24). Thus, a tensor T^{α}_{β} with one upper and one lower index will transform as follows:

$$T'^{\alpha}_{\beta}(x') = \frac{\partial x'^{\alpha}}{\partial x^{\gamma}} \frac{\partial x^{\delta}}{\partial x'^{\beta}} T^{\gamma}_{\delta}(x). \tag{2.25}$$

The above result can be extended straightforwardly to any number of upper and lower indices. It is easily seen that the metric $g_{\alpha\beta}$ transforms like a tensor with two covariant indices, i.e.,

$$g'_{\alpha\beta}(x') = g_{\gamma\delta}(x) \frac{\partial x^{\gamma}}{\partial x'^{\alpha}} \frac{\partial x^{\delta}}{\partial x'^{\beta}}, \tag{2.26}$$

while the tensor δ^{α}_{β} is invariant, i.e., $\delta'^{\alpha}_{\beta} = \delta^{\alpha}_{\beta}$. In addition to the tensors, for the case of general co-ordinate transformations, one also has tensor densities. Thus, consider the determinant of the metric tensor $g \equiv -det(g_{\alpha\beta})$. From the transformation property of $g_{\alpha\beta}$ given in Eq. (2.26), it is seen that

$$g'(x') = \left| \frac{\partial x}{\partial x'} \right|^2 g(x). \tag{2.27}$$

Since

$$\frac{\partial x'^{\alpha}}{\partial x^{\gamma}} \frac{\partial x^{\gamma}}{\partial x'^{\beta}} = \delta_{\beta}^{\alpha}, \tag{2.28}$$

one has $|\partial x'/\partial x||\partial x/\partial x'| = 1$, which implies that

$$g'(x') = \left|\frac{\partial x'}{\partial x}\right|^{-2} g(x), \tag{2.29}$$

where the quantity $|\partial x'/\partial x|$ is the Jacobian of the transformation. Because of the presence of the Jacobian, $g(x)$ in not a scalar but rather what one calls a scalar density, and it is, in fact, a scalar density of weight -2. More generally, one defines a tensor density $\mathbb{T}^{\alpha\cdots}_{\beta\cdots}$ of weight n such that

$$\mathbb{T}'^{\alpha\cdots}_{\beta\cdots} = \left|\frac{\partial x'}{\partial x}\right|^{n} \frac{\partial x'^{\alpha}}{\partial x^{\gamma}} \frac{\partial x^{\delta}}{\partial x'^{\beta}} \cdots \mathbb{T}^{\gamma\cdots}_{\delta\cdots}. \tag{2.30}$$

The tensor density arising from the determinant of the metric is useful because it allows one to write a volume element $d^4x\sqrt{g}$ which is invariant under the general co-ordinate transformations. Thus, the transformation

$$d^4x' = \left|\frac{\partial x'}{\partial x}\right| d^4x, \tag{2.31}$$

along with Eq. (2.27) gives

$$\sqrt{g'(x')}d^4x' = \sqrt{g(x)}d^4x. \tag{2.32}$$

It is also straightforward to show that the four-dimensional Levi–Civita symbol $\epsilon^{\alpha\beta\gamma\delta}$, with $\epsilon^{0123} = 1$, is a tensor density. To show this, let us write

$$J(x,x')\epsilon'^{\alpha\beta\gamma\delta} = \frac{\partial x'^{\alpha}}{\partial x^{\mu}} \frac{\partial x'^{\beta}}{\partial x^{\nu}} \frac{\partial x'^{\gamma}}{\partial x^{\lambda}} \frac{\partial x'^{\delta}}{\partial x^{\rho}} \epsilon^{\mu\nu\lambda\rho}. \tag{2.33}$$

Setting $\alpha = 0$, $\beta = 1$, $\gamma = 2$, and $\delta = 3$, we find Eq. (2.33) takes the form

$$J(x,x') = \frac{\partial x'^{0}}{\partial x^{\mu}} \frac{\partial x'^{1}}{\partial x^{\nu}} \frac{\partial x'^{2}}{\partial x^{\lambda}} \frac{\partial x'^{3}}{\partial x^{\rho}} \epsilon^{\mu\nu\lambda\rho} \tag{2.34}$$

$$= \left|\frac{\partial x'}{\partial x}\right|. \tag{2.35}$$

Thus, we have the transformation

$$\epsilon'^{\alpha\beta\gamma\delta} = \left|\frac{\partial x'}{\partial x}\right|^{-1} \frac{\partial x'^{\alpha}}{\partial x^{\mu}} \frac{\partial x'^{\beta}}{\partial x^{\nu}} \frac{\partial x'^{\gamma}}{\partial x^{\lambda}} \frac{\partial x'^{\delta}}{\partial x^{\rho}} \epsilon^{\mu\nu\lambda\rho}, \tag{2.36}$$

which tells us that $\epsilon^{\alpha\beta\gamma\delta}$ is a tensor density of weight -1, and consequently $g^{-\frac{1}{2}}\epsilon^{\alpha\beta\gamma\delta}$ is a tensor. We may also define $\epsilon_{\mu\nu\lambda\rho}$ so that

$$\epsilon_{\mu\nu\lambda\rho} = g_{\mu\alpha}g_{\nu\beta}g_{\lambda\gamma}g_{\rho\delta}\epsilon^{\alpha\beta\gamma\delta}. \tag{2.37}$$

We now discuss the transformation properties of the Christoffel symbol. Equation (2.16) defines the Christoffel symbol in some frame S, and we can similarly define it in another frame S' by the relation

$$\Gamma'^{\gamma}_{\alpha\beta} = \frac{\partial x'^{\gamma}}{\partial \bar{x}^{\delta}} \frac{\partial^2 \bar{x}^{\delta}}{\partial x'^{\alpha} \partial x'^{\beta}}. \tag{2.38}$$

Since $x' = x'(x)$, we can use the chain rule of differentiation to obtain a relation between the affinity in the primed and in the unprimed frames. Thus, we write

$$\begin{aligned}
\Gamma'^{\gamma}_{\alpha\beta}(x') &= \frac{\partial x'^{\gamma}}{\partial \bar{x}^{\delta}} \frac{\partial^2 \bar{x}^{\delta}}{\partial x'^{\alpha} \partial x'^{\beta}} \\
&= \frac{\partial x'^{\gamma}}{\partial x^{\mu}} \frac{\partial x^{\mu}}{\partial \bar{x}^{\delta}} \frac{\partial}{\partial x'^{\alpha}} \left[\frac{\partial x^{\nu}}{\partial x'^{\beta}} \frac{\partial \bar{x}^{\delta}}{\partial x^{\nu}} \right] \\
&= \frac{\partial x'^{\gamma}}{\partial x^{\mu}} \frac{\partial x^{\mu}}{\partial \bar{x}^{\delta}} \left[\frac{\partial^2 x^{\nu}}{\partial x'^{\alpha} \partial x'^{\beta}} \frac{\partial \bar{x}^{\delta}}{\partial x^{\nu}} + \frac{\partial x^{\nu}}{\partial x'^{\beta}} \frac{\partial^2 \bar{x}^{\delta}}{\partial x'^{\alpha} \partial x^{\nu}} \right]. \tag{2.39}
\end{aligned}$$

Next, we write the last term in Eq. (2.39) so that

$$\frac{\partial x^{\nu}}{\partial x'^{\beta}} \frac{\partial^2 \bar{x}^{\delta}}{\partial x'^{\alpha} \partial x^{\nu}} = \frac{\partial x^{\nu}}{\partial x'^{\beta}} \frac{\partial x^{\lambda}}{\partial x'^{\alpha}} \frac{\partial^2 \bar{x}^{\delta}}{\partial x^{\lambda} \partial x^{\nu}}. \tag{2.40}$$

Insertion of Eq. (2.40) into Eq. (2.39) allows us to write Eq. (2.39) in the form

$$\Gamma'^{\gamma}_{\alpha\beta}(x') = \frac{\partial x'^{\gamma}}{\partial x^{\mu}} \frac{\partial^2 x^{\mu}}{\partial x'^{\alpha} \partial x'^{\beta}} + \frac{\partial x'^{\gamma}}{\partial x^{\lambda}} \frac{\partial x^{\mu}}{\partial x'^{\alpha}} \frac{\partial x^{\nu}}{\partial x'^{\beta}} \Gamma^{\lambda}_{\mu\nu}. \tag{2.41}$$

From Eq. (2.41) we find that the Christoffel symbol does not transform like a tensor due to the presence of the first term on the right-hand side of Eq. (2.41).

The Christoffel symbol is needed in defining the covariant differentiation of tensors. Thus, using Eq. (2.23) the differentiation of a contravariant tensor gives

$$\frac{\partial V'^{\alpha}}{\partial x'^{\beta}} = \frac{\partial x^{\delta}}{\partial x'^{\beta}} \frac{\partial^2 x'^{\alpha}}{\partial x^{\delta} \partial x^{\gamma}} V^{\gamma} + \frac{\partial x'^{\alpha}}{\partial x^{\gamma}} \frac{\partial x^{\delta}}{\partial x'^{\beta}} \frac{\partial V^{\gamma}}{\partial x^{\delta}}. \tag{2.42}$$

From Eq. (2.42) we find that the ordinary derivative of a contravariant tensor does not transform like a tensor. However, the combination

$$V^{\alpha}_{;\beta} = \frac{\partial V^{\alpha}}{\partial x^{\beta}} + \Gamma^{\alpha}_{\beta\gamma} V^{\gamma} \tag{2.43}$$

does transform like a tensor, i.e., one has the transformation property

$$V'^{\alpha}_{;\beta} = \frac{\partial x'^{\alpha}}{\partial x^{\mu}} \frac{\partial x^{\nu}}{\partial x'^{\beta}} V'^{\mu}_{;\nu}. \tag{2.44}$$

Thus, while the ordinary derivative of a contravariant tensor by itself is not a tensor, a combination of the ordinary derivative and the Christoffel symbol in the combination of Eq. (2.43) does transform like a tensor, with one upper and one lower index. Similarly, one may define the covariant differentiation of a covariant vector so that

$$U_{\alpha;\beta} \equiv \frac{\partial U_\alpha}{\partial x^\beta} - \Gamma^\gamma_{\alpha\beta} U_\gamma. \tag{2.45}$$

Thus, $U_{\alpha;\beta}$ will transform like a tensor with two lower indices under the general co-ordinate transformation. In general, one can define a covariant differentiation of a tensor with mixed upper and lower indices so that

$$T^{\alpha\cdots}_{\mu\cdots;\rho} = T^{\alpha\cdots}_{\mu\cdots,\rho} + \Gamma^\alpha_{\rho\beta} T^{\beta\cdots}_{\mu\cdots} - \Gamma^\sigma_{\mu\rho} T^{\alpha\cdots}_{\sigma\cdots} + \cdots \tag{2.46}$$

With the above construction, $T^{\alpha\cdots}_{\mu\cdots;\rho}$ correctly transforms like a tensor with mixed upper and lower indices.

It is also instructive to look at the covariant differentiation of the metric tensor, which one can expand in the form

$$g_{\alpha\beta;\gamma} = g_{\alpha\beta,\gamma} - \Gamma^\delta_{\alpha\gamma} g_{\delta\beta} - \Gamma^\delta_{\beta\gamma} g_{\alpha\delta}. \tag{2.47}$$

On substitution of the affinity from Eq. (2.22), one finds that the covariant derivative of the metric tensor vanishes, i.e.,

$$g_{\alpha\beta;\gamma} = 0. \tag{2.48}$$

Now the affinity vanishes in a local inertial frame. Thus, the covariant derivative becomes the ordinary derivative, and one finds

$$g_{\alpha\beta,\gamma} = 0, \qquad \text{local inertial frame.} \tag{2.49}$$

It is also to be noted that the covariant differentiation of δ^α_β vanishes because it vanishes in the inertial frame, so that

$$\delta^\alpha_{\beta;\gamma} = 0. \tag{2.50}$$

Further carrying out a covariant differential of Eq. (2.14) and using Eqs. (2.48) and (2.50) one finds

$$g^{\alpha\beta}_{\;\;;\gamma} = 0. \tag{2.51}$$

Because of Eqs. (2.48) and (2.51), one can raise and lower the index of a tensor inside a covariant differentiation, i.e., one has

$$V^\alpha_{;\beta} = g^{\alpha\gamma} V_{\gamma;\beta}. \tag{2.52}$$

Covariant differentiation of a tensor density can also be defined. Assume we have a tensor density $\mathbb{T}^\alpha$ of weight n and we define T^α so that

$$T^\alpha = g^{n/2} \mathbb{T}^\alpha. \tag{2.53}$$

Since g is a tensor density of weight -2, T^α is an ordinary tensor, i.e., a tensor with zero weight. The covariant derivative of T^α is then given by

$$T^\alpha_{;\beta} = T^\alpha_{,\beta} + \Gamma^\alpha_{\beta\gamma} T^\gamma. \tag{2.54}$$

We define the covariant derivative of a tensor density $\mathbb{T}^{\alpha}_{;\beta}$ so that

$$
\begin{aligned}
\mathbb{T}^{\alpha}_{;\beta} &= g^{-n/2} T^{\alpha}_{;\beta} \\
&= g^{-n/2} \left[T^{\alpha}_{,\beta} + \Gamma^{\alpha}_{\beta\gamma} T^{\gamma} \right] \\
&= \mathbb{T}^{\alpha}_{,\beta} + \Gamma^{\alpha}_{\beta\gamma} \mathbb{T}^{\gamma} + \frac{n}{2} \frac{g_{,\beta}}{g} \mathbb{T}^{\alpha}.
\end{aligned}
\tag{2.55}
$$

The presence of the last term indicates that we are carrying out the covariant differentiation of a tensor density. We can use Eq. (2.55) to obtain the covariant differentiation of g. Here we set $n = -2$ and get

$$
g_{;\alpha} = 0.
\tag{2.56}
$$

We now discuss the covariant divergence of a contravariant vector, which is defined by

$$
V^{\alpha}_{;\alpha} = V^{\alpha}_{,\alpha} + \Gamma^{\beta}_{\beta\alpha} V^{\alpha}.
\tag{2.57}
$$

Here we need to evaluate $\Gamma^{\beta}_{\beta\alpha}$. Using Eq. (2.21) we find

$$
\begin{aligned}
\Gamma^{\beta}_{\beta\alpha} &= \frac{1}{2} g_{\beta\gamma,\alpha} g^{\beta\gamma} \\
&= \frac{1}{2} tr \left[(g)^{-1} \frac{\partial}{\partial x^{\alpha}} (g) \right] \\
&= \frac{\partial}{\partial x^{\alpha}} \log(\sqrt{g}).
\end{aligned}
\tag{2.58}
$$

where $(g) = (g_{\alpha\beta})$ and $g = -det(g)$. Thus, using Eqs. (2.57) and (2.58) we may write

$$
\begin{aligned}
V^{\alpha}_{;\alpha} &= V^{\alpha}_{,\alpha} + \frac{1}{\sqrt{g}} \left(\frac{\partial}{\partial x^{\alpha}} \sqrt{g} \right) V^{\alpha} \\
&= \frac{1}{\sqrt{g}} \frac{\partial}{\partial x^{\alpha}} (\sqrt{g} V^{\alpha}).
\end{aligned}
\tag{2.59}
$$

The above allows one to write the covariant form of Gauss's law so that

$$
\int d^4 x \, \sqrt{g} V^{\alpha}_{;\alpha} = \int d^4 x \frac{\partial}{\partial x^{\alpha}} (\sqrt{g} V^{\alpha}) = 0.
\tag{2.60}
$$

Thus, Eq. (2.60) is the correct generalization of Gauss's law in the presence of gravitation.

We now consider the covariant differentiation of a product of two tensors. Assume we have two tensors U^{α} and V_{β}, i.e, one contravariant and the other covariant. We form a product tensor $U^{\alpha} V_{\beta}$ and consider its covariant differentiation. It is straightforward to show that

$$
(U^{\alpha} V_{\beta})_{;\gamma} = U^{\alpha}_{;\gamma} V_{\beta} + U^{\alpha} V_{\beta;\gamma}.
\tag{2.61}
$$

Thus, the covariant differentiation obeys the Leibniz rule. We use the result of Eq. (2.61) to obtain the covariant differentiation of a scalar. Assume we then form a scalar from the product $U^\alpha V_\alpha$. In this case, using Eq. (2.61), we have

$$
\begin{aligned}
(U^\alpha V_\alpha)_{;\gamma} &= U^\alpha_{;\gamma} V_\alpha + U^\alpha V_{\alpha;\gamma} \\
&= (U^\alpha_{,\gamma} + \Gamma^\alpha_{\gamma\beta} U^\beta) V_\alpha + U^\alpha (V_{\alpha,\gamma} - \Gamma^\beta_{\alpha,\gamma} V_\beta) \\
&= U^\alpha_{,\gamma} V_\alpha + U^\alpha V_{\alpha,\gamma} + \left[\Gamma^\alpha_{\gamma\beta} U^\beta V_\alpha - \Gamma^\alpha_{\beta\gamma} U^\beta V_\alpha \right].
\end{aligned}
\tag{2.62}
$$

The last two terms in Eq. (2.62) vanish due to symmetry in the lower indices of the affinity $\Gamma^\alpha_{\beta\gamma}$, and one finds

$$
(U^\alpha V_\alpha)_{;\gamma} = (U^\alpha V_\alpha)_{,\gamma}.
\tag{2.63}
$$

Since $U^\alpha V_\alpha$ is a scalar, the result of Eq. (2.63) holds for any scalar field $\phi(x)$ so that

$$
\phi_{;\gamma} = \phi_{,\gamma}.
\tag{2.64}
$$

Thus, the covariant differentiation of a scalar field is the same as an ordinary differentiation. We can also define the covariant D'Alembertian of a scalar field. To do so we can first create a contravariant vector $\phi_{,}{}^\gamma$. Next, we take a covariant divergence of it, and so we define

$$
\begin{aligned}
\Box \phi &= (\phi_{,}{}^\gamma)_{;\gamma} \\
&= \frac{1}{\sqrt{g}} \frac{\partial}{\partial x^\gamma} (\sqrt{g}\, g^{\gamma\delta} \phi_{;\delta}) \\
&= \frac{1}{\sqrt{g}} \frac{\partial}{\partial x^\alpha} \left(\sqrt{g}\, g^{\alpha\beta} \frac{\partial \phi}{\partial x^\beta} \right).
\end{aligned}
\tag{2.65}
$$

In general, covariant differentiation does not commute, i.e., $V_{\alpha;\beta;\gamma} \neq V_{\alpha;\gamma;\beta}$. Their difference gives the so-called Riemann–Christoffel tensor. To compute the Riemann tensor, we carry out two covariant differentiations of the vector V_α in two different orderings, generating tensors $V_{\alpha;\beta;\gamma}$ and $V_{\alpha;\gamma;\beta}$. The difference between these two tensors tells us something about the curvature of space. Let us compute the difference explicitly. Let us consider $V_{\alpha;\beta;\gamma}$ and expand the outermost covariant derivative first so that

$$
V_{\alpha;\beta;\gamma} = V_{\alpha;\beta,\gamma} - \Gamma^\delta_{\alpha\gamma} V_{\delta;\beta} - \Gamma^\delta_{\beta\gamma} V_{\alpha;\delta}.
\tag{2.66}
$$

Next, we expand the remaining covariant derivative

$$
V_{\alpha;\beta} = V_{\alpha,\beta} - \Gamma^\epsilon_{\alpha\beta} V_\epsilon,
\tag{2.67}
$$

and similarly expand $V_{\delta;\beta}$ and $V_{\alpha;\delta}$. Substitution back into Eq. (2.66) gives

$$
\begin{aligned}
V_{\alpha;\beta;\gamma} = V_{\alpha,\beta\gamma} &- \Gamma^\epsilon_{\alpha\beta,\gamma} V_\epsilon - \Gamma^\epsilon_{\alpha\beta} V_{\epsilon,\gamma} \\
&- \Gamma^\delta_{\alpha\gamma} V_{\delta,\beta} + \Gamma^\delta_{\alpha\gamma} \Gamma^\epsilon_{\delta\beta} V_\epsilon \\
&- \Gamma^\delta_{\beta\gamma} V_{\alpha,\delta} - \Gamma^\delta_{\beta\gamma} \Gamma^\epsilon_{\alpha\delta} V_\epsilon.
\end{aligned}
\tag{2.68}
$$

Similarly, we compute $V_{\alpha;\gamma;\beta}$, which is obtained by simply interchanging β and γ. Now, we notice that terms in Eq. (2.68) involving single derivatives of V and involving two derivatives are symmetric under β and γ interchange, and they cancel in the difference between $V_{\alpha;\beta;\gamma}$ and $V_{\alpha;\gamma;\beta}$. Thus, we have

$$V_{\alpha;\beta;\gamma} - V_{\alpha;\gamma;\beta} = -V_\delta R^\delta_{\alpha\beta\gamma}, \tag{2.69}$$

where

$$R^\delta_{\alpha\beta\gamma} \equiv \Gamma^\delta_{\alpha\beta,\gamma} - \Gamma^\delta_{\alpha\gamma,\beta} + \Gamma^\epsilon_{\alpha\beta}\Gamma^\delta_{\epsilon\gamma} - \Gamma^\epsilon_{\alpha\gamma}\Gamma^\delta_{\epsilon\beta}. \tag{2.70}$$

Equation (2.70) shows that the Riemann tensor is anti-symmetric in the interchange of the last two indices. A similar result holds for the covariant differentiation of the vector V^α (see Eq. (2.218) in Problem 3). Using Eqs. (2.69) and (2.218), we can write the covariant commutator for the tensor t^α_β so that

$$t^\alpha_{\beta;\gamma;\delta} - t^\alpha_{\beta;\delta;\gamma} = t^\epsilon_\beta R^\alpha_{\epsilon\gamma\delta} - t^\alpha_\epsilon R^\epsilon_{\beta\gamma\delta}. \tag{2.71}$$

We note that the non-commutativity of the covariant differentiation arises due to the Riemann tensor. Obviously the covariant differentiations will commute if the Riemann tensor vanishes.

One can define a Riemann–Christoffel tensor with all lower indices by using the metric tensor, i.e.,

$$R_{\alpha\beta\gamma\delta} = g_{\alpha\epsilon}R^\epsilon_{\beta\gamma\delta}. \tag{2.72}$$

The Riemann–Christoffel tensor satisfies a number of symmetry properties. First, it is unchanged in the interchange of the first two and the last two indices:

$$R_{\alpha\beta\gamma\delta} = R_{\gamma\delta\alpha\beta}. \tag{2.73}$$

Second as noted after Eq. (2.70), the Riemann–Christoffel tensor is anti-symmetric in the last two indices and also because of Eq. (2.73) in the first two indices. Thus, we have

$$R_{\alpha\beta\gamma\delta} = -R_{\alpha\beta\delta\gamma} = -R_{\beta\alpha\gamma\delta}. \tag{2.74}$$

The Riemann–Christoffel tensor also satisfies a cyclicality condition, i.e.,

$$R_{\alpha\beta\gamma\delta} + R_{\alpha\delta\beta\gamma} + R_{\alpha\gamma\delta\beta} = 0. \tag{2.75}$$

The result of Eq. (2.75) follows by a simple expansion of the Riemann–Christoffel tensor in terms of the affinity as given by Eqs. (2.70) and (2.72). It is also seen that the identity

$$\epsilon^{\alpha\beta\gamma\delta}R_{\alpha\beta\gamma\delta} = 0 \tag{2.76}$$

holds. The proof of Eq. (2.76) follows by relabeling the last three indices, using $\epsilon^{\alpha\delta\beta\gamma} = \epsilon^{\alpha\gamma\delta\beta} = \epsilon^{\alpha\beta\gamma\delta}$, and Eq. (2.75). Finally, the Riemann–Christoffel tensor also satisfies a so-called Bianchi identity, which reads

$$R_{\alpha\beta\gamma\delta;\epsilon} + R_{\alpha\beta\epsilon\gamma;\delta} + R_{\alpha\beta\delta\epsilon;\gamma} = 0. \tag{2.77}$$

To establish this relation one can go to the inertial frame where the affinity vanishes. Thus, in the inertial frame

$$R_{\alpha\beta\gamma\delta;\epsilon} = \frac{1}{2}\left(g_{\alpha\gamma,\beta\delta\epsilon} - g_{\beta\gamma,\alpha\delta\epsilon} - g_{\alpha\delta,\beta\gamma\epsilon} + g_{\beta\delta,\alpha\gamma\epsilon}\right). \tag{2.78}$$

Substitution of Eq. (2.78) and its permutations leads directly to the result of Eq. (2.77) in the inertial frame. Since Eq. (2.77) holds in the inertial frame, it holds in general in all frames.

Let us now count the number of independent components of the Riemann–Christoffel tensor. $R_{\mu\nu\alpha\beta}$ is anti-symmetric in the first two indices and also in the last two indices, and a 4×4 matrix which is anti-symmetric has six independent components. Thus, we have a tensor of the form R_{xy} where x and y each take on six values. However, $R_{\mu\nu\alpha\beta}$ is symmetric under the interchange of the first two and the last two indices, which means that R_{xy} is symmetric under x and y interchange, which leaves us with $6 \times (6+1)/2 = 21$ components. However, from this set we must remove the possibility that $R_{\mu\nu\alpha\beta}$ could be totally anti-symmetric in all indices, since $R_{\mu\nu\alpha\beta}$ satisfies the identity Eq. (2.76), which imposes one constraint and thus we have 20 independent components of the Riemann–Christoffel tensor in four dimensions.

2.2 The Ricci Tensor and Curvature Scalar

From the Riemann–Christoffel tensor one can define a two-index tensor (Ricci tensor) by contraction, i.e.,

$$R_{\alpha\beta} = g^{\gamma\delta}R_{\gamma\alpha\delta\beta}. \tag{2.79}$$

Using Eq. (2.70), the explicit form of the Ricci tensor is given by

$$R_{\alpha\beta} = -\Gamma^{\gamma}_{\alpha\gamma,\beta} + \Gamma^{\gamma}_{\alpha\beta,\gamma} - \Gamma^{\gamma}_{\alpha\delta}\Gamma^{\delta}_{\gamma\beta} + \Gamma^{\gamma}_{\alpha\beta}\Gamma^{\delta}_{\gamma\delta}. \tag{2.80}$$

Similarly, one can define a curvature scalar by a further contraction of the Ricci tensor, i.e.,

$$R = g^{\alpha\beta}R_{\alpha\beta}. \tag{2.81}$$

From a contraction of the Bianchi identity of Eq. (2.77), one finds the Bianchi identity involving the Ricci tensor and the curvature scalar, i.e., one has

$$(R^{\alpha\beta} - \frac{1}{2}g^{\alpha\beta}R)_{;\beta} = 0. \tag{2.82}$$

2.3 The Einstein Field Equations

The Einstein field equations that describe gravity and its interactions with matter are given by

$$R^{\alpha\beta} - \frac{1}{2}g^{\alpha\beta}R = -8\pi G_N T^{\alpha\beta}, \tag{2.83}$$

where $T^{\alpha\beta}$ is the symmetric stress tensor and G_N is Newton's constant, as mentioned after Eq. (1.21). We will show shortly that the field equations (Eq. (2.83)), including the second term on the left-hand side with a front factor of $\frac{1}{2}$, arise naturally from an action principle. However, first we note that, as a consequence of the Bianchi identity of Eq. (2.82), Eq. (2.83) implies

$$T^{\alpha\beta}_{\;\;;\beta} = 0. \tag{2.84}$$

Thus, in the presence of gravitation, the quantity that is conserved in not the ordinary derivative of the stress tensor but rather the covariant derivative.

2.4 Variation of the Einstein Action

Next we turn to a derivation of Eq. (2.83) from an action principle, and write the total action including gravity and matter as follows:

$$I = I_E + I_M, \tag{2.85}$$

where I_E is the action for pure gravity and I_M is the action of the matter field and its interaction with gravity. We address the gravity part first. Here, we choose

$$I_E = -\frac{1}{16\pi G_N} \int d^4x \; \sqrt{g}R, \tag{2.86}$$

which is the simplest action one can write involving two derivatives and invariant under the general co-ordinate transformations. In order to vary the Einstein action we need to first write $R = g^{\alpha\beta}R_{\alpha\beta}$, so that

$$\begin{aligned}
\delta(\sqrt{g}R) &= \delta(\sqrt{g}g^{\alpha\beta}R_{\alpha\beta}) \\
&= \delta(\sqrt{g})\, g^{\alpha\beta}\, R_{\alpha\beta} + \sqrt{g}\, \delta g^{\alpha\beta}\, R_{\alpha\beta} + \sqrt{g}\, g^{\alpha\beta}\, \delta R_{\alpha\beta}.
\end{aligned} \tag{2.87}$$

The variation of the first and the second terms on the right-hand side of Eq. (2.87) can be obtained by writing

$$\delta(\sqrt{g}) = \frac{1}{2}\sqrt{g}g^{\alpha\beta}\,\delta g_{\alpha\beta} \tag{2.88}$$

$$\delta g^{\gamma\delta} = -g^{\gamma\alpha}\,\delta g_{\alpha\beta}g^{\beta\delta}, \tag{2.89}$$

which gives

$$\sqrt{g}\,\delta g^{\alpha\beta}R_{\alpha\beta} = -\sqrt{g}\,R^{\alpha\beta}\,\delta g_{\alpha\beta}. \tag{2.90}$$

Using Eq. (2.80), we obtain the variation of $R_{\alpha\beta}$ so that

$$\delta R_{\alpha\beta} = \delta\Gamma^{\gamma}_{\alpha\gamma,\beta} - \delta\Gamma^{\gamma}_{\alpha\beta,\gamma} + \delta\Gamma^{\gamma}_{\alpha\delta}\Gamma^{\delta}_{\gamma\beta} + \Gamma^{\gamma}_{\alpha\delta}\,\delta\Gamma^{\delta}_{\gamma\beta} - \delta\Gamma^{\gamma}_{\alpha\beta}\Gamma^{\delta}_{\gamma\delta} - \Gamma^{\gamma}_{\alpha\beta}\,\delta\Gamma^{\delta}_{\gamma\delta}. \quad (2.91)$$

We now note that

$$\delta\Gamma^{\gamma}_{\alpha\gamma;\beta} = \delta\Gamma^{\gamma}_{\alpha\gamma,\beta} - \Gamma^{\gamma}_{\alpha\beta}\,\delta\Gamma^{\delta}_{\gamma\delta} \quad (2.92)$$

$$\delta\Gamma^{\gamma}_{\alpha\beta;\gamma} = \delta\Gamma^{\gamma}_{\alpha\beta,\gamma} - \Gamma^{\gamma}_{\alpha\delta}\,\delta\Gamma^{\delta}_{\gamma\beta} - \Gamma^{\gamma}_{\beta\delta}\,\delta\Gamma^{\delta}_{\gamma\alpha} + \Gamma^{\delta}_{\gamma\delta}\,\delta\Gamma^{\delta}_{\alpha\beta}. \quad (2.93)$$

Using the above, we write

$$\delta R_{\mu\nu} = (\delta\Gamma^{\gamma}_{\mu\gamma})_{;\nu} - (\delta\Gamma^{\gamma}_{\mu\nu})_{;\gamma}. \quad (2.94)$$

Thus, the third term on the right-hand side of Eq. (2.87) may be written as

$$\sqrt{g}g^{\mu\nu}\,\delta R_{\mu\nu} = \sqrt{g}(g^{\mu\nu}\,\delta\Gamma^{\gamma}_{\mu\gamma})_{;\nu} - \sqrt{g}(g^{\mu\nu}\,\delta\Gamma^{\gamma}_{\mu\nu})_{;\gamma}. \quad (2.95)$$

Using the divergence formula (Eq. (2.59)), one may write Eq. (2.95) as follows:

$$\sqrt{g}g^{\mu\nu}\,\delta R_{\mu\nu} = (\sqrt{g}g^{\mu\nu}\,\delta\Gamma^{\gamma}_{\mu\gamma})_{,\nu} - (\sqrt{g}g^{\mu\nu}\,\delta\Gamma^{\gamma}_{\mu\nu})_{,\gamma}. \quad (2.96)$$

Since the right-hand side of Eq. (2.96) contains ordinary derivatives rather than covariant derivatives, one has

$$\int d^4x\,\sqrt{g}g^{\mu\nu}\,\delta R_{\mu\nu} = 0. \quad (2.97)$$

Consequently, the variation of the Einstein action is given by just the first two terms on the right-hand side of Eq. (2.87). Thus, using Eqs. (2.88) and (2.90), one finds

$$\delta I_E = \frac{1}{16\pi G_N}\int\sqrt{g}(R^{\mu\nu} - \frac{1}{2}g^{\mu\nu}R)\,\delta g_{\mu\nu}. \quad (2.98)$$

While the gravitational part of the action is completely specified by Eq. (2.86), the matter part of the action I_M is much more model dependent. In general, it would depend on our assumptions on the matter content of the theory, i.e., the number of spin 1 gauge fields, the number of quarks and leptons, and the Higgs fields in the theory. Thus, generally one can write the variation of the matter action I_M as follows:

$$\delta I_M = \frac{1}{2}\int d^4x\,\sqrt{g}T^{\mu\nu}\,\delta g_{\mu\nu}. \quad (2.99)$$

Equations (2.98) and (2.99) then give us Eq. (2.83), the field equation.

2.5 The Stress Tensor

As mentioned above, the stress tensor which acts as the source of the Einstein equations is highly model dependent. However, once the matter content is specified, one can calculate the corresponding stress tensor for the theory. We will consider two simple examples: the first is for Einstein gravity coupled to a scalar field ϕ and the second is coupling with the Maxwell field. For the scalar field in the absence of gravity, the free Lagrangian is given by

$$-\frac{1}{2}(\partial_\mu\phi\partial^\mu\phi + m^2\phi^2). \tag{2.100}$$

In the presence of gravity, the above expression assumes the form

$$-\frac{1}{2}\sqrt{g}(g^{\alpha\beta}\partial_\alpha\phi\partial_\beta\phi + m^2\phi^2). \tag{2.101}$$

Using Eq. (2.99), we can write the stress tensor for the scalar field so that

$$T_\phi^{\mu\nu} = \left[\phi^{;\mu}\phi^{;\nu} - \frac{1}{2}g^{\mu\nu}\left(\phi_{;\lambda}\phi^{;\lambda} + m^2\phi^2\right)\right]. \tag{2.102}$$

Next, we consider the case of Einstein gravity coupled to the Maxwell theory, which consists of a massless spin 1 field, described by the field A_μ. In the absence of gravity the Maxwell action is given by

$$\int d^4x \left[-\frac{1}{4}F_{\alpha\beta}F^{\alpha\beta}\right]. \tag{2.103}$$

However, to include the effect of gravity we need to make the action invariant under the general co-ordinate transformations. Thus, first we replace d^4x by $d^4x \sqrt{g}$. Second, we need to pay attention to how the indices on the Maxwell field strength are raised. We note that the field strength

$$F_{\alpha\beta} = \frac{\partial A_\beta}{\partial x^\alpha} - \frac{\partial A_\alpha}{\partial x^\beta} \tag{2.104}$$

has no dependence on the gravitational interaction. However, $F^{\alpha\beta}$ does depend on the gravitational interactions, since the indices are raised by using the inverse metric. Thus, the Maxwell action that exhibits explicitly the dependence on the gravitational interactions is given by

$$I_M = \int d^4x \sqrt{g} \left[-\frac{1}{4}g^{\alpha\gamma}g^{\beta\delta}F_{\alpha\beta}F_{\gamma\delta}\right]. \tag{2.105}$$

The Maxwell stress tensor can now be easily obtained by varying the metric. The variation of the integrand on the right-hand side of Eq. (2.105) gives

$$-\frac{1}{4}\left[\delta\sqrt{g}\, g^{\alpha\gamma}g^{\beta\delta}F_{\alpha\beta}F_{\gamma\delta} + \sqrt{g}\, \delta g^{\alpha\gamma}g^{\beta\delta}F_{\alpha\beta}F_{\gamma\delta} + \sqrt{g}\, g^{\alpha\gamma}\, \delta g^{\beta\delta}F_{\alpha\beta}F_{\gamma\delta}\right]. \tag{2.106}$$

Using Eqs. (2.88) and (2.89), one finds that

$$-\frac{1}{4}\delta\sqrt{g}\, g^{\alpha\gamma}g^{\beta\delta}F_{\alpha\beta}F_{\gamma\delta} = \frac{1}{2}\sqrt{g}[-\frac{1}{4}g^{\mu\nu}F_{\alpha\beta}F^{\alpha\beta}]\delta g_{\mu\nu} \tag{2.107}$$

$$-\frac{1}{4}\sqrt{g}\, \delta g^{\alpha\gamma}g^{\beta\delta}F_{\alpha\beta}F_{\gamma\delta} = \frac{1}{4}\sqrt{g}[F^{\mu\alpha}F_\alpha^\nu]\delta g_{\mu\nu}. \tag{2.108}$$

.

The third term in Eq. (2.106) equals the second term in Eq. (2.106), and thus using Eq. (2.107) and Eq. (2.108) we write

$$\delta I_M = \frac{1}{2} \int d^4x \; \sqrt{g} \left[F^{\mu\gamma} F^\nu_\gamma - \frac{1}{4} g^{\mu\nu} F_{\gamma\delta} F^{\gamma\delta} \right] \delta g_{\mu\nu}. \qquad (2.109)$$

A comparison of Eq. (2.109) with Eq. (2.99) gives us the Maxwell stress tensor, i.e.,

$$T^{\mu\nu}_{\text{Maxwell}} = F^{\mu\alpha} F^\nu_\alpha - \frac{1}{4} g^{\mu\nu} F_{\gamma\delta} F^{\gamma\delta}. \qquad (2.110)$$

Thus, the variation of the Maxwell action with respect to the metric tensor gives correctly the well-known form of the stress tensor for the Maxwell field.

2.6 The Linearized Field Equations of Einstein Gravity

In the presence of weak gravitational fields one may reduce the Einstein equations to a simple form. Thus, we consider Eq. (2.83) and eliminate first the curvature scalar by taking the trace of Eq. (2.83). Here we find

$$R = 8\pi G_N g_{\alpha\beta} T^{\alpha\beta}. \qquad (2.111)$$

Next, we use Eq. (2.111) in Eq. (2.83) and obtain $R^{\alpha\beta}$ in terms of the stress tensor, which gives

$$R^{\alpha\beta} = -8\pi G_N (T^{\alpha\beta} - \frac{1}{2} g^{\alpha\beta} g_{\gamma\delta} T^{\gamma\delta}), \qquad (2.112)$$

where

$$R^{\alpha\beta} = g^{\alpha\gamma} g^{\beta\delta} R_{\gamma\delta}. \qquad (2.113)$$

To obtain the field equations in the linearized form, we expand the metric tensor around the background so that

$$g_{\alpha\beta} = \eta_{\alpha\beta} + h_{\alpha\beta}, \qquad (2.114)$$

where $\eta_{\alpha\beta}$ is the metric in the Minkowski space and $h_{\alpha\beta}$ is the gravitational field, which we assume is weak. The inverse metric $g^{\alpha\beta}$ is in general an infinite-order expansion in the gravitational field, but to the lowest order in the gravitational field

$$g^{\alpha\beta} = \eta^{\alpha\beta} - h^{\alpha\beta} + \cdots . \qquad (2.115)$$

and to linear order in the gravitational field the affinity is

$$\Gamma^\gamma_{\alpha\beta} \simeq \frac{1}{2} \left[h^{\ \gamma}_{\alpha,\beta} + h^{\ \gamma}_{\beta,\alpha} - h_{\alpha\beta,}{}^\gamma \right]. \qquad (2.116)$$

Using the above, $R^{\alpha\beta}$ in the linearized approximation is

$$R^{\alpha\beta} = \frac{1}{2}\left(h^{\alpha\beta}{}^{,\gamma}_{,\gamma} - h^{\alpha\gamma}{}_{,\gamma}{}^{\beta} - h^{\beta\gamma}{}_{,\gamma}{}^{\alpha} + h^{\gamma}_{\gamma,}{}^{\alpha\beta}\right). \tag{2.117}$$

In order to simply it further, we can impose a gauge constraint which is similar to the gauge-fixing constraint. For the linearized theory, a constraint usually used is the so-called de Donder gauge constraint, which is

$$h^{\alpha\beta}{}_{,\beta} = \frac{1}{2}h^{\beta\alpha}_{\beta,}. \tag{2.118}$$

Under the constraint, of Eq. (2.118) the linearized $R^{\alpha\beta}$ is

$$R^{\alpha\beta} = \frac{1}{2}\Box h^{\alpha\beta}, \tag{2.119}$$

and the gravitational field equations in the linearized case take the form

$$\Box h^{\alpha\beta} = -16\pi G_N \left(T^{\alpha\beta} - \frac{1}{2}\eta^{\alpha\beta}\eta_{\gamma\delta}T^{\gamma\delta}\right). \tag{2.120}$$

2.7 Motion of a Particle in a Non-inertial Frame

Let us consider a free particle in an inertial frame. Let the co-ordinates of this particle be denoted by $\bar{x}^\mu$. The equation of motion of this particle is given by

$$\frac{d^2\bar{x}^\mu}{d\tau^2} = 0, \tag{2.121}$$

where $d\tau$ is the proper time defined by Eq. (2.11). We now make a transition to a non-inertial frame where the co-ordinates of the particle are given by x^μ. Here, x^μ is a function of $\bar{x}$, and reversibly $\bar{x}$ can be thought of as a function of x, i.e., $\bar{x} = \bar{x}(x)$. This allows us to write

$$\frac{d\bar{x}^\mu}{d\tau} = \frac{\partial\bar{x}^\mu}{\partial x^\lambda}\frac{dx^\lambda}{d\tau}. \tag{2.122}$$

We differentiate Eq. (2.122) once again and get

$$\frac{d^2\bar{x}^\mu}{d\tau^2} = \frac{\partial^2\bar{x}^\mu}{\partial x^\rho\partial x^\lambda}\frac{dx^\rho}{d\tau}\frac{dx^\lambda}{d\tau} + \frac{\partial\bar{x}^\mu}{\partial x^\lambda}\frac{d^2x^\lambda}{d\tau^2}. \tag{2.123}$$

Next, we multiply Eq. (2.123) by $\partial x^\sigma/\partial\bar{x}^\mu$,

$$\frac{d^2\bar{x}^\mu}{d\tau^2}\frac{\partial x^\sigma}{\partial\bar{x}^\mu} = \frac{\partial^2\bar{x}^\mu}{\partial x^\rho\partial x^\lambda}\frac{\partial x^\sigma}{\partial\bar{x}^\mu}\frac{dx^\rho}{d\tau}\frac{dx^\lambda}{d\tau} + \frac{\partial\bar{x}^\mu}{\partial x^\lambda}\frac{\partial x^\sigma}{\partial\bar{x}^\mu}\frac{d^2x^\lambda}{d\tau^2}$$
$$= \Gamma^\sigma_{\rho\lambda}\frac{dx^\rho}{d\tau}\frac{dx^\lambda}{d\tau} + \frac{d^2x^\sigma}{d\tau^2}, \tag{2.124}$$

where we used the relation $(\partial \bar{x}^\mu / \partial x^\lambda)(\partial x^\sigma / \partial \bar{x}^\mu) = \delta_\chi^\sigma$. Next, using Eq. (2.121) in Eq. (2.124) we find the relation

$$\frac{d^2 x^\sigma}{d\tau^2} + \Gamma_{\rho\lambda}^\sigma \frac{dx^\rho}{d\tau} \frac{dx^\lambda}{d\tau} = 0. \tag{2.125}$$

In order for Eq. (2.125) to be correct in all frames it must be a proper tensor equation. To check that this is indeed the case, we define the objects $T^{\prime\gamma}$ and T^λ so that

$$T^\lambda \equiv \frac{d^2 x^\lambda}{d\tau^2} + \Gamma_{\mu\nu}^\lambda \frac{dx^\mu}{d\tau} \frac{dx^\nu}{d\tau} \tag{2.126}$$

$$T^{\prime\gamma} \equiv \frac{d^2 x^{\prime\gamma}}{d\tau^2} + \Gamma_{\alpha\beta}^{\prime\gamma} \frac{dx^{\prime\alpha}}{d\tau} \frac{dx^{\prime\beta}}{d\tau}. \tag{2.127}$$

We then compute the transformations of the two parts in $T^{\prime\gamma}$. Thus, we have

$$\begin{aligned} \frac{d^2 x^{\prime\gamma}}{d\tau^2} &= \frac{d}{d\tau} \left(\frac{\partial x^{\prime\gamma}}{\partial x^\lambda} \frac{dx^\lambda}{d\tau} \right) \\ &= \frac{\partial^2 x^{\prime\gamma}}{\partial x^\nu \partial x^\lambda} \frac{dx^\nu}{d\tau} \frac{dx^\lambda}{d\tau} + \frac{\partial x^{\prime\gamma}}{\partial x^\lambda} \frac{d^2 x^\lambda}{d\tau^2} \end{aligned} \tag{2.128}$$

and

$$\Gamma_{\alpha\beta}^{\prime\gamma} \frac{dx^{\prime\alpha}}{d\tau} \frac{dx^{\prime\beta}}{d\tau} = \left[\frac{\partial x^{\prime\gamma}}{\partial x^\lambda} \frac{\partial x^\nu}{\partial x^{\prime\alpha}} \frac{\partial x^\nu}{\partial x^{\prime\beta}} \Gamma_{\mu\nu}^\lambda - \frac{\partial^2 x^{\prime\gamma}}{\partial x^\mu \partial x^\nu} \frac{\partial x^\mu}{\partial x^{\prime\alpha}} \frac{\partial x^\nu}{\partial x^{\prime\beta}} \right] \frac{dx^{\prime\alpha}}{d\tau} \frac{dx^{\prime\beta}}{d\tau}. \tag{2.129}$$

From Eqs. (2.126)–(2.129) we find

$$T^{\prime\gamma} = \frac{\partial x^{\prime\gamma}}{\partial x^\lambda} T^\lambda, \tag{2.130}$$

which shows that T^λ given by Eq. (2.126) is correctly a covariant tensor, and if it vanishes in one frame (i.e., in the inertial frame) it vanishes in all frames. The analysis above thus shows that Eq. (2.125) is correctly a tensor equation valid in all frames.

2.8 The Newtonian Limit

Next we demonstrate that the non-relativistic limit of Einstein's equations gives the Newtonian equations for gravitational interactions. To make the analysis concrete, we consider a particle moving along a trajectory $x^\mu(\tau)$. As discussed in the preceding section, the equations of motion of a particle in a gravitational potential are given by

$$\frac{d^2 x^\alpha}{d\tau^2} + \Gamma_{\beta\gamma}^\alpha \frac{dx^\beta}{d\tau} \frac{dx^\gamma}{d\tau} = 0. \tag{2.131}$$

To indicate the presence of gravity we expand $g_{\alpha\beta}$ around the flat metric

$$g_{\alpha\beta} = \eta_{\alpha\beta} + h_{\alpha\beta}. \tag{2.132}$$

For the Newtonian limit we assume $dx^i/d\tau \ll c$, gravitational fields to be weak, i.e., $h_{\mu\nu} \ll 1$, and time independent, i.e., $\partial h_{\mu\nu}/\partial t = 0$. In this approximation $dt/d\tau = 1$, and Eq. (2.131) gives

$$\frac{d^2 x^\alpha}{dt^2} + \Gamma^\alpha_{00} = 0. \tag{2.133}$$

In the same approximation where Eq. (2.133) is valid, one has

$$\Gamma^\alpha_{00} = \frac{1}{2} g^{\alpha\gamma} \left(2 g_{0\gamma,0} - g_{00,\gamma}\right)$$
$$= -\frac{1}{2} g^{\alpha\gamma} g_{00,\gamma}, \tag{2.134}$$

where we used the condition that $g_{0\gamma}$ are time independent. Next we compute Γ^α_{00} to linear order in the fields so that

$$\Gamma^0_{00} \sim -\frac{1}{2} \eta^{0\alpha} g_{00,\alpha} = 0 \tag{2.135}$$

$$\Gamma^i_{00} \sim -\frac{1}{2} \eta^{i\alpha} g_{00,\alpha}$$
$$= -\frac{1}{2} \nabla_i h_{00}(x). \tag{2.136}$$

From the above we obtain the relations

$$\frac{d^2 x^0}{dt^2} = 0 \tag{2.137}$$

$$\frac{d^2 x^i}{dt^2} = -\frac{1}{2} \nabla_i h_{00}(x). \tag{2.138}$$

On integration, the first equation gives $x^0 = at + b$, while the second relation can be compared with the motion of a particle in a Newtonian potential $\phi_N(x)$ which obeys the equation

$$\frac{d^2 x^i}{dt^2} = -\nabla_i \phi_N(x). \tag{2.139}$$

A comparison of Eq. (2.138) and Eq. (2.139) gives

$$\frac{1}{2} h_{00} = \phi_N(x) + \phi_0, \tag{2.140}$$

where ϕ_0 is a constant. The limit $h_{\alpha\beta} \to 0$ and $g_{\alpha\beta} \to \eta_{\alpha\beta}$ gives $\phi_0 = 0$, and thus

$$g_{00} = -1 + 2\phi_N. \tag{2.141}$$

Next we determine ϕ_N as given by the Einstein theory. To do so we first compute the stress tensor of a particle moving along a trajectory, for which the action

$$I_P = \int d\tau \frac{p^\alpha(\tau) p^\beta(\tau)}{2m} \eta_{\alpha\beta}. \tag{2.142}$$

Here, as before, τ is the proper time, and the four momentum $p^\alpha(\tau)$ is given by $p^\alpha(\tau) = m \, dx^\alpha(\tau)/d\tau$. When we switch on a gravitational field the particle moving along the trajectory interacts with this field, and Eq. (2.142) is modified to read

$$I_P = \frac{m}{2} \int d\tau \frac{dx^\alpha(\tau)}{d\tau} \frac{dx^\beta(\tau)}{d\tau} g_{\alpha\beta}(x(\tau)). \tag{2.143}$$

It is convenient to write the above in a form where we can freely vary $g_{\alpha\beta}(x)$. This is easily done by writing I_P so that

$$I_P = \frac{m}{2} \int d^4x \, g_{\alpha\beta}(x) \int d\tau \frac{dx^\alpha(\tau)}{d\tau} \frac{dx^\beta(\tau)}{d\tau} \delta^4(x - x(\tau)). \tag{2.144}$$

From the above we can determine the particle stress tensor $t^{\alpha\beta}$ so that

$$t^{\alpha\beta} = m \int d\tau \frac{dx^\alpha(\tau)}{d\tau} \frac{dx^\beta(\tau)}{d\tau} \delta^4(x - x(\tau)). \tag{2.145}$$

In the non-relativistic limit $dx^i/d\tau \to 0$ ($i = 1, 2, 3$), and thus $t^{ij} \to 0$ and $t^{i0} = t^{0i} \to 0$, and the only non-vanishing element is t^{00}. Then, integrating over $d\tau$ one finds

$$t^{00} = \rho(\vec{r}), \tag{2.146}$$

where $\rho(\vec{r}) == m\delta^3(\vec{r} - \vec{r}_p)$. Taking the non-relativistic limit of Eq. (2.120) and inserting Eq. (2.146), one finds

$$\nabla^2 h^{00} = -(8\pi G_N)\rho, \tag{2.147}$$

which on using Eq. (2.141) gives

$$\nabla^2 \phi_N(x) = -(4\pi G_N)\rho. \tag{2.148}$$

Using Eq. (2.148) we compute the force on an object at the surface of a sphere of radius r containing matter of mass density ρ. We assume that the mass density is spherically symmetric inside the sphere. From Gauss's theorem we have

$$\int d^3x \, \nabla^2 \phi_N = \int_S \vec{\nabla}\phi . \vec{ds} = -4\pi G_N \int d^3x \, \rho = -4\pi G_N M, \tag{2.149}$$

where $M = \int d^3x \rho$. Because of spherical symmetry the force on the surface will be in the radial direction and constant in magnitude on the surface of the sphere, so that

$$\vec{F} = -\hat{r} \frac{G_N M}{r^2}. \tag{2.150}$$

Thus, the force is directed inwards, i.e., it is attractive, and is exactly what one finds in Newtonian gravity.

2.9 Gauge Transformations of the Einstein Metric

So far we have considered the general co-ordinate transformations where $x^\alpha \to x'^\alpha = x'^\alpha(x)$ while the metric tensor also transforms so that $g_{\mu\nu}(x) \to g'_{\mu\nu}(x')$. We now consider a different situation where the metric tensor retains its form, i.e.,

$$g'_{\mu\nu}(x') = g_{\mu\nu}(x'). \tag{2.151}$$

Such a situation occurs when the system has certain symmetries present. To discuss the implications of Eq. (2.151), we consider the difference

$$\delta g_{\mu\nu}(x) = g'_{\mu\nu}(x) - g_{\mu\nu}(x). \tag{2.152}$$

We wish to compute $\delta g_{\mu\nu}(x)$ given by Eq. (2.152) under the co-ordinate transformation

$$x^\alpha = x'^\alpha + \xi^\alpha(x'), \tag{2.153}$$

where ξ^α is infinitesimal. In this case, one can compute $g'_{\alpha\beta}(x')$ using the relation

$$g'_{\alpha\beta}(x') = \frac{\partial x^\gamma}{\partial x'^\alpha} \frac{\partial x^\delta}{\partial x'^\beta} g_{\gamma\delta}(x), \tag{2.154}$$

which up to linear order in ξ^α gives

$$\begin{aligned} g'_{\alpha\beta}(x') &= (\delta^\gamma_\alpha + \xi^\gamma_{,\alpha})(\delta^\delta_\beta + \xi^\delta_{,\beta}) g_{\gamma\delta}(x) \\ &= g_{\alpha\beta}(x) + \xi^\gamma_{,\alpha} g_{\gamma\beta} + \xi^\gamma_{,\beta} g_{\gamma\alpha} + O(\xi^2). \end{aligned} \tag{2.155}$$

Further, up to linear order in ξ^α one also has $g'_{\alpha\beta}(x')$ given by

$$\begin{aligned} g'_{\alpha\beta}(x') &= g'_{\alpha\beta}(x - \xi) \\ &\simeq g'_{\alpha\beta}(x) - g'_{\alpha\beta,\gamma}\xi^\gamma + O(\xi^2). \end{aligned} \tag{2.156}$$

From Eqs. (2.152) and (2.154)–(2.156) we find to lowest order in ξ,

$$\delta g_{\alpha\beta} = \xi^\gamma_{,\alpha} g_{\gamma\beta} + \xi^\gamma_{,\beta} g_{\gamma\alpha} + g_{\alpha\beta,\gamma}\xi^\gamma. \tag{2.157}$$

One may view Eq. (2.157) as the gauge transformations of the Einstein metric. Next, we note that

$$\begin{aligned} \xi_{\alpha,\beta} &= (\xi^\gamma g_{\gamma\alpha})_{,\beta} \\ &= \xi^\gamma_{,\beta} g_{\gamma\alpha} + \xi^\gamma g_{\gamma\alpha,\beta}. \end{aligned} \tag{2.158}$$

From Eq. (2.158), interchanging α and β, we get

$$\xi^\gamma_{,\alpha} g_{\gamma\beta} + \xi^\gamma_{,\beta} g_{\gamma\alpha} = \xi_{\alpha,\beta} + \xi_{\beta,\alpha} - \xi^\gamma(g_{\gamma\alpha,\beta} + g_{\gamma\beta,\alpha}). \tag{2.159}$$

Using Eq. (2.159) in Eq. (2.157) we find

$$\begin{aligned} \delta g_{\alpha\beta} &= \xi_{\alpha,\beta} + \xi_{\beta,\alpha} - (g_{\gamma\alpha,\beta} + g_{\gamma\beta,\alpha} - g_{\alpha\beta,\gamma})\xi^\gamma \\ &= \xi_{\alpha,\beta} + \xi_{\beta,\alpha} - 2\Gamma^\gamma_{\alpha\beta}\xi_\gamma \\ &= \xi_{\alpha;\beta} + \xi_{\beta;\alpha}. \end{aligned} \tag{2.160}$$

The condition for form invariance is Eq. (2.151), which on use of Eq. (2.152) gives the constraint $\delta g_{\alpha\beta} = 0$, and which along with Eq. (2.160) implies

$$\delta g_{\alpha\beta} = \xi_{\alpha;\beta} + \xi_{\beta;\alpha} = 0. \tag{2.161}$$

Equation (2.161) determines the so-called isometries of the metric. The solutions to Eq. (2.161) are often referred to as the Killing vectors. Thus, one might consider the invariance in the inertial frame where $\Gamma^{\alpha}_{\beta\gamma} = 0$. In this case, Eq. (2.161) reduces to the relation

$$\xi_{\alpha,\beta} + \xi_{\beta,\alpha} = 0. \tag{2.162}$$

A solution to Eq. (2.162) gives

$$\xi_{\alpha} = \epsilon_{\alpha\beta} x^{\beta} + \epsilon_{\alpha}, \tag{2.163}$$

where $\epsilon_{\alpha\beta} = -\epsilon_{\beta\alpha}$. The transformations of Eq. (2.163) correspond to the Poincaré transformations, and correctly give the invariance of flat space.

2.10 The Palatini Variational Method

In obtaining equations of motion from the action of Eq. (2.86) we varied the metric tensor $g_{\alpha\beta}$. However, one may treat the metric $g_{\alpha\beta}$ and the connection $\Gamma^{\alpha}_{\beta\gamma}$ to be independent in the Einstein action. This method is known as the Palatini variational method[*]. Thus, here one writes the Einstein action (Eq. (2.86)), in the form

$$I_E = -\frac{1}{16\pi G_N} \int d^4x \, \sqrt{g} g^{\alpha\beta} R_{\alpha\beta}(\Gamma), \tag{2.164}$$

where $R_{\alpha\beta}$ is given in terms of the connection by Eq. (2.80), which we rewrite here for convenience:

$$R_{\alpha\beta} = \Gamma^{\gamma}_{\alpha\gamma,\beta} - \Gamma^{\gamma}_{\alpha\beta,\gamma} + \Gamma^{\gamma}_{\alpha\delta}\Gamma^{\delta}_{\gamma\beta} - \Gamma^{\gamma}_{\alpha\beta}\Gamma^{\delta}_{\gamma\delta}. \tag{2.165}$$

In varying I_E we treat $g_{\alpha\beta}$ and Γ as independent. Doing so, we have

$$\delta I_E = -\frac{1}{16\pi G_N} \int d^4x \left[\delta\sqrt{g} g^{\alpha\beta} R_{\alpha\beta} + \sqrt{g}\delta g^{\alpha\beta} R_{\alpha\beta} + \sqrt{g} g^{\alpha\beta} \delta R_{\alpha\beta} \right]. \tag{2.166}$$

Using Eqs. (2.88) and (2.89) in Eq. (2.166) and noting that $\delta R_{\alpha\beta}$ does not contain any $\delta g^{\alpha\beta}$ terms, one finds that the variation of $\delta g^{\alpha\beta}$ leads to the Einstein equations of motion in a vacuum:

$$R_{\alpha\beta} - \frac{1}{2}g_{\alpha\beta}R = 0. \tag{2.167}$$

---

[*] The Palatini method appears in a paper by Einstein in 1925 [8].

Next, we look at the variation of $\delta R_{\alpha\beta}$. Using Eq. (2.80) we have

$$\delta R_{\alpha\beta} = -\delta\Gamma^{\gamma}_{\alpha\gamma,\beta} + \delta\Gamma^{\gamma}_{\alpha\beta,\gamma} - \delta\Gamma^{\gamma}_{\alpha\delta}\Gamma^{\delta}_{\gamma\beta} + \delta\Gamma^{\gamma}_{\alpha\beta}\Gamma^{\delta}_{\gamma\delta} - \Gamma^{\gamma}_{\alpha\delta}\delta\Gamma^{\delta}_{\gamma\beta} + \Gamma^{\gamma}_{\alpha\beta}\delta\Gamma^{\delta}_{\gamma\delta}.$$
(2.168)

Using the result of Eq. (2.238), the variation of I_E with respect to the connection can be written as

$$\delta I_{\Gamma} = \int d^4x \, \sqrt{g}g^{\alpha\beta}\delta R_{\alpha\beta} = \int d^4x \left(\sqrt{g}g^{\alpha\gamma}\delta\Gamma^{\beta}_{\alpha\beta} - \sqrt{g}g^{\alpha\beta}\delta\Gamma^{\gamma}_{\alpha\beta} \right)_{,\gamma}$$
$$+ \int d^4x \left[(\sqrt{g}g^{\alpha\beta})_{;\gamma} - (\sqrt{g}g^{\alpha\delta})_{;\delta}\delta^{\beta}_{\gamma} \right] \delta\Gamma^{\gamma}_{\alpha\beta}. \quad (2.169)$$

The first term on the right-hand side of Eq. (2.169) vanishes by use of Gauss's theorem if we assume the appropriate boundary conditions on the surface, and for the remainder to be valid for arbitrary variations $\delta\Gamma^{\gamma}_{\alpha\beta}$ one must have the following constraint satisfied:

$$(\sqrt{g}g^{\alpha\beta})_{;\gamma} = (\sqrt{g}g^{\alpha\delta})_{;\delta}\delta^{\beta}_{\gamma}.$$
(2.170)

Setting $\beta = \gamma$ in Eq. (2.170) implies that $(\sqrt{g}g^{\alpha\beta})_{;\beta} = 0$, and its insertion back in Eq. (2.170) gives $(\sqrt{g}g^{\alpha\beta})_{;\gamma} = 0$, from which we can deduce

$$(g^{\alpha\beta})_{;\gamma} = 0.$$
(2.171)

Equation (2.171) implies that the connection is metric compatible, which allows us to determine the connection in terms of the metric, and we can follow the same procedure that leads to the deduction of Eq. (2.22). Conceptually, the Palatini method is more satisfying in that both the connection and the equations of motion are determined from the action principle.

2.11 The Vierbein Formalism

The general co-ordinate transformations have the group structure $GL(4)$, which, however, does not permit spinor representations. So, to introduce spinors in general relativity we have to turn to a local inertial frame. We can then at each point in the manifold define a tangent space which is a flat Minkowskian manifold which allows us to define spinors. The object that takes us from the global co-ordinate frame to the local co-ordinate frame is the vierbein. Thus, one defines the vierbein e^a_{α}, where a refers to local indices and α refers to global indices. Consequently, e^a_{α} has one "leg" in the global co-ordinate frame and the other in the local co-ordinate frame, where it transforms like a global vector according to the index α and like a local vector according to the index a. Further, e^a_{α} is related to the metric, so that

$$g_{\alpha\beta}(x) = e^a_{\alpha}\eta_{ab}e^b_{\beta},$$
(2.172)

where η_{ab} is the metric in the Minkowski space. We can define the inverse vierbein such that

$$e_\alpha^{\ a}(x)e_b^\alpha(x) = \delta_b^a$$
$$e_a^\alpha(x)e_\beta^{\ a}(x) = \delta_\beta^\alpha. \tag{2.173}$$

One may also check the property that

$$e_a^\alpha(x)g_{\alpha\beta}e_b^\beta(x) = \eta_{ab}. \tag{2.174}$$

We recall that the covariant derivative of a covariant global vector $V^\beta(x)$ is defined so that

$$D_\alpha V^\beta = \partial_\alpha V^\beta + \Gamma_{\alpha\gamma}^\beta V^\gamma. \tag{2.175}$$

We define the covariant derivative of a local vector V^a so that

$$D_\alpha V^\beta = e_a^\beta D_\alpha^{(\omega)} V^a, \tag{2.176}$$

where $D_\alpha^{(\omega)} V^a$ is defined as

$$D_\mu^{(\omega)} V^a(x) = \partial_\mu V^a(x) + \omega_\mu^{\ a}{}_b V^b. \tag{2.177}$$

Substitution gives us the following:

$$\begin{aligned} D_\alpha V^\beta &= e_a^\beta D_\alpha^{(\omega)} V^a \\ &= e_a^\beta [\partial_\alpha V^a(x) + \omega_\alpha^{\ a}{}_b V^b] \\ &= e_a^\beta [\partial_\alpha(e_{\ \gamma}^a V^\gamma) + \omega_\alpha^{\ a}{}_b V^b] \\ &= e_a^\beta [e_{\ \gamma}^a \partial_\alpha V^\gamma + (\partial_\alpha e_{\ \gamma}^a) V^\gamma + \omega_\alpha^{\ a}{}_b V^b] \\ &= \partial_\alpha V^\beta + e_a^\beta (\partial_\alpha e_{\ \gamma}^a) V^\gamma + \omega_\alpha^{\ a}{}_b e_{\ \gamma}^b V^\gamma]. \end{aligned} \tag{2.178}$$

Comparison of Eqs. (2.175) and (2.178) gives

$$\Gamma_{\alpha\gamma}^\beta = e_a^\beta (\partial_\alpha e_{\ \gamma}^a) + \omega_\alpha^{\ a}{}_b e_a^\beta e_{\ \gamma}^b. \tag{2.179}$$

Using Eq. (2.179), we can solve for the spin connection $\omega_\alpha^{\ a}{}_b$ so that we have

$$\omega_\alpha^{\ a}{}_b = e_\beta^{\ a} e_b^\gamma \Gamma_{\alpha\gamma}^\beta - e_b^\beta \partial_\alpha e_\beta^{\ a}. \tag{2.180}$$

Since a and b are local indices, the vectors V^a and V_a under Lorentz transformations transform so that

$$V^{a'} = \Lambda_b^{a'} V^b$$
$$V_{a'} = \Lambda_{a'}^b V_b, \tag{2.181}$$

where $\Lambda_a^{a'}$, etc., satisfy the constraint

$$\Lambda_a^{a'} \Lambda_b^{b'} \eta_{a'b'} = \eta_{ab}, \tag{2.182}$$

which leaves the line element invariant under local Lorentz transformations. Under Lorentz transformations the spin connection transforms as

$$\omega_\alpha{}^{a'}{}_{b'} = \Lambda_{b'}^b \Lambda_c^{a'} \omega_\alpha{}^c{}_b - \Lambda_{b'}^b \partial_\alpha \Lambda_b^{a'}. \tag{2.183}$$

Next, we express the Riemann–Christoffel tensor in terms of the spin connection. Thus, define

$$R^\alpha_{\beta\mu\nu} \equiv e_a^\alpha e_\beta^b R^a_{b\mu\nu}. \tag{2.184}$$

We can compute $R^a_{b\mu\nu}$ in terms of the spin connection by eliminating $\Gamma^\beta_{\alpha\gamma}$, etc., in terms of $\omega_\alpha{}^a{}_b$ using Eq. (2.179). The analysis gives

$$R^a_{b\mu\nu} = \partial_\mu \omega_\nu{}^a{}_b - \partial_\nu \omega_\mu{}^a{}_b + \omega_\mu{}^a{}_c \omega_\nu{}^c{}_b - \omega_\nu{}^a{}_c \omega_\mu{}^c{}_b. \tag{2.185}$$

We now discuss a Dirac fermion in the presence of a gravitational field. We note that in flat space-time one has the Dirac Lagrangian

$$I_D = -\int d^4x \bar\psi(x) \left(\frac{1}{i}\gamma^a \partial_a + m\right)\psi(x), \tag{2.186}$$

where γ^a satisfy the Clifford algebra so that

$$\gamma^a \gamma^b + \gamma^b \gamma^a = -2\eta^{ab}. \tag{2.187}$$

The Lagrangian of Eq. (2.186) is invariant under Lorentz transformations so that

$$\psi \to U\psi, \quad U = e^{-\frac{i}{4}\epsilon_{ab}\sigma_{ab}}. \tag{2.188}$$

where ϵ_{ab} are anti-symmetric under the interchange of a and b and σ_{ab} is defined so that

$$\sigma_{ab} = \frac{i}{2}[\gamma_a, \gamma_b] \tag{2.189}$$

and σ_{ab} satisfy the relation

$$[\sigma_{ab}, \sigma_{cd}] = 2i(\eta_{ac}\sigma_{bd} - \eta_{ad}\sigma_{bc} + \eta_{bd}\sigma_{ac} - \eta_{bc}\sigma_{ad}). \tag{2.190}$$

U is not unitary but instead obeys the property $\gamma^0 U^\dagger \gamma^0 = U^{-1}$. This property guarantees the Lorentz invariance of the Dirac action.

In order to obtain the Dirac equation in the curved space-time, we need to replace $\gamma^a \partial_a$ by their corresponding global counterparts. Thus, ∂_a is to be replaced by ∂_μ while γ^a is to be replaced by γ^μ, where γ^μ is defined by

$$\gamma^\mu = e_a^\mu \gamma^a. \tag{2.191}$$

It is easily checked that γ^μ satisfy the algebra

$$\gamma^\mu \gamma^\nu + \gamma^\nu \gamma^\mu = -2g^{\mu\nu}, \tag{2.192}$$

where

$$e_a^\mu e_b^\nu \eta^{ab} = g^{\mu\nu}. \tag{2.193}$$

To include the effect of gravitation we replace the ∂_μ in the Dirac equation by the covariant derivative D_μ, where

$$D_\mu \psi = (\partial_\mu - \frac{i}{2}\omega_\mu^{ab} s_{ab})\psi, \tag{2.194}$$

where $s_{ab} = \sigma_{ab}/2$. The full action of the Dirac field in curved space-time is given by

$$I_D = -\int d^4x \ \sqrt{g}(\frac{1}{i}\bar\psi\gamma^\mu D_\mu\psi + m\bar\psi\psi). \tag{2.195}$$

The action of Eq. (2.195) is invariant under general co-ordinate transformations. The action is also invariant under local Lorentz transformations. From Eq. (2.195) we directly obtain the Dirac equation in curved space-time, i.e.,

$$(\frac{1}{i}\gamma^\mu D_\mu + m)\psi = 0. \tag{2.196}$$

A more symmetrized way of writing the action is

$$I_D = -\int d^4x \ \sqrt{g}\left[\frac{1}{2i}g^{\mu\nu}(\bar\psi\gamma_\mu D_\nu\psi - D_\nu\bar\psi\gamma_\mu\psi) + m\bar\psi\psi\right], \tag{2.197}$$

where $D_\nu\bar\psi$ is given by

$$D_\nu\bar\psi = \partial_\nu\bar\psi + \frac{i}{2}\bar\psi\omega_\nu^{ab} s_{ab}, \tag{2.198}$$

where s_{ab} is defined after Eq. (2.194). We can also use variations of the vierbein in the action principle. Thus, in Eq. (2.98) we need to replace variation of the metric with variation of the vierbein. We then have

$$\delta g_{\alpha\beta} = \delta[e_\alpha{}^a e_\beta{}^b \eta_{ab}]$$
$$= (\delta e_\alpha{}^a e_\beta{}^b + \delta e_\alpha{}^a e_\beta{}^b)\eta_{ab}. \tag{2.199}$$

Next, using

$$e^{\alpha a} e_{\beta a} = \delta_\beta^\alpha \tag{2.200}$$

and setting the variation of the left-hand side of Eq. (2.200) to zero, we find that Eq. (2.199) can be written in the form

$$\delta g_{\alpha\beta} = -(g_{\alpha\gamma}e_\beta{}^a + g_{\beta\gamma}e_\alpha{}^a)\delta e_a^\gamma. \tag{2.201}$$

Substitution of Eq. (2.201) in Eq. (2.98) allows us to write Eq. (2.98) in the following form:

$$\delta I_E = -\frac{1}{16\pi G_N}\int \sqrt{g}(R_\beta^\alpha - \frac{1}{2}\delta_\beta^\alpha R)e_\alpha{}^a \ \delta e_a^\beta. \tag{2.202}$$

Setting $\delta I_E/\delta e_a^\beta = 0$ again leads us to Einstein equations in a vacuum.

2.12 Modifications of the Einstein Theory

Several possible modifications of the Einstein theory have been discussed. One possibility is to add a so-called cosmological constant to the field equations of Eq. (2.83), so that one has

$$R^{\alpha\beta} - \frac{1}{2}g^{\alpha\beta} - \Lambda g^{\alpha\beta} = -8\pi G_N T^{\alpha\beta}. \qquad (2.203)$$

This was a modification suggested by Einstein himself but later retracted. However, recent experimental observations indicate the existence of dark energy in the universe. Although the exact nature of this dark energy is unknown, it can be represented effectively by a Λ term as given by Eq. (2.203). Further, if one gives up the constraint of at most two derivatives, one may add higher-order corrections. Thus, one may include corrections to the Einstein action by adding terms of the type

$$\Delta I_E = \int dx \, \sqrt{g} \left(\alpha R^2 + \beta R_{\alpha\beta} R^{\alpha\beta} + \gamma R_{\alpha\beta\gamma\delta} R^{\alpha\beta\gamma\delta} + \cdots \right). \qquad (2.204)$$

Terms of this type do arise in effective gravitational corrections to Einstein gravity in string theory, where they appear in a so-called Gauss–Bonnet combination given by Eq. (2.205). For $d = 4$, this combination is a total derivative and does not contribute to the dynamics. However, for $d > 4$ it is dynamically relevant:

$$L_{GB} = R^2 - 4R_{\alpha\beta} R^{\alpha\beta} + R_{\alpha\beta\gamma\delta} R^{\alpha\beta\gamma\delta}. \qquad (2.205)$$

2.13 Parallel Transport

Suppose one has an object that is defined along a certain curve rather than over the entire manifold. Consider, for example, a particle moving along a trajectory $x^\alpha(\tau)$. The particle would carry certain properties with it such as energy–momentum. In this case, one would need to define differentiation along a curve. To be more concrete, consider two adjacent points along the curve x^α and $x^\alpha + dx^\alpha$. Let us also consider the transformation properties of a contravariant vector V^α defined along the trajectory so that at the adjacent points $A = x^\alpha$ and $B = x^\alpha + dx^\alpha$ the vectors are $V^\alpha(A)$ and $V^\alpha(B)$, respectively. Now, in general, the difference between them, i.e., $V^\alpha(B) - V^\alpha(A)$, is not a vector. Let us now try to generate a vector $\bar{V}^\alpha(B)$ by starting from the point x and transporting V^α to the point $x + dx$ along the trajectory. In this case, we expect the difference $\bar{V}^\alpha(x + dx) - V^\alpha(x)$ will be proportional to dx^β and V^γ so that we can write $[\bar{V}^\alpha(x + dx) - V^\alpha(x)] \propto dx^\beta V^\gamma$. To balance the indices we introduce a three-index quantity $G^\alpha_{\beta\gamma}$, and write

$$\bar{V}^\alpha(x + dx) = V^\alpha(x) - G^\alpha_{\beta\gamma}(x)V^\beta(x)dx^\gamma. \qquad (2.206)$$

Thus, at the point $x + dx$ we now have two vectors $V^\alpha(x + dx)$ and $\bar{V}^\alpha(x + dx)$, and we expect their difference $\bar{\Delta}V^\alpha = V^\alpha(x + dx) - \bar{V}^\alpha(x + dx)$ to be a vector. This gives

$$\begin{aligned}\bar{\Delta}V^\alpha &= V^\alpha(x + dx) - \bar{V}^\alpha(x + dx) \\ &= \left[V^\alpha(x)_{,\beta} + G^\alpha_{\gamma\beta}(x)V^\gamma(x) \right] dx^\beta.\end{aligned} \quad (2.207)$$

Eq. (2.207) implies that the quantity

$$D_\beta V^\alpha(x) = V^\alpha(x)_{,\beta} + G^\alpha_{\gamma\beta}(x)V^\gamma(x) \quad (2.208)$$

transforms like a second-rank tensor with one upper and one lower index. A comparison of Eq. (2.208) with Eq. (2.43) allows us to identify $D_\beta V^\alpha$ with $V^\alpha_{;\beta}$ and $G^\alpha_{\beta\gamma}$ with $\Gamma^\alpha_{\beta\gamma}$.

It is also straightforward to see how the Riemann tensor arises from the point of view of parallel transport. Assume we have a vector $V^\alpha(x)$ which we parallel transport to the point $x_1 = x + \delta_1 x$ and then parallel transport to the point $x_{12} = x_1 + \delta_2 x$ (Fig. 2.1). We label these points $A = x$, $B = x + \delta_1 x$, and $D = x + \delta_1 x + \delta_2 x_2$. However, we can also approach the point D starting from A by first going to the point $C = x_2 = x + \delta_2 x$ and then by parallel transport to the point $D = x_2 + \delta_1 x = x + \delta_2 x + \delta_1 x$. Let us work out the form of the transported vector by these two different paths. First we consider the path $A \to B \to D$. In this case we have

$$\begin{aligned}\bar{V}^\alpha(x + \delta_1 x + \delta_2 x) &= \bar{V}^\alpha(x_1 + \delta_2 x) \\ &= \bar{V}^\alpha(x_1) - \Gamma^\alpha_{\beta\delta}(x_1)\bar{V}^\beta(x_1)\delta_2 x^\delta\end{aligned} \quad (2.209)$$

$$\begin{aligned}\bar{V}^\alpha(x_1) &= V^\alpha(x) - \Gamma^\alpha_{\beta\gamma}V^\beta(x)\delta_1 x^\gamma \\ \Gamma^\alpha_{\beta\delta}(x_1) &= \Gamma^\alpha_{\beta\delta}(x) + \Gamma^\alpha_{\beta\delta,\gamma}(x)\delta_1 x^\gamma \\ \bar{V}^\beta(x_1) &= V^\beta(x) - \Gamma^\beta_{\epsilon\gamma}\delta_1 x^\gamma\, V^\epsilon(x).\end{aligned} \quad (2.210)$$

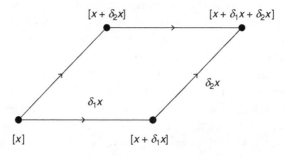

Figure 2.1. Parallel transport of a vector from x to $x + \delta_1 x + \delta_2 x$ along two different routes: counterclockwise, $x \to x + \delta_1 x \to x + \delta_1 x + \delta_2 x$, and clockwise, $x \to x + \delta_2 x \to x + \delta_2 x + \delta_1 x$.

Using Eq. (2.210) in Eq. (2.209), we find, keeping terms only up to quadratic order in the infinitesimals,

$$\bar{V}^\alpha(x + \delta_1 x + \delta_2 x) = V^\alpha(x) - \Gamma^\alpha_{\beta\gamma}V^\beta(x)\delta_1 x^\gamma - \Gamma^\alpha_{\beta\delta}(x)V^\beta(x)\delta_2 x^\delta$$
$$- \Gamma^\alpha_{\epsilon\delta,\gamma}V^\epsilon(x)\delta_1 x^\gamma\delta_2 x^\delta + \Gamma^\alpha_{\beta\delta}\Gamma^\beta_{\epsilon\gamma}V^\epsilon\delta_1 x^\gamma\delta_2 x^\delta. \quad (2.211)$$

Next, we carry out an identical analysis but for the path $A \to C \to D$. In this case, we have $x \to x + \delta_2 x \to x + \delta_2 x + \delta_1 x$. Thus, the roles of $\delta_1 x$ and $\delta_2 x$ are reversed. Making this change we have

$$\bar{V}^\alpha(x + \delta_2 x + \delta_1 x) = V^\alpha(x) - \Gamma^\alpha_{\beta\gamma}V^\beta(x)\delta_2 x^\gamma - \Gamma^\alpha_{\beta\delta}(x)V^\beta(x)\delta_1 x^\delta$$
$$- \Gamma^\alpha_{\epsilon\delta,\gamma}V^\epsilon(x)\delta_2 x^\gamma\delta_1 x^\delta + \Gamma^\alpha_{\beta\delta}\Gamma^\beta_{\epsilon\gamma}V^\epsilon\delta_2 x^\gamma\delta_1 x^\delta. \quad (2.212)$$

We now compute the quantity

$$\Delta V^\alpha \equiv \bar{V}^\alpha(x + \delta_1 x + \delta_2 x) - \bar{V}^\alpha(x + \delta_2 x + \delta_1 x). \quad (2.213)$$

Using Eqs. (2.211) and (2.212) in Eq. (2.213), we find that the terms symmetric under $\delta_1 x$ and $\delta_2 x$ interchange vanish, and rearranging indices we get

$$\Delta V^\alpha = [\Gamma^\alpha_{\epsilon\gamma,\delta} - \Gamma^\alpha_{\epsilon\delta,\gamma} + \Gamma^\alpha_{\beta\delta}\Gamma^\beta_{\epsilon\gamma} - \Gamma^\alpha_{\beta\gamma}\Gamma^\beta_{\epsilon\delta}]V^\epsilon\delta_1 x^\gamma\delta_2 x^\delta$$
$$= R^\alpha_{\epsilon\gamma\delta}V^\epsilon\delta_1 x^\gamma\delta_2 x^\delta, \quad (2.214)$$

where relabeling indices gives $R^\alpha_{\beta\gamma\delta}$ as

$$R^\alpha_{\beta\gamma\delta} = \Gamma^\alpha_{\beta\gamma,\delta} - \Gamma^\alpha_{\beta\delta,\gamma} + \Gamma^\alpha_{\epsilon\delta}\Gamma^\epsilon_{\beta\gamma} - \Gamma^\alpha_{\epsilon\gamma}\Gamma^\epsilon_{\beta\delta}. \quad (2.215)$$

Parallel transport is also useful in the analysis of Yang–Mills gauge theories, which are discussed in Chapter 3.

2.14 Problems

1. Show that for an anti-symmetric field $A_{\mu\nu}$ one has

$$A_{\mu\nu;\lambda} + A_{\lambda\mu;\nu} + A_{\nu\lambda;\mu} = A_{\mu\nu,\lambda} + A_{\lambda\mu,\nu} + A_{\nu\lambda,\mu}. \quad (2.216)$$

2. Show for an anti-symmetric tensor $A^{\alpha\beta}$ that

$$A^{\alpha\beta}_{;\beta} = \frac{1}{\sqrt{g}}\frac{\partial}{\partial x^\beta}(\sqrt{g}A^{\alpha\beta}), \quad (2.217)$$

where, as usual, $;\beta$ denotes a covariant derivative under the general coordinate transformations.

3. Show that, analogous to Eq. (2.69), the covariant differentiation of the contravariant vector V^α gives

$$V^\alpha_{;\beta;\gamma} - V^\alpha_{;\gamma;\beta} = V^\delta R^\alpha_{\delta\beta\gamma}. \quad (2.218)$$

4. Show that an alternative form for the transformation of the Christoffel symbol can be obtained by differentiating the relation

$$\frac{\partial x'^{\gamma}}{\partial x^{\mu}} \frac{\partial x^{\mu}}{\partial x'^{\beta}} = \delta^{\gamma}_{\beta}, \tag{2.219}$$

which reads

$$\Gamma'^{\gamma}_{\alpha\beta}(x') = -\frac{\partial^2 x'^{\gamma}}{\partial x^{\mu}\partial x^{\nu}} \frac{\partial x^{\mu}}{\partial x'^{\alpha}} \frac{\partial x^{\nu}}{\partial x'^{\beta}} + \frac{\partial x'^{\gamma}}{\partial x^{\lambda}} \frac{\partial x^{\mu}}{\partial x'^{\alpha}} \frac{\partial x^{\nu}}{\partial x'^{\beta}} \Gamma^{\lambda}_{\mu\nu}. \tag{2.220}$$

5. Show that the Riemann–Christoffel tensor satisfies a cyclicality condition, i.e.,

$$R_{\alpha\beta\gamma\delta} + R_{\alpha\delta\beta\gamma} + R_{\alpha\gamma\delta\beta} = 0, \tag{2.221}$$

and, further, that it also satisfies the identity

$$\epsilon^{\alpha\beta\gamma\delta} R_{\alpha\beta\gamma\delta} = 0. \tag{2.222}$$

6. Show that, in general, when one does not make the assumption of a symmetric affine connection, the anti-symmetric part of the affine connection enters into the analysis. Specifically, show that

$$V^{\alpha}_{;\nu\beta} - V^{\alpha}_{;\beta\nu} = V^{\mu} R^{\alpha}_{\mu\nu\beta} - (\Gamma^{\mu}_{\nu\beta} - \Gamma^{\mu}_{\beta\nu}) V^{\alpha}_{;\mu}. \tag{2.223}$$

7. Show that the Riemann–Christoffel tensor $R_{\rho\mu\nu\sigma} \equiv g_{\rho\lambda} R^{\lambda}_{\mu\nu\sigma}$ can be written in the form

$$R_{\sigma\mu\nu\rho} = \frac{1}{2} \left(g_{\sigma\nu,\mu\rho} - g_{\mu\nu,\sigma\rho} - g_{\sigma\rho,\mu\nu} + g_{\mu\rho,\sigma\nu} \right) - \Gamma^{\gamma}_{\rho\sigma} g_{\gamma\delta} \Gamma^{\delta}_{\mu\nu} + \Gamma^{\gamma}_{\nu\sigma} g_{\gamma\delta} \Gamma^{\delta}_{\mu\rho}. \tag{2.224}$$

Thus, establish the property that $R_{\sigma\mu\nu\rho} = R_{\nu\rho\sigma\mu}$.

8. Show that in d dimensions the Riemann–Christoffel tensor has $d^2(d^2 - 1)/12$ number of independent components.

9. Consider the line element for a 2-sphere in a three-dimensional space with radius ρ so that

$$ds^2 = \rho^2[(d\theta)^2 + \sin^2\theta (d\phi)^2]. \tag{2.225}$$

Show that the only non-vanishing components of the affinity are

$$\Gamma^{\theta}_{\phi\phi} = -\sin\theta\cos\theta, \quad \Gamma^{\phi}_{\theta\phi} = \Gamma^{\phi}_{\phi\theta} = \cot\theta. \tag{2.226}$$

Further, show that the only non-vanishing elements of the Ricci tensor are

$$R_{\theta\theta} = 1, \quad R_{\phi\phi} = \sin^2\theta \tag{2.227}$$

and that the Ricci scalar has the value $R = 2/\rho^2$.

10. The stress tensor for a perfect fluid model of the universe is given by

$$T^{\mu\nu} = pg^{\mu\nu} + (p + \rho)U^{\mu}U^{\nu}, \tag{2.228}$$

where U^μ is the local velocity of the fluid element and p and ρ are, respectively, the pressure and energy density measured in a local inertial frame moving with the fluid. Einstein equations require

$$T^{\mu\nu}_{;\nu} = 0.$$

Show that the above constraint leads to the hydrodynamic equation

$$\frac{\partial p}{\partial x^\nu} g^{\mu\nu} + g^{-1/2} \frac{\partial}{\partial x^\nu} (g^{1/2}(p+\rho)U^\mu U^\nu) + \Gamma^\mu_{\nu\lambda}(p+\rho)U^\nu U^\lambda = 0. \quad (2.229)$$

11. The proper time element $d\tau^2 = -dx^\mu \, g_{\mu\nu}(x)dx^\nu$ for a static, isotropic gravitational field can be written in the form

$$d\tau^2 = b(r)(dt)^2 - a(r)(dr)^2 - r^2(d\theta)^2 - r^2 \sin^2\theta(d\phi)^2. \quad (2.230)$$

Compute all the components of the affine connection and of the Ricci tensor. Show that

$$R_{\mu\nu} = 0, \quad \mu \neq \nu \quad (2.231)$$

and

$$R_{tt} = -\frac{b''}{2a(r)} + \frac{1}{4}\frac{b'}{a(r)}\left(\frac{a'(r)}{a(r)} + \frac{b'}{b(r)}\right) - \frac{1}{r}\frac{b'}{a(r)} \quad (2.232)$$

$$R_{rr} = \frac{b''}{2b} - \frac{1}{4}\frac{b'}{b}\left(\frac{a'}{a} + \frac{b'}{b}\right) - \frac{1}{r}\frac{a'}{a(r)} \quad (2.233)$$

$$R_{\theta\theta} = -1 + \frac{r}{2a}\left(-\frac{a'}{a} + \frac{b'}{b}\right) + \frac{1}{a} \quad (2.234)$$

$$R_{\phi\phi} = \sin^2\theta R_{\theta\theta}, \quad (2.235)$$

where $a' = da/dr$ and $a'' = d^2a/dr^2$, etc.

12. Show that the choice of $a(r)$ and $b(r)$ of the form

$$a(r) = \left(1 - \frac{2MG}{r}\right)^{-1}, \quad b(r) = \left(1 - \frac{2MG}{r}\right) \quad (2.236)$$

leads to a solution of the Einstein equation in empty space. This is the so-called Schwarzschild solution [9].

13. Show that the relation

$$\sqrt{g}g^{\alpha\beta}\delta R_{\alpha\beta} = -\sqrt{g}g^{\alpha\beta}\left[\delta\Gamma^\gamma_{\alpha\gamma,\beta} - \delta\Gamma^\gamma_{\alpha\beta,\gamma} + \delta\Gamma^\delta_{\alpha\gamma}\Gamma^\gamma_{\delta\beta}\right.$$

$$\left. + \Gamma^\delta_{\alpha\gamma}\delta\Gamma^\gamma_{\delta\beta} - \delta\Gamma^\delta_{\alpha\beta}\Gamma^\gamma_{\delta\gamma} - \Gamma^\delta_{\alpha\beta}\delta\Gamma^\gamma_{\delta\gamma}\right] \quad (2.237)$$

can be written in the form

$$\sqrt{g}g^{\alpha\beta}\delta R_{\alpha\beta} = -\left(\sqrt{g}g^{\alpha\gamma}\delta\Gamma^\beta_{\alpha\beta} - \sqrt{g}g^{\alpha\beta}\delta\Gamma^\gamma_{\alpha\beta}\right)_{,\gamma}$$

$$- \left[(\sqrt{g}g^{\alpha\beta})_{;\gamma} - (\sqrt{g}g^{\alpha\delta})_{;\delta}\delta^\beta_\gamma\right]\delta\Gamma^\gamma_{\alpha\beta}. \quad (2.238)$$

14. Show that in four dimensions a tensor which has the same symmetry properties as the Riemann tensor but with the extra constraint that it vanishes on contraction of any two indices is given by

$$C_{\alpha\beta\gamma\delta} = R_{\alpha\beta\gamma\delta} + \frac{1}{2}(g_{\alpha\delta}R_{\beta\gamma} - g_{\alpha\gamma}R_{\beta\delta} + g_{\beta\gamma}R_{\alpha\delta} - g_{\beta\delta}R_{\alpha\gamma})$$
$$+ \frac{1}{6}(g_{\alpha\gamma}g_{\beta\delta} - g_{\alpha\delta}g_{\beta\gamma})R. \tag{2.239}$$

15. Show that an extension of Eq. (2.239) to d dimensions gives

$$C_{\alpha\beta\gamma\delta} = R_{\alpha\beta\gamma\delta} + \frac{1}{d-2}(g_{\alpha\delta}R_{\beta\gamma} - g_{\alpha\gamma}R_{\beta\delta} + g_{\beta\gamma}R_{\alpha\delta} - g_{\beta\delta}R_{\alpha\gamma})$$
$$+ \frac{1}{(d-1)(d-2)}(g_{\alpha\gamma}g_{\beta\delta} - g_{\alpha\delta}g_{\beta\gamma})R. \tag{2.240}$$

16. Eliminating $\Gamma^{\beta}_{\alpha\gamma}$ in favor of $\omega_{\alpha}{}^{a}{}_{b}$ using Eq. (2.179) in Eq. (2.70) show that $R^{a}_{b\mu\nu}$ is given in terms of the spin connection so that

$$R^{a}_{b\mu\nu} = \partial_{\mu}\omega_{\nu}{}^{a}{}_{b} - \partial_{\nu}\omega_{\mu}{}^{a}{}_{b} + \omega_{\mu}{}^{a}{}_{c}\omega_{\nu}{}^{c}{}_{b} - \omega_{\nu}{}^{a}{}_{c}\omega_{\mu}{}^{c}{}_{b}. \tag{2.241}$$

17. Show that the stress tensor for a Dirac fermionic field is given by

$$T^{\alpha\beta}_{D} = \frac{i}{4}(\bar{\psi}\gamma^{\alpha}D^{\beta}\psi + \bar{\psi}\gamma^{\beta}D^{\alpha}\psi - D^{\alpha}\bar{\psi}\gamma^{\beta}\psi - D^{\beta}\bar{\psi}\gamma^{\alpha}\psi) - g^{\alpha\beta}L_{D}, \tag{2.242}$$

where the last term due to the variation of $\sqrt{g}$ actually vanishes as a consequence of equations of motion.

18. Given the spin connection $\omega_{\mu}{}^{b}{}_{c}$ one can define the $\omega_{a}{}^{b}{}_{c}$ by the relation

$$\omega_{a}{}^{b}{}_{c} = e^{\mu}_{a}\omega_{\mu}{}^{b}{}_{c}. \tag{2.243}$$

Show that

$$\omega_{b}{}^{b}{}_{a} = \frac{1}{\sqrt{g}}(\partial_{\mu}\sqrt{g}e^{\mu}_{a}). \tag{2.244}$$

19. Show that in the presence of gravitational interactions the Lagrangian for the Dirac fermion can be written in the form

$$\mathcal{L}_{D} = \sqrt{g}\left[-\frac{1}{2i}(\bar{\psi}\gamma^{\mu}\partial_{\mu}\psi - \partial_{\mu}\bar{\psi}\gamma^{\mu}\psi) - m\bar{\psi}\psi + \frac{i}{4}\omega_{[abc]}\bar{\psi}\gamma^{[a}\gamma^{b}\gamma^{c]}\psi\right], \tag{2.245}$$

where $\omega_{abc} = e^{\mu}_{a}\omega_{\mu bc}$.

References

[1] A. Einstein, *Preussische Akademie der Wissenschaften, Sitzungsberichte*, (1915), Part 2, 844–847; *Ann. Physik (ser. 4)*, **49**, 769–822.

[2] S. Weinberg, *Gravitation and Cosmology–Principles and Applications of the General Theory of Relativity*, Wiley (1972).

[3] R. V. Eötvös, D. Pek'ar, and E. Fekete, *Ann. Physik*, **68**, 11 (1922).

[4] T. A. Wagner, S. Schlamminger, J. H. Gundlach, and E. G. Adelberger, *Class. Quant. Grav.* **29**, 184002 (2012).

[5] S. Schlamminger, K.-Y. Choi, T. A. Wagner, J. H. Gundlach, and E. G. Adelberger, *Phys. Rev. Lett.* **100**, 041101 (2008).

[6] J. Hartwig, S. Abend, C. Schubert, D. Schlippert, H. Ahlers, K. Posso-Trujillo, N. Gaaloul, W. Ertmer *et al.*, *New J. Phys.* **17**(3), 035011 (2015).

[7] J. Berg *et al.* [MICROSCOPE Collaboration], *J. Phys. Conf. Ser.* **610**(1), 012009 (2015).

[8] A. Einstein, *S. B. preuss. Akad. Wiss.*, 414 (1925).

[9] K. Schwarzchild (translation and foreword by S.Antoci and A.Loinger), arXiv: physics/9905030 [physics.hist-ph] (1999).

Further Reading

J. L. Anderson, *Principles of Relativity Physics*, Academic Press, 1967.

S. Carrol, *Spacetime and Geometry: An Introduction to General Relativity*, Addison-Wesley (2003).

J. F. Donoghue, *Phys. Rev. D* **50**, 3874 (1994); *Phys. Rev. Lett.* **72**, 2996 (1994).

M. Gasperini and V. De Sabbata, *Introduction to Gravitation*, World Scientific (1986).

R. D'Inverno, *Introducing Einstein's Relativity*, Oxford University Press (1992).

T. W. B. Kibble, *J. Math. Phys.* **2**, 212 (1961).

C. W. Misner, K. S. Thorne, and J. A. Wheeler, *An Introduction to General Relativity and Cosmology*, W. H. Freeman (1973).

R. M. Wald, *General Relativity*, Chicago University Press (1984).

J. Yepez, arXiv:1106.2037 [gr-qc] (2011).

3

Non-abelian Gauge Theory

3.1 Yang–Mills Gauge Theory

A gauge theory for a non-abelian group was deduced in the works of Yang and Mills and others [1–3]. It is quite straightforward to extend the ideas of parallel transport which we discussed in the previous chapter in defining covariant derivatives in gravity theories to theories involving gauge fields. Thus, consider a matter field multiplet ϕ_i (where i is an internal symmetry index) belonging to a representation of a gauge group G. Such a field could be a multiplet of fermions or of Higgs bosons. Following the analysis as in the case with gravity, we consider the multiplet ϕ at two adjacent points, i.e., x and $x + dx$. Thus, at the point $x + dx$ the multiplet $\phi(x + dx)$ transforms according to the transformation properties appropriate for the representation to which ϕ belongs. We can, however, also create another multiplet $\tilde{\phi}(x + dx)$ at the point $x + dx$ by parallel transport so that

$$\tilde{\phi}_i(x + dx) = \phi_i(x) + ig A_{\alpha ij} \, dx^\alpha \, \phi_j(x). \tag{3.1}$$

Here, $A_{\mu ij} = A_\mu^a (T_a)_{ij}$, where T_a are the generators of the gauge group in the representation to which ϕ belongs. Here, $A_{\mu ij}$ play the same role that the connection $\Gamma^\alpha_{\beta\gamma}$ does for gravity. Since we have two quantities at the point $x + dx$ that transform like a multiplet of the gauge group G, the difference between these two must also transform in the same representation. Thus, expanding to first order in dx^α we get

$$\tilde{\Delta}\phi_i = \phi_i(x + dx) - \tilde{\phi}_i(x + dx)$$
$$= \partial_\alpha \phi_i \, dx^\alpha - ig A_{\alpha ij} \, dx^\alpha \, \phi_j. \tag{3.2}$$

Using Eq. (3.2) we can define a covariant derivative $D_\alpha \phi_i$ so that

$$D_\alpha \phi_i = \left(\frac{\tilde{\Delta} \phi_i}{dx^\alpha} \right)_{\text{lim}}$$
$$= (\delta_{ij} \partial_\alpha - ig A_\alpha^a (T_a)_{ij}) \phi_j. \tag{3.3}$$

The covariant derivative transforms as if the transformation was global, i.e.,

$$(D_\alpha \phi(x))' = U(g(x)) D_\alpha \phi(x). \tag{3.4}$$

Substitution of Eq. (3.3) into Eq. (3.4) allows us to compute the transformation of A_α under local gauge transformations. Thus, the left-hand side of Eq. (3.4) on expansion gives

$$(D_\alpha \phi(x))' = (\partial_\alpha - ig A_\alpha') U(g(x)) \phi(x)$$
$$= \partial_\alpha U \phi + U \partial_\alpha \phi - ig A_\alpha' U \phi, \tag{3.5}$$

while the right-hand side of Eq. (3.4) on expansion gives

$$U(g(x)) D_\alpha \phi(x) = U(g(x))(\partial_\alpha - ig A_\alpha) \phi$$
$$= U \partial_\alpha \phi - ig U A_\alpha \phi. \tag{3.6}$$

A comparison of Eqs. (3.5) and (3.6) gives the desired gauge transformations of A_α, i.e., we have

$$A_\alpha'(x) = U A_\alpha(x) U^{-1} + \frac{1}{ig} \partial_\alpha U U^{-1}. \tag{3.7}$$

For infinitesimal gauge transformations with transformation parameters ϵ^a we have

$$U(g(x)) = \mathbb{1} + i\epsilon^a(x) T_a. \tag{3.8}$$

Using Eq. (3.8) in Eq. (3.7) then gives the following:

$$A_\alpha'^{\,a} = A_\alpha^a + \frac{1}{g} \partial_\alpha \epsilon^a - C_{abc} \epsilon^b A_\alpha^c, \tag{3.9}$$

where C_{abc} are the structure constants of the gauge group G so that

$$[T_a, T_b] = i C_{abc} T_c. \tag{3.10}$$

Next, we compute the commutator of two covariant derivatives D_α and D_β:

$$[D_\alpha, D_\beta] = (\partial_\alpha - ig A_\alpha)(\partial_\beta - ig A_\beta) - (\alpha \leftrightarrow \beta)$$
$$= -ig(\partial_\alpha A_\beta - \partial_\beta A_\alpha) - g^2 [A_\alpha, A_\beta]$$
$$\equiv -ig F_{\alpha\beta}, \tag{3.11}$$

where

$$F_{\alpha\beta} = (\partial_\alpha A_\beta - \partial_\beta A_\alpha) - ig[A_\alpha, A_\beta]. \tag{3.12}$$

Here, $F_{\alpha\beta}$ is Lie valued, i.e., $F_{\alpha\beta} = F_{\alpha\beta}^a T_a$, and on using Eq. (3.12) we have

$$F_{\alpha\beta}^a = \partial_\alpha A_\beta^a - \partial_\beta A_\alpha^a + g C_{abc} A_\alpha^b A_\beta^c. \tag{3.13}$$

Further, from Eq. (3.4) we have

$$\begin{aligned} D_\alpha' \phi' &= D_\alpha' U \phi \\ &= U D_\alpha \phi, \end{aligned} \tag{3.14}$$

which leads to the relation

$$D_\alpha' = U D_\alpha U^{-1}. \tag{3.15}$$

We can use Eq. (3.15) to compute the transformation properties of $F_{\alpha\beta}$. Thus,

$$[D_\alpha', D_\beta'] = U[D_\alpha, D_\beta] U^{-1}, \tag{3.16}$$

which leads to the transformation

$$F_{\alpha\beta}' = U F_{\alpha\beta} U^{-1}. \tag{3.17}$$

For infinitesimal gauge transformations, Eq. (3.17) gives

$$\begin{aligned} F_{\alpha\beta}' &= U F_{\alpha\beta} U^{-1} \\ &= (1 + i\epsilon^b T_b) F_{\alpha\beta}^a T_a (1 - i\epsilon^c T_c) \\ &= F_{\alpha\beta}^a T_a + i[T_b, T_c] \epsilon^b F_{\alpha\beta}^c \\ &= \left[F_{\alpha\beta}^a - C_{abc} \epsilon^b F_{\alpha\beta}^c \right] T_a, \end{aligned} \tag{3.18}$$

which gives the transformation for $F_{\alpha\beta}$:

$$F_{\alpha\beta}'^{\,a} = F_{\alpha\beta}^a - C_{abc} \epsilon^b F_{\alpha\beta}^c. \tag{3.19}$$

Because of the absence of a derivative of the transformation parameter ϵ in Eq. (3.19) even under local gauge transformation, $F_{\alpha\beta}$ transforms as if the transformation was a global one. Thus, due to the transformation property of Eq. (3.19) the product with summed indices $F_{\alpha\beta}^a F^{a\alpha\beta}$ is invariant under the local gauge transformations. Using the above, we may now write the Lagrangian for spin 0 (ϕ), spin 1/2 (ψ), and massless spin 1 (A_α) fields so that the Lagrangian is invariant under the local gauge transformations, i.e.,

$$L_{\text{inv}} = -(D_\alpha^\phi \phi)^\dagger D^{\phi\alpha} \phi - \mu^2 \phi^\dagger \phi - \bar{\psi} \left(\frac{1}{i} \gamma^\alpha D_\alpha^\psi + m \right) \psi - \frac{1}{4} F_{\alpha\beta}^a F^{\alpha\beta a}, \tag{3.20}$$

where $D_\alpha^\phi = \partial_\alpha - ig A_\alpha^a T_a^\phi$, etc., and where T^ϕ (T^ψ) are the matrix representations of the generators in the representation to which ϕ (ψ) belong. From here on we will drop the superscripts since the dimension of the matrices will be clear from the context in which they are used.

3.2 Canonical Quantization of the Maxwell Theory

As a precursor to a discussion of the canonical quantization of a Yang–Mills theory, we discuss the canonical quantization of the Maxwell–Dirac theory. Here we start with the Lagrangian in the first-order formalism where we treat the Maxwell potential A_μ and the Maxwell field strength $F_{\mu\nu}$ as independent variables so that

$$\mathcal{L} = -\frac{1}{2}F^{\mu\nu}(\partial_\mu A_\nu - \partial_\nu A_\mu) + \frac{1}{4}F_{\mu\nu}F^{\mu\nu} + \mathcal{L}_\psi \tag{3.21}$$

$$\mathcal{L}_\psi = -\bar{\psi}\left(\frac{1}{i}\gamma^\mu\partial_\mu - e\gamma^\mu A_\mu + m\right)\psi. \tag{3.22}$$

Variation of Eq. (3.21) with respect to $F_{\mu\nu}$, i.e., $\delta\mathcal{L}/\delta F_{\mu\nu} = 0$, gives the field equation for $F_{\mu\nu}$,

$$F_{\mu\nu} = (\partial_\mu A_\nu - \partial_\nu A_\mu), \tag{3.23}$$

while the variation of $\mathcal{L}$ with A_μ, i.e., $\delta\mathcal{L}/\delta A_\mu$, gives

$$\partial_\nu F^{\mu\nu} = j^\mu \equiv \frac{\partial\mathcal{L}_\psi}{\partial A_\mu}. \tag{3.24}$$

Next, we analyze the constraint variables in comparison with the independent variables. This will allow us to choose a set of canonical variables which would be appropriate for quantizing the Maxwell–Dirac theory. First, we look at the components of Eq. (3.23). We choose two sets $\mu = i$, $\nu = j$ and $\mu = 0$, $\nu = j$. In this case, Eq. (3.23) reduces to the following set of relations:

$$\mu = i, \ \nu = j: \ F_{ij} = \partial_i A_j - \partial_j A_i, \tag{3.25}$$

$$\mu = 0, \ \nu = i: \ \partial_0 A_i = F_{0i} + \partial_i A_0. \tag{3.26}$$

For Eq. (3.24) the cases $\mu = 0$ and $\nu = i$ give the relations

$$\mu = 0: \ \partial_i F^{0i} = j^0 \tag{3.27}$$

$$\mu = k: \ \partial_0 F_{0k} = \partial_i F_{ik} + j_k. \tag{3.28}$$

It is now clear that Eqs. (3.25) and (3.27) are constraint equations while Eqs. (3.26) and (3.28) are dynamical equations, since they involve time. Let us examine the constraint equations first, to identity clearly the field variables that are the constraint ones. For convenience we introduce the notation

$$B_k \equiv \frac{1}{2}\epsilon_{ijk}F_{ij}; \ E^i \equiv F^{0i}. \tag{3.29}$$

In terms of $\vec{B}$ and $\vec{E}$ the constraints of Eqs. (3.25) and (3.27) can be expressed as follows:

$$\vec{B} = \vec{\nabla} \times \vec{A} \tag{3.30}$$

$$\vec{\nabla} \cdot \vec{E} = j^0. \tag{3.31}$$

We now see that F_{ij} are constraint variables, since they are determined in terms of A_i, as seen in Eqs. (3.25) or (3.30). On the other hand, the analysis of the constraint Eqs. (3.27) or (3.31) requires a more careful treatment. To accomplish that we carry out the following decomposition of $\vec{E}$:

$$\vec{E} = \vec{E}_T + \vec{E}_L. \tag{3.32}$$

Here, $\vec{E}_L$ and $\vec{E}_T$ satisfy the following constraints:

$$\vec{E}_L = \vec{\nabla} E_L, \quad \vec{\nabla} \cdot \vec{E}_T = 0. \tag{3.33}$$

Use of Eq. (3.33) in Eq. (3.31) gives $\nabla^2 E_L = j^0$, which leads to the relations

$$E_L = \nabla^{-2} j^0, \quad \vec{E}_L = \vec{\nabla} \nabla^{-2} j^0. \tag{3.34}$$

Thus, we can identity $\vec{E}_L$ and $\vec{B}$ as constraint variables. We now return to the two remaining relations given by Eqs. (3.26) and (3.28) so that

$$\partial_0 \vec{A} = -\vec{E} + \vec{\nabla} A_0 \tag{3.35}$$

$$\partial_0 \vec{E} = -\vec{\nabla} \times \vec{B} - \vec{j}. \tag{3.36}$$

We decompose the above into transverse and longitudinal parts as follows:

$$\partial_0 \vec{A}_T = -\vec{E}_T \tag{3.37}$$

$$\partial_0 \vec{A}_L = -\vec{\nabla}(\nabla^{-2} j^0) + \vec{\nabla} A_0 \tag{3.38}$$

$$\partial_0 \vec{E}_T = -\vec{\nabla} \times (\vec{\nabla} \times \vec{A}_T) - \vec{j}_T \tag{3.39}$$

$$\partial_0 \vec{E}_L = -\vec{j}_L, \tag{3.40}$$

where we have used $\vec{\nabla} \times \vec{A}_L = 0$, which gives $\vec{B} = \vec{\nabla} \times \vec{A} = \vec{\nabla} \times \vec{A}_T$. From Eqs. (3.37) and (3.39) we see that both $(\vec{A}_T, \vec{E}_T)$ are dynamical variables. From Eq. (3.34) we have already seen that $\vec{E}_L$ is a constraint variable. We are thus left with Eq. (3.38), which involves $\vec{A}_L$. On writing

$$\vec{A}_L = \vec{\nabla} A_L \tag{3.41}$$

Eq. (3.38) gives

$$A_0 = \partial_0 A_L + \nabla^{-2} j^0, \tag{3.42}$$

which determines A_0 in terms of A_L and j^0. All equations are now taken account of, and still A_L remains undetermined. The reason for the lack of determination of A_L is gauge invariance. Thus, Maxwell–Dirac theory is invariant under $U(1)$ gauge transformations. For the A_μ field it implies that

$$A'_\mu = A_\mu + \partial_\mu \Lambda. \tag{3.43}$$

For the spatial components of A_i it implies that $A'_i = A_i + \partial_i \Lambda$, which leads to $A'_{iT} = A_{iT}$ and

$$A'_L = A_L + \Lambda(x). \tag{3.44}$$

Thus, A_{iT} and also E_{iT} are gauge invariant, which is very desirable since they are dynamical variables. On the other hand, A_L is gauge variant. The underlying physics is unchanged by the choice of a gauge and thus by the choice of A_L. A common gauge choice is to set $A_L = 0$. Of course, once A_L is fixed, A_0 can be determined, and for the choice $A_L = 0$ one finds from Eq. (3.42) that

$$A_0 = \nabla^{-2} j^0. \tag{3.45}$$

Thus, in summary, we have the following. The quantities $\vec{A}_T$ and $\vec{E}_T$ are gauge invariant and are dynamical variables while A_L is gauge dependent. The quantity A_0 consists of two parts: one part is gauge dependent and the other part is dynamical, as it depends on j^0.

The Hamiltonian density for the Maxwell–Dirac theory can be written in a form given by Eq. (3.153). Here, the first term on the right-hand side is the energy density for the free photon field, the second term is the energy density for the free Dirac field, the third term is the interaction between the current and the radiation field, the fourth term is the Coulomb interaction energy, and the last term is the gauge part. From Eq. (3.153) the canonical variables for the radiation field are

$$A_{iT}, \pi_i = -E_{iT}. \tag{3.46}$$

In quantizing the fields we need to pay attention to the transverse nature of the canonical variables. Thus, we write

$$[A_{iT}(\vec{r}, t), E_{jT}(\vec{r}', t)] = -i\delta_{ij}^T(\vec{r} - \vec{r}'), \tag{3.47}$$

where $\delta_{ij}^T(\vec{r} - \vec{r}')$ satisfies the constraints

$$\partial_i \delta_{ij}^T(\vec{r} - \vec{r}') = 0 = \partial_j' \delta_{ij}^T(\vec{r} - \vec{r}'). \tag{3.48}$$

It is easily seen that $\delta_{ij}^T(\vec{r} - \vec{r}')$ has the form

$$\delta_{ij}^T(\vec{r} - \vec{r}') = \delta_{ij}\delta(\vec{r} - \vec{r}') - \nabla^{-2}\partial_i\partial_j\delta^3(\vec{r} - \vec{r}'). \tag{3.49}$$

A Fourier transform of $\delta_{ij}^T(\vec{r} - \vec{r}')$ gives

$$\delta_{ij}^T(\vec{r} - \vec{r}') = \frac{1}{(2\pi)^3} \int e^{i\vec{k}\cdot(\vec{r}-\vec{r}')}\delta_{ij}^T(k) d^3k \tag{3.50}$$

$$\delta_{ij}^T(k) = \frac{1}{k^2}\left(\delta_{ij} - \frac{k_i k_j}{k^2}\right), \tag{3.51}$$

where $\delta_{ij}^T(k)$ satisfies the constraint $k_i\delta_{ij}^T(k) = 0 = k_j\delta_{ij}^T(k)$.

3.3 Canonical Quantization of the Yang–Mills Theory

We now turn to the canonical quantization of the Yang–Mills theory. The Lagrangian of the Yang–Mills theory in a first-order formalism where A_μ^a and

$F^a_{\mu\nu}$ are treated as independent fields is given by

$$\mathcal{L}_{YM} = \frac{1}{4} F^a_{\mu\nu} F^a_{\mu\nu} - \frac{1}{2} F^{\mu\nu a} (\partial_\mu A^a_\nu - \partial_\nu A^a_\mu + g C_{abc} A^b_\mu A^c_\nu)$$
$$- \frac{1}{2} \bar{\psi} \left[\left(\frac{1}{i} \gamma^\mu \vec{D}_\mu + m \right) + \left(-\frac{1}{i} \gamma^\mu \overleftarrow{D}_\mu + m \right) \right] \psi, \qquad (3.52)$$

where

$$\vec{D}_\mu = \vec{\partial}_\mu - ig A^a_\mu T_a, \quad \overleftarrow{D}_\mu = \overleftarrow{\partial}_\mu + ig A^a_\mu T_a. \qquad (3.53)$$

Variation of $F^a_{\mu\nu}$ in Eq. (3.52) gives

$$F_{\mu\nu} = \partial_\mu A_\nu - \partial_\nu A_\mu - ig[A_\mu, A_\nu], \qquad (3.54)$$

where $A_\mu = A^a_\mu T_a$ and $F_{\mu\nu} = F^a_{\mu\nu} T_a$. Further, a variation of A^a_μ gives

$$\partial_\nu F^{\mu\nu}_a + g C_{abc} A^b_\nu F^{\mu\nu}_c = g j^\mu_a(x), \qquad (3.55)$$

where

$$j^\mu_a = \bar{\psi} \gamma^\mu T_a \psi. \qquad (3.56)$$

A variation of the Lagrangian with respect to $\bar{\psi}$ gives

$$\left(\frac{1}{i} \gamma^\mu D_\mu + m \right) \psi = 0. \qquad (3.57)$$

Taking a partial derivative ∂_μ of Eq. (3.55) we get

$$\partial_\mu \partial_\nu F^{\mu\nu}_a = 0 = g \partial_\mu J^\mu_{ta} \qquad (3.58)$$
$$j^\mu_{ta} = j^\mu_a - C_{abc} A^b_\nu F^{\mu\nu}_c, \qquad (3.59)$$

which tells us that the conserved current j^μ_{ta} has a matter part and a gauge part. We now look at the field equations (Eqs. (3.54) and (3.55)) in component form. From Eq. (3.54) we have the following two cases:

$$\mu = i, \nu = j : F_{ij} = \partial_i A_j - \partial_j A_i - ig[A_i, A_j] \qquad (3.60)$$

and

$$\mu = 0, \nu = i : -E_i = F_{0i} = \partial_0 A_i - \partial_i A_0 - ig[A_0, A_i]. \qquad (3.61)$$

Similarly, from Eq. (3.55) we have the following cases:

$$\mu = 0 : \vec{\nabla} \cdot \vec{E}_a + g C_{abc} \vec{A}_b \cdot \vec{E}_c = g j^0_a \qquad (3.62)$$

and

$$\mu = i : -\partial_o E^i_a + \partial_j F^{ij}_a + g C_{abc} \left(-A^b_0 E^i_c + A^b_j F^{ij}_c \right) = g j^i_a. \qquad (3.63)$$

As in the case for the Maxwell field A_μ, not all the components of the Yang–Mills gauge fields A^a_μ are dynamical: some are gauge dependent. In analogy to the Maxwell case we first consider the Coulomb gauge to fix the gauge-dependent components. The Yang–Mills analogue of the Coulomb gauge is given by

$$\vec{\nabla} \cdot A_a = 0, \tag{3.64}$$

which implies that $\vec{A}_{La} = 0$. Thus, $\vec{A}_{Ta}$ is one of the dynamical variables. It is also useful to decompose $\vec{E}_a$ into longitudinal and transverse parts so that $\vec{E}_a = \vec{E}_{La} + \vec{E}_{Ta}$. Also, from Eqs. (3.61) and (3.63) we get

$$\vec{E}_T = -\partial_0 \vec{A}_T + ig[A_0, \vec{A}_T]_T \tag{3.65}$$

and

$$-\partial_0 \vec{E}_{Ta} + \vec{\nabla} \times \vec{B}_a + gC_{abc}\left(-A_0^b E_c^i + A_j^b F_c^{ij}\right)_T = g\vec{j}_{Ta}, \tag{3.66}$$

where $B_{ia} = \frac{1}{2}\epsilon_{ijk}F_a^{jk}$. It is easy to check that A_i^{Ta} and $-E_{ia}^T$ are indeed canonical variables, and thus we assume for them the commutation relations

$$[A_i^{Ta}(\vec{r},t), E_{jb}^T(\vec{r}',t)] = -i\delta_{ij}^T(\vec{r}-\vec{r}')\delta_b^a. \tag{3.67}$$

3.4 Gribov Ambiguity

So far the analysis has been very similar to what we had for the Maxwell–Dirac theory. However, for the Yang–Mills case there is a new complexity, discovered by Gribov in 1977 [4] and often referred to as the Gribov ambiguity. We will discuss it first in the Coulomb gauge, which is defined by Eq. (3.64). Gribov pointed out that this gauge-fixing term does not fully fix the gauge. Assume that we want to check if the Coulomb gauge will fix the gauge uniquely. To do so we make a gauge transformation on A_i, which gives A_i', and for infinitesimal gauge transformation we have

$$\vec{A}_a' = \vec{A}_a + \vec{\nabla}\epsilon_a(x) + g\epsilon_{abc}\vec{A}_b\epsilon_c. \tag{3.68}$$

Let us assume that the Coulomb gauge is not unique and there are two vector potentials $\vec{A}_a$ and $\vec{A}_a'$, both of which satisfy the Coulomb gauge constraint of $\vec{\nabla} \cdot \vec{A}_a' = 0 = \vec{\nabla} \cdot \vec{A}_a$. Under this circumstance, taking the divergence of Eq. (3.68) and using the Coulomb gauge conditions, we get

$$\nabla^2\epsilon_a + gC_{abc}\vec{A}_b^T \cdot \vec{\nabla}\epsilon_c = 0. \tag{3.69}$$

Let us consider the case of an abelian gauge theory. In this case, Eq. (3.69) reduces to

$$\nabla^2\epsilon_a = 0. \tag{3.70}$$

Solutions to Eq. (3.70) which vanish at $|\vec{x}| \to \infty$ give $\epsilon_a = 0$. This means that for the abelian case the Coulomb gauge fixes the potential uniquely. However, for the non-abelian case we have Eq. (3.69). In this case, non-trivial solutions exist, and there can be different gauges consistent with the Coulomb gauge condition.

Such gauges are Gribov copies, and the number of such copies could be infinite. While we discussed the Gribov ambiguity in the context of a specific gauge, i.e., the Coulomb gauge, it has been shown by Singer [5] that Gribov ambiguity extends to all other gauge choices. We note that physical phenomena such as asymptotic freedom which can be understood by a perturbative approach are unlikely to be affected by the Gribov ambiguity. However, those phenomena which require non-perturbative dynamics such as quark confinement are likely to require taking account of this aspect of Yang–Mills theory. We discuss some further aspects of the Gribov ambiguity below.

In the above we discussed Gribov ambiguity in the context of the Coulomb gauge, which is a non-local gauge constraint. However, it is possible to make gauge choices where the gauge constraint on the components of A_a^μ is local. There are several such gauges such as

- the axial gauge [6], $A_3^a(x) = 0$;
- the temporal gauge, $A_0^a(x) = 0$;
- the null-plane gauge, $A_0^a + A_3^a = 0$.

We focus on the temporal gauge to see if Gribov ambiguity persists or if it can be removed in the temporal gauge. Now, consider the gauge transformation of an A_μ in an arbitrary gauge. Such a transformation is given by Eq. (3.7). We define a transformation U such that starting from an A_μ in any given gauge we have

$$A_0' = 0. \tag{3.71}$$

Using Eq. (3.71) in Eq. (3.7), we find the constraint on U so that

$$\partial_0 U = -ig U A_0(x). \tag{3.72}$$

The question then is if the temporal gauge is unique or not. To check this, let us investigate the constraint that is needed so that one is in a temporal gauge both before and after the transformation, i.e., that one has $A_0 = 0 = A_0'$. In this case, Eq. (3.72) reads

$$\partial_0 U = 0, \tag{3.73}$$

which implies that U is time independent. Returning to Eq. (3.7), we find that $\vec{A}$ and $\vec{A}'$ are related by

$$\vec{A}' = U\vec{A}U^{-1} + \frac{1}{ig}(\vec{\nabla}U)U^{-1}. \tag{3.74}$$

This mean that one can be in a temporal gauge with both $\vec{A}$ and $\vec{A}'$, and so the Gribov ambiguity still persists.

It is useful to examine the field equations in the temporal gauge. Here, one has

$$-E_i = \partial_0 A_i \tag{3.75}$$

$$F_{jk} = \partial_j A_k - \partial_k A_j - ig[A_j, A_k]. \tag{3.76}$$

Further, we have the following relations:

$$\vec{\nabla} \cdot \vec{E} - ig[A_i, E_i] = gj^0 \tag{3.77}$$

and

$$-\partial_0 E_a^i + \partial_j F_a^{ij} + gC_{abc}A_{jb}F_c^{ij} = gj_a^i. \tag{3.78}$$

Note that

$$\pi_{ia} = \frac{\delta \mathcal{L}}{\delta(\partial_0 A_i^a)} = -E_{ia}, \tag{3.79}$$

which implies that $(A_{ia}, -E_{ia})$ are canonical co-ordinates and momenta and satisfy the canonical commutation relations

$$[A_{ia}(\vec{r}, t), E_{jb}(\vec{r}', t)] = -i\delta_{ab}\delta^3(\vec{r} - \vec{r}'). \tag{3.80}$$

There are now two ways one can generate equations of motion. One is by use of the Lagrangian equations and the other by use of the Hamiltonian equations. An acceptable quantum theory requires the two ways to be consistent. So, we have the following Heisenberg equations of motion:

$$i\partial_0 A_{ia} = [A_{ia}, H] \tag{3.81}$$

$$i\partial_0 E_i^a = [E_i^a, H] \tag{3.82}$$

$$i\partial_0 \psi = [\psi, H]. \tag{3.83}$$

From Eqs. (3.156) and (3.157), we see that the Hamiltonian equations of motion reproduce the Lagrangian equations of motion given by Eqs. (3.61) and (3.63) in the temporal gauge. However, there is one missing piece, which is the analog of Gauss's law for the Yang–Mills case, i.e., Eq. (3.77).

To understand the difference between the two approaches, we can take the viewpoint that the theory defined by Eqs. (3.80), (3.154), and (3.155) is a larger theory, i.e., that the Hilbert space of this theory is larger than the true Yang–Mills theory. We can call it theory X, while theory Y is the true Yang–Mills theory which contains Gauss's law as given by Eq. (3.77). Thus, to go from theory X to the true Yang–Mills theory we must constrain the Hilbert space of X by imposing the constraint

$$G_a|X\rangle = 0, \tag{3.84}$$

where G_a is defined so that

$$G_a \equiv \vec{\nabla}.\vec{E}_a + gC_{abc}\vec{A}_b \cdot \vec{E}_c - gj^0. \tag{3.85}$$

It is interesting to ask if the constraint of Eq. (3.84) would hold as a function of time. To check this we can look at the Hamiltonian equation for G_a. Here, we get

$$i\frac{\partial G_a}{\partial t} = [G_a, H]. \tag{3.86}$$

Now, using Eqs. (3.80), (3.154), and (3.155) one finds that

$$[G_a, H] = 0. \tag{3.87}$$

Thus, it is indeed true that the Gauss's law constraint if imposed at one time will be maintained at other times. One can also define a unitary transformation $\mathcal{U}$ so that

$$\mathcal{U} = e^{i \int d^3x \, \theta_a(x) G_a(x)}, \tag{3.88}$$

and the action of $\mathcal{U}$ on physical states $|\text{phys}\rangle$ is such that

$$\mathcal{U}|\text{phys}\rangle = |\text{phys}\rangle. \tag{3.89}$$

Now, in addition to the gauge changes that can be obtained by continuous transformations starting from the identity, there are also other possible gauges that cannot be obtained by transformation starting from the identity. Thus, here one has a different set of operators $\mathcal{U}_n$ so that their action on states $|\text{phys}\rangle$ is given by

$$\mathcal{U}_n|\text{phys}\rangle = e^{\pm in\theta}|\text{phys}\rangle, \quad n = 0, 1, 2, 3, \cdots \tag{3.90}$$

Equation (3.90) points to the topological aspects of the Yang–Mills theory. This topic is discussed in further detail elsewhere [7,8].

3.5 Path Integral Quantization of Yang–Mills Theory

In this section we give a brief introduction to the path integral quantization of Yang–Mills theory. An extensive analysis of path integral quantization of non-abelian theories appears in a number of reviews in the literature (e.g., [9]). In this analysis we follow the approach of Abers and Lee [9], to which the reader is directed for a more thorough exposition. Path integral formulation offers a direct route to the quantization of Yang–Mills gauge theory. As in the operator formulation, gauge fixing needs to be addressed here as well. In the path integral formulation the vacuum-to-vacuum amplitude without gauge fixing is given by

$$Z[j] = \int [dA_\mu] e^{i \int d^4x [\mathcal{L}_{YM}(x) + A_\mu j^\mu(x)]}, \tag{3.91}$$

where j_a^μ are external sources. The Lagrangian $\mathcal{L}_{YM}$ in the second-order formalism is given by

$$\mathcal{L}_{YM}(x) = -\frac{1}{4}\vec{F}_{\mu\nu} \cdot \vec{F}^{\mu\nu}$$
$$\vec{F}_{\mu\nu} = \partial_\mu \vec{A}_\nu - \partial_\nu \vec{A}_\mu + g\vec{A}_\mu \times \vec{A}_\nu. \tag{3.92}$$

The quadratic part of $\mathcal{L}_{YM}$ can be written as

$$\mathcal{L}_{YM}^0(x) = \int d^4x \, d^4y \frac{1}{2} A_\mu(x) K^{\mu\nu}(x,y) A_\nu(y)$$
$$K^{\mu\nu}(x,y) = (-\eta^{\mu\nu}\Box + \partial^\mu \partial^\nu)\delta^4(x-y). \tag{3.93}$$

We now face the problem that the quadratic part cannot be inverted and a propagator corresponding to $K^{\mu\nu}$ cannot be defined. The origin of this problem resides in the gauge invariance of the action. Thus, the path integral is invariant under the gauge transformation

$$\vec{A}_\mu.\vec{T} \to \vec{A}_\mu^g.\vec{T} = U(g)\left[\vec{A}_\mu.\vec{T} + \frac{1}{ig}U^{-1}(g)\partial_\mu U(g)\right]U^{-1}(g), \qquad (3.94)$$

where g ranges over all allowed elements of the gauge group G. However, it was pointed out by Faddeev and Popov [10] that the path integral is to be performed over not all variations of the gauge fields but only over distinct orbits of $\vec{A}_\mu$ that arise from the action of the gauge group.

To implement this idea, we define an object $\Delta_c[A_\mu]$ so that

$$\Delta_c[A_\mu]\int \prod_x dg(x) \prod_{x,a} \delta(C_a(A_\mu^g(x)) = 1, \qquad (3.95)$$

where C_a is a gauge-fixing function and A^g are connected to other A' by gauge transformations and dg is a gauge group measure. We use Eq. (3.95) to write Eq. (3.91) in the following form:

$$Z = \int \prod_x dg(x)\ \Delta_c[A_\mu][dA_\mu]\prod_{x,a}\delta(C_a(A_\mu^g(x))e^{i\int d^4x\ \mathcal{L}_{YM}(x)}. \qquad (3.96)$$

Next, let us make the transformation so that $A_\mu(x) \to [A_\mu(x)]^{g^{-1}}$. Under this transformation we observe that the action and $[dA_\mu]$ are invariant. Further, using Eq. (3.159) one finds that $\Delta_c[A_\mu]$ is also invariant. This allows us to write Eq. (3.96) so that

$$Z = (\prod_x dg(x))\int \Delta_c[A_\mu][dA_\mu]\prod_{x,a}\delta(C_a(A_\mu(x))e^{i\int d^4x\mathcal{L}_{YM}(x)}, \qquad (3.97)$$

where we have factored out the infinite group factor. Dividing out the group factor we can write the gauge-fixed vacuum to vacuum amplitude Z_c, including an external source j^μ, so that

$$Z_c[j] = \int \Delta_c[A_\mu][dA_\mu]\prod_{x,a}\delta(C_a(A_\mu(x))e^{i\int d^4x[\mathcal{L}_{YM}(x)+j^\mu(x)A_\mu]}. \qquad (3.98)$$

It is the appearance of $\Delta_c(A_\mu)$ in the path integral that differentiates the non-abelian gauge theories from the abelian gauge theories [10]. We now wish to compute $\Delta_c[A_\mu]$. To do so we write

$$\Delta_c^{-1}[A_\mu] = \int \prod_{x,a} dg(x)\ \delta(C_a(A_\mu^g(x)), \qquad (3.99)$$

and expand $C_a(A^g_\mu(x))$ so that

$$C_a(A^g_\mu(x)) = C_a(A_\mu) + \int d^4y \sum_b (M(x,y))_{ab}\lambda_b(y) + O(\lambda^2), \qquad (3.100)$$

where

$$M_{ab}(x,y) = \frac{\partial C_a(x)}{\partial \lambda_b(y)}, \qquad (3.101)$$

and for the case when $C_a(A_\mu) = 0$ we get

$$\Delta^{-1}_c[A_\mu] = \int \prod_{x,a} d\lambda_a(x)\delta\left(\int d^4y(\sum_b M(x,y))_{ab}\lambda_b(y)\right). \qquad (3.102)$$

It is now easily seen that

$$\Delta_c(A_\mu)|_{C_a=0} = \det(M), \qquad (3.103)$$

and we can write

$$\det(M) = \int [d\xi][d\xi^*]e^{i\int d^4x\ \mathcal{L}_{FP}(x)}, \qquad (3.104)$$

where $\mathcal{L}_{FP}(x)$ is the Faddeev–Popov ghost Lagrangian given by

$$\mathcal{L}_{FP}(x) = \int d^4y\ \xi^*_a(x)(M(x,y))_{ab}\xi_b(y). \qquad (3.105)$$

where $\xi_a(x)$ and $\xi_a(x)^*$ are anti-commuting scalar fields called Faddeev–Popov ghost fields [10]. As an example, let us consider the case when

$$C_a(A_\mu) = g\partial_\mu A^\mu_a(x). \qquad (3.106)$$

where $\partial_\mu A^\mu_a = 0$ is the covariant Landau gauge condition. To compute the ghost Lagrangian, we consider the infinitesimal gauge transformation

$$\vec{A}_\mu \to \vec{A}_\mu + \vec{\lambda}(x) \times \vec{A}_\mu - \frac{1}{g}\partial_\mu\vec{\lambda}. \qquad (3.107)$$

In this case we get

$$g\partial_\mu\vec{A}^\mu \to g\partial_\mu\vec{A}^\mu - g\vec{A}^\mu \times \partial_\mu\vec{\lambda} - g\partial_\mu\vec{A}^\mu \times \vec{\lambda} - \Box\vec{\lambda}. \qquad (3.108)$$

Imposing the condition $C(A^\mu_a) = 0$ and using Eqs. (3.101) and (3.105) we get

$$\int d^4x\ \mathcal{L}_{FP} = \int d^4x \left(-\vec{\xi}^\dagger \cdot \Box\vec{\xi} - gC_{abc}\xi^\dagger_a A^\mu_b \partial_\mu\xi_c\right). \qquad (3.109)$$

Integrating by parts and again using the constraint $\partial_\mu A^\mu_a = 0$ and discarding the surface terms, one gets

$$\mathcal{L}_{FP} = \partial_\mu\vec{\xi}^\dagger \cdot \partial^\mu\vec{\xi} + gC_{abc}\partial_\mu\xi^\dagger_a A^\mu_b\xi_c. \qquad (3.110)$$

We now wish to cast the gauge-fixing condition in a different way. Let us write Eq. (3.98) in a more compact form:

$$Z_c = \int [dA] \Delta_c(A_\mu) \delta(C_a(A)) e^{i \int d^4x (L_{YM} + j.A)}, \qquad (3.111)$$

where $\Delta_c(A)$ defined by Eq. (3.95) is given in a more compact form by

$$\Delta_c(A) \int dg \, \delta(C_a(A^g)) = 1, \qquad (3.112)$$

where A^g is defined as in Eq. (3.94). As noted ealier, $\Delta_c(A)$ is gauge invariant (see Eq. (3.159)). Further, we note that $\Delta_c(A_\mu)$ appearing in Eq. (3.111) is multiplied by $\delta(C_a(A))$, i.e., it is evaluated at $C_a(A) = 0$. Thus, we write

$$\Delta_c(A)|_{C_a(A)=0} \equiv \Delta_c^0(A) = \det\left(\frac{\delta C_a}{\delta \lambda_b}\right). \qquad (3.113)$$

It is important to note that while $\Delta_c(A)$ is gauge invariant, $\Delta_c^0(A)$ is not, since $\Delta_c^0(A)$ is $\Delta_c(A)$ evaluated at $C_a(A) = 0$, and it is only $\Delta_c^0(A)$ that appears in the path integral. However, the following equality holds:

$$\Delta_c(A)\delta(C_a(A)) = \Delta_c^0(A)\delta(C_a(A)). \qquad (3.114)$$

Now, suppose we use a different type of gauge-fixing term where we replace $\delta(C_a(A))$ by $\delta(C_a(A) - \alpha_a)$. In this case, we define a new path integral so that

$$Z_\alpha = \int [dA] \Delta_{C(\alpha)}(A_\mu) \delta(C_a(A) - \alpha_a) e^{i \int d^4x (L_{YM} + j.A)}, \qquad (3.115)$$

where $\Delta_{C(\alpha)}(A)$ is defined by

$$\Delta_{C(\alpha)}(A) \int dg \, \delta(C_a(A) - \alpha_a) = 1, \qquad (3.116)$$

and we note that $\Delta_{C(\alpha)}(A)$ is again gauge invariant so that $\Delta_{C(\alpha)}(A^g) = \Delta_{C(\alpha)}(A)$. While $\Delta_{C(\alpha)}(A)$ depends on $\alpha_a(x)$, one may note that on multiplication by $\delta(C_a - \alpha_a)$ it is independent of α_a. This can be seen by noting

$$\Delta_{C(\alpha)}(A)|_{C_a(A)-\alpha_a=0} = \Delta_C^0(A). \qquad (3.117)$$

The above implies that the Faddeev–Popov ghost Lagrangian has no dependence on α, since it is determined by $\Delta_C^0(A)$. The independence of the Faddeev–Popov ghost Lagrangian on α is maintained if we multiply the path integral by a function of α_a and then integrate on α_a. Thus, we define a new path integral:

$$Z[j] = \int d\alpha_a \, e^{-\frac{i}{2} \int \alpha_a^2(x) dx} Z_\alpha[j]. \qquad (3.118)$$

Substitution of Eq. (3.115) into Eq. (3.118) gives us the following:

$$Z[j] = \int [dA] \Delta_C^0(A_\mu) e^{i \int d^4x \left(\mathcal{L}_{YM} - \frac{1}{2} C_a^2(x) + j \cdot A \right)}$$

$$= \int [dA] e^{i \int d^4x (\mathcal{L}_{YM} + \mathcal{L}_{GF} + \mathcal{L}_{FP} + j \cdot A)}. \tag{3.119}$$

3.6 BRST Transformations

Let us now summarize our analysis thus far. The effective Lagrangian is a sum of three parts:

$$\mathcal{L}_{YM}^{eff} = \mathcal{L}_{YM} + \mathcal{L}_{GF} + \mathcal{L}_{FP}, \tag{3.120}$$

where $\mathcal{L}_{YM}$ is the non-gauge-fixed Yang–Mills Lagrangian given by

$$\mathcal{L}_{YM} = -\frac{1}{4} F_{\mu\nu}^a F^{\mu\nu a}, \tag{3.121}$$

and $\mathcal{L}_{GF}$ is the gauge-fixing term, which for $C_a = \partial_\mu A_a^\mu$ is given by

$$\mathcal{L}_{GF} = -\frac{1}{2\alpha} (\partial_\mu A_a^\mu)^2, \tag{3.122}$$

where we have taken a more general form for $\mathcal{L}_{GF}$ by including a factor $1/\alpha$, and $\mathcal{L}_{FP}$ is the Faddeev–Popov ghost Lagrangian, which is given by

$$\mathcal{L}_{FP} = \left(\partial_\mu \xi_a^\dagger \partial^\mu \xi_a + g \epsilon_{abc} \partial^\mu \xi_a^\dagger A_\mu^b \xi_c \right). \tag{3.123}$$

The $\mathcal{L}_{FP}$ can be written as follows:

$$\mathcal{L}_{FP} = \partial_\mu \xi_a^\dagger D^\mu \xi_a \tag{3.124}$$

$$D^\mu \xi_a = \partial^\mu \xi_a + g C_{abc} A_b^\mu \xi_c. \tag{3.125}$$

BRST transformations [11,12] exhibit the gauge invariance of the theory even after the gauge fixing is carried out. The basic idea of the BRST theory is to relate the gauge transformation parameter $\epsilon_a(x)$ to the ghost field $\xi_a(x)$ by the relation

$$\epsilon_a(x) = g \xi_a(x) \delta\theta, \tag{3.126}$$

where $\delta\theta$ is an anti-commuting c-number. Specifically, the BRST [11,12] transformations are given by the following:

$$\delta A_{\mu a} = -(D_\mu \xi)_a \delta\theta \tag{3.127}$$

$$\delta \xi_a = -\frac{1}{2} g C_{abc} \xi_b \xi_c \delta\theta \tag{3.128}$$

$$\delta \xi_a^\dagger = -\frac{1}{\alpha} (\partial_\mu A_a^\mu) \delta\theta. \tag{3.129}$$

We will now show that $\mathcal{L}_{YM}^{eff}$ is invariant under the above transformations. We begin by noting that under the transformations Eqs. (3.127)–(3.129)

$$\delta\mathcal{L}_{YM} = 0. \qquad (3.130)$$

Next, we look at the variation of $\mathcal{L}_{GF}$. Here, one finds

$$\delta\mathcal{L}_{GF} = -\frac{1}{\alpha}(\partial_\nu A_a^\nu)(\partial_\mu \delta A_a^\mu) \qquad (3.131)$$

$$= \frac{1}{\alpha}(\partial_\nu A_a^\nu)(\partial_\mu D^\mu \xi_a)\delta\theta, \qquad (3.132)$$

where we have substituted the value of δA_μ^a from Eq. (3.127) in the last step. Next, the variation of $\mathcal{L}_{FP}$ given by Eq. (3.124) takes the form

$$\delta\mathcal{L}_{FP} = \partial_\mu(\delta\xi_a^\dagger)D_\mu\xi_a + \partial^\mu\xi_a^\dagger\Big[\partial_\mu\delta\xi_a + gC_{abc}\delta A_\mu^b\xi_c + gC_{abc}A_\mu^b\delta\xi_c\Big]. \qquad (3.133)$$

We compute the various terms appearing on the right-hand side of Eq. (3.133) separately. Thus, we have

$$\partial_\mu(\delta\xi_a^\dagger)D^\mu\xi_a = \partial_\mu\left(-\frac{1}{\alpha}\partial_\nu A_a^\nu\delta\theta\right)D^\mu\xi_a \qquad (3.134)$$

$$= \frac{1}{\alpha}\partial_\mu(\partial_\nu A_a^\nu)D^\mu\xi_a\delta\theta, \qquad (3.135)$$

where we have taken into account the fact that $\delta\theta$ and ξ_a anti-commute.

$$\partial_\mu\xi_a^\dagger\partial^\mu\delta\xi_a = \partial_\mu\xi_a^\dagger\partial^\mu\left[-\frac{1}{2}gC_{abc}\xi_b\xi_c\delta\theta\right]$$

$$= -\frac{g}{2}\partial^\mu\xi_a^\dagger(C_{abc}\partial^\mu\xi_b\xi_c + C_{abc}\xi_b\partial^\mu\xi_c)\delta\theta$$

$$= -g\partial_\mu\xi_a^\dagger C_{abc}\partial^\mu\xi_b\xi_c\delta\theta \qquad (3.136)$$

$$g\partial^\mu\xi_a^\dagger C_{abc}\delta A_\mu^b\xi_c = g\partial^\mu\xi_a^\dagger C_{abc}[-D_\mu\xi_b]\delta\theta\xi_c \qquad (3.137)$$

$$= -g\partial^\mu\xi_a^\dagger C_{abc}[\partial_\mu\xi_b + gC_{bed}A_{\mu e}\xi_d]\delta\theta\xi_c \qquad (3.138)$$

$$= gC_{abc}\partial^\mu\xi_a^\dagger\partial_\mu\xi_b\xi_c\delta\theta + g^2\partial^\mu\xi_a^\dagger C_{abc}C_{bb'c'}A_\mu^{b'}\xi_{c'}\xi_c\delta\theta \qquad (3.139)$$

$$\partial_\mu\xi_a^\dagger gC_{abc}A_b^\mu\delta\xi_c = \partial_\mu\xi_a^\dagger gC_{abc}A_b^\mu\left[-\frac{1}{2}gC_{cb'c'}\xi_{b'}\xi_{c'}\delta\theta\right] \qquad (3.140)$$

$$= -\frac{1}{2}g^2\partial_\mu\xi_a^\dagger C_{abc}C_{cb'c'}A_b^\mu\xi_{b'}\xi_{c'}\delta\theta. \qquad (3.141)$$

Thus, the variation of the $\mathcal{L}_{GF} + \mathcal{L}_{FP}$ can be written as

$$\delta\mathcal{L}_{GF} + \delta\mathcal{L}_{FP} = \left[\frac{1}{\alpha}(\partial_\nu A_a^\nu)(\partial_\mu D^\mu \xi_a)\delta\theta + \frac{1}{\alpha}\partial_\mu(\partial_\nu A_a^\nu)D^\mu \xi_a \delta\theta\right]$$

$$+ \left[-g\partial_\mu \xi_a^\dagger C_{abc}\partial^\mu \xi_b \xi_c \delta\theta + gC_{abc}\partial^\mu \xi_a^\dagger \partial_\mu \xi_b \xi_c \delta\theta\right]$$

$$+ [g^2 \partial^\mu \xi_a^\dagger C_{abc}C_{bb'c'}A_\mu^{b'}\xi_{c'}\xi_c \delta\theta - \frac{1}{2}g^2 \partial_\mu \xi_a^\dagger C_{abc}C_{cb'c'}A_b^\mu \xi_{b'}\xi_{c'}\delta\theta].$$

$$(3.142)$$

The terms in the second bracket on the right-hand side of Eq. (3.142) cancel. The cancellation of the terms in the last bracket in Eq. (3.142) comes about by use of the anti-commuting property of $\xi_{b'}$ and $\xi_{c'}$ and the Jacobi identity

$$C_{bcd}C_{ade} + C_{cad}C_{bde} + C_{abd}C_{cde} = 0. \qquad (3.143)$$

which can be obtained from the identity

$$[T_a, [T_b, T_c]] + [T_b, [T_c, T_a]] + [T_c, [T_a, T_b]] = 0. \qquad (3.144)$$

Thus, the variation of $\delta\mathcal{L}_{YM}^{eff}$ gives

$$\delta\mathcal{L}_{YM}^{eff} = \frac{1}{\alpha}(\partial_\nu A_a^\nu)(\partial_\mu D^\mu \xi_a)\delta\theta + \frac{1}{\alpha}\partial_\mu(\partial_\nu A_a^\nu)D^\mu \xi_a \delta\theta \qquad (3.145)$$

$$= \frac{1}{\alpha}\partial_\mu[\partial_\nu A^{\nu a}D^\mu \xi_a]\delta\theta. \qquad (3.146)$$

Since $\delta\mathcal{L}_{eff}$ is a total divergence, its contribution to the effective action vanishes under the usual boundary conditions. The above analysis shows that the effective quantum chronodynamics (QCD) action with inclusion of the gauge-invariant part, the gauge fixing part, and the Faddeev–Popov ghost part is invariant under the BRST transformations. The BRST transformations satisfy the property of nilpotency (see Eq. (3.160)).

3.7 Quantum Yang–Mills Theory

The properties of Yang–Mills theory at the quantum level are typically very different from those of classical Yang–Mills theory. Thus, for example, QCD based on an $SU(3)$ Yang–Mills theory coupled to quarks is meant to describe strong interactions. However, although the Yang–Mills field has no mass, the strong interactions are finite range, and thus there must exist a mass gap. Further, the quarks are not directly visible, which means that they are confined and must form bound states to produce mesons and baryons which are color neutral. Some of the desired aspects of the quantum Yang–Mills theory have been checked by lattice gauge theory analyses in computer simulations. However, a rigorous mathematical understanding of quantum Yang–Mills theory is lacking. This has led the Clay Mathematical Institute to propose a Millennium Problem related to a proper understanding of quantum Yang–Mills theory and the mass gap. Specifically, the Millennium Problem states [13]: "Prove that for any compact simple gauge group G, quantum Yang–Mills theory on $\mathbb{R}^4$ exists and has a mass gap $\Delta > 0$." A solution to this problem does not appear imminent and may require unconventional ways to formulate gauge theories.

3.8 Problems

1. Consider parallel transport around a closed loop with the corners defined by the co-ordinates x, $x + \delta_1 x$, $x + \delta_1 x + \delta_2 x$, $x + \delta_2 x$. We define a parallel transport operator from point x_1 to x_2 by

$$P(x_1, x_2) = e^{ig \int_{x_1}^{x_2} A_\mu(x) dx^\mu}. \tag{3.147}$$

Parallel transport along a closed loop can be constructed by the product $P(\text{loop}) = P_1 P_2 P_3 P_4$, where $P_1 = P(x, \ x + \delta_1 x)$, $P_2 = P(x + \delta_1 x, \ x + \delta_1 x + \delta_2 x)$, $P_3 = P(x + \delta_1 x + \delta_2 x, \ x + \delta_2 x)$, and $P_4 = P(x + \delta_2 x, \ x)$. Show that for infinitesimal $\delta_1 x$ and $\delta_2 x$

$$P(\text{loop}) = e^{ig F_{\mu\nu} \delta_1 x^\mu \delta_2 x^\nu}, \tag{3.148}$$

where

$$F_{\mu\nu} = \partial_\mu A_\nu - \partial_\nu A_\mu - ig[A_\mu, A_\nu]. \tag{3.149}$$

2. Show that the minimization of the pure Yang–Mills action in Euclidean space corresponds to either self-duality, i.e., $F_{\mu\nu} = \tilde{F}_{\mu\nu}$, or anti-self-duality, i.e., $F_{\mu\nu} = -\tilde{F}_{\mu\nu}$. To prove this, it may be helpful to first show that the Euclidean action can be written as

$$S_E = \frac{1}{2} \int (d^4 x)_E (\vec{E} \mp \vec{B})_E^2 \pm \int (\vec{E} \cdot \vec{B})_E (d^4 x)_E. \tag{3.150}$$

3. For Yang–Mills theory we can add a term of the following type to the Lagrangian density:

$$\mathcal{L}_\theta = \theta F_{\mu\nu}^a \tilde{F}^{\mu\nu a}. \tag{3.151}$$

Show that $\mathcal{L}_\theta$ is a total divergence, i.e.,

$$\mathcal{L}_\theta = 8\theta \partial_\alpha \left(\epsilon^{\mu\nu\alpha\beta} tr(A_\beta \partial_\mu A_\nu) - \frac{2i}{3} g tr(A_\beta A_\mu A_\nu) \right), \tag{3.152}$$

where $A_\mu = A_\mu^a T_a$ and $tr(T_a T_b) = \frac{1}{2}\delta_{ab}$.

4. Show that for Maxwell theory Eqs. (3.34) and (3.40) are compatible because of gauge invariance.

5. Show that the Hamiltonian density of the coupled Maxwell–Dirac theory can be written as

$$\mathcal{H} = \left[\vec{E}_T^2 + (\vec{\nabla} \times \vec{A}_T)^2\right] + \mathcal{H}_D^0 + \vec{j}_T \cdot \vec{A}_T - \frac{1}{2}j^0 \nabla^{-2} j^0$$
$$- (\vec{j}_L \cdot \vec{\nabla} A_L + j^0 \partial_0 A_L) \tag{3.153}$$

where $\mathcal{H}_D^0$ is the Hamiltonian for the free Dirac field.

6. Show that the Hamiltonian for the coupled Yang–Mills–Dirac fields in the temporal gauge is given by

$$H = \int d^3x \frac{1}{2}\left[\vec{E}_a \cdot \vec{E}_a + \vec{B}_a \cdot \vec{B}_a\right] + \int d^3x \; \psi^\dagger \left[\vec{\alpha} \cdot (\frac{1}{i}(\vec{\nabla} - ig\vec{A}) + \beta m\right]\psi, \tag{3.154}$$

where

$$\vec{B}_a = \vec{\nabla} \times \vec{A}_a + gC_{abc}\vec{A}_b \times \vec{A}_c. \tag{3.155}$$

7. Using canonical commutation relations, show the consistency of the Hamiltonian equations of motion arising from Eqs. (3.81)–(3.83) with the Lagrangian equations of motion in the temporal gauge. Specifically, show that the following equations of motion result:

$$\partial_0 A_{ia} = -E_{ia} \tag{3.156}$$

and

$$-\partial_0 E_a^i + \partial_j F_a^{ij} + gC_{abc}A_j^b F_c^{ij} = gj_a^i. \tag{3.157}$$

Show that these equations are consistent with the Lagrangian equations of motion (Eq. (3.61) and (3.63)) in the temporal gauge.

8. Using canonical commutation relations in the temporal gauge, show that G_a satisfy a Lie algebra, i.e.,

$$[G_a(\vec{r}, t), G_b(\vec{r}', t)] = iC_{abc}G_c(\vec{r}, t)\delta^3(\vec{r} - \vec{r}'), \tag{3.158}$$

where G_a are defined by Eq. (3.85).

9. Verify the validity of Eq. (3.87) using canonical commutation relations.

10. Using the definition of $\Delta_c[A_\mu]$ as given by Eq. (3.95) and the property that the group measure is group invariant [14], i.e., $d(gg') = d(g')$, show that $\Delta_c[A_\mu]$ is invariant under group transformations, i.e.,

$$\Delta_c[A_\mu] = \Delta_c[A_\mu^g]. \tag{3.159}$$

11. Show that the BRST transformations are nilpotent, i.e.,

$$\delta^2 A^\mu = 0, \quad \delta^2 \xi = 0, \quad \delta^2 \xi^\dagger = 0. \tag{3.160}$$

12. Couple the Yang–Mills field to the gravitational field, and by variation of the gravitational metric show that the stress tensor of the Yang–Mills field is given by

$$T_{\mu\nu} = F^{(a)}_{\mu\lambda} F^{\lambda}_{\nu(a)} - \frac{1}{4} g_{\mu\nu} F^{(a)}_{\lambda\rho} F^{\lambda\rho}_{(a)}. \tag{3.161}$$

References

[1] C.-N. Yang and R. L. Mills, *Phys. Rev.* **96**, 191 (1954).
[2] R. Shaw, Ph.D. thesis, Cambridge University (1955).
[3] R. Utiyama, *Phys. Rev.* **101**, 1597 (1956).
[4] V. N. Gribov, *Nucl. Phys. B* **139**, 1 (1978).
[5] I. M. Singer, *Commun. Math. Phys.* **60**, 7 (1978).
[6] R. L. Arnowitt and S. I. Fickler, *Phys. Rev.* **127**, 1821 (1962).
[7] R. Jackiw and C. Rebbi, *Phys. Rev. Lett.* **37**, 172 (1976).
[8] C. G. Callan, Jr., R. F. Dashen and D. J. Gross, *Phys. Lett. B* **63**, 334 (1976).
[9] E. S. Abers and B. W. Lee, *Phys. Rep.* **9**, 1 (1973).
[10] L. D. Faddeev and V. N. Popov, *Phys. Lett. B* **25**, 29 (1967).
[11] C. Becchi, A. Rouet and R. Stora, *Phys. Lett. B* **52**, 344 (1974); *Comm. Math. Phys.* **42**, 127 (1975); *Ann. Phys.* **98**(2), 287–321 (1976).
[12] I. V. Tyutin, *Gauge Invariance in Field Theory and Statistical Physics in Operator Formalism*, Lebedev Physics Institute (1975), preprint 39, arXiv:0812.0580.
[13] A. M. Jaffe and E. Witten, in J. Carlson, A. Jaffe, and A. Wiles, eds., *The Millennium Prize Problems*, American Mathematical Society (2000), pp. 129–152.
[14] A. Haar, *Ann. Math.* **34**(1), 147 (1933).

Further Reading

P. H. Frampton, *Gauge Field Theories*, Benjamin/Cummings (1987).
R. Jackiw, *Rev. Mod. Phys.* **52**, 661 (1980).
B. W. Lee and J. Zinn-Justin, *Phys. Rev. D* **5**, 3121 (1972); *Phys. Rev. D* **5**, 3137 (1972); *Phys. Rev. D* **5**, 3155 (1972); *Phys. Rev. D* **8**, 4654 (1973).
C. Quigg, *Gauge Theories of the Strong, Weak, and Electromagnetic Interactions*, 2nd edn, Princeton University Press (2013), p. 9–22.
S. Weinberg, *The Quantum Theory of Fields*, Cambridge University Press (2000).

4

Spontaneous Breaking of Global and Local Symmetries

As discussed in Chapter 1, symmetries play an important role in particle physics. But symmetries as observed in nature are often broken. They can be broken in an explicit way where one simply adds terms to the Lagrangian which do not obey the symmetry. However, such a breaking is not very interesting from a theoretical view point, as the addition of explicit breaking reduces the predictiveness of the theory and, further, may destroy the good properties of the theory at the loop level. More interesting is the case of spontaneous breaking of a symmetry where the Lagrangian possesses a minimum of the potential away from the symmetric point. Expansion of the Lagrangian around the new vacuum then generates terms in the Lagrangian which appear as symmetry-breaking terms. The Lagrangian, however, still possesses the original symmetry except that now it is in a hidden form.

4.1 Spontaneous Breaking of Abelian Symmetries

We will discuss first the simple case of a Lagrangian with a complex scalar field ϕ, i.e.,

$$\mathcal{L}(\phi) = -\partial_\mu \phi \partial^\mu \phi^* - V(\phi), \quad V(\phi) = \mu^2 \phi \phi^* + \lambda(\phi \phi^*)^2. \tag{4.1}$$

The Lagrangian of Eq. (4.1) is invariant under a global $U(1)$ symmetry:

$$\phi(x) \to e^{i\theta} \phi(x), \tag{4.2}$$

where θ is not a function of space-time. The Hamiltonian density $\mathcal{H}$ is given by

$$\mathcal{H} = \dot{\phi}\dot{\phi}^* + (\vec{\nabla}\phi).(\vec{\nabla}\phi^*) + V(\phi). \tag{4.3}$$

The first two terms on the right-hand side of Eq. (4.3) vanish for a constant ϕ, and the minimum is thus determined by the potential function $V(\phi)$. We would like the potential to be bounded from below for large $|\phi|$, which from Eq. (4.1)) implies that we must have $\lambda > 0$. The extrema condition by varying $V(\phi)$ with ϕ gives

$$\mu^2 \phi^* + 2\lambda(\phi\phi^*)\phi^* = 0. \tag{4.4}$$

There are two solutions to Eq. (4.4): $\phi = 0$ and $\phi \neq 0$. For the $\phi = 0$ case, $V(\phi) = 0$, and one has the symmetric vacuum, i.e., there is no spontaneous breaking of the global $U(1)$ symmetry. The case $\phi \neq 0$ gives

$$\phi_0 \equiv |\phi|_{SB} = \sqrt{-\frac{\mu^2}{2\lambda}}. \tag{4.5}$$

Of course, Eq. (4.5) makes sense only when $\mu^2 < 0$, i.e., we have a tachyonic mass term for the complex scalar field. In this case the potential at the extremum point ϕ_0 is

$$V(\phi_0) = -\frac{\mu^4}{4\lambda}. \tag{4.6}$$

Since $\lambda > 0$, $V(\phi_0) < 0$, and thus the vacuum solution $|\phi| = \phi_0$ lies lower than the symmetric vacuum $\phi = 0$. A check on stability conditions show that the point $\phi = 0$ is a point of instability while the point $|\phi| = \phi_0$ is a point of stability.

To see the particle spectrum around the new vacuum, we expand the complex scalar field ϕ around the new vacuum solution so that

$$\phi(x) = \frac{1}{\sqrt{2}}(v + \chi(x) + i\eta(x)), \tag{4.7}$$

where $\phi_0 = v/\sqrt{2}$. The Lagrangian around the new vacuum takes the form

$$\begin{aligned} -\mathcal{L}(\chi, \eta) = &\frac{1}{2}\partial_\mu\chi\partial^\mu\chi + \frac{1}{2}\partial_\mu\eta\partial^\mu\eta \\ &+ \lambda v^2\chi^2 + \lambda v\chi(\chi^2 + \eta^2) + \frac{\lambda}{4}(\chi^2 + \eta^2)^2. \end{aligned} \tag{4.8}$$

The Lagrangian of Eq. (4.8) exhibits the interesting feature that it has the following mass spectrum:

$$m_\chi = \sqrt{2\lambda}v, \quad m_\eta = 0. \tag{4.9}$$

Thus, the Lagrangian of Eq. (4.8) describes two interacting fields, one of which, i.e., the field χ, is massive while the other, i.e., the field η, is massless. The massless field η is the Goldstone boson corresponding to the breaking of the $U(1)$ symmetry. This is the simplest example of the Goldstone theorem which connects massless fields to broken global symmetries. (For early work on global symmetries and their breaking, see [1–3]; for a review, see Abers and Lee [4].) The presence of a massless scalar field is undesirable, and we need to remedy this situation.

A similar problem, i.e., the problem of massless modes, arises in the Bardeen–Cooper–Schrieffer (BCS) theory. There, it was shown by Anderson [5] that the undesirable modes can be eliminated by the introduction of a long-range force field such as the electromagnetic field. In the context of particle physics, there is a similar phenomenon. Thus, we consider the spontaneous breaking of the model discussed above but in the presence of a gauged $U(1)$ symmetry. There were three closely related works which addressed this problem around the mid-1960s: by Englert and Brout [6], Higgs [7,8], and Guralnik, Hagen, and Kibble [9,10] (for a comparison of the three approaches, see the review by Guralnik [11]). Here, we give an elementary discussion of the resolution of the massless mode problem. To discuss this possibility we extend the Lagrangian of Eq. (4.1) to possess a gauged $U(1)$ symmetry. This is easily done by converting the ordinary derivative to a covariant derivative so that

$$\partial_\mu \to D_\mu \equiv (\partial_\mu - igA_\mu),\tag{4.10}$$

and adding a kinetic term for the field A_μ so that the gauged $U(1)$ case has the Lagrangian

$$\mathcal{L}(\phi, A_\mu) = -(\partial_\mu - igA_\mu)\phi(\partial^\mu + igA^\mu)\phi^* - \frac{1}{4}F_{\mu\nu}F^{\mu\nu} - V(\phi).\tag{4.11}$$

The Lagrangian of Eq. (4.11) is invariant under the transformations

$$\phi(x) \to e^{i\theta(x)}\phi(x)$$

$$A_\mu \to A_\mu + \frac{1}{g}\partial_\mu\theta(x).\tag{4.12}$$

We now consider the spontaneous breaking in Eq. (4.11), assuming again that the potential $V(\phi)$ is given by Eq. (4.1) and $\mu^2 < 0$ and $\lambda > 0$. In this case, after spontaneous breaking and using the decomposition of $\phi(x)$ as given by Eq. (4.7), one finds that the kinetic terms assume the form

$$\mathcal{L}(\phi, A_\mu)_{K.E.} = -\frac{1}{4}F_{\mu\nu}F^{\mu\nu} - \frac{1}{2}\partial_\mu\chi\partial^\mu\chi - \frac{1}{2}\partial_\mu\eta\partial^\mu\eta + gvA^\mu\partial_\mu\eta.\tag{4.13}$$

From Eq. (4.13) we see that the kinetic terms are not diagonal due to the mixing term $A^\mu\partial_\mu\eta$, and thus the Lagrangian of Eq. (4.13) is not appropriate for interpretation in terms of particle states. We need to make a further transformation to eliminate the term $A^\mu\partial_\mu\eta$. To accomplish this instead of the transformation of Eq. (4.7), we make a different change of variables so that

$$\phi(x) = \frac{1}{\sqrt{2}}(v + h(x))e^{i\xi(x)/v}.\tag{4.14}$$

We also make the following change of variables for the spin 1 field so that

$$A_\mu(x) = B_\mu(x) + \beta\partial_\mu\xi(x).\tag{4.15}$$

It is easily checked that in order for $(\partial_\mu - igA_\mu)\phi$ to behave as a covariant derivative, one must have

$$\beta = \frac{1}{gv}.\tag{4.16}$$

In this case,

$$(\partial_\mu - igA_\mu)\phi = \frac{1}{\sqrt{2}}\left[\partial_\mu h - ig(v + h(x))B_\mu\right]e^{i\xi/v}, \tag{4.17}$$

and the spontaneously broken form of the Lagrangian takes the form

$$-\mathcal{L}(\phi, A_\mu) \to -\mathcal{L}_{\text{SB}} = \frac{1}{4}B_{\mu\nu}^2 + \frac{1}{2}m_B^2 B_\mu B^\mu + \frac{1}{2}\partial_\mu h \partial^u h + \frac{1}{2}m_h^2 h^2$$
$$+ (g^2 v)hB_\mu B^\mu + \frac{g^2}{2}h^2 B_\mu B^\mu + (\lambda v)h^3 + \frac{\lambda^4}{4}h^4, \tag{4.18}$$

where

$$m_B = gv, \quad m_h = \sqrt{2\lambda}v. \tag{4.19}$$

With the transformations of Eqs. (4.14) and (4.15) and the choice of Eq. (4.16), one finds that the kinetic terms in Eq. (4.18) have no mixing, and, further, that they are correctly canonically normalized.

Here, one finds that the massless vector field A_μ is replaced by a massive field B_μ. Thus, we started with a massless gauge field A_μ and a complex scalar field ϕ with a tachyonic mass term. After spontaneous breaking we find that the *physical* mass spectrum of the theory comprising of interacting fields consists of a massive vector field and a real scalar field, which is often referred to as the Higgs field. Before spontaneous breaking, the interacting fields had a total of four degrees of freedom, two from the massless gauge field A_μ and two from the tachyonic complex field ϕ. After spontaneous breaking, the *physical* spectrum consists of a massive field B_μ with three degrees of freedom and a massive real scalar field h with one degree of freedom, again with a total of four degrees of freedom. Thus, the process of spontaneous breaking has preserved the degrees of freedom in the physical sector consisting of interacting fields but has redistributed them between the spin 0 and the spin 1 sectors.

4.2 Spontaneous Breaking of Non-abelian Symmetries

Next, let us extend the analysis to the case when the underlying symmetry of the Lagrangian is non-abelian. Here, we first consider the simple case of an $O(n)$ symmetry where we have an n-plet of real fields ϕ_i ($i = 1, \ldots, n$) and a potential of the form

$$V(\phi) = \mu^2 \sum_{i=1}^{n} \phi_i^2 + \lambda \left(\sum_{i=1}^{n} \phi_i^2\right)^2, \tag{4.20}$$

where $\mu^2 < 0$ and $\lambda > 0$. This potential undergoes spontaneous breaking, and after spontaneous breaking one has $(n - 1)$ number of massless fields (see Problem 1 in Section 4.3). Spontaneous breaking reduces the symmetry of the Lagrangian from $O(n)$ to $O(n - 1)$. The group $O(n)$ has $n(n - 1)/2$ number of generators while the group $O(n-1)$ has $(n-1)(n-2)/2$, and the difference between the two is $n-1$, which precisely equals the number of massless degrees of freedom after spontaneous breaking. This equality of the number of broken generators and the number of massless degrees of freedom is by no means an accident. Rather, it is a consequence of a more general result which states that the number of massless degrees of freedom equals the number of broken generators after spontaneous breaking.

This result can be seen in the following: Let us assume that the potential $V(\phi)$ is invariant under a group G with the number of generators n_G and that ϕ belonging to an n-dimensional representation of G transforms so that

$$\delta\phi_i = -i\epsilon_a T_{ij}^a \phi_j, \tag{4.21}$$

where ϵ_a are the infinitesimal group transformation parameters. Let us vary the potential around the minimum of the potential so that

$$\left(\frac{\partial V}{\partial \phi_i}\right)_v = 0. \tag{4.22}$$

Two differentiations of the potential and use of Eq. (4.22) give us the relation

$$(M^2)_{ij} T_{jk}^a v_k = 0, \tag{4.23}$$

where $\langle \phi_k \rangle = v_k$ and where $(M^2)_{ij}$ is the mass squared matrix defined by $(M^2)_{ij} = (\partial^2 V/\partial\phi_i\partial\phi_j)_v$. For the case when $T_{jk}^a v_k$ are all zero, $(M^2)_{ij}$ are arbitrary. This is the case of no spontaneous symmetry breaking. Let us now assume that there is spontaneous breaking, which implies that one or more of the factors $T_{jk}^a v_k$ are non-zero. Let us suppose that there are K number of broken generators, which means that there are K independent combinations $T_{jk}^a v_k$ which are non-vanishing. For each of these non-vanishing combinations, Eq. (4.23) implies that the matrix $(M^2)_{ij}$ has a zero eigenvalue, which leads to K number of vanishing eigenvalues for the matrix $(M^2)_{ij}$ and thus K number of massless modes. This is a generalization of the $O(n)$ case so that for each broken generator we have one massless degree of freedom. Let the number of generators in the residual group H after symmetry breaking be n_H. This implies that $K = n_G - n_H$.

Analogous to the $U(1)$ case where the introduction of the Higgs mechanism removed the massless mode, which was absorbed by a massless gauge boson to become massive, one expects that such a phenomenon would extend to the non-abelian case. Thus, for the case when the Lagrangian has a global symmetry under the group G we can gauge this symmetry by the introduction of n_G number of gauge fields A_μ^a which belong to the adjoint representation of the gauge group, and the Lagrangian with the gauged symmetry has the form

$$\mathcal{L} = \mathcal{L}_{\text{kin}}^A + \mathcal{L}_{\text{kin}}^\phi - V(\phi). \tag{4.24}$$

where $\mathcal{L}_{\text{kin}}^A$ is the gauge-invariant kinetic energy term involving only the gauge fields and $\mathcal{L}_{\text{kin}}^\phi$ is the gauge-invariant kinetic term for the scalar fields, which is given by

$$\mathcal{L}^{\phi}_{\text{kin}} = -\frac{1}{2} D_{\mu}\phi (D^{\mu}\phi)^{\dagger}, \tag{4.25}$$

where $D_{\mu} = \partial_{\mu} - igT^a A^a_{\mu}$. Let us assume that there is spontaneous breaking which breaks $G \to H$ with $K = n_G - n_H$ number of massless fields. Weinberg [12] has shown that a gauge choice always exists, usually referred to as the unitary gauge, where the massless modes disappear from the spectrum and are absorbed by the vector gauge bosons to become massive [12]. We give a brief discussion of Weinberg's argument. The condition for the unitary gauge to exist is

$$(T^a v, \phi'(x)) = 0, \tag{4.26}$$

where ϕ' does not contain any massless spin zero components and can be obtained from $\phi(x)$ by a local unitary transformation. Let us assume that $\phi(x)$ is a real representation of the gauge group G. In this case, the desired local transformation is an orthogonal transformation. Thus, under the orthogonal transformation

$$\phi'(x) = O(x)\phi(x), \tag{4.27}$$

where

$$O(x) = e^{-i\xi_a(x)T^a}. \tag{4.28}$$

We may thus write Eq. (4.26) in the form

$$(T^a v, \, O_\phi \phi) = 0, \tag{4.29}$$

which is valid for all a where O_ϕ is the specific $O(x)$ that satisfies Eq. (4.29). To see how Eq. (4.29) comes about, let us consider the quantity Q defined by

$$Q = (v, O\phi). \tag{4.30}$$

Variation of O around O_ϕ, which is an extremum value, gives

$$\begin{aligned}
\delta Q = 0 &= (v, \delta O\phi)_{O=O_\phi} \\
&= -i\epsilon_a (v, T^a O_\phi \phi) \\
&= -i\epsilon_a (T^a v, O_\phi \phi).
\end{aligned} \tag{4.31}$$

Since Eq. (4.31) is valid for arbitrary ϵ_a, Eq. (4.29) follows. The vector boson masses in the unitary gauge are given by

$$\frac{1}{2}(gT^a v, gT^b v)A^a_\mu A^{\mu b}, \tag{4.32}$$

which gives a vector boson mass of

$$(M^2)^{ab} = g^2 (v, T^a T^b v). \tag{4.33}$$

The unitary gauge is the correct gauge to interpret the particle content of the theory, since in this case the massless fields are absent and the spectrum consists of only the physical fields. The massive vector bosons will now have a large p behavior of the type $p_\mu p_\nu / p^2 M^2$. This is to be contrasted with the propagator for the massless gauge boson, which is $(\eta_{\mu\nu} - p_\mu p_\nu / p^2)/p^2$. Thus, the analysis of renormalizability for the theory with massive vector bosons arising from spontaneously breaking is difficult in the unitary gauge [4]. A more appropriate gauge for the analysis of renormalizability, especially for the non-abelian case, is the R_ξ gauge (see e.g., [13]).

4.3 Problems

1. Consider the $O(n)$ invariant potential given by Eq. (4.20). Assuming $\mu^2 < 0$ and $\lambda > 0$, obtain the conditions for spontaneous breaking. Show that after spontaneous breaking there are $n-1$ massless fields left in the mass spectrum.
2. Write the Lagrangian of Eq. (4.11) after spontaneous breaking and expand the complex scalar field around the new vacuum using Eq. (4.7). Show that this Lagrangian, which involves the mixed term $A^\mu \partial_\mu \eta$ in the kinetic energy as seen in Eq. (4.13), is still invariant under gauge transformations so that

$$\chi \to \cos\theta(\chi + v) - \sin\theta\eta - v$$
$$\eta \to [\cos\theta\eta + \sin\theta(\chi + v)]$$
$$A_\mu \to \left[A_\mu + \frac{1}{g}\partial_\mu\theta\right].$$
(4.34)

References

[1] Y. Nambu and G. Jona-Lasinio, *Phys. Rev.* **122**, 345 (1961).
[2] J. Goldstone, *Nuovo Cim.* **19**, 154 (1961).
[3] J. Goldstone, A. Salam, and S. Weinberg, *Phys. Rev.* **127**, 965 (1962).
[4] E. S. Abers and B. W. Lee, *Phys. Rep.* **9**, 1 (1973).
[5] P. W. Anderson, *Phys. Rev.* **130**, 439 (1963).
[6] F. Englert and R. Brout, *Phys. Rev. Lett.* **13**, 321 (1964).
[7] P. W. Higgs, *Phys. Rev. Lett.* **13**, 508 (1964).
[8] P. W. Higgs, *Phys. Lett.* **12**, 132 (1964).
[9] G. Guralnik, C. Hagen, and T. Kibble, *Phys. Rev. Lett.* **13**, 585 (1964).
[10] T. W. B. Kibble, *Phys. Rev.* **155**, 1554 (1967).
[11] G. S. Guralnik, *Int. J. Mod. Phys. A* **24**, 2601 (2009).
[12] S. Weinberg, *Phys. Rev. D* **7**, 1068 (1973).
[13] K. Fujikawa, B. W. Lee, and A. I. Sanda, *Phys. Rev. D* **6**, 2923 (1972).

5

The Standard Model

Serious attempts to unify the electromagnetic and weak interactions can be traced to the early works by Schwinger [1] and by Salam and Ward [2, 3]. The successful unification of the electroweak interactions which gave rise to the current form of the standard model of electroweak interactions is due to the work of Glashow [4], Weinberg [5], and Salam [6], which is based on a gauged $SU(2)_L \otimes U(1)_Y$ theory which is spontaneously broken by the Higgs mechanism to a residual $U(1)_{em}$ gauge group. Further developments on the electroweak sector were carried out by Glashow, Iliopoulos, and Maiani [7], who resolved the problem of strangeness changing neutral current due to a rotation between the down and strange quarks given by the Cabibbo angle [8] and introduction of the CP phase in the quark sector of a three-generation model by Kobayashi and Maskawa [9]. The development of the strong interaction sector of the standard model involves extension of the gauge group structure of the theory from $SU(2)_L \otimes U(1)_Y$ to $SU(3)_C \otimes SU(2)_L \otimes U(1)_Y$, where the $SU(3)_C$ sector of the theory does not undergo spontaneous breaking and the $SU(3)_C$ gauge symmetry of color interactions is exact from the very high scales to the low scales. $SU(3)_C$ color interactions explain the phenomenon that the color interactions become feeble at high scales, which is known as asymptotic freedom [10–12], and strong at low scales, which leads to quark confinement.

5.1 Electroweak Sector of the Standard Model

As noted above, the electroweak sector of the standard model is based on the gauge group $SU(2)_L \otimes U(1)_Y$. It produces the correct V–A structure of weak interactions and also accommodates electromagnetism. Thus, the electroweak gauge sector of the model consists of a triplet of gauge vector bosons $A_{\mu a}$ ($a = 1, 2, 3$), which belong to the adjoint representation of the gauge group $SU(2)_L$, and B_μ, which is the gauge vector boson for the gauge group $U(1)_Y$. The kinetic energy for the gauge fields is given by

$$L_a = -\frac{1}{4}F_{\mu\nu a}F_a^{\mu\nu} - \frac{1}{4}B_{\mu\nu}B^{\mu\nu}, \tag{5.1}$$

where

$$F_{\mu\nu A} = \partial_\mu A_{\nu a} - \partial_\nu A_{\mu a} + gC_{abc}A_{\mu b}A_{\nu c}, \tag{5.2}$$

and where $C_{abc} = \epsilon_{abc}$ are the structure constants of the $SU(2)_L$ algebra and $B_{\mu\nu}$ is given by

$$B_{\mu\nu} = \partial_\mu B_\nu - \partial_\nu B_\mu. \tag{5.3}$$

In the electroweak model, leptons are placed in chiral left doublets of $SU(2)_L$ and right chiral singlets:

$$L_i = \begin{pmatrix} \nu_{Li} \\ e_{Li} \end{pmatrix}, \ e_{Ri}. \tag{5.4}$$

The gauge-covariant kinetic energy of the leptons can be written as

$$L_{\text{lep}}^{\text{kin}} = -\bar{L}_i \frac{1}{i}\gamma^\mu D_\mu L_i - \bar{e}_{R_i}\frac{1}{i}\gamma^\mu D_\mu e_{R_i}, \tag{5.5}$$

where the $SU(2)_L \otimes U(1)_Y$ gauge-covariant derivatives $D_\mu L_i$ and $D_\mu e_{R_i}$ are given by

$$D_\mu L_i = \left(\partial_\mu - iYg'B_\mu - ig\frac{\tau_a}{2}A_\mu^a\right)L_i \tag{5.6}$$

$$D_\mu e_{R_i} = (\partial_\mu - iYg'B_\mu)e_{R_i}, \tag{5.7}$$

where the hypercharge Y is related to the electrical charge by the relation Eq. (1.7). We now consider the lepton doublet $Y = -\frac{1}{2}$ and the right-handed charged lepton $Y = -1$. To break the gauge symmetry and also to give masses to the fermions, we introduce complex scalar fields Φ, which form a doublet of $SU(2)_L$ and carry a hypercharge $Y = \frac{1}{2}$:

$$\Phi = \begin{pmatrix} \phi^+ \\ \phi^0 \end{pmatrix}. \tag{5.8}$$

The gauge-covariant kinetic energy of the scalar fields is given by

$$-(D^\mu\Phi)^\dagger D_\mu\Phi, \tag{5.9}$$

where

$$D_\mu\Phi = \left(\partial_\mu - i\frac{g'}{2}B_\mu - ig\frac{\tau_a}{2}A_\mu^a\right)\Phi. \tag{5.10}$$

Finally, to obtain mass generation for the fermions we add a Yukawa coupling and a potential term, which breaks the $SU(2)_L \otimes U(1)_Y$ symmetry spontaneously, i.e.,

$$-(\bar{L}_i Y_{ij}\Phi e_{R_j} + \text{h.c.}) - V(\Phi^\dagger\Phi). \tag{5.11}$$

The scalar potential V is chosen to be of the simple form

$$V(\Phi^\dagger\Phi) = \mu^2(\Phi^\dagger\Phi) + \lambda(\Phi^\dagger\Phi)^2, \tag{5.12}$$

where $\lambda > 0$ and $\mu^2 < 0$ for spontaneous breaking of the group $SU(2)_L \otimes U(1)_Y$. With all the parts taken together, the electroweak Lagrangian including the gauge fields, the Higgs field, and the leptons takes the form

$$L = -\frac{1}{4}F_{\mu\nu a}F_a^{\mu\nu} - \frac{1}{4}B_{\mu\nu}B^{\mu\nu} - \bar{L}_i\frac{1}{i}\gamma^\mu D_\mu L_i - \bar{e}_{R_i}\frac{1}{i}\gamma^\mu D_\mu e_{R_i}$$
$$- (D^\mu\Phi)^\dagger D_\mu\Phi - (\bar{L}_i Y_{ij}\Phi e_{R_j} + \text{h.c.}) - V(\Phi^\dagger\Phi). \tag{5.13}$$

After spontaneous breaking, the group $SU(2)_L \otimes U(1)_Y$ will break to the group $U(1)_{em}$. If the group $SU(2)_L \otimes U(1)_Y$ was a global symmetry, the spontaneous breaking will produce three Goldstone bosons. However, we have here a gauged $SU(2)_L \otimes U(1)_Y$ group, and three Goldstone bosons are absorbed by three combinations of massless vector bosons of $SU(2)_L \otimes U(1)_Y$ to become massive. For the physical interpretation of the resultant theory, it is best to work in the unitary gauge where the Goldstone bosons have been absorbed and three of the four vector bosons are now massive. One may go to the unitary gauge by the following transformations:

$$\Phi(x) \to \Phi'(x) = U(\xi)\Phi(x)$$
$$L(x) \to L'(x) = U(\xi)L(x)$$
$$\vec{\tau}\cdot\vec{A}_\mu \to \vec{\tau}\cdot\vec{A}'_\mu = U(\xi)\left[\vec{\tau}\cdot\vec{A}_\mu - \frac{i}{g}U^{-1}(\xi)\partial_\mu U(\xi)\right]U^{-1}(\xi), \tag{5.14}$$

where

$$\Phi'(x) = \begin{pmatrix} 0 \\ \dfrac{v + h(x)}{\sqrt{2}} \end{pmatrix} \tag{5.15}$$

and where

$$U(\xi) = e^{-i\xi_a(x)\tau_a/2v}, \tag{5.16}$$

while e_R and B_μ are unchanged. In the above, we have replaced the complex scalar doublet $\Phi(x)$ with the fields $h(x)$ and $\xi_a(x)$ and defined a new lepton doublet L' and new fields A'^a_μ. In the unitary gauge, ξ_a disappear and we can drop the primes for convenience. With the new variables the scalar potential takes the form

$$V(H) = \frac{1}{2}m_h^2 h^2(x) + \frac{m_h^2}{2v}h^3(x) + \frac{m_h^2}{8v^2}h^4(x), \tag{5.17}$$

where

$$m_h = \sqrt{2\lambda}\,v. \tag{5.18}$$

The Yukawa couplings give rise to the mass growth for the leptons after the Φ field develops a vacuum expectation value (VEV). In terms of the new variables the Yukawa couplings have the expansion

$$-\mathcal{L}_Y = (\bar{L}_i Y_{ij} \Phi e_{R_j} + \text{h.c.})$$
$$= m_{e_i} \bar{e}_i(x) e_i(x) + 2^{1/4} \sqrt{G_F} m_{e_i} \bar{e}_i(x) e_i(x) h(x), \qquad (5.19)$$

where in the last step we have assumed the Yukawa matrices Y_{ij} to be diagonal and

$$m_{e_i} = \frac{1}{\sqrt{2}} Y_i v. \qquad (5.20)$$

From Eq. (5.19), it is seen that the Yukawa coupling of a charged lepton with the Higgs boson is proportional to the lepton mass. Thus, among the charged leptons the τ-lepton has the largest coupling to the Higgs boson while the electron has the smallest coupling.

Next, we look at the couplings of the charged vector bosons with the Higgs field. This interaction is given by Eq. (5.9). To get the mass terms we can retain just the vacuum expectation value of Φ, and obtain from $D_\mu \Phi$ the result

$$D_\mu \Phi \rightarrow -i \left[g \frac{\tau_a}{2} A_\mu^a + \frac{g'}{2} B_\mu \right] \begin{pmatrix} 0 \\ \frac{v}{\sqrt{2}} \end{pmatrix} \qquad (5.21)$$

$$= \frac{-iv}{2\sqrt{2}} \begin{pmatrix} g(A_{1\mu} - iA_{2\mu}) \\ -gA_{3\mu} + g'B_\mu \end{pmatrix}. \qquad (5.22)$$

Thus, the vector bosons mass term arising from $(D_\mu \Phi)^\dagger D^\mu \Phi$ is

$$(D_\mu \Phi)^\dagger D^\mu \Phi \rightarrow \frac{v^2}{8} \left[g^2 (A_{1\mu}^2 + A_{2\mu}^2) + (gA_{3\mu} - g'B_\mu)^2 \right]. \qquad (5.23)$$

We now introduce the normalized neutral vector boson fields

$$A_\mu = \frac{1}{\sqrt{g^2 + g'^2}} (g'A_{3\mu} + gB_\mu) \qquad (5.24)$$

$$Z_\mu = \frac{1}{\sqrt{g^2 + g'^2}} (gA_{3\mu} - g'B_\mu), \qquad (5.25)$$

where A_μ is a massless field and Z_μ is a massive neutral boson field. Defining

$$\tan \theta_W = \frac{g'}{g}, \qquad (5.26)$$

we may write A_μ and Z_μ in the following form:

$$A_\mu = \cos \theta_W B_\mu + \sin \theta_W A_{\mu 3} \qquad (5.27)$$
$$Z_\mu = -\sin \theta_W B_\mu + \cos \theta_W A_{\mu 3}, \qquad (5.28)$$

where θ_W is often referred to as the weak angle. Next, defining

$$M_W = \frac{v}{2} g \qquad (5.29)$$

$$M_Z = \frac{v}{2} \sqrt{g^2 + g'^2}, \qquad (5.30)$$

one finds the relation

$$\frac{M_W}{M_Z} = \frac{g}{\sqrt{g^2 + g'^2}} = \cos\theta_W.$$ (5.31)

A parameter used in precision electroweak fits of data is ρ defined by

$$\rho = \frac{M_W^2}{M_Z^2 \cos^2\theta_W}.$$ (5.32)

At the tree level, $\rho = 1$ in the standard model but small corrections to ρ arise at the loop level. The value $\rho = 1$ at the tree level reflects the fact that electroweak symmetry breaking is occurring from a Higgs doublet, and the value ρ will deviate significantly from unity if the electroweak symmetry was broken by another representation of the weak iso-spin. From Eq. (5.23) on using Eqs. (5.29) and (5.30), we can write the vector boson mass terms in the form

$$\frac{1}{2}M_W^2(A_{1\mu}A_1^\mu + A_{2\mu}A_2^\mu) + \frac{1}{2}M_Z^2 Z_\mu Z^\mu.$$ (5.33)

In Eq. (5.33) the M_W^2 term can be written as $M_W^2 W^+ W^-$, where

$$W_\mu^\pm(x) = (A_\mu^1 \mp iA_\mu^2)/\sqrt{2}.$$ (5.34)

Equation (5.31) implies $M_W < M_Z$, a result which is in conformity with experiment since experimentally $M_W \simeq 80.4$ GeV and $M_Z \sim 91.2$ GeV. We note that there is no mass term for the field A_μ. We will see shortly that A_μ can be identified as the photon field.

The Higgs interaction with $W^\pm$ is given by

$$\mathcal{L}_{WH} = -\frac{g^2}{4}(2vh + h^2)W_\mu^+ W^{\mu-},$$ (5.35)

and the interaction of the Z-boson with the Higgs field is given by

$$\mathcal{L}_{ZH} = -\frac{(g^2 + g'^2)}{8}(2vh + h^2)Z_\mu Z^\mu.$$ (5.36)

We now discuss the charged current interaction. For the leptons it is given by

$$\mathcal{L}_{CC} = \frac{g}{\sqrt{2}}\left[W_\mu^+ \bar{\nu}_{L_i}\gamma^\mu e_{L_i} + W_\mu^- \bar{e}_{L_i}\gamma^\mu \nu_{L_i}\right].$$ (5.37)

Correspondingly, the electromagnetic and the weak neutral current structure of leptons is given by

$$\mathcal{L}_{NC} = -\frac{1}{2}\bar{L}\gamma^\mu(g'B_\mu - g\tau_3 A_\mu^3)L - \bar{e}_R\gamma^\mu e_R g'B_\mu.$$ (5.38)

Substitution of B_μ and A_μ^3 in terms of A_μ and Z_μ gives us the following interaction:

$$\mathcal{L}_{NC} = - eA_\mu \bar{e}(x)\gamma^\mu e(x)$$
$$+ \frac{e}{4}(\sin\theta_W \cos\theta_W)^{-1} \left[\bar{\nu}\gamma^\mu(1-\gamma_5)\nu + \bar{e}_{\gamma^\mu}(-1 + 4\sin^2\theta_W + \gamma_5)e\right] Z_\mu,$$

$$(5.39)$$

where e, the electric charge, is defined by

$$e = \frac{gg'}{\sqrt{g^2 + g'^2}}. \tag{5.40}$$

using the relation

$$\sqrt{g^2 + g'^2} = e(\sin\theta_W \cos\theta_W)^{-1}. \tag{5.41}$$

From Eq. (5.39) we can identify A_μ to be indeed the photon field. Further, Eq. (5.39) indicates that the neutral current is not of the V–A form in the charged lepton sector.

We now discuss the low-energy energy limit of the weak interactions of the theory. As an illustration, let us consider the charged current interaction of the leptons given by Eq. (5.37). Taking the low-energy limit of this equation for the case when the energy involved in the scattering process is much smaller than M_W, one finds the effective interaction to be of the form

$$\frac{g^2}{8M_W^2} \bar{e}\gamma^\mu(1-\gamma_5)\nu_e \bar{\nu}_e\gamma^\mu(1-\gamma_5)e. \tag{5.42}$$

For a charged current interaction, the Fermi constant G_F is defined so that the four-fermion interaction has the form

$$\frac{G_F}{\sqrt{2}} \bar{e}\gamma^\mu(1-\gamma_5)\nu_e \bar{\nu}_e\gamma^\mu(1-\gamma_5)e. \tag{5.43}$$

A comparison of Eqs. (5.42) and (5.43) gives us the relation

$$\frac{G_F}{\sqrt{2}} = \frac{g^2}{8M_W^2}. \tag{5.44}$$

We next discuss the quark sector. Here, consider the first two generations of quarks where we introduce the following $SU(2)_L$ quark doublets and singlets:

$$q_1 = \begin{pmatrix} u_L \\ d_L^c \end{pmatrix}^{1/6}, \; u_R^{2/3}, \; d_R^{-1/3} \tag{5.45}$$

$$q_2 = \begin{pmatrix} c_L \\ s_L^c \end{pmatrix}^{1/6}, \; c_R^{2/3}, \; s_R^{-1/3}, \tag{5.46}$$

Here, u, d are the up and the down quarks while c, s are the charm and the strange quarks, respectively, and the superscripts are the values of Y. In the left doublets the objects d_L^c and s_L^c appear, which are the Cabibbo rotated quantities defined as follows:

$$\begin{pmatrix} d_L^c \\ s_L^c \end{pmatrix} = \begin{pmatrix} \cos\theta_C & \sin\theta_C \\ -\sin\theta_C & \cos\theta_C \end{pmatrix} \begin{pmatrix} d_L \\ s_L \end{pmatrix}. \tag{5.47}$$

The purpose of the Cabibbo rotated quantities is to allow for the strangeness changing charged current processes. With the iso-spin and hypercharge assignments of Eqs. (5.45) and (5.46), the gauge-covariant kinetic terms for the two quark generations are given by

$$\begin{aligned}
\mathcal{L}_{\text{KE}}^q = &-\bar{q}_1 \frac{1}{i}\gamma^\mu\left(\partial_\mu - igA_\mu^a\frac{\tau_a}{2} - i\frac{g'}{6}B_\mu\right)q_1 - \bar{q}_2\frac{1}{i}\gamma^\mu\left(\partial_\mu - igA_\mu^a\frac{\tau_a}{2} - i\frac{g'}{6}B_\mu\right)q_2 \\
&- \bar{u}_R\frac{1}{i}\gamma^\mu\left(\partial_\mu - i\frac{2g'}{3}B_\mu\right)u_R - \bar{c}_R\frac{1}{i}\gamma^\mu\left(\partial_\mu - i\frac{2g'}{3}B_\mu\right)c_R \\
&- \bar{d}_R\frac{1}{i}\gamma^\mu\left(\partial_\mu + i\frac{g'}{3}B_\mu\right)d_R - \bar{s}_R\frac{1}{i}\gamma^\mu\left(\partial_\mu + i\frac{g'}{3}B_\mu\right)s_R.
\end{aligned} \tag{5.48}$$

The charged current interaction arising from Eq. (5.48) is

$$\begin{aligned}
\mathcal{L}_{CC}^q = \frac{g}{\sqrt{2}}[(&\bar{u}_L\gamma^\mu d_L\cos\theta_C + \bar{u}_L\gamma^\mu s_L\sin\theta_C \\
&+ \bar{c}_L\gamma^\mu s_L\cos\theta_C - \bar{c}_L\gamma^\mu d_L\sin\theta_C)W_\mu^+ + \text{h.c.}].
\end{aligned} \tag{5.49}$$

The weak neutral current in Eq. (5.48) arises from the B_μ and A_μ^3 couplings. Equation (5.48) leads to the B_μ couplings so that

$$\mathcal{L}_{B_\mu} = \frac{g'}{3}\left[2\bar{u}_R\gamma^\mu u_R + 2\bar{c}_R\gamma^\mu c_R - (\bar{d}_R\gamma^\mu d_R + \bar{s}_R\gamma^\mu s_R) + \frac{1}{2}\bar{q}_1\gamma^\mu q_1 + \frac{1}{2}\bar{q}_2\gamma^\mu q_2\right]B_\mu. \tag{5.50}$$

Let us compute the last two terms of Eq. (5.50) separately. One finds

$$\begin{aligned}
\frac{1}{2}\bar{q}_1\gamma^\mu q_1 = \frac{1}{2}\Big(&\bar{u}_L\gamma^\mu u_L + \cos^2\theta_c\bar{d}_L\gamma^\mu d_L + \sin^2\theta_C\bar{s}_L\gamma^\mu s_L \\
&+ \sin\theta_C\cos\theta_C(\bar{d}_L\gamma^\mu s_L + \bar{s}_L\gamma^\mu d_L)\Big)
\end{aligned} \tag{5.51}$$

$$\begin{aligned}
\frac{1}{2}\bar{q}_2\gamma^\mu q_2 = \frac{1}{2}\Big(&\bar{c}_L\gamma^\mu c_L + \cos^2\theta_c\bar{s}_L\gamma^\mu s_L + \sin^2\theta_C\bar{d}_L\gamma^\mu d_L \\
&- \sin\theta_C\cos\theta_C(\bar{d}_L\gamma^\mu s_L + \bar{s}_L\gamma^\mu d_L)\Big).
\end{aligned} \tag{5.52}$$

We notice that in the absence of a charm quark to make up the doublet q_2 there will be a flavor-changing neutral current interaction, as can be seen from Eq. (5.51). Such a flavor-changing neutral current would produce a $K^0 \to \mu^+\mu^-$ decay at rates in violation of experiment. This problem was resolved by Glashow, Iliopoulos, and Maiani [7] by the introduction of the $SU(2)_L$ doublet q_2 defined by Eq. (5.46), where they introduced a new quark called the charm quark c for the second generation analogous to the up quark for the first generation.

With the introduction of the new doublet with s^c defined as in Eq. (5.46), one finds the remarkable result that there is a complete cancellation of the flavor-changing neutral current in the B_μ couplings arising from the sum of Eqs. (5.51) and (5.52). Thus, one has

$$\frac{1}{2}\bar{q}_1\gamma^\mu q_1 + \frac{1}{2}\bar{q}_2\gamma^\mu q_2 = \frac{1}{2}\left(\bar{u}_L\gamma^\mu u_L + \bar{d}_L\gamma^\mu d_L + \bar{c}_L\gamma^\mu c_L + \bar{s}_L\gamma^\mu s_L\right). \quad (5.53)$$

The cancellation of the flavor-changing currents in the neutral sector at the tree level arising from a cancellation between the first-generation and the second-generation contributions is referred to as the Glashow–Iliopoulos–Maiani (GIM) mechanism. Flavor-changing neutral current processes can arise at the loop level, but these are suppressed and pose no contradiction with experiments. We define the currents j^μ_{em} and j^3_μ as follows:

$$j^\mu_{em} = \frac{2}{3}(\bar{u}\gamma^\mu u + \bar{c}\gamma^\mu c) - \frac{1}{3}(\bar{d}\gamma^\mu d + \bar{s}\gamma^\mu s) \quad (5.54)$$

and

$$j^{\mu 3} = \frac{1}{2}\left(\bar{u}_L\gamma^\mu u_L - \bar{d}_L\gamma^\mu d_L + \bar{c}_L\gamma^\mu c_L - \bar{s}_L\gamma^\mu s_L\right). \quad (5.55)$$

One may now express the B_μ and A^3_μ couplings so that

$$\mathcal{L}^q_{NC} = g j^\mu_3 A^3_\mu + g'(j^\mu_{em} - j^\mu_3)B_\mu. \quad (5.56)$$

With substitution of B_μ and A^3_μ in terms of A_μ and Z_μ, one can write the neutral current interaction for the two generation of quarks in the form

$$\mathcal{L}^q_{NC} = e j^\mu_{em} A_\mu + \sqrt{g^2 + g'^2}(j^\mu_3 - \sin^2\theta_W j^\mu_{em})Z_\mu. \quad (5.57)$$

5.2 The Strong Interaction Sector of the Standard Model

Next, we consider the strong interaction sector of the theory. Here, the gauged group is color $SU(3)_C$, which has eight colored gluon fields A^μ_a, $a = 1, \cdots, 8$. Ignoring the $SU(2)_L \otimes SU(1)_Y$ gauge interactions, the color interactions between quarks and gluons and gluon self-interactions are described by the Lagrangian

$$\mathcal{L}_{SU(3)_C} = -\frac{1}{4}F_{\mu\nu a}F^{\mu\nu a} - \sum_i \bar{q}_i\frac{1}{i}\gamma^\mu(\partial_\mu - ig_s A^a_\mu T_a)q_i, \quad (5.58)$$

where the sum is over all the quark species i. Now, in a field theory the gauge coupling is a function of the renormalization group scale Q, and its evolution is governed by the renormalization group evolution equation. In a perturbative expansion, one has

$$Q\frac{d\alpha_s(Q)}{dQ} = \frac{\alpha_s^2}{\pi}b_1 + \left(\frac{\alpha_s^2}{\pi}\right)^2 b_2 + \cdots, \quad (5.59)$$

where $\alpha_s(Q) \equiv g_s^2/4\pi$. An explicit calculation of b_1 for a non-abelian gauge theory is given in the work of Politzer [10] and of Gross and Wilczek [11], and one has

$$b_1 = - \left[\frac{11}{6} C_A - \frac{2}{3} \sum_R n_R T_R \right], \tag{5.60}$$

where C_A is the quadratic Casimir for the adjoint representation in $SU(N)$, n_R is the number of fermions in the representation R of $SU(N)$, and T_R is the trace of the squares of generators in the R representation, i.e., $tr(T_a T_b) = T_R \delta_{ab}$. For $SU(N)$, $C_A = N$ and for n number of fermions in the fundamental representation of $SU(N)$ one has $T_F = \frac{1}{2}$. Thus, for a number n of fermions in the fundamental representation Eq. (5.60) reduces to

$$b_1 = -\frac{11}{6} N + \frac{1}{3} n. \tag{5.61}$$

When b_1 is negative the theory is asymptotically free, which means that the gauge coupling tends to become small for large Q while for small Q it tends to become large, leading to confinement for quarks. For the specific case $N = 3$, one finds from Eq. (5.61) that the theory is asymptotically free for up to $n = 16$ number of color triplets of quarks.

The standard model is anomaly free. This will be established in Chapter 6. The model has also been shown to be renormalizable [13]. Loop calculations in the model require the addition of gauge fixing and the inclusion of ghost terms, as discussed in Chapter 3.

5.3 Problems

1. Using the standard model interactions, show that the tree level decay widths of $W^- \to e^- \bar{\nu}$ and $Z^0 \to \bar{\nu}\nu$ are given by

$$\Gamma(W^- \to e^- \bar{\nu}) = \frac{G_F M_W^3}{6\pi\sqrt{2}} \tag{5.62}$$

$$\Gamma(Z^0 \to \nu\bar{\nu}) = \left(\frac{G_F M_Z^3}{12\pi\sqrt{2}} \right). \tag{5.63}$$

Also, show that the decay width $Z^0 \to e\bar{e}$ is related to the decay width of $Z^0 \to \nu\bar{\nu}$ so that

$$\Gamma(Z^0 \to e\bar{e}) = \Gamma(Z^0 \to \nu\bar{\nu}) \left[(2\sin^2\theta_W)^2 + (2\sin^2\theta_W - 1)^2 \right]. \tag{5.64}$$

2. In the standard model the first two generations of quarks consist of two left doublets $L_u^T = (u, d_c)_L$ and $L_c^T = (c, s_c)_L$ and four singlets u_R, d_R, c_R, s_R, where d_c and s_c are Cabibbo rotated, i.e., $d_c = d\cos\theta_c + s\sin\theta_c$ and $s_c = -d\sin\theta_c + s\cos\theta_c$. The Yukawa interaction for the above can be

written as

$$-L_{\text{Yukawa}} = C_1 \bar{L}_u \tilde{\Phi} u_R + C_2 \bar{L}_u \Phi d_R + C_3 \bar{L}_u \Phi s_R + C_4 \bar{L}_c \tilde{\Phi} c_R$$
$$+ C_5 \bar{L}_c \Phi d_R + C_6 \bar{L}_c \Phi s_R. \tag{5.65}$$

Imposing the condition that u, d, c, and s are mass eigenstates, obtain C_1–C_6 in terms of the four quark masses, m_u, m_d, m_c, and m_s.

3. In addition to the couplings of the Higgs and gauge fields with the fermions in the electroweak sector, one has couplings of Higgs and gauge fields. Show that the Lagrangian involving the Higgs fields and gauge fields is given by

$$\mathcal{L}_{HWZ} = - \left[M_W^2 W_\mu^+ W^{\mu-} + \frac{1}{2} M_Z^2 Z_\mu Z^\mu \right] \left(2 \frac{h(x)}{v} + \left(\frac{h(x)}{v} \right)^2 \right). \tag{5.66}$$

4. Show that the self-interaction of the gauge fields after spontaneous breaking is given by

$$\mathcal{L}_{AWZ}^{\text{int}} = ig(W_{\mu\nu}^+ W^{\mu-} - W_{\mu\nu}^- W^{\mu+})(\sin\theta_W A^\nu + \cos\theta_W Z^\nu)$$
$$+ igW^{\mu+} W^{\nu-}(\sin\theta_W F_{\mu\nu} + \cos\theta_W Z_{\mu\nu})$$
$$+ g^2 \left(\frac{1}{2} g^{\mu\alpha} g^{\nu\beta} + \frac{1}{2} g^{\mu\beta} g^{\nu\alpha} - g^{\mu\nu} g^{\alpha\beta} \right)$$
$$\times \left[W_\mu^+ W_\nu^- \left(A_\alpha A_\beta \sin^2\theta_W + Z_\alpha Z_\beta \cos^2\theta_W + 2A_\alpha Z_\beta \sin\theta_W \cos\theta_W \right) \right.$$
$$\left. - \frac{1}{2} W_\mu^+ W_\nu^+ W_\alpha^- W_\beta^- \right], \tag{5.67}$$

where

$$W_{\mu\nu}^\pm = \partial_\mu W_\nu^\pm - \partial_\nu W_\mu^\pm \tag{5.68}$$

and $F_{\mu\nu}$ and $Z_{\mu\nu}$ are similarly defined and the gauge kinetic energy terms are given by

$$\mathcal{L}_{AWZ}^0 = -\frac{1}{2} W_{\mu\nu}^+ W^{\mu\nu-} - \frac{1}{4} F_{\mu\nu} F^{\mu\nu} - \frac{1}{4} Z_{\mu\nu} Z^{\mu\nu}. \tag{5.69}$$

5. Consider a left–right symmetric Higgs sector where the covariant kinetic energy term is

$$\mathcal{L}_{kin} = -(D_\mu \phi_L)^\dagger D^\mu \phi_L - (D_\mu \phi_R)^\dagger D^\mu \phi_R \tag{5.70}$$

and where the covariant derivatives with respect to $SU(2)_L \otimes SU(2)_R \otimes U(1)$ are given by

$$D_\mu \phi_L = \partial_\mu \phi_L - ig \frac{\tau_a}{2} W_L^a \phi_L - \frac{i}{2} g' B_\mu \phi_L \tag{5.71}$$

$$D_\mu \phi_R = \partial_\mu \phi_R - ig \frac{\sigma_a}{2} W_R^a \phi_R - \frac{i}{2} g' B_\mu \phi_R. \tag{5.72}$$

Here, τ_a and σ_a $(a = 1, 2, 3)$ are the Pauli matrices. Assume that spontaneous symmetry breaking occurs so that

$$\langle \phi_L \rangle = \frac{1}{\sqrt{2}} \begin{pmatrix} 0 \\ v_L \end{pmatrix}, \quad \langle \phi_R \rangle = \frac{1}{\sqrt{2}} \begin{pmatrix} 0 \\ v_R \end{pmatrix}. \tag{5.73}$$

Compute masses for the $W_L^\pm$ and $W_R^\pm$ vector bosons. Also, obtain the mass-squared matrix for the neutral vector boson sector. Show that the eigenvalues contain a massless mode and two massive neutral vector bosons. Compute the masses for the massive neutrals.

References

[1] J. S. Schwinger, *Ann. Phys.* **2**, 407 (1957).
[2] A. Salam and J. C. Ward, *Nuovo Cim.* **11**, 568 (1959).
[3] A. Salam and J. C. Ward, *Phys. Lett.* **13**, 168 (1964).
[4] S. L. Glashow, *Nucl. Phys.* **22**, 579 (1961).
[5] S. Weinberg, *Phys. Rev. Lett.* **19**, 1264 (1967).
[6] A. Salam, in N. Svartholm, ed., *Elementary Particle Theory*, Almquist and Wiksells, (1969), p. 367.
[7] S. L. Glashow, J. Iliopoulos, and L. Maiani, *Phys. Rev. D* **2**, 1285 (1970).
[8] N. Cabibbo, *Phys. Rev. Lett.* **10**, 531 (1963).
[9] M. Kobayashi and T. Maskawa, *Prog. Theor. Phys.* **49**(2), 652 (1973).
[10] H. D. Politzer, *Phys. Rev. Lett.* **30**, 1346 (1973).
[11] D. J. Gross and F. Wilczek, *Phys. Rev. Lett.* **30**, 1343 (1973).
[12] D. J. Gross, in *Stanford 1992: The Rise of the Standard Model*, [hep-ph/9210207], pp. 199–232.
[13] G. 't Hooft, *Nucl. Phys. B* **35**, 167 (1971). See also *Nucl. Phys. B* **33**, 173 (1971); G. 't Hooft and M. J. G. Veltman, *Nucl. Phys. B* **44**, 189 (1972).

Further Reading

E. S. Abers and B. W. Lee, *Phys. Rep.* **9**, 1 (1973).
C. Quigg, *Front. Phys.* **56**, 1 (1983).

6

Anomalies

6.1 Introduction

The process $\pi^0 \to 2\gamma$ has been of theoretical interest for some time (for an early work, see Steinberger [1]). In the late 1960s this process played a central role in the discovery of anomalies. Thus, using current algebra techniques on the decay $\pi^0 \to 2\gamma$, Sutherland [2] in 1967 showed that the $\pi^0 \to 2\gamma$ decay amplitude vanished when one used the soft pion method when the divergence of the axial current is used as an interpolating field, i.e.,

$$\partial_\mu A_a^\mu = f_\pi m_\pi^2 \phi_a. \tag{6.1}$$

Here, A_a^μ ($a = 1, 2, 3$) is the axial vector current, ϕ_a is the pion field, f_π is the pion decay constant, and m_π is the pion mass. In the soft pion approximation it is assumed that the pion decay amplitude $T_\pi(p_\pi^2)$ is a smooth function of p_π^2 and $T_\pi(-m_\pi^2) \simeq T_\pi(0)$. Thus, one possible explanation for the vanishing of the $\pi^0 \to 2\gamma$ amplitude could be that the soft pion approximation breaks down in this case and the amplitude is a rapid function of p_π^2 vanishing at $p_\pi^\mu = 0$. However, if this were the case then an analysis which did not use the soft pion approximation should yield a non-null result. However, a current algebra calculation which did not use the soft pion approximation showed that the decay amplitude $\pi^0 \to 2\gamma$ again vanishes [3]. It was then proposed that the divergence of the axial current be modified with an additional term on the right-hand side of Eq. (6.1), which is independent of the pion mass [3]. This additional term survives in the limit of vanishing pion mass, and allows one to explain the two-photon decay of the π^0 and η^0 [3].

The above result is obtained in the framework of an effective Lagrangian. One can probe this question at the more fundamental level, which is at the level of quarks coupled to the electromagnetic field. In such a framework, the anomaly must arise at the quantum level. Two works which appeared in this period showed exactly that. In one work [4], the analysis done in the framework of the sigma model showed that the anomaly arises via a triangle-loop while the second one [5] investigated the axial-vector vertex in spinor electrodynamics. A modification of the partially conserved axial current (PCAC) condition was obtained, but this time with a front factor, which is determined. Specifically, even though the Lagrangian of the massless spinor electrodynamics is γ_5-invariant, the axial current

$$J_5^\mu(x) = \frac{\delta L}{\delta(\partial_\mu \theta)} = \bar{\psi}(x)\gamma^\mu \gamma_5 \psi(x), \tag{6.2}$$

associated with the γ_5 transformations

$$\psi(x) \to (1 + i\gamma_5\theta(x))\ \psi(x), \tag{6.3}$$

is not conserved. Assuming that the vector current is conserved, a perturbation calculation gives

$$\partial_\mu J_5^\mu(x) = -\frac{1}{16\pi^2}\epsilon^{\mu\nu\lambda\rho}F_{\mu\nu}(x)F_{\lambda\rho}(x). \tag{6.4}$$

More generally, consider the Lagrangian

$$L = -\bar{\psi}(x)\frac{1}{i}\gamma^\mu(\partial_\mu - iA_\mu^a T_a)\ \psi(x), \tag{6.5}$$

where T_a are the generators of some group G. In this case, a similar analysis gives

$$\partial_\mu J_5^\mu(x) = -\frac{1}{16\pi^2}\epsilon^{\mu\nu\lambda\rho}tr F_{\mu\nu}(x)F_{\lambda\rho}(x), \tag{6.6}$$

where $F_{\mu\nu} = F_{\mu\nu}^a T_a$, and $F_{\mu\nu}^a$ are non-abelian field strengths. The above result may be rewritten as

$$\partial_\mu J_5^\mu(x) = -\frac{1}{4\pi^2}\epsilon^{\mu\nu\lambda\rho}\ tr\ \partial_\mu(A_\nu \partial_\lambda A_\rho - i\frac{2}{3}A_\nu A_\lambda A_\rho). \tag{6.7}$$

We will refer to Eq. (6.7) as the abelian anomaly. Now assume we consider instead of Eq. (6.5) just the left chiral fields, i.e., the Lagrangian

$$L = -\bar{\psi}_L(x)\frac{1}{i}\gamma^\mu(\partial_\mu - iA_\mu^{La}T_a)\ \psi_L(x). \tag{6.8}$$

In this case we have the current

$$J_{L\alpha}^\mu(x) = \bar{\psi}_L(x)\gamma^\mu T_\alpha \psi_L(x). \tag{6.9}$$

Here, the analysis leads to [6]

$$\partial_\mu J_{L a}^\mu(x) = \frac{1}{24\pi^2}\epsilon_{\mu\nu\lambda\rho}\ tr\left[T_a \partial_\mu\left(A_\nu^L \partial_\lambda A_\rho^L - i\frac{1}{2}A_\nu^L A_\lambda^L A_\rho^L\right)\right]. \tag{6.10}$$

We could also consider the case with the left chiral fields replaced by the right chiral fields. In this case, Eq. (6.10) is replaced by

$$\partial_\mu J^\mu_{Ra}(x) = -\frac{1}{24\pi^2}\epsilon_{\mu\nu\lambda\rho}\, tr\left[T_a\partial_\mu\left(A^R_\nu\partial_\lambda A^R_\rho - i\frac{1}{2}A^R_\nu A^R_\lambda A^R_\rho\right)\right]. \qquad (6.11)$$

We will refer to Eqs. (6.10) and (6.11) as the non-abelian anomalies. It is to be noted that the terms cubic in the potential in the non-abelian anomalies have a factor of 1/2 compared with the factor of 2/3 for the case of the abelian anomaly of Eq. (6.7). One consequence of the above difference is that the non-abelian anomaly cannot be written in terms of the field strengths.

Since their discovery, the subject of anomalies has been further developed and applied in a variety of ways. Thus, aside from the Feynman diagram approach, other techniques in the investigations of anomalies include the path integral approach [7–11], the BRST approach [12–15] and the Wess–Zumino consistency conditions [16], and differential geometry techniques [17–20]. Non-renormalization of anomalies has been discussed [21–25] as well as the constraints on them in renormalizable gauge theories [26–30]. Anomalies also are found to play a role in the mesonic spectrum, specifically the mass of η' [31–37]. An investigation of anomalies in curved spaces appears in the literature [38–41]. The role of anomalies in instantons is also discussed [42–46], as are mathematical theorems related to anomalies [47–51]. Explicit evaluation of anomalies appears in a number of works [6, 52–55], and for the case with supersymmetry [56–58]. In the following we will discuss the Fujikawa approach, the Wess–Zumino consistency conditions and the BRST approach, and the differential geometric approach to anomalies.

6.2 The Path Integral Approach

We consider the case of abelian anomalies. The path integral describing the interactions of fermions ψ with gauge fields A_μ corresponding to a gauge group G is given by

$$Z = \int D\psi D\bar\psi\, e^{i\int S(A_\mu,\psi,\bar\psi)}. \qquad (6.12)$$

The central observation of Fujikawa [7–9] is that the fermionic measure is in general not invariant under local unitary chiral transformations, and this phenomenon leads to anomalies. Thus, we consider the change in the fermionic measure when we make a local unitary transformation:

$$\psi(x) \to \psi' = U(x)\psi(x), \quad \bar\psi(x) \to \bar\psi' = \bar\psi(x)\bar U(x), \qquad (6.13)$$

where $\bar{U}(x) = \gamma^0 U^\dagger(x)\gamma^0$. The transformation of the measure in this case will be given by

$$D\psi \to J^{-1}D\psi, \quad D\bar{\psi} \to \bar{J}^{-1}D\psi \tag{6.14}$$

and

$$D\psi D\bar{\psi} \to D\psi' D\bar{\psi}' = (J\bar{J})^{-1}D\psi D\bar{\psi}, \tag{6.15}$$

where $J = det(U(x))$ and $\bar{J} = det(\bar{U}(x))$ are the Jacobians of the respective transformations. The appearance of J^{-1} and $\bar{J}^{-1}$ is because the measure is for the fermionic fields. Let us first look at a non-chiral transformation where

$$\psi(x) \to \psi'(x) = e^{i\theta(x)}\psi(x)$$
$$\bar{\psi}(x) \to \bar{\psi}'(x) = \bar{\psi}(x)e^{-i\theta(x)}. \tag{6.16}$$

In this case $det\,\bar{U} = (det\,U)^{-1}$, which implies $J\bar{J} = 1$ and gives

$$D\psi D\bar{\psi} = D\psi' D\bar{\psi}'. \tag{6.17}$$

Here, the fermionic measure remains unchanged, and there is no anomaly. Next, we consider a local chiral transformations where

$$U(x) = e^{i\gamma_5\theta(x)}. \tag{6.18}$$

Here, $\gamma_5 = i\gamma^0\gamma^1\gamma^2\gamma^3$, where $\{\gamma^\mu, \gamma^\nu\} = -2\eta^{\mu\nu}\mathbb{1}$ with $\eta^{\mu\nu} = diag(-1, 1, 1, 1)$ and

$$\bar{U}(x) = \gamma^0 e^{-i\gamma_5\theta(x)}\gamma^0 = U(x), \tag{6.19}$$

where we have used $(\gamma^0)^2 = \mathbb{1}$ and Eq. (6.15) leads to

$$D\psi D\bar{\psi} = (det\,U)^{-2}D\psi' D\bar{\psi}'. \tag{6.20}$$

In this case the fermionic measure is no longer preserved under the chiral transformation. which points to the existence of an anomaly in this case. The rest of the analysis is to convert the factor of $(det\,U)^{-2}$ to an effective term in the Lagrangian. This requires regulating the theory in a careful way to extract the anomaly. We define the anomaly $\acute{a}(x)$ so that

$$(det\,U)^{-2} = e^{i\int d^4x\; \theta(x)\acute{a}(x)}. \tag{6.21}$$

Using the relation $\log det\,U = tr\;\log U$ we have

$$i\int d^4x\;\theta(x)\acute{a}(x) = -2\,tr\;\log\,U. \tag{6.22}$$

Since $\log U = i\theta(x)\gamma_5$, we need to compute the quantity $tr(i\theta(x)\gamma_5)$. This trace is over a complete set of fermionic functions, which we choose to be solutions of the gauge-covariant Dirac equation, i.e.,

$$\frac{1}{i}\gamma^\mu D_\mu f_n(x) = \lambda_n f_n(x). \tag{6.23}$$

Further, we choose these eigenfunctions to be properly normalized and to satisfy the completeness relation

$$\sum_n f_{n\alpha}(x) f_{n\beta}^\dagger(x') = \delta_{\alpha\beta}\delta^4(x - x'), \tag{6.24}$$

where α and β are Dirac indices. Using the basis functions $f_n(x)$, we write $tr \log U$ so that

$$tr \log U = tr(i\theta\gamma_5)$$
$$= i \int d^4x \sum_{n,\alpha,\beta} \theta(x) f_{n\alpha}^\dagger(x)(\gamma_5)_{\alpha\beta} f_{n\beta}(x). \tag{6.25}$$

This sum is divergent, and to control it we need to introduce a regulator, which we choose to be a function $G(t)$ with the properties that

$$G(0) = 1, \ [G^{(n)}(t)|_{\lim \ t\to\infty} = 0, \ n = 0, \ 1, \ 2, \tag{6.26}$$

where $G^{(n)}$ stands for the nth derivative. An example of such a regulator is e^{-t} but the details of the function do not concern us and the only property that we will need is Eq. (6.26). We also introduce the Fourier transform defined by

$$f_n(x) = \int d^4q \frac{e^{iq\cdot x}}{(2\pi)^4} \tilde{f}_n(q), \tag{6.27}$$

where the Fourier transform $\tilde{f}_n(q)$ satisfies the property

$$\sum_n \tilde{f}_{n\alpha}(q) \tilde{f}_{n\beta}^\dagger(q') = (2\pi)^4 \delta_{\alpha\beta}\delta^4(q - q'). \tag{6.28}$$

Using Eq. (6.27) and inserting the regulator function, we write Eq. (6.25) in the form

$$tr \ \log U = i \int d^4x \frac{d^4q}{(2\pi)^4} \frac{d^4q'}{(2\pi)^4} \theta(x) e^{iq\cdot x}$$
$$\sum_{n,\alpha,\beta} \tilde{f}_{n\alpha}^\dagger(q)(\gamma_5)_{\alpha\beta} G\left(\frac{\lambda_n^2}{\Lambda^2}\right) \tilde{f}_{n\beta}(q') e^{-iq'\cdot x}, \tag{6.29}$$

where the regulator function cuts off the integral for very large eigenvalues λ_n. Now, using Eq. (6.23) we can replace λ_n by the operator $\frac{1}{i}\gamma^\mu D_\mu$ and write

$$\left(\frac{1}{i}\gamma^\mu D_\mu\right)^2 = -\gamma^\mu \gamma^\nu D_\mu D_\nu$$
$$= -\frac{1}{2}(\gamma^\mu\gamma^\nu - \gamma^\nu\gamma^\mu)D_\mu D_\nu - \frac{1}{2}(\gamma^\mu\gamma^\nu + \gamma^\nu\gamma^\mu)D_\mu D_\nu$$
$$= -\frac{1}{2}\gamma^\mu\gamma^\nu[D_\mu, D_\nu] + D_\mu D^\mu$$
$$= \frac{i}{2}\gamma^\mu\gamma^\nu F_{\mu\nu} + D_\mu D^\mu. \tag{6.30}$$

Further, we have

$$\left(\frac{1}{i}\gamma^\mu D_\mu\right)^2 e^{-iq'\cdot x} = \left(\frac{i}{2}\gamma^\mu\gamma^\nu F_{\mu\nu} + D_\mu D^\mu\right) e^{-iq'\cdot x}$$

$$= e^{-q'\cdot x}\left[\frac{i}{2}\gamma^\mu\gamma^\nu F_{\mu\nu} + (-iq' + D)^2\right]. \qquad (6.31)$$

Using Eqs. (6.28) and (6.31) in Eq. (6.29), we have

$$tr \ \log \ U = i\int d^4x\frac{d^4q}{(2\pi)^4}tr\left[\theta\gamma_5 G(\frac{i}{2}\gamma^\mu\gamma^\nu F_{\mu\nu}/\Lambda^2 + (-iq + D)^2/\Lambda^2)\right]. \quad (6.32)$$

Next, we expand $G(q^2/\Lambda^2 + \cdots)$ around q^2/Λ^2 and compute the first non-vanishing term as $\Lambda \to \infty$. Here, we note that the following traces vanish: $tr(\gamma_5) = 0 = tr(\gamma_5\gamma^\mu\gamma^\nu) = 0$. The first non-vanishing term arises from the trace

$$tr(\gamma_5\gamma^\mu\gamma^\nu\gamma^\lambda\gamma^\rho) = -4i\epsilon^{\mu\nu\lambda\rho}. \qquad (6.33)$$

Thus, the anomaly is given by

$$\acute{a}(x) = -2\int\frac{d^4q}{(2\pi)^4}\frac{1}{\Lambda^4}\frac{1}{2}\left(-\frac{1}{4}\right)(-4i\epsilon^{\mu\nu\lambda\rho}F_{\mu\nu}F_{\lambda\rho})G''\left(\frac{q^2}{\Lambda^2}\right) + O(1/\Lambda^6). \quad (6.34)$$

Let us now make a change of variables and define $q^\mu/\Lambda \to q^\mu$ and take the $\Lambda \to \infty$ limit. In this case, Eq. (6.34) reads

$$\acute{a}(x) = -i\epsilon^{\mu\nu\lambda\rho}F_{\mu\nu}F_{\lambda\rho}\int\frac{d^4q}{(2\pi)^4}G''(q^2). \qquad (6.35)$$

To compute the integral we go to the Euclidean space by a Wick rotation so that $d^4q = id^4Q$ and

$$\int\frac{d^4q}{(2\pi)^4}G''(q^2) = \frac{i}{(2\pi)^4}(2\pi^2)\int_0^Q dQ \ Q^3G''(Q^2) = \frac{i}{16\pi^2}, \qquad (6.36)$$

where we have used Eq. (6.26). Using Eq. (6.36) in Eq. (6.35) we get

$$\acute{a}(x, A) = \frac{1}{16\pi^2}\epsilon^{\mu\nu\rho\lambda}tr(F_{\mu\nu}F_{\rho\lambda}). \qquad (6.37)$$

Let us now investigate the connection between the anomaly and the divergence of the axial current. This can be obtained by requiring that the path integral be invariant under the transformations of Eq. (6.13). Taking into account the non-invariance of the measure and the transformation property of the action, we find

$$Z = \int D\psi D\bar\psi e^{i\int S(A_\mu,\psi,\bar\psi)}e^{i\int d^4x \ \epsilon(x)\left(\partial_\mu J_5^\mu(x)+\acute{a}(x)\right)}. \qquad (6.38)$$

From the above, for the path integral to be invariant under the transformations of Eq. (6.13) with a fixed A_μ background, we must have

$$\acute{a}(x) = -\partial_\mu J_5^\mu(x). \tag{6.39}$$

Thus, we see that the axial current is not conserved, and the non-conservation of the axial current is given by the anomaly.

6.3 The Wess–Zumino Consistency Conditions

Next, we discuss the Wess–Zumino [16] consistency conditions, which are very powerful in determining the full form of the anomaly, including terms beyond those which are quadratic in the gauge fields. Let us assume that we are working only with the left chiral Lagrangian and we integrate out the fermionic fields. Thus, we will have an effective action which is a functional of the gauge fields. We define the effective action by

$$e^{i\Gamma[A_\mu]} = \int D\psi \, D\bar{\psi} \, \exp(i \int d^4x \, L). \tag{6.40}$$

We now look at the change in the functional with a variation of the gauge fields so that

$$\delta_\epsilon \Gamma[A] = \int d^4x \, \delta A_\mu^a(x) \frac{\delta\Gamma}{\delta A_\mu^a(x)} \tag{6.41}$$

$$= \int d^4x \, D_\mu \epsilon^a(x) \frac{\delta\Gamma}{\delta A_\mu^a(x)}, \tag{6.42}$$

where

$$D_\mu \epsilon^a = \partial_\mu \epsilon^a(x) + C_{abc} A_\mu^b \epsilon^c(x). \tag{6.43}$$

Integrating by parts we can write

$$\delta_\epsilon \Gamma[A] = \int d^4x \, \epsilon^a(x) X_a \Gamma[A], \tag{6.44}$$

where

$$X_a = \left(-D_\mu \frac{\delta}{\delta A_\mu(x)} \right)_a$$
$$= -\left[\frac{\partial}{\partial x^\mu} \frac{\delta}{\delta A_\mu^a(x)} + C_{abc} A_\mu^b(x) \frac{\delta}{\delta A_\mu^c(x)} \right]. \tag{6.45}$$

In the absence of an anomaly, the variation of the effective action will vanish under a variation of the gauge field but would be non-vanishing in the presence of the anomaly, and we define

$$\delta_\epsilon \Gamma[A] = \int d^4x \, \epsilon^a(x) \acute{a}_a(x). \tag{6.46}$$

Thus, the anomaly is given by

$$\acute{a}_a(A) = X_a\Gamma[A]$$

$$= \left(-D_\mu \frac{\delta}{\delta A_\mu(x)}\right)_a \Gamma[A]. \tag{6.47}$$

We now compute the commutator $[X_a(x), X_b(y)]$ so that

$$[X_a(x), X_b(y)] = C_{ab}^{(1)}(x,y) + C_{ab}^{(2)}(x,y), \tag{6.48}$$

where

$$C_{ab}^{(1)}(x,y) = -C_{abc}\left[\partial_\mu^x \delta^4(x-y)\frac{\delta}{\delta A_\mu^c(y)} + \partial_\mu^y \delta^4(x-y)\frac{\delta}{\delta A_\mu^c(x)}\right] \tag{6.49}$$

while

$$C_{ab}^{(2)}(x,y) = \delta^4(x-y)\left[-C_{cad}C_{bde} - C_{bcd}C_{ade}\right]A_\mu^c(x)\frac{\delta}{\delta A_\mu^e(y)}. \tag{6.50}$$

We note that $C_{ab}^{(1)}$ is effectively of the form (see Eq. (6.140))

$$C_{ab}^{(1)}(x,y) = -C_{abc}\delta^4(x-y)\partial_\mu^x \frac{\delta}{\delta A_\mu^c(x)}. \tag{6.51}$$

The $C_{ab}^{(2)}$ term on using the Jacobi identity on the product of $C's$ can be written as

$$C_{ab}^{(2)}(x,y) = -\delta^4(x-y)C_{abc}C_{cde}A_\mu^d(x)\frac{\delta}{\delta A_\mu^e(y)}. \tag{6.52}$$

Adding $C_{ab}^{(1)}$ and $C_{ab}^{(2)}$ one finds that

$$[X_a(x), X_b(y)] = \delta^4(x-y)C_{abc}X_c(x). \tag{6.53}$$

Multiplying the above equation by $\Gamma[A]$ from the right we find

$$X_a(x)\acute{a}_b(y) - X_b(y)\acute{a}_a(x) = \delta^4(x-y)C_{abc}\acute{a}_c(x). \tag{6.54}$$

Equation (6.54) is the Wess–Zumino consistency condition that the anomaly must satisfy. The consistency condition is powerful in the sense that a knowledge of the term in the anomaly which is quadratic in the gauge fields allows one to determine the full structure of the anomaly.

6.4 The BRST Approach and the Wess–Zumino Consistency Conditions

Here, we will show that from the nilpotency of the BRST transformations [12–15] one can deduce the Wess–Zumino consistency conditions. First, we introduce the

BRST operator s whose action on A_μ and the ghost fields is given by

$$sA_\mu^a = (D_\mu \xi)^a \tag{6.55}$$

$$s\xi^a = -\frac{1}{2}C_{abc}\xi^b \xi^c. \tag{6.56}$$

Let us look at the action of s on the effective action $\Gamma[A]$. Here, we have

$$\begin{aligned}
s\Gamma[A] &= \int d^4x (D_\mu \xi)^a \frac{\delta\Gamma}{\delta A_\mu^a(x)} \\
&= \int d^4x \, \xi^a \left[-D_\mu \frac{\delta}{\delta A_\mu(x)}\right]_a \Gamma[A] \\
&= \int d^4x \, \xi^a(x) X_a \Gamma \\
&= \int d^4x \, \xi^a(x) \acute{a}_a(A).
\end{aligned} \tag{6.57}$$

Equation (6.57) is similar to Eq. (6.46) except that ϵ^a in Eq. (6.46) is replaced by ξ^a in Eq. (6.57). It is thus useful to define a functional $\acute{a}[\xi, A]$ so that

$$\acute{a}[\xi, A] = \int d^4x \, \xi^a(x) \acute{a}_a(A). \tag{6.58}$$

Let us now operate by s once again and use the relation

$$s(\xi^a f(x)) = (s\xi^a)f(x) - \xi^a sf(x), \tag{6.59}$$

where we have used the anti-commuting property of s and ξ. Using the above we write

$$s^2\Gamma[A] = \int d^4x \left[(s\xi^a)\acute{a}_a(x, A) - \xi^a(s\acute{a}_a)\right]. \tag{6.60}$$

We compute the two terms on the right-hand side as follows:

$$\begin{aligned}
C_1 &\equiv \int d^4x (s\xi^a)\acute{a}_a \\
&= -\frac{1}{2}C_{abc}\int d^4x \, \xi^b(x)\xi^c(x)\acute{a}_a(x, A) \\
&= -\frac{1}{2}C_{abc}\int d^4x \, d^4y \, \xi^a(x)\xi^b(y)\acute{a}_c(x, A)\delta^4(x - y).
\end{aligned} \tag{6.61}$$

Next, we compute the second term:

$$\begin{aligned}
C_2 &= -\int d^4x \, \xi^a(s\acute{a}_a) \\
&= -\int d^4x \, \xi^a(x) \int d^4y \, \xi^b(y) X_b(y)\acute{a}_a(x) \\
&= -\frac{1}{2}\int d^4x \, d^4y \, \xi^a(x)\xi^b(y)[X_b(y)\acute{a}_a(x) - X_a(x)\acute{a}_b(y)].
\end{aligned} \tag{6.62}$$

Adding C_1 and C_2 we find

$$s^2\Gamma[A] = \int d^4x \, d^4y \left(-\frac{1}{2}\xi^a(x)\xi^b(y)\right)$$
$$\times \left[C_{abc}\delta^4(x-y)\acute{a}_c(x) + X_b(y)\acute{a}_a(x) - X_a(x)\acute{a}_b(y)\right]. \qquad (6.63)$$

Now, the use of the Wess–Zumino condition (Eq. (6.54)) in Eq. (6.63) leads to

$$s^2\Gamma[A] = 0. \qquad (6.64)$$

Alternatively, use of the nilpotency condition (Eq. (6.64)) leads to the Wess–Zumino consistency condition (Eq. (6.54)).

6.5 Application of Consistency Conditions to Determine Anomalies

In this section we demonstrate the utility of the Wess–Zumino consistency condition. In the preceding section we showed that the nilpotency of the BRST approach leads to the Wess–Zumino consistency conditions. In the analysis below, we will directly use the BRST consistency conditions to determine the missing parts of the anomaly equation. Consider the case of the anomaly arising from left chiral fermions coupling to a gauge fields. Using Eq. (6.10) we have (dropping the subscript L)

$$\acute{a}_a(x) = -\frac{1}{24\pi^2}\epsilon^{\mu\nu\lambda\rho} \, tr \left[T_a\partial_\mu\left(A_\nu\partial_\lambda A_\rho - i\frac{1}{2}A_\nu A_\lambda A_\rho\right)\right]. \qquad (6.65)$$

Next, using Eq. (6.65) in Eq. (6.46), we have

$$\delta_\epsilon\Gamma[A] = \int d^4x \, \epsilon^a(x)\acute{a}_a(x)$$
$$= -\frac{1}{24\pi^2}\int d^4x \, \epsilon^{\mu\nu\lambda\rho} \, Tr\left[\epsilon\partial_\mu\left(A_\nu\partial_\lambda A_\rho - i\frac{1}{2}A_\nu A_\lambda A_\rho\right)\right]. \qquad (6.66)$$

Next, we use the identity

$$\epsilon^{\mu\nu\lambda\rho}d^4x = dx^\mu \wedge dx^\nu \wedge dx^\lambda \wedge dx^\rho \qquad (6.67)$$

to write Eq. (6.66) in the form

$$\delta_\epsilon\Gamma[A] = -\frac{1}{24\pi^2}\int Tr\left[\epsilon \, dA \, dA + O(\epsilon A^3)\right]. \qquad (6.68)$$

To illustrate the power of the consistency conditions we will compute the $O(A^3)$ term from knowledge of the first term in Eq. (6.68). Thus, the constraint Eq. (6.64) reads

$$s\acute{a}[\xi, A] = 0. \qquad (6.69)$$

We use this constraint to determine the $O(A^3)$ terms in Eq. (6.68). For the case of Eq. (6.68), we can write the corresponding expression for $\acute{a}[\xi, A]$ by replacing ϵ by ξ inside the integral so that we have

$$\acute{a}[\xi, A] = \int d^4x \, \xi^a \acute{a}_a(x, \, A) \tag{6.70}$$

$$= -\frac{1}{24\pi^2} \int tr\left[\xi \, dA \, dA + O(\xi A^3)\right]. \tag{6.71}$$

For the analysis to follow it is more convenient to absorb factors of i in the gauge fields. Thus, we make the following field redefinitions:

$$\mathbf{A} = -\mathrm{i}A \tag{6.72}$$

$$\eta = -i\xi \tag{6.73}$$

$$\acute{a}[\eta, \, \mathbf{A}] \equiv \acute{a}[\xi, \, A]. \tag{6.74}$$

With these definitions, Eq. (6.71) can be written as follows:

$$\acute{a}[\eta, \mathbf{A}] = \frac{i}{24\pi^2} \int d^4x \, tr[\eta \, d\mathbf{A}d\mathbf{A} + \mathcal{O}(\eta\mathbf{A}^3)]. \tag{6.75}$$

We wish to use Eq. (6.69) to determine the $\mathcal{O}(\mathbf{A}^3)$ terms in Eq. (6.75).

To start off, we use dimensional analysis to guess the set of terms that can enter into $\acute{a}[\eta, \, \mathbf{A}]$. One finds straightforwardly that $s\acute{a}[\eta, \, \mathbf{A}]$ must have the general form

$$s\acute{a}[\eta, \mathbf{A}] = \frac{i}{24\pi^2} \int d^4x \, str[\eta \, (d\mathbf{A} \, d\mathbf{A} + a_1\mathbf{A}^2 \, d\mathbf{A}$$
$$+ a_2\mathbf{A} \, d\mathbf{A} \, \mathbf{A} + a_3 d\mathbf{A} \, \mathbf{A}^2 + a_4\mathbf{A}^4)], \tag{6.76}$$

where $a_1 - a_4$ are arbitrary parameters. We wish to use the constraint of Eq. (6.69) and the consistency condition Eq. (6.54) to determine the parameters. In the analysis below, the following relations involving the ghost field η and the BRST operator s will be found helpful:

$$d\eta = -\eta d$$
$$s\eta = -\eta\eta$$
$$sd = -ds$$
$$s\mathbf{A} = -d\eta - \{\mathbf{A}, \eta\}$$
$$s \, d\mathbf{A} = [d\mathbf{A}, \eta] - [\mathbf{A}, d\eta]. \tag{6.77}$$

The last part of Eq. (6.76) contains the term $str(\eta\mathbf{A}^4)$. We expand this term as below:

$$str(\eta\mathbf{A}^4) = tr[s\eta \, \mathbf{A}^4 - \eta s\mathbf{A}^4]$$
$$= -tr[\eta^2\mathbf{A}^4 + \eta s\mathbf{A}^4]. \tag{6.78}$$

We now use the following relation (see Eq. (6.142)):

$$s\mathbf{A}^4 = \mathbf{A}^3 \, d\eta - \mathbf{A}^2 \, d\eta \, \mathbf{A} + \mathbf{A} \, d\eta \, \mathbf{A}^2 - d\eta \, \mathbf{A}^3 + [\mathbf{A}^4, \eta]. \qquad (6.79)$$

Using the result of Eqs. (6.78) and (6.79), the terms with $\mathbf{A}^4$ in Eq. (6.76) are given by

$$a_4 \, str(\eta \mathbf{A}^4) = -a_4 \, tr \left(\eta^2 \mathbf{A}^4 + \eta [\mathbf{A}^4, \eta] \right) + \text{terms } O(\mathbf{A}^3) \text{ or lower}$$
$$= a_4 \, tr \, \eta^2 \mathbf{A}^4 + \cdots \qquad (6.80)$$

The BRST constraint demands that the right-hand side of Eq. (6.76) vanish (see Eq. (6.69)). Since Eq. (6.80) is the sum total of all $\mathbf{A}^4$ terms is Eq. (6.76), we deduce that

$$a_4 = 0. \qquad (6.81)$$

Looking at the middle three terms in Eq. (6.76), it can be seen that a significant simplification in the analysis will occur if these three terms arose from the derivative of a single term. The obvious such quantity is $d\mathbf{A}^3$, and since

$$d\mathbf{A}^3 = d\mathbf{A} \, \mathbf{A}^2 + \mathbf{A}^2 d\mathbf{A} - \mathbf{A} \, d\mathbf{A} \, \mathbf{A}, \qquad (6.82)$$

we find that a_1, a_2, and a_3 satisfy the following constraint:

$$a_1 = -a_2 = a_3. \qquad (6.83)$$

So, of the four arbitrary parameters a_1–a_4 in Eq. (6.76), we find that only one parameter is independent, which we may choose to be a_1, and our task now is to determine a_1.

In order to determine a_1 we need to evaluate explicitly the terms with coefficients a_1, a_2, and a_3. We first focus on the evaluation of the term in Eq. (6.76) with the coefficient a_1. Here, we have

$$a_1 \, str[\eta \mathbf{A}^2 \, d\mathbf{A}] = a_1 \, tr[s\eta A^2 \, d\mathbf{A} - \eta s \mathbf{A}^2 \, d\mathbf{A} - \eta \mathbf{A}^2 \, s \, d\mathbf{A}]$$
$$= a_1 \left(-\eta^2 \mathbf{A}^2 \, d\mathbf{A} \right) + a_1 \left(\eta \, d\eta \, \mathbf{A} \, d\mathbf{A} - \eta \mathbf{A} \, d\eta \, d\mathbf{A} - \eta [\mathbf{A}^2, \eta] d\mathbf{A} \right)$$
$$+ a_1 \left(-\eta \mathbf{A}^2 [\mathbf{A}^2, \eta] + \eta \mathbf{A}^2 [\mathbf{A}, d\eta] \right), \qquad (6.84)$$

where the three brackets on the right-hand side of Eq. (6.84) on the second and the third lines arise from the first, second, and third terms in the first line on the right-hand side of Eq. (6.84). From Eq. (6.84) we collect terms with one derivative and terms with two derivatives, which gives us

$$a_1 \, str[\eta \mathbf{A}^2 \, d\mathbf{A}] = a_1 \left(\eta^2 \mathbf{A}^2 \, d\mathbf{A} + d\eta \, \eta \mathbf{A}^3 + d\eta \, \mathbf{A}\eta \mathbf{A}^2 \right)$$
$$+ a_1 \left(\eta \, d\eta \, \mathbf{A} \, d\mathbf{A} - \eta \mathbf{A} \, d\eta \, d\mathbf{A} \right), \qquad (6.85)$$

where the first three terms on the right-hand side are the single -derivative terms and the remaining two terms are the terms involving two derivatives. Similar results for the a_2 and a_3 terms that appear in Eq. (6.76) can be obtained from

Eqs. (6.143) and (6.144). In the following we will use the results of Eqs. (6.85), (6.143), and (6.144). We first focus on terms with single derivatives. Here, we have

$$a_1 \; str(\eta \mathbf{A}^2 \; d\mathbf{A})_1 = a_1 \; tr \left(\eta^2 \mathbf{A}^2 \; d\mathbf{A} + d\eta \; \eta \mathbf{A}^3 + d\eta \; \mathbf{A} \eta \mathbf{A}^2\right) \tag{6.86}$$

$$a_2 s \; tr(\eta^2 \mathbf{A} \; d\mathbf{A} \; \mathbf{A})_1 = a_2 \; tr \left(\eta^2 \mathbf{A} \; d\mathbf{A} \; \mathbf{A} + \eta \mathbf{A} \; d\eta \; \mathbf{A}^2 + d\eta \; \mathbf{A} \eta \mathbf{A}^2\right) \tag{6.87}$$

$$a_3 \; str(\eta \; d\mathbf{A} \; \mathbf{A}^2)_1 = a_3 \; tr \left(\eta^2 \; d\mathbf{A} \; \mathbf{A}^2 - \eta \; d\eta \; \mathbf{A}^3 + \eta \mathbf{A} \; d\eta \; \mathbf{A}^2\right), \tag{6.88}$$

where the subscript on the left means that we are computing only the single-derivative terms. Using the constraint of Eq. (6.83) we find that the sum of the three terms of Eq. (6.88) gives

$$a_1 \; str[\eta \; (d\mathbf{A} \; \mathbf{A}^2 - \mathbf{A}^2 \; d\mathbf{A} + \mathbf{A} \; d\mathbf{A} \; \mathbf{A})]_1 = a_1 \; d(\eta^2 \mathbf{A}^3), \tag{6.89}$$

where the subscript means that we are computing only the single-derivative terms. Equation (6.89) shows that the single derivative terms add up to a total divergence, and thus on integration over $d^4 x$ the single-derivative terms do not contribute to the anomaly.

The absolute value of a_1 can be determined only by inclusion of the first term in Eq. (6.76). The constraint on a_1 would arise by imposing the condition that the combination of the first term along with the middle three terms lead to an $sá[\eta, \mathbf{A}]$, which is an integral of an exact form and would lead to satisfaction of the BRST constraint under the usual boundary conditions at the surface of a sphere in 4-space. To make progress, we notice that the first term in Eq. (6.76) contains two derivatives, and so we compute the two derivative terms arising from the three middle terms in Eq. (6.76). The two derivative terms arise from

$$(str \; \eta \mathbf{A}^2 \; d\mathbf{A})_2 = tr(\eta \; d\eta \; \mathbf{A} \; d\mathbf{A} - \eta \mathbf{A} \; d\eta \; d\mathbf{A})$$

$$(str \; \eta \mathbf{A} \; d\mathbf{A} \; \mathbf{A})_2 = tr(\eta \; d\eta \; d\mathbf{A} \; \mathbf{A} - d\eta \; \eta \mathbf{A} \; d\mathbf{A})$$

$$(str \; \eta \; d\mathbf{A} \; \mathbf{A}^2)_2 = tr(-d\eta \; \eta \; d\mathbf{A} \; \mathbf{A} + \eta \; d\mathbf{A} \; d\eta \; \mathbf{A}). \tag{6.90}$$

Adding them we get

$$X \equiv str \; \eta \left[a_1 \; d\mathbf{A} \; \mathbf{A}^2 + a_2 \; \mathbf{A}^2 \; d\mathbf{A} + a_3 \; \mathbf{A} \; d\mathbf{A} \; \mathbf{A}\right]_2$$

$$= a_1[(\eta \; d\eta \; \mathbf{A} \; d\mathbf{A} - \eta \mathbf{A} \; d\eta \; d\mathbf{A})$$

$$+ (-d\eta \; \eta \; d\mathbf{A} \; \mathbf{A} + \eta \; d\mathbf{A} \; d\eta \; \mathbf{A}) + (-\eta \; d\eta \; d\mathbf{A}\mathbf{A} + d\eta \; \eta \mathbf{A} \; d\mathbf{A})], \tag{6.91}$$

where we have used Eq. (6.83). Here, the first bracket on the right-hand side arises from the a_1 term in Eq. (6.76), the second from the a_2 term in Eq. (6.76), and the third from the a_3 term in Eq. (6.76). We can further simplify Eq. (6.91). Thus, we have

$$X = a_1 \; tr(\eta \; \overset{\mid}{d\mathbf{A}} \; d\eta \; \mathbf{A} - \eta \; d\eta \; d\mathbf{A}^2 - d\eta \; \eta \; d\mathbf{A}^2)$$

$$= a_1 \; tr(d\eta \; \mathbf{A} \; d\eta \; \mathbf{A} - 2 \; d\eta \; d\eta \; \mathbf{A}^2) + a_1 \; dG_1, \tag{6.92}$$

where G_1 is given in Eq. (6.149). Since dG_1 is closed, it does not contribute to the anomaly. Next, we note that $d\eta\,\mathbf{A}$ is an anti-commuting 2-form, so that

$$tr(d\eta\,\mathbf{A}\,d\eta\,\mathbf{A}) = 0. \tag{6.93}$$

Using Eq. (6.93) in Eq. (6.92), we find

$$X = -2a_1\,tr(d\eta\,d\eta\,\mathbf{A}^2) + a_1\,dG_1. \tag{6.94}$$

Next, we compute *str* $\eta d\mathbf{A}\,d\mathbf{A}$. We expand it as follows:

$$str\,\eta\,d\mathbf{A}\,d\mathbf{A} = tr[(s\eta)d\mathbf{A}\,d\mathbf{A} - \eta(s\,d\mathbf{A})d\mathbf{A} - \eta\,d\mathbf{A}(s\,d\mathbf{A})]. \tag{6.95}$$

In the computation of these, the following result is useful:

$$s\,d\mathbf{A} = [d\mathbf{A}, \eta] - [\mathbf{A}, d\eta]. \tag{6.96}$$

Using this result we can write the three terms on the right-hand side of Eq. (6.95) as follows:

$$tr[(s\eta)\,d\mathbf{A}\,d\mathbf{A}] = tr[-\eta^2\,d\mathbf{A}\,d\mathbf{A}]$$
$$tr[-\eta(s\,d\mathbf{A})d\mathbf{A}] = tr[-\eta\,[[d\mathbf{A}, \eta] - [\mathbf{A}, d\eta]]]$$
$$tr[-\eta\,d\mathbf{A}(s\,d\mathbf{A})] = tr[-\eta\,d\mathbf{A}\,[[d\mathbf{A}, \eta] - [\mathbf{A}, d\eta]]]. \tag{6.97}$$

The three parts in Eq. (6.97) provide a total of nine terms: one from the first, and four each from the second and the third. Of these, four terms cancel out, and we are left with five terms. Thus, using Eq. (6.97) in Eq. (6.95) one gets

$$str\,\eta\,d\mathbf{A}\,d\mathbf{A} = tr(\eta^2\,d\mathbf{A}\,d\mathbf{A} + \eta\mathbf{A}\,d\eta\,d\mathbf{A} + \eta\,d\mathbf{A}\,\mathbf{A}\,d\eta$$
$$- \eta\,d\eta\,\mathbf{A}\,d\mathbf{A} - \eta\,d\mathbf{A}\,d\eta\,\mathbf{A}). \tag{6.98}$$

Equation (6.98) can be reduced further:

$$str\,\eta\,d\mathbf{A}\,d\mathbf{A} = tr\left(d\eta\,d\eta\,\mathbf{A}^2 - d\eta\,\mathbf{A}\,d\eta\,\mathbf{A}\right) + dG_2$$
$$= tr\left(d\eta\,d\eta\,\mathbf{A}^2\right) + dG_2, \tag{6.99}$$

where G_2 is given in Eq. (6.150) and in the last step we have used Eq. (6.93). Adding the results of Eqs. (6.94) and (6.99), one finds that the terms with two derivatives cancel provided

$$a_1 = \frac{1}{2}. \tag{6.100}$$

Thus, we arrive at the conclusion that the BRST constraint of Eq. (6.69) determines the $O(\mathbf{A}^3)$ term, and $á[\eta, A]$ is determined to be

$$á[\eta, A] = \frac{i}{24\pi^2}\int tr\,\eta\left(d\mathbf{A}\,d\mathbf{A} + \frac{1}{2}d\mathbf{A}^3\right). \tag{6.101}$$

As mentioned earlier, we note that the $\mathbf{A}^3$ term in Eq. (6.101) has a 1/2 factor, which is different from the 2/3 factor that appears in the abelian anomaly case of Eq. (6.7). Further, the anomaly here cannot be expressed in terms of the $F\tilde{F}$ as was possible for the abelian case.

From Eq. (6.101), we deduce that the anomaly $\acute{a}_\alpha$ when written in normal notation and keeping in mind that the analysis is for the case of exchange of left chiral fermions, which we indicate by a superscript L, one has

$$
\begin{aligned}
\acute{a}_a^L &= -\frac{1}{24\pi^2}\epsilon^{\mu\nu\lambda\rho}\,tr(T_a\partial_\mu\left(A_\nu\partial_\lambda A_\rho - \frac{i}{4}A_\nu[A_\lambda, A_\rho]\right) \\
&= -\frac{1}{24\pi^2}\epsilon^{\mu\nu\lambda\rho}(\partial_\mu\left(A_\nu^b\partial_\lambda A_\rho^c - \frac{i}{4}A_\nu^b[A_\lambda, A_\rho]^c\right)d_{abc},
\end{aligned} \tag{6.102}
$$

where

$$
d_{abc} = tr(T_{(a}T_b T_{c)}). \tag{6.103}
$$

We can carry out a similar analysis for the exchange of right-handed fields in the triangle loop with an identical result except that the sign on γ_5 is reversed and as a consequence there is an overall minus sign relative to the anomaly for the left-handed fermion exchange case. Thus, we have

$$
\begin{aligned}
\acute{a}_a^R &= \frac{1}{24\pi^2}\epsilon^{\mu\nu\lambda\rho}\,tr(T_a\partial_\mu\left(A_\nu\partial_\lambda A_\rho - \frac{i}{4}A_\nu[A_\lambda, A_\rho]\right) \\
&= -\frac{1}{24\pi^2}\epsilon^{\mu\nu\lambda\rho}(\partial_\mu\left(A_\nu^b\partial_\lambda A_\rho^c - \frac{i}{4}A_\nu^b[A_\lambda, A_\rho]^c\right)d_{abc}.
\end{aligned} \tag{6.104}
$$

6.6 Mixed-gauge and Gravitational and Purely Gravitational Anomalies

We now discuss mixed-gauge and gravitational anomalies. Figures 6.1 and 6.2 show triangle diagrams where one has one vertex coupling to a $U(1)$ gauge field and two vertices coupling to gravitons with chiral fermions circulating in the loop. For the case of the standard model where $U(1)$ is $U(1)_L$, the anomaly is

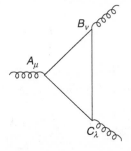

Figure 6.1. Anomaly diagram with gauge fields belong to three different gauge groups as external lines and with chiral fermions in the loop.

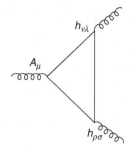

Figure 6.2. Anomaly diagram involving an abelian gauge field and two gravitons on external lines with chiral fermions in the loop.

proportional to

$$tr(Y)\epsilon^{\mu\nu\lambda\rho}R_{\mu\nu\alpha\beta}R_{\lambda\rho}{}^{\alpha\beta}. \tag{6.105}$$

For the case when $U(1)$ is replaced by $SU(N)$, $tr(Y)$ is replaced by $tr(T_a)$ in some given representation R. Since $tr(T_a) = 0$, $SU(N)$ gravitational anomalies all vanish, and thus the only mixed-gauge-gravitational anomaly one has to worry about is the $U(1)$ gravitational anomaly. In addition to the mixed-gauge and gravitational anomalies, there can also be pure gravitational anomalies. Here, one finds that such anomalies appear only in $4k + 2$ dimensions for k an integer, and are thus absent for the case of four dimensions.

6.7 Anomalies and Differential Geometry

In the preceding sections we have discussed a variety of different ways of looking at anomalies. Aside from the direct approach of Feynman diagrams, we discussed the path integral formulation of Fujikawa and the consistency conditions arising from the nilpotency of BRST transformations and the Wess–Zumino consistency conditions. The approach of differential geometry is very useful in obtaining solutions to the consistency equations [17–20]. Specifically, they are very useful for the analysis of anomalies in higher dimensions [59–62]. These techniques also extend to gravitational anomalies [20, 63–65]. Here, we discuss the case of the abelian anomaly in further detail. An object of geometric interest in the study of the abelian anomaly in four dimensions is the second Chern character $\Omega_4(A)$ defined by

$$\Omega_4(A) \equiv tr(F^2), \tag{6.106}$$

where a wedge product is understood, i.e., $tr(F^2) = tr(F \wedge F)$. [In $2n$ dimensional space-time the object of interest will be the $2n$ Chern character $\Omega_{2n} = tr(F^n)$.] The exterior derivative of Ω_4 gives

$$d\Omega_4 = 2 \, tr(dF \, F). \tag{6.107}$$

Next, let us consider the fields strength

$$F = dA + A^2. \tag{6.108}$$

The exterior derivative of the above gives

$$DF \equiv dF + [A, F] = 0. \tag{6.109}$$

which is the Bianchi identity. The use of Eq. (6.109) in Eq. (6.107) gives

$$\begin{aligned}
d\Omega_4 &= -2d\ tr([A, F]\ F) \\
&= -2d\ tr(AFF - FAF) \\
&= 0. \tag{6.110}
\end{aligned}$$

The above implies that Ω_4 is closed, and since Ω_4 is closed we can write it locally as an exterior derivative of a 3-form, i.e.,

$$\Omega_4(A) = d\omega_3^0(A). \tag{6.111}$$

Our next task is to determine $\omega_3^0(A)$. We proceed by making a variation on A so that $A \to A + \delta A$, which gives, on using Eq. (6.108),

$$\delta F = A\delta A + \delta AA + d(\delta A), \tag{6.112}$$

and leads to, on using Eq. (6.106),

$$\begin{aligned}
\delta\Omega_4 &= 2\ tr[A\delta AF + \delta AAF + d(\delta A)F] \\
&= 2\ tr[\delta A[A, F] + d(\delta A)F]. \tag{6.113}
\end{aligned}$$

Using the Bianchi identity (Eq. (6.109)) we have

$$\begin{aligned}
\delta\Omega_4 &= 2\ tr[-\delta AdF + d(\delta A)F] \\
&= 2d\ tr(\delta AF). \tag{6.114}
\end{aligned}$$

We need to integrate this equation to get Ω_4. To this end we consider the transformation

$$A \to A_\lambda = \lambda A \tag{6.115}$$

$$F \to F_\lambda = \lambda dA + \lambda^2 A^2, \tag{6.116}$$

where λ has the range (0, 1). Then,

$$\begin{aligned}
\frac{\delta\Omega_4^\lambda}{\delta\lambda} &= 2d\ tr\left(\frac{\delta A_\lambda}{\delta\lambda}F_\lambda\right) \\
&= 2d\ tr[A(\lambda dA + \lambda^2 A^2)]. \tag{6.117}
\end{aligned}$$

Integration of the above equation gives

$$\Omega_4 = \int_0^1 \delta\lambda \frac{\delta\Omega_4^\lambda}{\delta\lambda} = \int_0^1 \delta\lambda 2d\ tr[AdA\lambda + A^3\lambda^2]. \tag{6.118}$$

A comparison of Eqs. (6.118) and (6.111) leads to

$$\omega_3^0 = \int_0^1 d\lambda \; 2 \; tr[AdA\lambda + A^3\lambda^2]$$
$$= tr\left(AdA + \frac{2}{3}A^3\right). \tag{6.119}$$

The technique above extends easily to the analysis of anomalies in $2n$ dimensions. Thus, consider $tr(F^n)$. Using the same technique as in the deduction of Eq. (6.110) one has (see Eq. (6.151))

$$d \; tr(F^n) = 0. \tag{6.120}$$

Thus, $tr(F^n)$ is closed, and we can write

$$tr(F^n) = d\omega_{2n-1}. \tag{6.121}$$

Further, following the same procedure as before, we get

$$\omega_{2n-1} = n \; tr \int_0^1 d\lambda \; AF_\lambda^{n-1}. \tag{6.122}$$

This approach can be extended to the case of non-abelian anomalies but the analysis is more complex and the reader is directed to the literature. The differential geometric methods also provide an explanation of the factor of 2/3 versus 1/2 for the abelian versus non-abelian anomalies.

6.8 t'Hooft Matching Conditions

Anomalies provide constraints on preonic models [66–73]. Thus, t'Hooft has argued that the spectrum of bound states of a confining theory should be constrained by anomaly-matching conditions [66]. For instance, assume that we have a theory in terms of massless preonic particles. Next, assume that this theory confines but there is no spontaneous breaking. In that case the anomaly computed for the preonic theory must match the anomaly computed in terms of the composite spectrum. It has been argued that such a consistency condition could arise from the principles of analyticity and unitarity. The matching condition puts a significant constraint on building preonic models.

6.9 Anomaly Constraints for Unified Models

Model building on partially or fully unified groups must ensure the absence of anomalies for the model to be consistent at the quantum level. For some gauge groups this must be arranged by the choice of the matter representation while for other groups the absence of anomalies is automatic. Below is a brief summary of which groups are anomaly free and which are not, as this is helpful in the construction of unified models.

Anomalies in models using $U(1)$ and $SU(N)$ do not automatically vanish, and require constraints on matter representations for their cancellations. For example, for $SU(5)$ one needs a combination of $\bar{5} + 10$ representations of $SU(5)$ for the cancellation of the anomalies. For the $SU(N)$ gauge group, there will be matter in the anti-symmetric representations. For the kth rank anti-symmetric tensor the anomaly is given by [74]

$$A_N(k) = \frac{(N-3)!(N-2k)}{(N-k-1)!(k-1)!},$$
(6.123)

while the dimensionality of the kth anti-symmetric tensor is

$$d_N(k) = \frac{N!}{k!(N-k)!}.$$
(6.124)

Using the above, one can work out the anomaly-free combinations for any anti-symmetric representations of $SU(N)$. However, the absence of anomalies is automatic for many groups. We discuss this below.

- The following groups have only real and pseudo-real representations, and are thus automatically free of non-abelian anomalies [75]

$$SO(2N), \quad N \geq 2; \quad SO(2N+1), \quad N \geq 1$$
$$G_2, \ F_4, \ E_7, \ E_8$$
$$USp(2N), \quad N \geq 3$$
$$SU(2).$$
(6.125)

- There are groups where the representations are not real or pseudo-real but the group is anomaly free. Examples of these are

$$E_6; \quad SO(4N+2), \quad N \geq 1.$$
(6.126)

For model building we need chiral representations which are complex. Thus, the groups convenient for model building are $SO(4N+2), N \geq 1$; E_6; $SU(N)$; $U(1)$. Here, while $SO(2N+2), N \geq 1$, and E_6 are automatically anomaly free, for $SU(N)$ and $U(1)$ one needs to find combinations of matter representations which are anomaly free (e.g., see Eq. (6.154)).

6.9.1 Witten's Global Anomaly for SU(2)

It is known that an $SU(2)$ gauge theory with a single doublet of left-handed Weyl fermions is problematic since such a system does not allow fermion condensation. Consequently, the fermions would remain massless even in the presence of strong $SU(2)$ gauge interactions. It was shown by Witten [76] that such a theory is mathematically inconsistent and does not exist[*] The simple reason for this is that in a path integral formulation it is not possible to define a fermionic determinant which has a smooth gauge field dependence and is gauge invariant. Specifically, in a gauge theory with a doublet of Dirac fermions one has

$$\int d\psi \, d\bar{\psi} \, \exp(\bar{\psi}i\not{D}\psi) = det(i\not{D}). \tag{6.127}$$

For $SU(2)$ a doublet of Dirac fermions is equivalent of two doublets of Weyl fermions, and thus a gauge theory with a single doublet of Weyl fermions gives

$$\int d\psi \, d\bar{\psi} \, \exp(\bar{\psi}i\not{D}\psi) = (det(i\not{D}))^{1/2}. \tag{6.128}$$

However, the square root of the determinant has an ambiguous sign. To see the origin of the problem, one needs to look at the spectral flow of the eigenvalues of the Dirac operator. Thus, the determinant of Eq. (6.127) is a product of the eigenvalues of the Hermitian Dirac operator $i\not{D}$. One can look at the spectral flow of the eigenvalues as the gauge field $A_\mu(x, t)$ varies. As a specific example, we consider the case

$$A_\mu(x,t) = (1-t)A_\mu(x) + tA_\mu^g, \tag{6.129}$$

where A_μ^g is the gauge-transformed A_μ so that

$$A_\mu^g = g^{-1}A_\mu g - ig^{-1}\partial_\mu g, \tag{6.130}$$

where g refers to a large gauge transformation. Now, one may show that $(det(i\not{D}))^{1/2}$ has the property [76]

$$[det \, i\not{D}(A_\mu)]^{1/2} = -[det \, i\not{D}(A_\mu^g)]^{1/2}, \tag{6.131}$$

which implies that for the case of a single doublet of Weyl spinors the partition function vanishes when one includes contributions from both A_μ and A_μ^g. The same result holds for any odd number of doublets of Weyl spinors. In terms of the eigenvalues of the Dirac operators, one may write

$$[det \, i\not{D}]^{1/2} = \prod_{\lambda_k > 0} \lambda_k. \tag{6.132}$$

Now, using Eq. (6.129) one finds that n_j pairs of eigenvalues cross zero and switch places on moving from $t = 0$ to $t = 1$ so that

$$\lambda_j(0) = -\lambda_j(1), \quad j = 1, \ldots, n_j, \tag{6.133}$$

---

[*] Mathematically the Witten anomaly exists when one has a non-trivial fourth homotopy group of the gauge group. Specifically, this occurs only for the case $Sp(n)$ and $\pi_4[Sp(n)] = Z_2$. Since $SU(2) = Sp(1)$ one has $\pi_4[SU(2)] = Z_2$.

and this leads to the switch in sign shown in Eq. (6.131) when n_j is odd. The above analysis can be extended for $SU(2)$ representations where [77]

$$j = 2n + \frac{1}{2}, \quad n = 0, \, 1, \, \ldots. \tag{6.134}$$

Thus, aside from $j = 1/2$, the j values $5/2$, $9/2$, etc., corresponding to dimensionalities 2, 6, 10 etc., i.e., 2 mod 4, will also have a Witten anomaly. It is conjectured that the pair of eigenvalues n_j that cross zero is given by

$$n_j = j + \frac{1}{2}. \tag{6.135}$$

The number of zero crossings given by Eq. (6.135) has been numerically tested for $j = 1/2$, $3/2$, and $5/2$ for specific parametrizations, but the path independence of the result has not yet been established.

6.10 The Standard Model Anomalies

For the standard model with the gauge group $SU(3)_C \times SU(2)_L \times U(1)_Y$ the anomalies to be checked consist of

- $U_Y(1)^3$,
- $U_Y(1) \times (SU(2)_L)^2$;
- $U_Y(1) \times (SU(3)_C)^2$,
- $(SU(3)_C)^3$,
- $U_Y(1)$-gravitational anomalies.

Let us begin by recalling the quantum numbers of one generation of quarks and leptons in the standard model under the *SM* gauge group. These are listed in Table 6.1. The contributions to the various parts arise as follows for one generation of quarks and leptons:

- $U(1)^3$. Here, the anomaly is simply a sum over the cube of the hypercharges:

$$\sum_i Y_i^3 = \left[2 \times \left(-\frac{1}{2} \right)^3 + (1)^3 \right]$$
$$+ \left[3 \times 2 \times \left(\frac{1}{6} \right)^3 + 3 \times \left(-\frac{2}{3} \right)^3 + 3 \times \left(\frac{1}{3} \right)^3 \right] = 0, \tag{6.136}$$

where the first bracket on the right is a contribution due to leptons, and the second bracket is a contribution due to quarks. The factors of 3 inside the second brackets indicate the number of colors.

Table 6.1 *The $U(1)_Y$ charge assignments of the standard model particles in one generation where Y is normalized such that $Q = T_3 + Y$.*

Particles	$SU(3)$, $SU(2)$, $U(1)_Y$
(u_L, d_L)	3, 2, $\frac{1}{6}$
u^c	$\bar{3}$, 1, $-\frac{2}{3}$
d^c	$\bar{3}$, 1, $\frac{1}{3}$
(ν, e_L)	1, 2, $-\frac{1}{2}$
e^c	1, 1, 1

- $U(1) \times (SU(2)_L)^2$. Here, only the $SU(2)_L$ doublets contribute. The result is proportional to the sum over the hyper-charges of the doublets, which gives

$$\sum_i (Y_i) = -\frac{1}{2} + 3 \times \left(\frac{1}{6}\right) = 0. \tag{6.137}$$

- $U(1) \times (SU(3)_C)^2$. Here, the result is proportional to the sum of the hyper-charges over the color triplets and anti-triplets, which gives

$$\sum_i Y_i = 2 \times \left(\frac{1}{6}\right) - \frac{2}{3} + \frac{1}{3} = 0. \tag{6.138}$$

- $(SU(3)_C)^3$. Here, the representations that contribute are $2(3+\bar{3})$, which taken together are real and do not give the anomaly.
- $U(1)$ gravitatonal anomalies. Here, the anomaly is proportional to the sum of all the hyper-charges in the multiplet, which vanishes, i.e.,

$$\sum_i Y_i = 0. \tag{6.139}$$

The anomaly conditions are unaffected by higher-order loop corrections [21]. Suggested references for further reading are [78–81].

6.11 Problems

1. Show that the operator

$$C_{abc} \left[\partial_\mu^x \delta^4(x-y) \frac{\delta}{\delta A_\mu^c(y)} + \partial_\mu^y \delta^4(x-y) \frac{\delta}{\delta A_\mu^c(x)} \right] \tag{6.140}$$

is simulated by the operator $C_{abc}\delta^4(x-y)\partial^x_\mu(\delta/\delta A^c_\mu(x))$. To show the equality, multiply the operators by two functions, one a function of x and the other a function of y, and integrate over d^4x and d^4y, and show the equality of the integrals.

2. Show the following relations for the BRST operator s:

$$s[d\mathbf{A}] = [d\mathbf{A}, \eta] - [\mathbf{A}, d\eta] \tag{6.141}$$

$$s\mathbf{A}^4 = -d\eta\mathbf{A}^3 + \mathbf{A}d\eta\mathbf{A}^2 - \mathbf{A}^2 d\eta\mathbf{A} + \mathbf{A}^3 d\eta + [\mathbf{A}^4, \eta], \tag{6.142}$$

where A is defined in Eq. (6.72) and η is defined in Eq. (6.73).

3. Establish the following results for the action of the BRST operator involving the product of fields:

$$str[\eta\mathbf{A} \ d\mathbf{A} \ \mathbf{A}] = tr(\eta^2\mathbf{A} \ d\mathbf{A} \ \mathbf{A} + \eta\mathbf{A} \ d\eta \ \mathbf{A}^2 + d\eta \ \mathbf{A}\eta\mathbf{A}^2)$$
$$+ tr(\eta \ d\eta \ d\mathbf{A} \ \mathbf{A} - d\eta \ \eta\mathbf{A} \ d\mathbf{A}) \tag{6.143}$$

$$str(\eta \ d\mathbf{A} \ \mathbf{A}^2) = tr(\eta^2 \ d\mathbf{A} \ \mathbf{A}^2 - \eta \ d\eta \ \mathbf{A}^3 + \eta\mathbf{A} \ d\eta \ \mathbf{A}^2)$$
$$+ tr(-d\eta \ \eta \ d\mathbf{A} \ \mathbf{A} + \eta \ d\mathbf{A} \ d\eta \ \mathbf{A}). \tag{6.144}$$

4. Consider the following expressions:

$$C_1 = tr[-\eta \ d\eta \ d\mathbf{A}^2 - d\eta \ \eta \ d\mathbf{A}^2 - \eta\mathbf{A} \ d\eta \ d\mathbf{A}$$
$$+ \eta \ d\mathbf{A} \ d\eta \ \mathbf{A} - d\eta \ \mathbf{A} \ d\eta \ \mathbf{A} + 2 \ d\eta \ d\eta \ \mathbf{A}^2] \tag{6.145}$$

$$C_2 = tr[d\eta \ \mathbf{A} \ d\eta \ \mathbf{A} - d\eta \ d\eta\mathbf{A}^2 + \eta^2 \ d\mathbf{A}^2 \ d\mathbf{A} - \eta \ d\eta \ \mathbf{A} \ d\mathbf{A}$$
$$+ \eta\mathbf{A} \ d\eta \ d\mathbf{A} - \eta \ d\mathbf{A} \ d\eta \ \mathbf{A} + \eta \ d\mathbf{A} \ \mathbf{A} \ d\eta]. \tag{6.146}$$

Show that C_1 and C_2 are closed and can be written as

$$C_1 = dG_1 \tag{6.147}$$

$$C_2 = dG_2, \tag{6.148}$$

where G_1 and G_2 are given by

$$G_1 = tr[\eta \ d\eta \ \mathbf{A}^2 + d\eta \ \eta \ \mathbf{A}^2 - \eta\mathbf{A} \ d\eta \ \mathbf{A}] \tag{6.149}$$

$$G_2 = tr[\eta^2 \ d\mathbf{A} \ \mathbf{A} - \eta \ d\eta \ \mathbf{A}^2 + \eta\mathbf{A} \ d\eta \ \mathbf{A}]. \tag{6.150}$$

5. Show that in $2n$ dimensions $tr \ F^n$ satisfies the constraint

$$d \ tr \ F^n = 0, \tag{6.151}$$

and thus $tr \ F^n$ is closed and can be written as

$$tr \ F^n = d\omega_{2n-1}. \tag{6.152}$$

6. With reference to the preceding problem, show that for the case $n = 3$ we have

$$tr\, F^3 = d\, tr\left[A(dA)^2 + \frac{3}{5}A^5 + \frac{3}{2}A^3 dA\right].\qquad(6.153)$$

7. Consider, the irreducible representations $[N,\ m]$ for $SU(N)$ which are anti-symmetric in m indices. Obtain linear combinations in each of the following sets of representations, which are anomaly free:

$$SU(6):\ [6,1],[6,4]\,;\ SU(7):\ [7,2],[7,4],[7,6]$$
$$SU(8):\ [8,1],[8,2],[8,5]\,;\ SU(9,1):\ [9,3],[9,5],[9,7].\qquad(6.154)$$

References

[1]   J. Steinberger, *Phys. Rev.* **76**, 1180 (1949).

[2]   D. G. Sutherland, *Nucl. Phys. B* **2**, 433 (1967).

[3]   R. Arnowitt, M. H. Friedman, and P. Nath, *Phys. Lett. B* **27**, 657 (1968).

[4]   J. S. Bell and R. Jackiw, *Nuovo Cim. A* **60**, 47 (1969).

[5]   S. L. Adler, *Phys. Rev.* **177**, 2426 (1969).

[6]   W. A. Bardeen, *Phys. Rev.* **184**, 1848 (1969).

[7]   K. Fujikawa, *Phys. Rev. Lett.* **42**, 1195 (1979).

[8]   K. Fujikawa, *Phys. Rev. D* **29**, 285 (1984).

[9]   K. Fujikawa, *Phys. Rev. D* **21**, 2848 (1980), *Phys. Rev. D* **22**, 1499 (1980).

[10] A. P. Balachandran, G. Marmo, V. P. Nair, and C. G. Trahern, *Phys. Rev. D* **25**, 2713 (1982).

[11] M. B. Einhorn and D. R. T. Jones, *Phys. Rev. D* **29**, 331 (1984).

[12] C. Becchi, A. Rouet, and R. Stora, *Commun. Math. Phys.* **42**, 127 (1975).

[13] C. Becchi, A. Rouet, and R. Stora, *Phys. Lett. B* **52**, 344 (1974).

[14] C. Becchi, A. Rouet, and R. Stora, *Ann. Phys.* **98**, 287 (1976).

[15] I. V. Tyutin, arXiv:0812.0580 [hep-th] (19175).

[16] J. Wess and B. Zumino, *Phys. Lett. B* **37**, 95 (1971).

[17] B. Zumino, Y. S. Wu, and A. Zee, *Nucl. Phys. B* **239**, 477 (1984).

[18] B. Zumino, in S. B. Treiman, I. W. Jack, B, Zumino, and E. Witten, eds., *Current Algebra and Anomalies*, World Scientific (1985), p. 361.

[19] J. L. Manes, Lawrence Berkeley Laboratories, LBL-22304.

[20] W. A. Bardeen and B. Zumino, *Nucl. Phys. B* **244**, 421 (1984).

[21] S. L. Adler and W. A. Bardeen, *Phys. Rev.* **182**, 1517 (1969).

[22] A. Zee, *Phys. Rev. Lett.* **29**, 1198 (1972).

[23] J. H. Lowenstein and B. Schroer, *Phys. Rev. D* **7**, 1929 (1973).

[24] S. L. Adler, J. C. Collins, and A. Duncan, *Phys. Rev. D* **15**, 1712 (1977).

[25] K. Nishijima, *Prog. Theor. Phys.* **57**, 1409 (1977).

[26] D. J. Gross and R. Jackiw, *Phys. Rev. D* **6**, 477 (1972).

[27] C. Bouchiat, J. Iliopoulos, and P. Meyer, *Phys. Lett. B* **38**, 519 (1972).

[28] H. Georgi and S. L. Glashow, *Phys. Rev. D* **6**, 429 (1972).

[29] H. Georgi, *Nucl. Phys. B* **156**, 126 (1979).

[30] A. Zee, *Phys. Lett. B* **99**, 110 (1981).

[31] E. Witten, *Nucl. Phys. B* **156**, 269 (1979).

[32] E. Witten, *Ann. Phys.* **128**, 363 (1980).

[33] P. Di Vecchia and G. Veneziano, *Nucl. Phys. B* **171**, 253 (1980).

[34] C. Rosenzweig, J. Schechter, and C. G. Trahern, *Phys. Rev. D* **21**, 3388 (1980).

[35] P. Nath and R. L. Arnowitt, *Phys. Rev. D* **23**, 473 (1981).

[36] E. Witten, *Nucl. Phys. B* **223**, 433 (1983).

[37] E. Witten, *Nucl. Phys. B* **223**, 422 (1983).

[38] R. Delbourgo and A. Salam, *Phys. Lett. B* **40**, 381 (1972).

[39] S. W. Hawking, *Phys. Lett. A* **60**, 81 (1977).

[40] T. Eguchi and P. G. O. Freund, *Phys. Rev. Lett.* **37**, 1251 (1976).

[41] N. K. Nielsen, H. Romer and B. Schroer, *Nucl. Phys. B* **136**, 475 (1978).

[42] A. A. Belavin, A. M. Polyakov, A. S. Schwartz and Y. S. Tyupkin, *Phys. Lett. B* **59**, 85 (1975).

[43] G. 't Hooft, *Phys. Rev. Lett.* **37**, 8 (1976).

[44] G. 't Hooft, *Phys. Rev. D* **14**, 3432 (1976); *Phys. Rev. D* **18**, 2199 (1978).

[45] C. G. Callan, Jr., R. F. Dashen, and D. J. Gross, *Phys. Lett. B* **63**, 334 (1976).

[46] R. Jackiw and C. Rebbi, *Phys. Rev. Lett.* **37**, 172 (1976).

[47] M. F. Atiyah and I. M. Singer, *Ann. Math.* **87**, 484 (1968).

[48] M. Atiyah, R. Bott, and V. K. Patodi, *Invent. Math.* **19**, 279 (1973).

[49] M. F. Atiyah and R. S. Ward, *Commun. Math. Phys.* **55**, 117 (1977).

[50] L. S. Brown, R. D. Carlitz, and C. k. Lee, *Phys. Rev. D* **16**, 417 (1977).

[51] A. S. Schwarz, *Phys. Lett. B* **67**, 172 (1977).

[52] I. S. Gerstein and R. Jackiw, *Phys. Rev.* **181**, 1955 (1969).

[53] R. W. Brown, C. C. Shih, and B. L. Young, *Phys. Rev.* **186**, 1491 (1969).

[54] T. E. Clark and S. T. Love, *Nucl. Phys. B* **223**, 135 (1983).

[55] T. E. Clark and S. T. Love, *Nucl. Phys. B* **217**, 349 (1983).

[56] T. Curtright, *Phys. Lett. B* **71**, 185 (1977).

[57] D. R. T. Jones and J. P. Leveille, *Nucl. Phys. B* **206**, 473 (1982); *Nucl. Phys. B* **222**, 517 (1983).

[58] D. R. T. Jones and J. P. Leveille, *Phys. Lett. B* **109**, 449 (1982).

[59] P. H. Frampton and T. W. Kephart, *Phys. Rev. D* **28**, 1010 (1983).

[60] P. H. Frampton and T. W. Kephart, *Phys. Rev. Lett.* **50**, 1343 (1983); *Phys. Rev. Lett.* **51**, 232 (1983).

[61] P. H. Frampton and T. W. Kephart, *Phys. Rev. Lett.* **50**, 1347 (1983).

[62] P. K. Townsend and G. Sierra, *Nucl. Phys. B* **222**, 493 (1983).

[63] L. Alvarez-Gaume and E. Witten, *Nucl. Phys. B* **234**, 269 (1984).

[64] L. Alvarez-Gaume and P. H. Ginsparg, *Ann. Phys.* **161**, 423 (1985); *Ann. Phys.* **171**, 233 (1986).

[65] F. Langouche, T. Schucker, and R. Stora, *Phys. Lett. B* **145**, 342 (1984).

[66] G. 't Hooft, in *Recent Developments in Gauge Theories*, Springer (1980), p. 135.

[67] A. Zee, *Phys. Lett. B* **95**, 290 (1980).

[68] T. Banks, S. Yankielowicz, and A. Schwimmer, *Phys. Lett. B* **96**, 67 (1980).

[69] S. Dimopoulos, S. Raby, and L. Susskind, *Nucl. Phys. B* **173**, 208 (1980).

[70] S. Weinberg, *Phys. Lett. B* **102**, 401 (1981).

[71] J. Preskill and S. Weinberg, *Phys. Rev. D* **24**, 1059 (1981).

[72] I. Bars and S. Yankielowicz, *Phys. Lett. B* **101**, 159 (1981).

[73] S. R. Coleman and B. Grossman, *Nucl. Phys. B* **203**, 205 (1982).

[74] J. Banks and H. Georgi, *Phys. Rev. D* **14**, 1159 (1976).

[75] For constraints on $SU(2)$ gauge theories in the context of strings see: J. Halverson, *Phys. Rev. Lett.* **111** (26), 261 (2013).

[76] E. Witten, *Phys. Lett. B* **117**, 324 (1982).

[77] O. Bar, *Nucl. Phys. B* **650**, 522 (2003).

[78] Weinberg, *The Quantum Theory of Fields*, Cambridge University Press (1995).

[79] A. Bilal, arXiv:0802.0634 [hep-th] (2008).

[80] J. A. Harvey, hep-th/0509097 (2005).

[81] L. Alvarez-Gaume and M. A. Vazquez-Mozo, hep-th/9212006 (1992).

7

Effective Lagrangians

The use of effective Lagrangians is quite ancient. One example of its early use is in the paper by Heisenberg and Euler in 1935 [1], where they explored the implications of the Dirac theory. By integrating over the electron degrees of freedom they obtained a Lagrangian valid below the scale of the electron mass in terms of E and B fields. This was done more as a matter convenience since the fundamental theory in this case is known. A different motivation for the use of an effective Lagrangian is, however, to simulate physical phenomena in an energy domain where the fundamental Lagrangian is either unknown or, even if it is known, a direct use of it is highly non-trivial. An example of the latter is the quantum chromodynamics (QCD) Lagrangian which is given in terms of quarks and gluons where both the quarks and the gluons are confined and not directly visible. The manifestations of the quarks and the gluons in low-energy physics arise in the form of mesons and baryons, which are bound states of the quarks and the gluons. In this case the effective Lagrangians are a way to describe the properties for the observed particles in terms of some broad principles which we will discuss below. Indeed, it is shown by Witten [2] that the $1/N$ expansion of QCD leads to an effective Lagrangian although the $1/N$ expansion itself is not powerful enough to compute all the relevant properties of the states that appear in the effective Lagrangian such as their masses and the couplings. In some instances we do not know the fundamental Lagrangian, unlike the QCD case, and here the effective Lagrangian can be helpful in mapping the unknown regime of physics. The effective Lagrangian is, in general, scale dependent in that the spectrum of states that enter the effective Lagrangian could be different at different scales. Thus, each regime of energy would in general be described by an effective Lagrangian specific to that energy regime. From this viewpoint, even grand unified theories are also effective theories which could be the low-energy limit of a more fundamental theory such as string theory.

In what follows we discuss effective Lagrangians, but not in the sense of Heisenberg where we know the fundamental theory but for convenience we integrate over the heavy modes and generate an effective Lagrangian. Such an effective Lagrangian is only a matter of convenience and does not concern us here. Rather, we are interested in the exploration of energy domains where either the fundamental Lagrangian is not known or, if it is known, an explicit use of it in the description of physical phenomena is difficult. We pause here to mention phenomenological Lagrangians, where one starts with the assumption of a Lagrangian which obeys certain symmetries and then these symmetries are broken by hand to make the low-energy theory phenomenologically viable (e.g., [3–10]). In the analysis below, we do not *a priori* assume the existence of a Lagrangian. Rather, our aim is to develop a method for the computation of the T-product of fields which may populate a certain energy regime and ask how best to describe their interactions. We will assume certain broad principles in the analysis. These include single-particle saturation of T products along with the spectator approximation (which will be explained shortly), Lorentz invariance and crossing symmetry, and the assumptions that the fields are local operators and that the vertices have smooth dependence on momenta. It is then found that with these assumptions one is led to an effective Lagrangian, and such effective Lagrangians can be constructed for the computation of n-point functions [11–13]. These general principles are valid at any scale, and thus effective Lagrangians have a broad range of applicability from the low energy up to the Planck scale. In effective Lagrangian constructions, symmetry constraints are imposed only after the effective Lagrangian is in place. Although the techniques discussed below are valid in a broader context, we will illustrate this approach in the context of current algebra analyses.

In the determination of physical processes one encounters the necessity to compute the vacuum expectation of T-products of currents. The computation may involve, for example, the vacuum expectation value of, say, the T-product of three vector currents or the vacuum expectation value of the T-product of two axial and one vector current. Let us consider the case when one has $\langle T(A_a^\mu(x)V_c^\lambda(z)A_b^\nu(y))\rangle$, where A_a^μ and A_b^ν are the axial currents and V_c^λ is the vector current, and a, b, and c are $SU(2)$ indices. There are six time orderings in this T-product: let us consider one of them, i.e., $\langle A_a^\mu(x)V_c^\lambda(z)A_b^\nu(y)\rangle$. We expand it as follows:

$$F^{\mu\lambda\nu} \equiv \langle 0|A_a^\mu(x)V_c^\lambda(z)A_b^\nu(y)|0\rangle$$
$$= \sum_{n,m}\langle 0|A_a^\mu|n\rangle\langle n|V_c^\lambda(z)|m\rangle\langle m|A_b^\nu(y)|0\rangle. \qquad (7.1)$$

We now use the approximation of single-particle saturation, where we will treat the states n and m as single-particle "in" states. In addition, we must include those two-particle intermediated states which are dictated by Lorentz invariance,

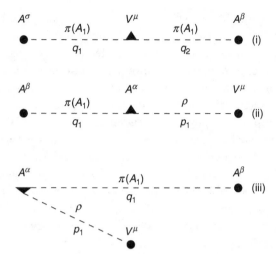

Figure 7.1. Illustration of single-particle saturation for the 3-point function. Diagram (i) gives the time ordering $x^0 > z^0 > y^0$ with π and A_1 intermediate states. The vacuum to one-particle matrix elements of the currents are represented by circles while the triangle is the one-particle matrix element of the current. Diagram (ii) gives the time ordering $x^0 > y^0 > z^0$ with π or A_1 in the first intermediate state and ρ in the second. Diagram (iii) gives the same time ordering as diagram (i) but with a two-particle intermediate state which is either $\pi\rho$ or $A_1\rho$ where π or A_1 is a spectator particle. (From Arnowitt *et al.* [12].)

and by crossing symmetry where one of the particles is a spectator. By "spectator," we mean that it moves freely across the vertex without interaction. As an illustration, if we limit ourselves to studying the dynamics of, say, the $\pi, \rho,$ and A_1 system, the first matrix element on the right-hand side of Eq. (7.1) would involve only matrix elements of the axial current between the vacuum and the pion state and the matrix element of the axial current between the vacuum and the A_1 state. This is illustrated by diagram (i) in Fig. 7.1 while diagram (ii) in Fig. 7.1 corresponds to another time ordering. However, as mentioned, one should also include two-particle states, which are required by Lorentz invariance and crossing symmetry where one of the particles is a spectator. Thus, for example, for the $\pi, \rho,$ and A_1 system for the time ordering of $x^0 > z^0 > y^0$, one should include terms with the following type of states:

$$F^{\mu\lambda\nu} = \sum_{a_1 a_2 a_3} \int d^3 q_1 d^3 q_2 d^3 p_1 \langle 0|A_a^\mu|\pi q_1 a_1, \rho p_1 a_3\rangle$$
$$\times \langle \pi q_1 a_1, \rho p_1 a_3|V_c^\lambda|\pi q_2 a_2\rangle \langle \pi q_2 a_2|A_b^\nu|0\rangle. \tag{7.2}$$

Here, the spectator approximation corresponds to the assumption that

$$\langle \pi q_1 a_1, \rho p_1 a_3|V_c^\lambda|\pi q_2 a_2\rangle = \delta^3(q_1 - q_2)\delta_{a_1 a_2}\langle \rho p_1 a_3|V_c^\lambda|0\rangle, \tag{7.3}$$

so that in this case the pion is a spectator. This is illustrated by diagram (iii) in Fig. 7.1. So, even though more than two particles are present in the intermediate sum in the matrix element of the vector current, the sum is only over the momentum of a single particle, which in this case is ρ while the pion is a spectator and its momentum is not summed over. Thus, by keeping one-particle states whose momenta are summed and all two-particle states where one of the particles is a spectator, one obtains a Lorentz covariant and crossing symmetric approximation.

For the specific ordering of Eq. (7.1) we have the matrix element $\langle \pi |V^\lambda| \pi \rangle$, which has a pole in the momentum transfer so we can write

$$\langle \pi qb|V_c^\lambda|\pi pa \rangle = iN\epsilon_{abc}\Delta_\rho^{\lambda\nu}(k)\Gamma_\nu(q,p), \qquad (7.4)$$

where N is a normalization and Δ_ρ is a propagator for the ρ meson and $\Gamma_\nu(q,\ p)$ is a vertex function which we assume is a smooth function of q, and p and can be expanded as

$$\Gamma_\nu(q,p) = (c_1 + c_2 k^2 + \cdots)(q+p)_\nu. \qquad (7.5)$$

Now, it is clear that for the computation of $\langle \pi qb|V_c^\lambda|\pi pa \rangle >$ one needs a replacement of V_c^λ as follows:

$$V_c^\lambda \rightarrow \epsilon_{abc}\int d^4y\, \Delta_\rho^{\lambda\nu}(x-y)\,[c_1 - c_2\Box + \cdots]\,\phi_a(y)\partial_\nu\phi_b(y). \qquad (7.6)$$

With the above prescription one can compute the vacuum expectation value of the T-products of three currents. However, we note from Eq. (7.6) that the expansion of the currents quadratic in the fields is a non-local operator. Now, locality can be achieved only if we assume that that V_a^μ is proportional to a Heisenberg ρ field v_a^μ and the ρ couples quadratically with the pion fields in a Lagrangian. Thus,

$$V_a^\mu = g_\rho v_a^\mu. \qquad (7.7)$$

In this case we require that the ρ meson field satisfies a Heisenberg field equation so that

$$K^\mu_{\ \nu}(x)v_c^\nu(x) = g_\rho^{-1}\epsilon_{abc}\,[c_1 - c_2\Box + \cdots]\,\phi_a(x)\partial^\mu\phi_b(x). \qquad (7.8)$$

In a similar fashion one is led to write the following field–current relation between the axial current and the pion field and the field of the axial vector meson A_1 so that

$$A_a^\mu = F_\pi\partial^\mu\phi_a + g_{A_1}a_a^\mu. \qquad (7.9)$$

Equations (7.7) and (7.9) are known as the field–current identities. They were introduced in the work of Lee *et al.* [14]. There and in the analysis by Arnowitt *et al.* [11] they arise as a consequence of single-particle saturation. Thus, the analysis above based on the single-particle saturation hypothesis, Lorentz invariance, crossing symmetry and the spectator approximation, locality, and the smoothness hypothesis dictates that we write an effective Lagrangian. As an example, we consider an explicit construction of an effective Lagrangian for the computation of a vertex function in the π, ρ, and A_1 system. The effective Lagrangian here is given by

$$\mathcal{L} = \mathcal{L}_0 + \mathcal{L}_3 \,, \tag{7.10}$$

where

$$\mathcal{L}_0 = \mathcal{L}_{0\pi} + \mathcal{L}_{0\rho} + \mathcal{L}_{0A} \tag{7.11}$$

and

$$\mathcal{L}_{0\pi} = -\phi_a^\mu \partial_\mu \phi_a + \frac{1}{2}(\phi_a^\mu \phi_{\mu a} - m_\pi^2 \phi_a^2)$$

$$\mathcal{L}_{0\rho} = -\frac{1}{2} G_a^{\mu\nu}(\partial_\mu v_{\nu a} - \partial_\nu v_{\mu a}) + \frac{1}{4} G_a^{\mu\nu} G_{\mu\nu a} - \frac{1}{2} m_\rho^2 v_a^\mu v_{\mu a}$$

$$\mathcal{L}_{0A} = -\frac{1}{2} H_a^{\mu\nu}(\partial_\mu a_{\nu a} - \partial_\nu a_{\mu a}) + \frac{1}{4} H_a^{\mu\nu} H_{\mu\nu a} - \frac{1}{2} m_A^2 a_a^\mu a_{\mu a}. \tag{7.12}$$

Here, we are using the first-order formalism where the pion fields ϕ_a and $\phi_{\mu a}$ are to be varied independently. Similarly, the ρ field v_a^μ and the ρ field strength $G_{\mu\nu a}$, and the A_1 field a_a^μ and the A_1 field strength $H_{\mu\nu a}$ are to be varied independently. Assuming there are no additional derivative couplings in the interaction Lagrangian, the canonically conjugate co-ordinate and momenta are given by (ϕ_a, ϕ_{0a}), (v_{ia}, G_{0ia}), and (a_{ia}, H_{0ia}). The interaction Lagrangian $\mathcal{L}_3$ dictated by the above considerations is given by

$$\begin{aligned}
\mathcal{L}_3 = \frac{1}{2}\epsilon_{abc}\big[& g_{\rho\rho\rho} v_{\mu a} v_{\nu b} G_c^{\nu\mu} + g_{\rho AA} v_{[\mu a} a_{\nu]b} H^{\nu\mu} + \lambda_{\rho AA} a_{\mu a} a_{\nu b} G_c^{\nu\mu} \\
& + 2g_{\pi\pi\rho} v_{\mu a}\phi_b^\mu \phi_c + \lambda_{\pi\pi\rho}\phi_{\mu a}\phi_{\nu b} G_c^{\nu\mu} + 2g_{\pi\rho A} v_{\mu a}\phi_b a_c^\mu \\
& + 2\mu_{\pi\rho A}\phi_a G_b^{\mu\nu} H_{\mu\nu c} + \lambda_{\pi\rho A} v_{[\mu a}\phi_{\nu]b} H_c^{\mu\nu} + \tilde{\lambda}_{\pi\rho A} a_{[\mu a}\phi_{\nu b} G_c^{\mu\nu}\big], \tag{7.13}
\end{aligned}$$

where $v_{[\mu a} a_{\nu]b} \equiv v_{\mu a} a_{\nu b} - a_{\nu b} a_{\mu a}$, etc. To compute the T-products of currents we express the T product in terms of the fields using the field–current identity of Eqs. (7.7) and (7.9) and we use the field equations to solve for the ρ, A_1, and pion fields in terms of the interactions, and then compute the T-product to first order in the coupling constants appearing in Eq. (7.13). We emphasize that the deduction of the effective Lagrangian of Eqs. (7.10)–(7.13) arose only as a consequence of single-particle saturation of the T-products of three field operators, Lorentz invariance and the spectator approximation, the assumption of the smoothness in the expansion of the vertex function in terms of momenta, and the constraint that the interactions involving the ρ, A_1, and the pion field operators be local [11, 12].

We can now extend the above analysis to higher-point functions. Consider a 4-point function given by a T-product of four currents consisting of two axial

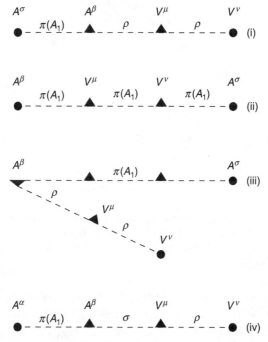

Figure 7.2. Single particle saturation with different time orderings of the 4-point function. Diagram (i) gives the time ordering $x^0 > y^0 > z^0 > w^0$ while diagram (ii) gives the time ordering $y^0 > z^0 > w^0 > x^0$. Diagram (iii) gives the time ordering $y^0 > z^0 > w^0 > x^0$ with two-particle intermediate state where one of the particles is a spectator. Diagram (iv) has the time ordering $x^0 > y^0 > z^0 > w^0$ with a sigma meson as an intermediate state. The σ field interpolates the σ-commutator, as discussed in the text relating to Eqs. (7.31) and (7.32). (From Arnowitt *et al.* [12].)

currents and two vector currents, i.e.,

$$F^{\mu\nu\lambda\rho}(x,y,z,w) \equiv \langle T(A_a^\mu(x)A_b^\nu(y)V^\lambda(z)V^\rho(w))\rangle. \tag{7.14}$$

Consider the time ordering $y^0 > z^0 > w^0 > x^0$, which gives

$$F^{\mu\nu\lambda\rho}(x,y,z,w) = \sum_{n,m,k} \langle 0|A^\nu|n\rangle\langle n|V^\lambda|m\rangle\langle m|V^\rho|k\rangle\langle k|A^\mu|0\rangle. \tag{7.15}$$

As in the preceding analysis, single-particle saturation requires that the intermediate states be single π, ρ, or A_1 mesons. This is illustrated in diagrams (i) and (ii) of Fig. 7.2. Additionally, Lorentz covariance and crossing symmetry requires, as discussed in the case of the 3-point functions, that one also includes all two-particle states where one of the particles is a spectator, i.e., that its momentum is not summed over. This is illustrated in diagram (iii) of Fig. 7.2. Thus, for $|n\rangle$ and $|m\rangle$ in Eq. (7.15) one could have a two-particle $\pi-\rho$ state while $|k\rangle$ is a single π state. This analysis is similar to the one for the case of the 3-point function.

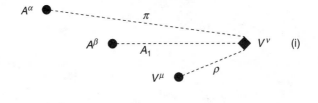

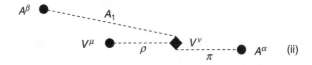

Figure 7.3. Diagram (i) illustrates single-particle saturation for a 4-point function when there are two particles (π and A_1) which are simultaneously spectators. Diagram (ii) is diagram (i) with the π meson crossed. (From Arnowitt *et al.* [12].)

However, in this case there exist additional diagrams that do not have an analog for the 3-point case. These are the cases where we have intermediate states with two particles as spectators. This is illustrated in Fig. 7.3. For example, we can expand (suppressing the iso-spin indices)

$$F^{\mu\nu\lambda\rho} = \sum \langle 0|A^\mu|\pi, q_1\rangle\langle\pi, q_1|A^\nu|\pi, q_2; A_1, q_3\rangle$$
$$\langle\pi q_2; A_1 q_3|V^\lambda|\pi, q_4; A_1, q_5; \rho, p_1\rangle\langle\pi, q_4; A_1, q_5; \rho, p_1|V^\rho|0\rangle. \quad (7.16)$$

Here we have up to two particles which are spectators. Thus, e.g.,

$$\langle\pi, q_2; A_1, q_3|V^\lambda|\pi, q_4; A_1, q_5; \rho, p_1\rangle = \delta^3(q_2 - q_4)\delta^3(q_3 - q_5)\langle 0|V^\lambda|\rho, p_1\rangle. \quad (7.17)$$

Equation (7.16) along with Eq. (7.17) shows that the assumption of single-particle saturation naturally leads us to a direct 4-point vertex $\langle\pi, q_1|A^\nu|\pi, q_2; A_1, q_3\rangle$. The above analysis implies that there can be a local contribution to the effective Lagrangian from a 4-point interaction so that $\mathcal{L} = \mathcal{L}_0 + \mathcal{L}_3 + \mathcal{L}_4$ where

$$\mathcal{L}_4 = c_1\phi_a^\mu\phi_{\mu a}v_b^\nu v_{\nu b} + c_2\phi_a\phi_{\mu a}G_b^{\mu\nu}v_{\nu b} + \cdots \quad (7.18)$$

in which we will assume that c_1, c_2 etc are second order in the (cubic) couplings. Thus, we can summarize the above analysis as follows. In the computation of a T-product of currents, replace the currents by fields according to the prescription of Eq. (7.7) and Eq. (7.9). Then, the conditions of single-particle saturation including the spectator approximation, smoothness of the vertices, and locality of the field operators leads us to the result that the T-products can be calculated from an effective Lagrangian which includes $\mathcal{L}_3$ and $\mathcal{L}_4$. The computation of a 4-point function is done to second order in the coupling constants where the interactions of $\mathcal{L}_4$ are considered to be second order in the couplings.

The analysis above can be extended to higher-point functions. For instance, for the computation of 5-point functions, the implementation of the single-particle saturation requires that we include the following set of contributions: (i) diagrams where we have a cascade of 3-point vertices; (ii) diagrams where we have 3- and 4-point vertices and (iii) diagrams where we have a 5-point vertex. Diagrams of type (i) arise from $\mathcal{L}_3$, diagrams of type (ii) arise from $\mathcal{L}_3$ and $\mathcal{L}_4$, while for diagrams of type (iii) we must include an additional $\mathcal{L}_5$ part in the Lagrangian which is quintic in the fields. We should then use the Lagrangian $\mathcal{L}_3 + \mathcal{L}_4 + \mathcal{L}_5$ to the first non-vanishing order, discarding disconnected diagrams to compute the T-product of five fields. It is now straightforward to extend the analysis to the computation of T-products of N-point functions. Thus, using the same principles as above, the analysis of N-point functions involves using an interaction Lagrangian

$$\mathcal{L}_I = g\mathcal{L}_3 + g^2\mathcal{L}_4 + \cdots + g^{N-2}\mathcal{L}_N, \tag{7.19}$$

where the order of the coupling in $\mathcal{L}_k$ will be viewed as $\mathcal{O}(g^{k-2})$. The computation of an N-point function is then done using the effective Lagrangian of Eq. (7.19) to order $N - 2$ in the couplings. We now summarize the basic principles that lead to the deduction of the effective Lagrangian for the computation of the T-product of N currents. These are:

- Single-particle saturation of T products. These include the "spectator" approximation and inclusion of higher vertices consistent with the spectator approximation.
- Lorentz invariance, which implies that diagrams that are related by crossing symmetry be included in the single-particle saturation.
- The assumption that the fields are local operators and the vertices have a smooth dependence on momenta.

These assumptions then lead us to the construction of an effective Lagrangian of Eq. (7.19) which is to be used to order $N - 2$ in the coupling constants in the computation of an N-point function where the order of the coupling in $\mathcal{L}_k$ will be viewed as $\mathcal{O}(g^{k-2})$

7.1 The Effective Lagrangian and Current Algebra

Next, we explore the implications of the current algebra constraints on the effective Lagrangian. We consider first the $SU(2) \times SU(2)$ current algebra constraints. Here, one has three vector currents $V_a^\mu(x)$ and three axial vector currents

$A_a^\mu(x)$, $a = 1$, 2, and 3. The current commutation relations are given by [15]

$$\delta(x^0 - y^0)[V_a^0(x), V_b^\mu(y)] = i\epsilon_{abc}V_c^\mu(x)\delta^4(x - y) + c\text{-No. S. T.} \qquad (7.20)$$

$$\delta(x^0 - y^0)[V_a^0(x), A_b^\mu(y)] = i\epsilon_{abc}A_c^\mu(x)\delta^4(x - y) + c\text{-No. S. T.} \qquad (7.21)$$

$$\delta(x^0 - y^0)[A_a^0(x), V_b^\mu(y)] = i\epsilon_{abc}A_c^\mu(x)\delta^4(x - y) + c\text{-No. S. T.} \qquad (7.22)$$

$$\delta(x^0 - y^0)[A_a^0(x), A_b^\mu(y)] = i\epsilon_{abc}V_c^\mu(x)\delta^4(x - y) + c\text{-No. S. T.}, \qquad (7.23)$$

where "c-No. S.T." stands for c-number Schwinger terms, which are terms proportional to derivatives of delta functions, i.e., we disallow q-number Schwinger terms. In addition, we impose the conservation of the vector current (CVC),

$$\partial_\mu V_a^\mu(x) = 0, \qquad (7.24)$$

and the partial conservation of the axial vector current condition (PCAC),

$$\partial_\mu A_a^\mu(x) = F_\pi m_\pi^2 \phi_a(x). \qquad (7.25)$$

In implementing the current algebra conditions (Eqs. (7.20)–(7.25)), it is convenient to expand the currents in terms of the Heisenberg canonical variables of the fields. As mentioned earlier, in the first-order formalism where the interaction Lagrangian has no explicit derivative terms, for the pion field the canonically conjugate pair of variables are (ϕ_a, ϕ_{0a}), and for the ρ and A_1 fields they are (v_a^i, G_{0ia}) and (A_a^i, H_{0ia}). The remaining variables ϕ_{ia}, v_a^0, a_a^0, G_{ija}, and H_{ija} are constraint variables. These constraint variables can be expressed in terms of the canonical variables using the field equations. Thus, for example, v_a^0 can be expanded as follows:

$$v_a^0(x) = m_\rho^{-2}\partial_i G_{0ia} + m_\rho^{-2}\delta\mathcal{L}_I/\delta v_{0a}. \qquad (7.26)$$

The first term on the right-hand side is linear in the canonical variables while the second term on the right-hand side involves non-linear terms in the canonical and the constraint variables. The constraint variables can be eliminated in terms of the canonical variables using the field equations. Elimination of the constraint variables then gives us v_a^0 as a power series expansion in terms of the canonical variables. An identical procedure can be carried out for the other constraint variables. Thus, using the field–current identities and the above procedure we can expand the currents J_a^μ in terms of the canonical variables in a power series expansion in the coupling constants. Since all the currents are now expressed in terms of the canonical variables, one can compute the current commutation relations in a straightforward fashion. The conservation of the vector current can be expressed as a constraint on the interaction Lagrangian so that it takes the form

$$g_\rho m_\rho^{-2}\partial_\mu(\delta\mathcal{L}_I/\delta v_{\mu a}) = 0, \qquad (7.27)$$

while the partial conservation of the axial vector current takes the form

$$g_{A_1} m_{A_1}^{-2} \partial_\mu (\delta \mathcal{L}_I / \delta a_{\mu a}) + F_\pi \partial_\mu (\delta \mathcal{L}_I / \delta \phi_{\mu a}) = F_\pi (\delta \mathcal{L}_I / \delta \phi_a). \tag{7.28}$$

As an illustration of this procedure, we will implement it in constraining the cubic interaction, i.e., we consider the case when $\mathcal{L}_I = \mathcal{L}_3$. In this case, V_a^0 and A_a^0 have expansions which are quadratic in the canonical variables, and the expansion is first order in the coupling constants. Using these forms and implementing the procedure outlined above, the current algebra commutation relations of Eqs. (7.20)–(7.23) produce constraints on the coupling constraints. Additional constraints are obtained from the conservation conditions of Eqs. (7.24) and (7.25). The condition that the q-number Schwinger term be absent produces the Weinberg sum rule [16]

$$\frac{g_{A_1}^2}{m_A^2} = F_\pi^2 + \frac{g_\rho^2}{m_\rho^2}. \tag{7.29}$$

In addition, if one assumes $g_{A_1} = g_\rho$ and uses the Kawarabayashi–Suzuki–Riazuddin–Fiazuddin (KSFR) relation [17] $g_\rho \simeq \sqrt{2} F_\pi m_\rho$, the following constraints on the coupling constants of the cubic Lagrangian $\mathcal{L}_3$ are obtained [11]:

$$g_{\pi\pi\rho} = g_{\rho\rho\rho} = g_{\rho AA} = m_\rho^2 g_\rho^{-1}$$

$$g_{\pi\rho A} = -m_A^2 \lambda_{\pi\rho A} = m_\rho^2 F_\pi^{-1}$$

$$\lambda_{\pi\pi\rho} = \frac{1}{2} \lambda_A g_\rho^{-1}, \ \mu_{\pi\rho A} = 0$$

$$\tilde{\lambda}_{\pi\rho A} = -\left(1 - \frac{1}{2} \lambda_A\right) F_\pi^{-1}$$

$$\lambda_A \equiv g_\rho m_\rho^{-2} \lambda_{\rho AA}, \tag{7.30}$$

where λ_A is the anomalous magnetic moment of the A_1 meson. A more general result is easily obtained in terms of the parameters $x \equiv \sqrt{2} m_\rho / m_{A_1}$, $y \equiv g_{A_1} / g_\rho$, and $z \equiv g_\rho / \sqrt{2} m_\rho F_\pi$, where the only constraint on these parameters is the condition that the q-number Schwinger term vanishes, which now reads $x^2 y^2 z^2 - 2z^2 + 1 = 0$. Equivalent results to the above have been obtained by the use of Ward identifies [18], phenomenological Lagrangians, and other techniques.

In addition to the PCAC relation there is another quantity which is needed to describe the breaking of the chiral symmetry, which is the so-called "σ commutator" and is defined by

$$\delta(x^0 - y^0)[\phi_a(x), A_b^0(y)] = i \delta^4(x - y) \sigma_{ab}(x). \tag{7.31}$$

The current algebra relations do not determine the σ commutator, and one needs additional assumptions to compute it. One possibility is to assume that σ_{ab} is dominated by the σ or the $f_0(500)$ meson where $\sigma(f_0(500))$ has the quantum numbers $J^{PC} = 0^{++}$ so that

$$\sigma_{ab} \sim \delta_{ab} \sigma(x). \tag{7.32}$$

In this case, we can extend the effective Lagrangian to include the terms

$$\mathcal{L} \supset \mathcal{L}_{0\sigma} + \mathcal{L}_{(3)\sigma}$$

$$\mathcal{L}_{0\sigma} = -\sigma^{\mu}\partial_{\mu}\sigma + \frac{1}{2}(\sigma^{\mu}\sigma_{\mu} - m_{\sigma}^{2}\sigma^{2})$$

$$\mathcal{L}_{(3)\sigma} = \frac{1}{2}g_{\sigma\pi\pi}\phi_{a}\phi_{a}\sigma + \frac{1}{2}\lambda_{\pi\pi\sigma}\phi_{a}^{\mu}\phi_{\mu a}\sigma + \lambda_{\sigma\pi A}\phi_{a}a_{a}^{\mu}\sigma_{\mu}$$

$$+ \tilde{\lambda}_{\sigma\pi A}a_{a}^{\mu}\phi_{\mu a}\sigma + \mu_{\sigma\pi\pi}\phi_{a}\phi_{a}^{\mu}\sigma_{\mu} + \cdots \qquad (7.33)$$

In the analysis thus far, V_{a}^{0} and A_{a}^{0} are quadratic in the canonical variables. Thus, V^{0} has the form

$$V_{a}^{0} = \epsilon_{abc}\left[v_{kb}G_{okc} + a_{kb}H_{0kc} + \phi_{b}\phi_{0c}\right] + \partial_{k}S^{k}. \qquad (7.34)$$

The integral of V^{0} over all space gives the iso-spin. The presence of the spatial divergence term $\partial_{k}S^{k}$ where $S^{k} = g_{\rho}m_{\rho}^{-2}G_{0ka} + F_{\pi}g_{A_{1}}^{-1}\epsilon_{abc}\phi_{b}H_{okc}$ is needed to ensure that the q-number Schwinger terms are absent in the current commutators. Similarly, A^{0} can be written as

$$A_{a}^{0} = -F_{\pi}\phi_{0a} + g_{A_{1}}m_{A}^{-2}\partial_{k}H_{0ka} + \epsilon_{abc}\left[v_{kb}H_{0kc} + a_{kb}G_{0kc}\right.$$

$$\left. + F_{\pi}g_{\rho}^{-1}\partial_{k}\phi_{b}G_{0kc}\right] + F_{\pi}g_{A_{1}}^{-1}\lambda_{1}\partial_{k}H_{0ka}\sigma + \lambda_{1}\phi_{0a}\sigma - \lambda_{3}\phi_{a}\sigma_{0}, \qquad (7.35)$$

where $\lambda_{1} \equiv (g_{A_{1}}m_{A}^{-2})\lambda_{\sigma\pi A}$, $\lambda_{2} \equiv (g_{A_{1}}m_{A}^{-2})\tilde{\lambda}_{\sigma\pi A}$, and $\lambda_{3} \equiv \lambda_{1} + F_{\pi}\mu_{\sigma\pi\pi}$. The condition that Eqs. (7.34) and (7.35) rigorously satisfy the current algebra constraint of Eqs. (7.20)–(7.23) requires that $\lambda_{1}\lambda_{3} = 1$. One can extend the above analysis straightforwardly to the analysis of higher-point functions. Thus, for 4-point functions, $\mathcal{L}_{3}$ as determined by the current algebra constraints of 3-point functions is used, and then current algebra conditions are used to determine $\mathcal{L}_{4}$. A simplifying assumption found useful in carrying out the extension is to treat Eqs. (7.34) and (7.35) as rigorous currents of the full theory.

7.2 The Chiral $U(3) \times U(3)$ Current Algebra

We now extend the previous analysis to chiral $U(3) \times U(3)$ current algebra. Such an analysis can describe the interactions of the nonets of pseudo-scalar mesons (π, K, η, η') and the nonets of vector mesons (ρ, K^{*}, ω, ϕ) and axial vector mesons (A_{1}, K_{A}, D, E) as well as the interactions involving scalar mesons (δ, κ, σ, η_{V}). We introduce the nonets of vector and axial vector currents

$$V_{A}^{\mu} \equiv (V_{a}^{\mu}, V_{\bar{a}}^{\mu} \equiv A_{a}^{\mu}), \qquad (7.36)$$

where $a = 1, \ldots, 9$ stand for the nonets of vector currents and $\bar{a} = \bar{1}, \ldots, \bar{9}$ stand for the nonets of axial vector currents. We assume the following set of current commutation relations among them:

$$\delta(x^0 - y^0)[V_A^0(x), V_B^\mu(y)] = iC_{ABC}\delta^4(x - y)V_C^\mu(y) + c\text{--No. S. T.}, \qquad (7.37)$$

where

$$C_{abc} = C_{a\bar{b}\bar{c}} = f_{abc}, \; C_{\bar{a}\bar{b}\bar{c}} = C_{ab\bar{c}} = 0, \quad a, b, c = 1, \cdots, 9,$$

where f_{abc} are the structure constants of $U(3)$ and are totally anti-symmetric in the indices a, b, and c. Additionally, we impose the conservation and the partial conservation of the current V_A^μ so that

$$\partial_\mu V_A^\mu = \mathcal{F}_{AB}\mu_B^2\chi_B, \qquad (7.38)$$

and assume the field–current identity as in the analysis of $SU(2)_L \times SU(2)_R$ current algebra so that

$$V_A^\mu = \mathsf{g}_{AB}v_B^\mu + \mathcal{F}_{AB}\partial^\mu\chi_B, \qquad (7.39)$$

where $\mathcal{F}_{AB}$ and g_{AB} are diagonal except possibly in the 8 and 9 channels.

A solution to the current algebra constraints leads to the following sum rules [12]:

$$(m_\eta F_{88})^2 + (M_{\eta'}F_{89})^2 = \frac{4}{3}(F_K^2 m_K^2 + F_\kappa^2 m_\kappa^2) - \frac{1}{3}F_\pi^2 m_\pi^2 \qquad (7.40)$$

$$F_{88}(F_{88} + \sqrt{2}F_{98})m_\eta^2 + F_{89}(F_{89} + \sqrt{2}F_{99})m_{\eta'}^2 = F_\pi^2 m_\pi^2 \qquad (7.41)$$

$$(F_{88} + \sqrt{2}F_{98})^2 m_\eta^2 + (F_{89} + \sqrt{2}F_{99})^2 m_{\eta'}^2 = 3F_\pi^2 m_\pi^2. \qquad (7.42)$$

Equation (7.42) puts a severe restriction on the mass of the η' meson. Thus, in the limit of no mixing, i.e., $F_{89} = 0$, and in the limit

$$F_{a9} \to \sqrt{\frac{N_\ell}{6}}F_\pi\delta_{a9}, \qquad (7.43)$$

where N_ℓ is the number of light quarks, substitution of Eq. (7.43) with $N_\ell = 3$ into Eq. (7.42) gives [19, 20]

$$m_{\eta'} < \sqrt{3}m_\pi. \qquad (7.44)$$

Specifically, this relation cannot accommodate the experimental mass of $\sim$958 GeV. A solution to this puzzle is given by the axial $U_A(1)$ anomaly, which is discussed next and gives an extra contribution to the mass of the η' meson.

7.3 The UA(1) Anomaly

A possible resolution to the η' puzzle is the suggestion that the axial $U(1)$ channel possesses an anomaly. Specifically, t'Hooft observed that the instanton effects generate an anomaly in this channel [21] which contributes to the η' mass. Thus, the PCAC condition is modified so that it reads[*]

$$\partial_\mu A_a^\mu = \bar{q} i \gamma_5 \{T_a, M\} q + \delta_{a9} \sqrt{\frac{2}{3}} N_\ell \partial_\mu \mathcal{K}^\mu, \tag{7.45}$$

where M is the light quark mass matrix so $M = diag(m_u, m_d, m_s)$ and $\mathcal{K}^\mu$ is a topological current density given by

$$\mathcal{K}^\mu(x) = \frac{g^2}{32\pi^2} \epsilon^{\mu\alpha\beta\gamma} A_\alpha^a \left(F_{\beta\gamma}^a - \frac{1}{3} f^{abc} A_\beta^b A_\gamma^c \right), \tag{7.46}$$

and the divergence of $\mathcal{K}^\mu$ is the topological charge density $Q(x)$ so that

$$Q(x) \equiv \partial_\mu \mathcal{K}^\mu = \frac{g^2}{32\pi^2} F^{\mu\nu a} \tilde{F}_{\mu\nu}^a. \tag{7.47}$$

We now include the effect of the $U_A(1)$ anomaly in the effective Lagrangian approach. This has been accomplished by a number of authors [22–26]. It is known that a necessary ingredient for the solution of the $U(1)$ anomaly is the existence of a singularity at $q^2 = 0$ in the matrix elements of non-gauge-invariant operators [27]. To accomplish this, we introduce an effective anomaly field K^μ so that the effective PCAC condition reads

$$\partial_\mu A_a^\mu = F_{ac}\mu_{cb}\phi_b + \delta_{a9} \sqrt{\frac{2}{3}} N_\ell \partial_\mu K^\mu. \tag{7.48}$$

Further, for the effective Lagrangian we assume the form

$$\mathcal{L} = \frac{1}{2C} (\partial_\mu K_\nu)^2 + G(\phi_a) \partial_\mu K^\mu - \theta \partial_\mu K^\mu + \mathcal{L}_{CA}, \tag{7.49}$$

where C is related to the topological charge and G is a function of the spin 0 and spin 1 fields and where $\mathcal{L}_{CA}$ is the usual current algebra effective Lagrangian without the anomaly. Now, the current algebra constraints determine G, and the effective Lagrangian with the inclusion of the field K^μ can be written as (see Eq. (7.114))

$$\mathcal{L} = \frac{1}{2C} (\partial_\mu K_\nu)^2 - \left(\frac{2}{3}\right)^{1/2} N_\ell F_{a9}^{-1} \phi_a \partial_\mu K^\mu - \theta \partial_\mu K^\mu + \mathcal{L}_{CA}. \tag{7.50}$$

We note that Eq. (7.50) automatically leads to Eq. (7.48) on using the field–current identity (Eq. (7.39)). With inclusion of the $U_A(1)$ anomaly, Eq. (7.42) is modified and the new relation including the $U_A(1)$ anomaly reads [24]

$$(F_{88} + \sqrt{2} F_{98})^2 m_\eta^2 + (F_{89} + \sqrt{2} F_{99})^2 m_{\eta'}^2 = 3F_\pi^2 m_\pi^2 + \frac{4}{3} N_\ell^2 C. \tag{7.51}$$

---

[*] The 9th axial current is normalized such that $A_9^\mu = \bar{q}\gamma^u \gamma_5 \frac{1}{2}\lambda_9 q$, where $\lambda_9 = \sqrt{\frac{2}{3}} I$.

It has been shown in QCD [28] that

$$\left(\frac{d^2E}{d\theta^2}\right)_{\theta=0} = i \int d^4x \; \partial_\mu \partial_\nu \langle 0|T(\mathcal{K}^\mu(x)\mathcal{K}^\nu(0))|0\rangle. \tag{7.52}$$

For the effective theory we define

$$\Delta_{\mu\nu}(x) \equiv i\langle 0|T(K_\mu(x)K_\nu(0))|0\rangle \tag{7.53}$$

and also define $\tau(x)$ by

$$\tau(x) \equiv i\langle 0|T(\partial_\mu K^\mu(x)\partial_\nu K^\nu(0))|0\rangle. \tag{7.54}$$

One then finds that

$$\Delta_{\mu\nu}(q) = -C\frac{\eta_{\mu\nu}}{q^2} + \frac{q_\mu q_\nu}{q^4}\tau(q), \tag{7.55}$$

where

$$\tau(q) = \frac{2}{3}N_\ell^2 C^2 \sum_a \frac{(F_{a9}^{-1})^2}{q^2 + m_a^2}. \tag{7.56}$$

From Eq. (7.55) it follows that

$$q_\mu q_\nu \Delta^{\mu\nu}(q) = \tau(q) - C. \tag{7.57}$$

From Eq. (7.56) we see that $\tau(0)$ vanishes in the limit $N_\ell = 0$, and thus C is related to the anomaly fields by the relation

$$C = \int d^4x \quad i\langle 0|T(\partial_\mu K^\mu(x)\partial_\nu K^\nu(0))|0\rangle_{CL}, \tag{7.58}$$

where CL indicates the chiral limit. C can also be related to the double derivative of the energy with respect to θ in the limit $\theta = 0$ and $N_\ell = 0$ [23,24], i.e.,

$$C = (d^2E/d\theta^2)_{\theta=0, N_\ell=0}. \tag{7.59}$$

Let us now consider the $\theta = 0$ and $N_\ell = 0$ limit of Eq. (7.51). In this case, we find on using Eq. (7.43) that

$$m_{\eta'}^2 \rightarrow \frac{4N_\ell}{F_\pi^2}(d^2E/d\theta^2)_{\theta=0, N_\ell=0}. \tag{7.60}$$

The result of Eq. (7.60) was first obtained by Witten [29] in the $1/N$ expansion and as indicated in the limit $\theta = 0$ and $N_\ell = 0$. Eq. (7.51) is the generalization of the Witten formula Eq. (7.60) including the chiral symmetry breaking effects arising from m_π and m_η. Further, Eq. (7.51) arises from an effective Lagrangian approach including the $U_A(1)$ anomaly. The fact that the two results match is no accident. This is because, as shown by Witten, the effective Lagrangians arise quite naturally in the $1/N$ expansion of QCD, and thus effective Lagrangians incorporate the $1/N$ QCD expansion in terms of the physical mesonic fields.

Consider an $SU(N)_C$ color group where N is the number of colors of the fundamental quark fields. In this case the matrix element of a bilinear quark operator between the vacuum and a one-meson state ϕ is of the form

$$\langle 0|\bar{q}\gamma_5\gamma^\mu\frac{\lambda_a}{2}q|\phi b\rangle = iq^\mu F_{ab}. \tag{7.61}$$

In the $1/N$ expansion, F_{ab} behaves like $\sqrt{N}$. Similarly, $\langle 0|V_A^\mu(0)|vb\lambda\rangle = g_{ABC}\epsilon^\mu(\lambda)$ $N_{vB}(q)$, where λ is the helicity, ϵ^μ is the polarization vector, and N_{vB} is the normalization factor. Again for large N, $g_{AB} \sim \sqrt{N}$. Now, in terms of the effective Lagrangian the PCAC condition reads

$$F_{AB}g_n\frac{\partial\mathcal{L}^{(n)}}{\partial\chi_B} \sim g_{n-1}\frac{\partial\mathcal{L}^{(n-1)}}{\partial\chi_C}\chi_C + \cdots \tag{7.62}$$

Using the above relation and other current algebra relations, one finds that $g_n \sim F_\pi^{2-n}$, and since in the large-N limit, $F_{AB} \sim \sqrt{N}$ one has $g_n \sim N^{1-(n/2)}$. This is identical to what one finds in the $1/N$ expansion in QCD. Thus, the $1/N$ expansion in QCD can be thought of as an $1/F_\pi$ expansion in the effective Lagrangian approach. The above exhibits the fact that the current algebra effective Lagrangians respect the $1/N$ expansion of QCD.

7.4 The θ Dependence of the Effective Lagrangian

Next, let us consider the quark Lagrangian with the effective anomaly term. Here, we have

$$\mathcal{L} = -\bar{q}\frac{1}{i}\gamma^\mu D_\mu q - \frac{1}{4}F_{\mu\nu a}F_a^{\mu\nu} - \sum_i m_i\bar{q}_i q_i + \frac{1}{2C}(\partial_\mu K_\nu)^2$$
$$+ G(q, q^\dagger)\partial_\mu K^\mu - \theta\partial_\mu K^\mu. \tag{7.63}$$

We require this Lagrangian to obey the PCAC condition

$$\partial_\mu A^\mu = 2i\sum_i m_i\bar{q}_i\gamma_5 q_i + 2N_\ell\partial_\mu K^\mu. \tag{7.64}$$

It is straightforward to check that a $G(q, q^\dagger)$ in Eq. (7.63) that satisfies the constraint Eq. (7.64) is given by

$$G = \frac{i}{2}(tr \ln \bar{q}_L q_R - tr \ln \bar{q}_R q_L). \tag{7.65}$$

Note that in Eq. (7.63) CP violation resides in the anomaly sector. We wish to transfer the CP violation to the quark sector. This can be done in a fashion similar to the work of Baluni [30] where one starts with CP violation in the fundamental QCD Lagrangian and then make a transformation to transfer it to the quark sector. To this end, we define a matrix $\chi_{ij} \equiv \langle\bar{q}_{Li}q_{Rj}\rangle$ and introduce a set of CP-violating phases in the vacuum expectation value of χ_{ij} so that

$$\langle 0|\bar{q}_{Li}q_{Rj}|0\rangle = \delta_{ij}e^{-i\beta_i}. \tag{7.66}$$

The vacuum expectation of the interaction Hamiltonian can be written in the form

$$-\langle H'\rangle_\beta = -2\sum_i m_i \cos(\beta_i + \omega_i) + \left(\sum(\beta_i + \omega_i) - \theta\right)\langle\partial_\mu K^\mu\rangle, \tag{7.67}$$

where we have used β_i as the minimum values and ω_i as the variations around the minimum. (This procedure is similar to Dashen's theorem [31] in finding the proper vacuum solution of the case of a broken chiral symmetry.) To make sure that the $\langle\partial_\mu K^\mu\rangle$ term is cancelled, we impose the conditions

$$\sum_i \beta_i = \theta$$

$$\sum_i \omega_i = 0. \tag{7.68}$$

Under these restrictions the minimization conditions give

$$m_i \sin\beta_i = m_j \sin\beta_j = m_k \sin\beta_k \equiv \lambda. \tag{7.69}$$

Assuming that β_i are small, Eq. (7.69) reads $m_i\beta_i \simeq m_j\beta_j \simeq m_k\beta_k$, and in this approximation one finds

$$\lambda = \frac{m_1 m_2 m_3}{m_1 m_2 + m_2 m_3 + m_3 m_1}\theta, \tag{7.70}$$

and the CP violation can now be exhibited in the form [30]

$$\delta\mathcal{L}_{CP} = i\theta\frac{m_1 m_2 m_3}{m_1 m_2 + m_2 m_3 + m_3 m_1}\bar{q}\gamma_5 q. \tag{7.71}$$

7.5 Appendix: The θ Dependence of the UA(1) Anomaly

Here, we give the derivation of the relation $d^2E/d\theta^2 = -q_\mu q_\nu \Delta^{\mu\nu}|_{q\to 0}$ for QCD. We take the fundamental QCD Lagrangian for the gluon field to be

$$\mathcal{L}_g = -\frac{1}{4}F^a_{\mu\nu}F^{\mu\nu}_a$$
$$F^a_{\mu\nu} = \partial_\mu A^a_\nu - \partial_\nu A^a_\mu + gC_{abc}A^b_\mu A^c_\nu. \tag{7.72}$$

We will work in the gauge defined by $A^0 = 0$. The Lagrangian can be written in the form

$$\mathcal{L}_g = \frac{1}{2}(\dot{A}^2_i - B^2_i), \tag{7.73}$$

where

$$B_i \equiv \frac{1}{2}\epsilon_{ijk}F^{jk}, \quad B^2_i = \frac{1}{2}F^{ij}F_{ij}. \tag{7.74}$$

We take for the color gluon anomaly

$$\partial_\mu \mathcal{K}^\mu = \frac{g^2}{32\pi^2} F^{\mu\nu} \tilde{F}_{\mu\nu}, \tag{7.75}$$

where $\tilde{F}_{\mu\nu} = \frac{1}{2}\epsilon_{\mu\nu\alpha\beta}F^{\alpha\beta}$ and $\epsilon_{0123} = +1$. $\mathcal{K}^\mu$ can be written in the form

$$\mathcal{K}^\mu = \frac{g^2}{32\pi^2} \epsilon^{\mu\alpha\beta\gamma} A_\alpha^a (F_{\beta\gamma}^a - \frac{1}{3} g C_{abc} A_\beta^b A_\gamma^c). \tag{7.76}$$

We define the quantity $d(q)$ so that

$$d(q) \equiv \int d^4x \left[e^{iq\cdot(x-y)} \partial_\mu^x \partial_\nu^y i\langle 0|T(\mathcal{K}^\mu(x)\mathcal{K}^\nu(y))|0\rangle \right]_{y=0}. \tag{7.77}$$

Now, it is easily seen that

$$\begin{aligned}
s &\equiv i\partial_\mu^x \partial_\nu^y \langle 0|T(\mathcal{K}^\mu(x)\mathcal{K}^\nu(y)) = \tau_1 + \tau_2 + \tau \\
\tau_1 &\equiv i\delta(x^0 - y^0)\langle 0|T[\mathcal{K}^0(x), \partial_\nu \mathcal{K}^\nu(y)]|0\rangle \\
\tau_2 &\equiv -i\partial_\mu^x \left[\delta(x^0 - y^0)\langle 0|T[\mathcal{K}^\mu(x), \mathcal{K}^0(y)]|0\rangle \right] \\
\tau &\equiv i\langle 0|T(\partial_\mu \mathcal{K}^\mu(x)\partial_\nu \mathcal{K}^\nu(0))|0\rangle.
\end{aligned} \tag{7.78}$$

We first compute the term τ_1. It involves $\mathcal{K}^0$ and $\partial_\nu \mathcal{K}^\nu$. The term $\partial_\nu \mathcal{K}^\nu$ can be written in the form

$$\partial_\nu \mathcal{K}^\nu = -\frac{g^2}{32\pi^2} \pi_a^i B_a^i(x), \tag{7.79}$$

where $\pi_a^i = \dot{A}_a^i$, and $\mathcal{K}^0$ can be written in the form

$$\mathcal{K}^0 = -\frac{g^2}{16\pi^2} A_i^a \left(B_{ia} - \frac{1}{6} g C_{abc} A_j^b A_k^c \right). \tag{7.80}$$

After doing some algebra the term τ_1 can be reduced to the form

$$\tau_1 = -\left(\frac{g^2}{8\pi^2} \right)^2 \langle (B_a^i(x))^2 \rangle \delta^4(x-y). \tag{7.81}$$

Next, we evaluate the term τ_2. This term involves the equal-time commutator of $\delta(x^0 - y^0)[\mathcal{K}^0(x), \mathcal{K}^0(y)]$ which vanishes because Eq. (7.80) contains only the canonical co-ordinates but no canonical momenta. Thus, $\tau_2 = 0$ and $d(q)$ take the form

$$d(q) = -\left(\frac{g^2}{8\pi^2} \right)^2 \langle (B_a^i)^2 \rangle + \tau(q). \tag{7.82}$$

Next, we show that $d(0)$ is related to $d^2E/d\theta^2$ where $E(\theta)$ is the vacuum energy for a QCD Lagrangian with a θ-term. The QCD Lagrangian with the theta dependence has the form

$$\mathcal{L}(\theta) = -\frac{1}{4}F_{\mu\nu}F^{\mu\nu} + \theta\Delta\mathcal{L}$$

$$\Delta\mathcal{L} \equiv \partial_\mu K^\mu = \frac{g^2}{32\pi^2}F_{\mu\nu}\tilde{F}^{\mu\nu}. \tag{7.83}$$

Following the same procedure as before, we can write

$$\mathcal{L}(\theta) = \frac{1}{2}(\dot{A}_{ia}^2 - B_{ia}^2) + \frac{\theta g^2}{8\pi^2}\dot{A}_a^i B_{ia}, \tag{7.84}$$

and here the canonical co-ordinates and momenta are (A_{ia}, π_{ia}), where

$$\pi_{ia} = \dot{A}_{ia} + \frac{\theta g^2}{8\pi^2}B_{ia}, \tag{7.85}$$

and the Hamiltonian density takes on the form [29]

$$\mathcal{H} = \frac{1}{2}(\dot{A}_{ia}^2 + B_{ia}^2) + \mathcal{H}_1(\theta) + \mathcal{H}_2(\theta), \tag{7.86}$$

where

$$\mathcal{H}_1(\theta) = -\frac{\theta g^2}{8\pi^2}\pi_{ia}B_{ia}$$

$$\mathcal{H}_2(\theta) = \frac{1}{2}\theta^2\left(\frac{g^2}{8\pi^2}\right)^2 B_{ia}^2. \tag{7.87}$$

From the $\mathcal{H}_2$ term in Eq. (7.87) we obtain a first-order perturbation contribution to $(d^2E/d\theta^2)$ so that

$$\left(\frac{d^2\langle 0|E|0\rangle}{d\theta^2}\right)^{(1)} = \left(\frac{g^2}{8\pi^2}\right)^2 \langle B_{ia}^2\rangle. \tag{7.88}$$

Regarding the contribution from the term $\mathcal{H}_1$, instead of using second-order perturbation theory one can carry out a variational analysis directly on $\mathcal{L}(\theta)$ [29]. So, we want to compute $d^2E/d\theta^2$ arising from $I = I_0 + I_\theta$ where $I_0 = -\int d^4x(\frac{1}{4})F_{\mu\nu}F^{\mu\nu}$ and

$$I_\theta = -\theta\int d^4x\frac{g^2}{32\pi^2}F_{\mu\nu a}\tilde{F}_a^{\mu\nu}. \tag{7.89}$$

To compute $d^2E/d\theta^2$ we define the vacuum-to-vacuum amplitude using the functional integral method so that

$$Z(\theta) = \langle 0+|0-\rangle_\theta = e^{iW(\theta)}. \tag{7.90}$$

One can view θ as an external source. For a proper treatment of the functional integral we can go to the Euclidian space by letting $ix^0 = x^4$ and $V_4 \equiv iL^3T$, where V_4 is the Euclidean volume. Further, in Euclidean space we may write $Z(\theta) \approx e^{-V_4E(\theta)}$, which gives us $Z^{-1}\,d^2Z/d\theta^2 = -V_4\,d^2E/d\theta^2$, where we have used $dE/d\theta = 0$. This result also holds in Minkowski space with V_4 replaced by iV_4, where $V_4 \equiv L^3 T$. It is then easily seen that

$$\left(\frac{d^2 E}{d\theta^2}\right)^{(2)}_{\theta=0} = -\left[\frac{1}{V_4 Z(\theta)}\frac{d^2 Z}{d\theta^2}\right]_{\theta=0}. \tag{7.91}$$

Now, we transform back to the Minkowski space and use Eq. (7.83) in computing the right-hand side of Eq. (7.91). This gives

$$\left(\frac{d^2 E}{d\theta^2}\right)^{(2)}_{\theta=0} = -i\int d^4x \langle 0|T(\partial_\mu(x)\mathcal{K}^\mu(x)\partial_\nu \mathcal{K}^\nu(0))|0\rangle = -\tau. \tag{7.92}$$

Combining Eqs. (7.88) and (7.92), we find that

$$\left(\frac{d^2 E}{d\theta^2}\right)_{\theta=0} = -\tau + \left(\frac{g^2}{8\pi^2}\right)^2 \langle 0|B_{ia}(0)B_{ia}(0)|0\rangle. \tag{7.93}$$

Thus, in pure QCD we have the following relation for $(d^2 E/d\theta^2)_{\theta=0}$ [28]:

$$\left(\frac{d^2 E}{d\theta^2}\right)_{\theta=0} = -\int d^4x \left[e^{iq\cdot(x-y)}\partial_\mu^x \partial_\nu^y i\langle 0|T(\mathcal{K}^\mu(x)\mathcal{K}^\nu(y))|0\rangle\right]_{y=0}. \tag{7.94}$$

Next, we show that Eq. (7.94) is also valid for the effective Lagrangian where the $U_A(1)$ anomaly is represented phenomenologically by the term $\partial_\mu \mathcal{K}^\mu$. Thus, using the effective Lagrangian of Eq. (7.50) we find the Hamiltonian to be of the form

$$H = \int d^4x \left[-\theta C \pi^0 + \frac{1}{2}C\theta^2 + \theta - \text{independent terms}\right]. \tag{7.95}$$

Here, π^μ is the momentum canonically conjugate to K_μ. As in the case of the fundamental QCD Lagrangian, here too we see that there is a term linear in θ and a term quadratic in θ. From the term quadratic in θ we can compute the contribution to $d^2 E/d\theta^2$ to first order in perturbation theory. This gives us the result

$$\left(\frac{d^2 E}{d\theta^2}\right)^{(1)}_{\theta=0} = C. \tag{7.96}$$

A comparison of Eqs. (7.88) and (7.96) shows that C is the value of the gluon condensate. The computation of the term which is second order in perturbation theory is done in the same way as for the case of the fundamental Lagrangian. Here, as in the analysis for the fundamental Lagrangian, we get

$$\left(\frac{d^2 E}{d\theta^2}\right)^{(2)}_{\theta=0} = -\tau. \tag{7.97}$$

Thus, from Eqs. (7.96) and (7.97),

$$\left(\frac{d^2 E}{d\theta^2}\right)_{\theta=0} = C - \tau. \tag{7.98}$$

Next, we use Eq. (7.78),

$$s = \tau_1 + \tau_2 + \tau, \tag{7.99}$$

and use the canonical commutation relations for the fields K^μ to evaluate τ_1 and τ_2. Thus, (K^μ, π_ν) are the canonically conjugate co-ordinates which obey the relations $\delta(x^0 - y^0)[K^\mu, \pi_\nu] = \delta^4(x - y)\delta^\mu_\nu$. Noting that $\pi_\nu = -(1/C)\dot{K}_\nu$, one has the following commutation relation between K^μ and $\dot{K}^\nu$:

$$\delta(x^0 - y^0)[K^\mu(x), \dot{K}^\nu(y)] = -iC\eta^{\mu\nu}\delta^4(x - y). \tag{7.100}$$

Using the above relation,

$$\tau_1 = -C$$
$$\tau_2 = 0, \tag{7.101}$$

which leads to the relation

$$\int d^4x \left[e^{iq\cdot(x-y)} \partial^x_\mu \partial^y_\nu i\langle 0|T(K^\mu(x)K^\nu(y))|0\rangle \right]_{y=0} = \tau - C. \tag{7.102}$$

From Eqs. (7.98) and (7.102) we find that Eq. (7.94) still holds, and thus the effective Lagrangian produces exactly the same result as the fundamental QCD Lagrangian.

Next, C, τ, and F_{ab} are evaluated. We consider the Lagrangian with the anomaly coupling with the pseudo-scalar fields. We have

$$\mathcal{L} = \left[\frac{1}{2C}(\partial_\mu K_\nu)^2 - \sqrt{\frac{2}{3}}N_\ell F^{-1}_{a9}\phi_a\partial_\mu K^\mu - \theta\partial_\mu K^\mu \right]$$
$$- \frac{1}{2}(\partial_\mu \phi_a)^2 - \frac{1}{2}\phi_a\mu_{ab}\phi_b + \cdots \tag{7.103}$$

The above Lagrangian contains a coupling between the pseudo-scalar sector and the anomaly. This coupling produces an additional mass term in the pseudo-scalar sector so that μ_{ab} is not the physical mass, as will be seen below. Now, using Eq. (7.103) we obtain for the propagator $\Delta_{ab}(x) \equiv i\langle 0|T(\phi_a(x)\phi_b(0))|0\rangle$ (a, b = 1, ..., 9) so that

$$\Delta_{ab}(q) = \delta_{ab}/(q^2 + m_a^2), \tag{7.104}$$

where (see Eq. (7.115))

$$m_a^2\delta_{ab} = \mu_{ab} + \frac{2}{3}N_\ell^2 CF^{-1}_{a9}F^{-1}_{b9}. \tag{7.105}$$

From Eq. (7.103) one also has the equation for K^μ so that

$$\Box K_\mu = C\sqrt{\frac{2}{3}}N_\ell F^{-1}_{a9}F^{-1}_{a9}\partial_\mu\phi_a. \tag{7.106}$$

Using the above and the interaction of Eq. (7.103), we obtain for the propagator of the anomaly field $\Delta_{\mu\nu}(x) = i\langle 0|T(K_\mu(x)K_\nu(0))|0\rangle$ the result

$$\Delta_{\mu\nu}(q) = -C\frac{\eta_{\mu\nu}}{q^2} + \frac{q^\mu q^\nu}{q^4}\tau(q). \tag{7.107}$$

From the above we can compute the quantity $q_\mu q_\nu \Delta^{\mu\nu}(q)$, which gives

$$q_\mu q_\nu \Delta^{\mu\nu}(q) = -C + \tau. \tag{7.108}$$

Using this result and Eq. (7.103) one finds

$$i\langle 0|T(\partial_\mu K^\mu(x)\phi_a(0))|0\rangle = C\sqrt{\frac{2}{3}}N_\ell F_{b9}^{-1}\Delta_{ab}. \tag{7.109}$$

7.6 Problems

1. Show that the condition that the current algebra commutation relations do not contain a q-number Schwinger term, i.e., the derivative of a delta function, leads to the Weinberg sum rule (Eq. (7.29)).
2. Use the effective Lagrangian of the $\pi\rho A_1$ system given by Eqs. (7.12) and (7.13) and the constraints of the current algebra commutation relations given by Eqs. (7.20)–(7.23), of CVC given by Eq. (7.24), and of PCAC given by Eq. (7.25) to deduce the constrains on the couplings given by Eq. (7.30).
3. Defining

$$x \equiv \sqrt{2}m_\rho/m_{A_1}, \ y \equiv g_{A_1}/g_\rho, \ z \equiv g_\rho/\sqrt{2}m_\rho F_\pi, \tag{7.110}$$

show that the current algebra constraints on the couplings given by Eq. (7.30) are modified to read

$$g_{\pi\pi\rho} = g_{\rho\rho\rho} = g_{\rho AA} = m_\rho^2 g_\rho^{-1}$$
$$g_{\pi\rho A} = -m_A^2\lambda_{\pi\rho A} = m_\rho^2(F_\pi x^2 yz^2)^{-1}$$
$$g_\rho\lambda_{\pi\pi\rho} = \frac{1}{2}x^4 y^2 z^2\lambda_A + 2(1 - z^2)$$
$$2F_\pi\mu_{\pi\rho A} = y - y^{-1}$$
$$F_\pi\tilde{\lambda}_{\pi\rho A} = -y\left(1 - \frac{1}{2}x^2\lambda_A\right). \tag{7.111}$$

Further, show that the condition that the q-number Schwinger terms vanish now reads $x^2 y^2 z^2 - 2z^2 + 1 = 0$. Note that the second Weinberg sum rule

$$g_A = g_\rho \tag{7.112}$$

arises from the condition that the coupling $\mu_{\pi\rho A}$ vanishes, where $\mu_{\pi\rho A}$ is the coupling of the term $\epsilon_{abc}\phi_a G_b^{\mu\nu} H_{\mu\nu c}$ and thus involves two derivatives. So, Eq. (7.111) tells us that the second Weinberg sum rule (Eq. (7.112)) may be viewed as a smoothness condition which constrains the number of derivatives in the cubic interaction.

4. Show that, using the field–current identity, the effective Lagrangian consistent with the modified PCAC condition including the effect of the anomaly, i.e.,

$$\partial_\mu A_a^\mu = F_{ac\mu}c_b\phi_b + \delta_{a9}\sqrt{\frac{2}{3}}N_\ell\partial_\mu K^\mu \tag{7.113}$$

is given by

$$\mathcal{L} = \frac{1}{2C}(\partial_\mu K_\nu)^2 - \left(\frac{2}{3}\right)^{1/2} N_\ell F_{a9}^{-1}\phi_a\partial_\mu K^\mu + \cdots \tag{7.114}$$

5. Using the effective Lagrangian, of Eq. (7.103) obtain the field equation for the propagating spin 0 fields ϕ_a including the effect of the anomaly. This is done by eliminating the $\partial_\mu K^\mu$ term in the field equation. Show that the mass of the propagating spin 0 field is given by

$$m_a^2\delta_{ab} = \mu_{ab} + \frac{2}{3}N_\ell^2 C F_{a9}^{-1} F_{b9}^{-1}, \tag{7.115}$$

where μ_{ab} are the masses without the anomaly effect, The second term on the right-hand side is the anomaly contribution and m_{ab} are the clothed masses, i.e., masses including the effect of the anomaly on the spin 0 masses. Show that in the limit of chiral symmetry, i.e., $\mu_{ab} \to 0$, and no mixing of the eightth and ninth components so that $F_{ab} = \delta_{ab}F_a$, the η' mass is given by

$$m_{\eta'}^2 = \frac{4N_\ell}{F_\pi^2}C. \tag{7.116}$$

References

[1] W. Heisenberg and H. Euler, *Z. Phys.* **98**, 714 (1936).

[2] E. Witten, *Nucl. Phys. B* **160**, 57 (1979).

[3] S. Weinberg, *Phys. Rev. Lett.* **17**, 616 (1966).

[4] S. L. Glashow and S. Weinberg, *Phys. Rev. Lett.* **20**, 224 (1968).

[5] J. S. Schwinger, *Phys. Lett. B* **24**, 473 (1967).

[6] S. Weinberg, *Phys. Rev.* **166**, 1568 (1968).

[7] J. Wess and B. Zumino, *Phys. Rev.* **163**, 1727 (1967).

[8] S. R. Coleman, J. Wess, and B. Zumino, *Phys. Rev.* **177**, 2239 (1969).

[9] C. G. Callan, Jr., S. R. Coleman, J. Wess, and B. Zumino, *Phys. Rev.* **177**, 2247 (1969).

[10] B. W. Lee and H. T. Nieh, *Phys. Rev.* **166**, 1507 (1968).

[11] R. Arnowitt, M. H. Friedman, and P. Nath, *Phys. Rev. Lett.* **19**, 1085 (1967).

[12] R. L. Arnowitt, M. H. Friedman, P. Nath, and R. Suitor, *Phys. Rev.* **175**, 1802 (1968).

[13] R. L. Arnowitt, M. H. Friedman, P. Nath, and R. Suitor, *Phys. Rev. D* **3**, 594 (1971).

[14] T. D. Lee, S. Weinberg, and B. Zumino, *Phys. Rev. Lett.* **18**, 1029 (1967).

[15] M. Gell-Mann, *Physics* **1**, 63 (1964).

[16] S. Weinberg, *Phys. Rev. Lett.* **18**, 507 (1967).

[17] K. Kawarabayashi and M. Suzuki, *Phys. Rev. Lett.* **16**, 255 (1966); Riazuddin and Fayyazuddin, *Phys. Rev.* **147**, 1071 (1966).

[18] H. J. Schnitzer and S. Weinberg, *Phys. Rev.* **164**, 1828 (1967).

[19] S. Weinberg, *Phys. Rev. D* **11**, 3583 (1975).

[20] S. L. Glashow, in *Hadrons and their Interactions*, Academic Press (1968), p. 83.

[21] G. 't Hooft, *Phys. Rev. Lett.* **37**, 8 (1976).

[22] C. Rosenzweig, J. Schechter, and C. G. Trahern, *Phys. Rev. D* **21**, 3388 (1980).

[23] P. Nath and R. L. Arnowitt, *Phys. Rev. D* **23**, 473 (1981).

[24] R. L. Arnowitt and P. Nath, *Nucl. Phys. B* **209**, 234 (1982).

[25] P. Di Vecchia and G. Veneziano, *Nucl. Phys. B* **171**, 253 (1980).

[26] E. Witten, *Ann. Phys.* **128**, 363 (1980).

[27] J. B. Kogut and L. Susskind, *Phys. Rev. D* **11**, 3594 (1975).

[28] R. J. Crewther, *Riv. Nuovo Cim.* **2N8**, 63 (1979).

[29] E. Witten, *Nucl. Phys. B* **156**, 269 (1979).

[30] V. Baluni, *Phys. Rev. D* **19**, 2227 (1979).

[31] R. F. Dashen, *Phys. Rev. D* **3**, 1879 (1971).

Further Reading

H. Georgi, *Ann. Rev. Nucl. Part. Sci.* **43**, 209 (1993).

J. Polchinski, *Nucl. Phys. B* **231**, 269 (1984).

S. Weinberg, *The Quantum Theory of Fields*, Vol. 2: *Modern Applications*, Cambridge University Press (2005).

8

Supersymmetry

Supersymmetry is a Fermi–Bose symmetry. Supersymmetry transforms bosons into fermions and vice versa. In two dimensions it appears in the work of Ramond [2] and in four dimensions in the works of Golfand and Likhtman [3], Volkov and Akulov [4], and of Wess and Zumino [5,6]. Further developments using superspace were carried out by Salam and Strathdee [7,8] and Arnowitt *et al.* [9]. Supersymmetry helps in resolving the hierarchy problem in particle physics models [10] and theories based on it are more well behaved at the quantum level [11] than the non-supersymmetric theories. The physics involving only bosonic operators is governed by the so-called Coleman–Mandula theorem [12], which we now briefly discuss.

8.1 The Coleman–Mandula Theorem

The Coleman–Mandula theorem [12] states that the bosonic symmetries of an S-matrix must be a direct product of Poincaré symmetry and internal symmetry, which means that the internal symmetries cannot change spin. The Poincaré group is described by the generators of translations P_μ and rotations $M_{\mu\nu}$ which obey the algebra

$$[P_\mu, P_\nu] = 0$$
$$[P_\mu, M_{\nu\lambda}] = i(\eta_{\mu\nu}P_\lambda - \eta_{\mu\lambda}P_\nu)$$
$$[M_{\mu\nu}, M_{\lambda\rho}] = i(\eta_{\mu\rho}M_{\nu\lambda} - \eta_{\nu\rho}M_{\mu\lambda} - \eta_{\mu\lambda}M_{\nu\rho} + \eta_{\nu\lambda}M_{\mu\rho}), \qquad (8.1)$$

---

[*] As quoted by Schwinger [1].

while the generators of the internal symmetry group, which we denote by F_a, obey the algebra

$$[F_a, F_b] = iC_{ab}{}^c F_c. \tag{8.2}$$

The generators of the Poincaré group and the internal symmetry group commute with each other so that

$$[F_a, P_\mu] = 0, \quad [F_a, M_{\mu\nu}] = 0. \tag{8.3}$$

We can define the Casimir operators

$$P^2 = P_\mu P^\mu, \quad U^2 = U_\mu U^\mu, \tag{8.4}$$

where U^μ is defined by

$$U^\mu = -\frac{1}{2}\epsilon^{\mu\nu\lambda\rho}P_\nu M_{\lambda\rho}. \tag{8.5}$$

The Casimirs commute with all the generators listed above, i..e., $[P^2, P_\mu] = 0 = [P^2, M_{\mu\nu}]$, $[U^2, P_\mu] = 0 = [U^2, M_{\mu\nu}]$, and

$$[F_a, P^2] = 0 \tag{8.6}$$

and

$$[F_a, U^2] = 0. \tag{8.7}$$

Since $M^2 = -P_\mu P^\mu$ is the mass squared operator, Eq. (8.6) implies that all elements of a massive irreducible representation of a group must have the same mass. To understand the meaning of Eq. (8.7) we can go to the rest frame of the massive particle, and here one finds that

$$U^2 = -m^2 L^2, \tag{8.8}$$

where

$$\vec{L} = [M_{23}, M_{31}, M_{12}]. \tag{8.9}$$

Equation (8.9) makes transparent that U^2 is the spin operator. Thus, Eq. (8.7) implies that all elements of a massive irreducible representation of a group must have the same spin.

8.2 Graded Lie Algebras

The analysis of Coleman and Mandula considered only the bosonic operators, and their conclusions are restricted to only this set. The situation changes drastically when one includes the possibility of both bosonic and fermionic operators. Thus, the simplest extension of the Poincaré algebra is the $N = 1$ supersymmetric extension, where N is the number of Majorana spinor generators. In general, we can for the case of $N = 1$ supersymmetry assign a grade g for each of the generators which is either 0 (mod 2) for bosonic generators or 1 (mod 2) for fermionic operators. For the present case, a graded commutator of two operators X and Y is defined so that

$$[X, Y\} = (-1)^{1 + g_X g_Y}[Y, X\}. \tag{8.10}$$

For both X and Y bosonic, the bracket $[X, Y\}$ in Eq. (8.10) is a commutator; for either X or Y bosonic and the other fermionic, the bracket is also a commutator; and with both X and Y fermionic, the bracket is an anti-commutator. Specifically, one may realize supersymmetry algebraically in the following way. Let Q_α be a Lorentz spinor obeying the following anti-commutation and commutation relations:

$$\{Q_\alpha, Q_\beta^\dagger\} = -2(P_L \gamma^\mu \gamma^0)_{\alpha\beta} P_\mu \tag{8.11}$$

$$\{Q_\alpha, Q_\beta\} = 0 \tag{8.12}$$

$$[Q_\alpha, P_\mu] = 0, \tag{8.13}$$

where $P_L = (1 - \gamma^5)/2$. Equations (8.11)–(8.13) constitute an $N = 1$ supersymmetry algebra.

For supersymmetry generators, the result of Eq. (8.7) does not hold since

$$[Q, \, U^2] \neq 0. \tag{8.14}$$

However, Eq. (8.6) still holds, i.e.,

$$[Q, \, P^2] = 0. \tag{8.15}$$

since $[Q, \, P_\mu] = 0$. Thus, our previous result that all elements of an irreducible representation have the same mass still holds. However, unlike the case of the Poincaré algebra, elements of the supermultiplets do not all have the same spin because of Eq. (8.14) but have spins differing by $1/2$. Thus, the $N = 1$ supermultiplets have spins of the following type: $(2, 3/2)$, $(3/2, 1)$, $(1, 1/2)$, and $(1/2, 0)$. Further, one can show that each supermultiplet would have an equal number of bosonic and fermionic degrees of freedom. This can be easily illustrated by use of the Witten index defined by $(-1)^{N_f}$, where N_f acting on a state counts the number of fermions. Thus, $(-1)^{N_f}$ is $+1$ for bosonic states and -1 for fermionic states. Thus,

$$tr[(-1)^{N_f}] = 0, \tag{8.16}$$

and implies that the number of bosonic degrees of freedom must equal the number of fermionic degrees of freedom for an irreducible multiplet of supersymmetry.

Another important property relates to the fact that the Hamiltonian is a positive definite operator in supersymmetry. This can be seen by taking the trace of Eq. (8.11). Here, $H \equiv P^0$ is given by

$$H = \frac{1}{4}(QQ^\dagger + Q^\dagger Q) \geq 0, \tag{8.17}$$

which shows that the Hamiltonian is a positive definite operator. Further, the supersymmetric vacuum state $|0\rangle$ satisfies the property that

$$Q_\alpha |0\rangle = 0$$
$$Q_\alpha^\dagger |0\rangle = 0. \tag{8.18}$$

Taking the expectation value of Eq. (8.17) in the vacuum state, we find

$$\langle 0|H|0\rangle = 0 \tag{8.19}$$

for a supersymmetric vacuum state. On the other hand, if there is spontaneous breaking of supersymmetry we have,

$$Q_\alpha |0\rangle \neq 0$$
$$Q_\alpha^\dagger |0\rangle \neq 0, \tag{8.20}$$

which leads to

$$\langle 0|H|0\rangle > 0. \tag{8.21}$$

From Eqs. (8.19) and (8.21), one finds that the supersymmetric vacuum state lies lower relative to the case when supersymmetry is broken. This result indicates that it would be difficult to break supersymmetry in an acceptable way. We will discuss this issue in greater detail later in this chapter.

A direct extension of Eqs. (8.11)–(8.13) can be obtained by giving fermionic charges an index i, i.e., the charge Q_α is extended to Q_α^i where i can take value $i = 1, \ldots, N$. However, N cannot exceed 8 since $N > 8$ would result in massless particles with spins greater than 2, which are undesirable [13, 14]. From the point of view of physics, one of the reasons supersymmetry becomes interesting is the following: for an S-matrix constructed from a local quantum field theory the supersymmtry algebra is the only graded algebra that is allowed. In this sense, supersymmetry constitutes a unique extension when constructing Lorentz-invariant theories [15].

8.3 Two-component Spinors

In the analysis of $N = 1$ supersymmetry, it is sometimes convenient to use two-component Weyl spinors, and there are two types: left handed and right handed. We will denote the left-handed Weyl spinor by ξ_α and the right handed by $\bar{\chi}^{\dot\alpha}$. Thus, for example, a Dirac field can be written as

$$\psi_D = \begin{pmatrix} \xi_\alpha \\ \bar{\chi}^{\dot\alpha} \end{pmatrix}, \tag{8.22}$$

and $\bar{\psi}_D$ is then given by

$$\bar{\psi}_D = (\chi^\alpha, \bar{\xi}_{\dot{\alpha}}). \qquad (8.23)$$

Now, from Eq. (8.22) we have in component form $\bar{\psi}_D = (\bar{\chi}^{\dot{1}*}, \bar{\chi}^{\dot{2}*}, \xi_1^*, \xi_2^*)$. Comparing it with Eq. (8.23) we have $\chi^\alpha = \bar{\chi}^{\dot{\alpha}*}$ and $\bar{\xi}_{\dot{\alpha}} = \chi_\alpha^*$ where for a quantized field the $*$ will indicate Hermitian conjugate. It is then seen that

$$\bar{\psi}_D \psi_D = \chi^\alpha \xi_\alpha + \bar{\xi}_{\dot{\alpha}} \bar{\chi}^{\dot{\alpha}}. \qquad (8.24)$$

So, each of the two parts on the right-hand side of Eq. (8.24) is separately Lorentz invariant. Thus, under Lorentz transformations the left-handed Weyl spinor ξ_α transforms like

$$\xi_\alpha \to \xi_\alpha' = L_\alpha{}^\beta \xi_\beta, \qquad (8.25)$$

where $L_\alpha{}^\beta$ ($\alpha, \beta = 1, 2$) is an element of an $SL(2, C)$ matrix which is complex and unimodular, i.e., $det(L) = 1$ and $L^\dagger L = 1$. Similarly, one can define a two-component spinor with an upper index ξ^α by the transformation

$$\xi^\alpha \to \xi^{\alpha'} = \xi^\beta (L^{-1})_\beta{}^\alpha, \qquad (8.26)$$

where L^{-1} is the inverse of the matrix L. From the above, it is clear that the product $\xi^\alpha \chi_\alpha$ is invariant under Lorentz transformations:

$$\xi^\alpha \chi_\alpha \to \xi^{\alpha'} \chi_{\alpha'} = \xi^\beta (L^{-1})_\beta{}^\alpha L_\alpha{}^\gamma \chi_\gamma = \xi^\beta (L^{-1}L)_\beta^\gamma \chi_\gamma = \xi^\beta \chi_\beta. \qquad (8.27)$$

We can also define higher-order tensors with lower and upper indices, so that we have tensors of the type $\psi_{\alpha\beta\cdots}$ and $\psi^{\alpha\beta\cdots}$ which have transformations of the type

$$\xi_{\alpha\beta\cdots} \to \xi_{\alpha\beta\cdots}' = L_\alpha{}^\gamma L_\beta{}^\delta \cdots . \xi_{\gamma\delta\cdots} \qquad (8.28)$$

Similarly, we have higher-order tensors $\xi^{\alpha\beta\cdots}$ which transform as follows:

$$\xi^{\alpha\beta} \to \xi^{\alpha\beta'} = \xi^{\gamma\delta\cdots} (L^{-1})_\gamma{}^\alpha . (L^{-1})_\delta{}^\beta \cdots \qquad (8.29)$$

We define the objects $\epsilon^{\alpha\beta}$ so that $\epsilon^{12} = 1$, and $\epsilon^{\alpha\beta} = -\epsilon^{\beta\alpha}$, and define $\epsilon_{12} = -1$ and $\epsilon_{\alpha\beta} = -\epsilon_{\beta\alpha}$. Thus, we have

$$\epsilon^{\alpha\beta} = \epsilon^{\dot{\alpha}\dot{\beta}} = \begin{pmatrix} 0 & 1 \\ -1 & 0 \end{pmatrix}, \quad \epsilon_{\alpha\beta} = \epsilon_{\dot{\alpha}\dot{\beta}} = \begin{pmatrix} 0 & -1 \\ 1 & 0 \end{pmatrix}. \qquad (8.30)$$

The epsilons then satisfy the relations

$$\epsilon^{\alpha\beta} \epsilon_{\beta\alpha} = \delta_\gamma^\alpha = \epsilon_{\alpha\beta} \epsilon^{\beta\gamma}, \qquad (8.31)$$

with identical relations holding for the case with dotted indices. The objects $\epsilon_{\alpha\beta}$, $\epsilon^{\alpha\beta}$, and $\epsilon_{\dot{\alpha}\dot{\beta}}$, $\epsilon^{\dot{\alpha}\dot{\beta}}$ are invariant tensors under the Lorentz transformation (see Problem 1 in Section 8.21). Further, the ϵ can be used to raise and lower the indices. so that

$$\xi^\alpha = \epsilon^{\alpha\beta} \xi_\beta, \quad \xi_\alpha = \epsilon_{\alpha\beta} \xi^\beta. \qquad (8.32)$$

Similarly, we have

$$\bar{\chi}^{\dot\alpha} = \epsilon^{\dot\alpha\dot\beta}\bar{\chi}_{\dot\beta}, \quad \bar{\chi}_{\dot\alpha} = \epsilon_{\dot\alpha\dot\beta}\bar{\chi}^{\dot\beta}. \tag{8.33}$$

We will now define the scalar product of two-component spinors using the epsilon tensor. Assume we have two spinors with undotted indices ξ^α and ξ'^α. We then define their scalar product so that

$$\xi\xi' \equiv \xi^\alpha\xi'_\alpha \tag{8.34}$$

$$= \xi^\alpha\epsilon_{\alpha\beta}\xi'^\beta \tag{8.35}$$

$$= -\xi_\alpha\epsilon^{\alpha\beta}\xi'_\beta. \tag{8.36}$$

In a similar fashion we define the scalar product of objects with dotted indices as follows:

$$\bar{\chi}\bar{\chi}' \equiv \bar{\chi}_{\dot\alpha}\bar{\chi}'^{\dot\alpha} \tag{8.37}$$

$$= -\bar{\chi}^{\dot\alpha}\epsilon_{\dot\alpha\dot\beta}\bar{\chi}'^{\dot\beta} \tag{8.38}$$

$$= \bar{\chi}_{\dot\alpha}\epsilon^{\dot\alpha\dot\beta}\bar{\chi}'_{\dot\beta}. \tag{8.39}$$

We may now check that using the fact that ξ and ξ' and χ and χ' are Grassmann that the following holds:

$$\xi\xi' = \xi'\xi, \quad \bar{\chi}\bar{\chi}' = \bar{\chi}'\bar{\chi}. \tag{8.40}$$

The transformations of a right-handed spinor $\bar{\chi}_{\dot\alpha}$ can be expressed as follows:

$$\bar{\chi}_{\dot\alpha} \to \bar{\chi}'_{\dot\alpha} = (L^*)_{\dot\alpha}^{\ \dot\beta}\bar{\chi}_{\dot\beta}, \tag{8.41}$$

while those of $\bar{\chi}^{\dot\alpha}$ can be expressed as

$$\bar{\chi}^{\dot\alpha} \to \bar{\chi}^{\dot\alpha'} = \bar{\chi}^{\dot\beta}(L^{-1})^*_{\dot\beta}{}^{\dot\alpha}. \tag{8.42}$$

As for the case of the undotted indices, one can check that the product $\bar{\chi}\bar{\chi}'$ is a Lorentz-invariant combination.

In order to write the Dirac equation in the two-component notation, we choose a representation of the gamma matrices in the two-component form:

$$\gamma^\mu = \begin{pmatrix} 0 & \sigma^\mu \\ \bar{\sigma}^\mu & 0 \end{pmatrix}, \quad \gamma_5 = \begin{pmatrix} -\mathbb{1} & O \\ 0 & \mathbb{1} \end{pmatrix}, \tag{8.43}$$

where $\mathbb{1}$ is a 2×2 identity matrix and σ^μ and $\bar{\sigma}^\mu$ are given by

$$\sigma^0 = \bar{\sigma}^0 = -\begin{pmatrix} 1 & 0 \\ 0 & 1 \end{pmatrix}, \quad \sigma^1 = -\bar{\sigma}^1 = \begin{pmatrix} 0 & 1 \\ 1 & 0 \end{pmatrix}$$

$$\sigma^2 = -\bar{\sigma}^2 = \begin{pmatrix} 0 & -i \\ i & 0 \end{pmatrix}, \quad \sigma^3 = -\bar{\sigma}^3 = \begin{pmatrix} 1 & 0 \\ 0 & -1 \end{pmatrix}. \tag{8.44}$$

The following properties of σ^μ and $\bar\sigma^\mu$ can be easily checked:

$$(\sigma^\mu\bar\sigma^\nu + \sigma^\nu\bar\sigma^\mu)_\alpha{}^\beta = -2\eta^{\mu\nu}\delta_\alpha{}^\beta \tag{8.45}$$

$$(\bar\sigma^\mu\sigma^\nu + \bar\sigma^\nu\sigma^\mu)^{\dot\alpha}{}_{\dot\beta} = -2\eta^{\mu\nu}\delta^{\dot\alpha}{}_{\dot\beta} \tag{8.46}$$

$$tr(\sigma^\mu\bar\sigma^\nu) = -2\eta^{\mu\nu}, \tag{8.47}$$

where the metric

$$\eta^{\mu\nu} = diag(-1,\,1,\,1,\,1). \tag{8.48}$$

The free Dirac Lagrangian in then expressed in the form

$$L_D = i\left[\xi_1^\dagger\bar\sigma^\mu\partial_\mu\xi_1 + \xi_2^\dagger\bar\sigma^\mu\partial_\mu\xi_2\right] - M\left(\xi_1\xi_2 + \xi_1^\dagger\xi_2^\dagger\right) - i\partial_\mu(\xi_2^\dagger\bar\sigma^\mu\xi_2). \tag{8.49}$$

In obtaining the above result we have used the following identity:

$$(\bar\sigma^\mu)^{\dot\alpha\alpha} = \epsilon^{\dot\alpha\dot\beta}\epsilon^{\alpha\beta}(\sigma^\mu)_{\beta\dot\beta}. \tag{8.50}$$

Using the above result, it is easy to see that

$$\xi_2^\dagger\bar\sigma_\mu\xi_2 = -\xi_2\sigma_\mu\xi_2^\dagger, \tag{8.51}$$

which is used in writing the Dirac Lagrangian in the two-component notation.

In the two-component notation the $N = 1$ supersymmetry algebra can be written as

$$\{Q_\alpha, \bar Q_{\dot\beta}\} = 2(\sigma^\mu)_{\alpha\dot\beta}P_\mu \tag{8.52}$$

$$\{Q^\alpha, \bar Q^{\dot\beta}\} = 2(\bar\sigma^\mu)^{\dot\beta\alpha}P_\mu \tag{8.53}$$

$$\{Q_\alpha, Q_{\dot\beta}\} = 0 \tag{8.54}$$

$$\{\bar Q_{\dot\alpha}, \bar Q_{\dot\beta}\} = 0. \tag{8.55}$$

The commutators of Q_α and $\bar Q^{\dot\alpha}$ with P_μ, and $M_{\mu\nu}$ are

$$[Q_\alpha, P_\mu] = 0 = [\bar Q^{\dot\alpha}, P_\mu] \tag{8.56}$$

$$[Q_\alpha, M_{\mu\nu}] = (\sigma_{\mu\nu})_\alpha{}^\beta Q_\beta \tag{8.57}$$

$$[\bar Q^{\dot\alpha}, M_{\mu\nu}] = (\sigma_{\mu\nu})^{\dot\alpha}{}_{\dot\beta}\bar Q^{\dot\beta}. \tag{8.58}$$

8.4 Left Chiral Superfields

We define a left chiral superfield so that it satisfies the condition

$$\bar D_{\dot\alpha}(x,\theta,\bar\theta)\Phi_L(x,\theta,\bar\theta) = 0, \tag{8.59}$$

where $\bar D_{\dot\alpha}(x,\theta,\bar\theta)$ is defined by

$$\bar D_{\dot\alpha}(x,\theta,\bar\theta) = -\frac{\partial}{\partial\bar\theta^{\dot\alpha}} - i\theta^\alpha(\sigma^\mu)_{\alpha\dot\alpha}\frac{\partial}{\partial x^\mu}. \tag{8.60}$$

Similarly, one defines a right chiral field $\Phi_R(x, \theta, \bar{\theta})$ so that it satisfies the condition

$$D_\alpha(x, \theta, \bar{\theta})\Phi_R(x, \theta, \bar{\theta}) = 0, \tag{8.61}$$

where $D_\alpha(x, \theta, \bar{\theta})$ is defined by

$$D_\alpha(x, \theta, \bar{\theta}) = \frac{\partial}{\partial \theta^\alpha} + i(\sigma^\mu)_{\alpha\dot{\alpha}}\bar{\theta}^{\dot{\alpha}}\frac{\partial}{\partial x^\mu}. \tag{8.62}$$

We will focus first on the left chiral field and later discuss the right chiral field. To exploit the constraint of Eq. (8.59), we introduce new variables:

$$y^\mu = x^\mu + i\theta(\sigma^\mu)\bar{\theta} \tag{8.63}$$

$$\theta' = \theta \tag{8.64}$$

$$\bar{\theta}' = \bar{\theta}. \tag{8.65}$$

It is then seen that

$$\bar{D}_{\dot{\alpha}}(x, \theta, \bar{\theta})y^\mu = -\frac{\partial y^\mu}{\partial \bar{\theta}^{\dot{\alpha}}} - i\theta^\alpha(\sigma^\mu)_{\alpha\dot{\alpha}}\frac{\partial y^\mu}{\partial x^\mu}$$

$$= i\theta^\alpha(\sigma^\mu)_{\alpha\dot{\alpha}} - i\theta^\alpha(\sigma^\mu)_{\alpha\dot{\alpha}} = 0. \tag{8.66}$$

We want to express the $\bar{D}_{\dot{\alpha}}(x, \theta, \bar{\theta})$ and $D_\alpha(x, \theta, \bar{\theta})$ in terms of the new variables $(y, \theta', \bar{\theta}')$. For this purpose the following results are useful:

$$\frac{\partial}{\partial \theta^\alpha} = \frac{\partial}{\partial \theta^{\alpha'}} + i(\sigma^\mu)_{\alpha\dot{\alpha}}\bar{\theta}'^{\dot{\alpha}}\frac{\partial}{\partial y^\mu} \tag{8.67}$$

$$\frac{\partial}{\partial \bar{\theta}^{\dot{\alpha}}} = \frac{\partial}{\partial \bar{\theta}^{\dot{\alpha}'}} - i\theta'^{\alpha}(\sigma^\mu)_{\alpha\dot{\alpha}}\frac{\partial}{\partial y^\mu}. \tag{8.68}$$

Now, let us define

$$D_\alpha^L(y, \theta', \bar{\theta}') \equiv D_\alpha(x, \theta, \bar{\theta}) \tag{8.69}$$

$$\bar{D}_{\dot{\alpha}}^L(y, \theta', \bar{\theta}') \equiv \bar{D}_{\dot{\alpha}}(x, \theta, \bar{\theta}). \tag{8.70}$$

Using Eqs. (8.60) and (8.62)–(8.65), one finds the following expressions for $D_{\dot{\alpha}}^L(y, \theta', \bar{\theta}')$ and $\bar{D}_{\dot{\alpha}}^L(y, \theta', \bar{\theta}')$:

$$\bar{D}_{\dot{\alpha}}^L(y, \theta', \bar{\theta}') = -\frac{\partial}{\partial \bar{\theta}^{\dot{\alpha}'}} \tag{8.71}$$

$$D_\alpha^L(y, \theta', \bar{\theta}') = \frac{\partial}{\partial \theta^{\alpha'}} + 2i(\sigma^\mu)_{\alpha\dot{\alpha}}\bar{\theta}'^{\dot{\alpha}}\frac{\partial}{\partial y^\mu}. \tag{8.72}$$

In terms of the new variables, the condition of Eq. (8.59) that the superfield be left chiral takes on the following form on using Eq. (8.70):

$$\bar{D}_{\dot{\alpha}}^L(y, \theta, \bar{\theta})\Phi_L(y, \theta, \bar{\theta}) = 0, \tag{8.73}$$

where we have written $\Phi_L(x, \theta, \bar{\theta})$ as $\Phi_L(y, \theta', \bar{\theta}')$ in order not to introduce excess notation. Equations (8.71) and (8.73) imply that

$$\frac{\partial}{\partial \bar{\theta}^{\dot{\alpha}}}\Phi_L(y, \theta, \bar{\theta}) = 0. \tag{8.74}$$

Thus, the field $\Phi_L(y, \theta, \bar{\theta})$ is independent of $\bar{\theta}$. It means that we can drop the dependence on $\bar{\theta}$ in $\Phi_L(y, \theta, \bar{\theta})$ since it is only a function of y and θ, i.e., one may write the left-handed field as $\Phi_L(y, \theta)$. Thus, in general, the left-handed field will have an expansion of the form

$$\Phi_L(y, \theta) = \phi(y) + \sqrt{2}\theta\psi(y) + \theta\theta F(y), \tag{8.75}$$

where the fields depend on $y^\mu = x^\mu + i\theta\sigma^\mu\bar{\theta}$. We want to write the left chiral field in terms of the x, θ, and $\bar{\theta}$. So, in this case we substitute for y and expand in θ and $\bar{\theta}$. We carry out the following expansions of the fields on the right-hand side of Eq. (8.75). We thus have

$$\phi(x + i\theta\sigma^\mu\bar{\theta}) = \phi(x) + i\theta\sigma^\mu\bar{\theta}\partial_\mu\phi(x) - \frac{1}{2}(\theta\sigma^\mu\bar{\theta})(\theta\sigma^\nu\bar{\theta})\partial_\mu\partial_\nu\phi(x) \tag{8.76}$$

$$\theta\psi(x + i\theta\sigma^\mu\bar{\theta}) = \theta\psi_\alpha(x) + i\theta\sigma^\mu\bar{\theta}\theta^\alpha\partial_\mu\psi_\alpha \tag{8.77}$$

$$\theta\theta F(x + i\theta\sigma^\mu\bar{\theta}) = \theta\theta F(x), \tag{8.78}$$

where we have used the fact that terms in the expansion which are θ^3 or $\bar{\theta}^3$ or higher vanish. Next, we use the identities

$$\theta^\alpha\theta^\beta = -\frac{1}{2}\epsilon^{\alpha\beta}\theta\theta \tag{8.79}$$

$$\bar{\theta}^{\dot{\alpha}}\bar{\theta}^{\dot{\beta}} = \frac{1}{2}\epsilon^{\dot{\alpha}\dot{\beta}}\bar{\theta}\bar{\theta} \tag{8.80}$$

$$\theta\sigma^\mu\bar{\theta}\theta\sigma^\nu\bar{\theta} = -\frac{1}{2}\eta^{\mu\nu}\theta\theta\bar{\theta}\bar{\theta}. \tag{8.81}$$

With the above we express the left chiral field so that

$$\begin{aligned}\Phi_L(x, \theta, \bar{\theta}) =&\phi(x) + \sqrt{2}\theta\psi(x) + \theta\theta F(x) + i\theta\sigma^\mu\bar{\theta}\partial_\mu\phi \\ &- \frac{i}{\sqrt{2}}\theta\theta\partial_\mu\psi(x)\sigma^\mu\bar{\theta} + \frac{1}{4}\theta^2\bar{\theta}^2\Box\phi(x).\end{aligned} \tag{8.82}$$

The transformation properties of the component fields under supersymmetry transformations are given by

$$\delta_\epsilon\phi(x) = \sqrt{2}\epsilon\psi(x) \tag{8.83}$$

$$\delta_\epsilon\psi_\alpha(x) = i\sqrt{2}(\sigma^\mu\bar{\epsilon})_\alpha\partial_\mu\phi(x) + \sqrt{2}\epsilon_\alpha F(x) \tag{8.84}$$

$$\delta_\epsilon F(x) = -i\sqrt{2}\partial_\mu\psi(x)\sigma^\mu\bar{\epsilon}, \tag{8.85}$$

where ϵ_α is an infinitesimal anti-commuting Grassmann parameter. Here, we find that under the supersymmetry transformation the fields of the left chiral multiplet rotate into themselves. To check that the algebra closes for the set of fields under consideration, we need to compute the commutator of two supersymmetry transformations. Thus, we consider the action of two supersymmetry transformations δ_{ϵ_1} and δ_{ϵ_2} in two different orderings. We first consider the action of

$\delta_{\epsilon_1}\delta_{\epsilon_2}$ and then the action of $\delta_{\epsilon_2}\delta_{\epsilon_1}$ on the components on the multiplet.

$$\delta_{\epsilon_1}\delta_{\epsilon_2}\phi(x) = \sqrt{2}\epsilon_2^\alpha \delta_{\epsilon_1}\psi_\alpha(x) \tag{8.86}$$

$$= \sqrt{2}\epsilon_2^\alpha \left[\sqrt{2}i(\sigma^\mu\bar{\epsilon}_1)_\alpha\partial_\mu\phi(x) + \sqrt{2}\epsilon_{1\alpha}F(x)) \right] \tag{8.87}$$

$$= 2i\epsilon_2\sigma^\mu\bar{\epsilon}_1\partial_\mu\phi(x) + 2\epsilon_2\epsilon_1 F(x). \tag{8.88}$$

Reversing the order, we compute the action of $\delta_{\epsilon_2}\delta_{\epsilon_1}$ on $\phi(x)$ to get

$$\delta_{\epsilon_2}\delta_{\epsilon_1}\phi(x) = 2i\epsilon_1\sigma^\mu\bar{\epsilon}_2\partial_\mu\phi(x) + 2\epsilon_1\epsilon_2 F(x). \tag{8.89}$$

Noting that $\epsilon_1\epsilon_2 = \epsilon_2\epsilon_1$, we have

$$[\delta_{\epsilon_1}, \delta_{\epsilon_2}]\phi(x) = 2i(\epsilon_2\sigma^\mu\bar{\epsilon}_1 - \epsilon_1\sigma^\mu\bar{\epsilon}_2)\partial_\mu\phi(x). \tag{8.90}$$

Thus, we see that under the action of the commutator of supersymmetry transformations the scalar component is rotated into itself. The partial derivative on the right-hand side of Eq. (8.90) tells us that the commutator of two supersymmetry transformations is a translation. A very similar analysis holds for the action of the commutator on $\psi_\alpha(x)$ and on $F(x)$.

8.5 Right Chiral Superfields

Next, we consider a right chiral superfield $\bar{\Phi}(x, \theta, \bar{\theta})$, which is defined so that

$$D_\alpha\bar{\Phi}(x, \theta, \bar{\theta}) = 0, \tag{8.91}$$

where D_α is as defined by Eq. (8.62). It is instructive to introduce a new variables $z^\mu = x^\mu - i\theta\sigma^\mu\bar{\theta}$, $\theta'_\alpha = \theta_\alpha$, and $\bar{\theta}'_{\dot{\alpha}} = \bar{\theta}_{\dot{\alpha}}$ The action of D_α on z^μ on the new variables is given by

$$D_\alpha z^\mu = (\partial_\alpha + i(\sigma^\lambda\bar{\theta})_\alpha\partial_\lambda)(x^\mu - i\theta\sigma^\mu\bar{\theta})$$
$$= -i\partial_\alpha(\theta\sigma^\mu\bar{\theta}) + i(\sigma^\lambda\bar{\theta})_\alpha\delta_\lambda^\mu = 0 \tag{8.92}$$

$$D_\alpha\bar{\theta}_{\dot{\alpha}} = 0. \tag{8.93}$$

Using the relations above, we can express $D_\alpha(x, \theta, \bar{\theta})$ given by Eq. (8.62) and $\bar{D}_{\dot{\alpha}}(x, \theta, \bar{\theta})$ given by Eq. (8.60) in terms of the new variables $(z, \theta', \bar{\theta}')$. For this purpose, the following results are useful:

$$\frac{\partial}{\partial\theta^\alpha} = \frac{\partial}{\partial\theta^{\alpha'}} - i(\sigma^\mu)_{\alpha\dot{\alpha}}\bar{\theta}'^{\dot{\alpha}}\frac{\partial}{\partial z^\mu} \tag{8.94}$$

$$\frac{\partial}{\partial\bar{\theta}^{\dot{\alpha}}} = \frac{\partial}{\partial\bar{\theta}^{\dot{\alpha}'}} + i\theta'^\alpha(\sigma^\mu)_{\alpha\dot{\alpha}}\frac{\partial}{\partial z^\mu}. \tag{8.95}$$

Next, define

$$D_\alpha^R(z, \theta', \bar{\theta}') \equiv D_\alpha(x, \theta, \bar{\theta}) \tag{8.96}$$

$$\bar{D}_{\dot{\alpha}}^R(z, \theta', \bar{\theta}') \equiv \bar{D}_{\dot{\alpha}}(x, \theta, \bar{\theta}). \tag{8.97}$$

Using Eqs. (8.60), (8.62) and (8.94), and (8.95), we find the following expressions for $D^R_\alpha(z, \theta', \bar{\theta}')$ and $\bar{D}^R_{\dot{\alpha}}(z, \theta', \bar{\theta}')$:

$$D^R_\alpha(z, \theta', \bar{\theta}') = \frac{\partial}{\partial \theta^{\alpha'}} \tag{8.98}$$

$$\bar{D}^R_{\dot{\alpha}}(z, \theta', \bar{\theta}') = -\frac{\partial}{\partial \bar{\theta}^{\dot{\alpha}'}} - 2i\theta'^\alpha (\sigma^\mu)_{\alpha\dot{\alpha}} \frac{\partial}{\partial z^\mu}. \tag{8.99}$$

Let us define

$$\Phi_R(z, \theta, \bar{\theta}) \equiv \bar{\Phi}(x, \theta, \bar{\theta}). \tag{8.100}$$

We may now write the condition for the right chirality condition (Eq. (8.91)) as follows:

$$\partial_\alpha \Phi_R(z, \theta, \bar{\theta}) = 0. \tag{8.101}$$

The above implies that $\Phi_R(z, \theta, \bar{\theta})$ is not a function of θ, and we can drop the dependence on θ and write the right chiral superfield only as a function of z and $\bar{\theta}$. Thus, the right chiral field can be expanded in component form up to powers of $\bar{\theta}^2$, i.e., one has the superfield expansion

$$\Phi_R(z, \bar{\theta}) = \bar{\phi}(z) + \sqrt{2}\bar{\theta}\bar{\psi}(z) + \bar{\theta}\bar{\theta}\bar{F}(z). \tag{8.102}$$

To express the superfield in terms of the standard variables x, θ, and $\bar{\theta}$, we expand $\bar{\phi}(z), \bar{\psi}(z)$, and $\bar{F}(z)$ as follows:

$$\bar{\phi}(z) = \bar{\phi}(x) - i\theta\sigma^\mu\bar{\theta}\partial_\mu\bar{\phi}(x) - \frac{1}{2}\theta\sigma^\mu\bar{\theta}\theta\sigma^\nu\bar{\theta}\partial_\mu\partial_\nu\bar{\phi}(x) \tag{8.103}$$

$$\bar{\theta}\bar{\psi}(z) = \bar{\theta}[\bar{\psi}(x) - i\theta\sigma^\mu\bar{\theta}\partial_\mu\bar{\psi}(x)] \tag{8.104}$$

$$\bar{\theta}\bar{\theta}\bar{F}(z) = \bar{\theta}\bar{\theta}\bar{F}(x), \tag{8.105}$$

where we have used the fact that θ^3 and $\bar{\theta}^3$ and higher powers vanish. An expansion of $\Phi_R(z, \bar{\theta})$ in terms of x, θ, and $\bar{\theta}$ gives

$$\Phi_R(z, \bar{\theta}) = \bar{\phi}(x) + \sqrt{2}\bar{\theta}\bar{\psi}(x) + \bar{\theta}\bar{\theta}\bar{F}(x) - i\theta\sigma^\mu\bar{\theta}\partial_\mu\bar{\phi}(x)$$
$$+ \frac{i}{\sqrt{2}}\bar{\theta}\bar{\theta}\theta\sigma^\mu\partial_\mu\bar{\psi} + \frac{1}{4}\theta\theta\bar{\theta}\bar{\theta}\Box\bar{\phi}(x). \tag{8.106}$$

Analogous to the case of the left chiral fields, the right chiral fields have the following transformations on the components $\bar{\phi}(x), \bar{\psi}(x)$, and $\bar{F}(x)$:

$$\delta_\epsilon \bar{\phi}(x) = \sqrt{2}\bar{\epsilon}\bar{\psi}(x)$$
$$\delta_\epsilon \bar{\psi}(x) = \sqrt{2}\bar{\epsilon}\bar{F}(x) - i\sqrt{2}(\epsilon\sigma^\mu)\partial_\mu\bar{\phi}(x)$$
$$\delta_\epsilon \bar{F}(x) = i\sqrt{2}\epsilon\sigma^\mu\partial_\mu\bar{\psi}(x). \tag{8.107}$$

8.6 Products of Chiral Superfields

In formulating particle physics theories based on supersymmetry, one needs products of chiral left superfields as well as products involving chiral left and chiral right superfields. Thus, we discuss now how such products of chiral superfields are formed. Let us consider first the product of left chiral superfields. Assume we have two superfields Φ_1 and Φ_2 which are left chiral and that they satisfy the constraint

$$\bar{D}_{\dot\alpha}\Phi_i = 0, \quad i = 1, 2. \tag{8.108}$$

We now consider the product $\Phi_{12} = \Phi_1\Phi_2$. Using Eq. (8.108) we find that Φ_{12} satisfies the constraint

$$\bar{D}_{\dot\alpha}\Phi_{12} = 0, \tag{8.109}$$

and so Φ_{12} is also a left chiral field. We can expand Φ_{12} so that

$$\Phi_{12}(y, \theta, \bar\theta) = \phi_{12}(y) + \sqrt{2}\theta\psi_{12}(y) + \theta\theta F_{12}(y). \tag{8.110}$$

The components of ϕ_{12} are given in terms of the components of the source fields so that

$$\phi_{12} = \phi_1\phi_2$$
$$\psi_{12} = \phi_1\psi_2 + \psi_1\phi_2$$
$$F_{12} = \phi_1 F_2 + F_1\phi_2 - \psi_1\psi_2. \tag{8.111}$$

One can extend the above analysis to a product of any number of chiral left superfields so that the product of any number of left chiral superfields is a left chiral superfield, i.e., we have

$$\bar{D}_{\dot\alpha}(\Phi_1\Phi_2\cdots\Phi_n) = 0. \tag{8.112}$$

A very similar analysis applies to products of right-handed chiral superfields $\bar\Phi_i$. Consider two chiral right superfields $\bar\Phi_1$ and $\bar\Phi_2$ which satisfy the relation

$$D_\alpha\bar\Phi_i = 0, \quad i = 1, 2, \tag{8.113}$$

where

$$\bar\Phi_i = \bar\phi_i(z) + \sqrt{2}\bar\theta\bar\psi_i(z) + \bar\theta\bar\theta\bar{F}_i(z). \tag{8.114}$$

We now define a product of two right-handed chiral fields

$$\bar\Phi_{12} = \bar\Phi_1\bar\Phi_2. \tag{8.115}$$

Using Eq. (8.113), we find that $\bar\Phi_{12}$ satisfies the relation

$$D_\alpha\bar\Phi_{12} = 0. \tag{8.116}$$

Thus, the product of two right chiral fields is also a right chiral field. We can expand $\bar\Phi_{12}$ in component form:

$$\bar\Phi_{12} = \bar\phi_{12} + \sqrt{2}\bar\theta\bar\psi_{12} + \bar\theta\bar\theta\bar{F}_{12}, \tag{8.117}$$

where the components of $\bar{\Phi}_{12}$ are given by

$$\bar{\phi}_{12} = \bar{\phi}_1 \bar{\phi}_2$$
$$\bar{\psi}_{12} = \bar{\phi}_1 \bar{\psi}_2 + \bar{\psi}_1 \bar{\phi}_2$$
$$\bar{F}_{12} = \bar{\phi}_1 \bar{F}_2 + \bar{F}_1 \bar{\phi}_2 - \bar{\psi}_1 \bar{\psi}_2. \tag{8.118}$$

One can extend the above analysis to a product of any number of chiral right superfields so that the product of any number of right chiral superfields is also a right chiral superfield since the product satisfies the constraint

$$D_\alpha(\bar{\Phi}_1 \bar{\Phi}_2 \cdots \bar{\Phi}_m) = 0. \tag{8.119}$$

8.7 Products of Chiral Left and Chiral Right Superfields

We now consider the product of a left chiral superfield $\Phi(y)$ and a right chiral superfield $\bar{\Phi}(z)$, and expand the product in powers of θ and $\bar{\theta}$, which gives the following expansion:

$$\Phi(y)\bar{\Phi}(z) = \phi(y)\bar{\phi}(z) + \sqrt{2}\theta\psi(y)\bar{\phi}(z) + \sqrt{2}\bar{\theta}\phi(y)\bar{\psi}(z) + \theta\theta F(y)\bar{\phi}(z)$$
$$+ \bar{\theta}\bar{\theta}\phi(y)\bar{F}(z) + 2\theta\psi(y)\bar{\theta}\bar{\psi}(z) + \sqrt{2}\theta\theta F(y)\bar{\theta}\bar{\psi}(z)$$
$$+ \sqrt{2}\bar{\theta}\bar{\theta}\theta\psi(y)\bar{F}(z) + \theta\theta\bar{\theta}\bar{\theta}F(y)\bar{F}(z). \tag{8.120}$$

We note that the left chiral and the right chiral fields are functions of different variables: the left chiral is a function of y and the right chiral is a function of z. It is useful to now express all the fields in terms of x, which requires an expansion. Thus, we have the following field expansions:

$$\phi(y) = (1 + i\theta\sigma^\mu\bar{\theta}\partial_\mu + \frac{1}{4}\theta\theta\bar{\theta}\bar{\theta}\Box)\phi(x), \tag{8.121}$$

and similar expansions hold for $\psi(y)$ and $F(y)$. For the bar fields we have

$$\bar{\phi}(z) = (1 - i\theta\sigma^\mu\bar{\theta}\partial_\mu + \frac{1}{4}\theta\theta\bar{\theta}\bar{\theta}\Box)\bar{\phi}(x), \tag{8.122}$$

and again similar formulae hold for $\bar{\psi}(z)$ and $\bar{F}(z)$. Using these results, we find for the product field the following result:

$$\Phi(y)\bar{\Phi}(z) = C(x) + \theta\chi_1(x) + \bar{\theta}\bar{\chi}_2(x) + \theta\theta p(x) + \bar{\theta}\bar{\theta}q(x)$$
$$+ \theta^\alpha\bar{\theta}^{\dot{\alpha}}v_{\alpha\dot{\alpha}}(x) + \theta\theta\bar{\theta}\bar{\lambda}(x) + \bar{\theta}\bar{\theta}\theta\psi(x) + \theta\theta\bar{\theta}\bar{\theta}D(x), \tag{8.123}$$

where

$$C(x) = \bar{\phi}(x)\phi(x)$$

$$\chi_1(x) = \sqrt{2}\bar{\phi}(x)\psi(x)$$

$$\bar{\chi}_2(x) = \sqrt{2}\phi(x)\bar{\psi}(x)$$

$$p(x) = \bar{\phi}(x)F(x)$$

$$q(x) = \phi(x)\bar{F}(x)$$

$$v_{2\alpha\dot\alpha} = i(\sigma^\mu)_{\alpha\dot\alpha}[\partial_\mu\phi(x)\bar{\phi}(x) - \partial_\mu\bar{\phi}(x)\phi(x)] - 2\bar{\psi}_{\dot\alpha}\psi_\alpha$$

$$\bar{\lambda}_{\dot\alpha}(x) = -\frac{i(\sigma^\mu)_{\alpha\dot\alpha}}{\sqrt{2}}[\bar{\phi}(x)\partial_\mu\psi^\alpha(x) + \psi^\alpha\partial_\mu\bar{\phi}(x)] + \sqrt{2}\bar{\psi}_{\dot\alpha}(x)F(x)$$

$$\psi_\alpha(x) = -\frac{i(\sigma^\mu)_{\alpha\dot\alpha}}{\sqrt{2}}[\bar{\psi}^{\dot\alpha}(x)\partial_\mu\phi(x) - \partial_\mu\bar{\psi}^{\dot\alpha}(x)\phi(x)] + \sqrt{2}\psi_\alpha(x)\bar{F}(x)$$

$$D(x) = \bar{F}(x)F(x) + \frac{1}{4}\bar{\phi}(x)\Box\phi(x) + \frac{1}{4}\phi(x)\Box\bar{\phi}(x)$$

$$- \frac{1}{2}\partial_\mu\bar{\phi}(x)\partial^\mu\phi(x) + \frac{i}{2}\partial_\mu\psi(x)\sigma^\mu\bar{\psi}(x) - \frac{i}{2}\psi(x)\sigma^\mu\partial_\mu\bar{\psi}(x). \qquad (8.124)$$

The product given by Eqs. (8.123) and (8.124) is a scalar superfield, which we denote by $\Phi(x, \theta, \bar{\theta})$.

8.8 Construction of Supersymmetric Action using Superfields

Supersymmetric actions can be constructed using superfields. This can be done by constructing products of superfields and then integrating over the Grassmann numbers θ and $\bar{\theta}$. There are two ways in which this can be accomplished. First, we could construct products of chiral left and chiral right fields as polynomials in θ and $\bar{\theta}$. The simplest such possibility is to consider the product of the fields $\Phi(z, \bar{\theta})$ and $\Phi(y, \theta)$ and integrate over θ and $\bar{\theta}$. In this case, only the highest component of the product field, i.e., the coefficient of $\theta^2\bar{\theta}^2$, survives. One can also have contributions where there are a product of several chiral left fields so that the product of these fields would again be a chiral left field. More generally, we could have sums of products of chiral left fields. We will call this combination a superpotential $W(y, \theta)$. In this case, we integrate on θ, which leaves us with the highest component of W in the action, i.e., the coefficient of θ^2. A very similar analysis holds for the case when we consider the sum of the products of right-handed fields. In this case, we integrate on $\bar{\theta}$, which leaves the highest component of $W^\dagger$ in the action, i..e, the coefficient of $\bar{\theta}^2$. Thus, the supersymmetric action takes the form

$$S = \int d^4x \left[\int d^2\theta\, d^2\bar{\theta}\ \Phi_a^\dagger(z, \bar{\theta})\Phi_a(y, \theta) + (\int d^2\theta\, d^2\bar{\theta}\ W(y, \theta)\delta(\bar{\theta}^2) + \text{h.c.}) \right],$$

$$(8.125)$$

where $d^2\theta \equiv \frac{1}{2}d\theta^1\, d\theta^2$ and $d^2\bar\theta = \frac{1}{2}d\bar\theta_1\, d\bar\theta_2$ and where we recall that $y^\mu = x^\mu + i\theta\sigma^\mu\bar\theta$ and $z^\mu = x^\mu - i\theta\sigma^\mu\bar\theta$. The engineering dimension of W is cubic in mass, and thus a typical expansion without introducing operators of dimension higher than 4 in the Lagrangian implies that one expands the superpotential up to cubic order in the fields. Thus, W will have the expansion

$$W(y,\theta) = c_a\Phi_a + \frac{1}{2}m_{ab}\Phi_a\Phi_b + \frac{1}{3}\lambda_{abc}\Phi_a\Phi_b\Phi_c. \tag{8.126}$$

An explicit computation gives

$$\int d^2\theta\, d^2\bar\theta\, \Phi_a^\dagger(z,\bar\theta)\Phi_a(y,\theta) = -\frac{1}{2}\partial_\mu\phi_a^*(x)\partial^\mu\phi_a(x) + \frac{1}{4}\phi_a(x)\Box\phi_a^*(x)$$
$$+ \frac{i}{2}\partial_\mu\psi_a(x)\sigma^\mu\bar\psi_a(x) + \frac{1}{4}\phi_a^*\Box\phi_a(x)$$
$$- \frac{i}{2}\psi_a(x)\sigma^\mu\partial_\mu\bar\psi_a + F_a^*(x)F_a(x). \tag{8.127}$$

Using the identity $\psi\sigma^\mu\bar\psi = -\bar\psi\bar\sigma^\mu\psi$, we can write the above in the form

$$\int d^2\theta\, d^2\bar\theta\, \Phi_a^\dagger(z,\bar\theta)\Phi_a(y,\theta) = \phi_a^*\Box\phi_a(x) + i\partial_\mu\bar\psi_a(x)\bar\sigma^\mu\psi_a(x)$$
$$+ F_a^*(x)F_a(x) + \partial_\mu D^\mu. \tag{8.128}$$

The total derivative term does not contribute in the action, and can be discarded.

Let us now check the invariance of the Lagrangian given by Eq. (8.128) under supersymmetry transformations. This requires that we compute the transformations of the various terms involved. We have

$$\delta_\epsilon(-\partial_\mu\phi^\dagger\partial^\mu\phi) = -\sqrt{2}\left[\partial_\mu\phi^\dagger\partial^\mu\epsilon\psi + \partial_\mu(\bar\psi\bar\epsilon)\partial^\mu\phi\right] \tag{8.129}$$

$$\delta_\epsilon\left(\frac{i}{2}\partial_\mu\psi\sigma^\mu\bar\psi\right) = \frac{i}{2}\partial_\mu\delta_\epsilon\psi\sigma^\mu\bar\psi + \frac{i}{2}\partial_\mu\psi\sigma^\mu\delta_\epsilon\bar\psi$$
$$= \frac{i}{2}\partial_\mu\left[\sqrt{2}(\epsilon F - i\bar\epsilon\bar\sigma^\nu\partial_\nu\phi(x))\right]\sigma^\mu\bar\psi$$
$$+ \frac{i}{2}\partial_\mu\psi\sigma^\mu\left[\sqrt{2}(\bar\epsilon F^\dagger + i\bar\sigma^\nu\epsilon\partial_\nu\phi^\dagger(x))\right]. \tag{8.130}$$

In a similar fashion, we have

$$\delta_\epsilon\left(-\frac{i}{2}\psi\sigma^\mu\partial_\mu\bar\psi\right) = -\frac{i}{2}\left[\sqrt{2}(\epsilon F - i\bar\epsilon\bar\sigma^\nu\partial_\nu\phi(x))\right]\sigma^\mu\partial_\mu\bar\psi$$
$$- \frac{i}{2}\psi\sigma^\mu\partial_\mu\left[\sqrt{2}(\bar\epsilon F^\dagger + i\bar\sigma^\nu\epsilon\partial_\nu\phi^\dagger(x))\right] \tag{8.131}$$

$$\delta_\epsilon(F^\dagger F) = \sqrt{2}i\left[\epsilon\sigma^\mu\partial_\mu\bar\psi(x)F - \partial_\mu\psi(x)\sigma^\mu\bar\epsilon F^\dagger\right]. \tag{8.132}$$

Adding them, we find that the kinetic energy Lagrangian under supersymmetry transformations transforms as a total divergence:

$$\delta_\epsilon L_{\Phi_L}^{kin} = \partial_\mu\left[-\sqrt{2}\partial^\mu\phi_a^\dagger\psi_a\epsilon - \frac{1}{\sqrt{2}}\psi_a\sigma^\mu\bar\sigma^\nu\epsilon\partial_\nu\phi_a^\dagger + \frac{i}{\sqrt{2}}\epsilon\sigma^\mu\psi_a^\dagger F_a + \text{h.c.}\right]. \tag{8.133}$$

Thus, the Lagrangian of Eq. (8.128) is invariant under the supersymmetry transformations up to a total divergence. Of course, our check was not necessary since the result of Eq. (8.128) was derived from a superspace formulation so the result was expected to be automatically supersymmetric. However, the analysis shows that the two approaches, i.e., the one in ordinary space and the other in superspace, agree.

Next, the superpotential term gives the following contribution:

$$\int d^2\theta \, d^2\bar{\theta} \; W(\theta, y)\delta(\bar{\theta}^2) = \int d^2\theta \, d^2\bar{\theta} \; W(\theta, x)\bar{\theta}\theta, \qquad (8.134)$$

where we have replaced $y^\mu = x^\mu + i\theta\bar{\sigma}^\mu\bar{\theta}$ with x since $\bar{\theta}^3 = 0$. One may write the above term in the form

$$\int d^2\theta \, d^2\bar{\theta} \; W(\theta, y)\delta(\bar{\theta}^2) = \frac{\partial W[\phi]}{\partial\phi_a}F_a(x) - \frac{1}{2}\frac{\partial^2 W[\phi]}{\partial\phi_a\partial\phi_b}\psi_a\psi_b. \qquad (8.135)$$

Thus, writing

$$S = \int d^4x \; L, \qquad (8.136)$$

and including all the terms, the supersymmetric Lagrangian constructed out of the chiral superfields Φ_a and $\Phi_a^\dagger$ is given by

$$L = \phi_a^*\Box\phi_a(x) + i\partial_\mu\bar{\psi}_a(x)\bar{\sigma}^\mu\psi_a(x) + F_a^*(x)F_a(x)$$
$$+ \left[\frac{\partial W[\phi]}{\partial\phi_a}F_a(x) - \frac{1}{2}\frac{\partial^2 W[\phi]}{\partial\phi_a\partial\phi_b}\psi_a\psi_b + \text{h.c.}\right]. \qquad (8.137)$$

This Lagrangian is an off-shell Lagrangian. It contains the fields F_a and $F_a^*(x)$, which have no kinetic energy terms and are not propagating fields. These are auxiliary fields, and can be eliminated by using field equations. Thus, the use of field equations gives

$$F_a^* + \left(\frac{\partial W_{\theta=0}}{\partial\phi_a}\right) = 0. \qquad (8.138)$$

Substituting for F_a we get the on-shell form of the Lagrangian:

$$L = \phi_a^*\Box\phi_a(x) + i\partial_\mu\bar{\psi}_a(x)\bar{\sigma}^\mu\psi_a(x)$$
$$- \left[\frac{1}{2}\frac{\partial^2 W[\phi]}{\partial\phi_a\partial\phi_b}\psi_a\psi_b + \text{h.c.}\right] - V(\phi, \phi^*), \qquad (8.139)$$

where $V(\phi, \phi^*)$ is the scalar potential, which is given by

$$V(\phi, \phi^*) = \sum_a F_a^* F_a = \sum_a \left|\frac{\partial W_{\theta=0}}{\partial\phi_a}\right|^2. \qquad (8.140)$$

Explicitly, for the assumed form of the superpotential we have

$$\left(\frac{\partial W_{\theta=0}}{\partial\phi_a}\right) = c_a + m_{ab}\phi_b(x) + \lambda_{abc}\phi_b(x)\phi_c(x), \qquad (8.141)$$

and here the Lagrangian takes the form

$$L = \phi_a^* \Box \phi_a(x) + i\partial_\mu \bar{\psi}_a(x)\bar{\sigma}^\mu \psi_a(x) - \frac{1}{2}m_{ab}\psi_a(x)\psi_b(x) - \frac{1}{2}m_{ab}^*\bar{\psi}_a(x)\bar{\psi}_b(x)$$
$$- \lambda_{abc}\psi_a(x)\psi_b(x)\phi_c(x) - \lambda_{abc}^*\bar{\psi}_a(x)\bar{\psi}_b(x)\phi_c^*(x) - V(\phi,\phi^*). \qquad (8.142)$$

The Lagrangian simplifies further for the case of a single superfield. Here, L reduces to

$$L = \phi^*(x)\Box\phi(x) + i\partial_\mu\bar{\psi}(x)\bar{\sigma}^\mu\psi(x) - \frac{1}{2}m(\psi(x)\psi(x) + \bar{\psi}(x)\bar{\psi}(x))$$
$$- \lambda\psi(x)\psi(x)\phi(x) - \lambda^*\bar{\psi}(x)\bar{\psi}(x)\phi^*(x) - V(\phi,\phi^*), \qquad (8.143)$$

where $V(\phi,\phi^*)$ has the explicit form

$$V(\phi,\phi^*) = c^2 + mc(\phi(x) + \phi^*(x)) + m^2|\phi(x)|^2 + c\lambda(\phi^2(x) + \phi^{*2}(x))$$
$$+ m\lambda(\phi(x) + \phi^*(x))|\phi(x)|^2 + \lambda^2|\phi(x)|^4. \qquad (8.144)$$

The field equations for this single superfield case are then given by

$$i\sigma^\mu\partial_\mu\bar{\psi}(x) + m\psi(x) + 2\lambda\phi(x)\psi(x) = 0$$
$$-i\partial_\mu\psi(x)\sigma^\mu + m\bar{\psi}(x) + 2\lambda\phi^*(x)\bar{\psi}(x) = 0$$
$$(-\Box + m^2)\phi(x) + 2c\lambda\phi^* + \lambda m\phi^2 + 2m\lambda|\phi|^2 + 2\lambda^2\phi|\phi|^2 + \lambda\bar{\psi}^2 + mc = 0. \qquad (8.145)$$

8.9 Vector Superfields

One can obtain a vector superfield from a scalar superfield $\Phi(x,\theta,\bar{\theta})$ by imposition of the constraint

$$\Phi(x,\theta,\bar{\theta}) = \Phi^\dagger(x,\theta,\bar{\theta}). \qquad (8.146)$$

We choose the vector field $\Phi(x,\theta,\bar{\theta})$ so that

$$\Phi(x,\theta,\bar{\theta}) = C(x) + \theta\chi(x) + \bar{\theta}\bar{\chi}(x) + \theta\theta p(x) + \bar{\theta}\bar{\theta}p^*(x)$$
$$- \theta\sigma^\mu\bar{\theta}v_\mu(x) - i\bar{\theta}\bar{\theta}\theta\lambda(x) + i\theta\theta\bar{\theta}\bar{\lambda}(x) + \frac{1}{2}\theta\theta\bar{\theta}\bar{\theta}D(x). \qquad (8.147)$$

Here, $C(x)$ and $D(x)$ are real scalars, $p(x)$ is a complex scalar, v_μ is a real vector, and χ and λ are spinors.

The transformation properties of the components of the vector superfield under the supersymmetry transformations are as follows:

$$\delta_s C(x) = \epsilon\chi(x) + \bar{\epsilon}\bar{\chi}(x)$$

$$\delta_s \chi(x) = 2\epsilon p + (\sigma^\mu\bar{\epsilon})(i\partial_\mu C(x) - v_\mu)$$

$$\delta_s p(x) = i\bar{\epsilon}\bar{\lambda} - \frac{i}{2}\partial_\mu\chi(x)\sigma^\mu\bar{\epsilon}$$

$$\delta_s v_\mu(x) = -i\epsilon\sigma_\mu\bar{\lambda}(x) + i\lambda(x)\sigma_\mu\bar{\epsilon} - \frac{i}{2}\left(\epsilon\partial_\mu\chi(x) + \partial_\mu\bar{\chi}(x)\bar{\epsilon}\right)$$

$$\delta_s \lambda(x) = i\epsilon D(x) - \frac{1}{2}\epsilon\partial^\mu v_\mu(x) - (\sigma^\mu\bar{\epsilon})\partial_\mu p^*$$

$$\delta_s D(x) = \left(\partial_\mu\lambda(x)\sigma^\mu\bar{\epsilon} + \partial_\mu\bar{\lambda}(x)\bar{\sigma}^\mu\epsilon\right). \tag{8.148}$$

In order to study the transformation properties of the component fields under gauge transformations it is useful to generate another vector superfield by the following substitutions:

$$\lambda(x) \to \lambda(x) + \frac{1}{2}\sigma^\mu\partial_\mu\bar{\chi}(x) \tag{8.149}$$

$$\bar{\lambda}(x) \to \bar{\lambda}(x) - \frac{1}{2}\bar{\sigma}^\mu\partial_\mu\chi(x) \tag{8.150}$$

$$D(x) \to D(x) - \frac{1}{2}\Box C(x). \tag{8.151}$$

With the above replacements, the vector superfield takes the form

$$\Phi(x,\theta,\bar{\theta}) = C(x) + \theta\chi(x) + \bar{\theta}\bar{\chi}(x) + \theta\theta p(x) + \bar{\theta}\bar{\theta}p^*(x)$$

$$- \theta\sigma^\mu\bar{\theta}v_\mu(x) - i\bar{\theta}\bar{\theta}\theta\left(\lambda(x) + \frac{1}{2}\sigma^\mu\partial_\mu\bar{\chi}\right)$$

$$+ i\theta\theta\bar{\theta}\left(\bar{\lambda}(x) - \frac{1}{2}\bar{\sigma}^\mu\partial_\mu\chi\right) + \frac{1}{2}\theta\theta\bar{\theta}\bar{\theta}\left(D(x) - \frac{1}{2}\Box C(x)\right). \tag{8.152}$$

We now discuss the transformation properties of the vector superfield under the abelian gauge transformation. Here, it is instructive to first consider a vector multiplet formed out of a chiral superfield. Consider the sum $\Phi(x,\theta,\bar{\theta}) + \Phi^\dagger(x,\theta,\bar{\theta})$ where we take for $\Phi(x,\theta,\bar{\theta})$ a left-handed chiral superfield which has the expansion

$$\Phi(x,\theta,\bar{\theta}) = \phi(x) + i\theta\sigma^\mu\bar{\theta}\partial_\mu\phi(x) - \frac{1}{4}\theta\theta\bar{\theta}\bar{\theta}\Box\phi(x) + \sqrt{2}\theta\psi(x)$$

$$- \frac{i}{\sqrt{2}}\theta\theta\partial_\mu\psi(x)\sigma^\mu\bar{\theta} + \theta\theta F(x). \tag{8.153}$$

Then, $\Phi^\dagger(x,\theta,\bar{\theta})$ is a right-handed chiral superfield, which is given by

$$\Phi^\dagger(x,\theta,\bar{\theta}) = \phi^*(x) - i\theta\sigma^\mu\bar{\theta}\partial_\mu\phi^*(x) - \frac{1}{4}\theta\theta\bar{\theta}\bar{\theta}\Box\phi^*(x) + \sqrt{2}\bar{\theta}\bar{\psi}(x)$$

$$+ \frac{i}{\sqrt{2}}\bar{\theta}\bar{\theta}\theta\sigma^\mu\partial_\mu\bar{\psi}(x) + \bar{\theta}\bar{\theta}F^*(x). \tag{8.154}$$

From the above, we can write the vector combination

$$\Phi + \Phi^\dagger = \phi(x) + \phi^*(x) + \sqrt{2}\theta\psi(x) + \sqrt{2}\bar{\theta}\bar{\psi}(x) + \theta\theta F(x)$$

$$+ \bar{\theta}\bar{\theta}F^*(x) + i\theta\sigma^\mu\bar{\theta}\partial_\mu(\phi(x) - \phi^*(x)) + \frac{i}{\sqrt{2}}\theta\theta\bar{\theta}\bar{\sigma}^\mu\partial_\mu\psi(x)$$

$$+ \frac{i}{\sqrt{2}}\bar{\theta}\bar{\theta}\theta\sigma^\mu\partial_\mu\bar{\psi}(x) - \frac{1}{4}\theta\theta\bar{\theta}\bar{\theta}\Box(\phi(x) + \phi^*(x))). \tag{8.155}$$

We note that the coefficient of the term $\theta\sigma^\mu\bar{\theta}$ is the quantity $\partial_\mu(\phi(x) - \phi^*(x))$, which is a total derivative. We also note that in the expansion of the vector superfield the coefficient of the term $\theta\sigma^\mu\bar{\theta}$ is the vector field v_μ. This leads us to believe that the combination $\Phi + \Phi^\dagger$ can be viewed as a gauge transformation of a vector superfield. Thus, we define the gauge transformation of a vector superfield as follows:

$$V \to V' = V + \Phi + \Phi^\dagger. \tag{8.156}$$

From the above, we obtain the following gauge transformations for the components of the redefined vector superfield given by Eq. (8.152) using Eqs. (8.155) and (8.156):

$$C'(x) = C(x) + \phi(x) + \phi^*(x)$$
$$\chi'(x) = \chi(x) + \sqrt{2}\psi(x)$$
$$\bar{\chi}'(x) = \bar{\chi}(x) + \sqrt{2}\bar{\psi}(x)$$
$$p'(x) = p(x) + F(x)$$
$$v'_\mu(x) = v_\mu(x) - i\partial_\mu(\phi(x) - \phi^*(x))$$
$$\lambda'(x) = \lambda(x)$$
$$D'(x) = D(x). \tag{8.157}$$

From Eq. (8.157) we note that $\lambda(x)$ and $D(x)$ are gauge invariant. The Wess–Zumino gauge is the one where the following are chosen:

$$C(x) + \phi(x) + \phi^*(x) = 0$$
$$\chi(x) + \sqrt{2}\psi(x) = 0$$
$$p(x) + F(x) = 0. \tag{8.158}$$

In this gauge the vector superfield contains only the components v_μ, $\lambda(x)$, $\bar{\lambda}(x)$, and $D(x)$. It is to be noted, however, that in the Wess–Zumino gauge the combination $v_\mu + \partial_\mu(\phi(x) - \phi^*(x))$ appears, and thus the gauge freedom associated with the conventional gauge theories is still present. Keeping this in mind, the vector superfield in the Wess–Zumino gauge is given by

$$V(x, \theta, \bar{\theta}) = -\theta\sigma^\mu\bar{\theta}v_\mu(x) - i\bar{\theta}\bar{\theta}\theta\lambda(x) + i\theta\theta\bar{\theta}\bar{\lambda}(x) + \frac{1}{2}\theta\theta\bar{\theta}\bar{\theta}D(x). \tag{8.159}$$

We note that the Wess–Zumino gauge is not supersymmetric. Thus, under the supersymmetry transformations we get

$$\delta_s C(x)|_{\text{WZ}} = 0$$
$$\delta_s \chi(x)|_{\text{WZ}} = -\sigma_\mu \bar{\epsilon} v_\mu(x)$$
$$\delta_s p(x)|_{\text{WZ}} = i\bar{\epsilon}\bar{\lambda}(x). \tag{8.160}$$

Thus, we see that Wess–Zumino gauge is not preserved under a supersymmetry transformation. However, the Wess–Zumino gauge can be restored by simultaneously making a gauge transformation.

8.10 Vector Field Strength

Next, we recall the covariant derivatives D_α and $\bar{D}_{\dot\alpha}$ introduced earlier, i.e.,

$$D_\alpha = \partial_\alpha + i(\sigma^\mu)_{\alpha\dot\beta}\bar{\theta}^{\dot\beta}\partial_\mu$$
$$\bar{D}_{\dot\alpha} = -\bar{\partial}_{\dot\alpha} - i\theta^\beta(\sigma^\mu)_{\beta\dot\alpha}\partial_\mu. \tag{8.161}$$

They satisfy the following algebra:

$$\{D_\alpha, \bar{D}_{\dot\beta}\} = -2i(\sigma^\mu)_{\alpha\dot\beta}\partial_\mu$$
$$\{D_\alpha, D_\beta\} = 0$$
$$\{\bar{D}_{\dot\alpha}, \bar{D}_{\dot\beta}\} = 0. \tag{8.162}$$

Given an unconstrained vector field $V(x,\theta,\bar{\theta})$, we can define the field strengths W_α and $\bar{W}_{\dot\alpha}$ so that

$$W_\alpha = -\frac{1}{4}\bar{D}\bar{D}D_\alpha V(x,\theta,\bar{\theta})$$
$$\bar{W}_{\dot\alpha} = -\frac{1}{4}DD\bar{D}_{\dot\alpha}V(x,\theta,\bar{\theta}). \tag{8.163}$$

We will show that the field strength W_α is invariant under the gauge transformation

$$V(x,\theta,\bar{\theta}) \to V'(x,\theta,\bar{\theta}) = V(x,\theta,\bar{\theta}) + \Phi + \Phi^\dagger. \tag{8.164}$$

To show this invariance, we need to consider the gauge transformation of W_α so that

$$W_\alpha \to W'_\alpha = -\frac{1}{4}\bar{D}\bar{D}D_\alpha[V(x,\theta,\bar{\theta}) + \Phi + \Phi^\dagger]. \tag{8.165}$$

First, we note that

$$\bar{D}\bar{D}D_\alpha\Phi^\dagger = 0, \tag{8.166}$$

since $D_\alpha\Phi^\dagger = 0$ for a right chiral field. Next, we consider the term $\bar{D}\bar{D}D_\alpha\Phi$. Here, using the fact that a left chiral field satisfies the relation $\bar{D}^{\dot\alpha}\Phi = 0$, we can write

$$\bar{D}_{\dot{\alpha}}\bar{D}^{\dot{\alpha}}D_{\alpha}\Phi = \bar{D}_{\dot{\alpha}}\{\bar{D}^{\dot{\alpha}}, D_{\alpha}\}\Phi. \tag{8.167}$$

The curly bracket $\{\bar{D}^{\dot{\alpha}}, D_{\alpha}\}$ can be rewritten in the following form:

$$\{\bar{D}^{\dot{\alpha}}, D_{\alpha}\} = \epsilon^{\dot{\alpha}\dot{\beta}}\{\bar{D}_{\dot{\beta}}, D_{\alpha}\} = -2i\epsilon^{\dot{\alpha}\dot{\beta}}(\sigma^{\mu})_{\alpha\dot{\beta}}\partial_{\mu}. \tag{8.168}$$

It then follows that

$$\bar{D}_{\dot{\alpha}}\bar{D}^{\dot{\alpha}}D_{\alpha}\Phi = -2i\epsilon^{\dot{\alpha}\dot{\beta}}(\sigma^{\mu})_{\alpha\dot{\beta}}\partial_{\mu}\bar{D}_{\dot{\alpha}}\Phi = 0, \tag{8.169}$$

where in the last step we have again used the fact that $\bar{D}_{\dot{\alpha}}\Phi = 0$. With the above, one finds that W_{α} transforms as follows:

$$W_{\alpha} \to W'_{\alpha} = W_{\alpha}. \tag{8.170}$$

Thus, the field strength W_{α} is gauge invariant. A very similar analysis holds in the proof of the gauge invariance of $\bar{W}_{\dot{\alpha}}$ to show that under a gauge transformation one has

$$\bar{W}_{\dot{\alpha}} \to \bar{W}'_{\dot{\alpha}} = \bar{W}_{\dot{\alpha}}. \tag{8.171}$$

We also note that W_{α} and $\bar{W}_{\dot{\alpha}}$ satisfy the following constraints:

$$\bar{D}_{\dot{\alpha}}W_{\beta} = 0, \quad D_{\alpha}\bar{W}_{\dot{\beta}} = 0. \tag{8.172}$$

We now want to compute explicitly the form of W_{α} and $\bar{W}_{\dot{\alpha}}$. We first consider the computation of W_{α}:

$$W_{\alpha} = -\frac{1}{4}\bar{D}\bar{D}D_{\alpha}\tilde{V}(y, \theta, \bar{\theta}), \tag{8.173}$$

where $\tilde{V}(y, \theta, \bar{\theta})$ is the vector multiplet in the Wess–Zumino gauge. We now recall that $\bar{D}_{\dot{\alpha}}$ and D_{α} are given by Eqs. (8.71) and (8.72), i.e.,

$$\bar{D}_{\dot{\alpha}}(y, \theta, \bar{\theta}) = -\frac{\partial}{\partial\bar{\theta}^{\dot{\alpha}}}$$

$$D_{\alpha}(y, \theta, \bar{\theta}) = \partial_{\alpha} + 2i(\sigma^{\mu})_{\alpha\dot{\alpha}}\bar{\theta}^{\dot{\alpha}}\frac{\partial}{\partial y^{\mu}}. \tag{8.174}$$

Using the above, we find

$$W_{\alpha} = -\frac{1}{4}\bar{D}_{\dot{\alpha}}\bar{D}^{\dot{\alpha}}[-(\sigma^{\mu})_{\alpha\dot{\beta}}\bar{\theta}^{\dot{\beta}}v_{\mu}(y) + 2i\theta_{\alpha}\bar{\theta}\bar{\lambda}(y) - i\bar{\theta}\bar{\theta}\lambda_{\alpha}(y)]$$
$$+ [\delta^{\beta}_{\alpha}D(y) - (\sigma^{\mu\nu})_{\alpha}{}^{\beta}v_{\mu\nu}(y)]\theta_{\beta}\bar{\theta}\bar{\theta} + \theta\theta\bar{\theta}\bar{\theta}(\sigma^{\mu})_{\alpha\dot{\beta}}\partial_{\mu}\bar{\lambda}^{\dot{\beta}}(y)]. \tag{8.175}$$

where

$$\sigma^{\mu\nu} = \frac{i}{4}(\sigma^{\mu}\bar{\sigma}^{\nu} - \sigma^{\nu}\bar{\sigma}^{\mu}), \quad v_{\mu\nu} = (\partial_{\mu}v_{\nu} - \partial_{\nu}v_{\mu}). \tag{8.176}$$

Next, we note that

$$\bar{D}_{\dot{\alpha}}\bar{D}^{\dot{\alpha}}\bar{\theta}\bar{\theta} = -4. \tag{8.177}$$

Explicit computation gives

$$W_\alpha = -(\sigma^{\mu\nu}\theta)_\alpha v_{\mu\nu}(y) - i\lambda_\alpha(y) + \theta\theta(\sigma^\mu\partial_\mu\bar\lambda(y))_\alpha + \theta_\alpha D(y). \qquad (8.178)$$

Similarly, for the field strength $\bar W_{\dot\alpha}$ we have

$$\bar W_\alpha = -\frac{1}{4}DD\bar D_{\dot\alpha}\tilde V(z,\theta,\bar\theta), \qquad (8.179)$$

where $\tilde V(z,\theta,\bar\theta)$ is the vector multiplet in the Wess–Zumino gauge and D_α and $\bar D_{\dot\alpha}$ are given by

$$D_\alpha(z,\theta,\bar\theta) = \frac{\partial}{\partial\theta^\alpha}$$
$$\bar D_{\dot\alpha} = -\partial_{\dot\alpha} - 2i\theta^\alpha(\sigma^\mu)_{\alpha\dot\alpha}\frac{\partial}{\partial z^\mu}, \qquad (8.180)$$

where $z^\mu = x^\mu - i\theta\sigma^\mu\bar\theta$. The field strength $\bar W_{\dot\alpha}$ in the Wess–Zumino gauge is then given by

$$\bar W_{\dot\alpha} = \epsilon_{\dot\alpha\dot\beta}(\bar\sigma^{\mu\nu}\bar\theta)^{\dot\beta}v_{\mu\nu}(z) + i\bar\lambda_{\dot\alpha}(z) + \bar\theta\bar\theta\partial_\mu\lambda^\beta(z)(\sigma^\mu)_{\beta\dot\alpha} + \bar\theta_{\dot\alpha}D(z). \qquad (8.181)$$

Next, we compute the quantity $W^\alpha W_\alpha$. A direct multiplication gives

$$W^\alpha W_\alpha = -\lambda^2(y) - 2i\lambda(y)\theta D(y) + 2i\lambda(y)\sigma^{\mu\nu}\theta v_{\mu\nu}(y)$$
$$+ \theta\theta\left[-\frac{1}{2}v_{\mu\nu}(y)v^{\mu\nu}(y) - 2i\lambda(y)\sigma^\mu\partial_\mu\bar\lambda(y) - \frac{i}{2}v_{\mu\nu}(y)\tilde v^{\mu\nu}(y) + D^2(y)\right], \qquad (8.182)$$

where $\tilde v^{\mu\nu} = \frac{1}{2}\epsilon^{\mu\nu\lambda\rho}v_{\lambda\rho}$. In arriving at the above result, we used the identity

$$tr(\sigma^{\mu\nu}\sigma^{\lambda\rho}) = \frac{1}{2}(\eta^{\mu\lambda}\eta^{\nu\rho} - \eta^{\mu\rho}\eta^{\nu\lambda}) - \frac{i}{2}\epsilon^{\mu\nu\lambda\rho}. \qquad (8.183)$$

In a similar fashion, one can show that

$$\bar W_{\dot\alpha}\bar W^{\dot\alpha} = -\bar\lambda^2(z) + 2i\bar\lambda(z)\bar\theta D(z) + 2i\bar\lambda(z)\bar\sigma^{\mu\nu}\bar\theta v_{\mu\nu}(z) + \theta\theta$$
$$\times\left[-\frac{1}{2}v_{\mu\nu}(z)v^{\mu\nu}(z) + 2i\partial_\mu\lambda(z)\sigma^\mu\bar\lambda(z) + \frac{i}{2}v_{\mu\nu}\tilde v^{\mu\nu}(z) + D^2(z)\right]. \qquad (8.184)$$

Thus, from the above we find

$$\int d^2\theta\frac{1}{4}W^\alpha W_\alpha + \int d^2\bar\theta\frac{1}{4}\bar W_{\dot\alpha}\bar W^{\dot\alpha} = -\frac{1}{4}v_{\mu\nu}(x)v^{\mu\nu}(x) + i\partial_\mu\lambda(x)\sigma^\mu\bar\lambda(x) + \frac{1}{2}D^2(x), \qquad (8.185)$$

where we have discarded the a total divergence term since it does not contribute in the action.

8.11 The Fayet–Iliopoulos D Term

For a $U(1)$ gauge theory it is possible to add a D term in the action consistent with gauge invariance and with supersymmetry invariance. Let us consider a term in the Lagrangian of the form [16]

$$\mathcal{L}_{FI} = \xi D. \tag{8.186}$$

$\mathcal{L}_{FI}$ transforms as a total divergence under supersymmetry transformations, and thus the addition of this term keeps the action supersymmetric. Further, from Eq. (8.157) we find that the D term is invariant under abelian $U(1)$ gauge transformations. Thus, in general, Eq. (8.185) can be enhanced to read

$$\mathcal{L}_V + \mathcal{L}_{FI} = -\frac{1}{4}v_{\mu\nu}(x)v^{\mu\nu}(x) + i\partial_\mu\lambda(x)\sigma^\mu\bar{\lambda}(x) + \frac{1}{2}D^2(x) + \xi D. \tag{8.187}$$

We will discuss the implications of the FI term in the context of supersymmetry breaking in Section 8.17.

8.12 Superfields for Non-abelian Gauge Theories

A general vector superfield $V_a(x,\theta,\bar{\theta})$ for a non-abelian case is given by

$$
\begin{aligned}
V_a(x,\theta,\bar{\theta}) = {} & C_a(x) + i\theta\chi_a(x) - i\bar{\theta}\bar{\chi}_a(x) + \frac{i}{2}\theta^2 p_a(x) - \frac{i}{2}\bar{\theta}^2 p_a^*(x) - \theta\sigma^\mu\bar{\theta}v_{\mu a} \\
& + i\theta^2\bar{\theta}\left[\bar{\lambda}_a(x) + \frac{i}{2}\bar{\sigma}^\mu\partial_\mu\chi_a(x)\right] - i\bar{\theta}^2\theta\left[\lambda_a(x) + \frac{i}{2}\sigma^\mu\partial_\mu\bar{\chi}_a(x)\right] \\
& + \frac{1}{2}\theta^2\bar{\theta}^2\left[D_a(x) - \frac{1}{2}\Box C_a(x)\right].
\end{aligned} \tag{8.188}
$$

Analogous to the abelian case, if we define the Wess–Zumino gauge by the conditions

$$C_a(x) = p_a(x) = p_a^*(x) = \chi_a(x) = \bar{\chi}_a(x) = 0, \tag{8.189}$$

the vector superfield takes the form

$$V_a(x,\theta,\bar{\theta}) = -\theta\sigma^\mu\bar{\theta}v_{\mu a} + i\theta^2\bar{\theta}\bar{\lambda}_a(x) - i\bar{\theta}^2\theta\lambda_a(x) + \frac{1}{2}\theta^2\bar{\theta}^2 D_a(x). \tag{8.190}$$

As in the abelian case, the Wess–Zumino gauge is not preserved under supersymmetry transformation. However, by a simultaneous gauge transformation one can maintain the Wess–Zumino gauge.

8.13 Supersymmetric Non-abelian Gauge Field Strengths and Lagrangians

We now discuss the field strengths for a supersymmetric non-abelian gauge theory, which will allow us to define a supersymmetric and gauge-invariant kinetic energy term for the vector multiplet. We begin by defining a chiral superfield W^α so that

$$W_\alpha = \frac{1}{4g}\bar{D}^2 e^{-gV} D_\alpha e^{gV},\qquad(8.191)$$

where $V(x,\theta,\bar{\theta})$ is a vector superfield which has the form $V(x,\theta,\bar{\theta}) = V_a(x,\theta,\bar{\theta})T_a$ where T_a are the generators in the adjoint representation of the gauge group. W_α satisfies the left chirality constraint

$$\bar{D}_{\dot{\alpha}}W_\alpha = \frac{1}{4g}\bar{D}_{\dot{\alpha}}\bar{D}^2 e^{-gV} D_\alpha e^{gV} = 0,\qquad(8.192)$$

where we used $\bar{D}_{\dot{\alpha}}\bar{D}^2 = 0$. Similarly, we define the anti-chiral field strength $\bar{W}_{\dot{\alpha}}$ so that

$$\bar{W}_{\dot{\alpha}} = \frac{1}{4g}D^2 e^{gV}\bar{D}_{\dot{\alpha}}e^{-gV},\qquad(8.193)$$

which satisfies the constraint

$$D_\alpha \bar{W}_{\dot{\alpha}} = 0.\qquad(8.194)$$

Again, we can check the satisfaction of the above constraint by noting that $D_\alpha D^2 = 0$. Under a gauge transformation

$$e^{gV'} = e^{-i\Lambda^\dagger}e^{gV}e^{i\Lambda},\qquad(8.195)$$

W_α transforms as

$$W_\alpha \to W'_\alpha = e^{-i\Lambda}W_\alpha e^{i\Lambda}.\qquad(8.196)$$

Let us now prove the transformation property (Eq. (8.196)).

$$\begin{aligned}
W'_\alpha &= \frac{1}{4g}\bar{D}^2 e^{-gV'} D_\alpha e^{gV'}\\
&= \frac{1}{4g}\bar{D}^2 e^{-i\Lambda}e^{-gV}e^{i\Lambda^\dagger} D_\alpha e^{-i\Lambda^\dagger}e^{gV}e^{i\Lambda}\\
&= \frac{1}{4g}e^{-i\Lambda}\bar{D}^2 e^{-gV}e^{i\Lambda^\dagger}\left[D_\alpha e^{-i\Lambda^\dagger}e^{gV}e^{i\Lambda}\right],
\end{aligned}\qquad(8.197)$$

where in the last step we have used the property that

$$\bar{D}_{\dot{\alpha}}\Lambda = 0.\qquad(8.198)$$

Further, the constraint $D_\alpha \Lambda^\dagger = 0$ gives

$$\left[D_\alpha e^{-i\Lambda^\dagger}e^{gV}e^{i\Lambda}\right] = e^{-i\Lambda^\dagger}D_\alpha e^{gV}e^{i\Lambda}.\qquad(8.199)$$

Thus, using Eq. (8.199) in Eq. (8.197), we have

$$
\begin{aligned}
W'_\alpha &= \frac{1}{4g} e^{-i\Lambda} \bar{D}^2 e^{-gV} D_\alpha e^{gV} e^{i\Lambda} \\
&= \frac{1}{4g} e^{-i\Lambda} \bar{D}^2 e^{-gV} \left[(D_\alpha e^{gV}) e^{i\Lambda} + e^{gV} D_\alpha e^{i\Lambda} \right] \\
&= e^{-i\Lambda} \left[\frac{1}{4g} \bar{D}^2 e^{-gV} (D_\alpha e^{gV}) \right] e^{i\Lambda} + \frac{1}{4g} e^{-i\Lambda} \bar{D}^2 D_\alpha e^{i\Lambda}.
\end{aligned}
\tag{8.200}
$$

Next, using $\bar{D}_{\dot\alpha}\Lambda = 0$ we have $\bar{D}D_\alpha \bar{D}e^{i\Lambda} = 0$, and adding this vanishing term we can write

$$
\begin{aligned}
\bar{D}^2 D_\alpha e^{i\Lambda} &= \bar{D}^2 D_\alpha e^{i\Lambda} + \bar{D}D_\alpha \bar{D}e^{i\Lambda} \\
&= \bar{D}_{\dot\alpha} \epsilon^{\dot\alpha\dot\beta} (\bar{D}_{\dot\beta} D_\alpha + D_\alpha \bar{D}_{\dot\beta}) e^{i\Lambda} \\
&= \bar{D}_{\dot\alpha} \epsilon^{\dot\alpha\dot\beta} \{\bar{D}_{\dot\beta}, D_\alpha\} e^{i\Lambda} \\
&= \bar{D}_{\dot\alpha} \epsilon^{\dot\alpha\dot\beta} [-2i(\sigma^\mu)_{\alpha\dot\beta} \partial_\mu] e^{i\Lambda} \\
&= 0,
\end{aligned}
\tag{8.201}
$$

where we have used once again the constraint $\bar{D}_{\dot\alpha}\Lambda = 0$. Using Eq. (8.201) in Eq. (8.200), we find that the last term in Eq. (8.200) vanishes, and Eq. (8.200) gives

$$
\begin{aligned}
W'_\alpha &= e^{-i\Lambda} \left[\frac{1}{4g} \bar{D}^2 e^{-gV} (D_\alpha e^{gV}) \right] e^{i\Lambda} \\
&= e^{-i\Lambda} W_\alpha e^{i\Lambda}.
\end{aligned}
\tag{8.202}
$$

Similarly, $\bar{W}_{\dot\alpha}$ transforms as follows:

$$
\bar{W}_{\dot\alpha} \to \bar{W}'_{\dot\alpha} = e^{-i\Lambda^\dagger} \bar{W}_{\dot\alpha} e^{i\Lambda^\dagger},
\tag{8.203}
$$

Next, we use the vector multiplet in the Wess–Zumino gauge as given by Eq. (8.190). We will show that in the Wess–Zumino gauge we can write W_α and $\bar{W}_{\dot\alpha}$ in the following forms:

$$
W_{\alpha a} = -\frac{1}{4} \left[\bar{D}^2 D_\alpha V_a - \frac{i}{2} g f_{abc} \bar{D}^2 V_b D_\alpha V_c \right]
\tag{8.204}
$$

$$
\bar{W}_{\dot\alpha a} = \frac{1}{4} \left[-D^2 \bar{D}_{\dot\alpha} V_a - \frac{i}{2} g f_{abc} D^2 V_b \bar{D}_{\dot\alpha} V_c \right],
\tag{8.205}
$$

where $W_\alpha = W_{\alpha a} T_a$ and $W_{\dot\alpha} = W_{\dot\alpha a} T_a$. Equations (8.204) and (8.205) have the correct abelian limits, i.e., setting $f_{abc} = 0$ and removing the group indices one finds $W_\alpha = \frac{1}{4} \bar{D}^2 D_\alpha V$ and $\bar{W}_{\dot\alpha} = -\frac{1}{4} D^2 \bar{D}_{\dot\alpha} V$, which are the correct expressions for W_α and $\bar{W}_{\dot\alpha}$ for the abelian gauge group case.

Let us now confirm the validity of Eq. (8.204) in the Wess–Zumino gauge. We start by considering the quantity $e^{-gV} D_\alpha (e^{gV} X)$, which we can expand in

the form

$$e^{-gV} D_\alpha(e^{gV} X) = e^{-gV}(D_\alpha e^{gV})X + e^{-gV}(e^{gV} D_\alpha X), \tag{8.206}$$

from which we deduce that

$$e^{-gV}(D_\alpha e^{gV}) = e^{-gV} D_\alpha e^{gV} - D_\alpha. \tag{8.207}$$

Next, we use the operator expansion

$$e^{-Q} P \, e^{Q} = P - [Q, P] + \frac{1}{2!}[Q, [Q, P]] - \cdots \tag{8.208}$$

to write Eq. (8.207) as follows:

$$e^{-gV}(D_\alpha e^{gV}) = \left(D_\alpha - [gV, D_\alpha] + \frac{1}{2!}[gV, [gV, D_\alpha]] - \cdots - D_\alpha \right)$$

$$= -g[V, D_\alpha] + \frac{g^2}{2!}[V, [V, D_\alpha]] - \cdots$$

$$= g(D_\alpha V) - \frac{g^2}{2}[V, (D_\alpha V)] - \cdots \tag{8.209}$$

The last step in the equation above is obtained by noting that $[V, D_\alpha]X = -(D_\alpha V)X$. Using Eq. (8.209) in Eq. (8.191) we write

$$W_\alpha = \frac{1}{4}\bar{D}^2(D_\alpha V) - \frac{g}{8}\bar{D}^2[V, (D_\alpha V)] - O(V^3). \tag{8.210}$$

We note that terms $O(V^3)$ and higher vanish in the Wess–Zumino gauge. This leads to the result

$$W_{\alpha a} = \frac{1}{4}\bar{D}^2(D_\alpha V_a) - \frac{i}{8}g f_{abc}\bar{D}^2 V_b(D_\alpha V_c), \tag{8.211}$$

where $W_\alpha = W_{\alpha a}T_a$ and $V = V_a T_a$ and $[T_a, T_b] = i f_{abc}T_c$. A very similar analysis shows that in the Wess–Zumino gauge one has

$$\bar{W}_{\dot\alpha a} = -\frac{1}{4}D^2(\bar{D}_{\dot\alpha} V_a) - \frac{i}{8}g f_{abc}D^2 V_b(\bar{D}_{\dot\alpha} V_c). \tag{8.212}$$

Since $W_{\alpha a}$ is a chiral multiplet, we expand it as follows:

$$W_{\alpha a}(y, \theta) = W_{\alpha a}^{(0)}(y) + \theta^\beta W_{\alpha\beta a}^{(1)}(y) + \theta^2 W_{\alpha a}^{(2)}(y). \tag{8.213}$$

We now determine $W^{(i)}$, with $i = 0$, 1, and 2. Thus, $W^{(0)}$ is given by

$$W_{\alpha a}^{(0)} = \left[\frac{1}{4}\bar{D}^2(D_\alpha V_a) \right]_0 - \left[\frac{i}{8}g f_{abc}\bar{D}^2 V_b(D_\alpha V_c) \right]_0. \tag{8.214}$$

Using the results of Eqs. (8.340) and (8.341), we find

$$W_{\alpha a}^{(0)} = i\lambda_{\alpha a}. \tag{8.215}$$

Next, we determine $W^{(1)}_{\alpha\beta a}$. Here, we have

$$W^{(1)}_{\alpha\beta a} = \left[\frac{1}{4}D_\beta\bar{D}^2(D_\alpha V_a)\right]_0 - \left[\frac{i}{8}gf_{abc}D_\beta\bar{D}^2 V_b(D_\alpha V_c)\right]_0. \qquad (8.216)$$

Using the results of Eq. (8.340) and Eq. (8.341) we find

$$W^{(1)}_{\alpha\beta a} = \epsilon_{\beta\gamma}\left[\delta^\gamma_\alpha D_a - \frac{i}{2}(\sigma^\mu\bar{\sigma}^\nu)_\alpha^\gamma(\partial_\mu v_{\nu a} - \partial_\nu v_{\mu a}) + \frac{i}{4}gf_{abc}(\sigma^\mu\bar{\sigma}^\nu)_\alpha^\gamma v_{\mu b}v_{\nu c}\right]$$

$$= \epsilon_{\beta\gamma}\left[\delta^\gamma_\alpha D_a - \frac{i}{2}(\sigma^\mu\bar{\sigma}^\nu)_\alpha^\gamma v_{\mu\nu a}\right], \qquad (8.217)$$

where $v_{\mu\nu a} = \partial_\mu v_{\nu a} - \partial_\nu v_{\mu a} - (g/2)f_{abc}v_{\mu b}v_{\nu b}$. Finally, we determine $W^{(2)}_{\alpha a}$. Here, we find that

$$-16W^{(2)}_{\alpha a} = [D^2\bar{D}^2(D_\alpha V_a)]_0 - \left[\frac{i}{2}gf_{abc}D^2\bar{D}^2 V_b(D_\alpha V_c)\right]_0. \qquad (8.218)$$

Next, noting that $D^2 D_\alpha = 0$, we write

$$[D^2\bar{D}^2(D_\alpha V_a)]_0 = [D^2\bar{D}^2(D_\alpha V_a) - D^2 D_\alpha\bar{D}^2 V_a]_0$$

$$= [D^2[\bar{D}^2, D_\alpha]V_a]_0$$

$$= -[D^2(4i(\sigma^\mu)_{\alpha\dot\alpha}\bar{D}^{\dot\alpha}\partial_\mu)V_a]_0$$

$$= -[4i(\sigma^\mu)_{\alpha\dot\alpha}\partial_\mu D^2\bar{D}^{\dot\alpha}V_a]_0$$

$$= 4i(\sigma^\mu)_{\alpha\dot\alpha}\partial_\mu(-4i\bar{\lambda}^{\dot\alpha}_a)$$

$$= 16(\sigma^\mu)_{\alpha\dot\alpha}\partial_\mu\bar{\lambda}^{\dot\alpha}_a. \qquad (8.219)$$

Further, using the results of Eqs. (8.340) and (8.341) we have

$$f_{abc}[D^2\bar{D}^2 V_b(D_\alpha V_c)]_0 = 16if_{abc}(\sigma^\mu)_{\alpha\dot\beta}\bar{\lambda}^{\dot\beta}_b v_{\mu c}. \qquad (8.220)$$

Using Eqs. (8.219) and (8.220) in Eq. (8.218) we have

$$-W^{(2)}_{\alpha a} = (\sigma^\mu)_{\alpha\dot\alpha}\left[\delta_{ab}\partial_\mu + \frac{g}{2}f_{abc}v_{\mu c}\right]\bar{\lambda}^{\dot\alpha}_b$$

$$= (\sigma^\mu)_{\alpha\dot\alpha}D_{\mu ab}\bar{\lambda}^{\dot\alpha}_b, \qquad (8.221)$$

where $D_{\mu ab} = \delta_{ab}\partial_\mu + (g/2)f_{abc}v_{\mu c}$. Collecting all the results, we find that $W_{\alpha a}$ in the Wess–Zumino gauge is given by

$$W_{\alpha a}(y,\theta) = i\lambda_{\alpha a}(y) - \theta_\beta\left[\delta^\beta_\alpha D_a(y) - \frac{i}{2}(\sigma^\mu\bar{\sigma}^\nu)_\alpha^\beta v_{\mu\nu a}(y)\right] - (\sigma^\mu)_{\alpha\dot\alpha}D_{\mu ab}\bar{\lambda}^{\dot\alpha}_b\theta^2. \qquad (8.222)$$

A very similar analysis shows that the $\bar{W}_{\alpha a}$ in the Wess–Zumino gauge is given by

$$\bar{W}_{\dot\alpha a}(z,\bar{\theta}) = i\bar{\lambda}_{\dot\alpha a}(z) + \bar{\theta}_{\dot\beta}\left[\delta^{\dot\beta}_{\dot\alpha}D_a(z) - \frac{i}{2}(\bar{\sigma}^\mu\sigma^\nu)^{\dot\beta}_{\dot\alpha}v_{\mu\nu a}(z)\right] + \bar{\theta}^2 D_{\mu ab}\lambda^\alpha_b(z)(\sigma^\mu)_{\alpha\dot\alpha}. \qquad (8.223)$$

Next, we use W_α and $\bar{W}_{\dot\alpha}$ to construct the kinetic energy terms for the gauge fields. Thus, we write

$$\mathcal{L}_G = \frac{1}{4}\left[\int d^2\theta\; W_a^\alpha W_{\alpha a} + \int d^2\bar\theta\; \bar{W}_{\dot\alpha a}\bar{W}^{\dot\alpha a}\right]. \qquad (8.224)$$

An explicit analysis then gives

$$\frac{1}{4}\int d^2\theta\; W_a^\alpha W_{\alpha a} = \left[-\frac{i}{2}\lambda_a\sigma^\mu D_\mu^{ab}\bar\lambda_b - \frac{1}{8}v_{\mu\nu a}v_a^{\mu\nu} - \frac{i}{8}v_{\mu\nu a}\tilde{v}_a^{\mu\nu} + \frac{1}{4}D_a D_a\right], \qquad (8.225)$$

where $\tilde{v}_a^{\mu\nu} = \frac{1}{2}\epsilon^{\mu\nu\lambda\rho}v_{\lambda\rho a}$. Similarly, we have

$$\frac{1}{4}\int d^2\bar\theta\; \bar{W}_{\dot\alpha a}\bar{W}_a^{\dot\alpha} = \frac{i}{2}D_{\mu ab}\lambda_b\sigma^\mu\bar\lambda_a - \frac{1}{8}v_{\mu\nu a}v^{\mu\nu a} + \frac{i}{8}v_{\mu\nu a}\tilde{v}^{\mu\nu a} + \frac{1}{4}D_a D_a. \qquad (8.226)$$

Adding the two contributions, one finds that the $v_{\mu\nu}\tilde{v}^{\mu\nu a}$ term cancels and the sum gives

$$\mathcal{L}_G = -\frac{1}{4}v_{\mu\nu a}v^{\mu\nu a} + \frac{i}{2}[D_{\mu ab}\lambda_b\sigma^\mu\bar\lambda_a - \lambda_a\sigma^\mu D_{\mu ab}\bar\lambda_b] + \frac{1}{2}D_a D_a. \qquad (8.227)$$

The sum $\frac{i}{2}[D_{\mu ab}\lambda_b\sigma^\mu\bar\lambda_a - \lambda_a\sigma^\mu D_{\mu ab}\bar\lambda_b]$ can now be written in the following way:

$$\frac{i}{2}[D_{\mu ab}\lambda_b\sigma^\mu\bar\lambda_a - \lambda_a\sigma^\mu D_{\mu ab}\bar\lambda_b] = -i\lambda_a\sigma^\mu D_{\mu ab}\bar\lambda_b + \frac{i}{2}\partial_\mu(\lambda_a\sigma^\mu\bar\lambda_a). \qquad (8.228)$$

Discarding the total derivative, as it makes no contribution in the action, we can write the kinetic energy for the non-abelian gauge multiplet in the Wess–Zumino gauge so that

$$\mathcal{L}_G = -\frac{1}{4}v_{\mu\nu a}v^{\mu\nu a} - i\lambda_a\sigma^\mu D_{\mu ab}\bar\lambda_b + \frac{1}{2}D_a D_a$$

$$v_{\mu\nu a} = \partial_\mu v_{\nu a} - \partial_\nu v_{\mu a} - \frac{1}{2}gf_{abc}v_{\mu b}v_{\nu c}$$

$$D_{\mu ab}\bar\lambda_b = \delta_{ab}\partial_\mu\bar\lambda_b + \frac{g}{2}f_{abc}v_{\mu b}\bar\lambda_c. \qquad (8.229)$$

To conform to the notation of Chapter 3 we need to let $-g/2 \to g$.

8.14 The Interaction between Matter and Gauge Fields

We now turn to a discussion of the interaction between matter and gauge fields. In general, matter will consist of quarks and leptons as well as Higgs fields. We will work in the Wess–Zumino gauge, so that the vector multiplet in this gauge denoted by $\tilde{V}$ is given by Eq. (8.159). Since θ^3, and $\bar\theta^3$ and higher powers of

θ and $\bar{\theta}$ vanish, we have

$$\tilde{V}\tilde{V} = -\frac{1}{2}\theta^2\bar{\theta}^2 v_\mu v^\mu \tag{8.230}$$

$$\tilde{V}^n = 0, \quad n > 2. \tag{8.231}$$

The gauge-invariant interactions of the matter fields Φ are given by

$$\mathcal{L}_M = \int d^2\theta\, d^2\bar{\theta}\; \Phi^\dagger e^{g\tilde{V}}\Phi \tag{8.232}$$

$$= \int d^2\theta\, d^2\bar{\theta}\left[\Phi^\dagger\Phi + \Phi^\dagger g\tilde{V}\Phi + \Phi^\dagger\frac{g^2}{2}\tilde{V}^2\Phi\right], \tag{8.233}$$

where we have used the expansion of $e^{g\tilde{V}}$ and the result of Eq. (8.231). We will now examine the various terms. The first term on the right-hand side of Eq. (8.233) is just the usual kinetic energy term for the chiral multiplet, so that we have, up to a total divergence,

$$\int d^4\theta\; \Phi_i^\dagger\Phi_i = -\partial_\mu\phi_i^\dagger\partial^\mu\phi_i - i\psi_i\sigma^\mu\partial_\mu\bar{\psi}_i + F_i^*F_i. \tag{8.234}$$

The second and the third terms in the brackets on the right-hand side of Eq. (8.233) give

$$\int d^4\theta\; \Phi^\dagger g\tilde{V}\Phi = \int d^4\theta\; g\Phi_i^\dagger\left[-\theta\sigma^\mu\bar{\theta}v_{\mu ij} - i\bar{\theta}^2\theta\lambda_{ij} + i\theta^2\bar{\theta}\bar{\lambda}_{ij} + \frac{1}{2}\theta^2\bar{\theta}^2 D_{ij}\right]\Phi_j$$

$$= -\frac{i}{2}g\left[\phi_i^*(v_\mu)_{ij}\partial^\mu\phi_j - \partial^\mu\phi_i^*(v_\mu)_{ij}\phi_j\right] + \frac{g}{2}\bar{\psi}_i\bar{\sigma}^\mu(v_\mu)_{ij}\psi_j$$

$$+ \frac{i}{\sqrt{2}}g\phi_i^*\lambda_{ij}\psi_j - \frac{i}{\sqrt{2}}g\bar{\psi}_i\bar{\lambda}_{ij}\phi_j + \frac{1}{2}g\phi_i^*D_{ij}\phi_j \tag{8.235}$$

$$\int d^4\theta\; \Phi^\dagger\frac{1}{2}g^2\tilde{V}^2\Phi = -\frac{1}{4}g^2\phi_i^*(v_\mu v^\mu)_{ij}\phi_j. \tag{8.236}$$

We now introduce the covariant derivatives

$$D_\mu^{ij} = \delta_{ij}\partial_\mu + \frac{i}{2}gv_{\mu a}(T_a)_{ij} \tag{8.237}$$

$$\bar{D}_\mu^{ij} = \delta_{ij}\partial_\mu - \frac{i}{2}gv_{\mu a}(T_a^*)_{ij}. \tag{8.238}$$

Using the covariant derivatives we can write the gauge-invariant kinetic energy for the matter fields so that

$$\mathcal{L}_M = -\bar{D}_{\mu ij}\phi_j^*D_{ik}^\mu\phi_k - i\psi_i\sigma^\mu\bar{D}_{\mu ij}\bar{\psi}_j + \frac{i}{\sqrt{2}}g(\phi_i^*\lambda_{ij}\psi_j - \bar{\psi}_i\bar{\lambda}_{ij}\phi_j)$$

$$+ F_i^*F_i + \frac{1}{2}g\phi_i^*D_{ij}\phi_j \tag{8.239}$$

where $\lambda_{ij} = \lambda_a(T_a)_{ij}$ and $D_{ij}(x) = D_a(x)(T_a)_{ij}$. As in Eq. (8.229) to conform to the notation of Chapter 3, we need to let $-g/2 \to g$. The Lagrangian of Eq. (8.239) gives the kinetic energy and the interactions of a chiral multiplet with the gauge multiplet within a supersymmetric and gauge invariant framework.

An important aspect of supersymmetry which makes it an attractive framework for model building is that quantum field theories based on supersymmetry have some remarkable properties. Thus, the F-type couplings in supersymmetry receive no renormalization corrections at the quantum level [11]. In fact, the only divergences in supersymmetric theories are logarithmic infinities of the wave function in the renomalization and renormalizations of the gauge couplings.

8.15 Supersymmetry Breaking

We begin by first reviewing the argument on the problem of breaking supersymmetry discussed in Section 8.2, but now in a two-component notation. In the two-component notation the commutator of the two supersymmetry generators Q_α and $\bar{Q}_{\dot{\beta}}$ is given by Eq. (8.52). Multiplying this relation by $(\bar{\sigma}^\nu)^{\dot{\beta}\alpha}$ and using the relation $tr(\sigma^\mu\bar{\sigma}^\nu) = -2\eta^{\mu\nu}$, we have

$$P^\mu = -\frac{1}{4}\{Q_\alpha, \bar{Q}_{\dot{\beta}}\}(\bar{\sigma}^\mu)^{\dot{\beta}\alpha}, \tag{8.240}$$

which allows us to write

$$H = \frac{1}{4}\sum_{\alpha,\dot{\alpha}}(Q_\alpha\bar{Q}_{\dot{\alpha}} + \bar{Q}_{\dot{\alpha}}Q_\alpha). \tag{8.241}$$

Noting that $\bar{Q}_{\dot{\alpha}} = Q_\alpha^\dagger$, the positive definite nature of H can again be seen. Taking the expectation value of Eq. (8.241) and using closure, we have

$$\langle 0|H|0\rangle = \frac{1}{4}\sum_{n,\alpha,\dot{\alpha}}\left[|\langle n|Q_\alpha|0\rangle|^2 + |\langle n|\bar{Q}_{\dot{\alpha}}|0\rangle|^2\right]. \tag{8.242}$$

This again shows that for the vacuum state which is supersymmetric, i.e.,

$$Q_\alpha|0\rangle = 0, \quad \bar{Q}_{\dot{\alpha}}|0\rangle = 0, \tag{8.243}$$

we have $\langle 0|H|0\rangle = 0$, and for states for which supersymmetry is spontaneously broken so that

$$Q_\alpha|0\rangle \neq 0, \quad \bar{Q}_{\dot{\alpha}}|0\rangle \neq 0, \tag{8.244}$$

we have $\langle 0|H|0\rangle > 0$. As we noted earlier in Section 8.2, the fact that the supersymmetric vacuum always lies lower than the one which is not supersymmetric creates a problem regarding the spontaneous breaking of supersymmetry. One possible way to achieve the breaking is to devise a way in which the supersymmetric vacuum is not an allowed solution of the extrema equations. Further, such an arrangement is needed at the tree level itself since it is known that if superysmmetry is not broken at the tree level it is not broken with quantum corrections, at least perturbatively [17, 18].

From a field-theoretical viewpoint there are two ways to break supersymmetry spontaneously. One of these is an F-type breaking and the other is a D-type breaking. To discuss F-type breaking, let us consider a single chiral superfield Φ with the components $\phi(x)$, $\psi(x)$, and $F(x)$ so that

$$\Phi = \phi + \sqrt{2}\theta\psi + \theta^2 F. \tag{8.245}$$

Here, $\phi(x)$ is a complex scalar field, $\psi(x)$ is a chiral fermion, and $F(x)$ is an auxiliary field. Under supersymmetry transformations they transform as follows:

$$\delta_s\phi(x) = \sqrt{2}\xi\psi(x)$$
$$\delta_s\psi_\alpha(x) = \sqrt{2}\left[\xi_\alpha F(x) + i(\sigma^\mu\bar{\xi})_\alpha\partial_\mu\phi(x)\right]$$
$$\delta_s F(x) = -i\sqrt{2}\partial_\mu\psi(x)\sigma^\mu\bar{\xi}. \tag{8.246}$$

If we take the vacuum expectation value (VEV) of these and use the conditions $\langle 0|\psi(x)|0\rangle = 0$ and the fact that $\langle 0|\phi(x)|0\rangle$ is a constant and hence the derivative of the VEV of ϕ vanishes, one finds the following:

$$\langle 0|\delta_s\phi(x)|0\rangle = 0$$
$$\langle 0|\delta_s\psi(x)|0\rangle = \sqrt{2}\xi_\alpha\langle 0|F(x)|0\rangle$$
$$\langle 0|\delta_s F(x)|0\rangle = 0. \tag{8.247}$$

It is clear that non-vanishing of $\langle 0|F(x)|0\rangle$ will lead to breaking of supersymmetry. The result can be extended to include any number of chiral fields, and more generally one can say that given a set of n-chiral superfields Φ_i with components $(\phi_i(x), \psi_i(x), F_i(x))$, at least one of the F_i must have a non-vanishing vacuum expectation value for supersymmetry to be broken. It is to be noted that since $F_i^* = -\partial W(\phi_i)/\partial\phi_i$, if there is a term in the superpotential which is linear in the chiral fields, then supersymmetry is broken.

Another generic type of breaking is D-type breaking. Here, one considers a $U(1)$ abelian gauge supermultiplet $V(x, \theta, \bar{\theta})$ where

$$V(x, \theta, \bar{\theta}) = \theta\sigma^\mu\bar{\theta}v_\mu(x) + \theta^2\bar{\theta}\bar{\lambda}(x) + \bar{\theta}^2\theta\lambda(x) + \theta^2\bar{\theta}^2 D(x). \tag{8.248}$$

The VEV of the transformations of the components under supersymmetry transformations gives us

$$\langle 0|\delta_s v_\mu(x)|0\rangle = 0$$
$$\langle 0|\delta_s\lambda_\alpha(x)|0\rangle = 2\xi_\alpha\langle 0|D(x)|0\rangle$$
$$\langle 0|\delta_s D(x)|0\rangle = 0. \tag{8.249}$$

Here, the non-vanishing of $\langle 0|D(x)|0\rangle$ leads to breaking of supersymmetry. We discuss these two cases in further detail below.

8.16 *F*-type Breaking of Supersymmetry

We now consider the Wess–Zumino model and the possibility of supersymmetry breaking in this model. Here, the superpotential is of the form

$$W[\phi] = \mu^2 \Phi + \frac{1}{2}m\Phi^2 + \frac{1}{3}\lambda\Phi^3, \tag{8.250}$$

and the scalar potential is

$$V(\phi) = |\mu^2 + m\phi(x) + \lambda\phi^2(x)|^2. \tag{8.251}$$

The condition that supersymmetry is not broken is $\langle V \rangle = 0$, and requires

$$\mu^2 + m\bar{\phi} + \lambda\bar{\phi}^2 = 0, \tag{8.252}$$

which gives

$$\bar{\phi} = -\frac{1}{2\lambda}\left[m \pm (m^2 - 4\lambda\mu^2)^{\frac{1}{2}}\right]. \tag{8.253}$$

Equations (8.253) tells us that a solution to Eq. (8.252), and thus to the condition $\langle V \rangle = 0$, always exists, and so in this case one has no breaking of supersymmetry. However, for the case when $m = 0 = \lambda$, a solution to Eq. (8.252) does not exist, $\langle V \rangle = |\mu|^4$, and thus supersymmetry is broken.

Next, we discuss the so-called O'Raifeartaigh breaking of supersymmetry [19]. Here, we consider three superfields and a superpotential of the form

$$W[\Phi_i] = \lambda\Phi_1(\Phi_3^2 - \mu^2) + m\Phi_2\Phi_3. \tag{8.254}$$

The fields Φ_i ($i = 1, 2, 3$) are chiral superfields and have the following expansion:

$$\Phi_i = \phi_i + \sqrt{2}\theta\psi_i + \theta^2 F_i. \tag{8.255}$$

Similarly,

$$\Phi_i^\dagger = \phi_i^* + \sqrt{2}\bar{\theta}\bar{\psi}_i + \bar{\theta}^2 F_i^*. \tag{8.256}$$

Let us define the lowest component of $\hat{W}(\Phi)$ to be $W(\phi)$. Then, the F_i and F_i^* are given by

$$F_i = -\frac{\partial W^*(\phi^*)}{\partial \phi_i^*}, \quad F_i^* = -\frac{\partial W(\phi)}{\partial \phi_i}, \tag{8.257}$$

so that

$$F_1^* = -\lambda(\phi_3^2 - \mu^2), \ F_2^* = -m\phi_3, \ F_3^* = -(2\lambda\phi_1\phi_3 + m\phi_2). \tag{8.258}$$

Now, the non-vanishing of one or more of the F_i implies that supersymmetry is broken, and it is easily seen from the expressions of F_i above that not all of them can vanish. Thus, the superpotential of Eq. (8.254) necessarily leads to the breaking of supersymmetry. Here, the scalar potential is

$$V = \sum_i F_i F_i^*. \tag{8.259}$$

The variation of this potential with respect to the three scalar fields ϕ_i gives the following three extrema equations:

$$\phi_3^*(m\phi_2 + 2\lambda\phi_1\phi_3) = 0 \tag{8.260}$$

$$m\phi_2 + 2\lambda\phi_1\phi_3 = 0 \tag{8.261}$$

$$2\lambda^2\phi_3^*(\phi_3^2 - \mu^2) + m^2\phi_3 + 2\lambda\phi_1^*(m\phi_2 + 2\lambda\phi_1\phi_3) = 0. \tag{8.262}$$

From Eq. (8.262) and its complex conjugate, one finds

$$(\phi_3^* - \phi_3)[2\lambda^2(|\phi_3|^2 + \mu^2) + m^2] = 0, \tag{8.263}$$

which implies $\phi_3^* = \phi_3$, i.e., that ϕ_3 is real. Next, the use of Eq. (8.261) in Eq. (8.262) gives

$$2\lambda^2\phi_3(\phi_3^2 - \mu^2) + m^2\phi_3 = 0. \tag{8.264}$$

We label the solutions to the extrema equations by a bar on the fields, i.e., $\bar{\phi}_i = \langle\phi_i\rangle$. Thus, by use of the extrema equations the vacuum energy takes the form

$$V_0 = m^2\bar{\phi}_3^2 + \lambda^2(\bar{\phi}_3^2 - \mu^2)^2. \tag{8.265}$$

Satisfaction of Eq. (8.264) requires one of the following two possibilities:

$$\text{(A)} \quad \bar{\phi}_3 = 0 \tag{8.266}$$

$$\text{(B)} \quad \bar{\phi}_3 = \pm\left(\mu^2 - \frac{m^2}{2\lambda^2}\right)^{1/2}. \tag{8.267}$$

Let us consider case (A). From Eq. (8.261) one also finds that $\bar{\phi}_2 = 0$ while $\bar{\phi}_1$ remains arbitrary. Here, one finds that $\langle F_1\rangle = \lambda\mu^2$, $\langle F_2\rangle = 0 = \langle F_3\rangle$, and the vacuum energy is

$$V_0^{(A)} = \lambda^2\mu^4 > 0, \tag{8.268}$$

so supersymmetry is broken. The Yukawa interactions for this case are given by

$$L_Y = -\left[2\lambda\phi_3\psi_1\psi_3 + m\psi_2\psi_3 + \lambda\phi_1\psi_3^2 + \text{h.c.}\right]. \tag{8.269}$$

In this case the mass matrix in the fermionic sector has the form

$$L_{mf}^{I} = -\frac{1}{2}(2\alpha\psi_3^2 + 2m\psi_2\psi_3 + \text{h.c.}), \tag{8.270}$$

where $\alpha = \lambda\bar{\phi}_1$. The eigenvalues of the fermion mass matrix are

$$0, \ \alpha \pm \gamma, \ \gamma \equiv \sqrt{\alpha^2 + m^2}. \tag{8.271}$$

Here, the zero eigenvalue implies that we have a massless chiral fermion, which is the Goldstino, while the remaining two eigenvalues $\alpha \pm \gamma$ are for Majorana states.

We now look at the mass terms in the bosonic sector. Here, noting that $\bar{\phi}_2 = \bar{\phi}_3 = 0$, the mass terms take the form

$$L_{mb} = m^2(|\phi_2|^2 + |\phi_3|^2) + 4\lambda^2\bar{\phi}_1^2|\phi_3|^2 - \lambda^2\mu^2(\phi_3^2 + \phi_3^{*2}) + 2\lambda m\bar{\phi}_1(\phi_2\phi_3^* + \phi_3\phi_2^*).$$

Decomposing the fields into real and imaginary parts, so that $\phi_i = \phi_{iR} + i\phi_{iI}$ ($i = 1, 2, 3$), the mass matrix for the bosonic fields takes on the form

$$L_{mb} = (m^2 + 2b)\phi_{3R}^2 + m^2\phi_{2R}^2 + 2a\phi_{2R}\phi_{3R} + (m^2 + 2c)\phi_{3I}^2 + m^2\phi_{2I}^2 + 2a\phi_{2I}\phi_{3I}.$$

In the above, we have defined

$$a = 2m\lambda\bar{\phi}_1, \ b = 2\lambda^2\bar{\phi}_1^2 - \lambda^2\mu^2, \ c = 2\lambda^2\bar{\phi}_1^2 + \lambda^2\mu^2. \tag{8.272}$$

Diagonalization of the above gives us two CP-even states with eigenvalues

$$m^2 + b \pm (a^2 + b^2)^{\frac{1}{2}}. \tag{8.273}$$

The diagonalization also gives us the eigenvalues for the two CP-odd states as follows:

$$m^2 + c \pm d, \ d \equiv (a^2 + c^2)^{\frac{1}{2}}. \tag{8.274}$$

Additionally, of course, the complex field ϕ_1 is massless. It is also straightforward to check that in this case the supertrace defined by

$$str \ M^2 = \sum_J (-1)^{2J}(2J+1)M_J^2 \tag{8.275}$$

vanishes when one inserts into the mass spectra obtained above. We note that the appearance of massless states in the spectrum is unrealistic, and for this reason this mechanism of supersymmetry breaking is not acceptable on phenomenological grounds.

Regarding case (B), it holds when $2\mu^2\lambda^2 > m^2$, and in this case the vacuum energy is given by

$$V_0^{(B)} = m^2\left(\mu^2 - \frac{m^2}{4\lambda^2}\right) > 0. \tag{8.276}$$

Since $V > 0$, supersymmetry is again broken. An analysis similar to case (A) can be carried out (see Section 8.21).

8.17 *D*-type Breaking of Supersymmetry

In addition to F-term breaking, one can also have breaking of supersymmetry by the D-term. Let us illustrate this by first considering the Lagrangian of Eq. (8.187). Here, since D is an auxiliary field, we can solve for it, which gives

$$D(x) = -\xi. \tag{8.277}$$

Substitution back into Eq. (8.187) gives us the potential

$$V = \frac{1}{2}\xi^2, \tag{8.278}$$

which implies that supersymmetry is broken spontaneously. We now consider a more realistic model where we have the interaction of a massless chiral super-field with a $U(1)$ abelian gauge field multiplet and we include an FI term. The Lagrangian for this system is given by

$$L = L_{kin} + L_1 + L_{FI}, \tag{8.279}$$

where L_{kin} is the covariant kinetic energy for the chiral field and the $U(1)$ field,

$$L_{kin} = -\frac{1}{4}v_{\mu\nu}v^{\mu\nu} + i\partial_\mu \lambda \sigma^\mu \bar{\lambda} - |D_\mu \phi(x)|^2 + iD_\mu^* \bar{\psi}\bar{\sigma}^\mu \psi(x), \tag{8.280}$$

where $D_\mu = (\partial_\mu - (i/2)gv_\mu)$, and where ($\phi(x)$ and $\psi(x)$) are the spin 0 and spin 1/2 components of the chiral multiplet and ($v_\mu(x)$ and $\lambda(x)$) are the spin 1 (vector field) and spin 1/2 (gaugino field) of the vector multiplet. The remainder of the Lagrangian is given by

$$L_1 + L_{FI} = -\frac{1}{2}\psi M \psi - \frac{1}{2}\bar{\psi}\bar{M}\bar{\psi} - \frac{g}{\sqrt{2}}\left[\lambda\psi\phi^* + \bar{\lambda}\bar{\psi}\phi\right] - V(\phi, \phi^*), \tag{8.281}$$

where $M = \partial^2 W/\partial\phi^2$ and $\bar{M} = \partial^2 \bar{W}/\partial\phi^{*2}$ and $\mathcal{L}_{FI}$ in Eq. (8.279) is given by Eq. (8.186). The scalar potential is given by

$$V(\phi, \phi^*) = \left|\frac{\partial W}{\partial\phi}\right|^2 + \frac{1}{2}\left(\xi + \frac{1}{2}g|\phi(x)|^2\right)^2. \tag{8.282}$$

Further, we assume that the chiral field is charged under the $U(1)$ gauge group and thus invariance under $U(1)$ transformations requires the superpotential to vanish. This implies that $M = \bar{M} = 0$, and the first two terms in Eq. (8.281) vanish. Additionally, the scalar potential in Eq. (8.282) contains only the D-term contribution. There now exist two possibilities depending on whether ξ/g is negative or positive. We first consider the case when ξ/g is negative. In this case it is clear that we can find solutions where the potential vanishes and there is no breaking of supersymmetry, i.e.,

$$V = 0, \quad |\phi| = \sqrt{-\frac{2\xi}{g}}, \quad \xi/g < 0. \tag{8.283}$$

Here, while supersymmetry is unbroken there is a breaking of the $U(1)$ gauge symmetry, which generates mass growth for some of the fields. Thus, using the VEV of ϕ as given by Eq. (8.283) in Eq. (8.281), we find the following mass terms for the fermionic fields:

$$L_{mf} = -(-g\xi)^{\frac{1}{2}}\left[\lambda\psi + \bar{\lambda}\bar{\psi}\right]. \tag{8.284}$$

It is more illuminating to define a four-component Dirac spinor so that

$$\psi_D = \begin{pmatrix} \lambda_\alpha \\ \bar{\psi}^{\dot\alpha} \end{pmatrix}$$

$$\bar{\psi}_D = (\psi^\alpha, \bar{\lambda}_{\dot\alpha}), \tag{8.285}$$

and the mass term Eq. (8.284) can be written as a mass term for the Dirac field,

$$L_{mf} = -m\bar{\psi}_D\psi_D$$

$$m \equiv (-g\xi)^{\frac{1}{2}}. \tag{8.286}$$

To determine the scalar mass spectrum, we expand around $\langle\phi\rangle = \bar{\phi}$ so that $\phi = \bar{\phi} + \phi'$, where we write $\phi' = \phi'_r + i\phi'_i$. We find that the expansion around the new vacuum gives a mass to ϕ'_r, so we have a mass term $\frac{1}{2}g^2\bar{\phi}^2\phi'^2_r = m^2\phi'^2_r$. Now, normalization of the kinetic energy of the residual scalar field requires that we define $s = \sqrt{2}\phi'_r$. Then, we have for the quadratic part of the s-field the terms $-\frac{1}{2}(\partial_\mu s)^2 - \frac{1}{2}m^2 s^2$. Additionally, the vector boson has a mass term $\frac{1}{4}g^2|\bar{\phi}|^2 v_\mu v^\mu = -\frac{1}{2}m^2 v_\mu v^\mu$. Collecting all the mass terms, we have

$$L_m = -\frac{1}{2}m^2 s^2 - m\bar{\psi}_D\psi_D - \frac{1}{2}m^2 v_\mu v^\mu. \tag{8.287}$$

Noting that the real scalar field has one degree of freedom, the massive vector boson has three degrees of freedom, and the Dirac field has 4 degrees of freedom, one finds that the supertrace $str\, M^2$ vanishes. These massive fields form a massive vector multiplet of supersymmetry with a common mass m.

We now consider the second possibility, i.e., $\xi/g > 0$. Here, the extrema equation $\partial V/\partial \phi = 0$ gives

$$\frac{\partial V}{\partial \phi} = 0 \; : \quad \left(\xi + \frac{g}{2}|\phi|^2\right)\phi^* = 0. \tag{8.288}$$

In this case the only allowed solution is $\phi = 0$, which gives

$$V = \frac{\xi^2}{2} > 0. \tag{8.289}$$

Here, the potential does not vanish, and supersymmetry is spontaneously broken.

Let us now consider the case of a $U(1)$ gauge theory with two chiral fields which are oppositely charged for anomaly cancellation. The Lagrangian for this model is

$$L = L_V + L_{\phi_+} + L_{\phi_-} + L_{\phi_+\phi_-} + \xi D, \tag{8.290}$$

where L_V is the Lagrangian of the $U(1)$ gauge multiplet given by Eq. (8.188),

$$L_V = -\frac{1}{4}v_{\mu\nu}v^{\mu\nu} + i\partial_\mu\lambda\sigma^\mu\bar{\lambda} + \frac{1}{2}D^2, \tag{8.291}$$

and $L_{\phi_\pm}$ is the Lagrangian for the chiral superfields $\Phi_\pm$ so that

$$L_{\Phi_+} = -|D_\mu\phi_+|^2 + iD_\mu^*\bar{\psi}_+\bar{\sigma}^\mu\psi_+ - \frac{g}{\sqrt{2}}(\lambda\psi_+\phi_+^* + \bar{\lambda}\bar{\psi}_+\phi_+) + |F_+|^2 + \frac{g}{2}D|\phi_+|^2.$$

L_{Φ_-} can be obtained from the above by letting $\phi_+ \to \phi_-$, $\psi_+ \to \psi_-$, $F_+ \to F_-$, $D_\mu \longleftrightarrow D_\mu^*$, and $g \to -g$. L_{+-} are terms that mix the chiral left and chiral right fields, and here one has

$$L_{\Phi_+\Phi_-} = -m(\psi_+\psi_- + \bar{\psi}_+\bar{\psi}_-) + m(\phi_+F_- + \phi_-F_+ + \text{h.c.}), \qquad (8.292)$$

and the last term in Eq. (8.290) is the Fayet–Iliopoulos D-term. Using the field equations to eliminate the auxiliary fields $F_\pm$ and D, the scalar potential reads

$$V = \left(m^2 + \frac{g\xi}{2}\right)|\phi_+|^2 + \left(m^2 - \frac{g\xi}{2}\right)|\phi_-|^2 + \frac{g^2}{8}(|\phi_+|^2 - |\phi_-|^2)^2 + \frac{\xi^2}{2}. \quad (8.293)$$

From the above, we get two extrema equations:

$$\left(m^2 + \frac{1}{2}g\xi\right)\phi_+ + \frac{g^2}{4}(|\phi_+|^2 - |\phi_-|^2)\phi_+ = 0$$

$$\left(m^2 - \frac{1}{2}g\xi\right)\phi_- - \frac{g^2}{4}(|\phi_+|^2 - |\phi_-|^2)\phi_- = 0. \qquad (8.294)$$

Multiplying the first equation in Eq. (8.294) by ϕ_- and the second equation by ϕ_+ and then adding the two, one finds

$$2m^2\phi_+\phi_- = 0. \qquad (8.295)$$

Thus, either ϕ_+ or ϕ_-, or both, must vanish. Let us assume that $\phi_+ = 0$. Then, inserting it into the second equation of Eq. (8.294), one gets

$$\left(m^2 - \frac{1}{2}g\xi + \frac{1}{4}g^2|\phi_-|^2\right)\phi_- = 0. \qquad (8.296)$$

Let us now consider two possibilities: (i) $m^2 > \frac{1}{2}g\xi$ and (ii) $m^2 < \frac{1}{2}g\xi$. For case (i) the factor multiplying ϕ_- in Eq. (8.296) is non-vanishing, and so one has $\phi_- = 0$. Here, there is breaking of supersymmetry but no breaking of the gauge symmetry. Thus, the mass terms in this case are given by

$$L_m^{(i)} = -\left(m^2 + \frac{g\xi}{2}\right)|\phi_+|^2 - \left(m^2 - \frac{g\xi}{2}\right)|\phi_-|^2 - m\bar{\psi}_D\psi_D, \qquad (8.297)$$

where the Dirac field is constructed out of the spin 1/2 components of the chiral superfields, i.e., $\psi_D^T = (\psi_{+\alpha}, \bar{\psi}_-^{\dot{\alpha}})$. Thus, in this case we have two massive complex scalar fields $\phi_\pm$, and one massive Dirac field. Additionally we have a massless vector field, and a massless spin 1/2 field $\lambda(x)$. The latter field acts like a Goldstino. So, while supersymmetry is broken in this model, the gauge symmetry is not broken.

Next, we consider the alternative possibility of case (ii). Here, the factor multiplying ϕ_- is not positive definite, and thus one obtains a non-vanishing VEV for ϕ_- so that

$$|\bar{\phi}_-|^2 = 2g^{-2}(g\xi - 2m^2). \tag{8.298}$$

In this case the local $U(1)$ gauge symmetry is broken. Also, $V(\phi_+ = 0, \phi_- = \bar{\phi}_-) = 2m^2g^{-2}(g\xi - m^2) > 0$, and thus supersymmetry is also broken. In the bosonic sector the mass terms for the scalar fields and for the vector are easily determined by expanding around the new vacuum, i.e., $\langle\phi_+\rangle = 0, \langle\phi_-\rangle = \bar{\phi}_-$. One finds

$$L_m^{(ii)}(\text{bose}) = -2m^2|\phi_+|^2 - \frac{1}{2}\bar{m}^2 s^2 - \frac{1}{2}\bar{m}^2 v_\mu v^\mu, \tag{8.299}$$

where $s = \sqrt{2}\,\mathrm{Re}\,\phi_-$ and $\bar{m} = g\bar{\phi}_-/\sqrt{2}$. Note that the imaginary part of ϕ_- is absorbed by the $U(1)$ vector field, to become massive. In the fermionic sector we obtain the following mass matrix in the basis $\psi_+, \psi_-,$ and λ. The matrix elements of the mass matrix $M^{(ii)}(\text{fermi})$ are $(12) = (21) = m$ and $(23) = (32) = \bar{m}$, and all other elements vanish. The eigenvalues of the mass matrix are $\pm(m^2 + \bar{m}^2)^{\frac{1}{2}}$ and 0. The two massive states can be combined into a Dirac fermion, while the massless state is the Goldstino. The Goldstino is a linear combination of ψ_+ and λ.

8.18 Spurion Fields

A general analysis of soft breaking in supersymmetric theories has been discussed in Girardello and Grisaru [20]. Spurion fields are sometimes used as a convenient device to characterize symmetry breaking. In the context of supersymmetry breaking, we can introduce a scalar superfield $S(\theta, \bar{\theta})$ which has the component expansion

$$S(\theta, \bar{\theta}) = S_0 + \theta^\alpha S_{1\alpha} + \bar{\theta}_{\dot{\alpha}} \bar{S}_1^{\dot{\alpha}} + \theta^\alpha \theta_\alpha S_2 + \bar{\theta}_{\dot{\alpha}} \bar{\theta}^{\dot{\alpha}} \bar{S}_2 + \theta^\alpha \sigma^\mu_{\alpha\dot{\beta}} \bar{\theta}^{\dot{\beta}} S_\mu$$
$$+ \bar{\theta}_{\dot{\alpha}} \bar{\theta}^{\dot{\alpha}} \theta^\alpha S_{3\alpha} + \theta^\alpha \theta_\alpha \bar{\theta}_{\dot{\alpha}} \bar{S}_3^{\dot{\alpha}} + \theta^\alpha \theta_\alpha \bar{\theta}_{\dot{\alpha}} \bar{\theta}^{\dot{\alpha}} S_4. \tag{8.300}$$

To break the symmetry, one makes the components $S_0, S_{1\alpha}$, etc., c-numbers. However, to preserve Lorentz invariance, some components must vanish, i.e.,

$$S_{1\alpha} = S_1^{\dot{\alpha}} = S_\mu = S_{3\alpha} = \bar{S}_3^{\dot{\alpha}} = 0. \tag{8.301}$$

Further, we assume $S(\theta, \bar{\theta})$ to be real so that $\bar{S}_2 = S_2^*$ and choose $S_0 = 1, S_2 = \xi_1$, and $S_4 = \xi_2$ so the spurion field takes the form

$$S = 1 + \theta^2 \xi_1 + \bar{\theta}^2 \xi_1^* + \theta^2 \bar{\theta}^2 \xi_2. \tag{8.302}$$

As an example of the utility of the spurion field, let us consider the Lagrangian term

$$\mathcal{L}_1 = \int d^2\theta\, d^2\bar{\theta}\, S(\theta, \bar{\theta}) \Phi_i^\dagger \Phi_i. \tag{8.303}$$

It is easily seen that the above Lagrangian gives a mass term for the scalar fields, i.e.,

$$\Delta \mathcal{L}_1^{SB} = \xi_2 \phi_i^\dagger \phi_i. \tag{8.304}$$

Next we consider a modified Lagrangian of the form

$$\mathcal{L}_2 = \int d^2\theta d^2\bar{\theta} S(\theta, \bar{\theta}) \left[\Phi_i^\dagger \Phi_i + \{ W \delta^2(\bar{\theta}) + \text{h.c.} \} \right], \tag{8.305}$$

where the superpotential has the form

$$W(\theta) = \frac{1}{2} m_{ij} \Phi_i \Phi_j + \frac{1}{3} h_{ijk} \Phi_i \Phi_j \Phi_k. \tag{8.306}$$

In this case the symmetry-breaking term is given by

$$\Delta \mathcal{L}_2^{SB} = (\xi_2 - |\xi_1|^2) |\phi_i|^2 - \left[\xi_1 \frac{1}{2} m_{ij} \phi_i \phi_j + \xi_1 \frac{2}{3} h_{ijk} \phi_i \phi_j \phi_k + \text{h.c.} \right]. \tag{8.307}$$

We note that the soft terms generated above are universal. Thus, $m_0^2 = (\xi_2 - |\xi_1|^2)^{1/2}$ is the universal scalar mass, $B_0 = \xi_1$ is the universal bilinear coefficient, and $A_0 = 2\xi_1$ is the universal trilinear coefficient. In this case, the bilinear and the trilinear coefficients are related so that $A_0 = 2B_0$. One can also generate a gaugino mass term by the spurion method. Since the mechanics of how the gaugino mass is generated are different from the way the scalar masses are generated, it is best to use a different spurion field in generating the gaugino masses. Thus, we use the spurion field $S'(\theta, \bar{\theta}) = (1 + \theta^2 \xi' + \bar{\theta}^2 \xi'^* + \theta^2 \bar{\theta}^2 \xi'')$. The gauge kinetic energy term arises from a spinor superfield W^α for a $U(1)$ gauge group. For an $SU(n)$ group, we promote the superfield to $W^{(i)\alpha}$, where $i = 1 - n$. Here, we focus on just the $U(1)$ case, as the extension to $SU(n)$ is straightforward. Consider the Lagrangian

$$L' = \frac{1}{4} \int d^2\theta \, d^2\bar{\theta} \, S'(\theta, \bar{\theta}) [W^{\alpha(i)} W_\alpha^{(i)} \delta^2(\bar{\theta}) + \bar{W}_{\dot{\alpha}}^{(i)} \bar{W}^{\dot{\alpha}(i)} \delta^2(\theta)]. \tag{8.308}$$

It is easily seen that Eq. (8.308) gives a gaugino mass term $\sim (\xi' \lambda^\alpha \lambda_\alpha + \xi'^* \bar{\lambda}_{\dot{\alpha}} \bar{\lambda}^{\dot{\alpha}})$. For further reading, see the literature [21, 22].

8.19 *R*-Symmetry

In superspace formulation of supersymmetry, R-symmetry corresponds to transformations on the θ-coordinates. It is imposed to enforce the baryon and lepton number [23–26]. Under R-transformations a superfield $\Phi(x, \theta, \bar{\theta})$ has the property that

$$R\Phi(x, \theta, \bar{\theta})R^{-1} = \Phi(x, \theta e^{-i\alpha}, \bar{\theta} e^{i\alpha}). \tag{8.309}$$

A superfield can carry an R-charge. Consider a left chiral superfield $\Phi(y, \theta)$ which we assume carries an R-charge of n. It has the expansion

$$\Phi(y, \theta) = \phi(y) + \theta\chi(y) + \theta^2 F(y). \tag{8.310}$$

Here, ϕ has the charge n, χ has the charge $n_\chi = n - 1$, and F has the charge $n_F = n - 2$. They transform under R-transformations as

$$\phi \to e^{in\alpha}\phi \tag{8.311}$$

$$\chi \to e^{i\gamma_5(n-1)\alpha}\chi \tag{8.312}$$

$$F \to e^{i(n-2)\alpha}F. \tag{8.313}$$

Next, consider a product of several left chiral fields Φ_i:

$$\Phi(y, \theta) = \prod_i \Phi_i(y, \theta) \tag{8.314}$$

$$\Phi_i \to e^{in_i\alpha}\Phi_i. \tag{8.315}$$

In this case, Φ will have an R-number of $n = \sum_i n_i$. Since each θ carries an R-charge of -1, the superpotential would have R-charge of $+2$ to have R invariance and contribute to the Lagrangian density. Let us assume that the superpotential has the following form:

$$W = c_1\Phi_1 + c_2\Phi_2^2 + c_3\Phi_3^3. \tag{8.316}$$

Since W has an R-charge of $+2$, Φ_1 has an R-charge of $+2$, Φ_2 has an R-charge of $+1$, and Φ_3 has an R-charge of $+2/3$. Then, expanding Φ_i as

$$\Phi_i = a_i\phi_i + b_i\theta\chi_i + c_i\theta^2 F_i, \tag{8.317}$$

one finds that ϕ_1, ϕ_2, and ϕ_3 have R-charges 2, 1, and 2/3; χ_1, χ_2, and χ_3 have R-charges of 1, 0, and $-1/3$; and F_1, F_2, and F_3 have R-charges of 0, -1, and $-4/3$. A vector superfield carries zero R-number. Its components in the Wess–Zumino gauge, i.e., V^μ, λ, and D, have the transformations

$$V^\mu \to V^\mu, \quad \lambda \to e^{i\gamma_5\alpha}\lambda, \quad D \to D. \tag{8.318}$$

For model building, a superpotential of the type in Eq. (8.316) with the assigned values of R quantum numbers will give R-invariant Lagrangians. R-invariance forbids gaugino masses, and must be broken to attain non-vanishing masses for the gauginos.

R-symmetry cannot be gauged in the framework of global supersymmetry but can be gauged within local supersymmetry. It is found that stringent anomaly cancellation conditions are needed which require inclusion of the Green–Schwarz [27] anomaly cancellation mechanism [28, 29].

8.20 Linear Multiplets

In $N = 1$ supersymmetry a linear multiplet satisfies the condition $L^\dagger = L$ and the additional constraints

$$\mathcal{D}^2 L = 0$$

$$\overline{\mathcal{D}}^2 L = 0, \tag{8.319}$$

where $\mathcal{D}^2 = \mathcal{D}^\alpha \mathcal{D}_\alpha$ and $\overline{\mathcal{D}}^2 = \overline{\mathcal{D}}_{\dot\alpha}\overline{\mathcal{D}}^{\dot\alpha}$, and $\mathcal{D}_\alpha$ and $\overline{\mathcal{D}}\dot\alpha$ are given by Eqs. (8.61) and (8.63).

$$\mathcal{D}_\alpha = \frac{\partial}{\partial \theta^\alpha} + i(\sigma^\mu \bar\theta)_\alpha \partial_\mu \tag{8.320}$$

$$\overline{\mathcal{D}}_{\dot\alpha} = -\frac{\partial}{\partial \bar\theta^{\dot\alpha}} - i(\theta \sigma^\mu)_{\dot\alpha} \partial_\mu. \tag{8.321}$$

With the above constraints, the linear multiplet can be expanded in component form so that

$$L = C(x) + i\theta\chi(x) - i\bar\theta\bar\chi(x) - \theta\sigma^\mu\bar\theta V_\mu(x) + \frac{1}{2}\theta\theta\bar\theta\bar\sigma^\mu\partial_\mu\chi(x)$$

$$- \frac{1}{2}\bar\theta\bar\theta\theta\sigma^\mu\partial_\mu\bar\chi(x) - \frac{1}{4}\theta\theta\bar\theta\bar\theta\,\Box C(x), \tag{8.322}$$

where the vector field V_μ is divergenceless, i..e,

$$\partial_\mu V^\mu(x) = 0. \tag{8.323}$$

Consequently, it can be expressed as

$$V^\mu = \frac{1}{\sqrt{2}}\epsilon^{\mu\nu\sigma\rho}\partial_\nu B_{\sigma\rho}. \tag{8.324}$$

The invariance of V^μ under the gauge transformation below is easily checked:

$$B_{\sigma\rho} \to B_{\sigma\rho} + \partial_\sigma\lambda_\rho - \partial_\rho\lambda_\sigma. \tag{8.325}$$

Thus, the fields in the linear multiplet consist of $C(x)$, which is a real scalar, $B_{\mu\nu}$, which is an anti-symmetric tensor, and χ, which is a Majorana spinor. From Eq. (8.322), we notice that there are no auxiliary fields in the multiplet.

Linear multiplets arise in superstring theory where, for example, the field $C(x)$ is essentially the dilation field. A linear multiplet, however, has a dual description. Thus, an anti-symmetric tensor with the gauge transformation of Eq. (8.325) has a dual description in terms of the pseudo-scalar field. The supersymmetric version of this is that the linear multiplet L has a dual description in terms of a chiral superfield. In the dual description of a chiral multiplet the components $C(x)$ and $B_{\mu\nu}$ of the linear multiplet are replaced by a complex chiral scalar field $\phi(x)$. Further, the chiral multiplet also has a Majorana spinor and, additionally, a complex auxiliary field. The chiral–linear duality holds in string perturbation theory, but is not on the same firm footing when non-perturbative effects such as gaugino condensation are included. For further reading on linear multiplets see the literature [30–32].

8.21 Problems

1. Show that $\epsilon_{\alpha\beta}$ and $\epsilon^{\alpha\beta}$ satisfy the following relations:

$$\epsilon_{\alpha\beta} = (L)_\alpha{}^\gamma (L)_\beta{}^\delta \epsilon_{\gamma\delta}, \epsilon^{\alpha\beta} = \epsilon^{\gamma\delta}(L^{-1})_\gamma{}^\alpha (L^{-1})_\delta{}^\beta, \qquad (8.325)$$

with similar relations holding for $\epsilon_{\dot\alpha\dot\beta}$ and $\epsilon^{\dot\alpha\dot\beta}$. Thus, $\epsilon_{\alpha\beta}$, $\epsilon^{\alpha\beta}$ $\epsilon_{\dot\alpha\dot\beta}$, and $\epsilon^{\dot\alpha\dot\beta}$ are invariant tensors under the $SL(2,C)$ transformations.

2. Establish the following identities:

$$(\sigma^\mu\bar\sigma^\nu + \sigma^\nu\bar\sigma^\mu)_\alpha{}^\beta = -2\eta^{\mu\nu}\delta_\alpha^\beta$$
$$\bar\sigma^\mu\sigma^\nu\bar\sigma^\lambda = -\eta^{\mu\nu}\bar\sigma^\lambda + \eta^{\mu\lambda}\bar\sigma^\nu - \eta^{\nu\lambda}\bar\sigma^\mu + i\epsilon^{\mu\nu\lambda\rho}\bar\sigma_\rho$$
$$\sigma^\mu\bar\sigma^\nu\sigma^\lambda = -\eta^{\mu\nu}\sigma^\lambda + \eta^{\mu\lambda}\sigma^\nu - \eta^{\nu\lambda}\sigma^\mu - i\epsilon^{\mu\nu\lambda\rho}\sigma_\rho.$$
$$tr[\sigma^\mu\bar\sigma^\nu\sigma^\lambda\bar\sigma^\rho - \sigma^\lambda\bar\sigma^\nu\sigma^\mu\bar\sigma^\rho] = 4i\epsilon^{\mu\nu\lambda\rho}. \qquad (8.326)$$

3. Given that $\sigma^{\mu\nu}$ and $\bar\sigma^{\mu\nu}$ are defined by

$$\sigma^{\mu\nu} = \frac{i}{4}(\sigma^\mu\bar\sigma^\nu - \sigma^\nu\bar\sigma^\mu) \qquad (8.327)$$

$$\bar\sigma^{\mu\nu} = \frac{i}{4}(\bar\sigma^\mu\sigma^\nu - \bar\sigma^\nu\sigma^\mu), \qquad (8.328)$$

show that $\sigma^{\mu\nu}$ ($\bar\sigma^{\mu\nu}$) satisfy the duality (anti-duality) property that

$$\sigma^{\mu\nu} = \frac{1}{2i}\epsilon^{\mu\nu\lambda\rho}\sigma_{\lambda\rho} \qquad (8.329)$$

$$\bar\sigma^{\mu\nu} = -\frac{1}{2i}\epsilon^{\mu\nu\lambda\rho}\bar\sigma_{\lambda\rho}. \qquad (8.330)$$

4. Show that the tensor $A_{\alpha\beta\dot\alpha\dot\beta}$ can be decomposed into its irreducible parts as follows:

$$A_{\alpha\beta\dot\alpha\dot\beta} = \epsilon_{\alpha\beta}A_{\{\dot\alpha\dot\beta\}} + \epsilon_{\dot\alpha\dot\beta}A_{\{\alpha\beta\}} + \epsilon_{\alpha\beta}\epsilon_{\dot\alpha\dot\beta}A + A_{\{\alpha\beta\}\{\dot\alpha\dot\beta\}}. \qquad (8.331)$$

Determine A, $A_{\{\alpha\beta\}}$, $A_{\{\dot\alpha\dot\beta\}}$, and $A_{\{\alpha\beta\}\{\dot\alpha\dot\beta\}}$ in terms of the components of the original tensor $A_{\alpha\beta\dot\alpha\dot\beta}$.

5. Consider the operators $D_\alpha(x,\theta,\bar\theta)$ and $\bar D_{\dot\alpha}(x,\theta,\bar\theta)$ defined by

$$D_\alpha(x,\theta,\bar\theta) = \frac{\partial}{\partial\theta^\alpha} + i(\sigma^\mu)_{\alpha\dot\alpha}\bar\theta^{\dot\alpha}\frac{\partial}{\partial x^\mu} \qquad (8.332)$$

$$\bar D_{\dot\alpha}(x,\theta,\bar\theta) = -\frac{\partial}{\partial\bar\theta^{\dot\alpha}} - i\theta^\alpha(\sigma^\mu)_{\alpha\dot\alpha}\frac{\partial}{\partial x^\mu}. \qquad (8.333)$$

Defining the variables $z^\mu = x^\mu - i\theta\sigma^\mu\bar\theta$, $\theta' = \theta$, and $\bar\theta' = \bar\theta$, show that $D_\alpha^R(z,\theta',\bar\theta') \equiv D_\alpha(x,\theta,\bar\theta)$ and $\bar D_{\dot\alpha}^R(z,\theta',\bar\theta') \equiv \bar D_{\dot\alpha}(x,\theta,\bar\theta)$ can be written as follows:

$$D_\alpha^R(z, \theta', \bar{\theta}') = \frac{\partial}{\partial\theta'^\alpha} \tag{8.334}$$

$$\bar{D}_{\dot{\alpha}}^R(z, \theta', \bar{\theta}') = -\frac{\partial}{\partial\bar{\theta}'^{\dot{\alpha}}} - 2i\theta'^\alpha(\sigma^\mu)_{\alpha\dot{\alpha}}\frac{\partial}{\partial z^\mu}. \tag{8.335}$$

6. Consider the free Lagrangian for the right chiral fields $\bar{\Phi}_R$, which we take to have the form

$$L_{\bar{\Phi}_R} = -\partial_\mu\bar{\phi}^\dagger\partial^\mu\bar{\phi} + \frac{i}{2}\partial_\mu\bar{\psi}(x)\sigma^\mu\bar{\psi}^\dagger(x) - \frac{i}{2}\bar{\psi}(x)\sigma^\mu\partial_\mu\bar{\psi}^\dagger(x) + \bar{F}^\dagger(x)\bar{F}(x). \tag{8.336}$$

Show that under supersymmetry transformations, variation of the Lagrangian is a total divergence.

7. Show that in the total derivative term $\partial_\mu D^\mu$ in Eq. (8.128) D^μ is given by

$$D^\mu = -\frac{1}{4}\phi_a(x)\partial^\mu\phi_a^*(x) + \frac{3}{4}\phi_a^*\partial^\mu\phi_a(x) - \frac{i}{2}\bar{\psi}_a(x)\bar{\sigma}^\mu\psi_a(x). \tag{8.337}$$

8. Show that under simultaneous supersymmetry transformations and gauge transformations which preserve the Wess–Zumino gauge, a chiral superfield with components $(\phi_i(x), \psi_i(x), F_i(x))$ has the following transformations:

$$\delta_{sg}\phi_i(x) = \sqrt{2}\epsilon\psi_i(x)$$
$$\delta_{sg}\psi_{\alpha i}(x) = \sqrt{2}\epsilon_\alpha F_i(x) + i\sqrt{2}(\sigma^\mu\bar{\epsilon})_\alpha D_{\mu ij}\phi_j(x)$$
$$\delta_{sg}F_i(x) = ig\epsilon\lambda_{ij}(x)\phi_j(x) - i\sqrt{2}D_{\mu ij}\psi_j\sigma^\mu\bar{\epsilon}. \tag{8.338}$$

Here, $D_{\mu ij} = \delta_{ij}\partial_\mu + (g/2)(t_a)_{ij}v_{\mu a}$ and $\delta_{gs} = \delta_s + \delta_g$, where δ_s is the supersymmetry transformation characterized by the parameter ξ and δ_g is the gauge transformation, which is chosen so that the Wess–Zumino gauge is preserved under the simultaneous supersymmetry and gauge transformations.

9. Prove the constraints of Eq. (8.172).

10. Consider the Wess–Zumino gauge where the vector multiplet is given by

$$V_a(WZ) = -(\theta\sigma^\mu\bar{\theta})v_{\mu a}(x) + i\theta^2(\bar{\theta}\bar{\lambda}_a(x)) - i\bar{\theta}^2(\theta\lambda_a(x)) + \frac{1}{2}\theta^2\bar{\theta}^2 D_a(x). \tag{8.339}$$

Show that the following results hold:

$$(\bar{D}^2 D_\alpha V_a)_0 = 4i\lambda_{\alpha a}$$
$$(D^\alpha\bar{D}^2 D_\alpha V_a) = 8D_a$$
$$[D_\alpha, \bar{D}_{\dot{\alpha}}]V_a = -2(\sigma^\mu)_{\alpha\dot{\alpha}}v_{\mu a}$$
$$(D^\alpha\bar{D}^2 D_\beta V_a)_0 = 4\delta_\beta^\alpha D_a - 2i(\sigma^\mu\bar{\sigma}^\nu)_\beta^\alpha(\partial_\mu v_{\nu a} - \partial_\nu v_{\mu a}), \tag{8.340}$$

where the subscript 0 on the brackets means that the brackets are evaluated at $\theta = 0 = \bar{\theta}$.

11. For the vector multiplet in the Wess–Zumino gauge, show the following:

$$[f_{abc}\bar{D}^2 V_b (D_\alpha V_c)]_0 = 0$$

$$[f_{abc} D_\alpha \bar{D}^2 V_b (D_\beta V_c)]_0 = -2 f_{abc} \epsilon_{\alpha\gamma} (\sigma^\mu \bar{\sigma}^\nu)^\gamma_\beta v_{\mu b} v_{\nu c}$$

$$[f_{abc} D^2 \bar{D}^2 V_b (D_\alpha V_c)]_0 = 16i f_{abc} (\sigma^\mu)_{\alpha\dot\beta} \bar\lambda_b^{\dot\beta} v_{\mu c} \tag{8.341}$$

12. Carry out a similar analysis for case (B) in Eq. (8.267), where $2\mu^2\lambda^2 > m^2$ and the fields ϕ_i have the following VEVs:

$$\bar\phi_2 = -\frac{2\lambda}{\mu}\bar\phi_1\bar\phi_3, \quad \bar\phi_3 = \left(\mu^2 - \frac{m^2}{2\lambda^2}\right)^{\frac{1}{2}}. \tag{8.342}$$

Show that in the fermionic sector there are two Majorana spinors with masses

$$\lambda\bar\phi_1 \pm (\lambda^2\phi_1^2 + 4\lambda^2\mu^2 - m^2)^{\frac{1}{2}}, \tag{8.343}$$

in addition to the massless Goldstino, which is a Weyl spinner. Show that in the bosonic sector one has two massive CP-even states, and determine their masses. Check if the supertrace relation $STr(M^2) = 0$ holds.

13. Verify the mass spectrum given in Eq. (8.299).

14. Prove the result of Eq. (8.307).

References

[1] J. Schwinger, *Ann. Phys.*, **2**, 407–434 (1967).

[2] P. Ramond, *Phys. Rev. D* **3**, 2415 (1971).

[3] Yu A. Golfand and E.P. Likhtman, *JETP Lett.* **13**, 323 (1971).

[4] D. Volkov and V.P. Akulov, *JETP Lett.* **16**, 438 (1972).

[5] J. Wess and B. Zumino, *Nucl. Phys. B* **78**, 1 (1974).

[6] J. Wess and B. Zumino, *Nucl. Phys. B* **70**, 39 (1974).

[7] A. Salam and J. A. Strathdee, *Phys. Rev. D* **11**, 1521 (1975).

[8] A. Salam and J. A. Strathdee, *Nucl. Phys. B* **86**, 142 (1975).

[9] R. L. Arnowitt, P. Nath, and B. Zumino, *Phys. Lett. B* **56**, 81 (1975).

[10] E. Gildener, *Phys. Rev. D* **14**, 1667 (1967); S. Weinberg, *Phys. Lett. B* **82**, 387 (1979).

[11] M. T. Grisaru, W. Siegel, and M. Rocek, *Nucl. Phys. B* **159**, 429 (1979).

[12] S. R. Coleman and J. Mandula, *Phys. Rev.* **159**, 1251 (1967).

[13] S. Weinberg and E. Witten, *Phys. Lett. B* **96**, 59 (1980).

[14] M. T. Grisaru, H. N. Pendleton, and P. van Nieuwenhuizen, *Phys. Rev. D* **15**, 996 (1977).

[15] R. Haag, J. T. Lopuszanski, and M. Sohnius, *Nucl. Phys. B* **88**, 257 (1975).

[16] P. Fayet and J. Iliopoulos, *Phys. Lett. B* **51**, 461 (1974).

[17] S. Weinberg, *Phys. Lett. B* **62**, 111 (1976).

[18] E. Witten, *Nucl. Phys. B* **202**, 253 (1982).

[19] L. O'Raifeartaigh, *Nucl. Phys. B* **96**, 331 (1975).

[20] L. Girardello and M. T. Grisaru, *Nucl. Phys. B* **194**, 65 (1982).

[21] J. Hirn and J. Stern, *Eur. Phys. J. C* **34**, 447 (2004).

[22] A. Kobakhidze, N. Pesor, and R. R. Volkas, *Phys. Rev. D* **79**, 075022 (2009).

[23] A. Salam and J. A. Strathdee, *Nucl. Phys. B* **87**, 85 (1975).

[24] P. Fayet, *Nucl. Phys. B* **90**, 104 (1975).

[25] G. R. Farrar and S. Weinberg, *Phys. Rev. D* **27**, 2732 (1983).

[26] P. Fayet and S. Ferrara, *Phys. Rep.* **32**, 249 (1977).

[27] M. B. Green and J. H. Schwarz, *Phys. Lett. B* **149**, 117 (1984).

[28] A. H. Chamseddine and H. K. Dreiner, *Nucl. Phys. B* **458**, 65 (1996).

[29] D. J. Castano, D. Z. Freedman, and C. Manuel, *Nucl. Phys. B* **461**, 50 (1996).

[30] W. Siegel, *Phys. Lett. B* **85**, 333 (1979).

[31] P. Binetruy, M. K. Gaillard and T. R. Taylor, *Nucl. Phys. B* **455**, 97 (1995).

[32] C. P. Burgess, J. P. Derendinger, F. Quevedo and M. Quiros, *Ann. Phys.* **250**, 193 (1996).

Further Reading

I. L. Buchbinder and S. Kuzenko, *Ideas and Methods of Supersymmetry and Supergravity*, Institute of Physics (1998).

S. J. Gates, M. T. Grisaru, M. Rocek, and W. Siegel, *Front. Phys.* **58**, 1 (1983).

S. Weinberg, *The Quantum Theory of Fields*, Cambridge University Press (2000).

J. Wess and J. Bagger, *Supersymmetry and Supergravity*, Princeton University Press (1992).

9

Grand Unification

As discussed in Chapter 5, the standard model is described by the gauge group $SU(3)_C \otimes SU(2)_L \otimes U(1)_Y$. It has three coupling constants, g_3, g, and g', where g_3 is the gauge coupling for $SU(3)_C$, g is for $SU(2)_L$, and g' is for $U(1)_Y$. The $SU(2)_L \otimes U(1)_Y$ gauge group unifies the electroweak interactions while $SU(3)_C$ describes the strong interactions. It is reasonable to ask if the standard model is a remnant of a more unified theory valid at a high scale. Thus, just as the group $SU(2)_L \otimes U(1)_Y$ produces the electromagnetic and the weak interactions after spontaneous breaking, it is possible that a larger theory does the same, i.e., that there is a more unified theory which breaks at a higher scale producing the electroweak and the strong interactions. If the standard model gauge group arises from a unified theory, the three coupling constants could be replaced by just one coupling, reducing the number of arbitrary parameters. Further, the more unified framework could resolve another puzzle which relates to the neutrality of ordinary matter: specifically, the charge of the proton is equal and opposite to the charge of the electron to a high degree of accuracy

$$\left| 1 + \frac{Q_e}{Q_p} \right| < 10^{-21}. \tag{9.1}$$

Since the proton is composed of three quarks uud, it provides a strong hint that the quarks and the electron have a common ancestry, which in particle physics implies that they likely arise from a common multiplet.

9.1 $SU(4) \otimes SU(2)_L \otimes SU(2)_R$ Unification

An important step towards unification beyond the standard model was taken by Pati and Salam [1, 2], who proposed the fundamental gauge group to be $G(4, 2, 2) \equiv SU(4) \otimes SU(2)_L \otimes SU(2)_R$. One of the major implications of this proposal is the unification of quarks and leptons. Thus, the quarks and leptons of one generation arise from the multiplets $(4, 2, 1)$ and $(\bar{4}, 1, 2)$ as follows:

$$(4, 2, 1) = \begin{pmatrix} u_1 & u_2 & u_3 & \nu \\ d_1 & d_2 & d_3 & e^- \end{pmatrix} \tag{9.2}$$

$$(\bar{4}, 1, 2) = \begin{pmatrix} d_1^c & d_2^c & d_3^c & e^c \\ u_1^c & u_2^c & u_3^c & \nu^c \end{pmatrix}, \tag{9.3}$$

where u_1, u_2, and u_3 refer to the quark colors. In this model, the leptons appear as the fourth color. Assuming that the unification of quarks and leptons appears at the grand unified theory (GUT) scale, we need to break the symmetry at the GUT scale so that

$$SU(4) \otimes SU(2)_L \otimes SU(2)_R \to SU(3)_C \otimes SU(2)_L \otimes U(1)_Y. \tag{9.4}$$

To break the symmetry at the GUT scale, we need to introduce heavy Higgs representations $(4, 2, 1)$ and $(\bar{4}, 1, 2)$ such that

$$(4, 2, 1)_H = \begin{pmatrix} u_{H1} & u_{H2} & u_{H3} & \nu_H \\ d_{H1} & d_{H2} & d_{H3} & e_H^- \end{pmatrix} \tag{9.5}$$

$$(\bar{4}, 1, 2)_H = \begin{pmatrix} d_{H1}^c & d_{H2}^c & d_{H3}^c & e_H^c \\ u_{H1}^c & u_{H2}^c & u_{H3}^c & \nu_H^c \end{pmatrix}. \tag{9.6}$$

The GUT group $G(4, 2, 2)$ breaks to the standard model gauge group when ν_H and ν_H^c develop a vacuum expectation value (VEV) so that

$$\langle \nu_H \rangle = \langle \nu_H^c \rangle = M_{\nu_H}, \tag{9.7}$$

where M_{ν_H} is the size of the grand unification scale. Under the breaking pattern Eq. (9.4) the $G(4, 2, 2)$ multiplets can be decomposed as follows:

$$(4, 2, 1) = \left(3, 2, \frac{1}{6}\right) \oplus \left(1, 2, -\frac{1}{2}\right) \tag{9.8}$$

$$(\bar{4}, 1, 2) = \left(\bar{3}, 1, \frac{1}{3}\right) \oplus \left(\bar{3}, 1. -\frac{2}{3}\right) \oplus (1, 1, 1) + \oplus(1, 1, 0), \tag{9.9}$$

where Eq. (9.8) gives the $SU(2)_L$ doublets of quarks u_L and d_L and of leptons e_L, and ν_L, and Eq. (9.9) gives the quark–lepton singlets (d^c, u^c, e^c, ν^c). Of course, additional Higgs fields are needed to break the symmetry down to the residual gauge group $SU(3)_C \otimes U(1)_{em}$:

$$SU(3)_C \otimes SU(2)_L \otimes U(1)_Y \to SU(3)_C \otimes U(1)_{em}. \tag{9.10}$$

This is accomplished by introducing a Higgs representation, which is $(1, 2, 2)$, so that we can accommodate the two minimal supersymmetry standard model (MSSM) Higgs doublets in it. Thus, we have

$$(1, 2, 2) = \begin{pmatrix} H_2^+ & H_1^0 \\ H_2^0 & H_1^- \end{pmatrix}. \tag{9.11}$$

The symmetry breaking of Eq. (9.10) is accomplished when H_1^0 and H_2^0 develop VEVs. In $G(4, 2, 2)$ the following charge formula holds:

$$Q_{em} = T_{3L} + T_{3R} + \frac{B - L}{2}. \tag{9.12}$$

From Eq. (9.12) we see that the standard model hypercharge Y is given by

$$Y = T_{3R} + \frac{B - L}{2}. \tag{9.13}$$

The group $G(4, 2, 2)$ can arise from the breaking of $SO(10)$. Thus, the minimal set of Higgs representations needed to break the $SO(10)$ symmetry down to the standard model gauge group and then to the residual gauge group $SU(3)_C \otimes U(1)_Y$ is the set

$$210_H \oplus 126_H \oplus \overline{126}_H \oplus 10_H. \tag{9.14}$$

Here, the breaking pattern is as follows:

$$
\begin{aligned}
SO(10) &\to SU(4)_C \otimes SU(2)_L \otimes SU2)_R && \langle 210_H \rangle \\
&\to SU(3)_C \otimes SU(2)_L \otimes U(1)_Y && \langle 126_H \rangle = \langle \overline{126}_H \rangle \\
&\to SU(3)_C \otimes U(1)_{em} && \langle 10_H \rangle.
\end{aligned} \tag{9.15}
$$

Thus, the breaking of $SO(10)$ on the top line of Eq. (9.15) occurs via VEV formation for 210_H, which breaks the group $SO(10)$ giving $SU(4)_C \times SU(2)_L \times SU(2)_R$. The VEV formation of 126_H and $\overline{126}_H$ breaks it down further to the standard model gauge group, as shown in the middle line of Eq. (9.15). Finally, the VEV formation for 10_H breaks the standard model gauge group to the residual gauge group $SU(3)_C \otimes U(1)_{em}$, as shown in the bottom line of Eq. (9.15). Aside from Eq. (9.14) there are other Higgs representations which can be used for the breaking of $SO(10)$ down to the standard model gauge group.

9.2 $SU(5)$ Grand Unification

The Pati–Salam group $G(4, 2, 2)$ is a product group and thus not fully unified. The simplest fully unified group that can accommodate the standard model gauge group is $SU(5)$ proposed by Georgi and Glashow [3]. The two lowest representations of this group are 5 and 10 and their conjugate representations $\bar{5}$ and $\overline{10}$. One combination of five-dimensional and 10-dimensional representations will produce 15 components, which is the same as the number of left chiral fields of the standard model which we specify below in terms of their $SU(3)_C$, $SU(2)_L$, and $U(1)_Y$ quantum numbers. For the leptons we have

$$(\nu_L, e_L) : \quad \left(1, 2, -\frac{1}{2}\right); \quad e_L^c : \quad (1, 1, 1), \tag{9.16}$$

while for the quarks we have

$$(u_L, d_L): \quad \left(3, 2, \frac{1}{6}\right); \quad u_L^c: \quad \left(\bar{3}, 1, -\frac{2}{3}\right); \quad d_L^c: \quad \left(\bar{3}, 1, \frac{1}{3}\right). \quad (9.17)$$

A decomposition of the 5- and 10-plets under $SU(3)_C \otimes SU(2)_L \otimes U(1)_Y$ gives the following:

$$5: \quad \left(3, 1, -\frac{1}{3}\right) \oplus \left(1, 2, \frac{1}{2}\right)$$

$$10: \quad \left(3, 2, \frac{1}{6}\right) \oplus \left(\bar{3}, 1, -\frac{2}{3}\right) + (1, 1, 1). \quad (9.18)$$

A comparison with the standard model field content shows that all of the standard model fields in one generation can be accommodated in $\bar{5} + 10$ such that the $\bar{5}$-plet contains the fields (d_L^c, ν_L, e_L) and the 10-plet contains the fields (u_L, d_L, u_L^c, e_L^c). We need three copies to have three generations. Additionally, we need Higgs fields to break the $SU(5)$ gauge symmetry down to the gauge symmetry $SU(3)_C \otimes SU(2)_L \otimes U(1)_Y$ and to further break down $SU(2)_L \otimes U(1)_Y$ to $U(1)_{em}$. The breaking of the grand unified symmetry $SU(5)$ can be accomplished by a scalar field Σ_j^i ($i, j = 1, 2, \cdots, 5$) which is 24-dimensional and belongs to the adjoint representation of the gauge group. To break the $SU(2)_L \otimes U(1)_Y$ symmetry, one needs a 5-plet Higgs field H^i ($i = 1, 2, \cdots, 5$). The unified $SU(5)$ model then contains the following set of matter and Higgs fields:

$$\bar{\psi}_{i\acute{a}}[\bar{5}_{\acute{a}}], \quad \psi_{\acute{a}}^{ij}[10_{\acute{a}}], \quad H^i[5], \quad \Sigma_j^i[24], \quad (9.19)$$

where $\psi_{\acute{a}}^{ij}$ in an anti-symmetric tensor $\psi_{\acute{a}}^{ij} = -\psi_{\acute{a}}^{ji}$ and $\acute{a}$ is a generation index and assumes values $\acute{a} = 1, 2,$ and 3. The charge quantization follows simply. Using $Q = T_3 + Y$ one finds that $Q(e) = -1$ and $Q(d^c) = \frac{1}{3}$. The above result can be understood in another way. For a semi-simple group the trace of any generator acting on a representation of the group vanishes. Since Q is a linear combination of the generators of the group, the same result holds for the charge operator. This means that the charges in any given representation must sum up to zero, and our explicit example demonstrates that. Closely related to the quantization of charge in GUTs is the existence of magnetic monopoles. Thus, in the context of Maxwell electromagnetism, it was shown by Dirac [4, 5] that the existence of magnetic monopoles implies the quantization of charge, i.e., one has the relation

$$eg = \frac{1}{2}n\hbar, \quad (9.20)$$

where g is the magnetic charge of the monopole and n is an integer. The above implies that if monopoles exist the charge would be quantized, i.e., it would be a multiple of $1/2g$ which acts as the basic unit of charge. It is interesting to speculate if the converse is true, i.e., that the existence of charge quantization necessarily implies the existence of monopoles. A framework to check this would be unified models. Thus, while the standard electroweak model $SU(2) \times U(1)$ does not have monopoles, the Georgi–Glashow $SO(3)$ electroweak model [6] leads to monopoles [7,8]. Without going into details, these monopoles arise as solutions to the field equations and lead to the relation between e and g of the form

$$eg = n\hbar, \tag{9.21}$$

which is the Schwinger quantization [9]. The Georgi–Glashow electroweak model is ruled out by experiments on neutral current interactions. Nonetheless, it provides the simplest example of how a monopole arises in non-abelian gauge theories. The $SU(5)$ grand unification also contains monopoles [10]. The monopoles arising in GUTs are expected to be on the grand unification scale in size. The monopoles, if they exist, would be produced in the early universe and could overclose the universe. In Chapter 17, we will see that one needs an inflationary universe to resolve this problem. Currently there is no experimental evidence for the existence of monopoles.

The $SU(5)$ symmetry holds at the grand unification scale, and it must be broken down to the standard model gauge group at low energy and further down to the residual gauge group $SU(3)_C \times U(1)_{em}$. Thus, consider the scalar potential

$$V = M^2 \, tr \, \Sigma^2 + \frac{\lambda_1}{2} tr(\Sigma^4) + \frac{\lambda_2}{2}(tr \, \Sigma^2)^2 + \mu \, tr(\Sigma^3) + \frac{1}{2}\lambda_3 H^\dagger H \, tr(\Sigma^2)$$

$$+ \frac{1}{2}\lambda_4 H^\dagger \Sigma^2 H + \frac{\lambda}{4}(H^\dagger H)^2. \tag{9.22}$$

In order that the GUT group breaks to the standard model gauge group, the VEV formation of Σ should have the form

$$\langle \Sigma \rangle = diag\,(2, 2, 2, -3, -3)v. \tag{9.23}$$

If we were to take a VEV formation of the form

$$\langle \Sigma \rangle = diag\,(1, 1, 1, 1, -4)v, \tag{9.24}$$

it would break $SU(5)$ to only $SU(4) \otimes U(1)$. Thus, we consider the first case, and here the spontaneous breaking of the symmetry leads to the condition

$$M^2 + (7\lambda_1 + 30\lambda_2)v^2 - \frac{3}{2}\mu v = 0. \tag{9.25}$$

A deduction of Eq. (9.25) is given in Section 9.8.3, where the technique of projection operators to reduce $SU(5)$-invariant couplings in terms of $SU(3)_C \otimes SU(2)_L \otimes U(1)_Y$ irreducible tensors is discussed. One serious problem, however, is that the breaking leads to a super-massive Higgs doublet. Now, of course, we need a light Higgs doublet to achieve an electroweak symmetry breaking. This can be achieved by setting $10\lambda_3 + 3\lambda_4 = 0$, and this relation must be satisfied to approximately one part in 10^{28} in order that the Higgs doublet field has a mass in the $\mathcal{O}(100)$ GeV range while the color Higgs triplets, which have the

$SU(3)$, $SU(2)$, and $U(1)$ quantum numbers $(3, 1, 0) + (\bar{3}, 1, 0)$, are left super-heavy. Additionally, one has $(3, 2, 5/3) + (\bar{3}, 2, -5/3)$ vector bosons, which are supermassive with mass $M_V = 5\sqrt{2}gv$, and $(8, 1, 0)$ and $(3, 1, 0)$ Σ-fields, which are supermassive with mass $M_\Sigma = 5\lambda_1 v/2$, as well as a singlet Σ- field of mass $M_\Sigma/5$, where v is as defined in Eq. (9.23).

In $SU(5)$ we can define the analogue of the Gell–Mann matrices for $SU(3)$. We label these as λ_A so that $\lambda_A/2$ are the five-dimensional representation of the $SU(5)$ generators. We will discuss them in detail later. Here, of special interest to us is λ_{24}, which is given by

$$\lambda_{24} = \frac{1}{\sqrt{15}} \, diag\,(2, 2, 2, -3, -3). \tag{9.26}$$

The hypercharge generator Y commutes with all the $SU(3) \otimes SU(2)$ generators, and can be defined so that

$$Y = diag\left(-\frac{1}{3}, -\frac{1}{3}, -\frac{1}{3}, \frac{1}{2}, \frac{1}{2}\right). \tag{9.27}$$

Thus, the charge operator in $SU(5)$ reads

$$Q = \frac{1}{2}\lambda_{23} - \frac{1}{2}\sqrt{\frac{5}{3}}\lambda_{24}, \tag{9.28}$$

which is consistent with the definition $Q = T_3 + Y$ where λ_{23} is defined in Eq. (9.50). Now, the hypercharge couplings are defined by

$$g_Y \bar{f}\gamma^\mu Y f B_\mu. \tag{9.29}$$

Identifying V_{24}^μ with the field B_μ gives us the connection between g_Y and g_5 so that

$$g_Y = \sqrt{\frac{3}{5}}g_5. \tag{9.30}$$

The weak angle θ_W is defined in the electroweak theory in terms of the $SU(2)_L$ gauge coupling constant g_2 and the hypercharge gauge coupling constant g_Y so that

$$\sin^2\theta_W = \frac{g_Y^2}{g_2^2 + g_Y^2}. \tag{9.31}$$

In the $SU(5)$ unified theory, $g_2 = g_5$, and along with Eq. (9.30) the above immediately gives

$$\sin^2\theta_W = \frac{3}{8}. \tag{9.32}$$

However, this is the value of $\sin^2\theta_W$ at the GUT scale. It must be renormalized down to obtain its value at the electroweak scale to compare it with its experimental value, which is $\sim$0.23. The fit of the supersymmetric theory (to be discussed later) to this quantity is better than for the standard model.

We now discuss a related issue, which is the unification of the gauge coupling constants. Very accurate measurements of the coupling constants now exist owing

to the Large Electron–Positron Collider (LEP) experiment. The current evalua-
tions at the M_Z scale are

$$\alpha_1(M_Z) = 0.016985 \pm 0.000020$$
$$\alpha_2(M_Z) = 0.03358 \pm 0.00011$$
$$\alpha_3(M_Z) = 0.118 \pm 0.007, \tag{9.33}$$

where $\alpha_1 \equiv (5/3)\alpha_Y$. The values of the gauge coupling constants can be obtained
at other mass scales using the renormalization group (RG) equations. In this way,
one can verify if they meet at a high scale. The evolution of the gauge coupling
constants is of the form

$$\mu \frac{dg_i(\mu)}{d\mu} = \beta_i(g_i(\mu)), \tag{9.34}$$

where i takes on the values 1, 2, and 3 for the gauge groups $U(1)_1$, $SU(2)$, and
$SU(3)_C$, respectively. At the one-loop level, β_i are given by [11–13]

$$\beta_i(g_i(\mu)) = \frac{g_i^3}{16\pi^2} \left(-\frac{11}{3} C_2(G_i) + \frac{2}{3} T_i(F)d(F) + \frac{1}{3} T(S)d(S) \right), \tag{9.35}$$

where we assume that the fermion representations (F) are chiral and the scalar
representations (S) are complex. In the above, the first term in the bracket on the
right-hand side is the contribution to the beta function from the gauge bosons
corresponding to the gauge group G_i, the second term is the contribution arising
from fermions in the representation F, and the last term is the contribution from
scalars in the representation S. The quantities $d(F)$ and $d(S)$ are the dimensionalities for the representations F and S. In the above, $C_2(R)$ is the quadratic
Casimir for the representation R, and $T(R)$ is defined so that

$$tr(T^a T^b) = T(R)\delta^{ab}, \tag{9.36}$$

where T^a are the matrix representations of the generators in the representation
R with dimensionality $d(R)$. $T(R)$ and $C_2(R)$ are related by the following:

$$C_2(R)d(R) = rT(R), \tag{9.37}$$

where r is the number of generators of the group. Now, for the group $SU(N)$ for
the fundamental representation, which is of dimensionality N, $T(R) = 1/2$, and
for the adjoint representation, which is of dimensionality $N^2 - 1$, $T(R) = N$.
For the $U(1)_Y$ case, $C_2 = 0$ and $T(R) = Y^2$, where Y is normalized such that
the charge operator is given by $Q = T_3 + Y$ where T_3 is the third component of
$SU(2)_L$.

For comparison with the LEP data, it is important to take into account the
two-loop corrections in the gauge-coupling constant evolution. Thus, up to two-
loop order the RG equations for $\alpha_i(\mu) \equiv g_i^2/(4\pi)$ ($i = 1, 2, 3$) are given by

$$\mu \frac{d\alpha_i(\mu)}{d\mu} = -\frac{1}{2\pi} \left[b_i + \frac{1}{4\pi} \sum_j b_{ij}\alpha_j(\mu) \right] \alpha_i^2(\mu), \tag{9.38}$$

where b_i are related to β_i by $\beta_i = \alpha_i b_i/(4\pi)$, and b_{ij} are the two-loop beta functions. First, let us consider the non-supersymmetric case. Here, b_i are given by

$$b_i = (0, -22/3, -11) + N_G(4/3, 4/3, 4/3) + N_H(1/10, 1/6, 0), \qquad (9.39)$$

where N_G is the number of generations and N_H is the number of Higgs doublets. For the standard model case, $N_G = 3$ and $n_H = 1$, which gives

$$b_i = \left(\frac{41}{10}, -\frac{19}{6}, -7\right). \qquad (9.40)$$

The two-loop contribution is found to be significant, and the relevant b_{ij} are

$$b_{ij} = \begin{pmatrix} 0 & 0 & 0 \\ 0 & -136/3 & 0 \\ 0 & 0 & -102 \end{pmatrix} + N_G \begin{pmatrix} 19/15 & 3/5 & 44/15 \\ 1/5 & 49/3 & 4 \\ 11/30 & 3/2 & 76/3 \end{pmatrix}$$

$$+ N_H \begin{pmatrix} 9/50 & 9/10 & 0 \\ 3/10 & 13/6 & 0 \\ 0 & 0 & 0 \end{pmatrix}. \qquad (9.41)$$

Including the two-loop corrections for the case of the standard model, one finds that an upward extrapolation of the gauge couplings does not lead to a simultaneous meeting of all the three constants. This is shown in Fig. 9.1.

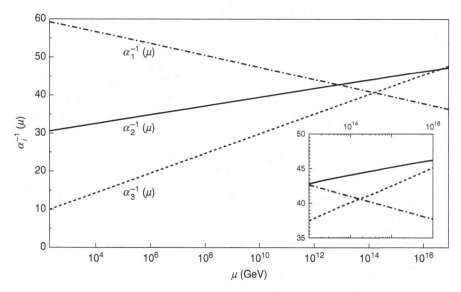

Figure 9.1. $\alpha_i(\mu)$ for the standard model with one Higgs doublet showing that the coupling constants do not unify. (From Amaldi *et al.* [14]).

We now consider Yukawa couplings for the matter fields with the 5-plet of the Higgs field. We write the interaction in the form

$$\mathcal{L}_Y = \lambda_d \bar{\psi}_i^c \psi^{ij} H_j^\dagger + \lambda_u \epsilon_{ijklm} \bar{\psi}^{cij} \psi^{kl} H^m + \text{h.c.}, \qquad (9.42)$$

where ψ_i^c is the $\bar{5}$-plet and ψ^{ij} is the 10-plet of fermions, where we have suppressed the generation indices, and H^i is the 5-plet of Higgs. Applying the VEV to the neutral component of the Higgs field, one finds several mass relations:

$$m_e = m_d, \ m_\mu = m_s, \ m_b = m_\tau. \qquad (9.43)$$

These relations hold at the GUT scale, and must be evolved down to low energies by the renormalization group equations. However, the ratio

$$\frac{m_e}{m_\mu} = \frac{m_d}{m_s} \qquad (9.44)$$

is to one-loop order independent of the scale and can be tested at low energies. It is obvious that this relation is invalid by a large amount. Thus, the minimal $SU(5)$ as a theory of flavor is inadequate.

We now discuss the $SU(5)$ generators in greater detail. Here, one procedure is to start with the generators of $U(5)$ and then extract the $SU(5)$ generators. We can illustrate this procedure more generally for $SU(n)$. First, consider the generators of $U(n)$. This can be done by defining a set of generators consisting of the following set:

$$U_a(a = 1 - n^2): \quad U_j^i + U_j^{i\dagger}, -i(U_j^i - U_j^{i\dagger}), i, j = 1 - n, i \neq j$$
$$U_i^i, \quad i = 1 - n, \qquad (9.45)$$

where

$$(U_j^i)_m^k = \delta_{ik}\delta_{jm}. \qquad (9.46)$$

The U satisfies the property that $U_j^{i\dagger} = U_i^j, tr(U_i^i) = 1$. It is easy to see that each of the n^2 generators U_a $(a = 1 - n^2)$ are Hermitian. The $n \times n$ matrix representation of the $SU(n)$ generators can be obtained from the above by making the matrices traceless. This can be done by defining

$$T_j^i = U_j^i - \frac{\delta_j^i}{n}\mathbb{1}. \qquad (9.47)$$

All the T_j^i are now traceless. For $SU(5)$ it is convenient to define the set T_A $(A = 1 - 24)$ such that

$$T_A = \frac{1}{2}\lambda_A, \quad A = 1 - 24. \qquad (9.48)$$

We define λ_A, $A = 1 - 8$ so that

$$\lambda_A = \begin{pmatrix} (\lambda_A)_{3\times 3} & (0)_{3\times 2} \\ (0)_{2\times 3} & (0)_{2\times 2} \end{pmatrix}, \qquad (9.49)$$

where $(\lambda_A)_{3\times3}$ $(A = 1, \ldots, 8)$ are the Gell–Mann matrices, and we define the remaining λ as follows:

$$\lambda_9 = U_4^1 + U_1^4, \ \lambda_{10} = -i(U_4^1 - U_1^4), \lambda_{11} = (U_5^1 + U_1^5), \ \lambda_{12} = -i(U_5^1 - U_1^5)$$
$$\lambda_{13} = U_4^2 + U_2^4, \ \lambda_{14} = -i(U_4^2 - U_2^4), \lambda_{15} = (U_5^2 + U_2^5), \ \lambda_{16} = -i(U_5^2 - U_2^5)$$
$$\lambda_{17} = U_4^3 + U_3^4, \ \lambda_{18} = -i(U_4^3 - U_3^4), \lambda_{19} = (U_5^3 + U_3^5), \ \lambda_{20} = -i(U_5^3 - U_3^5)$$
$$\lambda_{21} = U_5^4 + U_4^5, \ \lambda_{22} = -i(U_5^4 - U_4^5), \lambda_{23} = (U_4^4 - U_5^5). \tag{9.50}$$

with λ_{24} defined so that

$$\lambda_{24} = \frac{1}{\sqrt{15}} \begin{pmatrix} 2 & & & & \\ & 2 & & & \\ & & 2 & & \\ & & & -3 & \\ & & & & -3 \end{pmatrix}, \tag{9.51}$$

and all the off-diagonal elements are trivially traceless. The T_A satisfy the $SU(5)$ algebra, i.e.,

$$[T_A, T_B] = if_{ABC}T_C, \tag{9.52}$$

where f_{ABC} are the structure constants of $SU(5)$. Thus, for $SU(5)$ we have 24 vector gauge bosons in the adjoint representation, and one can define a Lie-valued object,

$$V = \frac{1}{\sqrt{2}} \sum_{1}^{24} \lambda_A V_A, \tag{9.53}$$

which can be expressed in a matrix form so that

$$V = \left[\begin{array}{c|c} V_p & \bar{V}_r \\ \hline V_r & V_q \end{array} \right], \tag{9.54}$$

where V_p is given by

$$V_p = \frac{1}{\sqrt{2}} \sum_{A=1}^{8} (\lambda_A)_{3\times3} G_A + \frac{\sqrt{2}}{\sqrt{15}} B(\mathbb{1})_{3\times3}, \tag{9.55}$$

where G_A $(A = 1-8)$ are the eight gluon fields and $B = V_{24}$. Similarly, V_q are given by

$$V_q = \frac{1}{\sqrt{2}} \sum_{a=1}^{3} \tau_a W_a - \sqrt{\frac{3}{10}} B(\mathbb{1})_{2\times2}, \tag{9.56}$$

where $W_1 = V_{21}$, $W_2 = V_{22}$, and $W_3 = V_{23}$ are the $SU(2)$ gauge bosons, τ_a $(a = 1-3)$ are the Pauli matrices, and $\bar{V}_r$ and V_r are given by

$$V_r = \begin{pmatrix} X_1 & X_2 & X_3 \\ Y_1 & Y_2 & Y_3 \end{pmatrix}, \ \bar{V}_r = \left[\begin{array}{cc} \bar{X}_1 & \bar{Y}_1 \\ \bar{X}_2 & \bar{Y}_2 \\ \bar{X}_3 & \bar{Y}_3 \end{array} \right]. \tag{9.57}$$

The fields X and Y are lepto-quarks and they carry charges.

We can compute the charges of the various components of a representation by the action of the charge operator Q. Assume ϕ^i is a 5-plet representation of $SU(5)$. The charge of the component ϕ^i is then given by

$$Q\phi^i = q^i\phi^i, \tag{9.58}$$

where from Eq. (9.28) one finds

$$\{q^i\} = (-1/3, -1/3, -1/3, 1, 0). \tag{9.59}$$

Further, the action of Q on the tensor ϕ_i is

$$Q\phi_i = q_i\phi_i, \quad q_i = -q^i. \tag{9.60}$$

It is easy to see that the action of Q on a tensor $\phi_{j_1 j_2 \cdots}^{i_1 i_2 \cdots}$ is given by

$$Q\phi_{j_1 j_2 \cdots}^{i_1 i_2 \cdots} = (q^{i_1} + q^{i_2} + \cdots + q_{j_1} + q_{j_2} + \cdots)\phi_{j_1 j_2 \cdots}^{i_1 i_2 \cdots} \tag{9.61}$$

From Eqs. (9.59)–(9.61) we find the following charge assignments for $\bar{X}$ and $\bar{Y}$ and X and Y:

$$V_r = \begin{pmatrix} X_1^{4/3} & X_2^{4/3} & X_3^{4/3} \\ Y_1^{1/3} & Y_2^{1/3} & Y_3^{1/3} \end{pmatrix}, \quad \bar{V}_r = \begin{bmatrix} \bar{X}_1^{-4/3} & \bar{Y}_1^{-1/3} \\ \bar{X}_2^{-4/3} & \bar{Y}_2^{-1/3} \\ \bar{X}_3^{-4/3} & \bar{Y}_3^{-1/3} \end{bmatrix}. \tag{9.62}$$

We now discuss details of how one may write $SU(5)$ gauge-covariant derivatives for matter fields in different representations of $SU(5)$. First, let us consider the fundamental representation ξ^i of $SU(5)$. Under $SU(5)$ gauge transformations it transforms as

$$\xi'^i = U^i{}_j\xi^j, \tag{9.63}$$

where $U = \exp(-ig\epsilon_A T_A)$ and T_A are the $SU(5)$ generators. The derivative of ξ' transforms so that

$$\partial_\mu \xi' = -ig\partial_\mu \epsilon_A T_A U\xi + U\partial_\mu \xi. \tag{9.64}$$

As discussed in Chapter 3, one may now define a covariant derivative $D_\mu \xi$ so that

$$D_\mu \xi = (\partial_\mu + igA_\mu)\xi, \tag{9.65}$$

where with an appropriate transformation for A_μ, $D_\mu \xi$ transforms as a global tensor, i.e., provided

$$A'_\mu = U(A_\mu - ig^{-1}\partial_\mu)U^\dagger \tag{9.66}$$

one has

$$(D_\mu \xi)' = U(D_\mu \xi). \tag{9.67}$$

A $\bar{5}$-plet $\bar{\xi}_i$ of $SU(5)$ will transform as

$$\xi'_i = \xi_j(U^\dagger)^j{}_i, \tag{9.68}$$

and the covariant derivative here is defined so that

$$D_\mu \bar{\xi} = (\partial_\mu - ig A_\mu^T)\bar{\xi}. \tag{9.69}$$

Next, we want to determine the covariant derivative of a tensor with two upper indices. Assume we have a tensor ϕ^{ij} given by

$$\phi^{ij} = \xi^i \eta^j. \tag{9.70}$$

In this case, ϕ^{ij} will transform as follows:

$$\phi^{ij'} = \xi^{i'} \eta^{j'}. \tag{9.71}$$

Using the transformation for ξ and η, we have

$$
\begin{aligned}
\phi^{ij'} &= U^i{}_k \xi^k U^j{}_l \eta^l \\
&= U^i{}_k \xi^k \eta^l (U^T)_l{}^j \\
&= U^i{}_k \phi^{kl} (U^T)_l{}^j.
\end{aligned} \tag{9.72}
$$

It is now straightforward to show that the covariant derivative of ϕ^{ij} is given by

$$(D_\mu \phi)^{ij} = \partial_\mu \phi^{ij} + ig \left(A_{\mu k}^i \phi^{kl} + \phi^{il} (A^T)_l{}^j \right). \tag{9.73}$$

Using the above, we can write the $SU(5)$ gauge-invariant kinetic energy term for the $\bar{5} + 10$ of matter fields. Let $\psi_{\bar{5}}$ be the $\bar{5}$-plet of matter fields and ψ_{10} is a 10-plet of matter fields. The gauge-invariant kinetic terms are given by

$$\mathcal{L}_{\bar{5}+10}^{kin} = -\frac{1}{i} \bar{\psi}_{\bar{5}} \gamma^\mu D_\mu \psi_{\bar{5}} - \frac{1}{i} \bar{\psi}_{10} \gamma^\mu D_\mu \psi_{10}. \tag{9.74}$$

The interaction terms involving the above fields then have the form

$$\mathcal{L}_{\bar{5}+10}^{int} = g \bar{\psi}_{\bar{5}} \gamma^\mu A_\mu^T \psi_{\bar{5}} - g \bar{\psi}_{10} \gamma^\mu A_\mu \psi_{10}. \tag{9.75}$$

Eq. (9.75) can be expanded to give all the relevant gauge interaction for the $\bar{5} + 10$ of matter fields. The gauge-covariant derivative of the 24-plet scalar field of $SU(5)$ can be obtained from Eq. (9.301). Thus, the gauge-covariant kinetic term for the 24-plet is given by

$$\mathcal{L}_{24}^{kin} = -\frac{1}{2} tr((D_\mu \Sigma)^\dagger (D^\mu \Sigma)), \tag{9.76}$$

which leads to the quartic interaction

$$\mathcal{L}_{24}^{int} = -\frac{g_5^2}{2} tr \left([A_\mu, \Sigma]^\dagger [A^\mu, \Sigma] \right). \tag{9.77}$$

Now, let us assume that Σ develops a VEV such that $\langle \Sigma \rangle = v(2, 2, 2, -3, -3)$. In this case the X- and Y-fields develop mass terms so that

$$\mathcal{L}_{X,Y}^m = -(M_X^2 \bar{X}_\mu^i X^{i\mu} + M_Y^2 \bar{Y}_\mu^i Y^{i\mu}), \tag{9.78}$$

where

$$M_X = M_Y = \frac{5}{\sqrt{2}} gv. \tag{9.79}$$

9.3 Supersymmetric Grand Unification

In the grand unification discussed thus far we noted that there were some problems. One of these concerns the non-unification of gauge coupling constants when we extrapolate the low-energy data on the couplings constants to the grand unification scale. This problem is overcome in supersymmetric grand unification, the simplest version of which is supersymmetric $SU(5)$ grand unification. Here, as in the non-supersymmetric case, we have three generations of quarks and leptons and a Higgs field $\Sigma_j^i[24]$ to break the $SU(5)$ GUT symmetry. However, we need to introduce a $H_{1i}[\bar{5}] \oplus H_2^i[5]$ of Higgs fields to break the electroweak symmetry. The reason for this is twofold: first, in the supersymmetric framework, Higgs multiplets have Higgsino fields which are Weyl fermions and contribute to anomalies. These anomalies are automatically cancelled when one chooses a vector-like pair $[\bar{5}] \oplus [5]$. Second, we will see shortly that it is necessary to have a pair of Higgs fields to give masses to the up quarks, the down quarks, and the leptons. Thus, the minimal supersymmetric $SU(5)$ model has the following field content [1, 5, 16]:

$$\hat{\bar{\psi}}_{i\acute{a}}[\bar{5}_{\acute{a}}], \quad \hat{\psi}_{\acute{a}}^{ij}[10_{\acute{a}}], \quad \hat{H}_{1i}[\bar{5}] \quad \hat{H}_2^i[5], \quad \hat{\Sigma}_j^i[24], \tag{9.80}$$

where the "hat" indicates that they are superfields. However, to keep the notation as simple as possible we will not display the hats when there is no possibility of confusion. The field content of the Higgs $H_1[\bar{5}]$ and $H_2[5]$ multiplets is as follows: $H_1[\bar{5}] = (H_{1a}, H_{1\alpha})$, and $H_2[5] = (H_2^a, H_2^\alpha)$, where H_{1a}, H_2^a ($a = 1, 2, 3$) are the Higgs color anti-triplets and triplets and $H_{1\alpha}, H_2^\alpha$ ($\alpha = 4, 5$) are the Higgs doublets which enter into the breaking of the electroweak symmetry.

The superpotential for the minimal supersymmetric $SU(5)$ model involving matter and Higgs fields is given by

$$W = W_G + W_Y, \tag{9.81}$$

where W_G enters in the breaking of the GUT symmetry and has the form

$$W_G = \lambda_1 \left[\frac{1}{3} \, tr \, \Sigma^3 + \frac{1}{2} M \, tr \, \Sigma^2 \right] + \lambda_2 \bar{H}_{1i} [\Sigma_j^i + 2M' \delta_j^i] H_2^j. \tag{9.82}$$

The GUT symmetry breaks when the scalar field Σ develops a VEV, just as in the non-supersymmetric case. Further, a fine tuning is necessary in order that the Higgs doublets be light. This requires the condition $M = M'$, which, of course, leaves the color Higgs triplets and anti-triplets heavy, which is desirable in order to avoid too rapid a proton decay. The W_Y is the superpotential that gives the

Yukawa couplings where

$$W_Y = f_{1\acute{a}\acute{b}}H_{1i}\psi_{\acute{a}j}\psi_{\acute{b}}^{ij} + f_{2\acute{a}\acute{b}}\varepsilon_{ijklm}H_2^i\psi_{\acute{a}}^{jk}\psi_{\acute{b}}^{lm}. \tag{9.83}$$

We could have, in principle, written additional terms in the superpotential, i.e.,

$$f_{3\acute{a}}H_2^i\psi_{i\acute{a}} + f_{4\acute{a}\acute{b}\acute{c}}\psi_{i\acute{a}}\psi_{j\acute{b}}\psi_{\acute{c}}^{ij}. \tag{9.84}$$

Each of these terms violates R-parity, and thus under the assumption of R-party conservation they are not allowed.

Before proceeding further, it is useful to ask how good supersymmetric grand unification is regarding the unification of gauge couplings. For the supersymmetric case there are super partners which contribute to the running of the coupling constants. This modifies b_i and b_{ij} in the renormalization evolution equations. Assuming that all of the super particles are degenerate, b_i and b_{ij} for the supersymmetric case are given by

$$b_i = (0, -6, -9) + n_G(2, 2, 2) + n_H(3/10, 1/2, 0) \tag{9.85}$$

$$b_{ij} = \begin{pmatrix} 0 & 0 & 0 \\ 0 & -24 & 0 \\ 0 & 0 & -54 \end{pmatrix} + n_G \begin{pmatrix} 38/15 & 6/5 & 88/15 \\ 2/5 & 14 & 8 \\ 11/15 & 3 & 68/3 \end{pmatrix}$$

$$+ n_H \begin{pmatrix} 9/50 & 9/10 & 0 \\ 3/10 & 7/2 & 0 \\ 0 & 0 & 0 \end{pmatrix}, \tag{9.86}$$

where $n_G = 3$ and $n_H = 2$ for the case of the minimal supersymmetric $SU(5)$ model. Here, the extrapolation of the gauge-coupling constants shows that unification does occur for the gauge-coupling constants at a common scale of $Q \sim 2 \times 10^{16}$ GeV. It should be emphasized that the unification of the gauge couplings using the precision data occurs for the couplings α_1, α_2, and α_3 where $\alpha_1 = \frac{5}{3}\alpha_Y$, i.e., it is α_1 that unifies with α_2 and α_3 and not α_Y. The reason for this is easily understood since in $SU(5)$ it is α_1 along with α_2 and α_3 that assume the value α_G at the unification scale. There are, of course, several fine points to be taken into account for a more accurate analysis. The first is that the masses of the superparticles are not all the same. There can be a wide dispersion. The low-energy region then consists of different domains, each with its own set of renormalization evolution coefficients b_i, etc. The second issue relates to the high scale. Here, too, one may have different heavy thresholds, and the renormalization group evolution is affected by the heavy thresholds.

We now discuss the inclusion of heavy thresholds for the $SU(5)$ theory with the 24-plet of the gauge vector multiplet and with a matter content of 5, $\bar{5}$, and 24. At the one-loop level, we use a step function approximation, which gives

$$b_1 = \frac{33}{5} - (\Delta b_1)\Theta_S - 10\Theta_V + \frac{2}{5}\Theta_{H_3} \qquad (9.87)$$

$$b_2 = 1 - (\Delta b_2)\Theta_S + 2\Theta_\Sigma - 6\Theta_V \qquad (9.88)$$

$$b_3 = -3 - (\Delta b_3)\Theta_S + 3\Theta_\Sigma - 4\Theta_V + \Theta_{H_3}, \qquad (9.89)$$

where H_3 is the color Higgs triplet and where Δb_i are given by

$$\Delta b_1 = \frac{5}{2}, \ \Delta b_2 = \frac{25}{6}, \ \Delta b_3 = 4. \qquad (9.90)$$

In Eq. (9.87) Θ_S is defined by $\Theta_S \equiv \Theta(M_S - \mu)$, where we have assumed all low-lying sparticles to have the common mass scale M_S. If we are at a renormalization group scale μ which is above the common sparticle mass scale M_S, then $(\Delta b_i)\Theta(M_S - \mu) = 0$ and the one-loop evolution of α_i is controlled by $(b_1, b_2, b_3) = (33/5, 1, -3)$, as can be seen from Eqs. (9.87)–(9.89). On the other hand, if we are below the scale M_S, then we must subtract the supersymmetry contributions which are given by Eq. (9.90). In this case, we recover the result of Eq. (9.40). The b_i given by Eqs. (9.87)–(9.89) also contain corrections from heavy thresholds. Thus, Θ_{H_3} refers to the heavy Higgs triplet threshold so that $\Theta_{H_3} = \Theta(\mu - M_{H_3})$, Θ_Σ is for the heavy Σ threshold, and Θ_V is the threshold for the heavy lepto-quark fields. In the RG evolution one uses the so-called match and run technique between different thresholds. An integration of the one-loop equations gives

$$\alpha_1^{-1} = \alpha_G^{-1} + \frac{1}{2\pi}\left[\frac{33}{5}\ln\frac{M_G}{M_Z} - \frac{5}{2}\ln\frac{M_S}{M_Z} - 10\ln\frac{M_G}{M_V} + \frac{2}{5}\ln\frac{M_G}{M_{H_3}}\right]$$

$$\alpha_2^{-1} = \alpha_G^{-1} + \frac{1}{2\pi}\left[\ln\frac{M_G}{M_Z} - \frac{25}{6}\ln\frac{M_S}{M_Z} - 6\ln\frac{M_G}{M_V} + 2\ln\frac{M_G}{M_\Sigma}\right]$$

$$\alpha_3^{-1} = \alpha_G^{-1} + \frac{1}{2\pi}\left[-3\ln\frac{M_G}{M_Z} - 4\ln\frac{M_S}{M_Z} - 4\ln\frac{M_G}{M_V} + 3\ln\frac{M_G}{M_\Sigma} + \ln\frac{M_G}{M_{H_3}}\right].$$

$$(9.91)$$

From Eq. (9.91) one can derive the relation [17]

$$3\alpha_2^{-1}(M_Z) - 2\alpha_3^{-1}(M_Z) - \alpha_1^{-1}(M_Z) = \frac{1}{2\pi}\left[\frac{12}{5}\ln\frac{M_{H_3}}{M_Z} - 2\ln\frac{M_S}{M_Z}\right]. \qquad (9.92)$$

This relation is interesting, since a knowledge of low-energy data on the gauge-coupling constants along with a knowledge of M_S will give a determination of M_{H_3}, and thus low-energy data become a direct probe of a high scale. Further, M_{H_3} itself is an important high-scale parameter, since it enters into proton decay via dimension-5 operators. Another relation also follows from Eq. (9.91), which reads

$$5\alpha_1^{-1}(M_Z) - 3\alpha_2^{-1}(M_Z) - 2\alpha_3^{-1}(M_Z) = \frac{1}{2\pi}\left[12\ln\frac{M_V^2 M_\Sigma}{M_Z^3} + 8\ln\frac{M_S}{M_Z}\right]. \qquad (9.93)$$

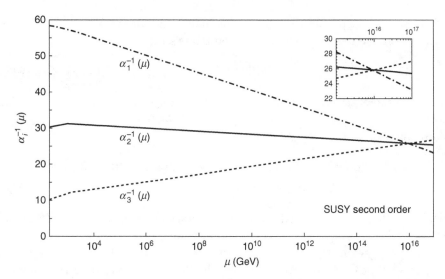

Figure 9.2. $\alpha_i(\mu)$ for MSSM/minimal supergravity model with a pair of Higgs doublets showing that the coupling constants do unify. (From Amaldi *et al.* in [14]).

An exhibition of the unification of the gauge couplings in supersymmetric $SU(5)$ is given in Fig. 9.2. Compared with the analysis of Fig. 9.1, here one finds that the three gauge couplings come together to a very good accuracy [18–22].

9.3.1 Higgs Doublet–Triplet Splitting

As discussed after Eq. (9.25) and Eq. (9.82), one needs significant fine tuning to make the Higgs doublets light and keep the Higgs triplets heavy. While fine tuning is needed in the minimal $SU(5)$ model to get light Higgs doublets, it is possible to achieve this without fine tuning in non-minimal or extended $SU(5)$ models. We discuss two of these extended $SU(5)$ models, one being the so-called missing partner mechanism [23, 24] and the other the flipped $SU(5) \otimes U(1)$ model [25, 26].

The Missing Partner Mechanism

In the missing partner mechanism, one uses a 75-plet representation to break the GUT symmetry and, additionally, a 50-plet and $\overline{50}$ representations to give masses to the color triplets [23, 24]. In this scheme, the Higgs doublets naturally remain light. We now give the details. The 50-, $\overline{50}$- and 75-plets have the following irreducible tensors associated with them:

$$50 = \Phi_{lm}^{ijk}, \quad \Phi_{lm}^{ijm} \equiv 0$$
$$\overline{50} = \overline{\Phi}_{ijk}^{lm}, \quad \overline{\Phi}_{ijm}^{lm} \equiv 0$$
$$75 = \Sigma_{lm}^{ij}, \quad \Sigma_{lj}^{ij} \equiv 0. \tag{9.94}$$

where the subscript and the superscript indices are all anti-symmetric. In this case, we choose the superpotential in the Higgs sector to be of the form

$$W_G = \lambda_1 \overline{\Phi}^{lm}_{ijk} \Sigma^{ij}_{lm} H^k_2 + \lambda_2 \Phi^{ijk}_{lm} \Sigma^{lm}_{ij} \overline{H}_{1k} + M \overline{\Phi}^{lm}_{ijk} \Phi^{ijk}_{lm} + W_\Sigma(\Sigma). \qquad (9.95)$$

The superpotential $W_\Sigma(\Sigma)$ is chosen to produce a GUT symmetry breaking when Σ^{ij}_{kl} develops a non-vanishing VEV of $O(M)$. The $SU(3) \otimes SU(2) \otimes U(1)$ decomposition of the 75-plet is as follows:

$$75 = (1,1)(0) \oplus (3,1)(10) \oplus (3,2)(-5) \oplus (\bar{3},1)(-10) \oplus (\bar{3},2)(5)$$
$$\oplus (\bar{6},2)(-5) \oplus (6,2)(5) \oplus (8,1)(0) \oplus (8,3)(0), \qquad (9.96)$$

where the first bracket specifies the representations under $SU(3) \otimes SU(2)$ and the second bracket gives the $U(1)$ quantum number.

The GUT symmetry breaking occurs when the $(1,1)(0)$ component of the 75-plet develops a VEV. We note that the decomposition above contains the $SU(3)$ Higgs triplets $(3,1)(10)$ and the $SU(3)$ Higgs anti-triplets $(\bar{3},1)(-10)$ but no $SU(2)$ Higgs doublets. Further, the $SU(3) \otimes SU(2) \otimes U(1)$ content of the 5-plet field and the 50-field Φ^{ijk}_{lm} are as follows:

$$5 = (1,2)(3) \oplus (3,1)(-2) \qquad (9.97)$$
$$50 = (1,1)(-12) \oplus (3,1)(-2) \oplus (\bar{3},2)(-7)$$
$$\oplus (\bar{6},3)(-2) \oplus (6,1)(8) \oplus (8,2)(3). \qquad (9.98)$$

It is interesting to note that the 50-plet contains a color triplet field $(3,1)$, and similarly the $\overline{50}$-plet contains a color anti-triplet field $(\bar{3},1)$. These can join to become superheavy via the $M50\overline{50}$ term. However, we note that there is no Higgs doublet field in the 50-plet and similarly no Higgs doublet field in the $\overline{50}$-plet. Consequently, the Higgs doublets in H_1 and H_2 cannot tie up with the heavy Higgs doublets and remain light. We note that after spontaneous breaking when the $(1,1)(0)$ component of the 75-plet develops a VEV, the Higgs color triplets in the 5-plet tie up with the Higgs color anti-triplets in the $\overline{50}$-plet, and the Higgs color anti-triplets in the $\bar{5}$-plet of the Higgs tie up with the Higgs color-triplets in the 50-plet, each becoming supermassive. Similarly, the color triplets and the anti-triplets in the 75-plet become supermassive. However, the Higgs doublets in the 5-plet and $\bar{5}$-plet cannot tie up, and remain massless.

The Flipped $SU(5) \otimes U(1)$

Another method of having light Higgs doublets while keeping the Higgs triplets heavy is to use the flipped $SU(5) \otimes U(1)$. As in the case of $SU(5)$, one uses $\bar{5} \oplus 10$-plet representations of $SU(5)$ for each generation of quarks and leptons. However, the u and the d quarks as well as e and ν leptons are interchanged, the right-handed neutrino ν^c replaces e^c in the 10-plet representation, and e^c appears in the singlet representation. The particle content of the model is given by

$$
\bar{5} = \begin{pmatrix} u^c_{La} \\ e^-_L \\ -\nu_{eL} \end{pmatrix}, \quad 10 = \begin{pmatrix} 0 & d^c_3 & -d^c_2 & -u^1 & -d^1 \\ -d^c_3 & 0 & d^c_1 & -u^2 & -d^2 \\ d^c_2 & -d^c_1 & 0 & -u^3 & -d^3 \\ u^1 & u^2 & u^3 & 0 & \nu^c \\ d^1 & d^2 & d^3 & -\nu^c & 0 \end{pmatrix}_L, \quad 1 = e^c_L. \quad (9.99)
$$

To break the GUT symmetry, one uses $\overline{10}$-fields rather than the 24-plet as in $SU(5)$. For the breaking of the electroweak symmetry, one introduces a $\bar{5} \oplus 5$, i.e., H_{1_i} and H^i_2 as for the standard $SU(5)$ case. The superpotential for the Higgs sector is then

$$
W_{\text{flipped}} = W_0(10) + \lambda_1 \epsilon_{ijklm} H^{ij} H^{kl} H^m_2 + \lambda_2 \epsilon^{ijklm} \bar{H}_{ij} \bar{H}_{kl} \bar{H}_{1m}. \quad (9.100)
$$

The $SU(3) \otimes SU(2) \otimes U(1)$ branchings of the 10-plet of $SU(5)$ are given by

$$
10 = (\bar{3}, 1)(-4) \oplus (3, 2)(1) \oplus (1, 1)(6). \quad (9.101)
$$

In the 10-plet, it is the singlet field $(1,1)$ that grows a VEV, and one finds that the color triplet in the 5-plet combines with the color anti-triplet in 10-plet to become supermassive, while the $SU(2)$ doublet and color singlet $(2,1)$ in the 5-plet have no partner in the 10-plet. Thus, we have a natural missing partner mechanism in this case. The X and Y gauge fields of $SU(5)$ have the $SU(3)_C \otimes SU(2)_L \otimes U(1)_Y$ quantum numbers $(X, Y) = (3, 2, 5/6)$, where the hypercharge is normalized so that $Q = T_3 + Y$, and the charges for them are $Q_X = 4/3, Q_Y = 1/3$. In contrast, the (X', Y') gauge bosons of the flipped $SU(5) \otimes U(1)$ have the quantum numbers $(X', Y') = (3, 2, -1/6)$ so that charges for X' and Y' are given by $Q_{X'} = 1/3$ and $Q_{Y'} = -2/3$, respectively. The unusual charge assignment in this case requires that the hypercharge be a linear combination of $U(1)$ and of generators in $SU(5)$. One drawback of the model is that we do not have a fully unified theory since the underlying structure is based on a product group.

9.4 *SO(10)* Grand Unification [27, 28]

The $O(N)$ group consist of $N \times N$ real orthogonal transformations that preserve the scalar product $x^T y$ of two N-dimensional vectors. Thus, under $O(N)$ transformations a vector x_μ will transform as

$$
x_\mu = O_{\mu\nu} x_\nu, \quad \mu, \nu = 1, \dots, N, \quad (9.102)
$$

where

$$
O^T O = OO^T = \mathbb{1}. \quad (9.103)
$$

Since $det(O) = \pm 1$, we have two disconnected sets of transformations. The case $det(O) = +1$ forms the $SO(N)$ group, and we will focus on this case. The $N \times N$ matrix O can be represented by

$$O = e^{\frac{i}{2} a_{\mu\nu} A_{\mu\nu}}, \tag{9.104}$$

where $a_{\mu\nu}$ is anti-symmetric, i.e., $a_{\mu\nu} = -a_{\nu\mu}$, and has only $N(N-1)/2$ non-vanishing components. The $N \times N$ matrices $A_{\mu\nu}$ are given by

$$(A_{\mu\nu})_{\sigma\rho} = -i(\delta_{\mu\sigma}\delta_{\nu\beta} - \delta_{\nu\sigma}\delta_{\mu\rho}), \quad \mu < \nu, \tag{9.105}$$

and satisfy the following algebra:

$$[A_{\mu\nu}, A_{\sigma\rho}] = -i\,[\delta_{\mu\rho}A_{\nu\sigma} + \delta_{\nu\sigma}A_{\mu\rho} - \delta_{\mu\sigma}A_{\nu\rho} - \delta_{\nu\rho}A_{\mu\sigma}]. \tag{9.106}$$

We can also construct more general tensors $t_{\mu\nu\lambda\cdots}$ which transform as follows:

$$t'_{\mu\nu\lambda\cdots} = O_{\mu\alpha}O_{\nu\beta}O_{\lambda\gamma}\cdots t_{\alpha\beta\gamma\cdots} \tag{9.107}$$

We now consider $SO(2N)$ groups. In addition to the tensor representations, an $SO(2N)$ group also has spinor representations. This can be easily seen by writing the length in the $2N$-dimensional vector space as follows:

$$\sum_{\mu=1}^{2N} x_\mu^2 = \left(\sum_{i=1}^{2N} \Gamma_\mu x_\mu\right)^2, \tag{9.108}$$

under the constraints

$$\Gamma_\mu \Gamma_\nu + \Gamma_\nu \Gamma_\mu = 2\delta_{\mu\nu}\mathbb{1}, \quad \mu,\,\nu = 1,\,\cdots,2N. \tag{9.109}$$

Here, $\Gamma_\mu^2 = \mathbb{1}$ and $tr(\Gamma_\mu) = 0$, which implies that the Γ_μ should have an even dimensionality.

The explicit representation of the Clifford algebra of Eq. (9.109) for $SO(2N)$ can be obtained in terms of $2^N \times 2^N$ dimensional matrices. This can be done by using the two-dimensional unit matrix $\mathbb{1}$ and the Pauli matrices

$$\mathbb{1} = \begin{pmatrix} 1 & 0 \\ 0 & 1 \end{pmatrix}, \quad \sigma_1 = \begin{pmatrix} 0 & 1 \\ 1 & 0 \end{pmatrix}, \quad \sigma_2 = \begin{pmatrix} 0 & -i \\ i & 0 \end{pmatrix}, \quad \sigma_3 = \begin{pmatrix} 1 & 0 \\ 0 & -1 \end{pmatrix},$$
$$\tag{9.110}$$

where $\{\sigma_a, \sigma_b\} = 2\delta_{ab}$. An explicit representation of the gamma matrices for the case $SO(2N)$ can be constructed by multiplication of N number of 2×2 matrices as follows:

$$\gamma_{2k-1}^{(N)} = (\mathbb{1} \otimes \mathbb{1} \otimes \cdots \mathbb{1})_{k-1} \otimes \sigma_1 \otimes (\sigma_3 \otimes \sigma_3 \otimes \cdots \otimes \sigma_3)_{N-k}$$
$$\gamma_{2k}^{(N)} = (\mathbb{1} \otimes \mathbb{1} \otimes \cdots \mathbb{1})_{k-1} \otimes \sigma_2 \otimes (\sigma_3 \otimes \sigma_3 \otimes \cdots \otimes \sigma_3)_{N-k}$$
$$k = 1, 2, \cdots, N, \tag{9.111}$$

where the subscript indicates the number of factors. From the above, one finds that

$$\sigma^{(N)} = \frac{i}{2}[\gamma_{2k-1}^{(N)}, \gamma_{2k}^{(N)}] = -(1 \otimes 1 \otimes \cdots 1)_{k-1} \otimes \sigma_3 \otimes (1 \otimes 1 \otimes \cdots 1)_{n-k}. \tag{9.112}$$

An apparent weakness of this approach is that there are many different choices one can make in forming products such as Eq. (9.111).

The 2^N dimensional spinor representation $\psi(\xi)$ transforms as follows:

$$\psi'(x') = S(O)\psi(x). \tag{9.113}$$

It is then easily checked that $S(O)$ is given by

$$S = e^{-\frac{i}{4}a_{\mu\nu}\Sigma_{\mu\nu}}. \tag{9.114}$$

Now, under rotations Γ_μ transforms so that

$$S(O)\Gamma_\mu S^{-1}(O) = O_{\mu\nu}\Gamma_\nu. \tag{9.115}$$

Expansion of this equation for infinitesimal rotations then determines $\Sigma_{\mu\nu}$ so that

$$\Sigma_{\mu\nu} = \frac{i}{2}[\Gamma_\mu, \Gamma_\nu], \tag{9.116}$$

where $\Sigma_{\mu\nu}$ obey the algebra

$$[\Sigma_{\mu\nu}, \Sigma_{\sigma\rho}] = -2i\left(\delta_{\mu\sigma}\Sigma_{\nu\rho} - \delta_{\mu\rho}\Sigma_{\nu\sigma} - \delta_{\nu\sigma}\Sigma_{\mu\rho} + \delta_{\nu\rho}\Sigma_{\mu\sigma}\right). \tag{9.117}$$

This is the same algebra as satisfied by $A_{\mu\nu}$ in Eq. (9.106). The set of generators defined by

$$\Sigma_{2i+1.2i+2}, \quad i = 0, 1, \cdots, (N-1) \tag{9.118}$$

gives a set of N generators which are diagonal and commuting and thus form the Cartan subalgebra. This also tells us that the rank of $SO(2N)$ is N. Now, $U(N)$ can be shown to be a subgroup of $SO(2N)$. To show this, we look at the generators $\Sigma_{\mu\nu}$ in the fermionic creation and destruction operators $b_i^\dagger$ and b_i, where $i = 1, \ldots, N$, which we now introduce. The use of fermionic creation and destruction operators in this context was introduced by Mohapatra and Sakita [29]. Thus, $b_i^\dagger$ and b_i ($i = 1, N$) are defined so that

$$b_i|0\rangle = 0, \quad \langle 0|b_i^\dagger = 0, \tag{9.119}$$

where $b_i^\dagger$ and b_i satisfy the following anti-commutation relations:

$$\{b_i, b_j^\dagger\} = \delta_i^j, \quad \{b_i, b_j\} = 0, \quad \{b_i^\dagger, b_j^\dagger\} = 0. \tag{9.120}$$

We now introduce the quantities

$$\Gamma_{2i} = (b_i + b_i^\dagger) \tag{9.121}$$

$$\Gamma_{2i-1} = -i(b_i - b_i^\dagger) \tag{9.122}$$

and define Γ_μ so that

$$\Gamma_\mu = \{\Gamma_{2i}, \Gamma_{2i-1}\}. \tag{9.123}$$

Since $\Gamma_{2i}^\dagger = \Gamma_{2i}$ and $\Gamma_{2i-1}^\dagger = \Gamma_{2i-1}$, Γ_μ are Hermitian, i.e., $\Gamma_\mu^\dagger = \Gamma_\mu$.

In terms of the creation and the destruction operators, we can write the following relations:

$$\Sigma_{2i-1,2j-1} = \frac{i}{2}\left[b_i, b_j^\dagger\right] - \frac{i}{2}\left[b_j, b_i^\dagger\right] - i\left(b_i b_j + b_i^\dagger b_j^\dagger\right)$$

$$\Sigma_{2i,2j-1} = \frac{1}{2}\left[b_i, b_j^\dagger\right] + \frac{1}{2}\left[b_j, b_i^\dagger\right] - \left(b_i b_j - b_i^\dagger b_j^\dagger\right)$$

$$\Sigma_{2i,2j} = \frac{i}{2}\left[b_i, b_j^\dagger\right] - \frac{i}{2}\left[b_j, b_i^\dagger\right] + i\left(b_i b_j + b_i^\dagger b_j^\dagger\right). \tag{9.124}$$

Next, we define a set of generators x_j^i so that

$$x_j^i = \frac{1}{4}\left[\Sigma_{2i2j-1} + \Sigma_{2j2i-1} + i\Sigma_{2i2j} + i\Sigma_{2i-12j-1}\right]. \tag{9.125}$$

Substitution of Eq. (9.124) gives x_j^i in terms of b_i and $b_j^\dagger$ so that

$$x_j^i \equiv -\frac{1}{2}\left[b_i^\dagger, b_j\right]. \tag{9.126}$$

The quantities x_j^i $(i, j = 1 - N)$ are N^2 quantities, and it can be easily seen that they satisfy the relation

$$[x_j^i, x_l^k] = \delta_{il} x_j^k - \delta_{jk} x_l^i, \tag{9.127}$$

which is the algebra satisfied by the generators of the group $U(N)$. It then follows that since x_j^i can be written as a linear combination of the generators of $SO(2N)$, $U(N)$ is a subgroup of $SO(2N)$. The number of diagonal generators of $U(N)$ is N, which is also the rank of $SO(2N)$. We have already seen that $U(N)$ can decompose into $SU(N) \otimes U(1)$. Thus, $SO(2N)$ representations can be decomposed in representations of $SU(N) \otimes U(1)$. The generator of $U(1)$, which we label as X, commutes with the generators of $SU(5)$, and is given by

$$X = \frac{1}{2}(\Sigma_{1\,2} + \Sigma_{3\,4} + \Sigma_{5\,6} + \Sigma_{7\,8} + \Sigma_{9\,10}). \tag{9.128}$$

In terms of fermionic creation and destruction operators, it is given by

$$X = \frac{5}{2} - \sum_{i=1}^{5} b_i^\dagger b_i. \tag{9.129}$$

It is also useful to record here the forms of $B - L$, T_{3L}, T_{3R}, Y, and Q:

$$B - L = \frac{1}{3}(\Sigma_{12} + \Sigma_{34} + \Sigma_{56}) \qquad (9.130)$$

$$T_{3L} = \frac{1}{4}(\Sigma_{78} - \Sigma_{9\ 10}) \qquad (9.131)$$

$$T_{3R} = -\frac{1}{4}(\Sigma_{78} + \Sigma_{9\ 10}). \qquad (9.132)$$

In terms of the fermionic creation and destruction operators, they take the following forms:

$$B - L = 1 - \frac{2}{3}\sum_{i=1}^{3} b_i^\dagger b_i \qquad (9.133)$$

$$T_{3L} = \frac{1}{2}(b_4^\dagger b_4 - b_5^\dagger b_5) \qquad (9.134)$$

$$T_{3R} = \frac{1}{2}(b_4^\dagger b_4 + b_5^\dagger b_5 - 1). \qquad (9.135)$$

Next, we compute the hypercharge defined by

$$Y = T_{3R} + \frac{1}{2}(B - L). \qquad (9.136)$$

Substitution of $B - L$ and T_{3R} using Eqs. (9.130), (9.132), (9.133), and (9.135) gives

$$Y = \frac{1}{6}(\Sigma_{12} + \Sigma_{34} + \Sigma_{56}) - \frac{1}{4}(\Sigma_{78} + \Sigma_{9\ 10}) \qquad (9.137)$$

$$= -\frac{1}{3}\sum_{i=1}^{3} b_i^\dagger b_i + \frac{1}{2}\sum_{j=4}^{5} b_j^\dagger b_j. \qquad (9.138)$$

Further, the charge operator is defined so that

$$Q = T_{3L} + Y, \qquad (9.139)$$

which on using the result of Eqs. (9.134), (9.137) and (9.138) gives

$$Q = \frac{1}{6}(\Sigma_{12} + \Sigma_{34} + \Sigma_{56}) - \frac{1}{2}\Sigma_{9\ 10} \qquad (9.140)$$

$$= -\frac{1}{3}\sum_{i=1}^{3} b_i^\dagger b_i + b_4^\dagger b_4. \qquad (9.141)$$

An alternative form for the generators X, $B - L$, T_{3L}, T_{3R}, Y, and Q in the 10-dimensional representation is given in Section 9.8.2.

It is useful to define a chirality operator for $SO(2N)$. Let us consider the matrix Γ_0 defined by

$$\Gamma_0 = (-i)^N \Gamma_1 \Gamma_2 \cdots \Gamma_{2N}, \qquad (9.142)$$

which satisfies the conditions $tr(\Gamma_0) = 0$, $\Gamma_0^2 = 1$, and $\Gamma_0^\dagger = \Gamma_0$, where Γ_0 anti-commutes with all the Γ_μ, i.e.,

$$\Gamma_0\Gamma_\mu = -\Gamma_\mu\Gamma_0, \quad \mu = 1, 2, \cdots, 2N. \tag{9.143}$$

It is also seen that Γ_0 commutes with all the generators of $SO(2N)$, i.e.,

$$[\Gamma_0, \Sigma_{\mu\nu}] = 0. \tag{9.144}$$

In terms of the creation and the destruction operators, we can write Γ_0 in the form

$$\Gamma_0 = \prod_{i=1}^{N}(1 - 2b_i^\dagger b_i), \tag{9.145}$$

and it is useful to note that

$$\Gamma_0 b_i^\dagger = -b_i^\dagger \Gamma_0. \tag{9.146}$$

Using Γ_0 we define projection operators such that

$$P_\pm = \frac{1}{2}(1 \pm \Gamma_0), \tag{9.147}$$

where $P_\pm$ satisfy the relations $P_+ + P_- = 1$, $P_\pm^2 = P_\pm$, and $P_+P_- = 0 = P_-P_+$. The $P_\pm$ and Γ_0 satisfy the following relations:

$$\Gamma_0 P_\pm = \pm P_\pm$$
$$P_\pm \Gamma_\mu = \Gamma_\mu P_\mp, \quad \mu = 1, 2, \cdots, 2N. \tag{9.148}$$

The projection operators allow one to reduce the 2^N-dimensional spinor representation ψ_N of $SO(2N)$ into two irreducible 2^{N-1}-dimensional representations. Thus, we can define

$$\psi_N^\pm = P_\pm \psi_N, \tag{9.149}$$

and the action of Γ_0 on $\psi_N^\pm$ gives

$$\Gamma_0 \psi_N^\pm = \pm \psi_N^\pm, \tag{9.150}$$

where $\psi_N^\pm$ are chiral projections of ψ_N. The $\psi_N^\pm$ can be expanded in terms of the oscillators as follows:

$$|\psi_N^+\rangle = \sum_{k=0,2,\cdots}^{N-1} \frac{1}{k!}\psi^{i_1\cdots i_k}\prod_{j=1}^{k} b_{i_j}^\dagger |0\rangle$$

$$|\psi_N^-\rangle = \sum_{k=1,3,\cdots}^{N-1} \frac{1}{k!}\chi^{i_1\cdots i_k}\prod_{j=1}^{k} b_{i_j}^\dagger |0\rangle. \tag{9.151}$$

Let us introduce a charge conjugation operator $\mathbb{B}$ for $SO(2N)$ and use it to define a charge conjugate state Ψ^c given a state Ψ so that

$$\Psi^c \equiv \mathbb{B}^{-1}\Psi^*, \tag{9.152}$$

where $\mathbb{B}^{-1}\mathbb{B} = \mathbb{B}\mathbb{B}^{-1} = 1$. For the $SO(10)$, case, the charge conjugation operator takes the form

$$\mathbb{B} = \prod_{\mu=\text{odd}} \Gamma_\mu = -i \prod_{k=1}^{5} (b_k - b_k^\dagger). \tag{9.153}$$

Under $SO(10)$ rotations, Ψ transforms as

$$\delta\Psi = -\frac{i}{4}\epsilon_{\mu\nu}\Sigma_{\mu\nu}\Psi. \tag{9.154}$$

Using the fact that $\Sigma_{\mu\nu}$ are Hermitian, we can write the transformation of Ψ^c so that

$$\delta\Psi^c = \frac{i}{4}\epsilon_{\mu\nu}\mathbb{B}^{-1}\Sigma_{\mu\nu}^T\mathbb{B}\Psi^c. \tag{9.155}$$

If we want Ψ^c to transform the same way that Ψ does, then we require $\mathbb{B}^{-1}\Sigma_{\mu\nu}^T\mathbb{B} = -\Sigma_{\mu\nu}$, which can be achieved if Γ' satisfy the following relation:

$$\mathbb{B}^{-1}\Gamma_\mu^T\mathbb{B} = \pm\Gamma_\mu. \tag{9.156}$$

We will make the choice of a minus sign on the left-hand side. Using the above, we can show that B has the property that

$$\mathbb{B}^{-1}\Gamma_0^T\mathbb{B} = (-1)^N\Gamma_0, \tag{9.157}$$

where Γ_0 is defined by Eq. (9.142). It is then easily seen that one has the relations

$$\mathbb{B}P_\pm = P_\pm^T\mathbb{B} \quad (N \text{ even}) \tag{9.158}$$

$$\mathbb{B}P_\mp = P_\pm^T\mathbb{B} \quad (N \text{ odd}), \tag{9.159}$$

where $P_\pm$ are the projection operators defined in Eq. (9.147).

9.4.1 SO(10) Model Building

In this case, $N = 5$, and the 2^N dimensional spinor representation can be decomposed into $16 \oplus \overline{16}$, where we denote the 16-plet by $\Psi_{(+)}$ and the $\overline{16}$-plet by $\Psi_{(-)}$. Further, the 16-plet under $SU(5) \otimes U(1)$ has the decomposition

$$16 = [1] \oplus [\overline{5}] \oplus [10]. \tag{9.160}$$

As noted earlier, the $SO(10)$ model was introduced by Georgi, Fritzsch, and Minkowski [27, 28], and further early work was done by Harvey and others

[30–34]. The 16-plet can be expanded in terms of $SU(5)$ creation and destruction operators as follows:

$$|16_á\rangle = |0\rangle\psi_á + \frac{1}{2}b_i^\dagger b_j^\dagger|0\rangle\psi_á^{ij} + \frac{1}{24}\epsilon^{ijklm}b_i^\dagger b_j^\dagger b_k^\dagger b_l^\dagger|0\rangle\bar{\psi}_{ám}. \qquad (9.161)$$

As already noted, the combination $[\bar{5}] \oplus [10]$ is anomaly free, and, further, the $[1]$ in Eq. (9.160) is a right-handed neutrino field which is a singlet of the standard model gauge group, and thus the 16-plet of the matter field in $SO(10)$ is anomaly free. Such a cancellation is the result of a more general result for $SO(N)$ gauge theories. Thus, in general, anomalies arise due to the non-vanishing of the trace over the product of three group generators in some given group representation

$$tr\left(\{T_a, T_b\}T_c\right). \qquad (9.162)$$

For $SO(10)$, one will have the following trace:

$$tr\left(\{\Sigma_{\mu\nu}, \Sigma_{\alpha\beta}\}\Sigma_{\lambda\rho}\right). \qquad (9.163)$$

However, there is no invariant tensor to which the above quantity can be proportional, which then automatically guarantees vanishing of the anomaly for $SO(10)$. This analysis extends to other $SO(N)$ groups. One exception is $SO(6)$, where there a six-index-invariant tensor $\epsilon_{\mu\nu\alpha\beta\lambda\rho}$ exists, and so in this case vanishing of the anomaly is not automatic.

Analogous to Eq. (9.161), the $\overline{16}$-plet can be expanded so that

$$|\overline{16}_á\rangle = b_i^\dagger|0\rangle\bar{\psi}_á^i + b_1^\dagger b_2^\dagger b_3^\dagger b_4^\dagger b_5^\dagger|0\rangle\bar{\psi}_á + \frac{1}{12}\epsilon^{ijklm}b_k^\dagger a_l^\dagger b_m^\dagger|0\rangle\bar{\psi}_{áij}. \qquad (9.164)$$

The particle identification of the various components of the 16-plet field are as follows:

$$\psi_{áa} = d_{Láa}^c, \quad \psi_á^{ab} = \epsilon^{abc}u_{Lác}^c, \quad \psi_{á4} = \overset{-}{e}_{Lá}, \quad \psi_á = \nu_{Lá}^c$$

$$\psi_á^{4a} = u_{Lá}^a, \quad \psi_á^{5a} = d_{Lá}^a, \quad \psi_{á5} = \nu_{Lá}, \quad \psi_á^{45} = \overset{+}{e}_{Lá}. \qquad (9.165)$$

We show below the $|16\rangle$-plet in terms of the expansion in $b^\dagger$ with the particle content as follows:

$$|16\rangle = |0\rangle\nu_L^c + \frac{1}{2}b_5^\dagger b_a^\dagger|0\rangle d_L^a + \frac{1}{2}b_4^\dagger b_a^\dagger|0\rangle u_L^a$$

$$+ \frac{1}{2}b_4^\dagger b_5^\dagger|0\rangle e_L^+ + \frac{1}{2}\epsilon^{abc}b_a^\dagger b_b^\dagger|0\rangle u_{Lc}^c$$

$$+ \frac{1}{2}\epsilon^{abc}b_4^\dagger b_5^\dagger b_a^\dagger b_b^\dagger|0\rangle d_{Lc}^c + b_5^\dagger b_1^\dagger b_2^\dagger b_3^\dagger|0\rangle e_L^- + b_1^\dagger b_2^\dagger b_3^\dagger b_4^\dagger|0\rangle\nu_L. \qquad (9.166)$$

9.4.2 The Variety of SO(10) Grand Unified Models

The Yukawa couplings involve couplings of two 16-plets with Higgs fields. To see which Higgs fields couple, we expand the product $16 \otimes 16$ as a sum over the irreducible representations of $SO(10)$. Here, we have

$$16 \otimes 16 = 10_s \oplus 120_a \oplus 126_s, \tag{9.167}$$

where the $s(a)$ refer to symmetric (anti-symmetric) under the interchange of the two 16-plets. In building $SO(10)$ models, one chooses for the matter sector three generations of quarks and leptons that belong to three copies of the 16-plet representation of $SO(10)$. The gauge vector bosons are chosen to belong to the adjoint representation of $SO(10)$, i.e., the 45-plet representation of $SO(10)$. This leaves us with the choice of picking the Higgs sector. Here, we need to pick a Higgs sector which reduces the $SO(10)$ gauge symmetry first to the standard model gauge group, i.e., $SU(3)_C \otimes SU(2)_L \otimes U(1)_Y$, and then to $SU(3)_C \times U(1)_{em}$. We note that $SO(10)$ has rank 5 while the gauge group $SU(3)_C \otimes SU(2)_L \otimes U(1)_Y$ has rank 4. Thus, we must not only break $SO(10)$, which contains the above subgroups, but also reduce the rank in the process of breaking. In analogy to $SU(5)$ where we chose a scalar multiplet belonging to the adjoint representation of $SU(5)$, i.e., the 24-plet of $SU(5)$, to break $SU(5)$ down to the standard model gauge group, here too we can start with a scalar multiplet in the adjoint representation of $SO(10)$, i.e., the 45 representation of $S0(10)$, to break the symmetry. The branchings of 45 under $SU(5) \otimes U(1)$ are

$$45 = 1(0) \oplus 10(4) \oplus \overline{10}(-4) \oplus 24(0). \tag{9.168}$$

A spontaneous breaking of the GUT symmetry can take place when the fields $1(0)$ and $24(0)$ develop a VEV. However, while these VEVs do break the symmetry, they do not reduce the rank, since both the singlet and the 24-plet fields carry a vanishing $U(1)$ quantum number. To reduce the rank we need an additional field which carries a non-vanishing $U(1)$ quantum number. We could try to break the gauge group symmetry with a 54-plet or a 210 of Higgs. To check if this will work, we again look at branchings under $SU(5) \otimes U(1)$. Now, the 54-plet has the branchings

$$54 = 15(4) \oplus \overline{15}(-4) \oplus 24(0). \tag{9.169}$$

Symmetry breaking can occur with the $24(0)$ developing a VEV. However, here again one has the same problem, i.e., the 24-plet carries a vanishing $U(1)$ quantum number and thus cannot reduce the rank. We do the same analysis on the 210 representation. Here, $SU(5) \otimes U(1)$ branchings are

$$210 = 1(0) \oplus 5(-8) \oplus \bar{5}(8) \oplus 10(4) \oplus \overline{10}(-4) \oplus 24(0) \oplus 40(-4) \oplus \overline{40}(4) \oplus 75(0). \tag{9.170}$$

The possible candidates for developing VEV which would preserve the standard model gauge group are the components $1(0)$, $24[0]$, and $75(0)$. However, here again one has a vanishing $U(1)$ quantum number for the fields, and thus giving a

VEV to these fields will not reduce the rank. A possible candidate to reduce the rank is the $16 \oplus \overline{16}$ of Higgs. This is easily seen by looking at the $SU(5) \otimes U(1)$ branchings of 16:

$$16 = 1(-5) \oplus \bar{5}(3) \oplus 10(-1). \tag{9.171}$$

Here, since the singlet field carries a non-vanishing $U(1)$ quantum number, a VEV for the singlet field will reduce the rank. Thus, a set of Higgs fields that can reduce the $SO(10)$ gauge symmetry down to the symmetry of the standard model gauge group is the combination $45 \oplus 16 \oplus \overline{16}$. There is another possibility where one can replace $16 + \overline{16}$ with $126 \oplus \overline{126}$. This can be seen by looking at the $SU(5) \otimes U(1)$ branchings of 126, i.e.,

$$126 = 1(-10) \oplus \bar{5}(-2) \oplus 10(-6) \oplus \overline{15}(6) \oplus 45(2) \oplus \overline{50}(-2). \tag{9.172}$$

Since the 126 contains the component $1(-10)$, a VEV for this singlet field will reduce the rank of the group after breaking. Thus, using the combination $45 \oplus 16 \oplus \overline{16}$ or the combination $45 \oplus 126 \oplus \overline{126}$, the $SO(10)$ will break to the standard model gauge group $SU(3)_C \otimes SU(2)_L \otimes U(1)_Y$. However, we still need to break it down further to the residual gauge group $SU(3)_C \otimes U(1)_Y$, and for this we need a pair of Higgs doublets. These can arise from a 10-plet representation of $SO(10)$ where under $SU(5) \otimes U(1)$ the 10-plet has the branchings $10 = 5(2) \oplus \bar{5}(-2)$.

Thus, the breaking of $SO(10)$ down to $SU(3)_C \otimes U(1)_{em}$ requires at least three sets of Higgs representations: one to reduce the rank, the second to break the rest of the gauge group to the standard model gauge group, and then at least one 10-plet to break the electroweak symmetry. We can do this by a combination of various fields from the set: 10, $16 \oplus \overline{16}$, 45, 54, $126 \oplus \overline{126}$, and 210. As is obvious from the above, one drawback of this method of breaking the $SO(10)$ gauge group is that we need several Higgs fields. It would be interesting to see if the breaking could come about with just one Higgs field. One possibility where the $SO(10)$ symmetry can break all the way to $SU(3)_C \otimes U(1)$ arises from using the combination $144 + \overline{144}$ [35, 36]. To see how this can happen, we look at the branchings of 144 under $SU(5) \otimes U(1)$:

$$144 = \bar{5}(3) \oplus 5(7) \oplus 10(-1) \oplus 15(-1) \oplus 24(-5) \oplus 40(-1) \oplus \overline{45}(3). \tag{9.173}$$

Here, we find that, quite remarkably, the $24(-5)$-plet carries a $U(1)$ charge. This means that once the standard model singlet in it acquires a VEV, there will be a reduction in the rank of the group, and the residual subgroup after breaking will be just the standard model gauge group. Now, the branchings of 144 in Eq. (9.173) contain the $SU(5)$ multiplets $5(3), 5(7)$, and $45(3)$. All of these have Higgs doublets which have the same quantum numbers as the Higgs doublets that enter into the electroweak symmetry breaking. The spontaneous breaking of the electroweak symmetry can occur via the Higgs self-interactions given by

$$W_H = M\overline{144} \otimes 144 + \sum_i \frac{\lambda_1}{M'} (\overline{144} \otimes 144)_i \otimes (\overline{144} \otimes 144)_i$$

$$+ \sum_j \frac{\lambda_2}{M'} (144 \otimes 144)_j \otimes (\overline{144} \otimes 144)_j. \tag{9.174}$$

A light pair of Higgs doublet as in the minimal supersymmetric standard model case can be obtained by fine tuning. While this is not ideal, it is technically natural, and perhaps a justification of this could be given in the context of a string landscape. It would, of course, be much better if a large fine tuning of the size needed to achieve a light Higgs doublet could be circumvented while keeping the nice feature of the $144 + \overline{144}$ representation. This can be be achieved if one uses a $560 + \overline{560}$ representation along with a missing partner mechanism [37]. Here, the $SU(5) \otimes U(1)$ branchings of 560 are given by

$$560 = 1(-5) \oplus \overline{5}(3) + \overline{10}(-9) \oplus 10(-1)_1 \oplus 10(-1)_2 + 24(-5) \oplus 40(-1)$$
$$\oplus 45(7) \oplus \overline{45}(3) \oplus \overline{50}(3) \oplus \overline{70}(3) \oplus 75(-5) \oplus 175(-1). \tag{9.175}$$

After spontaneous symmetry breaking, the VEV of $1(-5)$ and $75(-5)$ reduce the rank and break the symmetry down to the standard model gauge group. An interesting aspect of $560 \oplus \overline{560}$ is that with the inclusion of a light Higgs sector consisting of $SO(10)$ multiplets $320 \oplus 2 \times 10$, one can achieve a missing partner mechanism [37, 38]. This offers an alternative to the conventional mechanism of VEV alignment to achieve doublet–triplet splitting [39].

An interesting issue relates to R-parity in $SO(10)$. In $SO(10)$, R-parity can be automatic for certain choices of the Higgs and matter representations. Thus, if we assume that the Higgs representations are always chosen to be tensor representations such as 10, 45, 54, 120, 126, $\overline{126}$, 210, etc., and the matter representations always in the spinor representations, e.g., 16, 144, 560, then R-parity is automatic. This is rather obvious in that $SO(10)$ invariance must always involve spinor interactions appearing in pairs.

Spontaneous breaking of $SO(10)$ has been discussed by a number of authors (see, for example, [40–47]). In addition to the breaking of $SO(10)$ via the chain $SO(10) \rightarrow SU(5) \otimes U(1)$ and $SU(5) \rightarrow SU(3)_C \otimes SU(2)_L \otimes U(1)_Y$, there are other chains such as the one discussed at the beginning of this section and given in Eq. (9.15). Another possible chain is $SO(10) \rightarrow SO(8) \otimes U(1)$. However, the group $SO(8)$ is too small to accommodate $SU(3)_C$ and $SU(2)_L$ together, and thus this chain of breaking is not useful.

9.4.3 Techniques for Computation of SO(10) Vertices

We discuss now a technique for the computation of $SO(10)$ vertices, which is a natural extension of the work of Mohapatra and Sakita [29], which introduced the oscillator expansion in the analysis of $SO(2N)$ interactions. One may also use completely group theoretical methods to determine the couplings. However, the technique we discuss here is field theoretical and more straightforward. A result which is very useful in the computation of $SO(10)$ vertices is the basic theorem, which we discuss below [48, 49, 50]. The quantity that enters in the determination of any vertex is the contraction of an $SO(10)$ tensor with a string of Γ'_μ such as

$$\Gamma_\mu \Gamma_\nu \Gamma_\lambda \cdots \phi_{\mu\nu\lambda\ldots} \tag{9.176}$$

We wish to decompose this contraction in terms of $SU(5)$ representations. Let us consider the simplest vertex $\Gamma_\mu \phi_\mu$. It can be decomposed as follows:

$$\Gamma_\mu \phi_\mu = b_i \phi_{\bar{c}_i} + b_i^\dagger \phi_{c_i}, \tag{9.177}$$

where

$$\phi_{c_i} \equiv \phi_{2i} + i\phi_{2i-1}$$
$$\phi_{\bar{c}_i} = \phi_{2i} - i\phi_{2i-1}. \tag{9.178}$$

As the next example, we consider the expansion of $\Gamma_\mu \Gamma_\nu \phi_{\mu\nu}$. Here, one has

$$\Gamma_\mu \Gamma_\nu \phi_{\mu\nu} = b_i b_j \phi_{\bar{c}_i \bar{c}_j} + b_i b_j^\dagger \phi_{\bar{c}_i c_j} + b_i^\dagger b_j \phi_{c_i \bar{c}_j} + b_i^\dagger b_j^\dagger \phi_{c_i c_j}, \tag{9.179}$$

where each of the tensors are the sum of four terms when expanded in $SO(10)$ indices. Thus, e.g.,

$$\phi_{\bar{c}_i c_j} = \phi_{2ic_j} - i\phi_{2i-1c_j}$$
$$= \phi_{2i2j} + i\phi_{2i(2j-1)} - i\phi_{(2i-1)2j} + \phi_{(2i-1)(2j-1)}. \tag{9.180}$$

More generally, a reducible $SU(5)$ representation $\phi_{c_i \tilde{c}_j \ldots c_\ell \tilde{c}_m}$ with n number of c_i or $\tilde{c}_j$ subscripts will have 2^n number of terms in an expansion using b_i and $b_i^\dagger$ such as is given in Eq. (9.179). We notice that the $SU(5)$ destruction operator b_i goes with the index $\bar{c}_i$ and the creation operator $b_j^\dagger$ goes with the index c_j. A generalization to an arbitrary vertex is now clear. Thus, an $SO(10)$ vertex with an arbitrary number of Γ matrices and contracted with an $SO(10)$ tensor with an equal number of indices can be expanded in an $SU(5)$ decomposition so that one writes all possible permutations using b_i and $b_j^\dagger$ with each term multiplied by an appropriate $SU(5)$ tensor, which is in general a reducible representation of $SU(5)$. The $SU(5)$ decomposition of an arbitrary $SO(10)$ vertex $\Gamma_\mu \Gamma_\nu \Gamma_\lambda \ldots \Gamma_\sigma \phi_{\mu\nu\lambda\ldots\sigma}$ can then be expanded in the following form [48]:

$$\begin{aligned}
\Gamma_\mu \Gamma_\nu \Gamma_\lambda .. \Gamma_\sigma \phi_{\mu\nu\lambda\ldots\sigma} =& b_i^\dagger b_j^\dagger b_k^\dagger \ldots b_n^\dagger \phi_{c_i c_j c_k \ldots c_n} + (b_i b_j^\dagger b_k^\dagger \ldots b_n^\dagger \phi_{\bar{c}_i c_j c_k \ldots c_n} + perms) \\
&+ (b_i b_j b_k^\dagger \ldots b_n^\dagger \phi_{\bar{c}_i \bar{c}_j c_k \ldots c_n} + perms) + \cdots \\
&+ (b_i b_j b_k \ldots b_{n-1} b_n^\dagger \phi_{\bar{c}_i \bar{c}_j \bar{c}_k \ldots \bar{c}_{n-1} c_n} + perms) + \\
&+ b_i b_j b_k \ldots b_n \phi_{\bar{c}_i \bar{c}_j \bar{c}_k \ldots \bar{c}_n}. \tag{9.181}
\end{aligned}$$

In the above, $\phi_{c_i c_j c_k \cdots c_n}$, etc., are reducible $SU(5)$ tensors which must be decomposed into irreducible ones, and, further, must be rescaled so that the fields have their kinetic terms correctly normalized.

9.4.4 16 · 16 · 10 Couplings

We illustrate the technique for the decomposition of $SO(10)$ vertices in representations of $SU(5)$ by examining the simplest non-trivial case of the coupling $16 \cdot 16 \cdot 10$ so that the $SO(10)$ invariant coupling has the form

$$W_{16 \cdot 16 \cdot 10} = f^{(10)}_{\acute{a}\acute{b}} \tilde{\psi}_{\acute{a}} \mathbb{B} \Gamma_\mu \psi_{\acute{b}} \phi_\mu, \tag{9.182}$$

where $\mathbb{B}$ is the charge conjugation and matrix for $SO(10)$ defined in Eq. (9.153) and $\acute{a}$, and $\acute{b}$ are the generation indices. Using the property that $\tilde{\mathbb{B}} = -\mathbb{B}$ and $\mathbb{B}^{-1} \tilde{\Gamma}_\mu \mathbb{B} = -\Gamma_\mu$, we can write

$$f^{(10)}_{\acute{a}\acute{b}} \tilde{\psi}_{\acute{b}} \tilde{\Gamma}_\mu \tilde{\mathbb{B}} \psi_{\acute{a}} \phi_\mu = -f^{(10)}_{\acute{a}\acute{b}} \tilde{\psi}_{\acute{b}} \mathbb{B} \mathbb{B}^{-1} \tilde{\Gamma}_\mu \mathbb{B} \psi_{\acute{a}} \phi_\mu \tag{9.183}$$

$$= f^{(10)}_{\acute{b}\acute{a}} \tilde{\psi}_{\acute{a}} \mathbb{B} \Gamma_\mu \psi_{\acute{b}} \phi_\mu, \tag{9.184}$$

which shows that $f^{(10)}_{\acute{a}\acute{b}} = f^{(10)}_{\acute{b}\acute{a}}$. For the explicit determination, we display $\mathbb{B}$ in terms of b_i and $b_j^\dagger$ so that

$$\mathbb{B} = \Pi_{\mu \text{ odd}} \Gamma_\mu = \Gamma_1 \Gamma_3 \Gamma_5 \Gamma_7 \Gamma_9$$
$$= (-i)^5 (b_1 - b_1^\dagger)(b_2 - b_2^\dagger)(b_3 - b_3^\dagger)(b_4 - b_4^\dagger)(b_5 - b_5^\dagger). \tag{9.185}$$

We expand $\Gamma_\mu \phi_\mu$ in terms of the familiar $\bar{5}$ and 5 Higgs multiplets H_i and H^i of $SU(5)$ so that

$$\Gamma_\mu \phi_\mu = b_i \phi_{\bar{c}_i} + b_i^\dagger \phi_{c_i}. \tag{9.186}$$

Next, we expand the 16-plet of fermion fields denoted by the spinor $|\psi_{\acute{b}}\rangle$ in $SU(5)$ decomposition so that

$$\psi_{\acute{b}} \rightarrow |\psi_{\acute{b}}\rangle = \psi_{0\acute{b}}|0\rangle + \psi_{\acute{b}}^{ij} b_i^\dagger b_j^\dagger |0\rangle + \psi_{\acute{b}}^{ijk\ell} b_i^\dagger b_j^\dagger b_k^\dagger b_\ell^\dagger |0\rangle, \tag{9.187}$$

where $\phi_{0\acute{b}}$ is the singlet, $\psi_{\acute{b}}^{ij} = -\psi_{\acute{b}}^{ji}$ is the 10-plet, and $\psi_{\acute{b}}^{ijk\ell}$ is the $\bar{5}$-plet of $SU(5)$. We interpret $\tilde{\psi}_{\acute{a}}$ so that

$$\tilde{\psi}_{\acute{a}} \rightarrow \langle \psi_{\acute{a}}^*| = \psi_{0\acute{a}}^T \langle 0| + \psi_{\acute{a}}^{ijT} \langle 0|b_j b_i + \psi_{\acute{a}}^{ijk\ell T} \langle 0|b_\ell b_k b_j b_i. \tag{9.188}$$

Suppressing the generation index, the $SO(10)$ vertex can be written in the form

$$\langle \psi^*|\mathbb{B}\Gamma_\mu|\psi\rangle \phi_\mu = A^n \phi_{\bar{c}_n} + B_n \phi_{c_n}, \tag{9.189}$$

where

$$A_n \equiv \langle \psi^*|\mathbb{B}b_n|\psi\rangle \tag{9.190}$$

$$B_n \equiv \langle \psi^*|\mathbb{B}b_n^\dagger|\psi\rangle. \tag{9.191}$$

We will show a few steps for the explicit computation of A^n and B_n. An expansion A^n gives us two terms:

$$A_n = A_{1n} + A_{2n}$$
$$A_{1n} = \psi^{ij}\langle 0|b_j b_i \mathbb{B} b_n b_i^\dagger b_j^\dagger b_k^\dagger b_l^\dagger|0\rangle\psi^{ijkl}$$
$$A_{2n} = \psi^{ijkl}\langle 0|b_\ell b_k b_j b_i \mathbb{B} b_n b_i^\dagger b_j^\dagger|0\rangle\psi^{ij}. \tag{9.192}$$

To compute A_{1n} we use $SU(5)$ invariance and the relation

$$\psi^{ijk\ell} = \epsilon^{ijk\ell m}\psi_m, \tag{9.193}$$

and write

$$\psi^{pq}\langle 0|b_q b_p \mathbb{B} b_n b_i^\dagger b_j^\dagger b_k^\dagger b_\ell^\dagger \epsilon^{ijk\ell m}|0\rangle = \lambda_1\psi^{mn} \tag{9.194}$$

A simple computation gives $\lambda_1 = 48i$, and

$$A_{1n} = 48i\psi^{mn}\psi_m, \tag{9.195}$$

and a similar computation gives

$$A_{2n} = 48i\psi^{mn}\psi_m. \tag{9.196}$$

We now turn to the computation of B_n. Here, using $SU(5)$ invariance we write

$$\langle 0|b_j b_i \mathbb{B} b_m^\dagger b_k^\dagger b_\ell^\dagger|0\rangle = \lambda_2\epsilon_{ijk\ell m}. \tag{9.197}$$

A simple procedure gives $\lambda_2 = i$, and

$$B_n = i\epsilon_{ijk\ell n}\psi^{ij}\psi^{k\ell}. \tag{9.198}$$

Thus, collecting the results and inserting in the generation indices we find that the $16 \cdot 16 \cdot 10$ interaction takes the form

$$W_{16\cdot16\cdot10} = if^{(10)}_{\acute{a}\acute{b}}\left(48\psi^{ij}_{\acute{a}}\psi_{\acute{b}i}\phi_{\bar{c}_j} + 48\psi_{\acute{a}i}\psi^{ij}_{\acute{b}}\phi_{\bar{c}_j} + \epsilon_{ijk\ell m}\psi^{ij}_{\acute{a}}\psi^{k\ell}_{\acute{b}}\phi_{c_m}\right). \tag{9.199}$$

The fields appearing in Eq. (9.199) are not yet properly normalized. To do this we compute the kinetic energy terms which have a $16^\dagger \cdot 16$ form in the Lagrangian. The computation of this term is straightforward, and a rescaling of the fields is needed so that

$$\psi'_0 = \psi_0$$
$$\psi^{ij'} = 2\psi^{ij}$$
$$\psi'_i = 24\psi_i, \tag{9.200}$$

where the prime fields are correctly normalized. Further, a term in the superpotential of the form $10 \cdot 10$ for the Higgs mass will give $\phi_{\bar{c}_i}\phi_{c_i}$, and we define $H_{1i} = \phi_{\bar{c}_i}/\sqrt{2}$ and $H_2^i = \phi_{c_i}/\sqrt{2}$ to get the normalized mass term. With the above change of variables the superpotential for the $16 \cdot 16 \cdot 10$ couplings in $SU(5)$ decomposition takes the form [48]

$$W_{16 \cdot 16 \cdot 10} = (\sqrt{2} i f^{(10+)}_{\acute{a}\acute{b}}) \left(\psi^{ij}_{\acute{a}} \psi_{i\acute{b}} H_{1j} + \frac{1}{8} \epsilon_{ijk\ell m} \psi^{ij}_{\acute{a}} \psi^{k\ell}_{\acute{b}} H^m_2 - \psi_{0\acute{a}} \psi_{i\acute{b}} H^i_2 \right),$$

(9.201)

where we have dropped the primes on the matter fields ψ_0, ψ_i, and ψ^{ij} to avoid an overabundance of symbols, and it will be understood that the fields are canonically normalized and $f^{(\pm)}_{\acute{a}\acute{b}}$ are in general defined so that

$$f^{(\pm)}_{\acute{a}\acute{b}} = \frac{1}{2}(f_{\acute{a}\acute{b}} \pm f_{\acute{b}\acute{a}}).$$

(9.202)

Thus, the front factor $f^{(10+)}_{\acute{a}\acute{b}}$ in Eq. (9.201) implies that the interaction is symmetric in the generation indices $\acute{a}\acute{b}$. Eq. (9.201) is the usual form of the $SU(5)$ interaction. Here, the relative factor of $1/8$ arises naturally from the underlying $SO(10)$ invariance of the interaction. Similar procedures can be applied to other interactions in the superpotential, i.e., $16 \cdot 16 \cdot 120$ and $16 \cdot 16 \cdot \overline{126}$, and in the derivation of terms in the Lagrangian of the form $16^\dagger \cdot 16 \cdot 45$ and $16^\dagger \cdot 16 \cdot 210$.

9.4.5 $16 \cdot 16 \cdot 120$ *Couplings*

Next, we discuss the $16 \cdot 16 \cdot 120$ coupling, which is given by

$$\frac{1}{3!} f^{(120)}_{\acute{a}\acute{b}} \tilde{\psi}_{\acute{a}} \mathbb{B} \Gamma_\mu \Gamma_\nu \Gamma_\lambda \psi_{\acute{b}} \phi_{\mu\nu\lambda}.$$

(9.203)

As for the case of the $16 \cdot 16 \cdot 10$ coupling, here also using the property that $\tilde{\mathbb{B}} = -\mathbb{B}$ and $\mathbb{B}^{-1}\tilde{\Gamma}_\mu \mathbb{B} = -\Gamma_\mu$ as well as the total anti-symmetry of $\phi_{\mu\nu\lambda}$ so that $\phi_{\lambda\nu\mu} = -\phi_{\mu\nu\lambda}$ it is straightforward to see that $f^{(120)}_{\acute{a}\acute{b}}$ is anti-symmetric in the generation indices so that $f^{(120)}_{\acute{a}\acute{b}} = -f^{(120)}_{\acute{b}\acute{a}}$. In this case an expansion of the vertex in Eq. (9.203) gives

$$\Gamma_\mu \Gamma_\nu \Gamma_\lambda \phi_{\mu\nu\lambda} = b_i b_j b_k \phi_{\bar{c}_i \bar{c}_j \bar{c}_k} + b^\dagger_i b^\dagger_j b^\dagger_k \phi_{c_i c_j c_k} + 3(b^\dagger_i b_j b_k \phi_{c_i \bar{c}_j \bar{c}_k} + b^\dagger_i b^\dagger_j b_k \phi_{c_i c_j \bar{c}_k})$$
$$+ (3b_i \phi_{\bar{c}_n c_n \bar{c}_i} + 3b^\dagger_i \phi_{\bar{c}_n c_n c_i}).$$

(9.204)

The 120-plet of $SO(10)$ has the $SU(5)$ decomposition

$$120 = 5(h^i) + \bar{5}(h_i) + 10(h^{ij}) + \overline{10}(h_{ij}) + 45(h^{ij}_k) + \overline{45}(h^i_{jk}).$$

(9.205)

Using Eq. (9.203), Eq. (9.204), and the technique of Section 9.4.4, one can obtain in a straightforward fashion a decomposition of the interaction of Eq. (9.203) to give [48]

$$W^{(120)} = i\frac{2}{\sqrt{3}} f^{(120-)}_{\acute{a}\acute{b}} (2\psi_{0\acute{a}} \psi_{i\acute{b}} h^i + \psi^{ij}_{\acute{a}} \psi_{0\acute{b}} h_{ij} + \psi_{i\acute{a}} \psi_{j\acute{b}} h^{ij} - \psi^{ij}_{\acute{a}} \psi_{i\acute{b}} h_j$$
$$+ \psi_{i\acute{a}} \psi^{jk}_{\acute{b}} h^i_{jk} - \frac{1}{4} \epsilon_{ijklm} \psi^{ij}_{\acute{a}} \psi^{mn}_{\acute{b}} h^{kl}_n),$$

(9.206)

where the fields $\psi_0, \psi_i,$ and ψ^{ij} are correctly normalized. The interactions in Eq. (9.206) are in $SU(5)$ representations, and they include the following set of $SU(5)$-invariant couplings: $1_M \cdot \bar{5}_M \cdot 5_H$, $1_M \cdot 10_M \cdot \overline{10}_H$, $\bar{5}_M \cdot \bar{5}_M \cdot 10_H$, $10_M \cdot \bar{5}_M \cdot \bar{5}_H$, $\bar{5}_M \cdot 10_M \cdot \overline{45}_H$, and $10_M \cdot 10_M \cdot 45_H$, where the subscript M refers to matter and H refers to Higgs. Further, as noted already, the couplings are anti-symmetric in the generation indices. This is reflected in the front factor $f_{\acute{a}b}^{(-)}$.

9.4.6 $16 \cdot 16 \cdot \overline{126}$ *Couplings*

We now turn to the most difficult of the three cases, i.e., the $16 \cdot 16 \cdot \overline{126}$ coupling, which is given by

$$\frac{1}{5!} f_{\acute{a}b}^{(126)} \tilde{\psi}_{\acute{a}} \mathbb{B} \Gamma_\mu \Gamma_\nu \Gamma_\lambda \Gamma_\rho \Gamma_\sigma \psi_{\acute{b}} \Delta_{\mu\nu\lambda\rho\sigma}, \tag{9.207}$$

where $\Delta_{\mu\nu\lambda\rho\sigma}$ is 252 dimensional and can be decomposed so that

$$\Delta_{\mu\nu\lambda\rho\sigma} = \bar{\phi}_{\mu\nu\lambda\rho\sigma} + \phi_{\mu\nu\lambda\rho\sigma}, \tag{9.208}$$

where $\phi_{\mu\nu\lambda\rho\sigma}$ is the 126-plet and is self-dual, and $\bar{\phi}_{\mu\nu\lambda\rho\sigma}$ is the $\overline{126}$-plet and is anti-self-dual. They can be expressed in terms of the 252-plet representation by

$$\begin{pmatrix} \bar{\phi}_{\mu\nu\lambda\rho\sigma} \\ \phi_{\mu\nu\lambda\rho\sigma} \end{pmatrix} = \frac{1}{2} (\delta_{\mu\alpha}\delta_{\nu\beta}\delta_{\rho\gamma}\delta_{\lambda\delta}\delta_{\sigma\theta} \pm \frac{i}{5!}\epsilon_{\mu\nu\rho\lambda\sigma\alpha\beta\gamma\delta\theta}) \Delta_{\alpha\beta\gamma\delta\theta}. \tag{9.209}$$

It is to be noted that only the $\overline{126}$-plet couples in Eq. (9.207). It turns out that in the analysis of the couplings it is more convenient to keep the full 252-plet representation in Eq. (9.207), and in the end only the couplings involving $\overline{126}$ will survive. Following the same procedure as for the $16 \cdot 16 \cdot 10$ and $16 \cdot 16 \cdot 120$ couplings, we use once again the property that $\tilde{\mathbb{B}} = -\mathbb{B}$ and $\mathbb{B}^{-1}\tilde{\Gamma}_\mu\mathbb{B} = -\Gamma_\mu$. It is then straightforward to see that $f_{\acute{a}b}^{(126)}$ is symmetric in the interchange of the generation indices so that $f_{\acute{a}b}^{(126)} = f_{\acute{b}a}^{(126)}$.

As in the previous two cases, here too one begins by expanding the vertex $\Gamma_\mu\Gamma_\nu\Gamma_\lambda\Gamma_\rho\Gamma_\sigma\Delta_{\mu\nu\lambda\rho\sigma}$ in terms of the fermionic creation and destruction operators. One then follows the same steps as in the previous two analyses. In terms of $SU(5)$ components, $\overline{126}$ has the following expansion:

$$\overline{126} = 1(h) + 5(h^i) + \overline{10}(h_{ij}) + 15 h_S^{ij}) + \overline{45}(h_{jk}^i) + 50(h_{rs}^{ijk}), \tag{9.210}$$

and an explicit analysis gives [48]

$$\begin{aligned} W^{(\overline{126})} = i f_{\acute{a}b}^{(126+)} \frac{\sqrt{2}}{\sqrt{15}} \Big[&- \sqrt{2}\psi_{0\acute{a}}\psi_{0\acute{b}}h - \sqrt{3}\psi_{0\acute{a}}\psi'_{i\acute{b}}h^i + \psi_{0\acute{a}}\psi_{\acute{b}}^{ij}h_{ij} \\ &- \frac{1}{8\sqrt{3}}\psi_{\acute{a}}^{ij}\psi_{\acute{b}}^{kl}h^m\epsilon_{ijklm} - h_S^{ij}\psi'_{i\acute{a}}\psi'_{j\acute{b}} + \psi_{\acute{a}}^{ij}\psi'_{\acute{b}k}h_{ij}^k \\ &- \frac{1}{12\sqrt{2}}\epsilon_{ijklm}\psi_{\acute{a}}^{lm}\psi_{\acute{b}}^{rs}h_{rs}^{ijk} \Big]. \end{aligned} \tag{9.211}$$

The $SU(5)$-invariant couplings in Eq. (9.211) consist of $1_M \cdot 1_M \cdot 1_H$, $1_M \cdot \bar{5}_M \cdot 5_H$, $1_M \cdot 10_M \cdot \overline{10}_H$, $10_M \cdot 10_M \cdot 5_H$, $\bar{5}_M \cdot \bar{5}_M \cdot 15_H$, $10_M \cdot \bar{5}_M \cdot \overline{45}_H$, and $10_M \cdot 10_M \cdot 50_M$. As already noted, the front factor $f_{ab}^{(126+)}$ is symmetric in the generation indices, just as in the couplings for the 10-plet of Higgs.

9.4.7 Couplings with Vector Fields in SO(10)

The couplings with vector fields involve the product of $16^\dagger \otimes 16$, which has the following decomposition:

$$16^\dagger \otimes 16 = 1 \oplus 45 \oplus 210. \tag{9.212}$$

We now wish to couple the 16-plets with vector fields in the adjoint representation of $SO(10)$, where vector fields arise in the process of gauging the $SO(10)$ symmetry. This coupling in the Lagrangian can be written as

$$\int d^4\theta \ \ tr(\hat{\phi}^\dagger e^{g\hat{V}} \hat{\phi}), \tag{9.213}$$

where $\hat{V}$ is the Lie-valued vector superfield. More explicitly, we have

$$\mathsf{L}^{(45)} = \frac{1}{i}\frac{1}{2!}g^{(45)} \langle \Psi_{(+)\acute{a}}|\gamma^0\gamma^A\Sigma_{\mu\nu}|\Psi_{(+)\acute{a}}\rangle \Phi_{A\mu\nu}, \tag{9.214}$$

where $\psi_{(+)\acute{a}} = 16_{\acute{a}}$ and $\gamma^A(A = 0 - 3)$ spans the Clifford algebra associated with the Lorentz group, g is the gauge coupling constant, and $\Phi_{A\mu\nu}$ are gauge vector fields of dimensionality 45. The gauge fields have an $SO(10)$-invariant gauge kinetic energy term of the form

$$-\frac{1}{4}\mathcal{F}_{\mu\nu}^{AB}\mathcal{F}_{AB\mu\nu}, \tag{9.215}$$

where $\mathcal{F}_{\mu\nu}^{AB}$ is the 45 of the SO(10) field strength tensor. Now, the 45 of $SO(10)$ has the following set of irreducible $SU(5)$ tensors: $1 + 10 + \overline{10} + 24$. We can thus break the 45-plet into these $SU(5)$ invariant tensors so that

$$\Phi_{Ac_n\bar{c}_n} = \phi_A, \ \ \Phi_{Ac_i\bar{c}_j} = \phi_{Aj}^i + \frac{1}{5}\delta_j^i\phi_A, \ \ \Phi_{Ac_ic_j} = \phi_A^{ij}, \ \ \Phi_{A\bar{c}_i\bar{c}_j} = \phi_{Aij}. \tag{9.216}$$

Next, we normalize them so that

$$\phi_A = 2\sqrt{5}\mathsf{G}_A, \ \ \phi_A^{ij} = \sqrt{2}\mathsf{G}_A^{ij}, \ \ \phi_{Aij} = \sqrt{2}\mathsf{G}_{Aij}, \ \ \phi_{Aj}^i = \sqrt{2}\mathsf{G}_{Aj}^i, \tag{9.217}$$

and the normalized kinetic energy terms take the form

$$\mathsf{L}_{kin}^{45-gauge} = -\frac{1}{2}\mathcal{G}_{AB}\mathcal{G}^{AB\dagger} - \frac{1}{4}\mathcal{G}^{ABij}\mathcal{G}_{AB}^{ij\dagger} - \frac{1}{4}\mathcal{G}_{ij}^{AB}\mathcal{G}_{ABij}^\dagger - \frac{1}{4}\mathcal{G}_j^{ABi}\mathcal{G}_{ABi}^j. \tag{9.218}$$

Using the above normalizations and the technique of the basic theorem, we find the couplings of the 16-plet of matter to the 45-plet of vector gauge bosons so that [49]

$$
\mathsf{L}^{(45)} = \tilde{g}^{(45)} \left[\sqrt{5} \left(-\frac{3}{5} \overline{\psi}_{\acute{a}}^{i} \gamma^A \psi_{\acute{a}i} + \frac{1}{10} \overline{\psi}_{\acute{a}ij} \gamma^A \psi_{\acute{a}}^{ij} + \overline{\psi}_a \gamma^A \psi_{\acute{a}} \right) \mathsf{G}_A \right.
$$

$$
+ \frac{1}{\sqrt{2}} \left(\overline{\psi}_a \gamma^A \psi_{\acute{a}}^{lm} + \frac{1}{2} \epsilon^{ijklm} \overline{\psi}_{\acute{a}ij} \gamma^A \psi_{\acute{a}k} \right) \mathsf{G}_{Alm}
$$

$$
- \frac{1}{\sqrt{2}} \left(\overline{\psi}_{alm} \gamma^A \psi_{\acute{a}} + \frac{1}{2} \epsilon_{ijklm} \overline{\psi}_{\acute{a}}^{i} \gamma^A \psi_{\acute{a}}^{jk} \right) \mathsf{G}_A^{lm}
$$

$$
\left. + \sqrt{2} \left(\overline{\psi}_{\acute{a}ik} \gamma^A \psi_{\acute{a}}^{kj} + \overline{\psi}_{\acute{a}}^{j} \gamma^A \psi_{\acute{a}i} \right) \mathsf{G}_{Aj}^{i} \right]. \tag{9.219}
$$

We are primarily interested in the couplings of the 16-plets with the 24-plet of $SU(5)$ gauge fields since they contain the standard model gauge fields corresponding to the gauge group $SU(3)_C \otimes SU(2)_L \otimes U(1)_Y$. This leads to the couplings

$$
\mathcal{L}_{24/45} = g \left(\overline{\psi}_{\acute{a}ik} \gamma^A \psi_a^{kj} + \overline{\psi}_{\acute{a}}^{j} \gamma^A \psi_{\acute{a}i} \right) \mathsf{G}_{Aj}^{i}. \tag{9.220}
$$

Using the projection operator method given in Section 9.8.3, Eq. (9.220) gives [49]

$$
\mathcal{L}_{24/45} = g \sum_{x=1}^{8} \left[\overline{u}_{\acute{a}} \gamma^A V_A^x \frac{\lambda_x}{2} u_{\acute{a}} + \overline{d}_{\acute{a}} \gamma^A V_A^x \frac{\lambda_x}{2} d_{\acute{a}} \right]
$$

$$
+ g \sum_{y=1}^{3} \left[(\overline{\nu} \quad \overline{e}^-)_{\acute{a}L} \gamma^A W_A^y \frac{\tau_y}{2} \begin{pmatrix} \nu \\ e^- \end{pmatrix}_{\acute{a}L} + (\overline{u} \quad \overline{d})_{\acute{a}L} \gamma^A W_A^y \frac{\tau_y}{2} \begin{pmatrix} u \\ d \end{pmatrix}_{\acute{a}L} \right]
$$

$$
+ g^{(45)} \sqrt{\frac{3}{5}} \left[-\frac{1}{2} \left(\overline{e}_{\acute{a}L}^- \gamma^A B_A e_{\acute{a}L}^- + \overline{\nu}_{\acute{a}L} \gamma^A B_A \nu_{\acute{a}L} \right) \right.
$$

$$
+ \frac{1}{6} \left(\overline{u}_{\acute{a}L} \gamma^A B_A u_{\acute{a}L} + \overline{d}_{\acute{a}L} \gamma^A B_A d_{\acute{a}L} \right) + \frac{2}{3} \overline{u}_{\acute{a}R} \gamma^A B_A u_{\acute{a}R}
$$

$$
\left. - \frac{1}{3} \overline{d}_{\acute{a}R} \gamma^A B_A d_{\acute{a}R} - \overline{e}_{\acute{a}R}^- \gamma^A B_A e_{\acute{a}R}^- \right] + \cdots \tag{9.221}
$$

In the above, V_A^x is an $SU(3)$ octet of gluons, W_A^y is an $SU(2)$ isospin vector multiplet of gauge bosons, and B_A is the hypercharge boson. Further, in Eq. (9.221) λ_x are the Gell–Mann matrices, τ_y are the Pauli matrices, and the dots stand for terms not displayed for the couplings of the lepto-quark/diquark bosons to fermions. For alternative techniques for the analysis of couplings in $SO(10)$, see the literature [51, 52].

The low-energy theory below the GUT scale can also be affected by corrections arising from the Planck scale [22, 53–56]. Thus, e.g., one may have corrections at low energy of the type

$$
\frac{1}{M_{Pl}} tr(FF\Phi). \tag{9.222}
$$

After Φ develops a VEV, one will have corrections to the kinetic energy of size $\langle \Phi \rangle / M_{Pl}$. Such a correction will affect gauge couplings after one has normalized the gauge kinetic energy function. For $SO(10)$, a 45-plet of scalar fields for

Φ gives a vanishing contribution due to anti-symmetry, and the non-vanishing contributions arise from Φ either as a 54-plet or a 210-plet representation of $SO(10)$.

9.5 Unification with Exceptional Groups

There are five exceptional groups: G_2, F_4, E_6, E_7, and E_8. Among these, the possible candidates for unification are E_6, E_7, and E_8. However, E_7 and E_8 do not have chiral representations and are not suitable for model building. Thus, the two lowest representations of E_7 are 56 and 133 , and have the following branchings into representations of E_6:

$$56 = 1 \oplus 1 \oplus 27 \oplus \overline{27}$$
$$133 = 1 \oplus 27 \oplus \overline{27} + \oplus 78. \tag{9.223}$$

One notices that 27 and $\overline{27}$ are conjugate representations and they appear in pairs, while 78 is vector-like. A very similar situation occurs for E_8. Thus, the lowest representation of E_8 is 248 dimensional, and its branchings under $SU(5) \otimes SU(5)$ are

$$248 = (1, 24) \oplus (24, 1) \oplus (5, \overline{10}) + (\overline{5}, 10) \oplus (10, 5) \oplus (\overline{10}, \overline{5}). \tag{9.224}$$

All the branchings in $SU(5) \otimes SU(5)$ appear in conjugate pairs, and thus it is difficult to get chiral families in this case. However, E_6 does possess non-real representations. The lowest representation of E_6 is 27-plet, and has the branchings under $SO(10) \otimes U(1)$ as follows:

$$27 = 1(4) \oplus 10(-2) \oplus 16(1), \tag{9.225}$$

where the 16-plet can accommodate one full generation of quarks and leptons and a right-handed neutrino, and the 10-plet can accommodate a pair of Higgs fields needed for the breaking of the electroweak symmetry. A more convenient branching to consider is

$$E_6 \supset SU(3)_C \otimes SU(3)_L \otimes SU(3)_R, \tag{9.226}$$

and here the 27-plet decomposes into the sum

$$27 = (1, 3, \bar{3}) \oplus (3, \bar{3}, 1) \oplus (\bar{3}, 1, 3). \tag{9.227}$$

It is useful to display the particle spectrum for one generation of 27-plet in this case. A single generation of 27-plet has the following particle spectrum:

$$L_{\acute{a}}(1, 3, \bar{3}) = \ell_{\acute{a}}(\nu_{\acute{a}}, e_{\acute{a}}); \ e^c_{\acute{a}}; \ H_{\acute{a}}, \ H'_{\acute{a}}; \ \nu^c_{\acute{a}}, N_{\acute{a}}$$
$$Q_{\acute{a}}(3, \bar{3}, 1) = q^a_{\acute{a}}(u^a_{\acute{a}}, d^a_{\acute{a}}), D^a_{\acute{a}}$$
$$Q^c_{\acute{a}}(\bar{3}, 1, 3) = (u^c)^a_{\acute{a}}, (d^c)^a_{\acute{a}}, (D^c)^a_{\acute{a}}. \tag{9.228}$$

Here, $\ell_{\acute{a}}$, $H_{\acute{a}}$, $H'_{\acute{a}}$, and $q^a_{\acute{a}}$ are $SU(2)_L$ doublets, $D^a_{\acute{a}}$ $(D^c)^a_{\acute{a}}$ are color triplets, $\nu^c_{\acute{a}}$ are $SU(5)$ singlets, and $N_{\acute{a}}$ are $SO(10)$ singlets. So, $27 \otimes \overline{27} = 1 \oplus 78 \oplus 650$, where 78 is the adjoint representation of E_6, and thus a model based on E_6 contains a 78-plet of gauge bosons. Now, the product of two 27-plets gives

$$27 \otimes 27 = \overline{27}_s + \overline{351}_a + \overline{351}'_s, \tag{9.229}$$

which allows one to write an invariant superpotential of the form

$$W_{27} = \lambda \; 27 \otimes 27 \otimes 27. \tag{9.230}$$

Thus, quite remarkably, this single-interaction type controls the interactions of all the quarks, leptons, and Higgs bosons. As the reader may have noticed, we have three pairs of Higgs doublets in this model, and two pairs must be removed to get back to the standard one pair of Higgs doublets of the minimal supersymmetric standard model.

9.6 Quark-Lepton Masses

In $SU(5)$ the lepton and down quark Yukawa couplings are equal at the grand unification scale. These give rise to incorrect relations among quark and lepton masses at low energy. On the other hand, an extrapolation of experimental values of the quark and lepton masses from the low energy to the grand unification scale implies a relation between Yukawa couplings of the Georgi–Jarlskog form [57]:

$$m_b \simeq m_\tau, \; m_\mu \simeq 3m_s, \; m_e \simeq \frac{1}{3}m_d. \tag{9.231}$$

One notices an interesting factor of 3, pointing to a possible group theoretical origin of this factor. A specific $SU(5)$ model with 5-plets and a 45-plet of Higgs can now be constructed to obtain up, down, and charged lepton mass matrices of the form

$$M_d = \begin{bmatrix} 0 & a' & 0 \\ a & c & 0 \\ 0 & 0 & b \end{bmatrix}, \; M_\ell = \begin{bmatrix} 0 & a' & 0 \\ a & -3c & 0 \\ 0 & 0 & b \end{bmatrix}, \; M_u = \begin{bmatrix} 0 & d & 0 \\ d & 0 & f \\ 0 & f & e \end{bmatrix}. \tag{9.232}$$

The factor -3 in the 22-element of M_ℓ arises from the VEV of the 45-plet of the Higgs field. Under the assumption, $a' = a$ and $b \gg c \gg a$, $e \gg f \gg d$, Eq. (9.232) on diagonalization reproduces the result of Eq. (9.231).

The above features are naturally reproduced in $SO(10)$. As shown in Eq. (9.160) in $SO(10)$, a full generation of fermions belong to a 16-plet of $SO(10)$, and the $16 \otimes 16$ of fermions has the decomposition as given in Eq. (9.167). Thus, as discussed in Section 9.4.4, 9.4.5, and 9.4.6, the Yukawa coupling involving the

16-plets has the decomposition

$$W_Y = Y_{\acute{a}\acute{b}}^{10(s)} \ 16_{\acute{a}} \otimes 16_{\acute{b}} \otimes 10 + Y_{\acute{a}\acute{b}}^{120(a)} \ 16_{\acute{a}} \otimes 16_{\acute{b}} \otimes 120 + Y_{\acute{a}\acute{b}}^{\overline{126}(s)} \ 16_{\acute{a}} \otimes 16_{\acute{b}} \otimes \overline{126},$$
(9.233)

where $s(a)$ refer to the symmetry (anti-symmetry) in the generation indices. After spontaneous breaking, the mass matrices for the quarks and the leptons can be written in the following form:

$$M_u = v_u Y^{10(s)} + v_u' Y^{\overline{126}(s)} + v_u'' Y^{120(a)}$$

$$M_d = v_d Y^{10(s)} + v_d' Y^{\overline{126}(s)} + v_d'' Y^{120(a)}$$

$$M_e = v_e Y^{10(s)} + v_e' Y^{\overline{126}(s)} + v_e'' Y^{120(a)}$$

$$M_D = v_D Y^{10(s)} + v_D' Y^{\overline{126}(s)} + v_D'' Y^{120(a)},$$
(9.234)

where M' and Y' are matrices in the generation space. The $SO(10)$ couplings relate the coefficients of the Yukawa couplings, which are the VEVs of Higgs fields such as $v_e = v_d$, $v_e' = -3v_d'$, and $v_e'' = -3v_d''$. The factor of -3 is easily understood. The reason for this is that 120 contains in its $SU(5)$ components a $45 + \overline{45}$ of SU(5) while $\overline{126}$ contains a 45 of $SU(5)$, and it is the VEV of these 45-plets of $SU(5)$ in the decompositions of 120 and $\overline{126}$ that produce the desired -3 factor. A similar situation occurs for the relation between u and d couplings. Thus, with the mass matrices of Eq. (9.234), one can once again reproduce the result of Eq. (9.231). Several other approaches to quark-lepton textures exist (e.g., [58]). Textures are found to be sensitive to Planck scale corrections (e.g., [59]).

Additionally one has Majorana mass terms for the right-handed and the left-handed neutrinos given by

$$M_{RR} = v_R Y^{126(s)}, \quad M_{LL} = v_L Y^{126(s)}.$$
(9.235)

The right-handed neutrino fields are assumed to be heavy. Integration over them leads to see-saw neutrino masses [60]. The effective neutrino masses then take the form

$$M_\nu = M_{LL} - M_D M_{RR}^{-1} M_D.$$
(9.236)

To obtain neutrino masses in the sub-eV range, the Majorana mass M_{RR} must be in the vicinity of 10^{14} GeV. Further, the term M_{LL} should either vanish or be of size $O(eV)$. Now, M_{LL} arises from the VEV of the 15-plet of the $SU(5)$ component in $\overline{126}$. Thus, either the VEV of this 15-plet component of $SU(5)$ should have vanishing VEV or it should be highly suppressed.

To explain neutrino oscillations, one needs flavor mixing so that the flavor states are linear combinations of mass eigenstates:

$$\begin{bmatrix} \nu_e \\ \nu_\mu \\ \nu_\tau \end{bmatrix} = \begin{bmatrix} U_{e1} & U_{e2} & U_{e3} \\ U_{\mu 1} & U_{\mu 2} & U_{\mu 3} \\ U_{\tau 1} & U_{\tau 2} & U_{\tau 3} \end{bmatrix} \begin{bmatrix} \nu_1 \\ \nu_2 \\ \nu_3 \end{bmatrix},$$
(9.237)

where ν_1, ν_2, and ν_3 are neutrino mass eigenstates. The matrix U is often referred to as the Pontecorvo–Maki–Nakagawa–Sakata (PMNS) matrix [61,62]. The usual parametrization of the PMNS matrix is in terms of three mixing angles θ_{12}, θ_{13}, and θ_{23} and a phase δ. In terms of these, the PMNS matrix takes the form

$$\begin{pmatrix} 1 & 0 & 0 \\ 0 & c_{23} & s_{23} \\ 0 & -s_{23} & c_{23} \end{pmatrix} \times \begin{pmatrix} c_{13} & 0 & s_{13}e^{-i\delta} \\ 0 & 1 & 0 \\ -s_{13}e^{i\delta} & 0 & c_{13} \end{pmatrix} \times \begin{pmatrix} c_{12} & s_{12} & 0 \\ -s_{12} & c_{12} & 0 \\ 0 & 0 & 1 \end{pmatrix}, \quad (9.238)$$

where $c_{ij} = \cos\theta_{ij}$ and $s_{ij} = \sin\theta_{ij}$. The product has the familiar form of the Cabibbo–Kobayashi–Maskawa (CKM) matrix, so that

$$V_{\text{PMNS}} = \begin{pmatrix} c_{12}c_{13}, & s_{12}c_{13}, & s_{13}e^{-i\delta}, \\ -s_{12}c_{23} - c_{12}s_{23}s_{13}e^{i\delta}, & c_{12}c_{23} - s_{12}s_{23}s_{13}e^{i\delta} & s_{23}c_{13} \\ s_{12}s_{23} - c_{12}c_{23}s_{13}e^{i\delta}, & -c_{12}s_{23} - s_{12}c_{23}s_{13}e^{i\delta}. & c_{23}c_{13} \end{pmatrix}.$$
$$(9.239)$$

There are two main scenarios for neutrino mass orderings, known as normal hierarchy (NH), where $m_1 < m_2 < m_3$, and inverted hierarchy (IH), where $m_3 < m_1 < m_2$. The PMNS matrices for the NH and for the IH [63] are given by

$$\sin^2_{12} \simeq 0.307, \ s^2_{23} \simeq 0.386, \ s^2_{13} \simeq 0.0241, \delta \sim 1.08\pi \ \text{(NH)} \quad (9.240)$$
$$\sin^2_{12} \simeq 0.307, \ s^2_{23} \simeq 0.392, \ s^2_{13} \simeq 0.0244, \delta \sim 1.09\pi \ \text{(IH)}. \quad (9.241)$$

9.6.1 See-saw Neutrino Masses

In the above, we showed how see-saw masses arise in $SO(10)$ grand unification. However, it is useful to discuss see-saw masses in a more general context [64–72]. We begin by noting that small neutrino masses can arise from the Weinberg operator given by [64]

$$O_5 = \frac{\lambda_{ij}}{M} L_i L_j H H, \quad (9.242)$$

where L_i ($i = 1, 2, 3$) are the leptonic doublets and H are the Higgs doublets. The operator of Eq. (9.242) violates $B - L$, and a VEV formation for the Higgs doublets will lead to the generation of Majorana masses for the neutrinos. Thus, after spontaneous breaking when $\langle H \rangle = v$, one has neutrino masses so that [64]

$$\lambda_{ij} \frac{v^2}{M} L_i L_j. \quad (9.243)$$

There are many ways in which an operator of type in Eq. (9.242) can arise. First, we note that the lepton multiplets are iso-spin 1/2 and so is the Higgs multiplet. Since $\frac{1}{2} \otimes \frac{1}{2} = 0 \oplus 1$, one may have exchanges of heavy fields which are iso-singlet or iso-triplet that lead to the operator in Eq. (9.242). The analysis above points to three mechanisms for the generation of small neutrino masses, as discussed below:

1. *Type I see-saw.* Here, one has exchange of a heavy fermion singlet so that $LH \to$ fermion singlet $\to LH$. An interaction of the following type is needed:

$$W_I = \lambda_a \epsilon_{\alpha\beta} \nu^{cT} L_a^\alpha H_2^\beta + \frac{1}{2} M \nu^{cT} \nu^c, \qquad (9.244)$$

where ν^c is the fermionic singlet. An integration of the heavy singlet fermion ν^c gives the type I see-saw operator.

2. *Type II see-saw.* Here, one has exchange of a heavy scalar triplet so that $LL \to$ scalar triplet $\to HH$. An interaction of the following type in the superpotential is needed:

$$W_{II} = \lambda_1 L \vec{\tau}.\vec{\Delta} L + \lambda_2 H \vec{\tau}.\vec{\Delta} H, \qquad (9.245)$$

and the integration of the heavy scalar triplet $\vec{\Delta}$ leads to the type II see-saw.

3. *Type III see-saw.* Here, one has exchange of a heavy fermion triplet so that $LH \to$ fermion triplet $\to LH$. An interaction of the following type in the superpotential is needed:

$$W_{III} = \lambda_1 L \vec{\tau}.\vec{\Sigma} H_2 + M_\Sigma \, tr(\Sigma^2), \qquad (9.246)$$

where $\vec{\Sigma}$ is the heavy fermionic triplet. An integration of $\vec{\Sigma}$ leads to the type III see-saw. In the $SU(5)$ model the interaction given by Eq. (9.246) can arise from interactions of the type

$$\bar{M}_x.24_y^x H_2^y + M_{24} 24 \cdot 24. \qquad (9.247)$$

9.7 Vector Multiplets

Vector multiplets carrying particle quantum numbers and their mirrors exist in GUTs, and some of them can survive at low scales in vector- like combinations [73–77]. Let us discuss the transformation properties of mirror particles under $SU(3)_C \otimes SU(2)_L \otimes U(1)_Y$. In $SU(5)$ matter resides in $\bar{5}$- and 10-plet representations of $SU(5)$. Assume there was additional matter which existed in vector-like combinations such as in $5 + \bar{5}$ and $10 + \overline{10}$. The $\bar{5} + 10$ would count as an extra sequential generation while the $5 + \overline{10}$ would be a mirror generation. Thus, 5 is the mirror of $\bar{5}$ and $\overline{10}$ is the mirror of 10. It is instructive to decompose the mirrors in terms of $SU(3)_C \otimes SU(2)_L \otimes U(1)_Y$ and contrast them with the components of a sequential generation. If such vector-like multiplets exist, they could mix with the third-generation leptons and quarks. Thus, we consider an ordinary lepton generation. Here, one has

$$\psi_L \equiv \begin{pmatrix} \nu_L \\ e_L \end{pmatrix} \left(1, 2, -\frac{1}{2} \right), \ e_L^c(1,1,1), \ \nu_L^c(1,1,0), \qquad (9.248)$$

where the numbers in each bracket give the $SU(3)_C$ and $SU(2)_L$ representations, and the last entry is the value of the hypercharge Y. The corresponding mirror generation is

$$\chi^c \equiv \left(\begin{array}{c} E_L^c \\ N_L^c \end{array} \right) \left(1, 2, \frac{1}{2}\right), \; E_L(1, 1, -1), \; N_L(1, 1, 0). \tag{9.249}$$

For quarks, one has analogous relations. Thus, an ordinary quark generation is

$$q \equiv \left(\begin{array}{c} u_L \\ d_L \end{array} \right) \left(3, 2, \frac{1}{6}\right), \; u_L^c \left(3^*, 1, -\frac{2}{3}\right), \; d_L^c \left(3^*, 1, \frac{1}{3}\right), \tag{9.250}$$

while for the mirror quarks one has

$$Q^c \equiv \left(\begin{array}{c} D_L^c \\ U_L^c \end{array} \right) \left(3^*, 2, -\frac{1}{6}\right), \; U_L \left(3, 1, \frac{2}{3}\right), \; D_L \left(3^*, 1, -\frac{1}{3}\right). \tag{9.251}$$

It is now easily seen that while the ordinary quarks and leptons have $V - A$ interactions because of their left-handed chiralities, the mirror quarks and leptons have $V + A$ interactions. Thus, the charged current interaction of an ordinary lepton multiplet and a mirror lepton multiplet with a W-boson are given by

$$\mathcal{L}_{CC} = \frac{g}{2\sqrt{2}} W_\mu^+ \left[\bar{\nu}\gamma^\mu(1 - \gamma_5)e + \bar{N}\gamma^\mu(1 + \gamma_5)E \right] + \text{h.c.} \tag{9.252}$$

In a similar fashion, the neutral current interactions of the ordinary lepton multiplet and of the mirror lepton multiplet are given by

$$\mathcal{L}_{NC} = \frac{g}{4\cos\theta_W} Z_\mu \left[\bar{e}\gamma^\mu(4x - 1 + \gamma_5)e + \bar{E}\gamma^\mu(4x - 1 - \gamma_5)E \right], \tag{9.253}$$

where $x = \sin^2\theta_W$. New effects on low-energy physics arise because of the possible mixing of the ordinary lepton and quark multiplets with mirrors. Thus, for example, mixings can arise from non-renormalizable interactions of the type

$$\frac{1}{M_{Pl}} \nu_L^c N_L \Phi_1 \Phi_2. \tag{9.254}$$

Here, if the Higgs fields Φ_1 and Φ_2 develop VEVs, a mixing term bilinear in the fields will appear. Mixings of this type can generate a large magnetic moment for the leptons. Thus, if the mixings involved the third-generation leptons, one could generate a large magnetic moment for the τ-lepton [77].

9.8 Appendix

In this appendix we give further details of some of the items discussed in this chapter.

9.8.1 Yukawa Coupling Evolution of Quarks and of Leptons

For the standard model the evolution of the Yukawa couplings up to one loop
are given by [11–13]

$$\frac{dY_u}{dt} = \frac{1}{16\pi^2}\Big[\frac{3}{2}Y_uY_u^\dagger - \frac{3}{2}Y_dY_d^\dagger + Tr(3Y_uY_u^\dagger + 3Y_dY_d^\dagger + Y_eY_e^\dagger + Y_\nu Y_\nu^\dagger)$$
$$-\frac{17}{20}g_1^2 - \frac{9}{4}g_2^2 - 8g_3^2\Big]Y_u$$

$$\frac{dY_d}{dt} = \frac{1}{16\pi^2}\Big[\frac{3}{2}Y_dY_d^\dagger - \frac{3}{2}Y_uY_u^\dagger + Tr(3Y_uY_u^\dagger + 3Y_dY_d^\dagger + Y_eY_e^\dagger + Y_\nu Y_\nu^\dagger)$$
$$-\frac{1}{4}g_1^2 - \frac{9}{4}g_2^2 - 8g_3^2\Big]Y_d$$

$$\frac{dY_e}{dt} = \frac{1}{16\pi^2}\Big[\frac{3}{2}Y_eY_e^\dagger - \frac{3}{2}Y_\nu Y_\nu^\dagger + Tr(3Y_uY_u^\dagger + 3Y_dY_d^\dagger + Y_eY_e^\dagger + Y_\nu Y_\nu^\dagger)$$
$$-\frac{9}{4}g_1^2 - \frac{9}{4}g_2^2\Big]Y_e$$

$$\frac{dY_\nu}{dt} = \frac{1}{16\pi^2}\Big[\frac{3}{2}Y_\nu Y_\nu^\dagger - \frac{3}{2}Y_eY_e^\dagger + Tr(3Y_uY_u^\dagger + 3Y_dY_d^\dagger + Y_eY_e^\dagger + Y_\nu Y_\nu^\dagger)$$
$$-\frac{9}{20}g_1^2 - \frac{9}{4}g_2^2\Big]Y_\nu, \tag{9.255}$$

while the evolution of the gauge-coupling constants up to two loops is given by

$$\frac{dg_i}{dt} = \frac{g_i}{16\pi^2}\left[b_i'g_i^2 + \frac{1}{16\pi^2}\left(\sum_{j=1}^{3}b_{ij}'g_i^2g_j^2 - \sum_{j=u,d,e}a_{ij}'g_i^2Tr[Y_jY_j^\dagger]\right)\right], \tag{9.256}$$

where

$$b_i' = \left(\frac{41}{10}, -\frac{19}{6}, -7\right), \quad a_{ij}' = \begin{pmatrix} \frac{17}{10} & \frac{1}{2} & \frac{3}{2} \\ \frac{3}{2} & \frac{3}{2} & \frac{1}{2} \\ 2 & 2 & 0 \end{pmatrix}, \quad b_{ij}' = \begin{pmatrix} \frac{199}{50} & \frac{27}{10} & \frac{44}{5} \\ \frac{9}{10} & \frac{35}{6} & 12 \\ \frac{11}{10} & \frac{9}{2} & -26 \end{pmatrix}. \tag{9.257}$$

For the case of the MSSM the evolution equations for the Yukawa couplings
of quarks and leptons at one loop are given by [11–13]

$$\frac{dY_u}{dt} = \frac{1}{16\pi^2}\Big[3Y_uY_u^\dagger + Y_dY_d^\dagger + Tr(3Y_uY_u^\dagger + Y_\nu Y_\nu^\dagger) - \frac{13}{15}g_1^2 - 3g_2^2 - \frac{16}{3}g_3^2\Big]Y_u$$

$$\frac{dY_d}{dt} = \frac{1}{16\pi^2}\Big[Y_uY_u^\dagger + 3Y_dY_d^\dagger + Tr(3Y_dY_d^\dagger + Y_eY_e^\dagger) - \frac{7}{15}g_1^2 - 3g_2^2 - \frac{16}{3}g_3^2\Big]Y_d$$

$$\frac{dY_e}{dt} = \frac{1}{16\pi^2}\Big[3Y_eY_e^\dagger + 3Y_\nu Y_\nu^\dagger + Tr(3Y_dY_d^\dagger + Y_eY_e^\dagger) - \frac{9}{5}g_1^2 - 3g_2^2\Big]Y_e$$

$$\frac{dY_\nu}{dt} = \frac{1}{16\pi^2}\Big[3Y_\nu Y_\nu^\dagger + 3Y_eY_e^\dagger + Tr(3Y_uY_u^\dagger + Y_\nu Y_\nu^\dagger) - \frac{3}{5}g_1^2 - 3g_2^2\Big]Y_\nu, \tag{9.258}$$

while the evolution of the gauge-coupling constants up to two loops is given by

$$
\frac{dg_i}{dt} = \frac{g_i}{16\pi^2} \left[b_i g_i^2 + \frac{1}{16\pi^2} \left(\sum_{j=1}^{3} b_{ij} g_i^2 g_j^2 - \sum_{j=u,d,e} a_{ij} g_i^2 Tr[Y_j Y_j^\dagger] \right) \right], \quad (9.259)
$$

where

$$
b_i = \left(\frac{33}{5}, 1, -3 \right), \quad
a_{ij} = \begin{pmatrix} \frac{26}{5} & \frac{14}{5} & \frac{18}{5} \\ 6 & 6 & 2 \\ 4 & 4 & 0 \end{pmatrix}, \quad
b_{ij} = \begin{pmatrix} \frac{199}{25} & \frac{27}{5} & \frac{88}{5} \\ \frac{9}{5} & 25 & 24 \\ \frac{11}{5} & 9 & 14 \end{pmatrix}. \quad (9.260)
$$

The renormalization group evolution equations given above are valid up to the electroweak scale. Below the electroweak scale only the quantum electrodynamic and the quantum chromodynamic couplings in the renormalization group running need be included.

9.8.2 The Generators X, $B - L$, T_{3L}, and T_{3R}

We have discussed the generators X, $B-L$, T_{3L}, and T_{3R} in terms of the fermionic creation and destruction operators. It is also useful to express them in the 10-dimensional representation in which the generators $A_{\mu\nu}$ are given in Eq. (9.105). Notice that the generators X, $B - L$, etc., involve $\Sigma_{\mu\nu}$, which are of the form $\Sigma_{i,i+1}$. Thus, X involves $\Sigma_{12}, \Sigma_{34}, \ldots, \Sigma_{9\,10}$. Using the result of Eq. (9.105), we can write

$$
\Sigma_{12} = \sigma_2 \otimes \mathrm{diag}(1, 0, 0, 0, 0) \quad (9.261)
$$

$$
\vdots
$$

$$
\Sigma_{9\,10} = \sigma_2 \otimes \mathrm{diag}(0, 0, 0, 0, 1), \quad (9.262)
$$

where the 10-dimensional matrices have been reduced to an outer product of two-dimensional and five-dimensional matrices. Using the above, we can write Eq. (9.128) so that

$$
X = \frac{1}{2} \sigma_2 \otimes \mathrm{diag}(1, 1, 1, 1, 1). \quad (9.263)
$$

Similarly, $B - L$, T_{3L}, and T_{3R} can be written as

$$
(B - L) = \frac{1}{3} \sigma_2 \otimes \mathrm{diag}(1, 1, 1, 0, 0)
$$

$$
T_{3L} = \frac{1}{4} \sigma_2 \otimes \mathrm{diag}(0, 0, 0, 1, -1)
$$

$$
T_{3R} = -\frac{1}{4} \sigma_2 \otimes \mathrm{diag}(0, 0, 0, 1, 1). \quad (9.264)
$$

The hypercharge Y and the charge operator Q are then given by

$$Y = \frac{1}{4}\sigma_2 \otimes diag\left(\frac{2}{3}, \frac{2}{3}, \frac{2}{3}, -1, -1\right)$$

$$Q = \frac{1}{4}\sigma_2 \otimes diag\left(\frac{2}{3}, \frac{2}{3}, \frac{2}{3}, 0, -2\right). \tag{9.265}$$

9.8.3 Decomposition of SU(5) Tensors in Terms of Standard Model Ones

For phenomenological analyses, $SU(5)$-invariant vertices need to be further decomposed into components which are irreducible tensors under $SU(3)_C \otimes SU(2)_L \otimes U(1)_Y$. For this purpose, one convenient procedure is to use projection operators. Assume we have a 5-plet of the $SU(5)$ field ϕ^i. In this case, we can write

$$\phi^i = P^i_\alpha \phi^\alpha + P^i_a \phi^a, \tag{9.266}$$

where $a = 1, 2$, and 3 are the $SU(3)_C$ color indices and $\alpha = 4$, and 5 are the $SU(2)_L$ indices and where $P^i_a = \delta^i_a$ and $P^i_\alpha = \delta^i_\alpha$ are the projection operators. We can also express in a similar way the $\bar{5}$-field

$$\phi_i = P^\alpha_i \phi_\alpha + P^a_i \phi_a, \tag{9.267}$$

where $P^\alpha_i = \delta^\alpha_i$ and $P^a_i = \delta^a_i$. The projection operators satisfy the condition

$$\sum_{i=1}^{5} P^i_b P^a_i = \delta^a_b, \quad \sum_{i=1}^{5} P^i_\beta P^\alpha_i = \delta^\alpha_\beta \tag{9.268}$$

$$\sum_{i=1}^{5} P^i_\alpha P^a_i = 0, \quad \sum_{i=1}^{5} P^i_a P^\alpha_i = 0. \tag{9.269}$$

Using this technique we can decompose any $SU(5)$-invariant tensor into tensors irreducible under $SU(3)_C \otimes SU(2)_L \otimes U(1)_Y$.

Next, let us consider the field Σ^i_j, which is a 24-plet of $SU(5)$. This field is traceless. Here, a short-cut exists in effecting an $SU(3)_C \otimes SU(2)_L \otimes U(1)_Y$ decomposition in that we can expand it using the $SU(5)$ generators so that

$$\Sigma^i_j = \sum_{u=1}^{24} (\lambda_u)^i_j \sigma_u. \tag{9.270}$$

In this case, using the property that $tr(\lambda_u \lambda_w) = 2\delta_{uw}$, we can determine σ_u. Now, the fields σ_u consist of a singlet field σ which appears with the hypercharge generator λ_{24}, fields σ_A ($A = 1, 2, 3$) in the adjoint representation of the group $SU(2)_L$, and the fields σ_x ($x = 1, 2, \ldots, 8$) in the adjoint representation of the color group $SU(3)_C$. These account for 12 of the 24 fields of Σ. The remaining fields are lepto-quark scalars. Assume that we now wish to break the global $SU(5)$ symmetry using the field Σ.

For this purpose we consider a scalar potential which is the most general renormalizable potential, which is

$$V = M^2 \, Tr(\Sigma^2) + \frac{\lambda_1}{2} tr(\Sigma^4) + \frac{\lambda_2}{2} (tr(\Sigma^2))^2 + \mu \, tr(\Sigma^3). \qquad (9.271)$$

We expand Σ^i_j in terms of the irreducible tensors of $SU(3)_C \otimes SU(2)_L \otimes U(1)_Y$. We then have

$$\Sigma^i_j = P^i_{1j}\sigma + P^i_{3Aj}\sigma_A + P^i_{8xj}\sigma_x + P^{ia}_{ja}\sigma^\alpha_a + P^{i\alpha}_{ja}\sigma^a_\alpha, \qquad (9.272)$$

where P's are the projection operators, and σ^α_a and σ^a_α are lepto-quark scalars. We now focus on just the fields σ, σ_A, and σ_x as we want to discuss spontaneous breaking as well as determine the masses of the fields after spontaneous breaking. The projection operators for the fields σ, σ_A, and σ_x are

$$P^i_{1j} = \sqrt{\frac{2}{15}} \left(\delta^i_a \delta^a_j - \frac{3}{2} \delta^i_\alpha \delta^\alpha_j \right)$$

$$P^i_{3Aj} = \frac{1}{\sqrt{2}} \delta^i_\alpha \delta^\beta_j (T_A)^\alpha_\beta$$

$$P^i_{8xj} = \frac{1}{\sqrt{2}} \delta^i_a \delta^b_j (T_x)^a_b. \qquad (9.273)$$

First, we determine the potential involving σ. Here, the following relations are useful:

$$tr(P^2_1) = 1, \ tr(P^3_1) = -\frac{1}{\sqrt{30}}, \ tr(P^4_1) = \frac{7}{30}. \qquad (9.274)$$

Using these, the scalar potential takes the form

$$V(\sigma) = M^2 \sigma^2 + \left(\frac{7\lambda_1}{60} + \frac{\lambda_2}{2} \right) \sigma^4 - \mu \frac{1}{\sqrt{30}} \sigma^3. \qquad (9.275)$$

Variation of this potential gives

$$M^2 + \left(\frac{7\lambda_1}{30} + \lambda_2 \right) \sigma^2_0 - \frac{3\mu\sigma_0}{2\sqrt{30}} = 0, \qquad (9.276)$$

where $\sigma_0 \equiv \langle \sigma \rangle$. Conventionally, one uses

$$\langle \Sigma \rangle = v \, diag(2, 2, 2, -3, -3), \qquad (9.277)$$

which leads to

$$\sigma_0 = \sqrt{30}v, \qquad (9.278)$$

and the condition for spontaneous breaking (Eq. (9.276)) becomes

$$M^2 + (7\lambda_1 + 30\lambda_2)v^2 - \frac{3}{2}\mu v = 0. \qquad (9.279)$$

Next, we determine the masses of the fields after spontaneous breaking. We will determine the contributions from the various terms in the scalar potential.

For the $tr(\Sigma^2)$ term we use the following properties of the projection operators:

$$tr(P_{3A}P_{3B}) = \delta_{AB}$$
$$tr(P_{8x}P_{8y}) = \delta_{xy}$$
$$tr(P_1 P_{3A}) = 0, \ tr(P_1 P_{8x}) = 0, \ tr(P_{3A}P_{8x}) = 0. \tag{9.280}$$

Using these, we get

$$M^2 \ tr(\Sigma^2) \supset M^2(\sigma^2 + \sigma_A^2 + \sigma_x^2). \tag{9.281}$$

The $(\lambda_2/2)(tr(\Sigma^2))^2$ term can be computed by simply squaring the previous result, which gives

$$\frac{\lambda_2}{2}(tr(\Sigma^2))^2 \supset 30\lambda_2^2 v^2(\sigma_A^2 + \sigma_x^2) + 90\lambda_2^2 v^2 \sigma_1^2, \tag{9.282}$$

where $\sigma = \sigma_0 + \sigma_1$. We now look at the $\mu\Sigma^3$ term. Here, we have

$$\mu \ tr\Sigma^3 \supset \mu \ tr(P_1\sigma + P_3\sigma_3 + P_8\sigma_8)^3. \tag{9.283}$$

Expanding, we find the following relevant terms:

$$\mu \ tr\Sigma^3 \supset \mu\sigma^3 \ tr(P_1^3) + 3\sigma\sigma_A\sigma_B \ tr(P_1 P_{3A}P_{3B}) + 3\sigma\sigma_x\sigma_y \ tr(P_1 P_{8x}P_{8y}). \tag{9.284}$$

Next, using the identities

$$tr(P_1 P_{3A}P_{3B}) = -\frac{3}{2}\sqrt{\frac{2}{15}}\delta_{AB}$$
$$tr(P_1 P_{8x}P_{8y}) = \sqrt{\frac{2}{15}}\delta_{xy}. \tag{9.285}$$

we get

$$\mu \ tr(\Sigma^3) \supset -3\mu v\sigma_1^2 - 9\mu v\sigma_A^2 + 6\mu v\sigma_x^2. \tag{9.286}$$

Finally let us look at the $(\lambda_1/2)tr\Sigma^4$ term. The relevant terms in the expansion are

$$\frac{\lambda_1}{2} \ tr\Sigma^4 \supset \frac{\lambda_1}{2}\big[\sigma^4 \ tr(P_1^4) + \sigma^2\sigma_{3A}\sigma_{3B} \ tr(P_1^2 P_{3A}P_{3B})$$
$$+\sigma^2\sigma_{8x}\sigma_{8y}tr(P_1^2 P_{8x}P_{8y})\big]. \tag{9.287}$$

The following identities are helpful in arriving at Eq. (9.287):

$$tr(P_1^2 P_{3A}P_{3B}) = \frac{3}{10}\delta_{AB}$$
$$tr(P_1^2 P_{8x}P_{8y}) = \frac{2}{15}\delta_{xy}. \tag{9.288}$$

An expansion around the vacuum state gives

$$\frac{\lambda_1}{2} tr \Sigma^4 \supset 21\lambda_1 v^2 \sigma_1^2 + 27\lambda_1 v^2 \sigma_{3A}^2 + 12\lambda_1 v^2 \sigma_{8x}^2. \qquad (9.289)$$

Using the result of Eqs. (9.281)–(9.289) and the evaluation of M^2 as given by Eq. (9.279), we obtain the following for the mass terms for the fields σ_1, σ_A, and σ_x:

$$V \supset \frac{1}{2} m_1^2 \sigma_1^2 + \frac{1}{2} m_3^2 \sigma_A^2 + \frac{1}{2} m_8^2 \sigma_x^2, \qquad (9.290)$$

where

$$\frac{1}{2} m_1^2 = 14\lambda_1 v^2 - \frac{3}{2}\mu v + 60\lambda_2^2 v^2$$

$$\frac{1}{2} m_3^2 = 20\lambda_1 v^2 - \frac{15}{2}\mu v$$

$$\frac{1}{2} m_8^2 = 5\lambda_1 v^2 + \frac{15}{2}\mu v. \qquad (9.291)$$

The example given above illustrates the usefulness of the projection operators in the $SU(3)_C \otimes SU(2)_L \otimes U(1)_Y$ decomposition of the $SU(5)$-invariant vertices.

9.9 Problems

1. Show that the $\Delta B = \Delta L = 0$ interactions arising from the couplings of the $SU(5)$ gauge bosons with the $\bar{5}$-plet of matter are given by

$$\mathcal{L}_{int-5}(\Delta B = \Delta L = 0) = g_5 \sum_{x=1}^{8} \bar{d}_R \gamma^\mu g_\mu^x \frac{\lambda_x}{2} d_R + g_5 \sum_{A=1}^{3} \bar{L} \gamma^\mu W_\mu^A \frac{\tau_A}{2} L$$

$$- \sqrt{\frac{3}{5}} \left(\frac{1}{3} \bar{d}_R \gamma^\mu d_R + \frac{1}{2}(\bar{e}_L \gamma^\mu e_L + \bar{\nu}_L \gamma^\mu \nu_L) \right) B_\mu \qquad (9.292)$$

and with the 10-plet of matter are given by

$$\mathcal{L}_{int-10}(\Delta B = \Delta L = 0) = g_5 \sum_{x=1}^{8} \left[\bar{u} \gamma^\mu \frac{\lambda_x}{2} u + \bar{d}_L \gamma^\mu \frac{\lambda_x}{2} d_L \right] g_\mu^x$$

$$+ g_5 \sum_{A=1}^{3} \bar{q}_L \gamma^\mu \frac{\tau_A}{2} q_L W_\mu^A$$

$$+ g_5 \sqrt{\frac{3}{5}} \left[\frac{2}{3} \bar{u}_R \gamma^\mu u_R \right.$$

$$\left. + \frac{1}{6}(\bar{u}_L \gamma^\mu u_L + \bar{d}_L \gamma^\mu d_L) - \bar{e}_R \gamma^\mu e_R \right] B_\mu. \qquad (9.293)$$

2. For $SU(5)$ the conventional kinetic energy term for the gauge fields is $-(1/2)F_{\mu\nu}F^{\mu\nu}$, where $F_{\mu\nu}$ is the Lie-valued field strength in the adjoint, i.e., the 24-plet representation of the gauge group. Assume one adds a correction to the Lagrangian which might arise from Planck scale physics of the form

$$\mathcal{L}_{Pl} = \frac{c}{2M_{Pl}}Tr(FF\Phi), \tag{9.294}$$

where Φ is the 24-plet of heavy scalar fields whose VEV formation breaks $SU(5) \rightarrow SU(3) \otimes SU(2) \otimes U(1)_Y$. Show that the proper normalization of the kinetic energy terms will lead to a splitting of the fine structure constant at the GUT scale for the $SU(3) \otimes SU(2) \otimes U(1)$ sectors as follows:

$$\alpha_G^{-1}(M_X) \rightarrow \alpha_G^{-1}(1 + \frac{cM}{2M_{Pl}}r_i). \tag{9.295}$$

Here M is defined so that $\langle\Phi\rangle = M \, diag(2,2,2,-3,-3)$ and $r_i = (2,-3,-1)$ for $i = SU(3), SU(2),$ and $U(1)$.

3. The 24-plet of $SU(5)$ can be decomposed into irreducible tensors of $SU(3)_C \times SU(2)_L \times U(1)_Y$ as follows:

$$24 = (1,1,0) + (1,3,0) + (8,1,0) + (3,2,-5/3) + (\bar{3},2,5/3). \tag{9.296}$$

To obtain the $SU(3)_C \times SU(2)_L \times U(1)_Y$ irreducible tensors explicitly, one may use the tensorial decomposition

$$\delta_j^i = \sum_{a=1}^{3} \delta_a^i \delta_j^a + \sum_{\alpha=4}^{5} \delta_\alpha^i \delta_j^\alpha. \tag{9.297}$$

Here, $a = 1, 2,$ and 3 is the color index and $\alpha = 4$ and 5 is the $SU(2)$ index. Show that the $SU(3)_C \times SU(2)_L \times U(1)_Y$ irreducible part $(1,1,0)$ in Eq. (9.296) has the following projection in $SU(5)$:

$$24_j^i(1,1,0) = \sqrt{\frac{2}{15}} \left(\sum_{a=1}^{3} \delta_a^i \delta_j^a - \frac{3}{2}\sum_{\alpha=4}^{5} \delta_\alpha^i \delta_j^\alpha\right)\sigma_{(110)}, \tag{9.298}$$

where $\sigma_{(110)}$ is the $SU(3)_C \times SU(2)_L \times U(1)$ singlet field.

4. Show that for the $SU(3)_C \otimes SU(2)_L \otimes U(1)$ singlet field in Σ the potential of Eq. (9.271) reduces to the potential given in Eq. (9.275).

5. Consider the coupling of a $\overline{45}$-plet of the Higgs field with $\bar{5} + 10$ of the matter fields so that the superpotential has the coupling

$$\psi^{ij}h_{ij}^k\psi_k', \tag{9.299}$$

where h_{ij}^k is a 45-plet representation which satisfies the constraints $h_{ij}^k = -h_{ji}^k$ and $\sum_k h_{ik}^k = 0$. Assume that a VEV growth for h_{ij}^k occurs so that

$$\langle h_{5j}^k \rangle = v\left(\frac{1}{4}\delta_j^k - \delta_4^k\delta_j^4\right) \tag{9.300}$$

and assume that this term contributes only to the second generation of quarks and leptons. Show that $\delta m_\mu : \delta m_s = (-3) : 1$, where the Δm_μ and Δm_s are the contributions to the muon and the s-squark mass from the 45-plet VEV formation.

6. Show that for $SU(n)$ the covariant derivative of a scalar field in the adjoint representation, i.e., the field Σ^i_j where $i, j = 1, \ldots, n$ is given by

$$D_\mu \Sigma = \partial_\mu \Sigma + ig[A_\mu, \Sigma], \tag{9.301}$$

where Σ and A_μ are Lie valued.

7. The kinetic energy of the 120-plet field $\phi_{\mu\nu\lambda}$, which is totally anti-symmetric in all the $SO(10)$ indices μ, ν, and λ which run from 1 to 10 is given by

$$\mathcal{L}^{120}_{\text{KE}} = -\partial_\alpha \phi_{\mu\nu\lambda} \partial^\alpha \phi^\dagger_{\mu\nu\lambda}, \tag{9.302}$$

where α are the space-time indices. Show that in terms of the reducible $SU(5)$ tensors $\phi_{c_i c_j c_k}\cdot$, etc., the $\mathcal{L}^{120}_{\text{KE}}$ takes the form

$$\mathcal{L}^{120}_{\text{KE}} = -\frac{1}{8}\left[\partial_\alpha \phi_{c_i c_j c_k} \partial \phi^\dagger_{c_i c_j c_k} + 3\partial_\alpha \phi_{c_i c_j \bar{c}_k} \partial \phi^\dagger_{c_i c_j \bar{c}_k} \right.$$
$$\left. + 3\partial_\alpha \phi_{c_i \bar{c}_j \bar{c}_k} \partial \phi^\dagger_{c_i \bar{c}_j \bar{c}_k} + \partial_\alpha \phi_{\bar{c}_i \bar{c}_j \bar{c}_k} \partial \phi^\dagger_{\bar{c}_i \bar{c}_j \bar{c}_k} \right]. \tag{9.303}$$

Define the irreducible $SU(5)$ tensors h^i, h^{ij}_k, etc., by

$$\phi_{c_i c_j \bar{c}_k} = \frac{2}{\sqrt{3}} h^{ij}_k + \frac{1}{\sqrt{3}}(\delta^i_k h^j - \delta^j_k h^i),$$

$$\phi_{c_i \bar{c}_j \bar{c}_k} = \frac{2}{\sqrt{3}} h^i_{jk} + \frac{1}{\sqrt{3}}(\delta^i_j h_k - \delta^i_k h_j)$$

$$\phi_{c_i c_j c_k} = \frac{1}{\sqrt{3}} \epsilon^{ijklm} h_{lm}, \quad \phi_{\bar{c}_i \bar{c}_j \bar{c}_k} = \frac{1}{\sqrt{3}} \epsilon_{ijklm} h^{lm},$$

$$\phi_{\bar{c}_n c_n c_i} = \frac{4}{\sqrt{3}} h^i, \phi_{\bar{c}_n c_n \bar{c}_i} = \frac{4}{\sqrt{3}} h_i,$$

where the fields $h^i, h_i, h^{ij}, h_{ij}, h^{ij}_k$, and h^k_{ij} are the 5, $\bar{5}$, 10, $\overline{10}$, 45, $\overline{45}$ of Higgs, respectively. Show that in terms of the irreducible tensor fields the kinetic energy terms for the components of the 120-plet take the form

$$L^{(120)}_{\text{kin}} = -\left(\frac{1}{2}\partial_\alpha h^{ij}\partial^\alpha h^{ij\dagger} + \frac{1}{2}\partial_\alpha h_{ij}\partial^\alpha h^\dagger_{ij} + \frac{1}{2}\partial_\alpha h^{ij}_k \partial^\alpha h^{ij\dagger}_k \right.$$
$$\left. + \frac{1}{2}\partial_\alpha h^i_{jk}\partial^\alpha h^{i\dagger}_{jk} + \partial_\alpha h^i \partial^\alpha h^{i\dagger} + \partial_\alpha h_i \partial^\alpha h^\dagger_i \right), \tag{9.304}$$

where the factors of 1/2 account for (ij) permutations.

References

[1] J. C. Pati and A. Salam, *Phys. Rev. D* **8**, 1240 (1973).

[2] J. C. Pati and A. Salam, *Phys. Rev. D* **10**, 275 (1974) [erratum, *ibid.* D **11**, 703 (1975)].

[3] H. Georgi and S. L. Glashow, *Phys. Rev. Lett.* **32**, 438 (1974).

[4] P. A. M. Dirac, *Proc. R. Soc. London, Ser. A* **133**, 60 (1931).

[5] P. A. Dirac, *Phys. Rev.* **74**, 817 (1948).

[6] M. Georgi and S. L. Glashow, *Phys. Rev. Lett.* **28**, 1494 (1972).

[7] G. 't Hooft, *Nucl. Phys. B* **79**, 276 (1974);

[8] A. M. Polyakov, *JETP Lett.* **20**, 194 (1974).

[9] J. S. Schwinger, *Phys. Rev.* **144**, 1087 (1966).

[10] M. Daniel, G. Lazarides, and Q. Shafi, *Nucl. Phys. B* **170**, 156 (1980).

[11] M. B. Einhorn and D. R. T. Jones, *Nucl. Phys. B* **196**, 475 (1982).

[12] M. E. Machacek and M. T. Vaughn, *Nucl. Phys. B* **249**, 70 (1985); *Nucl. Phys. B* **236**, 221 (1984); *Nucl. Phys. B* **222**, 83 (1983).

[13] S. P. Martin and M. T. Vaughn, *Phys. Rev. D* **50**, 2282 (1994) [Erratum, **78**, 039903 (2008)]; *Phys. Lett. B* **318**, 331 (1993).

[14] U. Amaldi, W. de Boer, and H. Furstenau, *Phys. Lett. B* **260**, 447 (1991).

[15] S. Dimopoulos and H. Georgi, *Nucl. Phys. B* **193**, 150 (1981).

[16] N. Sakai, *Z. Phys. C* **11**, 153 (1981).

[17] J. Hisano, H. Murayama, and T. Yanagida, *Nucl. Phys. B* **402**, 46 (1993).

[18] S. Dimopoulos, S. Raby, and F. Wilczek, *Phys. Rev. D* **24**, 1681 (1981).

[19] J. Ellis, S. Kelley, and D. V. Nanopoulos, *Phys. Lett. B* **249**, 441 (1990); **260**, 131 (1991).

[20] P. Langacker and M. x. Luo, *Phys. Rev. D* **44**, 817 (1991).

[21] F. Anselmo, L. Cifarelli, A. Peterman, and A. Zichichi, *Nuov. Cim.* **104A**, 1817 (1991); **115A**, 581 (1992).

[22] T. Dasgupta, P. Mamales, and P. Nath, *Phys. Rev. D* **52**, 5366 (1995).

[23] A. Masiero, D. V. Nanopoulos, K. Tamvakis, and T. Yanagida, *Phys. Lett. B* **115**, 380 (1982).

[24] B. Grinstein, *Nucl. Phys. B* **206**, 387 (1982).

[25] S. M. Barr, *Phys. Lett. B* **112**, 219 (1982).

[26] J. P. Derendinger, J. E. Kim, and D. V. Nanopoulos, *Phys. Lett. B* **139**, 170 (1984).

[27] H. Georgi, in C. E. Carlson, *Particles and Fields* AIP (1975).

[28] H. Fritzsch and P. Minkowski, *Ann. Phys.* **93**, 193 (1975).

[29] J. A. Harvey, P. Ramond, and D. B. Reiss, *Phys. Lett. B* **92**, 309 (1980).

[30] R. N. Mohapatra and B. Sakita, *Phys. Rev. D* **21**, 1062 (1980).

[31] J. A. Harvey, D. B. Reiss, and P. Ramond, *Nucl. Phys. B* **199**, 223 (1982).

[32] T. E. Clark, T. K. Kuo, and N. Nakagawa, *Phys. Lett. B* **115**, 26 (1982).

[33] F. Wilczek and A. Zee, *Phys. Rev. D* **25**, 553 (1982).

[34] C. S. Aulakh and R. N. Mohapatra, *Phys. Rev. D* **28**, 217 (1983).

[35] K. S. Babu, I. Gogoladze, P. Nath, and R. M. Syed, *Phys. Rev. D* **72**, 095011 (2005).

[36] K. S. Babu, I. Gogoladze, P. Nath, and R. M. Syed, *Phys. Rev. D* **74**, 075004 (2006).

[37] K. S. Babu, I. Gogoladze, P. Nath, and R. M. Syed, *Phys. Rev. D* **85**, 075002 (2012).

[38] K. S. Babu, I. Gogoladze, and Z. Tavartkiladze, *Phys. Lett. B* **650**, 49 (2007).

[39] S. Dimopoulos and F. Wilczek, Report. No. NSF-ITP-82-07 (1981 unpublished).

[40] M. Abud, F. Buccella, A. Della Selva, A. Sciarrino, R. Fiore, and G. Immirzi, *Nucl. Phys. B* **263**, 336 (1986).

[41] J. Basecq, S. Meljanac, and L. O'Raifeartaigh, *Phys. Rev. D* **39**, 3110 (1989).

[42] X. G. He and S. Meljanac, *Phys. Rev. D* **40**, 2098 (1989). doi:10.1103/PhysRevD.40.2098

[43] C. S. Aulakh, B. Bajc, A. Melfo, G. Senjanovic, and F. Vissani, *Phys. Lett. B* **588**, 196 (2004).

[44] S. Bertolini, L. Di Luzio, and M. Malinsky, *Phys. Rev. D* **81**, 035015 (2010).

[45] L. Di Luzio, arXiv:1110.3210 [hep-ph] (unpublished).

[46] S. Bertolini, L. Di Luzio, and M. Malinsky, *Phys. Rev. D* **85**, 095014 (2012).

[47] H. S. Goh, R. N. Mohapatra, and S. Nasri, *Phys. Rev. D* **70**, 075022 (2004).

[48] P. Nath and R. M. Syed, *Phys. Lett. B* **506**, 68 (2001).

[49] P. Nath and R. M. Syed, *Nucl. Phys. B* **618**, 138 (2001); *Nucl. Phys. B* **676**, 64 (2004).

[50] P. Nath and R. M. Syed, *JHEP* **0602**, 022 (2006).

[51] C. S. Aulakh and A. Girdhar, *Int. J. Mod. Phys. A* **20**, 865 (2005); *Nucl. Phys. B* **711**, 275 (2005).

[52] T. Fukuyama, A. Ilakovac, T. Kikuchi, S. Meljanac, and N. Okada, *J. Math. Phys.* **46**, 033505 (2005).

[53] J. R. Ellis and M. K. Gaillard, *Phys. Lett. B* **88** (1979) 315;

[54] C. T. Hill, *Phys. Lett. B* 135 **47** (1984).

[55] Q. Shafi and C. Wetterich, *Phys. Rev. Lett.* **52**, 875 (1984).

[56] L. J. Hall and U. Sarid, *Phys. Rev. Lett.* **70**, 2673 (1993).

[57] H. Georgi and C. Jarlskog, *Phys. Lett. B* **86**, 297 (1979).

[58] P. Ramond, R. G. Roberts, and G. G. Ross, *Nucl. Phys. B* **406**, 19 (1993).

[59] P. Nath, *Phys. Rev. Lett.* **76**, 2218 (1996); *Phys. Lett. B* **381**, 147 (1996).

[60] P. Minkowski, *Phys. Lett. B* **67**, 421 (1977); M. Gell-Mann, P. Ramond and R. Slansky, in *Supergravity*, P. van Nieuwenhuizen *et al.* (eds.), North-Holland (1979), p. 315.

[61] B. Pontecorvo, *Sov. Phys. JETP* **7**, 172 (1958).

[62] Z. Maki, M. Nakagawa, and S. Sakata, *Prog. Theor. Phys.* **28**, 870 (1962).

[63] G.L. Fogli *et al.*, *Phys. Rev. D*, **86**, 013012 (2012).

[64] S. Weinberg, *Phys. Rev. Lett.* **43**, 1566 (1979).

[65] T. Yanagida, *KEK Report 79-18*, Tsukuba (1979).

[66] S. L. Glashow, in M. Lévy *et al.* (eds), *Cargése Summer Institute: Quarks and Leptons*, Plenum Press (1980), p. 707.

[67] R. N. Mohapatra and G. Senjanovic, *Phys. Rev. Lett.* **44**, 912 (1980).

[68] M. Magg, C. Wetterich, *Phys. Lett. B* **94**, 61 (1980).

[69] J. Schechter and J. W. F. Valle, *Phys. Rev. D* **22**, 2227 (1980).

[70] G. Lazarides, Q. Shafi, and C. Wetterich, *Nucl. Phys. B* **181**, 287 (1981).

[71] R. N. Mohapatra and G. Senjanovic, *Phys. Rev. D* **23**, 165 (1981).

[72] R. Foot, H. Lew, X.-G. He, and G. C. Joshi, *Z. Phys. C* **44**, 441 (1989).

[73] G. Senjanovic, F. Wilczek, and A. Zee, *Phys. Lett. B* **141**, 389 (1984).

[74] J. Maalampi and M. Roos, *Phys. Rep.* **186**, 53 (1990).

[75] J. Maalampi, J.T. Peltoniemi, and M. Roos, *PLB* **220**, 441 (1989).

[76] T. Ibrahim and P. Nath, *Phys. Rev. D* **78**, 075013 (2008).

[77] T. Ibrahim and P. Nath, *Phys. Rev. D* **81**(3), 033007 (2010).

Further Reading

R. Cahn, *Semi-Simple Lie Algebras and Their Representatins*, Benjamin Cummings (1984).

H. Georgi, *Lie Groups in Particle Physics*, Benjamin Cummings (1982).

G. G. Ross, "Grand Unified Theories", Benjamin Cummings (1984).

R. Slansky, *Phys. Rep.* **79**, 1 (1981).

R. M. Syed, "Couplings in SO(10) grand unification," Ph.D. thesis, Northeastern University, hep-ph/0508153 (unpublished).

B. G. Wybourne, *Classical Groups for Physicists* Wiley (1974).

10

The MSSM Lagrangian

The foundation of the minimal supersymmetric standard model (MSSM) can be traced to the pioneering works by Golfand and Likhtman [1], Volkov and Akulov [2], and by Wess and Zumino [3–5]. Specifically, the realization of supersymmetry in four dimensions and the construction of a supersymmetric Lagrangian by Wess and Zumino [3–5] opened the way for rapid developments [6–8]. A further impetus for building supersymmetric models came from the work of Haag *et al.* [9], who showed that, aside from minor generalizations, the only graded algebra for an S-matrix constructed from a local relativistic quantum field theory is the supersymmetry algebra. Thus, supersymmetry turns out to be the only graded extension of the Lorentz covariant field theory. Supersymmetry also exhibits the remarkable renormalization properties [10, 11], where the F-terms are not renormalized. The first detailed model building was initiated by the work of Fayet [12–14], and this was followed by works on how supersymmetry may in part resolve the gauge hierarchy problem [15–17] via the no-renormalization theorems [10, 11]. However, since supersymmetry is not an exact symmetry of nature, it must be broken, but spontaneous breaking of supersymmetry is problematic [18], and to obtain realistic models one must add soft breaking terms by hand [19]. However, the number of such terms is rather large [20]. Thus, most of the progress in supersymmetry phenomenology occurred only after the formulation of supergravity models [21,22] described in later chapters. However, we discuss the globally supersymmetric models in this chapter as a prelude to the discussion in later chapters.

The spectrum of MSSM is based on the gauge group $SU(3)_C \otimes SU(2)_L \otimes U(1)_Y$, and its matter spectrum consists of all the particles in the standard model along with additional particles. Thus, all the particles now come in supermultiplets. This means that the gauge bosons are now accompanied by gauginos, and fermions are accompanied by sfermions. Specifically, for every quark there is a corresponding scalar quark or squark with all the same $SU(3) \otimes SU(2) \otimes U(1)_Y$ properties as the quark except for its spin. Similarly, for every lepton there is a corresponding scalar lepton or slepton, again with all the same $SU(3)_C \otimes SU(2)_L \otimes U(1)_Y$ properties as the lepton. In the supersymmetry extension, the Higgs boson doublet is now part of a supersymmetry multiplet with a supersymmetric partner – the Higgsino doublet, which is the spin $1/2$ partner of the Higgs boson doublet. However, unlike the case for the standard model the Higgs sector now has two doublets, H_1 and H_2, and, correspondingly, two Higgsino doublets. There are at least two reasons for requiring a pair of doublets in the Higgs sector. One reason is anomaly cancellation. The fact that the Higgs boson now comes with its fermionic partner which couples to gauge bosons implies that there is an additional contribution to the gauge anomaly from the fermionic partner of the Higgs boson. This anomaly can be cancelled by inclusion of a second Higgs multiplet which carries the opposite hypercharge, as shown in Table 10.1. The second reason relates to Yukawa couplings. As we have seen in the discussion of Chapter 8, in the supersymmetric model the superpotential contains the Yukawa coupling, and it is a holomorphic function. Since only the Higgs field and not its Hermitian conjugate is allowed in the superpotential, one needs two Higgs doublets, one to couple to the down quarks and the leptons and the other to couple to the up quarks. In the notation of Table 10.1 it is the field H_1 which couples to the down quarks and the leptons while the doublet H_2 couples to the up quarks. Thus, after spontaneous breaking when the neutral components of the Higgs bosons H_1 and H_2 develop vacuum expectation values (VEVs), the down quark and lepton masses arise from the VEV of H_1 and the up quark mass arises from the VEV of H_2.

The MSSM Lagrangian consists of many parts. It includes the kinetic energy terms for the gauge fields and their self-interactions for the case of non-abelian gauge fields, kinetic energy terms for the matter fields and the Higgs fields, interactions of the matter fields with the gauge fields, and the Yukawa interactions arising from the superpotential. Further, one adds in an *ad hoc* manner soft terms for the sfermions and for the gauginos. In addition, one has gauge fixing and ghost terms in the Lagrangian. Thus, the total MSSM Lagrangian is given by

$$\mathcal{L}_{MSSM} = \mathcal{L}_1 + \mathcal{L}_2 + \mathcal{L}_3 + \mathcal{L}_4 + \mathcal{L}_5. \tag{10.1}$$

Here, $\mathcal{L}_1$ is the Lagrangian for the kinetic energy terms for the gauge fields and their self-interactions, $\mathcal{L}_2$ is the kinetic energy terms for the matter fields and their interactions with the gauge fields, $\mathcal{L}_3$ gives the Yukawa couplings, and $\mathcal{L}_4$ gives the soft breaking terms for the sfermions and for the gauginos, which are included in an *ad hoc* fashion. Finally, $\mathcal{L}_5$ contains the gauge-fixing and ghost terms in the Lagrangian. We give further details below of the various parts in the MSSM Lagrangian.

The Lagrangian for the vector multiples takes the form

$$\mathcal{L}_1 = \int d^2\theta \left[\frac{1}{4} W_1^\alpha W_{1\alpha} + \frac{1}{4} Tr(W_{2A}^\alpha W_{2A\alpha}) + \frac{1}{4} Tr(W_{3a}^\alpha W_{3a\alpha}) \right] + \text{h.c.}, \tag{10.2}$$

Table 10.1 $SU(3)_C$, $SU(2)_L$, and $U(1)_Y$ *properties of the matter fields in the MSSM. The first two entries in the bracket in the second column refer to the* $SU(3)_C$ *and* $SU(2)_L$ *representations while the third entry refers to the value of the hypercharge* Y *where* Y *is defined so that* $Q = T_3 + Y$.

Superfields	$(SU(3), SU(2), U(1)^Y$	Spin 0,	spin 1/2
$\hat{q}$	$\left(3, 2, \frac{1}{6}\right)$	$\begin{pmatrix} \tilde{u}_L \\ \tilde{d}_L \end{pmatrix}$,	$\begin{pmatrix} u_L \\ d_L \end{pmatrix}$
$\hat{u}^c$	$\left(3, 1, -\frac{2}{3}\right)$	$\tilde{u}_R^*$,	u_L^c
$\hat{d}^c$	$\left(3, 1, \frac{1}{3}\right)$	$\tilde{d}_R^*$,	d_L^c
$\hat{\ell}$	$\left(1, 2, -\frac{1}{2}\right)$	$\begin{pmatrix} \tilde{\nu}_L \\ \tilde{e}_L^- \end{pmatrix}$,	$\begin{pmatrix} \nu_L \\ e_L^- \end{pmatrix}$
$\hat{e}^c$	$(1, 1, 1)$	$\tilde{e}_R^*$,	e_L^c
$\hat{H}_1$	$\left(1, 2, -\frac{1}{2}\right)$	$\begin{pmatrix} H_1^0 \\ H_1^- \end{pmatrix}$,	$\begin{pmatrix} \tilde{H}_1^0 \\ \tilde{H}_1^- \end{pmatrix}$
$\hat{H}_2$	$\left(1, 2, \frac{1}{2}\right)$	$\begin{pmatrix} H_2^+ \\ H_2^0 \end{pmatrix}$,	$\begin{pmatrix} \tilde{H}_2^+ \\ \tilde{H}_2^0 \end{pmatrix}$

where W^α are chiral superfields for the field strengths of the vector multiplet where we work in the Wess–Zumino gauge. Thus, the gauge kinetic energy for the abelian $U(1)$ case is given by W_1 so that integration over $d^2\theta$ gives

$$\mathcal{L}_1(U(1)) = -\frac{1}{4}v_{\mu\nu}v^{\mu\nu} - i\lambda(x)\sigma^\mu\partial_\mu\bar{\lambda}(x) + \frac{1}{2}D(x)D(x). \tag{10.3}$$

For the non-abelian case of $SU(2)_L$ gauge bosons, one finds

$$\mathcal{L}_1(SU(2)) = -\frac{1}{4}W_{\mu\nu A}W_A^{\mu\nu} - i\lambda_A(x)\sigma^\mu\partial_\mu\bar{\lambda}_A$$
$$+ ig\epsilon_{ABC}\lambda_A\sigma^\mu\bar{\lambda}_B W_{\mu C} + \frac{1}{2}D_A D_A, \tag{10.4}$$

where A, B, and C are 1–3 and

$$W_{\mu\nu A} = \partial_\mu W_{\nu A} - \partial_\nu W_{\mu A} + g\epsilon_{ABC}W_{\mu B}W_{\nu C}. \tag{10.5}$$

For $SU(3)_C$ we have

$$\mathcal{L}_1(SU(3)) = -\frac{1}{4}G_{\mu\nu a}G_a^{\mu\nu} - i\lambda_a(x)\sigma^\mu\partial_\mu\bar{\lambda}_a + ig_s f_{abc}\lambda_a\sigma^\mu\bar{\lambda}_b G_{\mu c} + \frac{1}{2}D_a D_a, \tag{10.6}$$

where a, b, and c are 1–8 and

$$G_{\mu\nu a} = \partial_\mu G_{\nu a} - \partial_\nu G_{\mu a} + g_s f_{abc}G_{\mu b}G_{\nu c}. \tag{10.7}$$

Next, we look at the kinetic energy terms for the matter fields and their couplings with the gauge fields. The Lagrangian for the interactions of the lepton, quark, and Higgs fields with the gauge fields are given by

$$\mathcal{L}_2 = \int d^2\,\theta\, d^2\bar{\theta}\,[C_q + C_l + C_H],$$ (10.8)

where

$$C_q = q^\dagger\,\exp\left[2\left(g'Y_q V + g\frac{\tau_A}{2}V^A + g_3\frac{\lambda_a}{2}V^a\right)\right]q$$

$$+\, u^\dagger\exp\left[2\left(g'Y_u V - g_3\frac{\lambda_a^*}{2}V^a\right)\right]u$$ (10.9)

$$+\, d^\dagger\,\exp\left[2\left(g'Y_d V - g_3\frac{\lambda_a^*}{2}V^a\right)\right]d$$ (10.10)

$$C_\ell = \ell^\dagger\,\exp\left[2\left(g'Y_\ell V + g\frac{\tau_A}{2}V^A\right)\right]\ell + e^\dagger\,\exp\left[2g'Y_e V\right]e$$ (10.11)

$$C_H = H_1^\dagger\,\exp\left[2\left(g'Y_{H_1}V + g\frac{\tau_A}{2}V^A\right)\right]H_1$$

$$+\, H_2^\dagger\,\exp\left[2\left(g'Y_{H_2}V + g\frac{\tau_A}{2}V^A\right)\right]H_2,$$ (10.12)

where $A = 1, \ldots, 3$ and $a = 1, \ldots, 8$. The expansion of each of the terms above can be done using Eqs. (8.234), (8.236), and (8.237) in Chapter 8.

The Lagrangian $\mathcal{L}_3$ arises from the superpotential term, i.e.,

$$\mathcal{L}_3 = \int d^2\theta\, W + \int d^2\bar{\theta}\,\overline{W}$$

$$= \frac{\partial W}{\partial\phi_i}F_i - \frac{1}{2}M_{ij}\psi_i\psi_j + \text{h.c.}$$ (10.13)

For the MSSM we can write the superpotential to cubic order so that

$$W = \lambda_d H_1 q d^c - \lambda_u H_2 q u^c + \lambda_e H_1 \ell e^c - \mu\epsilon_{ij}H_1^i H_2^j,$$ (10.14)

where we have suppressed the generational indices. The reason for the negative sign in front of the λ_u relative to the other terms is the following:

$$H_1 q d^c = \epsilon_{ij}H_1^i q^j d^c = (H_1^0 d - H_1^- u)d^c$$
$$H_2 q u^c = \epsilon_{ij}H_2^i q^j d^c = (H_2^+ d - H_2^0 u)u^c.$$ (10.15)

Thus, with an extra minus in the λ_u term in the superpotential, Eq. (10.14) will give all mass terms coming with a positive sign after the Higgs fields develop VEVs. One may add to Eq. (10.14) R-parity-violating interactions. Such gauge-invariant but R-parity-violating terms up to cubic order in the matter fields can be written as

$$W_{\not R} = \lambda_1 LQd^c + \lambda_2 LLe^c + \lambda_3 u^c d^c d^c + \mu' LH_2,$$ (10.16)

where again we have suppressed the generation indices. R-parity-violating terms violate lepton number and/or baryon number, and can give rise to fast proton decay unless additional constraints are taken into account.

In the MSSM, one supplements the supersymmetric Lagrangian in an *ad hoc* fashion with soft supersymmetry breaking terms. As mentioned earlier the number of such terms one can add is rather large [20]. Thus, under the constraint of gauge invariance, one may add over 100 parameters including soft parameters and μ. Limiting to a subset of the hundred or so parameters leads to constrained MSSM or CMSSM models. However, the landscape of such models is enormous. For example a simplified version of soft terms consists of

$$
\begin{aligned}
-\mathcal{L}_4 =\ & m_1^2 |H_1|^2 + m_2^2 |H_2|^2 - [m_{12}^2 \epsilon_{ij} H_1^i H_2^j + \text{h.c.}] \\
& + M_{\tilde{Q}}^2 [\tilde{u}_L^* \tilde{u}_L + \tilde{d}_L^* \tilde{d}_L] + M_{\tilde{U}}^2 \tilde{u}_R^* \tilde{u}_R + M_{\tilde{D}}^2 \tilde{d}_R^* \tilde{d}_R \\
& + M_{\tilde{L}}^2 [\tilde{\nu}_e^* \tilde{\nu}_e + \tilde{e}_L^* \tilde{e}_L] + M_{\tilde{E}}^2 \tilde{e}_R^* \tilde{e}_R \\
& + \epsilon_{ij} [\lambda_e A_e H_1^i \tilde{l}_L^j \tilde{e}_R^* + \lambda_d A_d H_1^i \tilde{q}_L^j \tilde{d}_R^* - \lambda_u A_u H_2^i \tilde{q}_L^j \tilde{u}_R^* + \text{h.c.}] \\
& + \frac{1}{2} [\tilde{M}_3 \lambda_a \lambda_a + \tilde{M}_2 \lambda_A \lambda_A + \tilde{M}_1 \lambda_0 \lambda_0],
\end{aligned}
\tag{10.17}
$$

where we have suppressed the generational indices and λ_0 denotes the bino. In supergravity models the number of parameters is significantly reduced [21, 22]. In addition to the above, we would need to add to the Lagrangian gauge-fixing and the ghost terms so that

$$
\mathcal{L}_5 = \mathcal{L}_{\text{gauge-fixing}} + \mathcal{L}_{\text{ghost}}.
\tag{10.18}
$$

10.1 The Scalar Potential

The scalar potential of the theory arises after elimination of the auxiliary fields in the Lagrangian, which are the F- and D-terms. We consider the F-terms first. They arise from Eqs. (8.129) and (10.13). Together they give

$$
\mathcal{L}_{aux}^F = \sum_i \left[\frac{\partial W}{\partial \phi_i} F_i + \frac{\partial \bar{W}}{\partial \phi_i^*} F_i^* + F_i F_i^* \right].
\tag{10.19}
$$

Varying $\mathcal{L}_{aux}^F$ with respect to F_i^* and F_i, one gets

$$
F_i = -\frac{\partial \bar{W}}{\partial \phi_i^*}
$$

$$
F_i^* = -\frac{\partial W}{\partial \phi_i}.
\tag{10.20}
$$

Substitution back into $\mathcal{L}_{aux}^F$ gives the result

$$
\begin{aligned}
\mathcal{L}_{aux}^F &= -\sum_i F_i F_i^* \\
&= -\sum_i \frac{\partial W}{\partial \phi_i} \frac{\partial \bar{W}}{\partial \phi_i^*}.
\end{aligned}
\tag{10.21}
$$

Similarly, parts of the Lagrangian containing the auxiliary fields D_A are

$$\mathcal{L}_{aux}^D = \sum_A \left[J_A D_A + \frac{1}{2} D_A D_A \right]. \tag{10.22}$$

Here, the first set of terms arises from the couplings of the matter fields with the gauge fields (e.g., see Eqs. (8.234), (8.236), and (8.237)) while the last set of terms arise purely from the gauge sector (see Eqs. (10.3), (10.4), and (10.6)). Since the D-terms are auxiliary fields, we can eliminate them by using field equations. Thus, varying Eq. (10.22) with respect to D_A one finds

$$D_A = -J_A \equiv -\frac{\partial \mathcal{L}_{\text{matter-gauge}}}{\partial D_A}, \tag{10.23}$$

where $\mathcal{L}_{\text{matter-gauge}}$ is the Lagrangian which contains the coupling of the matter fields with the gauge fields. Substitution of Eq. (10.23) into Eq. (10.22) gives

$$\mathcal{L}_{aux}^D = -\frac{1}{2} \sum_A J_A J_A = -\frac{1}{2} \sum_A D_A D_A. \tag{10.24}$$

The scalar potential is given by Eqs. (10.21), (10.24), and (10.17). Thus, one finds

$$V = V_F + V_D + V_{soft}, \tag{10.25}$$

where

$$V_F = -\mathcal{L}_{aux}^F = \sum_i F_i F_i^* = \sum_i \left| \frac{\partial W}{\partial \phi_i} \right|^2 \tag{10.26}$$

$$V_D = -\mathcal{L}_{aux}^D = \frac{1}{2} \sum_A D_A D_A = \frac{1}{2} \sum_A \left| \frac{\partial \mathcal{L}_{\text{matter-gauge}}}{\partial D_A} \right|^2 \tag{10.27}$$

$$V_{soft} = -\mathcal{L}_4. \tag{10.28}$$

In Eq. (10.26), ϕ_i stand for all the scalar fields in the theory, and the sum on A in Eq. (10.27) runs over all the gauge groups.

10.2 The Fermion Sector

Let us look at the masses for the quarks and the leptons. They arise from Yukawa couplings. In the two-component notation we have

$$\mathcal{L}_{Yuk}^f = -\lambda_e \epsilon_{ij} H_1^i L^j e_L^c - \lambda_d \epsilon_{ij} H_1^i Q^j d_L^c + \lambda_u \epsilon_{ij} H_2^i Q^j u_L^c + \text{h.c.} \tag{10.29}$$

After the VEV formation for the Higgs we have

$$v_1 = \sqrt{2}\langle H_1^1 \rangle = v\cos\beta$$
$$v_2 = \sqrt{2}\langle H_2^2 \rangle = v\sin\beta$$
$$v = \sqrt{v_1^2 + v_2^2}$$
$$\langle H_1^2 \rangle = 0 = \langle H_2^1 \rangle. \tag{10.30}$$

Thus, the fermion mass terms are given by

$$\mathcal{L}_m^f = -\frac{1}{\sqrt{2}}\lambda_e v_1(e_L e_L^c + \bar{e}_L \bar{e}_L^c) - \frac{1}{\sqrt{2}}\lambda_d v_1(d_L d_L^c + \bar{d}_L \bar{d}_L^c) - \frac{1}{\sqrt{2}}\lambda_u v_2(u_L u_L^c + \bar{u}_L \bar{u}_L^c). \tag{10.31}$$

We define Dirac spinors so that

$$\psi_e = \begin{pmatrix} e_\alpha \\ \bar{e}^{c\dot\alpha} \end{pmatrix}, \quad \psi_d = \begin{pmatrix} d_\alpha \\ \bar{d}^{c\dot\alpha} \end{pmatrix}, \quad \psi_u = \begin{pmatrix} u_\alpha \\ \bar{u}^{c\dot\alpha} \end{pmatrix}, \tag{10.32}$$

In terms of Dirac spinors, the fermion masses have the conventional form

$$\mathcal{L}_m^f = -m_e \bar{\psi}_e \psi_e - m_d \bar{\psi}_d \psi_d - m_u \bar{\psi}_u \psi_u, \tag{10.33}$$

where

$$m_e = \frac{\sqrt{2}\lambda_e}{g} M_W \cos\beta$$

$$m_d = \frac{\sqrt{2}\lambda_d}{g} M_W \cos\beta$$

$$m_u = \frac{\sqrt{2}\lambda_u}{g} M_W \sin\beta \tag{10.34}$$

and where we have used the relations $v_1 = 2M_W \cos\beta/g$ and $v_2 = 2M_W \sin\beta/g$.

10.3 Gaugino–Higgsino Sector

From the expansion of Eq. (10.12) the gaugino–Higgsino couplings are given by

$$\mathcal{L}_{\text{gaugino-Higgsino}} = \frac{ig}{\sqrt{2}}\left[H_1^* \tau_A \lambda_A \tilde{H}_1 - H_1 \tau_A \bar{\lambda}_A \overline{\tilde{H}}_1\right]$$
$$+ \frac{ig}{\sqrt{2}}\left[H_2 \tau_A \lambda_A \tilde{H}_2 - H_2 \tau_A \bar{\lambda}_A \overline{\tilde{H}}_2\right]$$

$$+ \sqrt{2} \left[H_1 Y_{H_1} \lambda_0 \tilde{H}_1 - H_1 Y_{H_1} \bar{\lambda}_0 \overline{\tilde{H}}_1 \right]$$
$$+ \sqrt{2} \left[H_2 Y_{H_2} \lambda_0 \tilde{H}_2 - H_2 Y_{H_2} \bar{\lambda}_0 \overline{\tilde{H}}_2 \right]$$
$$+ \mu [\tilde{H}_1 \epsilon \tilde{H}_2 + \text{h.c.}] + \frac{1}{2} \left[M_1 \lambda_0 \lambda_0 + M_2 \lambda_A \lambda_A + M_3 \lambda_a \lambda_a + \text{h.c.} \right],$$

$$(10.35)$$

where λ_0, λ_A and λ_a are $U(1)$, $SU(2)$ and $SU(3)$ gauginos and where $Y_{H_1} = -1$ and $Y_{H_2} = 1$. After spontaneous breaking of the electroweak symmetry, we can write the mass terms in the Lagrangian for the gauginos and the Higgsinos as follows:

$$\mathcal{L}^m_{\text{gaugino–Higgsino}} = \mathcal{L}^{\tilde{\chi}^\pm}_m + \mathcal{L}^{\tilde{\chi}^0}_m + \mathcal{L}^{\text{gluinos}}_m. \qquad (10.36)$$

The various parts are as follows:

$$\mathcal{L}^{\tilde{\chi}^\pm}_m = i\sqrt{2} M_W \cos\beta \lambda^+ \tilde{H}_1^2 + i\sqrt{2} M_W \sin\beta \lambda^- \tilde{H}_2^1 - \mu \tilde{H}_1^2 \tilde{H}_2^1 + M\lambda^+ \lambda^- + \text{h.c.}$$
$$\mathcal{L}^{\tilde{\chi}^0}_m = i M_Z \cos\beta \cos\theta_W \lambda_3 \tilde{H}_1^1 - i M_Z \sin\beta \cos\theta_W \lambda_3 \tilde{H}_2^2$$
$$- i M_Z \cos\beta \sin\theta_W \lambda_0 \tilde{H}_1^1 + i M_Z \sin\beta \sin\theta_W \lambda_0 \tilde{H}_2^2$$
$$+ \mu \tilde{H}_1^1 \tilde{H}_2^2 + \frac{1}{2} M_1 \lambda_0 \lambda_0 + \text{h.c.}$$
$$\mathcal{L}^{\text{gluinos}}_m = \frac{M_3}{2} (\lambda_a \lambda_a + \bar{\lambda}_a \bar{\lambda}_a). \qquad (10.37)$$

We can express the gluino mass part in terms of the Majorana spinors. To this end, we introduce the Majorana spinor ψ_a where

$$\psi_a = \begin{pmatrix} -i\lambda_{a\sigma} \\ i\bar{\lambda}_a^{\dot{\sigma}} \end{pmatrix}, \quad \bar{\psi}_a = \psi^\dagger \gamma^0 = (-i\lambda_a^\sigma, \ i\bar{\lambda}_{a\dot{\sigma}}), \qquad (10.38)$$

and where γ^0 is in the Weyl representation (see Chapter 22). The gluino mass terms can now be written in terms of Majorana spinors:

$$\mathcal{L}^{\text{gluinos}}_m = -\frac{M_3}{2} \bar{\psi}_a \psi_a, \quad a = 1, \ldots, 8. \qquad (10.39)$$

In the charged gaugino–Higgsino sector, the mass terms take the form

$$\mathcal{L}^{\tilde{\chi}^\pm}_m = \left[\sqrt{2} M_W \cos\beta i \lambda^+ \tilde{H}_1^2 + \sqrt{2} M_W \sin\beta i \lambda^- \tilde{H}_2^1 \right.$$
$$\left. + \mu \tilde{H}_1^2 \tilde{H}_2^1 - M_2 \lambda^+ \lambda^- + \text{h.c.} \right] \qquad (10.40)$$

To construct a mass matrix we define

$$\zeta^+ = \begin{pmatrix} -i\lambda^+ \\ \tilde{H}_2^1 \end{pmatrix}, \quad \zeta^- = \begin{pmatrix} -i\lambda^- \\ \tilde{H}_1^2 \end{pmatrix}. \qquad (10.41)$$

In the ζ^+ and ζ^- basis we can write the mass terms in the charged gaugino–Higgsino sector in the following form:

$$\mathcal{L}_m^{\tilde{\chi}^\pm} = -\frac{1}{2} \left(\zeta^{+T} \zeta^{-T} \right) \begin{pmatrix} 0 & M_C^T \\ M_C & 0 \end{pmatrix} \begin{pmatrix} \zeta^+ \\ \zeta^- \end{pmatrix} + \text{h.c.}, \tag{10.42}$$

where M_C is given by

$$M_C = \begin{pmatrix} M_2 & \sqrt{2} M_W \sin\beta \\ \sqrt{2} M_W \cos\beta & \mu \end{pmatrix}. \tag{10.43}$$

Since M_C is not a symmetric matrix, its diagonalization requires a biunitary transformation so that

$$M_{\chi^\pm} = U^* M_C V^\dagger, \tag{10.44}$$

where U and V are unitary matrices and $M_{\chi^\pm}$ is the matrix with diagonal elements. The two-component mass eigenstates are $\zeta_{Di}^\pm$, where

$$\zeta_{Di}^- = U_{ij} \zeta_j^-, \quad \zeta_{Di}^+ = V_{ij} \zeta_j^+. \tag{10.45}$$

Let us assume that the eigenvalues are $m_{\tilde{\chi}_1^\pm}$ and $m_{\tilde{\chi}_2^\pm}$, and define Dirac fields so that

$$\tilde{\chi}_1 = \begin{pmatrix} \zeta_{D1\sigma}^+ \\ \bar{\zeta}_{D1}^{-\dot{\sigma}} \end{pmatrix}, \quad \tilde{\chi}_2 = \begin{pmatrix} \zeta_{D2\sigma}^+ \\ \bar{\zeta}_{D2}^{-\dot{\sigma}} \end{pmatrix}. \tag{10.46}$$

Further, using

$$\overline{\tilde{\chi}}_i = \bar{\zeta}_{Di}^\dagger \gamma^0 = (\zeta_{Di}^{-\sigma}, \overline{\zeta^+}_{i\dot{\sigma}}), \quad i = 1, 2 \tag{10.47}$$

the chargino mass terms take the form

$$\mathcal{L}_m^{\text{chargino}} = -m_{\tilde{\chi}_1^\pm} \overline{\tilde{\chi}}_1^\pm \tilde{\chi}_1^\pm - m_{\tilde{\chi}_2^\pm} \overline{\tilde{\chi}}_2^\pm \tilde{\chi}_2^\pm. \tag{10.48}$$

The eigenvalues of the chargino mass-squared matrix are given by

$$m_{\chi_{1,2}^\pm} = \frac{1}{2} \Big[(M_2^2 + \mu^2 + 2M_W^2) \\ \pm \left((M_2^2 + \mu^2 + 2M_W^2)^2 - 4(M_2\mu - M_W^2 \sin 2\beta)^2 \right)^{1/2} \Big]. \tag{10.49}$$

Note that $\det(M_C) = M_2\mu - M_W^2 \sin 2\beta$. When $(M_2\mu - M_W^2 \sin 2\beta) < 0$ we need to make a further transformation so that positivity of the eigenvalues will be maintained. This can be done by replacing V by PV, where

$$P = \begin{pmatrix} 1 & 0 \\ 0 & \epsilon \end{pmatrix} \tag{10.50}$$

and $\epsilon = sign(M_2\mu - M_W^2 \sin 2\beta)$.

Next, we look at the mass terms in the neutral-gaugino–Higgsino sector:

$$\mathcal{L}_m^{\chi^0} = iM_Z \cos\theta_W \cos\beta \lambda_3 \tilde{H}_1^1 - iM_Z \cos\theta_W \sin\beta \lambda_3 \tilde{H}_2^2$$
$$- iM_Z \sin\theta_W \cos\beta \lambda_B \tilde{H}_1^1 + iM_Z \sin\theta_W \sin\beta \lambda_B \tilde{H}_2^2$$
$$+ \mu \tilde{H}_1^1 \tilde{H}_2^2 + M_1 \lambda_B \lambda_B + \frac{1}{2} M_2 \lambda_3 \lambda_3 + \text{h.c.} \qquad (10.51)$$

Here, we define the basis states so that

$$\psi_N = \begin{pmatrix} -i\lambda_B \\ -i\lambda_3 \\ \tilde{H}_1^1 \\ \tilde{H}_2^2 \end{pmatrix}. \qquad (10.52)$$

In this basis we can write the neutralino mass matrix in the following form:

$$M_N = \begin{pmatrix} M_1 & 0 & -M_Z \sin\theta_W \cos\beta & M_Z \sin\theta_W \sin\beta \\ 0 & M_2 & M_Z \cos\theta_W \cos\beta & -M_Z \cos\theta_W \sin\beta \\ -M_Z \sin\theta_W \cos\beta & M_Z \cos\theta_W \cos\beta & 0 & -\mu \\ M_Z \sin\theta_W \sin\beta & -M_Z \cos\theta_W \sin\beta & -\mu & 0 \end{pmatrix}.$$

$$(10.53)$$

M_N is a real symmetric matrix. We can diagonalize it by a unitary transformation as follows:

$$X^* M_N X^{-1} = M_{\chi^0}, \qquad (10.54)$$

and the mass eigenstates in the two-component notation are ζ_i^0 where

$$\zeta_i^0 = X_{ij} \psi_{Nj}. \qquad (10.55)$$

The mass terms are then given in the form

$$\mathcal{L}_m^{\chi^0} = -\frac{1}{2} m_{\chi_i^0} (\zeta_i^0 \zeta_i^0 + \overline{\zeta_i^0 \zeta_i^0}). \qquad (10.56)$$

We define Majorana spinors so that

$$\tilde{\chi}_i^0 = \begin{pmatrix} \zeta_{i\sigma}^0 \\ \overline{\zeta_i^0}^{\dot{\sigma}} \end{pmatrix}. \qquad (10.57)$$

In terms of the Majorana spinors, the neutralino mass terms take the form

$$\mathcal{L}_m^{\chi^0} = -\frac{1}{2} m_{\chi_i^0} \overline{\tilde{\chi}_i^0} \tilde{\chi}_i^0. \qquad (10.58)$$

Further details of the interactions in the MSSM are given in Section 21.7.

References

[1] Y. A. Golfand and E. P. Likhtman, *JETP Lett.* **13**, 323 (1971).
[2] D. V. Volkov and V. P. Akulov, *Phys. Lett. B* **46**, 109 (1973).
[3] J. Wess and B. Zumino, *Phys. Lett. B* **49**, 52 (1974).
[4] J. Wess and B. Zumino, *Nucl. Phys. B* **78**, 1 (1974).
[5] J. Wess and B. Zumino, *Nucl. Phys. B* **70**, 39 (1974).
[6] S. Ferrara, J. Wess, and B. Zumino, *Phys. Lett. B* **51**, 239 (1974).
[7] A. Salam and J. A. Strathdee, *Phys. Rev. D* **11**, 1521 (1975).
[8] A. Salam and J. A. Strathdee, *Nucl. Phys. B* **86**, 142 (1975).
[9] R. Haag, J. T. Lopuszanski, and M. Sohnius, *Nucl. Phys. B* **88**, 257 (1975).
[10] J. Iliopoulos and B. Zumino, *Nucl. Phys. B* **76**, 310 (1974).
[11] M. T. Grisaru, W. Siegel, and M. Rocek, *Nucl. Phys. B* **159**, 429 (1979).
[12] P. Fayet, *Nucl. Phys. B* **90**, 104 (1975); *Phys. Lett. B* **64**, 159 (1976).
[13] G. R. Farrar and P. Fayet, *Phys. Lett. B* **79**, 442 (1978).
[14] P. Fayet, *Phys. Lett. B* **86**, 272 (1979).
[15] S. Dimopoulos and H. Georgi, *Nucl. Phys. B* **193**, 150 (1981).
[16] N. Sakai, *Z. Phys. C* **11**, 153 (1981).
[17] R. K. Kaul, *Phys. Lett. B* **109**, 19 (1982).
[18] L. O'Raifeartaigh, *Nucl. Phys. B* **96**, 331 (1975).
[19] L. Girardello and M. T. Grisaru, *Nucl. Phys. B* **194**, 65 (1982).
[20] S. Dimopoulos and D. W. Sutter, *Nucl. Phys. B* **452**, 496 (1995).
[21] A. H. Chamseddine, R. L. Arnowitt, and P. Nath, *Phys. Rev. Lett.* **49**, 970 (1982).
[22] R. Barbieri, S. Ferrara, and C. A. Savoy, *Phys. Lett. B* **119**, 343 (1982).

Further Reading

H. Baer and X. Tata, *Weak Scale Supersymmetry: From Superfields to Scattering Events*, Cambridge University Press (2012).

M. Drees, R. Godbole, and P. Roy, *Theory and Phenomenology of Sparticles: An Account of Four-dimensional N = 1 Supersymmetry in High Energy Physics*, World Scientific (2004).

H. E. Haber and G. L. Kane, *Phys. Rep.* **117**, 75 (1985).

S. P. Martin, *Adv. Ser. Direct. High Energy Phys.* **21**, 1 (2010) [*Adv. Ser. Direct. High Energy Phys.* **18**, 1 (1998)] [hep-ph/9709356].

S. Weinberg, *The Quantum Theory of Fields, vol. 3, Modern Applications*, Cambridge University Press (1996).

11

$N = 1$ Supergravity

Supersymmetry as formulated by Golfand and Likhtman, Volkov and Akulov, and Wess and Zumino discussed in Chapter 8 is a global symmetry. However, if supersymmetry is to be a fundamental symmetry of nature, it ought to be a local symmetry [1,2]. However, it was soon realized that any effort to gauge supersymmetry brings in gravity [1,2], and since gravity is described by the Einstein theory, a direct approach to making supersymmetry a local symmetry was to extend the geometry of Einstein gravity to superspace. That is, one extends the ordinary space-time with co-ordinates x^μ to superspace with co-ordinates $z^A = (x^\mu, \theta^\alpha)$, where θ^α are the anti-commuting Grassmann co-ordinates. Further, the invariance of the line element $ds^2 = dx^\mu \, g_{\mu\nu}(x) \, dx^\nu$ under general co-ordinate transformations is extended to invariance of the line element $ds^2 = dz^A \, g_{AB}(z) \, dz^B$ under the general co-ordinate transformations in superspace $z^\Lambda = z'^\Lambda + \xi^\Lambda(z)$. The extended transformations of the metric contain in it Einsten transformations as well as transformations of local supersymmetry. Further, the metric now contains in addition to the vierbein also spin $3/2$ and other fields. Thus, for example, for the case $A = \mu$ and $B = \alpha$, one may expand $g_{\mu\alpha}$ so that

$$g_{\mu\alpha}(z) = -i(\bar{\theta}\gamma_m)_\alpha e_\mu^m + (\bar{\psi}_\mu\gamma_m\theta)(\bar{\theta}\gamma^m)_\alpha + \cdots, \qquad (11.1)$$

where e_μ^m is the vierbein of Einstein gravity and ψ^μ is a spin $3/2$ field [3] and the $\cdots$ indicates additional terms in the metric. This theory was referred to as gauge supersymmetry, and was the first supergravity theory. The tangent space group of this theory is $OSp(3,1|4)$. Details on this approach are given in Section 21.4. Subsequent to the formulation of gauge supersymmetry the modern form of supergravity was formulated [4,5], and this theory was originally formulated in ordinary space. The connection of gauge supersymmtry to the current form of supergravity is the following: gauge supersymmetry reduces to supergravity in a group theoretical contraction, as discussed in [6–8] and in Section 21.4. Below, we discuss further aspects of $N = 1$ supergravity.

We have already seen, as discussed above in the framework of gauge supersymmetry in superspace, that local supersymmetry requires gravity to appear [1]. We can show this phenomenon in a non-superspace formulation by considering the case of a free scalar supermultiplet consisting of the components ($\phi(x)$ and $\psi(x)$) for which the free Lagrangian is given by[*]

$$\mathcal{L} = -\partial_\mu \phi^\dagger \partial^\mu \phi - \overline{\psi} \frac{1}{i} \gamma^\mu \partial_\mu \psi. \tag{11.2}$$

In global supersymmetry, the Lagrangian of Eq. (11.2) is invariant under the following supersymmetry transformations:

$$\delta\phi(x) = \overline{\varepsilon}\psi(x) \tag{11.3}$$

$$\delta\psi(x) = -i\gamma^\mu [\partial_\mu \phi(x)]\varepsilon, \tag{11.4}$$

where ϵ is an anti-commuting parameter independent of space-time co-ordinates. To make the Lagrangian invariant under a local supersymmetry transformation, we make ϵ a function of space-time co-ordinates, i.e., we promote ϵ to $\epsilon(x)$. It is then easily seen that the Lagrangian of Eq. (11.2) in this case is no longer invariant, and one is left with a residual term, which is

$$\delta\mathcal{L} = \partial_\mu \overline{\varepsilon}(x) \gamma^\mu \gamma^\nu \psi(x) \partial_\nu \phi(x) + \text{total div.}, \tag{11.5}$$

where "total div." indicates total divergence, which vanishes in the action. Now, to achieve an invariance we proceed as in constructing an ordinary gauge theory. In ordinary gauge theories, one introduces a compensating field to restore gauge invariance. To do this here requires that we introduce a spin 3/2 field $\psi_\alpha^\mu(x)$, where α is the Dirac spinor index with the interaction

$$\Delta\mathcal{L} = -\kappa \overline{\psi}_\mu(x) \gamma^\mu \gamma^\nu \psi(x) \partial_\nu \phi(x). \tag{11.6}$$

Here, κ is a dimensioned parameter with the dimension of inverse mass. We assume that ψ_α^μ has the transformation property under local supersymmetry gauge transformations so that

$$\delta\psi^\mu = \kappa^{-1}\partial^\mu \varepsilon(x). \tag{11.7}$$

It is easily seen that the addition of Eq. (11.6) cancels the term in Eq. (11.5). Now, the addition of a spin 3/2 term leads us to introduce the kinetic energy term for a spin 3/2 particle. Such a kinetic energy term has been known since the work Rarita and Schwinger [3], and it reads

$$\mathcal{L}_{RS} = -\frac{1}{2}\epsilon^{\mu\nu\rho\sigma}\overline{\psi}_\mu \gamma_5 \gamma_\nu \partial_\rho \psi_\sigma. \tag{11.8}$$

In supersymmetry, the spin 3/2 field should belong to a supermultiplet, and the simplest such multiplet consists of spins $(3/2, 2)$. Thus, we are led to introduce a

[*] In this chapter and in further discussion of supergravity-based analyses we use four-component spinors.

massless spin 2 field, and the natural candidate for such a field is gravity, and so we have gravity appearing in a natural way in gauging of supersymmetry. The Lagrangian for gravity is, of course, known, and it is the Einstein Lagrangian, but since we have spinors around, we use vierbein e_μ^a so that

$$g_{\mu\nu} = e_\mu{}^a(x)\eta_{ab}e_\nu{}^b(x), \tag{11.9}$$

and write

$$\mathcal{L}_E = -\frac{1}{2}\kappa^{-2}eR(e_\mu{}^a, \omega_\mu{}^{ab}), \tag{11.10}$$

where e is the determinant of the vierbein, i.e., $e = det[e_\mu{}^a]$, R is the curvature scalar, and $\omega_\mu{}^{ab}$ is the spin connection. The Einstein Lagrangian, of course, has general covariance, and to make the full Lagrangian generally covariant we require that the spin 3/2 Lagrangian be generally covariant as well. This can be easily done by replacing ∂_ρ by the covariant derivative D_ρ, where

$$D_\rho = \partial_\rho + \frac{1}{2}\omega_\rho{}^{ab}s_{ab}, \tag{11.11}$$

where for notation see Chapter 2. This leads us to write the Lagrangian for the spin 3/2–spin 2 system so that [4, 5]

$$\mathcal{L}_{3/2-2} = -\frac{1}{2}\kappa^{-2}eR(e_\mu{}^a, \omega_\mu{}^{ab}) - \frac{1}{2}\epsilon^{\mu\nu\rho\sigma}\bar{\psi}_\mu\gamma_5\gamma_\nu D_\rho\psi_\sigma. \tag{11.12}$$

In the above, we have given a very heuristic argument to construct Eq. (11.12). However, there are several subtle points to establish to ensure that Eq. (11.12) is indeed invariant under local supersymmetry transformations. If one works within the second-order formalism, then $e_\mu{}^a$ and ψ_μ are treated as independent fields and $\omega_\mu{}^{ab}$ is taken to be $\omega_\mu{}^{ab}(e)$, where $\omega_\mu{}^{ab}$ is given in terms of the vierbein so that

$$\omega_{\mu ab}(e) = \frac{1}{2}e_a{}^\nu(\partial_\mu e_{b\nu} - \partial_\nu e_{b\mu}) - \frac{1}{2}e_b{}^\nu(\partial_\mu e_{a\nu} - \partial_\nu e_{a\mu})$$
$$- \frac{1}{2}e_a{}^\rho e_b{}^\sigma(\partial_\rho e_{c\sigma} - \partial_\sigma e_{c\rho})e^c{}_\mu. \tag{11.13}$$

Then, one corrects Eq. (11.12) by additional terms, specifically terms quartic in spin 3/2 fields, and corrects the transformation rules so that consistency is reached. One consequence is that the transformation laws one finds in the second-order formalism suggests that $\omega_{\mu ab}(e)$ is replaced by $\omega_{\mu ab}$, where

$$\omega_{\mu ab} = \omega_{\mu ab}(e) + \frac{\kappa^2}{4}(a_{\mu ba} - a_{\mu ab} + a_{ab\mu}) \tag{11.14}$$

and

$$a_{\mu ba} = e_b{}^\nu\bar{\psi}_\mu\gamma_a\psi_\nu, \text{ etc.} \tag{11.15}$$

Matters simplify if one works within the first-order formalism [9], where $e_\mu{}^a$, ψ_μ and $\omega_{\mu ab}$ are treated as independent variables. and $\omega_{\mu ab}$ is determined by the field equations. Now, the $\omega_{\mu ab}$ obtained this way gives the result of Eq. (11.14). In fact, the two approaches, i.e., the second order and the first order, are equivalent on the mass shell. However, the simplest approach is what is sometimes referred to as the 1.5 formalism [10]. To see this, let us look at the variation of the action so that

$$\delta I = \int \left(\frac{\delta I}{\delta e_\mu{}^a} \delta e_\mu{}^a + \frac{\delta I}{\delta \psi_\mu} \delta \psi_\mu + \frac{\delta I}{\delta \omega_{\mu ab}} \delta \omega_{\mu ab} \right) d^4 x = 0. \tag{11.16}$$

Here, we use the idea that the spin connection obeys the condition

$$\frac{\delta I}{\delta \omega_{\mu ab}} = 0, \tag{11.17}$$

and thus the last term can be dropped provided we use $\omega_{\mu ab}(e, \psi)$. This is equivalent to saying that one can set $\delta \omega_{\mu ab} = 0$. Now, one can establish the invariance of Eq. (11.12) under local supersymmetry transformations so that

$$\delta e_\mu{}^m = \frac{1}{2} \, \kappa \bar{\varepsilon}(x) \gamma^m \psi_\mu \tag{11.18}$$

$$\delta \psi_\mu = \kappa^{-1} D_\mu \varepsilon(x), \tag{11.19}$$

and $\delta w_{\mu ab} = 0$. A more detailed discussion of this can be found in the literature [10].

We now recall that global supersymmetry has the problem that spontaneous breaking produces a massless Goldstino field, which is undesirable. As we will discuss in Chapter 12 and 13, this problem is resolved in supergravity in that the massless spin 3/2 field of the supergravity multiplet absorbs the massess Goldstino and becomes massive. This phenomenon is the supersymmetric analog of the Higgs mechanism where a massless gauge field absorbs a Goldstone boson and becomes massive. We will discuss this phenomenon in greater length in the following chapters.

11.1 Supergravity with Auxiliary Fields

The supergravity that we have discussed above consisting of spin 2 and spin 3/2 fields closes only on shell, i.e., one needs to use field equations to establish the closure of the algebra. However, auxiliary fields are needed for closure of the algebra off shell. We are already familiar with the presence of auxiliary fields for the case of global supersymmetry, where, for example, for the chiral scalar multiplet one has, in addition to the spin 0 and spin 1/2 fields, the auxiliary field F. For the case of supergravity, auxiliary fields are also needed when we couple supergravity with the matter fields, i.e., if we have a different set of matter fields which we couple to supergravity we should be able to do so without changing the supergravity transformation laws. For $N = 1$ supergravity consisting of the spin 2 field e_μ^a and the spin 3/2 field ψ_μ, we seek the minimum set of auxiliary fields that can close the supergravity algebra off shell. It turns out that this minimal set consists of a scalar field S, a pseudo-scalar field P, and an axial vector A_μ [11, 12]. Thus, the supergravity Lagrangian with the minimal set of auxiliary fields is given by

$$\mathcal{L}_{SG} = -\frac{e}{2\kappa^2} R(e, \omega) - \frac{1}{2} \bar{\psi}_\mu R^\mu - \frac{e}{3} |u|^2 + \frac{e}{3} A_\mu A^\mu, \qquad (11.20)$$

where u is defined by

$$u = S - iP \qquad (11.21)$$

and R^μ is defined by

$$R^\mu = \epsilon^{\mu\nu\rho\sigma} \gamma_5 \gamma_\nu D_\rho(\omega) \psi_\sigma, \qquad (11.22)$$

where $D_\mu(\omega)$ is the covariant derivative defined in Eq. (11.11). $R(e, \omega)$ in Eq. (11.20) is given by

$$R = e_a{}^\mu e_b{}^\nu R_{\mu\nu}{}^{ab}, \qquad (11.23)$$

where $R_{\mu\nu}{}^{ab}$ is defined by

$$R_{\mu\nu}{}^{ab} = (\partial_\mu \omega_\nu{}^{ab} + \omega_\mu{}^{ac} \omega_{\nu c}{}^b) - (\mu \longleftrightarrow \nu). \qquad (11.24)$$

The supergravity Lagrangian of Eq. (11.20) with the minimal set of auxiliary fields S, P and A_μ is invariant under the following set of supersymmetry transformations [11]:

$$\delta_s e_\mu^a = \kappa \bar{\epsilon} \gamma^a \psi_\mu$$

$$\delta_s \psi_\mu = 2\kappa^{-1} D_\mu[\omega(e, \psi)]\epsilon + \frac{1}{3} \gamma_\mu (S - i\gamma_5 P)\epsilon + i\gamma_5 \left(\delta_\mu^\nu - \frac{1}{3} \gamma_\mu \gamma^\nu \right) \epsilon A_\nu, \quad (11.25)$$

while the auxiliary fields S, P, and A_μ transform as

$$\delta_s S = \frac{1}{2} e^{-1} \bar{\epsilon} \gamma_\mu R^\mu - \frac{\kappa}{2} \bar{\epsilon}(S + i\gamma_5 P)\gamma^\mu \psi_\mu + i\frac{\kappa}{2} \bar{\epsilon} \gamma_5 \psi_\nu A^\nu$$

$$\delta_s P = -\frac{i}{2} e^{-1} \bar{\epsilon} \gamma_5 \gamma_\mu R^\mu + i\frac{\kappa}{2} \bar{\epsilon} \gamma_5 (S + i\gamma_5 P)\gamma^\mu \psi_\mu + \frac{\kappa}{2} \bar{\epsilon} \psi_\nu A^\nu$$

$$\delta_s A_\mu = i\frac{3}{2} e^{-1} \bar{\epsilon} \gamma_5 \left(\delta_\mu^\nu - \frac{1}{3} \gamma_\mu \gamma^\nu \right) R_\nu - \frac{e}{4} \kappa \epsilon_{\mu\nu\rho\sigma} \bar{\epsilon} \gamma_5 \gamma^\rho \psi^\sigma A^\nu$$

$$+ \frac{i}{2} \kappa \bar{\epsilon} \gamma_5 (S - i\gamma_5 P)\psi_\mu + \kappa \bar{\epsilon} \gamma^\nu \left(\psi_\mu A_\nu - \frac{1}{2} \psi_\nu A_\mu \right), \qquad (11.26)$$

where $\epsilon(x)$ is the transformation parameter.

It should be noted that the choice of auxiliary fields is not unique, and other sets of auxiliary field choices exist such as the new minimal set [13] and the Breitenlohner set [14,15] as well as other possibilities [16–18]. Further, it is known that these various versions of $N = 1$ Poincaré supergravity theories with different sets of auxiliary fields can be derived from a unique superconformal theory. A good account of conformal supergravity and its reduction to Poincaré supergravities can be found in the papers of Kugo and Uehara [19,20] and the references quoted therein. It is found that superconformal tensor calculus simplifies in a significant way the derivation of couplings of supergravity with matter fields [21, 22]. Further, as discussed at the beginning of this chapter, the connection between gauge supersymmetry [1] and supergravity is established by noting that supergravity of [4, 5] can be obtained as a limiting case of gauge supersymmetry [6]. Supergravity can also be formulated in superspace [6–8, 23–28].

11.2 Tensor Calculus for Scalar and Vector Multiplets in Supergravity

Tensor calculus techniques are useful for obtaining products of chiral multiplets and of vector multiplets [29,30]. We shall make use of these techniques. Consider left chiral multiplets Γ^a with components (z^a, χ^a, h^a), where z^a is a complex scalar, χ^a is a left-handed Weyl spinor, and h^a are auxiliary fields. The components of these multiplets have the following supersymmetry transformations in the presence of couplings with supergravity:

$$\delta_s z^a = 2\bar{\epsilon}_R \chi^a$$

$$\delta_s \chi^a = h^a \epsilon_L + \hat{\slashed{D}} z^a \epsilon_R$$

$$\delta_s h^a = 2\bar{\epsilon}_R \hat{\slashed{D}} \chi^a - 2\kappa \bar{\eta}_R \chi^a, \tag{11.27}$$

where $\hat{D}_\mu z^a$, $\hat{D}_\mu \chi^a$, and η_R are defined as follows:

$$\hat{D}_\mu z^a = \partial_\mu z^a - \kappa \bar{\psi}_\mu \chi^a$$

$$\hat{D}_\mu \chi^a = D_\mu \omega(e,\psi)\chi^a - \frac{i}{2}\kappa A_\mu \chi^a - \frac{\kappa}{2}\slashed{D} z^a \psi_{\mu R} - \frac{\kappa}{2} h^a \psi_{\mu L}$$

$$\eta_R = \frac{1}{3}(u^* \epsilon_R - i\slashed{A}\epsilon_L). \tag{11.28}$$

Now, assume we are given a multiplet Γ^a of definite chirality. From Γ^a we can generate a multiplet Γ_a of opposite chirality which is given by the relation

$$\Gamma_a \equiv (\Gamma^a)^\dagger = (z_a, \chi_a, h_a), \tag{11.29}$$

where the components (z_a, χ_a, h_a) are related to the components (z^a, χ^a, h^a) as follows:

$$z_a = z^{a\dagger} = \phi_1^a - i\phi_2^a \tag{11.30}$$

$$\chi_a = C^{-1}\overline{\chi^a} \equiv (\chi^a)^c \tag{11.31}$$

$$h_a = h^{a\dagger} = F^a - iG^a, \tag{11.32}$$

where C is the charge conjugation matrix and χ_a is the right-handed spinor. We will use the upper indices for the left-handed chiral multiplets and the lower indices for the right-handed chiral multiplets.

We now discuss the rules for the multiplication of chiral fields. First, we consider the case when two multiplets of the same chirality are multiplied. These rules can be obtained from the superfield multiplication. Let Γ_1 and Γ_2 be two such multiplets so that $\Gamma_i = (z_i, \chi_i, h_i)(i = 1, 2)$. In this case, we define the product as follows [31]:

$$\Gamma_1 \cdot \Gamma_2 = (z_1 z_2, z_1 \chi_2 + z_2 \chi_1, z_1 h_2 + z_2 h_1 - 2\bar{\chi}_1^c \chi_2). \tag{11.33}$$

Next, we consider two multiplets of opposite chirality. Let Γ_1 and Γ_2 be two such multiplets. These two multiplied symmetrically then produce a vector multiplet:

$$\Gamma_1 \times \Gamma_2 \equiv \frac{1}{2}(\Gamma_1^\dagger \Gamma_2 + \Gamma_2^\dagger \Gamma_1). \tag{11.34}$$

The components of this vector multiplet are given by

$$\Gamma_1 \times \Gamma_2 = \Big(\frac{1}{2}z_1^* z_2, \; i(z_1^* \chi_2 - z_1 \chi_2^c), \; -h_i z_2^*, \frac{i}{2}(z_1^* \hat{D}_\mu z_2 - z_1 \hat{D}_\mu z_2^*) - 2\bar{\chi}_1 \gamma_\mu \chi_2$$
$$- ih_1^* \chi_2 + ih_1 \chi_2^c - i\gamma^\mu \hat{D}_\mu z_1^* \chi_2 + i\gamma^\mu \hat{D}_\mu z_1 \chi_2^c$$
$$h_1^* h_2 - \hat{D}_\mu z_1^* \hat{D}_\mu z_2 - \bar{\chi}_1 \overleftrightarrow{\not{D}} \chi_2\Big) + 1 \longleftrightarrow 2. \tag{11.35}$$

It is to be noted that the spinor components in the resultant vector multiplet of Eq. (11.35) are Majorana, as expected for a vector multiplet.

Next, assume that we have Π_1, Π_2, Π_1', and Π_2', which are left-handed Majorana multiplets, and

$$\Gamma_1 = \Pi_1 + i\Pi_2, \; \Gamma_2 = \Pi_1' + i\Pi_2'. \tag{11.36}$$

In this case, we have

$$\frac{1}{2}(\Gamma_1 \Gamma_2^\dagger + \Gamma_2 \Gamma_1^\dagger) = (\Pi_1 \Pi_1' + \Pi_2 \Pi_2') - i(\Pi_1 \Pi_2' - \Pi_1' \Pi_2). \tag{11.37}$$

The anti-symmetric product rule for multiplets with Majorana spinors is given by

$$\Pi_1 \wedge \Pi_2 \equiv -\frac{i}{2}(\Pi_1^\dagger \Pi_2 - \Pi_1 \Pi_2^\dagger)$$
$$= \Big(\frac{i}{2}z_1^* z_2, \; z_1^* \chi_{2L} + z_1 \chi_{2R}, \; iz_1^* h_2, \; \frac{1}{2}(z_1^* D_\mu z_2 + z_1^* D_\mu z_2 - \bar{\chi}_{1L} \gamma_\mu \chi_{2L})$$
$$h_2^* \chi_{1L} + \gamma^\mu \hat{D}_\mu z_2^* \chi_{1L}, \; i(h_1^* h_2 - D_\mu z_1^* D_\mu z_2 - \bar{\chi}_2 \gamma^\mu D_\mu \chi_1)\Big) - 1 \leftrightarrow 2. \tag{11.38}$$

A general vector multiplet V (where $V^\dagger = V$) has components

$$V = (C, \xi, H, K, V_\mu, \lambda, D). \tag{11.39}$$

where C, H, and K are scalar fields, ξ and λ are Majorana fields, and D is an auxiliary field. One may define

$$v = K + i\gamma_5 H. \tag{11.40}$$

so that the vector multiplet may be written as

$$V = (C, \xi, v, V_\mu, \lambda, D). \tag{11.41}$$

The product of two-vector multiplets is a vector multiplet. The components of the product of two-vector multiplets are given by

$$V_1 \cdot V_2 = (C_1, \xi_1, v_1, V_{\mu 1}, \lambda_1, D_1) \cdot (C_2, \xi_2, v_2, V_{\mu 2}, \lambda_2, D_2)$$
$$= (C_{12}, \xi_{12}, v_{12}, V_{\mu 12}, \lambda_{12}, D_{12}), \tag{11.42}$$

where

$$C_{12} = C_1 C_2$$

$$\xi_{12} = C_1 \xi_2 + C_2 \xi_1$$

$$v_{12} = C_1 v_2 - \frac{1}{2}\bar{\xi}_{1R}\xi_{2L} + C_2 v_1 - \frac{1}{2}\bar{\xi}_{2R}\xi_{1L}$$

$$V_{\mu 12} = \left(C_1 V_{\mu 2} - \frac{i}{4}\bar{\xi}_1\gamma_\mu\gamma_5\xi_2\right) + 1 \leftrightarrow 2$$

$$\lambda_{12} = \left(C_1\lambda_2 - \frac{1}{2}\gamma^\mu\hat{D}_\mu C_1\xi_2 + \frac{1}{2}v_1^*\xi_2 + \frac{i}{2}\gamma_5\gamma^\mu\xi_2 V_{\mu 1}\right) + 1 \leftrightarrow 2$$

$$D_{12} = \left(C_1 D_2 - \frac{1}{2}\hat{D}_\mu C_1 \hat{D}^\mu C_2 - \frac{1}{2}V_{\mu 1}V_2^\mu + \frac{1}{2}v_1^* v_2 - \bar{\xi}_1\lambda_2\right.$$
$$\left. - \frac{1}{2}\bar{\xi}_2\gamma^\mu\hat{D}_\mu\xi_1\right) + 1 \leftrightarrow 2, \tag{11.43}$$

where $\hat{D}_\mu C$ and $\hat{D}_\mu\xi$ are defined by

$$\hat{D}_\mu C = \partial_\mu C + \frac{i}{2}\kappa\bar{\psi}_\mu\gamma_5\xi$$

$$\hat{D}_\mu\xi = D_\mu(\omega(e,\psi))\xi - \frac{\kappa}{2}(\gamma^\mu V_\mu + i\gamma_5\gamma^\mu\hat{D}_\mu C - i\gamma_5 v)\psi_\mu - \frac{i}{2}\kappa A_\mu\gamma_5\xi. \tag{11.44}$$

The vector multiplet kinetic energy in the Wess–Zumino gauge can be written as

$$e^{-1}\mathcal{L}_V = \frac{1}{2}f_{\alpha\beta}\left(-\frac{1}{4}F_{\mu\nu}^\alpha F^{\mu\nu\beta} - \frac{1}{2}\bar{\lambda}^\alpha\slashed{D}\lambda^\beta + \frac{1}{2}D^\alpha D^\beta\right.$$
$$\left. + \frac{1}{4}F_{\mu\nu}^\alpha\tilde{F}_{\mu\nu}^\beta - \frac{1}{2}D_\mu(\bar{\lambda}^\alpha\gamma^\mu\lambda_R^\beta)\right) + \text{h.c.} \tag{11.45}$$

$$F_{\mu\nu}^\alpha = \partial_\mu V_\nu^\alpha - \partial_\nu V_\mu^\alpha + g_\alpha f^{\alpha\beta\gamma}V_\mu^\beta V_\nu^\gamma$$

$$D_\mu\lambda^\alpha = \partial_\mu\lambda^\alpha + g_\alpha f^{\alpha\beta\gamma}V_\mu^\beta\lambda^\gamma + \frac{1}{2}\omega_{\mu ab}s^{ab}\lambda^\alpha + \frac{i}{2}\kappa A_\mu\gamma_5\lambda^\alpha, \tag{11.46}$$

where $f_{\alpha\beta}$ is the gauge kinetic energy function, g_α are the gauge-coupling constants, and s^{ab} is defined as after Eq. (11.11). Further, in the above we have

$V = V^\alpha T^\alpha$ and $[T^\alpha/2, T^\beta/2] = if^{\alpha\beta\gamma}T^\gamma/2$. For simplicity, we will discuss here the case $f_{\alpha\beta} = \delta_{\alpha\beta}$, and the inclusion of an arbitrary $f_{\alpha\beta}$ is straightforward. The Lagrangian of Eq. (11.45) is invariant under the supersymmetry transformations

$$\delta_s V_\mu^\alpha = \bar{\epsilon}\gamma_\mu\lambda^\alpha$$
$$\delta_s \lambda^\alpha = -\sigma^{\mu\nu}\bar{\epsilon}\hat{F}_{\mu\nu}^\alpha - i\gamma_5\epsilon D^\alpha$$
$$\delta_s D^\alpha = -i\bar{\epsilon}\gamma_5\hat{\slashed{D}}\lambda^\alpha. \tag{11.47}$$

Here, $\hat{F}_{\mu\nu}^\alpha$ and $\hat{D}_\mu\lambda^\alpha$ are super-covariant objects, and they are related to the usual covariant quantities $F_{\mu\nu}^\alpha$ and $D_\mu\lambda^\alpha$ as follows:

$$\hat{F}_{\mu\nu}^\alpha = F_{\mu\nu}^\alpha - \frac{\kappa}{2}(\bar{\psi}_\mu\gamma_\nu\lambda^\alpha - \bar{\psi}_\nu\gamma_\mu\lambda^\alpha)$$
$$\hat{D}_\mu\lambda^\alpha = D_\mu\lambda^\alpha + \frac{\kappa}{2}(\sigma^{\mu\nu}\hat{F}_{\mu\nu}^\alpha + i\gamma_5 D^\alpha)\psi_\mu. \tag{11.48}$$

For further exposition, the reader is referred to several sources [10, 31–33].

References

[1] P. Nath and R. L. Arnowitt, *Phys. Lett. B* **56**, 177 (1975).

[2] R. L. Arnowitt, P. Nath, and B. Zumino, *Phys. Lett. B* **56**, 81 (1975).

[3] W. Rarita and J. Schwinger, *Phys. Rev.* **60**, 61 (1941).

[4] D. Z. Freedman, P. van Nieuwenhuizen, and S. Ferrara, *Phys. Rev. D* **13**, 3214 (1976).

[5] S. Deser and B. Zumino, *Phys. Lett. B* **62**, 335 (1976).

[6] P. Nath and R. L. Arnowitt, *Phys. Lett. B* **65**, 73 (1976).

[7] R. L. Arnowitt and P. Nath, *Phys. Lett. B* **78**, 581 (1978).

[8] P. Nath and R. L. Arnowitt, *Nucl. Phys. B* **165**, 462 (1980).

[9] T. W. B. Kibble, *J. Math. Phys.* **2**, 212 (1961).

[10] P. van Nieuwenhuizen, *Phys. Rep.* **68**, 189 (1981).

[11] K. S. Stelle and P. C. West, *Phys. Lett. B* **74**, 330 (1978).

[12] S. Ferrara and P. van Nieuwenhuizen, *Phys. Lett. B* **74**, 333 (1978).

[13] M. F. Sohnius and P. C. West, *Phys. Lett. B* **105**, 353 (1981).

[14] P. Breitenlohner, *Phys. Lett. B* **67**, 49 (1977).

[15] P. Breitenlohner, *Nucl. Phys. B* **124**, 500 (1977).

[16] B. de Wit and P. van Nieuwenhuizen, *Nucl. Phys. B* **139**, 216 (1978).

[17] V. O. Rivelles and J. G. Taylor, *Phys. Lett. B* **113**, 467 (1982).

[18] V. O. Rivelles and J. G. Taylor, *Nucl. Phys. B* **212**, 173 (1983).

[19] T. Kugo and S. Uehara, *Nucl. Phys. B* **226**, 49 (1983).

[20] T. Kugo and S. Uehara, *Nucl. Phys. B* **226**, 93 (1983).

[21] T. Kugo and S. Uehara, *Nucl. Phys. B* **222**, 125 (1983).

[22] S. Ferrara, L. Girardello, T. Kugo, and A. Van Proeyen, *Nucl. Phys. B* **223**, 191 (1983).

[23] J. Wess and B. Zumino, *Phys. Lett. B* **66**, 361 (1977).

[24] J. Wess and B. Zumino, *Phys. Lett. B* **74**, 51 (1978).

[25] L. Brink, M. Gell-Mann, P. Ramond, and J. H. Schwarz, *Phys. Lett. B* **76**, 417 (1978).

[26] L. Brink, M. Gell-Mann, P. Ramond, and J. H. Schwarz, *Phys. Lett. B* **74**, 336 (1978).

[27] S. J. Gates, Jr. and W. Siegel, *Nucl. Phys. B* **163**, 519 (1980).

[28] W. Siegel and S. J. Gates, Jr., *Nucl. Phys. B* **147**, 77 (1979).

[29] K. S. Stelle and P. C. West, *Phys. Lett. B* **77**, 376 (1978).

[30] S. Ferrara and P. van Nieuwenhuizen, *Phys. Lett. B* **76**, 404 (1978).

[31] P. Nath, R. Arnowitt, and A. H. Chamseddine, *Applied N = 1 Supergravity*, Trieste Lecture Series vol. I, World Scientific Singapore, (1984).

[32] D. Z. Freedman and A. Van Proeyen, *Supergravity*, Cambridge University Press (2012).

[33] S. Weinberg, *The Quantum Theory of Fields*. Vol. 3: *Supersymmetry*, Cambridge University Press (2005).

12

Coupling of Supergravity with Matter and Gauge Fields

The analysis of this section is based on the results given in Chapter 11, and it is useful to summarize these results before going into the details of the supergravity–matter–gauge fields couplings. In building models with supergravity, we need to couple supergravity with matter fields and gauge fields. Regarding matter fields, we consider matter belonging to an irreducible representation of a gauge group G. Let Γ^a be a left chiral multiplet with components given by

$$\Gamma^a = (z^a,\ \chi_L^a,\ h^a), \qquad (12.1)$$

where z^a denotes complex scalars given by $z^a = A^a + iB^a$, χ_L^a are left-handed Weyl spinor fields, and h^a are auxiliary fields which are complex while the superscript "a" indicates that the fields belong to an irreducible representation of the gauge group G. The coupling of a single chiral multiplet to supergravity is given by

$$e^{-1}\mathcal{L}_F = Re\left[h + uz + \bar{\psi}_\mu\gamma^\mu\chi + \bar{\psi}_\mu\sigma^{\mu\nu}\psi_{\nu R}z\right], \qquad (12.2)$$

where $u = S - iP$. In addition to matter fields, we need gauge fields in building particle physics models. Thus, we consider a vector multiplet V which belongs to the adjoint representation of the gauge group G and satisfies the condition $V^\dagger = V$, and has components

$$V = (C,\ \xi,\ H,\ K,\ V_\mu,\ \lambda,\ D). \qquad (12.3)$$

Here C, H, and K are scalar fields, ξ and λ are Majorana spinors, V^μ is a vector field, and D is an auxiliary scalar field as discussed in Chapter 8. The coupling of a single vector multiplet to supergravity is given by

$$e^{-1}\mathcal{L}_D = D - \frac{i\kappa}{2}\bar{\psi}_\mu\gamma^5\gamma^\mu\lambda - \frac{2}{3}\kappa V_\mu\left(A^\mu + \frac{3}{8}ie^{-1}\epsilon^{\mu\rho\sigma\tau}\bar{\psi}_\rho\gamma_\tau\psi_\sigma\right) - \frac{2}{3}(SK - PH)$$

$$+ i\frac{k}{3}e^{-1}\bar{\xi}\gamma_5\gamma_\mu R^\mu + \frac{ik^2}{8}\epsilon^{\mu\nu\rho\sigma}\bar{\psi}_\mu\gamma_\nu\psi_\rho\bar{\xi}\psi_\sigma - \frac{2}{3}e^{-1}\kappa^2 C\mathcal{L}_{SG}. \qquad (12.4)$$

Couplings of one chiral field to supergravity were determined by Cremmer *et al.* [1]. However, for the construction of models, one needs to couple an arbitrary number of chiral multiplets charged under a gauge group and a vector multiplet belonging to the adjoint representation of the gauge group to supergravity. This has been accomplished [2–5]. Below, we follow the procedure discussed in the literature [2,4], which consists of the following steps:

1. Produce the most general F- and D-multiplets out of the chiral multiplets Γ and the vector multiplet V using the rules of tensor calculus while maintaining gauge invariance under a desired gauge group.
2. Couple the F- and the D-multiplets to supergravity using Eqs. (12.2) and (12.4).
3. Carry out an elimination of the auxiliary fields in the chiral matter sector, gauge sector, and the supergravity sector.
4. Make point transformations to write the kinetic energy of the dynamical fields in canonical form.

The resulting Lagrangian will consist of only dynamical fields with their proper kinetic energies, and is the desired final form of the Lagrangian. It depends on three arbitrary functions, which are [2–5].

$$W(z),\ K(z, z^\dagger),\ f_{\alpha\beta}(z). \tag{12.5}$$

These functions are defined as follows, The function $W(z)$ is the lowest element of the most general F-multiplet one can form from a set of scalar components of chiral fields z^a so that

$$W(z) = \sum C_{a_1 \cdots a_k} z^{a_1} \cdots z^{a_k}, \tag{12.6}$$

where $C_{a_1 \cdots a_k}$ are arbitrary parameters which are chosen so that $W(z)$ is gauge invariant. Similarly, from the chiral multiplets we can form a D-multiplet, and its lowest element can be written so that

$$\phi(z, z^\dagger) = \sum A^{a_1 \cdots a_m}{}_{b_1 \cdots b_n} z_{a_1} \cdots z_{a_m} z^{b_1} \cdots z^{b_n}, \tag{12.7}$$

where again the coefficients $A^{a_1 \cdots a_m}{}_{b_1 \cdots b_n}$ are arbitrary but are chosen so that they maintain gauge invariance. It is found useful to introduce a function $K(z, z^\dagger)$, which is related to $\phi(z, z^\dagger)$ as follows:

$$K(z, z^\dagger) = -\frac{6}{\kappa^2} \ln \left[-\frac{\kappa^2}{3} \phi(z, z^\dagger) \right]. \tag{12.8}$$

The function $f_{\alpha\beta}(z)$ enters in the Yang–Mills Lagrangian as given by Eq. (11.45). Supergravity constraints do not determine $f_{\alpha\beta}$, and additional model-dependent assumptions are needed to specify it.

At the tree level, the other two functions $W(z)$ and $\phi(z, z^\dagger)$ enter the Lagrangian in a specific combination, which is given by the functions $\mathcal{G}(z, z^\dagger)$ so that

$$\mathcal{G}(z, z^\dagger) = -\frac{\kappa^2}{2} K(z, z^\dagger) - \ln\left(\frac{\kappa^6}{4}|W(z)|^2\right). \tag{12.9}$$

The quantity $\mathcal{G}^a{}_{,b} \equiv \partial^2\mathcal{G}/\partial z^b \partial z_a$ acts as the metric in the Kahler manifold, which is defined by the co-ordinates z_a, and z^b. The total Lagrangian consists of the following parts:

$$\mathcal{L} = \mathcal{L}_V + \mathcal{L}_F + \mathcal{L}_D, \tag{12.10}$$

where $\mathcal{L}_V$ is given by Eq. (11.45), $\mathcal{L}_F$ is the Lagrangian that arises from the self-couplings of the chiral fields, i.e., from $W(z)$, and $\mathcal{L}_D$ is the Lagrangian that arises from the couplings of the chiral fields to the vector multiplet and to supergravity. A discussion of $\mathcal{L}_F$ and $\mathcal{L}_D$ is given in Section 21.6. The full Lagrangian involving the couplings of the matter and gauge fields to themselves and to supergravity can be written as

$$\mathcal{L} = \mathcal{L}_{\text{Bose}} + \mathcal{L}_{\text{Fermi}}. \tag{12.11}$$

We discuss the bosonic part first, and for simplicity we give below the result under the assumption that $f_{\alpha\beta} = \delta_{\alpha\beta}$ so that [2–4]

$$\mathcal{L}_{\text{Bose}} = -\frac{e}{2\kappa^2} R(e, \omega) + \frac{e}{\kappa^2}\mathcal{G}^a{}_{,b}\mathcal{D}_\mu z_a \mathcal{D}^\mu z^b + \frac{e}{\kappa^4}e^{-\mathcal{G}}\left[3 + (\mathcal{G}^{-1})^a{}_{,b}\mathcal{G}_{,a}\mathcal{G}^b_{,}\right]$$
$$- \frac{1}{4}eF^\alpha_{\mu\nu}F^{\mu\nu\alpha} - \frac{e}{8\kappa^4}|g_\alpha\mathcal{G}_{,a}(T^\alpha z)^a|^2. \tag{12.12}$$

Alternatively, we may write Eq. (12.12) so that

$$\mathcal{L}_{\text{Bose}} = -\frac{e}{2\kappa^2} R(e, \omega) + \frac{e}{2}K^a{}_{,b}\mathcal{D}_\mu z_a \mathcal{D}^\mu z^b - \frac{1}{4}eF^\alpha_{\mu\nu}F^{\mu\nu\alpha} - e^{-1}V, \tag{12.13}$$

where V is the scalar potential given by

$$e^{-1}V = -\frac{1}{\kappa^4}e^{-\mathcal{G}}\left[3 + (\mathcal{G}^{-1})^a{}_{,b}\mathcal{G}_{,a}\mathcal{G}^b_{,}\right] + \frac{1}{8\kappa^4}|g_\alpha\mathcal{G}_{,a}(T^\alpha Z)^a|^2$$
$$= \frac{1}{2}e^{\frac{\kappa^2}{2}K}\left[(K^{-1})^a{}_{,b}\left(\frac{\partial W}{\partial z^a} + \frac{\kappa^2}{2}K_{,a}W\right)\left(\frac{\partial W^*}{\partial z_b} + \frac{\kappa^2}{2}K^{,b}W^*\right) - \frac{3}{2}\kappa^2|W|^2\right]$$
$$+ \frac{1}{8\kappa^4}|g_\alpha\mathcal{G}_{,a}(T^\alpha z)^a|^2, \tag{12.14}$$

where $(K^{-1})^a{}_b$ is the inverse of the matrix $(K)^b{}_{,a}$ and K is the Kahler potential defined by Eq. (12.8).

We note that Eq. (12.9) is invariant under the transformation

$$W \to e^{-f(\phi)}W \tag{12.15}$$

$$K \to K + \frac{2}{\kappa^2}(f(\phi) + f^\dagger(\phi^\dagger)). \tag{12.16}$$

However, this invariance is valid only at the classical level and is not valid at the quantum level because the transformation is anomalous [6–8].

Next, we give the result for the fermionic part of the Lagrangian, again for the case $f_{\alpha\beta} = \delta_{\alpha\beta}$ for simplicity. More details of the construction are given in Section 21.6. Here, we state the result [3,4]

$$
\begin{aligned}
\mathcal{L}_{\text{Fermi}} = &-\frac{1}{2}\epsilon^{\mu\nu\rho\sigma}\bar{\psi}_\mu\gamma_5\gamma_\nu D_\rho(\omega(e,\psi))\psi_\sigma + \frac{e}{\kappa^2}\mathcal{G}_{,b}^{\ a}\overline{\chi^a}\overleftrightarrow{\not{D}}\chi^b - \frac{e}{2}\overline{\lambda^\alpha}\gamma^\mu D_\mu\lambda^\alpha \\
&- e\frac{\kappa}{2}\bar{\psi}_\mu\sigma^{\kappa\lambda}\gamma^\mu\lambda^\alpha F_{\kappa\lambda}^\alpha + \frac{1}{8}\epsilon^{\mu\nu\rho\sigma}\bar{\psi}_\mu\gamma_\nu\psi_\rho\left(\mathcal{G}_{,a}\mathcal{D}_\sigma z^a - \mathcal{G}^a_{,}\mathcal{D}_\sigma z_a\right) \\
&- \frac{e}{\kappa}\mathcal{G}_{,b}^{\ a}\left(\mathcal{D}_\nu z^b\overline{\chi^a}\gamma^\mu\gamma^\nu\psi_\mu + \bar{\psi}_\mu\gamma^\nu\gamma^\mu\chi^b\mathcal{D}_\nu z_a\right) \\
&+ \frac{e}{\kappa^2}\overline{\chi^a}\gamma^\mu\chi^b\left[\left(\mathcal{G}^a_{,\ bc} - \frac{\kappa^2}{4}\mathcal{G}^a_{,\ b}\mathcal{G}_{,c}\right)\mathcal{D}_\mu z^c - \left(\mathcal{G}^{ac}_{,\ b} - \frac{\kappa^2}{4}\mathcal{G}^a_{,\ b}\mathcal{G}^c_{,}\right)\mathcal{D}_\mu z_c\right] \\
&+ \frac{e}{8}\overline{\lambda^\alpha}\gamma^\mu\gamma_5\lambda^\alpha\left(\mathcal{G}_{,a}\mathcal{D}_\mu z^a - \mathcal{G}^a_{,}\mathcal{D}_\mu z_a\right) \\
&+ \frac{e}{\kappa^3}e^{-\frac{\mathcal{G}}{2}}\Big[\left(\mathcal{G}_{,ab} - \mathcal{G}_{,a}\mathcal{G}_{,b} - (\mathcal{G}^{-1})^c_{\ d}\mathcal{G}_{,c}\mathcal{G}^d_{,\ ab}\right)\bar{\chi}_a\chi^b \\
&\quad - \kappa\mathcal{G}_{,a}\bar{\psi}_\mu\gamma^\mu\chi^a + \kappa^2\bar{\psi}_\mu\sigma^{\mu\nu}\psi_{\nu R} + h.c.\Big] \\
&+ ie\frac{g_\alpha}{4k}(\mathcal{G}_{,a}(T^\alpha z)^a\bar{\psi}_\mu\gamma_5\gamma^\mu\lambda^\alpha + ei\frac{g_\alpha}{\kappa^2}\mathcal{G}^a_{,\ b}\left((T^\alpha z)_a\overline{\lambda^\alpha}\chi^b - \overline{\chi^a}\lambda^\alpha(T^\alpha z)^b\right) \\
&- \frac{e}{\kappa^2}\left(\mathcal{G}_{,ab}^{\ \ cd} - (\mathcal{G}^{-1})^e_{\ f}\mathcal{G}_{,ab}^{\ \ e}\mathcal{G}^{cd}_{,\ f} + \frac{1}{2}\mathcal{G}^c_{,\ a}\mathcal{G}^d_{,\ b}\right)\bar{\chi}_a\chi^b\overline{\chi^c}\chi_d \\
&+ e\frac{\kappa^2}{4}\overline{\lambda^\alpha}\gamma^\mu\sigma^{\nu\rho}\psi_\mu\overline{\psi_\nu}\gamma_\rho\lambda^\alpha + \frac{e}{8}\mathcal{G}^a_{,\ b}\overline{\chi^a}\gamma^\mu\chi^b\overline{\lambda^\alpha}\gamma_\mu\gamma_5\lambda^\alpha \\
&+ 3e\frac{\kappa^2}{64}\overline{\lambda^\alpha}\gamma^\mu\gamma_5\lambda^\alpha\overline{\lambda^\beta}\gamma_\mu\gamma_5\lambda^\beta + \frac{1}{4}\mathcal{G}^a_{,\ b}\overline{\chi^a}\gamma_\sigma\chi^b\left(\epsilon^{\mu\nu\rho\sigma}\overline{\psi_\mu}\gamma_\nu\psi_\rho - e\bar{\psi}_\mu\gamma_5\gamma^\sigma\psi^\mu\right).
\end{aligned}
$$

$$(12.17)$$

In order to arrive at Eq. (12.17), one arranges terms with W and K with similar structures and then combines them into the function $\mathcal{G}$. Thus, the Lagrangian of Eq. (12.17) depends only the single function $\mathcal{G}$. Since the auxiliary fields have now been eliminated, the Lagrangian of Eq. (12.17) is invariant under supersymmetry transformations only on the mass shell. The supersymmetry transformations on the mass shell can be obtained by the substitution of the auxiliary fields using Eq. (21.232) into the general supersymmetry transformations. Further, one uses Eq. (21.230) to redefine the variables. With these substitutions, one has the following set of transformations [2–4]:

$$
\delta_s e_\mu^{\ r} = \kappa\bar{e}\gamma^r\psi_\mu
$$

$$
\delta_s\psi_{\mu L} = 2\kappa^{-1}D_\mu(e,\psi_\nu)\epsilon_L + \kappa^{-2}e^{-\frac{\mathcal{G}}{2}}\gamma_\mu\epsilon_R + \frac{1}{2}\left(\mathcal{G}_{,a}\bar{\epsilon}\chi^a - \mathcal{G}^a_{,}\overline{\chi^a}\epsilon\right)\psi_{\mu L}
$$

$$
\qquad + \kappa^{-1}(\sigma_{\mu\nu}\epsilon_L)\mathcal{G}^a_{,\ b}\overline{\chi^a}\gamma^\nu\chi^b - \frac{\kappa}{8}(\delta_\mu^{\ \nu} + \gamma^\nu\gamma_\mu)\epsilon_L\overline{\lambda^\alpha}\gamma_\nu\gamma_5\gamma^\alpha
$$

$$- \frac{\kappa^{-1}}{2} \left(\mathcal{G}_{,a} \mathcal{D}_\mu z^a - \mathcal{G}_{,}{}^a \mathcal{D}_\mu z_a \right)$$

$$\delta_s V_\mu{}^\alpha = -\bar{\epsilon} \gamma_\mu \lambda^\alpha$$

$$\delta_s \lambda^\alpha = -\frac{1}{2} \left(\mathcal{G}_{,a} \bar{\epsilon} \chi^a - \mathcal{G}_{,}{}^a \overline{\chi^a} \epsilon \right) \lambda^\alpha - \sigma^{\mu\nu} \epsilon \hat{F}_{\mu\nu}{}^\alpha - i \frac{g_\alpha}{2\kappa^2} (\gamma_5 \epsilon) \left(\mathcal{G}_{,a} (T^\alpha z)^a \right)$$

$$\delta_s z^a = 2 \bar{\epsilon} \chi^a$$

$$\delta_s \chi^a = \gamma^\mu \epsilon_R \hat{\mathcal{D}}_\mu z^a - \frac{1}{2} \left(\mathcal{G}_{,b} \bar{\epsilon} \chi^b - \mathcal{G}_{,}{}^b \overline{\chi^b} \epsilon \right) \chi^a$$

$$+ (\mathcal{G}^{-1})^a{}_b \mathcal{G}_{,}{}^b{}_{cd} \overline{\chi c} \chi^d - \kappa^{-1} (\mathcal{G}^{-1})^a{}_b \mathcal{G}_{,}{}^b e^{-\frac{\mathcal{G}}{2}} \epsilon_L . \tag{12.18}$$

A further discussion of the details of the construction are given in Section 21.6.

12.1 Problems

1. Show that for the case of a single chiral multiplet ϕ the F-part of the scalar potential can be written in the form

$$V = -\frac{e}{\kappa^4} 9 e^{\frac{-4}{3} \mathcal{G}} \mathcal{G}_{\phi\phi^\dagger}^{-1} \partial_\phi \partial_{\phi^\dagger} e^{\frac{1}{3} \mathcal{G}} . \tag{12.19}$$

Further, show that for a $\mathcal{G}$ of the form

$$\mathcal{G} = \frac{3}{2} \log(f(\phi) + f^\dagger(\phi^\dagger))^2 \tag{12.20}$$

the potential of Eq. (12.19) vanishes. Extend the analysis to the case of N chiral superfields and show that the F-part of the scalar potential takes the form

$$V = -\frac{9e}{N^2} e^{-\frac{N+3}{3} \mathcal{G}} (\mathcal{G}^{-1})^a_b \partial_a \partial^b e^{\frac{N}{3} \mathcal{G}} . \tag{12.21}$$

Find the form of $\mathcal{G}$ for which V given by Eq. (12.21) vanishes.

References

[1] E. Cremmer, B. Julia, J. Scherk, P. van Nieuwenhuizen, S. Ferrara, and L. Girardello, *Phys. Lett. B* **79**, 231 (1978).

[2] A. H. Chamseddine, R. L. Arnowitt, and P. Nath, *Phys. Rev. Lett.* **49**, 970 (1982).

[3] E. Cremmer, S. Ferrara, L. Girardello, and A. Van Proeyen, *Phys. Lett. B* **116**, 231 (1982); *Nucl. Phys. B* **212**, 413 (1983).

[4] P. Nath, R. Arnowitt, and A. H. Chamseddine, *Applied N = 1 Supergravity*, Trieste Lecture Series vol. I, World Scientific (1984).

[5] J. Bagger and E. Witten, *Phys. Lett. B* **118** (1982) 103.

[6] G. Lopes Cardoso and B. A. Ovrut, *Nucl. Phys. B* **418**, 535 (1994).

[7] J. A. Bagger, T. Moroi, and E. Poppitz, *JHEP* **0004**, 009 (2000); *Nucl. Phys. B* **594**, 354 (2001).

[8] M. K. Gaillard and B. Nelson, *Nucl. Phys. B* **588**, 197 (2000).

13

Supergravity Grand Unification

We now consider the phenomenon of spontaneous breaking of supersymmetry within the supergravity grand unified models. We recall that one of the major problems in the globally supersymmetric models arose from the appearance of a massless spin 1/2 field, the Goldstino, after the spontaneous breaking of supersymmetry. This is the analog of a similar problem that arose in non-supersymmetric theories regarding the breaking of a global symmetry which gave rise to a massless Goldstone boson. In that case the problem was resolved through the Higgs mechanism where the vector field absorbs the massless Goldstone and becomes massive. A very similar super-Higgs phenomenon occurs in local supersymmetry where the massless Goldstino is absorbed by a massless spin 3/2 field, the gravitino, to become massive. We give further details below of how this comes about. To accomplish this we first discuss the master equations which control the breaking of symmetry, both ordinary symmetry as well as supersymmetry, in the framework of supergravity. We will focus on flat Kahler manifolds for simplicity. Here, variation of the supergravity scalar potential leads to the following extrema equations [1, 2]:

$$\left(\frac{\partial G_b}{\partial Z^a} + \frac{\kappa^2}{2} Z_a G_b \right) G^b - \kappa^2 W G_a = 0, \tag{13.1}$$

where G_a is defined by

$$G_a = \frac{\partial W}{\partial Z^a} + \frac{\kappa^2}{2} K_{,a} W. \tag{13.2}$$

We now limit our analysis to real manifolds, in which case Eq. (13.1) may be written in the form

$$T_{ab} G_b = 0, \tag{13.3}$$

where T_{ab} is given by

$$T_{ab} = \frac{\partial^2 W}{\partial Z^a \partial Z^b} + \frac{\kappa^2}{2} \left(Z_a \frac{\partial W}{\partial Z^b} + Z_b \frac{\partial W}{\partial Z^a} \right) + \frac{\kappa^4}{4} Z_a Z_b W - \kappa^2 \delta_{ab} W. \tag{13.4}$$

13.1 Spontaneous Breaking of Gauge Symmetry

As mentioned above, Eq. (13.3) governs the spontaneous breaking of gauge symmetry and also the spontaneous breaking of supersymmetry. We examine first the case in which Eq. (13.3) is satisfied through the condition

$$G_a = 0. \tag{13.5}$$

The constraint of Eq. (13.5) does not break supersymmetry but may break gauge symmetry. As an illustration, let us consider an $SU(5)$ grand unification where one has a 24-plet of scalar fields $\phi^a_{\ b}$ with a superpotential W_ϕ given by

$$W_\phi = \left[\frac{1}{2} M \ Tr \ \phi^2 + \frac{1}{3} \ Tr \ \phi^3 \right]. \tag{13.6}$$

Now, the satisfaction of the condition Eq. (13.5) leads to the following three possible solutions:

1. $\phi^a_{\ b} = 0$;
2. $\phi^a_{\ b} = \frac{1}{3} M \ [\delta^a_{\ b} - 5\delta^a_{\ 5}\delta^5_{\ b}]$;
3. $\phi^a_{\ b} = M \ [2\delta^a_{\ b} - 5(\delta^a_{\ 5}\delta^5_{\ b} + \delta^a_{\ 4}\delta^4_{\ b})]$.

The three solutions correspond to no breaking of the gauge symmetry for case 1, breaking of the gauge symmetry to $SU(4) \times U(1)$ for case 2, and the breaking of the gauge symmetry to $SU(3) \times SU(2) \times U(1)$ for case 3. Now, it is well known that the breaking of the gauge symmetry for the case of global supersymmetry does not break the degeneracy of the vacuum, i.e., for the case of global supersymmetry the vacuum expectation value (VEV) of the scalar potential vanishes for the cases 1, 2 and 3 discussed above. However, this is not necessarily the case for breaking in supergravity. Here, one finds that the vacuum energy may break the degeneracy. Thus, the potential energy at the minimum for the supergravity case is given by [1]

$$V_{min}(\phi_0, \phi_0^*) = -\frac{3}{4} \kappa^2 \ e^{(\frac{1}{2}\kappa^2 \phi_0 \phi_0^*)} \ |W(\phi_0)|^2. \tag{13.7}$$

Now, by adding a constant term to the superpotential we can make the vacuum energy vanish for one case, but then the vacuum energy for the other cases will be negative and we will have an anti-de Sitter vacuum. Normally, it would appear that the Minkowskian vacuum would be unstable. However, stability could be helped by gravity [3]. In fact, it has been argued that even though the Minkowskian vacuum would not have the lowest energy, it could be stable for any finite-size perturbation [4]. However, even in supergravity, vacuum degeneracy is not lifted in all cases when a gauge symmetry breaks producing multiple vacua. See, e.g., Section 13.5.

13.2 Spontaneous Breaking of Supersymmetry

We now discuss the conditions under which supersymmetry is broken [1,2,5,6]. Here, it is instructive to look at the transformation properties of the spin 1/2 Weyl spinor fields χ^a and the spin 3/2 field ψ_μ as given in Eq. (12.18). Thus, for the spin 1/2 field Eq. (12.18) gives the transformation

$$\delta\chi^a = (\gamma^\mu \epsilon_R)\hat{\mathcal{D}}_\mu Z^a - \frac{1}{2}(\mathcal{G}_{,b}\bar{\epsilon}\chi^b - \mathcal{G}_{,b}\bar{\chi}^b\epsilon)\chi^a + (\mathcal{G}^{-1})^a{}_b\mathcal{G}_{,\,cd}^{\ b}\bar{\chi}^c\chi^d\epsilon_L$$
$$- \kappa^{-1}e^{-\mathcal{G}/2}(\mathcal{G}^{-1})^a{}_b\mathcal{G}_{,}^{\ b}\epsilon_L, \tag{13.8}$$

while for the spin 3/2 field Eq. (12.18) gives the transformation

$$\delta\psi_{\mu L} = 2\kappa^{-1}D_\mu(e,\psi_\nu)\epsilon_L + \kappa^{-2}e^{-\mathcal{G}/2}\gamma_\mu\epsilon_R + \frac{1}{2}(\mathcal{G}_{,a}\bar{\epsilon}\chi^a - \mathcal{G}_{,}^{\ a}\bar{\chi}^a\epsilon)\psi_{\mu L}$$
$$+ \kappa^{-1}\sigma_{\mu\nu}\epsilon_{\mu\nu}\epsilon_L\mathcal{G}_{,\,b}^a\bar{\chi}^a\gamma^\nu\chi^b - \frac{\kappa}{8}(\delta_\mu^\nu + \gamma^\nu\gamma_\mu)\epsilon_L\bar{\lambda}^\alpha\gamma_\nu\gamma_5\lambda^\alpha$$
$$- \frac{1}{2}\kappa^{-1}\left(\mathcal{G}_{,a}\mathcal{D}_\mu Z^a - \mathcal{G}_{,}^a\mathcal{D}_\mu Z_a\right)\epsilon_L. \tag{13.9}$$

Let us consider the flat Kahler limit and take the VEVs of Eqs. (13.8) and (13.9), which give

$$\langle\delta\chi^a\rangle_0 = G_a(Z_0)\epsilon_L e^{(\frac{\kappa^2}{2}Z_0^a Z_{0a})} \tag{13.10}$$

and

$$\langle\delta\psi_{\mu L}\rangle_0 = \frac{\kappa}{2}G(Z_0)\gamma_\mu\epsilon_R e^{(\frac{\kappa^2}{2}Z_0^a Z_{0a})} + 2\kappa^{-1}\partial_\mu\epsilon_L. \tag{13.11}$$

Now, the non-vanishing of $\langle\delta\chi^a\rangle_0$ is an indication that supersymmetry is spontaneously broken, and from Eqs. (13.10) and (13.11) we find that this is possible only if at least one of the $G_a(Z_0)$ is non-vanishing. Further, in order that Eq. (13.3) be satisfied, we also have the condition that one of T_{ab} must be non-zero in the basis where T_{ab} is diagonal. The spontaneous breaking of supersymmetry produces a Goldstino which is absorbed by the gravitino to become massive. We will see later (Eq. (13.19)) that the mass of the gravitino after absorption of the Goldstino is given by

$$m_{3/2} = \frac{1}{\kappa}e^{-\mathcal{G}/2}. \tag{13.12}$$

Using Eq. (12.9) for a flat Kahler potential, one may also write the mass of the gravitino as

$$m_{3/2} = \frac{1}{2}\kappa^2|W(Z_0)|e^{(\frac{\kappa^2}{2}Z_0^\dagger Z_0)}. \tag{13.13}$$

A concrete example of the super-Higgs is given by a superpotential of the form

$$W_{SH}(z) = m^2\kappa^{-1}f_s(\kappa Z), \tag{13.14}$$

where $f_s(\kappa Z)$ is a function of the dimensionless combination κZ. Now, supersymmetry is broken by imposing the condition $T_{ZZ} = 0$, which gives solutions of the form

$$\langle Z \rangle \sim O(\kappa^{-1}) \tag{13.15}$$

so that $f_s(\kappa Z) \sim O(1)$. It then follows that $m_{3/2} \sim O(m_s)$ where $m_s = \kappa m^2$. If one chooses m to be an intermediate scale of size $O(10^{10-11})$ GeV, it would imply that $m_{3/2}$ would lie in the 1–10 TeV range. A simple choice often made for W_{SH} is $W_{SH} = m^2(z + B)$. The more general form, of course, is given by Eq. (13.14)). However, not all forms consistent with Eq. (13.14) are desirable.

13.3 Mass Matrices

Now, using the Lagrangian of Eqs. (12.12) and (12.17), the mass matrices for the physical fields can be obtained. These fields are

$$\psi_\mu, \chi^a, z^a, \lambda^\alpha, V_\mu^\alpha. \tag{13.16}$$

We begin by providing the mass terms for the fields ψ_μ and χ^a for the case of the flat Kähler manifold case [2]:

$$\mathcal{L}_{\text{mass}} = \frac{e}{\kappa^3} e^{-\frac{\mathcal{G}}{2}} \left[(\mathcal{G}_{ab} - \mathcal{G}_{,a}\mathcal{G}_{,b}) \overline{\chi_a} \chi^b - \kappa \mathcal{G}_{,a} \bar{\psi}_\mu \gamma^\mu \chi^a + \kappa^2 \bar{\psi}_\mu \sigma^{\mu\nu} \psi_{\nu R} + \text{h.c.} \right]$$
$$- ie \frac{g_\alpha}{8} \kappa (z_a (T^\alpha z)^a) \bar{\psi}_\mu \gamma_5 \gamma^\mu \lambda^\alpha - ie \frac{g_\alpha}{2} \left((T^\alpha z)_a \bar{\lambda}^\alpha \chi^a - \bar{\chi}^a \lambda^\alpha (T^\alpha z)^a \right). \tag{13.17}$$

After breaking of supersymmetry the gravitino field is defined by ψ'_μ, where

$$\psi'_{\mu R} = \psi_{\mu R} - \frac{1}{3\kappa} \gamma_\mu \left(\mathcal{G}_{,a} \chi^a + \frac{i}{4} g_\alpha \kappa^3 e^{-\frac{\mathcal{G}}{2}} (z_a (T^\alpha z)^a) \lambda_L^\alpha \right). \tag{13.18}$$

Substitution of Eq. (13.18) into Eq. (13.17) gives

$$\mathcal{L}_{\text{mass}} = \frac{e}{\kappa} e^{-\frac{\mathcal{G}}{2}} (\overline{\psi'_\mu} \sigma^{\mu\nu} \psi'_{\nu L}) - \frac{e}{2} (m_{ab} \overline{\chi_a} \chi^b) - em_a^\alpha \overline{\lambda^\alpha} \chi^a - \frac{e}{2} m^{\alpha\beta} \overline{\lambda}_L^\alpha \lambda_L^\beta + \text{h.c.}, \tag{13.19}$$

where the matrices appearing in Eq. (13.19) are defined by

$$m_{ab} = \frac{2}{\kappa^3} e^{-\frac{\mathcal{G}}{2}} (\mathcal{G}_{,ab} - \frac{1}{3} \mathcal{G}_{,a} \mathcal{G}_{,b}) \tag{13.20}$$

$$m_a^\alpha = \frac{i}{2} g_\alpha z_b (T^\alpha)^b_{\ a} \tag{13.21}$$

$$m^{\alpha\beta} = -\kappa^3 e^{-\frac{\mathcal{G}}{2}} \frac{g_\alpha^2}{24} (z_a (T^\alpha z)^a)(z_b (T^\beta z)^b). \tag{13.22}$$

From Eq. (13.19) we see that the gravitino mass is given by Eq. (13.12). Next, we use the minimization condition $V_{,a} = 0$, as well as the condition that the vacuum energy vanishes, i.e., $V = 0$. These conditions are equivalent to the following set of constraints:

$$\mathcal{G}_{,ab} \mathcal{G}^b_{,} = \frac{\kappa^2}{2} \mathcal{G}_{,a}, \ (z_a (T^\alpha z)^a) = 0, \ \mathcal{G}_{,a} \mathcal{G}^{\ a}_{,} = \frac{3}{2} \kappa^2. \tag{13.23}$$

We now determine the supertrace for the case of N number of chiral multiplets where the supertrace is given by

$$STr\ M^2 = \sum_{J=0}^{3/2}(-1)^{2J}(2J+1)m_J^2.\tag{13.24}$$

The masses of the vector gauge bosons are the same as for global supersymmetry. Thus, we will set $g_\alpha = 0$ for the determination of the supertrace. We now discuss the mass formula for the complex fields z^a and z_b. We follow here the analysis by Nath *et al.* [2]. The expansion of the potential V in terms of the complex fields z^a and z_b is given by

$$V(z^a, z_b) = \frac{1}{2}(V_{,ab})_0 z^a z^b + \frac{1}{2}(V^{ab})_0 z_a z_b + (V^a_{,b})_0 z_a z^b + \cdots\tag{13.25}$$

Instead of using the complex fields z^a, we now use the two real fields ϕ_1^a and ϕ_2^a, since after spontaneous breaking, their masses will be split. Thus, we determine the mass matrices for the fields ϕ_1^a and ϕ_2^a so that

$$V(z^a, z_b) = \frac{1}{2}\left((M_{\phi_1}^2)_{ab}\phi_1^a\phi_1^b + (M_{\phi_2}^2)_{ab}\phi_2^a\phi_2^b\right) + \cdots\tag{13.26}$$

The matrices $(M_{\phi_1}^2)_{ab}$ and $(M_{\phi_2}^2)_{ab}$ can be written in terms of the derivatives of the potential so that

$$\begin{aligned}(M_{\phi_1}^2)_{ab} &= \left(V^a_{,b} + V^b_{,a} + 2V_{,ab}\right)_0 \\ (M_{\phi_2}^2)_{ab} &= \left(V^a_{,b} + V^b_{,a} - 2V_{,ab}\right)_0.\end{aligned}\tag{13.27}$$

For the case of a flat Kähler manifold, the matrices can also be expressed in terms of $\mathcal{G}$, and one has for V_{ab} and $V^a_{,b}$ [2] setting $g_\alpha = 0$ the result

$$V_{ab} = \frac{2e}{\kappa^4}e^{-\mathcal{G}}\left[\frac{1}{2}(\mathcal{G}_{,ab} - \mathcal{G}_{,a}\mathcal{G}_{,b}) + \frac{1}{\kappa^2}\left(\mathcal{G}_{,abc} - 3\mathcal{G}_{,(a}\mathcal{G}_{,bc)} + \mathcal{G}_{,a}\mathcal{G}_{,b}\mathcal{G}_{,c}\right)\mathcal{G}^c_{,}\right]\tag{13.28}$$

$$\begin{aligned}V^a_{,b} = \frac{2e}{\kappa^4}e^{-\mathcal{G}}&\left[-\frac{1}{2}\mathcal{G}^a_{,}\mathcal{G}_{,b} + \frac{1}{\kappa^2}\left(\mathcal{G}_{,bc} - \mathcal{G}_{,b}\mathcal{G}_{,c}\right)\left(\mathcal{G}^{,ac} - \mathcal{G}^a_{,}\mathcal{G}^c_{,}\right)\right.\\ &\left. - \frac{\kappa^2}{2}\delta^a_{\ b} + \frac{1}{2}\delta^a_{\ b}\mathcal{G}_{,c}\mathcal{G}^c_{,}\right].\end{aligned}\tag{13.29}$$

The supertrace can now be determined, and gives

$$\begin{aligned}Tr(M_{\phi_1}^2) + Tr(M_{\phi_2}^2) - 2\,Tr(m_{ab}^2) - 4m_{3/2}^2 = 4V^a_{,a} - &\frac{8}{\kappa^6}e^{-\mathcal{G}}\left(\mathcal{G}_{,ab} - \frac{1}{3}\mathcal{G}_{,a}\mathcal{G}_{,b}\right)\\ &\times\left(\mathcal{G}^{,ab} - \frac{1}{3}\mathcal{G}^a_{,}\mathcal{G}^b_{,}\right) - \frac{4}{\kappa^2}e^{-\mathcal{G}}.\end{aligned}\tag{13.30}$$

After using Eqs. (13.29) and (13.23), one finds that the above result gives

$$STr\ M^2 = 2(N-1)\frac{1}{\kappa^2}e^{-\mathcal{G}},\tag{13.31}$$

where N is the number of chiral multiplets coupled to supergravity and which immediately leads to the sum rule for masses in supergravity [2,5]:

$$\sum_{J=0}^{3/2}(-1)^{2J}(2J+1)m_J^2 = 2(N-1)m_{3/2}^2. \tag{13.32}$$

13.4 Generation of Soft Terms in Supergravity Models

Having broken supersymmetry we now want to communicate this breaking to the physical sector of the theory, which contains the supersymmetric partners of the matter and gauge fields. This requires that the manifolds of fields be extended to include the super-Higgs field, i.e., the full set of fields consists of the set $Z_A = (Z_a, Z)$, where Z_a are the physical set of fields and Z is the super-Higgs field. Since the VEV of the Z-field is $O(M_{Pl})$, any direct interaction of dimension 3 or less in the superpotential that involves the direct coupling of the super-Higgs field and the physical fields would be disastrous after spontaneous breaking of supersymmetry. Thus, for example, a term such as ZZ_aZ_b in the superpotential would lead to a mass term of $\sim \langle Z \rangle Z_a Z_b$ where $\langle Z \rangle \sim O(M_{Pl})$. In order to maintain the hierarchy of scales, it is imperative that we do not allow direct terms of this type in the superpotential, which leads us to postulate a separation of the super-Higgs and the physical fields in the superpotential, i.e., we write that total superpotential to be of the form [1,7]

$$W_{tot} = W(Z_a) + W_{SH}(Z), \tag{13.33}$$

where $W(Z_a)$ is the superpotential for all the fields in the physical sector of the theory, i.e., the quarks, the leptons, and the Higgs fields, while W_{SH} is the superpotential for the super-Higgs fields given by Eq. (13.14). Now, in the limit when $\kappa = 0$ the two sectors would be completely decoupled so there is no influence of one sector on another. However, with inclusion of gravitational interactions there is an effect of the super-Higgs sector on the physical sector, and this effect appears in the form of soft terms in the physical sector.

In the preceding discussion we showed that the assumption of Eq. (13.33) resolves the Planck scale hierarchy problem, i.e., that soft terms in the physical sector proportional to the Planck mass are avoided by separating the superpotential terms of the super-Higgs fields from the superpotential which depends on the fields in the visible sector. However, in grand unified theories (GUTs) there is another scale that appears, which is M_G, which although not as large as the Planck mass is much larger than the electroweak scale. This scale can create a new hierarchy problem, as it can generate mass terms in the low-energy regions of the type [8]

$$m_s M_G, \ m_s M_G(\kappa M_G), \cdots, m_s M_G(\kappa M_G)^k, \tag{13.34}$$

which can destroy the gauge hierarchy when $k \leq 6$. To discuss this problem further, we segregate the physical sector fields in two classes by writing $Z_a = (Z_\alpha, Z_i)$, where Z_α are the light fields such as quarks, leptons, and Higgs particles and their super-partners and Z_i are the fields with GUT size masses. Additionally, we may allow the superpotential to depend on the combination $\bar{Z} = \kappa Z$. Thus, in this case the total superpotential has the field dependence

$$W = W(\bar{Z}, Z_\alpha, Z_i). \tag{13.35}$$

In order to protect the low-energy theory from GUT scale masses, it is necessary to impose the following constraints:

$$W_{,\alpha i} = O(m_s), \quad W_{,\alpha\beta} = O(m_s). \tag{13.36}$$

The constraints of Eq. (13.36) kill the terms of the type given in Eq. (13.34) [8] so that the low-energy theory after integration over the heavy fields is protected both from the Planck size corrections as well as corrections of the type in Eq. (13.34). The fact that one can cancel or suppress such terms allows one to integrate on the heavy fields and discuss a low-energy theory at scales $E < M_G$. The procedure for generating a low-energy theory below the scale M_G is to consider the field equations for the super-Higgs field Z and for the heavy fields Z_i, i.e.,

$$\frac{\partial V}{\partial Z} = 0, \quad \frac{\partial V}{\partial Z_i} = 0. \tag{13.37}$$

The procedure consists of using Eq. (13.37) to solve for the Z-field and the fields Z_i in terms of the light fields Z_α. Let us now illustrate how this procedure works. A direct way to generate the low-energy theory is to expand the VEVs of the super-Higgs field, the heavy field, and, the light fields in powers of $1/M_{Pl}$, i.e., in powers of κ. Thus, we have the expansions

$$Z = Z^{-1} + Z^{(0)} + \cdots$$
$$Z_i = Z_i^{(0)} + Z_i^{(1)} + Z_i^{(2)} + \cdots$$
$$Z_\alpha = Z_\alpha^{(1)} + Z_\alpha^{(2)} + \cdots \tag{13.38}$$

Here, the fields $Z^{(m)}$, $Z_i^{(m)}$, and $Z_\alpha^{(m)}$ are of order the of κ^m. Elimination of the VEVs of Z and Z_i gives rise to an effective potential at scales $E < M_G$, which depends only on the fields Z_α, i.e., one has

$$V_{eff}(Z_\alpha) = V[Z_i(Z_\alpha); \ Z_\alpha; \ Z(Z_\alpha)]. \tag{13.39}$$

What one needs to show is that Eq. (13.39) leads to an effective potential which is independent of terms of the type Eq. (13.34). To proceed further, it is convenient to use the dimensionless lower-case variables z, z_i, and z_α, which are related to the dimensioned fields Z, Z_i, and Z_α so that

$$Z = M_{Pl}z, \quad Z_i = M_G z_i, \quad Z_\alpha = m_s z_\alpha. \tag{13.40}$$

Additionally, we introduce the following rescaling:

$$W = m^2 M_{Pl} \overline{W}, \quad G_\alpha = m_s^2 \overline{G}_\alpha, \quad G_i = m_s^2 \overline{G}_i, \quad G_Z = m^2 \overline{G}_Z. \tag{13.41}$$

Here, we have introduced the barred quantities $\overline{W}$, $\bar{G}_\alpha$, $\bar{G}_i$, and $\bar{G}_Z$, which are all dimensionless. It is also useful to introduce two parameters of smallness, ϵ and δ_s, which are defined so that

$$\epsilon = \kappa M_G \sim 10^{-2}, \quad \delta_s = \kappa m_s \sim 10^{-16}. \tag{13.42}$$

Next, from Eq. (13.3) we deduce the following three equations. For the case $A = Z$ we find [8,9]

$$\left[\frac{1}{m_s} W_{SH,ZZ} + (z\bar{G}_Z - \overline{W} - \frac{1}{4}z^2\bar{W}) + \frac{1}{2}(\epsilon\delta_s z_i \overline{G}_i + \delta_s^2 z_\alpha \overline{G}_\alpha) \right] \overline{G}_Z$$

$$+ \frac{1}{4}\epsilon\delta_s z z_i \overline{G}_i \overline{W} + \frac{1}{2}\delta_S^2 (z\overline{G}_i^2 + z\overline{G}_\alpha^2 - \frac{1}{2}z z_\alpha \overline{G}_\alpha \overline{W} = 0. \tag{13.43}$$

The case $A = Z_i$ gives [8,9]

$$[W_{,ij} + \frac{M_G}{2} \left\{ \delta_s^2(z_i\bar{G}_j + z_j\bar{G}_i) - \frac{1}{2}\epsilon\delta_s z_i z_j \bar{W} \right\}$$

$$+ m_s\delta_{ij} \left\{ \frac{1}{2}z\bar{G}_z - \bar{W} + \frac{1}{2}\delta_s^2 z_\alpha \bar{G}_\alpha \right\}]\overline{G}_j$$

$$+ M_G \left[\frac{1}{2}z_i\overline{G}_Z^2 - \frac{1}{4}z_i z \overline{W} G_Z + \frac{1}{2}\delta_s^2 z_i(\overline{G}_\alpha^2 - \frac{1}{2}\overline{G}_\alpha \overline{W}) \right] = 0, \tag{13.44}$$

while the case $A = Z_\alpha$ gives [8,9]

$$\left(\frac{1}{m_s} W_{,\alpha\beta} + \delta_{\alpha\beta}(\frac{1}{2}z\overline{G}_Z - \overline{W}) + \frac{1}{2}\epsilon\delta_s\delta_{\alpha\beta}z_i\overline{G}_i + \frac{1}{2}\delta_s^2(z_\alpha\overline{G}_\beta \right.$$

$$\left. + z_\beta\overline{G}_\alpha) - \frac{1}{4}z_\alpha z_\beta\overline{W} \right) \overline{G}_\beta$$

$$+ \left(\overline{W}_{,\alpha i}\overline{G}_i + \frac{1}{2}z_\alpha\overline{G}_Z^2 - \frac{1}{4}z_\alpha z\overline{W}\ \overline{G}_z \right) - \frac{1}{4}\epsilon\delta_s z_\alpha z_i\overline{W}\ \overline{G}_i + \frac{1}{2}\delta_s^2 z_\alpha\overline{G}_i^2 = 0. \tag{13.45}$$

Now, $W_{,ij} \sim M_G$. As a consequence, it is easily seen that all the barred quantities are $O(1)$. To appreciate how remarkable this result is, one may note that on dimensional grounds one expects $G_i \sim M_G^2$ while, since $\bar{G}_i$ is $O(1)$, Eq. (13.41) implies that $G_i \sim O(m_s^2)$. Thus, the proof of independence of the low energy from M_G proceeds in solving for z, and z_i in terms of z_α using Eqs. (13.43) and (13.44), and using them in Eq. (13.45) to determine z_α, which is then clearly seen to be independent of M_G. The dependence on the GUT mass at low-energy surfaces is only via the correction $\epsilon\delta_s$, which is very small. A further analysis shows that Eq. (13.45) can be deduced from an effective low-energy potential of

the form

$$V_{eff} = \left| \frac{\partial \tilde{W}}{\partial Z_\alpha} \right|^2 + m_0^2 Z_\alpha^\dagger Z_\alpha + (B_0 W^{(2)} + A_0 W^{(3)} + \text{h.c}) + \frac{1}{2}[g_\sigma \kappa^{-2}(G_\alpha (T^\sigma Z)_\alpha)]^2,$$

(13.46)

where m_0, A_0, and B_0 are dimensioned parameters of size m_s. Thus, the scales M_G and M_{Pl} are absent in the low-energy domain [8, 10]. It is the absence of these scales at low energies which makes supergravity grand unification a viable model of particle interactions valid at the scale $E < M_G$. As already mentioned, M_G appears only via the correction $\epsilon \delta_s$, which is very small, and has been neglected along with the δ_s^2 correction in going from Eq. (13.45) to Eq. (13.46). Here, we note that the lack of positive definiteness of the supergravity scalar potential allows one to set the cosmological constant to zero, which is not possible in global supersymmetry, so the supergravity model is consistent with what is observed except for minor corrections. We note, however, that supergravity unified models exist where the effective Lagrangian at the minimum has a large negative cosmological constant [11]. Thus, these models require explicit additional contributions to cancel such a term. Further, it is to be noted that Eq. (13.36) plays an important role in maintaining the hierarchy. It constrains the couplings of the light fields with the heavy fields. Thus, for example, consider the coupling $h_{\alpha\beta i} Z_\alpha Z_\beta Z_i$ in the superpotential. Here, one finds that $W_{,\alpha i}$ is proportional to Z_β. Since the VEV of Z_β is $O(m_s)$, this coupling is consistent with the first constraint of Eq. (13.36). However, the satisfaction of the second constraint of Eq. (13.36) would require that the VEV of the field Z_i should vanish. Another illustration is the coupling $h_{\alpha i j} Z_\alpha Z_i Z_j$. Here, the first constraint of Eq. (13.36) would forbid the coupling unless the VEV of Z_i and Z_j vanishes. Such couplings can destabilize the gauge hierarchy at the loop level [12] if Z_α also has couplings to light fields [12, 13].

We now give a more transparent and physical explanation of the origin of the soft terms of Eq. (13.46). For this purpose we simplify the scalar potential by letting $K/2 \to K$ and $W/2 \to W$. In this case the scalar potential takes the form

$$V = e^{\kappa^2 K} \left[(K^{-1})_j^i \left(\frac{\partial W}{\partial z_i} + \kappa^2 K_{,i} W \right) \left(\frac{\partial W}{\partial z_i} + \kappa^2 K_{,i} W \right)^\dagger - 3\kappa^2 |W|^2 \right] + V_D.$$

(13.47)

For illustration, we will consider the flat Kahler case where $K = Z_i Z_i^\dagger$. We note in passing that for the flat Kahler case can recover the F term contribution to the scalar potential in the global limit by letting kappa go to zero. Thus, the term $((K^{-1})_j^i \kappa^2 K_{,i} W)((K^{-1})_j^i \kappa^2 K_{,i} W)^\dagger$ generates the term $(\kappa^2 |W|^2) Z_i Z_i^\dagger$, which leads to the soft term $m_0^2 Z_i Z_i^\dagger$, where $m_0 = \kappa^2 \langle W_{SH} \rangle$. Next, we look at the cross-term $(K^{-1})_j^i \kappa^2 K_i W (\partial W/\partial Z_j)$. This generates the term $Z_i (\partial W^{(2)}/\partial Z_i) \kappa^2 \langle W_{SH} \rangle$, which gives rise to the $B_0 W^{(2)}$ part in the V_{eff} of Eq. (13.46). In a similar fashion, the term $Z_i(\partial W^{(3)}/\partial Z_i)\kappa^2 \langle W_{SH}\rangle$ gives rise to the term $A_0 W^{(3)}$ in the superpotential.

In addition to the above, we need to generate a bilinear term with the two Higgs doublets of the minimal supersymmetric standard model, i.e., a term of the form $\mu\epsilon_{ij}H_1^i H_2^j$ in the superpotential, and thus the μ-term is supersymmetric. Further, for phenomenological reasons we need this term to be of electroweak size. This can happen if the μ-term arises as a consequence of supersymmetry breaking [1, 2, 14, 15]. A variety of possibilities exist for generating such a term. If one starts with a cubic interaction UH_1H_2 in the superpotential where U is a scalar, the spontaneous supersymmetry breaking also produces a VEV for U where $\langle U\rangle$ is of electroweak size, and thus a quadratic term of the form $\mu H_1 H_2$ will emerge, and this term will be typically of the same size as the soft terms [1]. A more satisfactory possibility that was found later is to generate this term through a Kahler transformation, where one may start with a Kahler potential of the form cH_1H_2+h.c. and transfer the term through a Kahler transformation to the superpotential [14, 15].

In supergravity models the gauginos can gain mass at the loop level through their gauge interactions with the fields in the heavy sector. The loop contributions give masses of size [16, 17]

$$\tilde{m}_i = \frac{\alpha_i}{4\pi}m_{3/2}C\frac{D(R)}{D(A)}, \tag{13.48}$$

where $\alpha_i(i = 1, 2, 3)$ are the fine structure constants of the $U(1)$, $SU(2)$, and $SU(3)$ gauge groups, $D(R)$ is the dimensionality of the exchanged representation, $D(A)$ is the dimensionality of the adjoint representation, and C is the quadratic Casimir of the fields that contribute. One can also generate gaugino masses at the tree level when the supergravity Lagrangian coupled with matter and gauge fields has a field-dependent kinetic energy function $f_{\alpha\beta}$. Thus, the supergravity Lagrangian coupled with matter and gauge fields has a term which is quadratic in the gaugino fields of the form [6]

$$\left[\frac{1}{4}\,\kappa^{-1}e^{-\mathcal{G}/2}(\mathcal{G}^{-1})^i_j\mathcal{G}^{,j}f^\dagger_{\alpha\beta,i}\right]\,\overline{\lambda}^\alpha\lambda^\beta. \tag{13.49}$$

After the breaking of supersymmetry when the bracket in Eq. (13.49) develops a VEV, there is generation of a gaugingo mass term, which we may write as $\mathcal{L}^\lambda_m = -(1/2)m_{\alpha\beta}\overline{\lambda}^\alpha\lambda^\beta$. Now, $f_{\alpha\beta}$ transforms like the symmetric product of adjoint representations. If one assumes that it transforms like a singlet in the product representation, the gaugino mass term becomes universal, i.e.,

$$\mathcal{L}^\lambda_m = -\frac{1}{2}m_{1/2}\overline{\lambda}\lambda. \tag{13.50}$$

In addition to the supersymmetry breaking discussed here by the VEV formation of scalar fields, supersymmetry can also be broken by gaugino condensation with $\langle\lambda\gamma^0\lambda\rangle \neq 0$ [18]. This type of breaking is often invoked in supersymmetry breaking in the string context. Since the breaking of supersymmetry by gravity mediation was discovered, other mechanisms for supersymmetry breaking have been proposed. Chief among these are gauge mediation [19, 20] and anomaly mediation [21–24]. The reader is directed to the original papers for these.

13.5 Problems

1. Consider the case where one has two scalar fields ϕ_1 and ϕ_2 in the adjoint representation of $SU(5)$ along with two singlets ξ and η and with a superpotential given by

$$W = \lambda\xi(\mu^2 - Tr\phi_1^2) + \lambda_1 Tr(\phi_1^2\phi_2) + \lambda_2\eta Tr(\phi_1\phi_2). \quad (13.51)$$

Show that the minimization of the potential leads to solutions such that $\xi = 0 = \phi_2$ and η and ϕ_1 have the following solutions:

(a) $\phi_{1}^{a}{}_{b} = \frac{\mu}{\sqrt{20}}(\delta^a{}_b - 5\delta^a{}_5\delta^5{}_b), \eta = \frac{3}{\sqrt{20}}\mu$

(b) $\phi_{1}^{a}{}_{b} = \frac{\mu}{\sqrt{30}}(2\delta^a{}_b - 5(\delta^a{}_4\delta^4{}_b + \delta^a{}_5\delta^5{}_b)), \eta = \frac{1}{\sqrt{30}}\mu\frac{\lambda_1}{\lambda_2}.$

Here, case (a) gives a residual gauge symmetry which is $SU(4) \times U(1)$ while case (b) gives a residual symmetry which is $SU(3) \times SU(2) \times U(1)$. Show that cases (a) and (b) both have a vanishing W at the minimum, and thus gravity does not lift the degeneracy in this case.

2. Consider a superpotential of the form

$$W(Z) = m^2(Z + B). \quad (13.52)$$

Obtain a spontaneous supersymmetry breaking solution that also has a vanishing vacuum energy. Compute the gravitino mass in this case and show that it is proportional to κm^2.

3. Consider the case when

$$\mathcal{G}(z, z^*) = -\frac{1}{2}zz^* - \log|z + B|^2, \quad (13.53)$$

with B given by $B = 2\sqrt{2} - \sqrt{6}$. Discuss the super-Higgs effect in this case. Replacing z by two real fields ϕ_1 and ϕ_2, show that their masses are given by

$$m_{\phi_1}^2 = 2\sqrt{3}m_{3/2}^2$$
$$m_{\phi_2}^2 = 2(2 - \sqrt{3})m_{3/2}^2. \quad (13.54)$$

References

[1] A. H. Chamseddine, R. L. Arnowitt, and P. Nath, *Phys. Rev. Lett.* **49**, 970 (1982).
[2] P. Nath, R. L. Arnowitt, and A. H. Chamseddine *Applied N = 1 Supergravity*, World Scientific (1984); *Trieste Particle Phys.* **1** (1983) (QCD161:W626:1983).

[3] S. R. Coleman and F. De Luccia, *Phys. Rev. D* **21**, 3305 (1980).

[4] S. Weinberg, *Phys. Rev. Lett.* **48**, 1776 (1982).

[5] E. Cremmer, B. Julia, J. Scherk, P. van Nieuwenhuizen, S. Ferrara, and L. Girardello, *Phys. Lett. B* **79**, 231 (1978).

[6] E. Cremmer, S. Ferrara, L. Girardello, and A. Van Proeyen, *Nucl. Phys. B* **212**, 413 (1983).

[7] R. Barbieri, S. Ferrara, and C. A. Savoy, *Phys. Lett. B* **119**, 343 (1982).

[8] P. Nath, R. L. Arnowitt, and A. H. Chamseddine, *Nucl. Phys. B* **227**, 121 (1983).

[9] R. Arnowitt, A. H. Chamseddine, and P. Nath, *Int. J. Mod. Phys. A* **27**, 1230028 (2012) [*Int. J. Mod. Phys. A* **27**, 1292009 (2012)].

[10] L. J. Hall, J. D. Lykken, and S. Weinberg, *Phys. Rev. D* **27**, 2359 (1983).

[11] L. E. Ibanez, *Phys. Lett. B* **118**, 73 (1982).

[12] H. P. Nilles, M. Srednicki, and D. Wyler, *Phys. Lett. B* **124**, 337 (1983); *Phys. Lett. B* **120**, 346 (1983).

[13] A. Sen, *Phys. Rev. D* **30**, 2608 (1984).

[14] S. K. Soni and H. A. Weldon, *Phys. Lett. B* **126**, 215 (1983).

[15] G. F. Giudice and A. Masiero, *Phys. Lett. B* **206**, 480 (1988).

[16] L. Alvarez-Gaume, J. Polchinski, and M. B. Wise, *Nucl. Phys. B* **221**, 495 (1983).

[17] R. Arnowitt, A.H. Chamseddine, and P. Nath, in *Proceedings of Workshop on Problems in Unfication and Supergravity*, La Jolla Institute, La Jolla, CA, USA (1983).

[18] H. P. Nilles, *Phys. Lett. B* **115**, 193 (1981); S. Ferrara, L. Girardello, and H. P. Nilles, *Phys. Lett. B* **125**, 457 (1983).

[19] M. Dine and A. E. Nelson, *Phys. Rev. D* **48**, 1277 (1993); M. Dine, A.E. Nelson, and Y. Shirman, *Phys. Rev. D* **51**, 1362 (1995); M. Dine, A.E. Nelson, Y. Nir, and Y. Shirman, *Phys. Rev. D* **53**, 2658 (1996).

[20] G. F. Giudice and R. Rattazzi, *Phys. Rep.* **322**, 419 (1999).

[21] L. Randall and R. Sundrum, *Nucl. Phys. B* **557**, 79 (1999).

[22] G. F. Giudice, M. A. Luty, H. Murayama, and R. Rattazzi, *JHEP* **9812**, 027 (1998).

[23] J. A. Bagger, T. Moroi, and E. Poppitz, *JHEP* **0004**, 009 (2000).

[24] P. Binetruy, M. K. Gaillard, and B. D. Nelson, *Nucl. Phys. B* **604**, 32 (2001).

14

Phenomenology of Supergravity Grand Unification

Supergravity grand unification [1–3] provides a framework for the calculation of low-energy parameters such as sparticle masses and coupling constants in terms of a few parameters defined at a high scale, usually taken to be the grand unification scale $M_G \sim 2 \times 10^{16}$ GeV. Supergravity unified theories can accommodate both universal and as well as non-universal high-scale boundary conditions. Further supergravity models can accommodate low-energy limits of string and D-brane models since they correspond to specific choices of the Kahler potential, the superpotential, and the gauge kinetic energy function, which are the three arbitrary functions that define a supergravity model.

We now discuss the details of how a supergravity model makes predictions at low scales, starting with data specified at the grand unification scale. We will discuss the renormalization group evolution [4–6] of the various parameters that enter into the effective low-energy supergravity Lagrangian and the role that the renormalization group plays in the breaking of the electroweak symmetry [7,8]. For simplicity, we will discuss the renormalization group evolution only at one loop level. The two-loop renormalization group equations can be found in the literature [5,6]. Further, again for simplicity, we include only the top-quark Yukawa coupling in the evolution, which is a good approximation for a relatively small $\tan \beta$, i.e. $\tan \beta < 10$. We begin with a discussion of the evolution of the gauge couplings. Defining $\tilde{\alpha}_i = g_i^2/(4\pi)^2$ ($i = 1, 2, 3$) where g_1 is related to the hypercharge gauge coupling so that $g_1 = \sqrt{5/3} g_Y$, the evolution for $\tilde{\alpha}_i$ is given by

$$\frac{d\tilde{\alpha}_i}{dt} = -b_i \tilde{\alpha}_i^2(t), \quad b_i = (33/5, 1, -3). \tag{14.1}$$

Here

$$t \equiv \ln(M_G^2/Q^2) \tag{14.2}$$

where Q is the renormalization group scale and where the boundary conditions on the coupling constants at the grand unified theory (GUT) scale are $\alpha_i(0) = \alpha_G$.

As mentioned above, among the Yukawas it is the top quark Yukawa that makes the most significant contribution, at least for low values of $\tan\beta$, i.e., $\tan\beta < 10$. The renormalization group evolution for the t-quark Yukawa coupling is

$$\frac{dY_t}{dt} = \left(\frac{16}{3}\tilde{\alpha}_3 + 3\tilde{\alpha}_2 + \frac{13}{15}\tilde{\alpha}_1 - 6Y_t\right)Y_t, \tag{14.3}$$

where $Y_t = h_t^2/(4\pi)^2$. Next, we discuss the evolution of μ, B, and A_t. Here, we have [8]

$$\frac{d\mu^2}{dt} = (3\tilde{\alpha}_2 + (3/5)\tilde{\alpha}_1 - 3Y_t)\mu^2 \tag{14.4}$$

$$\frac{dB}{dt} = (3\tilde{\alpha}_2\tilde{m}_2 + (3/5)\tilde{\alpha}_1\tilde{m}_1) - 3Y_t A_t \tag{14.5}$$

$$\frac{dA_t}{dt} = (16/3)\tilde{\alpha}_3\tilde{m}_3 + 3\tilde{\alpha}_2\tilde{m}_2 + (13/15)\tilde{\alpha}_1\tilde{m}_1) - 6Y_t A_t, \tag{14.6}$$

where $\mu(0) = \mu_0$, $B(0) = B_0$, and $A_t(0) = A_0$ (see Eq. (13.46)). Regarding the gaugino masses $\tilde{m}_i(t)$ ($i = 1$–3), at the one-loop level they obey

$$\frac{d\tilde{m}_i}{dt} = -b_i\tilde{\alpha}_i(t)\tilde{m}_i(t). \tag{14.7}$$

Under the assumption of universal boundary conditions for the gaugino masses at the GUT scale, one has $\tilde{m}_i(0) = m_{1/2}$. We note the strong similarity of Eq. (14.1) and Eq. (14.7), which shows that the evolution of $\tilde{m}_i$ is identical to the evolution of $\tilde{\alpha}_i$ at one loop.

We now look at the evolution of the sparticle masses, beginning with sleptons. The evolution equations for the three generations of sleptons

$$L_i \equiv (\tilde{\nu}_{Li},\ \tilde{e}_{Li}),\ \tilde{E}_i \equiv \tilde{e}_{Ri}\ i = 1,\ 2,\ 3 \tag{14.8}$$

at the one-loop level neglecting Yukawa couplings are given by

$$\frac{dm_{Li}^2}{dt} = 3\alpha_2\tilde{m}_2^2 + (3/5)\tilde{\alpha}_1\tilde{m}_1^2$$

$$\frac{dm_{Ei}}{dt} = (12/5)\tilde{\alpha}_1\tilde{m}_1^2. \tag{14.9}$$

Assuming universal boundary conditions at the GUT scale for the three generations, one has $m_{Li}(0) = m_{Ei}(0) = m_0$.

Regarding the squark masses, we must treat the third generation separately from the first two because of the large Yukawa couplings for the third-generation squarks. The masses for the first two generations of squarks

$$Q_i \equiv (\tilde{u}_{iL},\ \tilde{d}_{iL}),\quad U_i \equiv \tilde{u}_{Ri},\quad D_i \equiv \tilde{d}_{Ri},\quad i = 1, 2, \tag{14.10}$$

neglecting Yukawa couplings, obey the evolution

$$\frac{dm_{Q_i}^2}{dt} = (16/3)\tilde{\alpha}_3\tilde{m}_3^2 + 3\tilde{\alpha}_2\tilde{m}_2^2 + (1/15)\tilde{\alpha}_1\tilde{m}_1^2$$

$$\frac{dm_{U_i}^2}{dt} = (16/3)\tilde{\alpha}_3\tilde{m}_3^2 + (16/15)\tilde{\alpha}_1\tilde{m}_1^2$$

$$\frac{dm_{D_i}^2}{dt} = (16/3)\tilde{\alpha}_3\tilde{m}_3^2 + (4/15)\tilde{\alpha}_1\tilde{m}_1^2), \tag{14.11}$$

and, under universal boundary conditions, $m_{Q_i}(0) = m_{U_i}(0) = m_{D_i}(0) = m_0$. For the third-generation squarks (i.e., $i = 3$),

$$Q_3 \equiv (\tilde{u}_{3L}, \tilde{d}_{3L}), \ U_3 \equiv \tilde{u}_{3R}, D_3 \equiv \tilde{d}_{3R}. \tag{14.12}$$

The evolution of the third generation masses must include at least the top Yukawa coupling, and thus we have

$$\frac{dm_{Q_3}^2}{dt} = \frac{16}{3}\tilde{\alpha}_3\tilde{m}_3^2 + 3\tilde{\alpha}_2\tilde{m}_2^2 + \frac{1}{15}\tilde{\alpha}_1\tilde{m}_1^2 - Y_t(m_{H_2}^2 + m_{Q_3}^2 + m_{U_3}^2 + A_t^2) \tag{14.13}$$

$$\frac{dm_{U_3}^2}{dt} = \frac{16}{3}\tilde{\alpha}_3\tilde{m}_3^2 + \frac{16}{15}\tilde{\alpha}_1\tilde{m}_1^2 - 2Y_t(m_{H_2}^2 + m_{Q_3}^2 + m_{U_3}^2 + A_t^2) \tag{14.14}$$

$$\frac{dm_{D_3}^2}{dt} = \frac{16}{3}\tilde{\alpha}_3\tilde{m}_3^2 + \frac{4}{15}\tilde{\alpha}_1\tilde{m}_1^2, \tag{14.15}$$

where, under the assumption of universal boundary conditions, $m_{Q_3}(0) = m_{U_3}(0) = m_{D_3}(0) = m_0$. Next, we consider the Higgs masses M_{H_1} and M_{H_2}, which obey the evolution equations

$$\frac{dm_{H_1}^2}{dt} = 3\tilde{\alpha}_2\tilde{m}_2^2 + \frac{3}{5}\tilde{\alpha}_1\tilde{m}_1^2 \tag{14.16}$$

$$\frac{dm_{H_2}^2}{dt} = (3\tilde{\alpha}_2\tilde{m}_2^2 + \frac{3}{5}\tilde{\alpha}_1\tilde{m}_1^2) - 3Y_t(m_{H_2}^2 + m_{Q_3}^2 + m_{U_3}^2 + A_t^2), \tag{14.17}$$

and, under the assumption of universal boundary conditions at the GUT scale, $m_{H_i}(0) = m_0$. From Eqs. (14.13), (14.14), and (14.17), it is seen that the evolution equations for M_{Q_3}, m_{U_3}, and m_{H_2} are highly coupled due to the large top quark Yukawa coupling, and one may write these three equations in a matrix form:

$$\frac{d}{dt}\begin{bmatrix} m_{H_2}^2 \\ m_{U_3}^2 \\ m_{Q_3}^2 \end{bmatrix} = -Y_t \begin{bmatrix} 3 & 3 & 3 \\ 2 & 2 & 2 \\ 1 & 1 & 1 \end{bmatrix} \begin{bmatrix} m_{H_2}^2 \\ m_{U_3}^2 \\ m_{Q_3}^2 \end{bmatrix} - \begin{bmatrix} 3 \\ 2 \\ 1 \end{bmatrix} Y_t A_t^2$$

$$+ \begin{bmatrix} 3\tilde{\alpha}_2\tilde{m}_2^2 + \frac{3}{5}\tilde{\alpha}_1\tilde{m}_1^2 \\ \frac{16}{3}\tilde{\alpha}_3\tilde{m}_3^2 + \frac{16}{15}\tilde{\alpha}_1\tilde{m}_1^2 \\ \frac{16}{3}\tilde{\alpha}_3\tilde{m}_3^2 + 3\tilde{\alpha}_2\tilde{m}_2^2 + \frac{1}{15}\tilde{\alpha}_1\tilde{m}_1^2 \end{bmatrix}. \tag{14.18}$$

The above coupled set of equations plays a central role in the spontaneous breaking of the electroweak symmetry, which we discuss next.

14.1 Radiative Breaking of $SU(2)_L \times U(1)_Y$

The $SU(2) \otimes U(1)$ invariance in the standard model is broken spontaneously by the introduction of a tachyonic mass term for the Higgs boson. Of course, the introduction of a tachyonic mass term is done in an *ad hoc* way, and points to a missing piece of underlying physics. One of the remarkable aspects of supergravity grand unification is that it provides a rational explanation for the Higgs mass turning tachyonic [1,7,8]. In supergravity grand unification the Higgs mass turns tachyonic as a consequence of soft terms. More concretely, there are two Higgs boson fields in supergravity grand unification, H_1 and H_2. The renormalization group equations for H_2 are coupled with the third-generation squarks, as shown in Eq. (14.18). The evolution of m_{H_2} is strongly governed by the first term on the right-hand side of Eq. (14.18), which has a negative sign. Thus, as we go from the GUT scale to low scales the H_2 mass turns tachyonic, which triggers the breaking of the electroweak symmetry. It is then clear that the breaking of the electroweak symmetry is intimately tied to the breaking of supersymmetry, and in the gravity-mediated breaking of supersymmetry the soft terms are governed in part by the Planck mass. Thus, we have the remarkable result that the spontaneous breaking of the electroweak symmetry and the spontaneous breaking of supersymmetry are linked, and, further, that the electroweak symmetry breaking is triggered by the soft breaking terms. In addition, the same linkage of two spontaneous breakings, one for $SU(2) \times U(1)$ and the other for supersymmetry, also connects the electroweak scale to the Planck scale.

We now give further details of how the spontaneous breaking of the electroweak symmetry comes about. The scalar potential which generates spontaneous breaking of the electroweak symmetry depends on the scalar fields in the theory which includes matter, i.e., sleptons and squarks, as well as Higgs fields. We can decompose the scalar potential into two parts, one of which depends on just the Higgs bosons fields and the other includes squark and slepton fields as well as Higgs fields. Thus, we may write

$$V = V_H^0 + V_{\tilde{f}}^0, \tag{14.19}$$

where V_H is part of the potential that depends only on the Higgs fields and $V_{\tilde{f}}$ is the potential that includes sfermion fields as well as Higgs fields. Since we wish to preserve R parity conservation, we ensure that the vacuum expectation values (VEVs) of the squark and the slepton fields vanish, and thus we focus on V_H^0. A simple derivation gives

$$V_H^0 = \frac{g'^2}{8} \left(|H_2|^2 - |H_1|^2 \right)^2 + \frac{g^2}{8} \left(H_1^* \tau_a H_1 + H_2^* \tau_a H_2 \right)^2$$
$$+ (\mu^2 + m_{H_1}^2)|H_1|^2 + (\mu^2 + m_{H_2}^2)|H_2|^2 - (B\mu\epsilon_{ij}H_1^i H_2^j + \text{h.c.}). \tag{14.20}$$

Using the identity given in Eq. (14.111), one can write the Higgs potential in the following alternative form:

$$V_H^0 = (\mu^2 + m_{H_1}^2)\left(|H_1^0|^2 + |H_1^-|^2\right)^2 + (\mu^2 + m_{H_2}^2)\left(|H_2^+|^2 + |H_2^0|^2\right)^2$$

$$- B\mu(H_1^0 H_2^0 - H_1^- H_2^+ + \text{h.c.}) + \frac{g^2}{2}|H_1^{0*}H_2^+ + H_1^{-*}H_2^0|^2$$

$$+ \frac{g'^2 + g^2}{8}\left(|H_1^0|^2 + |H_1^-|^2 - |H_2^+|^2 - |H_2^0|^2\right)^2. \qquad (14.21)$$

In the minimization of the potential, the condition that charge (and color) be preserved requires that the vacuum expectation value of the charged fields vanish. For this reason, for the purposes of electroweak symmetry breaking one sets to zero the VEVs of the charged fields when minimizing the potential, and the relevant part of the Higgs potential that enters into the symmetry breaking can be taken to be of the following form:

$$V_H^0 = (\mu^2 + m_{H_1}^2)|H_1^0|^2 + (\mu^2 + m_{H_2}^2)|H_2^0|^2 - B\mu(H_1^0 H_2^0 + \text{h.c.})$$

$$+ \frac{g'^2 + g^2}{8}\left(|H_1^0|^2 - |H_2^0|^2\right)^2. \qquad (14.22)$$

In addition, we take into account one loop correction to the Higgs potential so that the total potential to minimize is

$$V = V_H^0 + \Delta V, \qquad (14.23)$$

where ΔV is the loop correction given by [9]

$$\Delta V = \frac{1}{64\pi^2}\sum_a (-1)^{2J_a}(2J_a + 1)M_a^4 \ln\left[\frac{M_a^2}{e^{3/2}Q^2}\right], \qquad (14.24)$$

where J_a is the spin of particle a and M_a are the tree level masses that enter the loop and are functions of neutral Higgs VEVs.

The potential used in the analysis is renormalization group improved, which means that all the coupling constants and the soft parameters appearing in the potential are evaluated by renormalization group evolution from the grand unification scale to the electroweak scale. Specifically, the quantities $g_2(t)$, $g_Y(t)$, $m_i(t)$, and $M_a(t)$ are all evaluated at the scale Q, where Q is related to t by Eq. (14.2). Thus, we define

$$m_i^2(t) = m_{H_i}^2(t) + \mu^2(t), \ i = 1, 2$$
$$m_3^2(t) = -B(t)\mu(t), \qquad (14.25)$$

with the boundary conditions at $Q = M_G$ ($t = 0$) of

$$m_i^2(0) = m_0^2 + \mu_0^2, \ i = 1, 2, \quad B(0) = B_0. \qquad (14.26)$$

The renormalization group evolution allows us to derive all the sparticle masses in terms of the parameters at the GUT scale, i.e., in terms of m_0, $m_{1/2}$, A_0, B_0, and μ_0. The scalar mass squares are all positive at the GUT scale. As we evolve the potential to lower scales, if one of the scalar mass square terms turns negative, then one has spontaneous breaking. One hidden danger in this evolution is that the mass squares of a squark or a charged lepton might turn negative, which would imply spontaneous violation of color and charge. To ward against such a situation, we must restrict the parameter space that none of the squark or charged slepton mass-squared terms turns negative. (The reader is referred to the literature for early [10] and more recent [11] work on color and charge conservation constraints.) For the case of the potential given by Eqs. (14.22) and (14.25), it is easy to check that spontaneous breaking of $SU(2)_L \otimes U(1)_Y$ occurs when

$$m_1^2 m_2^2 - m_3^4 < 0. \tag{14.27}$$

Further, in order that one has a stable minimum, the potential must be bounded from below, which is achieved with the condition

$$m_1^2 + m_2^2 - 2|m_3^2| > 0. \tag{14.28}$$

Under the above conditions the minimization of the potential gives two relations, i.e., $\partial V_H / \partial v_i = 0$ for $i = 1$ and 2. One of these relations gives

$$\frac{1}{2} M_Z^2 = \frac{\mu_1^2 - \mu_2^2 \tan^2 \beta}{\tan^2 \beta - 1}, \tag{14.29}$$

where $\tan \beta \equiv v_2/v_1$ and $\mu_i^2 = m_i^2 + \Sigma_i$, where Σ_i is the loop correction where

$$\Sigma_i = \frac{1}{32\pi^2} \Sigma_a (-1)^{2s_a} n_a M_a^2 ln[M_a^2/e^{1/2}Q^2] \, (\partial M_a^2/\partial v_i). \tag{14.30}$$

and n_a counts the number of degrees of freedom for the particle a. The second relation arising from the minimization of the potential is

$$\sin 2\beta = \frac{2|m_3^2|}{\mu_1^2 + \mu_2^2}. \tag{14.31}$$

The radiative breaking equations Eqs. (14.29) and (14.31) can be used to determine the VEVs v_i of the Higgs fields in terms of the input parameters. A more common practice is to use Eq. (14.29) to determine μ^2, which determines its magnitude but not its sign, and use Eq. (14.31) to determine B_0 in terms of $\tan \beta$. Thus, instead of the five input parameters $m_0, m_{1/2}, A_0, B_0$, and μ_0, one now has the new set, which consists of four parameters and the sign, of μ, i.e., one has

$$m_0, \ m_{1/2}, \ A_0, \ \tan \beta = v_2/v_1, \ \text{sign}(\mu). \tag{14.32}$$

The model of Eq. (14.32) is referred to in the literature as minimal supergravity (mSUGRA),[*] and has been the subject of much analysis. The masses of the

---

[*] Sometimes the name constrained minimal supersymmetric standard model (CMSSM) is used for the parameter set of Eq. (14.32). This usage, however, is inappropriate since as discussed in Chapter 10 the choice of any subset of over 100 parameters in MSSM gives a constrained

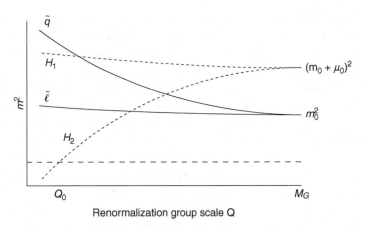

Renormalization group scale Q

Figure 14.1. A schematic diagram of the running of masses from the GUT scale M_G to the electroweak scale. The heavy top quark bends $m_{H_2}^2$ downwards as Q decreases and turns $m_{H_2}^2$ negative at the scale Q_0, breaking electroweak symmetry.

sparticles as well as of the Higgs bosons are determined in terms of these four parameters and the sign of μ [12, 13]. Consequently, there are many sum rules connecting the sparticle masses [14].

We now discuss some further details of radiative breaking. Fig. 14.1 shows the evolution of the scalar mass squares as a function of the renormalization group scale Q as we move from the GUT scale down to the electroweak scale. We note that the mass square of the field H_2 shows the most rapid decrease as a function of Q, and turns negative at low scales, triggering breaking of the electroweak symmetry. As stated earlier the reason for this is that H_2 couples with the top quark, which has a large Yukawa coupling and is largely responsible for driving the Higgs mass square negative. More precisely, three conditions are needed for radiative breaking of the electroweak symmetry to occur. These are: (1) there be non-vanishing soft breaking terms; (2) a non-vanishing μ-term should exist; (3) the top quark should be heavy. Condition (1) is needed since the electroweak symmetry breaking is governed by soft terms. Condition (2) is required for sensible phenomenology. Finally, condition (3), i.e., the requirement of a heavy top, is needed since in the renormalization group evolution it is the largeness of the top Yukawa coupling that forces the Higgs H_2 mass square to turn negative. Thus, the radiative breaking of the electroweak symmetry is the result

MSSM. Thus, the soft parameter sets arising not only in gravity-mediated breaking with its various variations aside from mSUGRA (such as supergravity models with non-universalities in the Higgs sector, in the gaugino sector, and in the sfermion flavor sector) but also via other mechanisms such as by gauge mediation, anomaly mediation, and other forms of mediation are all constrained sets, and are thus different shades of constrained MSSM. Further, the use of constrained MSSM to label Eq. (14.32) hides the fact that the set in Eq. (14.32) first originated from gravity mediation while other mediations are differently parametrized.

of the three conditions working together. The observation that the top must be heavy, i.e., $m_t > 90$ GeV, was made before the experimental measurement of the top quark mass, and thus it was one of the first predictions of supergravity GUTs.

Another relevant issue regarding radiative breaking relates to the scale Q at which minimization is to be performed. Let us define a scale Q_0 where the determinant of the Higgs mass-squared matrix first turns negative, i.e., where Eq. (14.27) is first realized, and let Q_1 be the scale below which Eq. (14.28) is no longer valid. In this case, the scale Q must lie in the interval between Q_0 and Q_1, i.e., $Q_1 < Q < Q_0$. The relevant question is, what is the value of Q where we should stop the renormalization group evolution and minimize the potential? It turns out that if one takes into account the loop correction, the minimization is rather insensitive to what value of Q is chosen. However, the effects of the loop corrections are significantly reduced if one chooses $Q \simeq (m_{\tilde{t}_1} m_{\tilde{t}_2})^{1/2}$. We also need to emphasize that the analysis rests on the fact that there is no color or charge breaking. Thus, sufficient constraints must be applied on the soft parameters to eliminate such regions of the parameter space where color or charge breakings can occur. Remarkably, in supergravity grand unification this is easily accomplished.

We now discuss the solutions to the one-loop renormalization group equations for the sparticle masses given at the beginning of this chapter. The results are for mSUGRA with the parameter space defined by Eq. (14.32). We list the results below.

Gaugino masses

At the one-loop level the renormalization group evolution of the gaugino masses parallels that for the corresponding gauge-coupling fine structure parameter $\alpha_i(t)$. Thus, from Eqs. (14.1) and (14.7) one has

$$\tilde{m}_i(t) = \frac{\alpha_i(t)}{\alpha_G} m_{1/2}, \tag{14.33}$$

where $\alpha_i(t) = \alpha_G/(1 + \beta_i t)$ and $\beta_i = b_i \tilde{\alpha}_G$, where $\tilde{\alpha}_G$ and b_i are as defined in Eq. (22.19).

$$\tilde{m}_1 : \tilde{m}_2 : \tilde{m}_3 = \alpha_1 : \alpha_2 : \alpha_3. \tag{14.34}$$

The preceding result is valid only at the one-loop level, and there can be significant deviations from this relation at the two-loop level.

Slepton masses

An integration of the renormalization group evolution equations gives the slepton masses at the electroweak scale so that

$$m_{\tilde{e}_{iL}}^2 = m_0^2 + m_{ei}^2 + \tilde{\alpha}_G \left(\frac{3}{2} f_2 + \frac{3}{10} f_1 \right) m_{1/2}^2 + \left(-\frac{1}{2} + \sin^2 \theta_W \right) M_Z^2 \cos 2\beta$$
(14.35)

$$m_{\tilde{\nu}_{iL}}^2 = m_0^2 + \tilde{\alpha}_G \left(\frac{3}{2} f_2 + \frac{3}{10} f_1 \right) m_{1/2}^2 + \frac{1}{2} M_Z^2 \cos 2\beta \qquad (14.36)$$

$$m_{\tilde{e}iR}^2 = m_0^2 + m_{e_i}^2 + \tilde{\alpha}_G (6/5) f_1 m_{1/2}^2 - \sin^2 \theta_W M_Z^2 \cos 2\beta, \qquad (14.37)$$

where $f_k(t) = t(2 + b_k \tilde{\alpha}_G t)/(1 + b_k \tilde{\alpha}_G t)^2$. In the analysis above, we have ignored the left–right mixing for the charged leptons. Including this term, the mass square matrix becomes

$$\begin{pmatrix} m_{\tilde{e}_{iL}}^2 & m_{e_i}(A_{e_i} - \mu \tan \beta) \\ \\ m_t(A_{e_i} - \mu \tan \beta) & m_{\tilde{e}_{iR}}^2 \end{pmatrix}.$$
(14.38)

The contributions of the off-diagonal terms are proportional to the charged lepton masses and are small except for the third generation because of the relative largeness of the tau mass.

Squark masses for the first two generations

For the first two generations ($i = 1, 2$) of up squarks, integration of the renormalization group equations gives

$$m_{\tilde{u}_{iL}}^2 = m_0^2 + m_{u_i}^2 + \tilde{\alpha}_G \left(\frac{8}{3} f_3 + \frac{3}{2} f_2 + \frac{1}{30} f_1 \right) m_{1/2}^2$$
$$+ \left(\frac{1}{2} - \frac{2}{3} \sin^2 \theta_W \right) M_Z^2 \cos 2\beta \qquad (14.39)$$

$$m_{\tilde{u}_{iR}}^2 = m_0^2 + m_{u_i}^2 + \tilde{\alpha}_G \left(\frac{8}{3} f_3 + \frac{8}{15} f_1 \right) m_{1/2}^2 + \frac{2}{3} \sin^2 \theta_W M_Z^2 \cos 2\beta. \quad (14.40)$$

Similarly, for the down squarks one has the relations

$$m_{\tilde{d}_{iL}}^2 = m_0^2 + m_{d_i}^2 + \tilde{\alpha}_G \left[\frac{8}{3} f_3 + \frac{3}{2} f_2 + \frac{1}{30} f_1 \right] m_{1/2}^2$$
$$- \left(\frac{1}{2} - \frac{1}{3} \sin^2 \theta_W \right) M_Z^2 \cos 2\beta \qquad (14.41)$$

$$m_{\tilde{d}_{iR}}^2 = m_0^2 + m_{d_i}^2 + \tilde{\alpha}_G \left(\frac{8}{3} f_3 + \frac{2}{15} f_1 \right) m_{1/2}^2 - \frac{1}{3} \sin^2 \theta_W M_Z^2 \cos 2\beta, \quad (14.42)$$

where m_{u_i} and m_{d_i} are quark masses. In the above, we have neglected the mixings between the left chiral and the right chiral squarks. Including these, one has the following mass matrices for the first two generations of up and down squarks. For the up squarks one has

$$\begin{pmatrix} m_{\tilde{u}_{iL}}^2 & m_{u_i}(A_{u_i} - \mu \cot \beta) \\ \\ m_{u_i}(A_{u_i} - \mu \cot \beta) & m_{\tilde{u}_{iR}}^2 \end{pmatrix},$$
(14.43)

and for the down squarks one has

$$
\begin{pmatrix}
m^2_{\tilde{d}_{iL}} & m_{d_i}(A_{d_i} - \mu \tan \beta) \\
m_{d_i}(A_{d_i} - \mu \tan \beta) & m^2_{\tilde{d}_{iR}}
\end{pmatrix}. \tag{14.44}
$$

Squark masses for the third generation.
For the third generation the evolution of the mass squares for the sleptons and $\tilde{b}_R$ will be similar to the case for the first two generations. However, the evolution of the mass squares for $\tilde{t}_L$, $\tilde{b}_L$, $\tilde{t}_R$, and H_2 are highly coupled and strongly affected by the top quark Yukawa coupling, as can be seen from the evolution equation (Eq. (14.18)). The analytic form for these is rather complicated, and we do not display it here. As for the first two generations, the mass-squared matrix for the top squarks will have the form

$$
\begin{pmatrix}
m^2_{\tilde{t}_L} & m_t(A_t - \mu \cot \beta) \\
m_t(A_t - \mu \cot \beta) & m^2_{\tilde{t}_R}
\end{pmatrix}. \tag{14.45}
$$

In this case, the largeness of the Yukawa coupling can produce tachyonic states when the mass squares of one of the top squarks (stops) turns negative. This can happen for the case of a large trilinear coupling A_t, and thus the constraints that there be no tachyons puts limits on the allowed range of A_t. The sfermion mass squares matrix can be diagonalized by a unitary transformation. Thus, the generic sfermion mass squares matrix

$$
(M^2_{\tilde{f}}) = \begin{pmatrix}
m^2_{\tilde{f}_L} & m^2_{\tilde{f}_{LR}} \\
m^{2*}_{\tilde{f}_{LR}} & m^2_{\tilde{f}_R}
\end{pmatrix} \tag{14.46}
$$

can be diagonalized so that

$$
U^\dagger_{\tilde{f}}(M^2_{\tilde{f}})U_{\tilde{f}} = diag\,(m^2_{\tilde{f}_1}, m^2_{\tilde{f}_2}). \tag{14.47}
$$

For sfermion masses with no phases, $U_{\tilde{f}}$ is an orthogonal matrix:

$$
U_{\tilde{f}} = \begin{pmatrix}
\cos \theta_f & -\sin \theta_f \\
\sin \theta_f & \cos \theta_f
\end{pmatrix}. \tag{14.48}
$$

The left chiral and the right chiral sfermion states can be expressed in terms of the mass eigenstates so that

$$
\begin{pmatrix}
\tilde{f}_L \\
\tilde{f}_R
\end{pmatrix} = \begin{pmatrix}
\cos \theta_f & -\sin \theta_f \\
\sin \theta_f & \cos \theta_f
\end{pmatrix} \begin{pmatrix}
\tilde{f}_1 \\
\tilde{f}_2
\end{pmatrix}. \tag{14.49}
$$

The elements of the mass matrix $(M_{\tilde{f}}^2)$ could be expressed in terms of the diagonal elements so that

$$m_{\tilde{f}_L}^2 = \cos^2 \theta_f m_{\tilde{f}_1}^2 + \sin^2 \theta_f m_{\tilde{f}_2}^2$$
$$m_{\tilde{f}_R}^2 = \sin^2 \theta_f m_{\tilde{f}_1}^2 + \cos^2 \theta_f m_{\tilde{f}_2}^2$$
$$m_{\tilde{f}_{LR}}^2 = \cos \theta_f \sin \theta_f (m_{\tilde{f}_1}^2 - m_{\tilde{f}_2}^2). \tag{14.50}$$

For the case when the sfermion mass-squared matrix is complex, one needs to include the dependence on CP phases, and this analysis is given in Chapter 15.

14.2 The Super-GIM Mechanism

One of the strong constraints on model building is the suppression of the flavor-changing neutral-current (FCNC) processes. A prime example of this is the process $K_L^0 \to \mu^+ \mu^-$, which experimentally is highly suppressed. In the standard model this process proceeds by a box diagram with the exchange of $W^\pm, u(c)$, and ν_μ in the loop, as shown in the left diagram in Fig. 14.2. Consistency with the FCNC constraint is obtained because in the standard model the left diagram is of size $(m_c^2 - m_u^2)/M_W^2$. This suppression is referred to as the Glashow–Iliopoulos–Maini (GIM) mechanism. In supersymmetry, the $K_L^0 \to \mu^+ \mu^-$ process proceeds via the right diagram of Fig. 14.2 with the exchanges of charginos, squarks, and sneutrinos. In global supersymmetry, the assumption of universality of the first two-generation squark masses except for the quark mass corrections lead to the same result [15]. In supergravity unification with universal boundary conditions at the grand unification scale, a super-GIM mechanism exists, i.e., one finds that at the electroweak scale because of Eqs. (14.39) and (14.40) one has the relation

$$m_{\tilde{c}}^2 - m_{\tilde{u}}^2 = m_c^2 - m_u^2. \tag{14.51}$$

Thus, in supergravity unification the corrections to the $K_L^0 \to \mu^+ \mu^-$ process from supersymmetry are of size $(m_{\tilde{c}}^2 - m_{\tilde{u}}^2)/M_{\chi^\pm}^2$ or $(m_{\tilde{c}}^2 - m_{\tilde{u}}^2)/M_{\tilde{q}}^2$, which are of the same size or less than in the standard model as a consequence of Eq. (14.51).

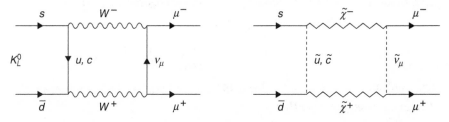

Figure 14.2. One loop contributions to $K_L \to \mu^+ \mu^-$. The left diagram (the usual GIM term) is of size $(m_c^2 - m_u^2)/M_W^2$ while the right diagram (the additional supersymmetry term) is of size $(m_{\tilde{c}}^2 - m_{\tilde{u}}^2)/M_{\chi^\pm}^2$ or $(m_{\tilde{c}}^2 - m_{\tilde{u}}^2)/(m_{\tilde{q}}^2)$.

14.3 Higgs Bosons

Let us now look at the mass spectrum in the Higgs boson sector after spontaneous breaking of the electroweak symmetry. Using $H_1^1 = (u_1 + \phi_1^0 + i\psi_1)/\sqrt{2}$ and $H_2^2 = (u_2 + \phi_2^0 + i\psi_2)/\sqrt{2}$ the Higgs boson mass-squared matrix in the neutral sector takes the form

$$\mathcal{L}^m_{\text{neutral Higgs}} = -\frac{1}{2}(\phi_1^0, \phi_2^0)(\mathcal{M}_H^2)\begin{pmatrix} \phi_1^0 \\ \phi_2^0 \end{pmatrix} - \frac{1}{2}(\psi_1, \psi_2)(\mathcal{M}_A^2)\begin{pmatrix} \psi_1 \\ \psi_2 \end{pmatrix}, \quad (14.52)$$

where ϕ_i^0 and ψ_i are the real and the imaginary parts of the neutral Higgs boson fields H_i^0 ($i = 1, 2$), respectively. The matrix $(\mathcal{M}_A^2)$ in the CP odd sector is given by

$$(\mathcal{M}_A^2) = \frac{2m_3^2}{\sin 2\beta}\begin{pmatrix} \sin^2\beta & -\sin\beta\cos\beta \\ -\sin\beta\cos\beta & \cos^2\beta \end{pmatrix}, \quad (14.53)$$

so the CP-odd mass-squared matrix $\mathcal{M}_A^2$ has eigenvalues $(0, M_A^2)$ where the zero eigenvalue corresponds to the Goldstone mode and m_A^2 is given by

$$M_A^2 = \frac{2m_3^2}{\sin 2\beta}. \quad (14.54)$$

In the neutral CP-even sector the mass-squared matrix $(\mathcal{M}_H^2)$ is given by

$$(\mathcal{M}_H^2) = \begin{pmatrix} M_A^2\sin^2\beta + M_Z^2\cos^2\beta & -(M_A^2 + M_Z^2)\sin\beta\cos\beta \\ -(M_A^2 + M_Z^2)\sin\beta\cos\beta & M_A^2\cos^2\beta + M_Z^2\sin^2\beta \end{pmatrix}. \quad (14.55)$$

The mass eigenstates h_0 and H^0 are related to ϕ_1^0 and ϕ_2^0 by the relation

$$\begin{pmatrix} H^0 \\ h^0 \end{pmatrix} = \begin{pmatrix} \cos\alpha & \sin\alpha \\ -\sin\alpha & \cos\alpha \end{pmatrix}\begin{pmatrix} \phi_1^0 \\ \phi_2^0 \end{pmatrix}, \quad (14.56)$$

where

$$\sin 2\alpha = -\sin 2\beta\left(\frac{m_{H^0}^2 + m_{h^0}^2}{m_{H^0}^2 - m_{h^0}^2}\right)$$

$$\cos 2\alpha = -\cos 2\beta\left(\frac{m_{A^0}^2 - M_Z^2}{m_{H^0}^2 - m_{h^0}^2}\right). \quad (14.57)$$

Here, $M_{h^0}^2$ and $M_{H^0}^2$ are eigenvalues of the CP even mass-squared matrix Eq. (14.55) so that

$$M_{h^0, H^0}^2 = \frac{1}{2}\left[M_A^2 + M_Z^2 \mp ((M_A^2 + M_Z^2)^2 - 4M_A^2 M_Z^2\cos^2\beta)^{1/2}\right]. \quad (14.58)$$

The corresponding eigenstates are

$$H^0 = \phi_1^0\cos\alpha + \phi_2^0\sin\alpha$$
$$h^0 = -\phi_1^0\sin\alpha + \phi_2^0\cos\alpha. \quad (14.59)$$

Next, we look into the charged Higgs boson sector. Here, the mass-squared matrix is given by

$$\mathcal{L}^m_{\text{charged Higgs}} = -(M_A^2 + M_W^2)(\phi_1^+, \phi_2^+) \begin{pmatrix} \sin^2\beta & -\sin\beta\cos\beta \\ -\sin\beta\cos\beta & \cos^2\beta \end{pmatrix} \begin{pmatrix} \phi_1^- \\ \phi_2^- \end{pmatrix}.$$

(14.60)

The eigenstates of the above matrix are $(G^\pm, H^\pm)$, where

$$\begin{pmatrix} G^\pm \\ H^\pm \end{pmatrix} = \begin{pmatrix} \cos\beta & \sin\beta \\ -\sin\beta & \cos\beta \end{pmatrix} \begin{pmatrix} \phi_1^\pm \\ \phi_2^\pm \end{pmatrix}.$$

(14.61)

They correspond to the eigenvalues $(0, M_{H^\pm}^2)$, where the zero eigenvalue corresponds to the Goldstone boson and $M_{H^\pm}^2$ is the mass square of the charged Higgs boson, where

$$M_{H^\pm}^2 = M_A^2 + M_W^2.$$

(14.62)

Thus, after spontaneous breaking of the electroweak symmetry the massive Higgs boson sector has the following particle spectrum:

$$h^0, H^0, A^0, H^\pm.$$

(14.63)

14.4 Loop Corrections to the Higgs Boson Mass

The light CP-even Higgs boson mass m_h is less than M_Z at the tree level, and loop corrections are needed to lift its mass above M_Z. The loop correction to the Higgs boson mass can be obtained from the loop-corrected effective potential, i.e.,

$$V(H_1, H_2) = V_0 + \Delta V,$$

(14.64)

where V_0 is the renormalization group improved tree level potential and ΔV is the loop correction. At the one-loop level the correction ΔV to the effective potential is [9, 15, 16]

$$\Delta V = \frac{1}{64\pi^2} str \left[M_i^4(H_1, H_2) \left(\ln \frac{M_i^2(H_1, H_2)}{Q^2} - \frac{3}{2} \right) \right].$$

(14.65)

Here, str stands for the supertrace $\sum_i c_i(2J_i+1)(-1)^{2J_i}$ where $c_i(2J_i+1)$ counts the degrees of freedom and the sum runs over both bosonic and fermionic particles i which enter into the self-energy diagram for the Higgs boson. The computation of the CP-even Higgs boson mass-squared matrix is then given by

$$(M_H)^2_{\alpha\beta} = \frac{\partial^2 V}{\partial v_\alpha \partial v_\beta} = (M_H^2)^0_{\alpha\beta} + \Delta M_{H\alpha\beta}^2,$$

(14.66)

where $(\alpha, \beta) = (1, 2)$ and $v_1 \equiv \sqrt{2}\langle H_1^1 \rangle$, $v_2 \equiv \sqrt{2}\langle H_2^2 \rangle$. Here, $(M_H^2)^0_{\alpha\beta}$ is the tree level contribution from V_0 and $\Delta M_{H\alpha\beta}^2$ is the one-loop contribution from ΔV, and $\Delta M_{H\alpha\beta}^2$ is given by

$$\Delta M_{H\alpha\beta}^2 = \frac{1}{32\pi^2} str \left[\frac{\partial M_i^2}{\partial v_\alpha} \frac{\partial M_i^2}{\partial v_\beta} \ln \frac{M_i^2}{Q^2} + M_i^2 \frac{\partial^2 M_i^2}{\partial v_\alpha \partial v_\beta} \left(\ln \frac{M_i^2}{Q^2} - 1 \right) \right]. \quad (14.67)$$

The variations of the potential with respect to v_1 and v_2 give the constraints

$$m_1^2 + \frac{g_2^2 + g_1^2}{8} (v_1^2 - v_2^2) + |B\mu|^2 \tan\beta + \frac{1}{v_1} \frac{\partial \Delta V}{\partial v_1} = 0 \quad (14.68)$$

$$m_2^2 - \frac{g_2^2 + g_1^2}{8} (v_1^2 - v_2^2) + |B\mu|^2 \cot\beta + \frac{1}{v_2} \frac{\partial \Delta V}{\partial v_2} = 0. \quad (14.69)$$

In the analysis of the Higgs mass-squared matrix, we can use Eq. (14.68) and Eq. (14.69) to write the CP-even mass-squared matrix in the following form:

$$M_H^2 = \begin{pmatrix} M_Z^2 c_\beta^2 + M_A^2 s_\beta^2 + \Delta_{11} & -(M_Z^2 + M_A^2) s_\beta c_\beta + \Delta_{12} \\ -(M_Z^2 + M_A^2) s_\beta c_\beta + \Delta_{12} & M_Z^2 s_\beta^2 + M_A^2 c_\beta^2 + \Delta_{22} \end{pmatrix}, \quad (14.70)$$

where the corrections Δ_{ij} are given by Eq. (14.113) (e.g., [17,18]). We are specifically interested in the correction to the light CP-even Higgs mass h^0. The dominant contribution to it arises from the stop masses. Their contribution is given by [19,20].

$$\Delta m_h^2 \simeq \frac{3m_t^4}{2\pi^2 v^2} \ln \frac{M_S^2}{m_t^2} + \frac{3m_t^4}{2\pi^2 v^2} \left(\frac{X_t^2}{M_S^2} - \frac{X_t^4}{12M_S^4} \right), \quad (14.71)$$

where $v = 246$ GeV, M_S is an average stop mass, and X_t is given by

$$X_t \equiv A_t - \mu \cot\beta. \quad (14.72)$$

From Eq. (14.71) one finds that the loop correction is maximized at $X_t \sim \sqrt{6} M_S$. Recently, experiments at CERN [21,22] have confirmed the existence of the Higgs boson with mass ~ 126 GeV by using data from the diphoton decay of the Higgs boson.

14.5 Low-energy Higgs Theorems

There are low-energy theorems for the Higgs bosons which are often useful, and we discuss these here (for some of the works on low-energy Higgs boson theorems, see the literature, e.g., [23–28]). In the standard model one can write the Higgs interaction after spontaneous breaking in the fermion sector so that

$$L_{hff} = -\left(1 + \frac{h}{v} \right) \sum_i m_{f_i} \bar{f}_i f_i, \quad (14.73)$$

where $v \simeq 246$ GeV. The above implies that if a mass term appears in the Lagrangian where the mass is such that it originates from spontaneous breaking of the electroweak symmetry and depends linearly on the Higgs VEV, then the

Lagrangian including the interaction with the Higgs field can be obtained by the replacement

$$m \rightarrow m \left(1 + \frac{h}{v} \right). \tag{14.74}$$

Thus, the Higgs interactions with the W and Z bosons have the form

$$L_{hVV} = - \left(1 + \frac{h}{v} \right)^2 \left(M_W^2 W_\mu^+ W_\mu^- + \frac{1}{2} M_Z^2 Z_\mu Z^\mu \right). \tag{14.75}$$

The above leads to the following low-energy theorem in the limit of the vanishing four-momentum p_h of the Higgs boson h:

$$\mathcal{M}(X \rightarrow Y + h)_{lim \; p_h \rightarrow 0} = \sum_i \frac{m_i}{v} \frac{\partial}{\partial m_i} \mathcal{M}(X \rightarrow Y), \tag{14.76}$$

where i runs over both the fermionic and the vector boson modes.

Equation (14.76) is useful in deriving the low-energy interactions of the Higgs with other fields. Thus, for example, one can derive the low-energy Higgs–gluon–gluon and the Higgs–gamma–gamma interactions by considering the corrections to the gluon and to the photon self-energies. For the case of the gluon, the self-energy arises from the top quark exchange, and the effective Higgs–gluon–gluon interaction is given by

$$L_{hgg} = \frac{\alpha_s}{12\pi v} h G_{\mu\nu}^a G^{\mu\nu a}. \tag{14.77}$$

A very similar analysis can be done to obtain the effective Higgs–gamma–gamma coupling. Here, the dominant contributions to the effective gauge kinetic energy term for the photon include one-loop corrections from the top exchange and from the W boson exchange. In this case, the effective photon–photon–Higgs interaction is

$$L_{h\gamma\gamma} = \frac{\alpha_{em}}{16\pi} \sum_i \left[b_i Q_i^2 \frac{\partial}{\partial v} \log \; m_i^2(v) \right] h F_{\mu\nu} F^{\mu\nu}, \tag{14.78}$$

where $b_1 = -7$ for a vector boson, i..e, the W-boson, and $b_{1/2} = 4/3$ for a Dirac fermion, i.e., the top quark, and Q_i is the charge of the heavy particle in the loop and where m_i is the mass of the particle circulating in the loop. The above result is valid when the particles running in the loop are much heavier than the Higgs boson. When the particle masses in the loop are light, they cannot be integrated out, and in this case one must include the full loop functions. When there are many particles with the same charge, $\log m_i^2$ is replaced by $\log(det \; M^2)$, where M^2 is the mass-squared matrix of all the particles that are circulating in the loop producing the decay of the Higgs bosons. For the supersymmetric case, we have two Higgs doublets, and instead of one Higgs VEV we have two, v_u and v_d.

In this case, Eq. (14.78) is modified to read

$$\mathcal{L}^{\text{SUSY}}_{h\gamma\gamma} = \frac{\alpha_{em}}{16\pi} \sum_i b_i Q_i^2 \left[\cos\alpha \frac{\partial}{\partial v_2} \log m_i^2(v_2) - \sin\alpha \frac{\partial}{\partial v_1} \log m_i^2(v_1) \right] h F_{\mu\nu} F^{\mu\nu},$$

(14.79)

where α is the mixing angle between the two CP-even Higgs in the MSSM given by Eq. (14.57).

14.6 Diphoton Decay of the Higgs Boson

Let us consider the case of one Higgs doublet with $H^T = (H^+, H^0)$ where after spontaneous breaking $\langle Re\ H^0 \rangle = (v + h)\sqrt{2}$ (where in the standard model $v = 246$ GeV). Let us assume a coupling of h to spins 1, 1/2, and 0 of the form

$$- \mathcal{L}_{int} = g_{hVV} h V_\mu^+ V^{-\mu} + g_{hff} h f \bar{f} + g_{hSS} h S \bar{S}, \tag{14.80}$$

where g_{hVV} is the coupling to spin 1 fields, g_{hff} is the coupling to spin 1/2 fields, and g_{hSS} is the coupling to spin 0 fields. In this case, the decay of the Higgs boson into two photons goes through a loop, involving charge vector bosons, charged fermions, and charged scalars. The loop correction is given by

$$\Gamma(h \to \gamma\gamma) = \frac{\alpha^2 m_h^3}{1024\pi^3} \left| \frac{g_{hVV}}{m_V^2} Q_V^2 A_1(\tau_V) + \frac{2g_{hff}}{m_f} N_{c,f} Q_f^2 A_{\frac{1}{2}}(\tau_f) \right.$$
$$\left. + \frac{g_{hSS}}{m_S^2} N_{c,S} Q_S^2 A_0(\tau_S) \right|^2, \tag{14.81}$$

where N is the number of colors and Q is the charge, while A are loop functions which are defined as follows:

$$A_0(\tau) = -\tau[1 - \tau f(\tau)] \tag{14.82}$$

$$A_{\frac{1}{2}}(\tau) = 2\tau[1 + (1 - \tau)f(\tau)] \tag{14.83}$$

$$A_1(\tau) = -[2 + 3\tau + 3\tau(2 - \tau)f(\tau)]. \tag{14.84}$$

The function $f(\tau)$ appearing above is defined by

$$f(\tau) = \begin{cases} \left(\arcsin \dfrac{1}{\sqrt{\tau}} \right)^2, & \tau \geq 1 \\ -\dfrac{1}{4} \left[\ln \dfrac{\eta_+}{\eta_-} - i\pi \right]^2, & \tau < 1, \end{cases} \tag{14.85}$$

where $\eta_\pm \equiv (1 \pm \sqrt{1 - \tau})$ and $\tau = 4m^2/m_h^2$ for a particle running in the loop with mass m. For the case when the particles running in the loop are much heavier that $m_h/2$, so that $\tau \gg 1$, one has

$$f(\tau) \to \frac{1}{\tau} \left(1 + \frac{1}{3\tau} + \cdots \right). \tag{14.86}$$

In this limit $A_1 \to -7$, $A_{\frac{1}{2}} \to 4/3$, and $A_0 \to 1/3$.

Let us now consider the case of the standard model. Here, the vector bosons in the loop are the $W^{\pm}$ bosons and the fermions are the quarks and the leptons, so that we have $g_{hWW} = g_2 M_W$ and $g_{hff} = g_2 m_f/(2M_W)$, where g_2 is the $SU(2)$ gauge coupling. Thus, $g_{hWW}/M_W^2 = 2g_{hff}/m_f = 2/v$. Among the fermions, the largest contribution comes from the top quark exchange because of the large coupling of the top with the Higgs boson. The dominant contribution to the diphoton rate arises from the W-exchange diagrams, which is partially cancelled by the top quark exchange. Thus, for the standard model Eq. (14.81) gives

$$\Gamma_{SM}(h \to \gamma\gamma) \approx \frac{\alpha_{em}^2 m_h^3}{256v^2\pi^3} \Big| A_1(\tau_W) + N_c Q_t^2 A_{\frac{1}{2}}(\tau_t) \Big|^2 \to \frac{\alpha_{em}^2 m_h^3}{256v^2\pi^3} |\mathcal{A}_{SM}|^2, \quad (14.87)$$

where $\mathcal{A}_{SM} \approx -6.49$.

Let us now consider the supersymmetric extension of Eq. (14.87). Keeping in mind that for the MSSM case one has two Higgs doublets with two mixing angles α and β, a direct extension of Eq. (14.87) gives

$$\Gamma_{SUSY}(h \to \gamma\gamma) \approx \frac{\alpha_{em}^2 m_h^3}{256v^2\pi^3} \Big| \sin(\beta - \alpha) Q_W^2 A_1(\tau_W) + \frac{\cos\alpha}{\sin\beta} N_t Q_t^2 A_{\frac{1}{2}}(\tau_t) \Big|^2. \quad (14.88)$$

A comparison of the decay width in Eq. (14.88) with Eq. (14.87) shows that for the case of the MSSM the Higgs couplings to the W-boson and to the top quark are modified by the factors $\sin(\beta - \alpha)$ and $\cos\alpha/\sin\beta$ compared with the standard model case. For the supersymmetric case, there would be, in addition, contributions from the exchange of charginos and from the exchange of squarks, sleptons, and the charged Higgs fields.

14.7 Hyperbolic Geometry: The Focus Point, Focal Curves, and Focal Surfaces

We now discuss an interesting region of the parameter space of SUGRA models where TeV-size scalars appear quite naturally. To illustrate this point we write the radiative breaking equation to one-loop order in the following form [29, 30]:

$$\mu^2 = -\frac{1}{2}M_Z^2 + m_0^2 C_1 + A_0^2 C_2 + m_{1/2}^2 C_3 + m_{1/2}A_0 C_4 + \Delta\mu_{\text{loop}}^2, \quad (14.89)$$

where

$$C_1 = \frac{1}{\tan^2\beta - 1}\Big(1 - \frac{3D_0 - 1}{2}\tan^2\beta\Big), \quad (14.90)$$

and C_2, C_3, and C_4 are as defined in the literature [29, 30]. Here, $D_0(t)$ is defined by

$$D_0(t) = (1 + 6Y_0 K(t))^{-1}, \quad (14.91)$$

where Y_0 is related to the top Yukawa coupling $h_t(0)$ at the GUT scale so that $Y_0 = h_t^2(0)/(4\pi^2)$, and $K(t)$ is defined by

$$K(t) \equiv \int_0^t dt' \, (1 + \beta_3 t')^{16/3b_3} (1 + \beta_2 t')^{3/b_2} (1 + \beta_1 t')^{13/15b_1}. \qquad (14.92)$$

Here, $\beta_i = \alpha_i(0)b_i/(4\pi)$ and $b_i = \left(\frac{33}{5}, 1, -3\right)$ for $U(1)$, $SU(2)$, $SU(3)$, and $\alpha_3(0) = \alpha_2(0) = \alpha_1(0) = \alpha_G$. For large $\tan\beta$, i.e., $\tan\beta \gg 1$, Eq. (14.90) reduces to

$$C_1 \simeq -\frac{3D_0 - 1}{2}. \qquad (14.93)$$

μ^2 as given by Eq. (14.89) consists of two parts, one arising from the tree level potential V_0 and the other arising from the one-loop correction $\Delta\mu_{\text{loop}}^2$. Typically, each of them separately could have a strong dependence on the renormalization group scale Q. However, together the dependence is significantly reduced, and μ has only a very mild dependence on Q. It turns out that the one-loop contribution $\Delta\mu_{\text{loop}}^2$ is typically minimized at $Q_m \sim \mathcal{O}\left(\sqrt{M_{\tilde{t}_1} M_{\tilde{t}_2}}\right)$, where $M_{\tilde{t}_1}$ and $M_{\tilde{t}_2}$ are the stop masses. Thus, at the scale Q_m the geometrical interpretation of Eq. (14.89) is much simpler. Here, we may write Eq. (14.89) in the form

$$\mu^2 + \frac{1}{2}M_Z^2 = m_0^2 C_1 + \bar{A}_0^2 C_2 + m_{1/2}^2 \bar{C}_3, \qquad (14.94)$$

where $\bar{A}_0 \equiv A_0 + \frac{1}{2}M_{1/2}C_4/\sqrt{C_2}$ and $\bar{C}_3 = C_3 - C_4^2/4$. For μ fixed there are now two possibilities. The first possibility corresponds to the case when C_1, C_2, and $\bar{C}_3$ are all positive. In this case, the soft parameters m_0, $m_{1/2}$, and $\bar{A}_0$ lie on the surface of an ellipsoid whose radii are determined by μ, the top mass, and $\tan\beta$. On this branch there is an upper limit on how large the soft parameters m_0, $m_{1/2}$, and $\bar{A}_0$ can get. One may call this the ellipsoidal branch (EB). Next, assume that one of the functions C_1, C_2, and $\bar{C}_3$ turns negative. In this case, for a fixed μ the soft parameters m_0, $m_{1/2}$, and $\bar{A}_0$ lie on the surface of a hyperboloid, which implies that two or all three soft parameters can become large while μ remains small. Specifically, here we have TeV-size scalars with no fine tuning needed to keep μ small. One may call this the hyperbolic branch (HB) [30]. Numerical analyses show that C_1 does turn negative for certain ranges of the parameters. The interpretation of the point $C_1 = 0$ is rather simple. It is seen from Eq. (14.94) that when $C_1 = 0$, μ becomes independent of m_0, and thus we can have TeV-size scalars consistent with small μ. As can be seen from Eq. (14.93), this is the limit point of the hyperbolic branch which occurs when $D_0 \sim 1/3$, and is often referred to as the focus point [31,32].). For the case when $C_1 < 0$, we have several possibilities [33]. m_0 and $\bar{A}_0$, or m_0 and $m_{1/2}$, may become large while μ remains small. One may label these as focal curves. More generally, for the case when $C_1 < 0$, one has the possibility that $m_0, m_{1/2}$, and $\bar{A}_0$ all become large while μ remains small. One may call this the focal surface region. Data from the Large Hadron Collider is expected to provide hints regarding which branch of radiative breaking of the electroweak symmetry one is on. An extreme limit of the hyperbolic branch is the case when the universal scalar mass m_0 becomes superheavy but the gaugino masses remain light. This is the split supersymmetry limit [34] of the hyperbolic branch. The literature for works on fine tuning and naturalness in supersymmetry and supergravity models can be found in [35–41].

14.8 Non-universalities in Soft Breaking

Since the nature of Planck scale physics is still largely unknown, it is useful to consider the possibility of non-universalities of soft breaking parameters at the GUT scale consistent with the flavor-changing neutral-current constraints. These considerations allow for non-universalities in the gaugino sector, the Higgs sector, and the third-generation sector (non-universalities have been discussed in a number of works and a small sample is [42–46]). Non-universalities appear quite naturally in supergravity models with more general gauge kinetic energy functions for the gaugino masses and with non-minimal Kähler potential. Such non-universalities are also generic in string and D-brane models. Non-universalities for the gaugino masses at the tree level can arise from non-universalities of the gauge kinetic energy function, so that in general for the gauge group $SU(3)_C \times SU(2)_L \times U(1)$ on using Eq. (13.49) the gaugino masses M_a are given by

$$\mathcal{L}_\lambda^m = -\frac{1}{2} M_a \bar{\lambda}_a \lambda_a. \tag{14.95}$$

We may write $M_a = m_{1/2}(1 + \delta_a)(a = 1\text{--}3)$, where δ_a parametrize the non-universalities. The gaugino sector non-universalities are consistent with the FCNC constraints. Similarly, one can introduce non-universalities in the Higgs sector, i.e., $M_{H_u} = m_0(1 + \delta_{H_u})$ and $M_{H_d} = m_0(1 + \delta_{H_d})$, which are also consistent with the FCNC constraints. Regarding non-universalities in the sfermion sector, the FCNC constraints are much weaker for the third generation compared with the case for the first two generations. Thus, consistent with the FCNC constraint, one may introduce non-universalities for the case of the third-generation sfermions so that $M_{\tilde{q}3} = m_0(1 + \delta_{\tilde{q}3})$, $M_{\tilde{t}_R} = m_0(1 + \delta_{\tilde{t}_R})$, and $M_{\tilde{b}_R} = m_0(1 + \delta_{\tilde{b}_R})$, where the δ' parametrize the non-universalities.

In the discussion above, we have limited ourselves to soft parameters on real manifolds. A display of the sparticle spectrum for one set of parameters on the real manifold for mSUGRA is given in Fig. 14.3. However, in general one has CP phases that enter into the gauge kinetic energy function and the Kähler potential as well into the superpotential. Consequently, the soft parameters would, in general, possess phases. Some of the phases can be absorbed in the redefinition of fields. For the minimal supergravity case, mSUGRA one will have two phases in the soft sector, which one can choose to be the phase of μ and the phase of A_0. With the inclusion of non-universalities, additional phases will enter. For instance, one will have phases in the gaugino sector and in the third-generation sector, as well as additional phases for trilinear parameters. These phases are severely constrained by the experimental data on the electric dipole moments (EDMs) of the electron, the neutron, and atoms. This topic is discussed in Chapter 15.

14.9 Yukawa Unification

In addition to the gauge-coupling unification, in grand unified models one also encounters unification of Yukawa couplings. Thus, in $SU(5)$ unified models the b-quark Yukawa coupling and the Yukawa coupling of the τ-lepton are equal at the grand unification scale. While this is also the case for the first two generations, one expects significant corrections from textures for the first two generations, but which are expected to be small for the third generation. Thus, as a rough approximation one could treat the third-generation Yukawas independently and ask if such a simple picture fits the experimental data. To discuss $b-\tau$ unification one uses renormalization group analysis to evolve the Yukawa couplings from the GUT scale down to one common scale M_{SUSY}. In the renormalization group evolution at least two-loop order is necessary to achieve good accuracy. At the

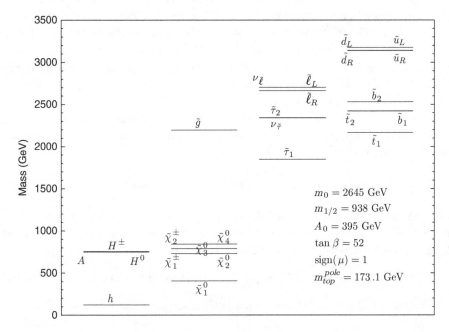

Figure 14.3. An illustration of the prediction of 32 sparticle masses for supergravity grand unification with universal boundary conditions so that $m_0 = 2645$ GeV, $m_{1/2} = 938$ GeV, $A_0 = 395$ GeV, $\tan \beta = 52$, and $\text{sign}(\mu) = 1$ when the top pole mass is 173.1 GeV. The sparticle masses show a mass hierarchy.

scale M_{SUSY}, one then matches the solutions above and below the supersymmetry scale so that

$$\lambda_i(M_{SUSY}^-) = \lambda_i(M_{SUSY}^+)\cos\beta \quad (i = b, \tau)$$
$$\lambda_t(M_{SUSY}^-) = \lambda_t(M_{SUSY}^+)\sin\beta. \qquad (14.96)$$

Below this scale, one evolves the b/τ masses down to the b/τ mass scales by use of renormalization group evolution up to three-loop order in quantum chromodynamics (QCD) and one-loop order in quantum electrodynamics. The physical mass of the b-quark is taken to be the pole mass M_b, and is related to the running mass $m_b(M_b)$ as follows:

$$M_b = \left(1 + \frac{4\alpha_3(M_b)}{3\pi} + 12.4\frac{\alpha_3(M_b)^2}{\pi^2}\right)m_b(M_b), \qquad (14.97)$$

where $m_b(M_b)$ is determined from $m_b(M_Z)$ by a renormalization group running from the scale M_Z to the scale M_b. Interestingly, it is difficult to achieve b/τ unification in non-supersymmetric grand unification but it can occur in supersymmetric unification. However, supersymmetric analyses show that b/τ unification is sensitive to small changes in α_3, $\tan\beta$, and the top quark mass, and is sensitive to GUT threshold corrections and to gravitational smearing effects [47–50]. Further, it is known that b/τ unification prefers a negative sign of μ [51,52] in the conventional sign convention for μ, while the Brookhaven data indicate a positive sign [53–59]. Several resolutions of this exist in the literature [60,61]. One solution within supersymmetric $SO(10)$ unification suggests a resolution by inclusion of D-term corrections to the sfermion and Higgs masses after the breaking of the $SO(10)$ gauge symmetry [60]. Another possibility suggested is via the use of non-universalities [61].

In further analyses on b/τ unification it is important to also take into account supersymmetric QCD and supersymmetric electroweak corrections to $b\,\tau$ masses [62,63]. Thus, $m_b(M_Z)$ is given by

$$m_b(M_Z) = \frac{v}{\sqrt{2}}h_b(M_Z)\cos\beta(1 + \Delta_b). \qquad (14.98)$$

Here, Δ_b is the supersymmetric QCD and supersymmetric electroweak loop correction to the running mass of the b-quark at the Z-scale, and can be obtained from the loop corrections to the Yukawa couplings

$$-L_{bbH^0} = (m_b + \delta h_b)\bar{b}_R b_L H_1^0 + \Delta h_b \bar{b}_R b_L H_2^{0*} + \text{H.c.} \qquad (14.99)$$

The correction Δ_b to the τ-upon mass is then given by the relation

$$\Delta_b = \left[\frac{Re(\Delta h_b)}{h_b}\tan\beta + \frac{Re(\delta h_b)}{h_b}\right]. \qquad (14.100)$$

The diagrams that contribute to Δ_b are shown in Fig. 14.4, which include the one-loop corrections arising from the chargino, neutralino, and gluino exchanges.

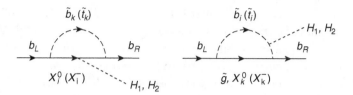

Figure 14.4. One-loop correction from the exchange of charginos, neutralinos, and the gluino to the bottom quark mass.

In a similar fashion, one has corrections to the τ-lepton mass so that

$$m_\tau(M_Z) = \frac{v}{\sqrt{2}} h_\tau(M_Z) \cos\beta (1 + \Delta_\tau), \qquad (14.101)$$

where Δ_τ is the supersymmetric electroweak loop corrections to the running mass of the τ-lepton at the Z-scale. Here, one has

$$- L_{\tau\tau H^0} = (h_\tau + \delta h_\tau)\bar{\tau}_R \tau_L H_1^0 + \Delta h_\tau \bar{\tau}_R \tau_L H_2^{0*} + \text{H.c.} \qquad (14.102)$$

The correction Δ_τ to the τ-lipton mass is then given by the relation

$$\Delta_\tau = \left[\frac{Re(\Delta h_\tau)}{h_\tau} \tan\beta + \frac{Re(\delta h_\tau)}{h_\tau} \right]. \qquad (14.103)$$

The supersymmetric QCD and the supersymmetric electroweak correction to the b-quark mass can be as large as 20% or more while the loop correction to the τ-mass is significantly smaller, i.e., less than 1%. Regarding the supersymmetric loop correction to the top quark mass, it is found to be typically small, i.e., less than 1% at best. Let us now consider the possibility of a $b-t-\tau$ unification. Reverting to the tree level relations, one has

$$(m_t, m_b, m_\tau) = \frac{v}{\sqrt{2}} (h_t \sin\beta, h_b \cos\beta, h_\tau \cos\beta). \qquad (14.104)$$

In supersymmetric $SO(10)$ models a $b-t-\tau$ unification can occur for large $\tan\beta$ as large as $\tan\beta = 50$ in models where the couplings to the matter fields arise from 10-plets of $SO(10)$ [64]. However, in $SO(10)$ models with different Higgs structures it is also possible to get a unification of Yukawas with low values of $\tan\beta$ [65].

14.10 Supersymmetry at the Large Hadron Collider

Models based on supergravity unification have a large number of signatures which can be detected in a variety of settings, including colliders, underground and satellite-based experiments. Here, we discuss collider-based signatures which arise from the decay of sparticles once they are produced at colliders (See the literature for discussion of some of the early work [66–71] and for reviews [72, 73].)

Currently, the Large Hadron Collider is the largest particle physics collider, and the initial planning for it was to run at a center of mass energy of 14 TeV and collect 10^3 fb^{-1} of data. The machine is now commissioned to run at a center of mass energy of 13 TeV, is currently running and collecting data. As discussed earlier under the assumption that R-parity is conserved, the least massive R-parity odd supersymmetric particle (LSP) will be stable. Thus, at colliders, supersymmetric particles will be produced in pairs such as two squarks ($\tilde{q}_1\tilde{q}_2$), two gluinos ($\tilde{g}\tilde{g}$), or a combination of a squark and a gluino ($\tilde{q}\tilde{g}$). One would also produce pairs of electroweak gauginos, i.e., combinations of charginos $\tilde{\chi}^{\pm}$ and neutralinos $\tilde{\chi}_i^0$, so that the overall R-parity is even. Except for the LSP, the other odd R-parity particles will decay, producing at the end of the chain at least two LPSs at colliders. Remarkably, in supergravity unified models over most of the parameter space of models consistent with electroweak symmetry breaking, the LSP turns out to be the lightest neutralino which is neutral and thus escapes the detector without detection. As a consequence, supersymmetric decays are characterized by a significant amount of missing energy.

One possible supersymmetric signature is opposite-sign same-flavor (OSSF) dileptonic events. Such events can arise from the decay of a neutral supersymmetric particle. Thus, for example, $\tilde{\chi}_2^0$ can have leptonic decays so that $\tilde{\chi}_2^0 \to \tilde{\chi}_1^0 \ell^+ \ell^-$. Further, such events can arise also from two independent sources such as from the decays of two final states $\tilde{\chi}_1^+ \tilde{\chi}_1^-$. Additionally, of course, one has standard model background events which come from the decay of $t\bar{t}$ and other standard model final states. Such decays give us not only the OSSF events but also opposite-sign different-flavor events such as $e^{\pm}\mu^{\mp}$. One can remove such events by considering the flavor-subtracted combination, i.e., $e^+e^- + \mu^+\mu^- - e^+\mu^- - e^-\mu^+$. The above combination gives a clearer signal for supersymmetry. Another possible signature is an excess of trileptonic events. In pp collisions one can produce $W^{\pm}$-bosons, which can decay into $\tilde{\chi}_1^{\pm}\tilde{\chi}_2^0$. Initially, the analyses were for the on-shell decays of the W- and Z-bosons [66, 74, 75]. However, later these analyses were extended to include off-shell W- and Z-decays [76]. Thus, in a $\tilde{\chi}_1^{\pm}\tilde{\chi}_2^0$ production the chargino may decay into a charged lepton, a neutrino or an antineutrino, and a neutralino, while $\tilde{\chi}_2^0$ will decay into a $\ell^+\ell^-$ and χ_1^0. Thus, a trileptonic signal can come from the process

$$pp \to W^{-*} \to \chi^- \chi_2^0 \to (\tilde{\chi}_1^- \to \ell_1^- \nu \tilde{\chi}_1^0) + (\chi_2^0 \to \ell_2^+ \ell_2^- \tilde{\chi}_1^0). \qquad (14.105)$$

Here, the off-shell production of $W^{\pm*}$ will lead to a trileptonic signature with two neutralinos in the final state which will escape detection, and along with the unobserved neutrino constitute missing energy, i.e.,

$$p + p \to l_1^{\pm} l_2^{\pm} l_2^{\mp} + \text{jets} + \text{ missing energy}. \qquad (14.106)$$

The background for these events in the standard model arises from the decay of the $W + Z$ final states where, for example, $W^- \to \ell_1^- \nu_{\ell_1}$ and $Z \to \ell_2 \bar{\ell}_2$, which together produce a trileptonic state. However, an excess of such events would be an indication of supersymmetry.

Other sets of signatures consist of a combination of jets and leptons. Thus, for example, one may have the production of gluino pairs or chargino pairs or a combination of a chargino and a neutralino. Their decays will produce signatures involving jets and leptons and missing energy. Thus, for the gluino pair production, one has

$$pp \to \tilde{g}\tilde{g} \to (\tilde{g} \to q\bar{q}\tilde{\chi}_i^0, q\bar{q}'\tilde{\chi}_j^{\pm})(\tilde{g} \to q\bar{q}\tilde{\chi}_i^0, q\bar{q}'\tilde{\chi}_j^{\pm}). \tag{14.107}$$

This is followed by further decays of the chargino and the non-LSP neutralinos such as in Eq. (14.105). Thus, a gluino will have decays of the type

$$\begin{aligned}\tilde{g} &\to q\bar{q}'\ell^{\pm} + E_T^{miss} \\ &\to q\bar{q}\ell^+\ell^- + E_T^{miss}. \end{aligned} \tag{14.108}$$

The final quarks will hadronise after they are produced, and would appear as jets. The gluino pair production will generate a variety of signatures where the signatures would consist of many jets and jets and leptons all with E_T^{miss}. One interesting signature that arises is due to the Majorana nature of the gluinos in which one can get an excess of the same-sign leptons. This occurs because two gluinos can each decay, producing the same-sign lepton. This signal is an important supersymmetry signal. In addition, kinematical edges in invariant mass distributions provide useful information regarding mass differences among sparticles. Thus, for example, in the decay $\tilde{\chi}_2^0 \to \ell^+\ell^-\tilde{\chi}_1^0$ the dilepton-invariant mass has an upper limit of $m_{\tilde{\chi}_2^0} - m_{\tilde{\chi}_1^0}$. So, a plot of the event distribution for the process as a function of the dilepton-invariant mass $M_{\ell\ell}$ must have an edge at $M_{\ell\ell} = m_{\tilde{\chi}_2^0} - m_{\tilde{\chi}_1^0}$.

The two main detectors at the Large Hadron Collider, i.e., the ATLAS and the CMS, collect data using a set of trigger level cuts which are imposed on the transverse momentum P_T of jets and leptons, on missing energy and on pseudorapidity η, which is defined in terms of an angle θ which is the angle of the particle momentum $\vec{p}$ relative to the beam axis so that

$$\eta = -\ln\left[\tan\left(\frac{\theta}{2}\right)\right]. \tag{14.109}$$

To enhance the prospects for the discovery of supersymmetry, one imposes further cuts on the measured quantities. These are the so-called post-trigger level cuts. The choice of such cuts can vary widely and is geared to enhance the signal and reduce the background. Of course, the cuts are also dependent on how well the experiment can measure the specific parameter on which the cut is placed. In the analysis of any of the supersymmetry signatures, an accurate knowledge of the standard model background is essential.

As is obvious from the discussion above, the observability of a supersymmetry signature depends critically on the mass difference between the decaying sparticle and the LSP, and this mass difference varies widely as one moves in the parameter space of supergravity models. Thus, a signature which is dominant for one model point in the parameter space is not necessarily the dominant signature for other model points. For this reason all discovery channels need be investigated. Further, one expects that a signal would be visible in more than one channel, providing corroborating evidence that one has observed a supersymmetric signal. All the signals discussed above are for the case when one assumes that R-parity is preserved. Supersymmetric signatures would be very different if R-parity was violated. Further, it is possible that even though R-parity conservation may be valid at high scales, the renormalization group evolution may lead to spontaneous breaking of R-parity (see the literature for early work on violations of R-parity [77] and also for more recent work [78]), although this is not a generic feature of most of the parameter space of supergravity unified models.

As discussed above, in supergravity grand unified models the soft parameters are defined at the grand unification scale and evolved down to the low-energy scale by a renormalization group. Such an evolution creates a dispersion in the sparticle masses at low scale. The precise mass hierarchy of the sparticles at low scale contains in it information about its GUT boundary conditions. Thus, knowledge of the sparticle mass hierarchies can allow one to partially resolve the inverse problem, i.e., how one may trace back the model from experimental knowledge of the sparticle masses (see the literature for works on the inverse problem [79–83].). Since there are 32 sparticle masses including the Higgs bosons, their masses could, in principle, generate a large number of hierarchical patterns. In supergravity models, the number of possibilities is drastically reduced, and, further, the hierarchical patterns are strongly correlated with the parameters at high scale. So, from a knowledge of sparticle mass hierarchies that are observed, one will be able to at least partially trace back the high-scale models and thus the nature of supersymmetry breaking.

14.11 Problems

1. Show that for the minimal supersymmetric standard model using the renormalization group evolution between M_Z and M_S where M_S is the common sparticle mass scale and between M_S and M_G, one has the following relation:

$$\ln\left(\frac{M_S}{M_Z}\right) = \frac{4\pi}{19}(5\alpha_1^{-1} - 12\alpha_2^{-1} + 7\alpha_3^{-1}), \qquad (14.110)$$

where $\alpha_1 = \frac{5}{3}\alpha_Y$.

2. Show that the contribution from the D-term of the $SU(2)$ sector to the Higgs potential can be written in the following alternative form:

$$\frac{g^2}{8} |H_1^* \tau_a H_1 + H_2^* \tau_a H_2|^2 = \frac{g^2}{8} \left(|H_1|^2 - |H_2|^2 \right)^2 + \frac{g^2}{2} |H_1^* H_2|^2. \qquad (14.111)$$

The identity $(\tau_a)_{ij}(\tau_a)_{kl} = 2\delta_{il}\delta_{jk} - \delta_{ij}\delta_{kl}$ is useful in showing the equivalence of the two forms.

3. Using the potential of Eqs. (14.22)–(14.24), show that the conditions needed for spontaneous electroweak symmetry breaking give Eq. (14.29).

4. Using the potential of Eqs. (14.22)–(14.24), show that after spontaneous electroweak symmetry breaking the mass-squared matrix in the charged Higgs sector is given by

$$(M_A^2 + M_W^2) \begin{pmatrix} \sin^2 \beta & -\sin \beta \cos \beta \\ \\ -\sin \beta \cos \beta & \cos^2 \beta \end{pmatrix}. \qquad (14.112)$$

5. Show that the loop correction to the tree Higgs boson mass square matrix given by Eq. (14.67) may be written as

$$\Delta M_H^2 = \begin{pmatrix} \left(-\dfrac{1}{v_1}\dfrac{\partial}{\partial v_1} + \dfrac{\partial^2}{\partial v_1^2} \right) & \dfrac{\partial^2}{\partial v_2 \partial v_1} \\ \\ \dfrac{\partial^2}{\partial v_2 \partial v_1} & \left(-\dfrac{1}{v_2}\dfrac{\partial}{\partial v_2} + \dfrac{\partial^2}{\partial v_2^2} \right) \end{pmatrix} \Delta V, \qquad (14.113)$$

where ΔV is the loop correction to the potential.

References

[1] A. H. Chamseddine, R. L. Arnowitt, and P. Nath, *Phys. Rev. Lett.* **49**, 970 (1982); P. Nath, R. L. Arnowitt, and A. H. Chamseddine, *Phys. Lett. B* **121**, 33 (1983); *Nucl. Phys. B* **227**, 121 (1983).

[2] L. J. Hall, J. D. Lykken, and S. Weinberg, *Phys. Rev. D* **27**, 2359 (1983).

[3] For a more complete list and description of works related to the development of supergravity unified models see: R. Arnowitt, A. H. Chamseddine, and P. Nath, *Int. J. Mod. Phys. A* **27**, 1230028 (2012) [*Int. J. Mod. Phys. A* **27**, 1292009 (2012)].

[4] M. B. Einhorn and D. R. T. Jones, *Nucl. Phys. B* **196**, 475 (1982).

[5] M. E. Machacek and M. T. Vaughn, *Nucl. Phys. B* **222**, 83 (1983); *Nucl. Phys. B* **236**, 221 (1984); *Nucl. Phys. B* **249**, 70 (1985).

[6] S. P. Martin and M. T. Vaughn, *Phys. Lett. B* **318**, 331 (1993); *Phys. Rev. D* **50**, 2282 (1994); I. Jack, D. R. Jones, S. P. Martin, M. T. Vaughn, and Y. Yamada, *Phys. Rev. D* **50**, 5481 (1994).

[7] K. Inoue, A. Kakuto, H. Komatsu and S. Takeshita, *Prog. Theor. Phys.* **68**, 927 (1982); L. Ibañez and G. G. Ross, *Phys. Lett. B* **110**, 227 (1982); L. E. Ibanez, *Phys. Lett. B* **118**, 73 (1982); J. Ellis, J. Hagelin, D. V. Nanopoulos, and K. Tamvakis, *Phys. Lett. B* **125**, 2275 (1983).

[8] L. E. Ibañez and C. Lopez, *Phys. Lett. B* **128**, 54 (1983); L. Alvarez-Gaumé, J. Polchinski, and M. B. Wise, *Nucl. Phys. B* **250**, 495 (1983); L. E. Ibanez and C. Lopez, *Nucl. Phys. B* **233**, 511 (1984); L. E. Ibañez, C. Lopez, and C. Muños, *Nucl. Phys. B* **256**, 218 (1985); J. Ellis and F. Zwirner, *Nucl. Phys. B* **338**, 317 (1990).

[9] S. R. Coleman and E. J. Weinberg, *Phys. Rev. D* **7**, 1888 (1973).

[10] J.-M. Fr'ere, D.R.T. Jones, and S. Raby, *Nucl. Phys. B* **222**, 11 (1983); M. Claudson, L. Hall, and I. Hinchcliffe, *Nucl. Phys. B* **228**, 501 (1983); H.-P. Nilles, M. Srednicki, and D. Wyler, *Phys. Lett. B* **120**, 346 (1983);M. Drees, M. Glück, and K. Grassie, *Phys. Lett. B* **157**, 164 (1985).

[11] H. Baer, M. Brhlik, and D. Castano, *Phys. Rev. D* **54**, 6944 (1996); J. E. Camargo-Molina, B. O'Leary, W. Porod, and F. Staub, *JHEP* **1312**, 103 (2013).

[12] G. G. Ross and R. G. Roberts, *Nucl. Phys. B* **377**, 571 (1992).

[13] R. L. Arnowitt and P. Nath, *Phys. Rev. Lett.* **69**, 725 (1992); M. Drees and M. M. Nojiri, *Phys. Rev. D* **47**, 376 (1993); V. D. Barger, M. S. Berger, and P. Ohmann, *Phys. Rev. D* **49**, 4908 (1994); G. L. Kane, C. F. Kolda, L. Roszkowski, and J. D. Wells, *Phys. Rev. D* **49**, 6173 (1994).

[14] S. P. Martin and P. Ramond, *Phys. Rev. D* **48**, 5365 (1993).

[15] S. Dimopoulos and H. Georgi, *Nucl. Phys. B* **193**, 150 (1981).

[16] R. L. Arnowitt and P. Nath, *Phys. Rev. D* **46**, 3981 (1992).

[17] T. Ibrahim and P. Nath, *Phys. Rev. D* **63**, 035009 (2001).

[18] S. P. Martin, *Adv. Ser. Direct. High Energy Phys.* **21**, 1 (2010).

[19] M.S. Berger, *Phys. Rev. D* **41**, 225 (1990); Y. Okada, M. Yamaguchi, and T. Yanagida, *Prog. Theor. Phys.* **85**, 1 (1991); J. R. Ellis, G. Ridolfi, and F. Zwirner, *Phys. Lett. B* **257**, 83 (1991); H. E. Haber and R. Hempfling, *Phys. Rev. Lett.* **66**, 1815 (1991).

[20] J. A. Casas, J. R. Espinosa, M. Quiros, and A. Riotto, *Nucl. Phys. B* **436**, 3 (1995) [*Nucl. Phys. B* **439**, 466 (1995)]; M. Carena, M. Quiros, and C. E. M. Wagner, *Nucl. Phys. B* **461**, 407 (1996); J. R. Espinosa and R. J. Zhang, *Nucl. Phys. B* **586**, 3 (2000).

[21] G. Aad, T. Abajan, B. Abbott, *et al.* [ATLAS Collaboration], *Phys. Lett. B* **716**, 1 (2012).

[22] S. Chatrchyan, V. Khachatryan, A. M. Sirunyan, *et al.* [CMS Collaboration], *Phys. Lett. B* **716**, 30 (2012).

[23] J. R. Ellis, M. K. Gaillard, and D. V. Nanopoulos, *Nucl. Phys. B* **106**, 292 (1976).

[24] M. A. Shifman, A. I. Vainshtein, M. B. Voloshin, and V. I. Zakharov, *Sov. J. Nucl. Phys.* **30**, 711 (1979).

[25] J. F. Gunion, H. E. Haber, G. L. Kane, and S. Dawson, *Front. Phys.* **80**, 1 (2000).

[26] B. A. Kniehl and M. Spira, *Z. Phys. C* **69**, 77 (1995).

[27] A. Djouadi, *Phys. Rep.* **459**, 1 (2008).

[28] S. Dawson and H. E. Haber, *Int. J. Mod. Phys. A* **7**, 107 (1992).

[29] P. Nath and R. L. Arnowitt, *Phys. Rev. D* **56**, 2820 (1997).

[30] K. L. Chan, U. Chattopadhyay, and P. Nath, *Phys. Rev. D* **58**, 096004 (1998).

[31] J. L. Feng, K. T. Matchev, and T. Moroi, *Phys. Rev. Lett.* **84**, 2322 (2000).

[32] J. L. Feng, K. T. Matchev, and F. Wilczek, *Phys. Lett. B* **482**, 388 (2000).

[33] S. Akula, M. Liu, P. Nath, and G. Peim, *Phys. Lett. B* **709**, 192 (2012).

[34] N. Arkani-Hamed and S. Dimopoulos, *JHEP* **0506**, 073 (2005).

[35] R. Barbieri and G. F. Giudice, *Nucl. Phys. B* **306**, 63 (1988).

[36] G. Bhattacharyya and A. Romanino, *Phys. Rev. D* **55**, 7015 (1997).

[37] P. H. Chankowski, J. R. Ellis, and S. Pokorski, *Phys. Lett. B* **423**, 327 (1998)

[38] G. L. Kane and S. F. King, *Phys. Lett. B* **451**, 113 (1999).

[39] D. Feldman, G. Kane, E. Kuflik, and R. Lu, *Phys. Lett. B* **704**, 56 (2011).

[40] G. Belanger, D. Ghosh, R. Godbole, M. Guchait, and D. Sengupta, *Phys. Rev. D* **89**, 015003 (2014).

[41] H. Baer, V. Barger, P. Huang, D. Mickelson, A. Mustafayev, and X. Tata, *Phys. Rev. D* **87**, no. 11, 115028 (2013).

[42] S. K. Soni and H. A. Weldon, *Phys. Lett. B* **126**, 215 (1983); D. Matalliotakis and H. P. Nilles, *Nucl. Phys. B* **435**, 115 (1995); M. Olechowski and S. Pokorski, *Phys. Lett. B* **344**, 201 (1995).

[43] P. Nath and R. Arnowitt, *Phys. Rev. D* **56**, 2820 (1997); E. Accomando, R. Arnowitt, B. Dutta, and Y. Santoso, *Nucl. Phys.* **B585**, 124 (2000).

[44] J. Ellis, K. Enqvist, D. V. Nanopoulos, and K. Tamvakis, *Phys. Lett. B* **155**, 381 (1985); M. Drees, *Phys. Lett. B* **158**, 409 (1985).

[45] N. Polonsky and A. Pomarol, *Phys. Rev. D* **51**, 6532 (1995); G. Anderson, C. H. Chen, J. F. Gunion, J. Lykken, T. Moroi, and Y. Yamada, hep-ph/9609457 (1996); G. Anderson, H. Baer, C. h. Chen, and X. Tata, *Phys. Rev. D* **61**, 095005 (2000); H. Baer, M. A. Diaz, P. Quintana, and X. Tata, *JHEP* **0004**, 016 (2000).

[46] A. Corsetti and P. Nath, *Phys. Rev. D* **64**, 125010 (2001); A. Birkedal-Hansen and B. D. Nelson, *Phys. Rev. D* **67**, 095006 (2003); N. Chamoun, C. S. Huang, C. Liu, and X. H. Wu, *Nucl. Phys. B* **624**, 81 (2002); U. Chattopadhyay and P. Nath, *Phys. Rev. D* **65**, 075009 (2002); G. Belanger, F. Boudjema, A. Cottrant, A. Pukhov, and A. Semenov, *Nucl. Phys. B* **706**, 411 (2005); Y. Mambrini and C. Munoz, *Astropart. Phys.* **24**, 208 (2005); K. Huitu, R. Kinnunen, J. Laamanen, S. Lehti, S. Roy, and T. Salminen, *Eur. Phys. J. C* **58**, 591 (2008); S. P. Martin, Phys. Rev. D **79**, 095019 (2009); D. Feldman, Z. Liu, and P. Nath, *Phys. Rev. D* **80**, 015007 (2009); I. Gogoladze, F. Nasir, and Q. Shafi, *Int. J. Mod. Phys. A* **28**, 1350046 (2013).

[47] H. Arason, D.J. Castano, B.E. Kesthelyi, S. Mikaelian, E.J. Piard, P. Ramond, and B.D. Wright, *Phys. Rev. Lett.* **67**, 2933(1991).

[48] V. Barger, M.S. Berger, and P. Ohman, *Phys. Lett. B* **314**, 3511 (1993); *Phys. Rev. D* **47**, 1093 (1993).

[49] P. Langacker and N. Polonsky, *Phys. Rev. D* **50**, 2199 (1994).

[50] T. Dasgupta, P. Mamales, and P. Nath, *Phys. Rev. D* **52**, 5366 (1995).

[51] W. de Boer, M. Huber, A. V. Gladyshev, and D. I. Kazakov, *Eur. Phys. J. C* **20**, 689 (2001).

[52] M. E. Gomez, T. Ibrahim, P. Nath, and S. Skadhauge, *Phys. Rev. D* **72**, 095008 (2005).

[53] G. W. Bennett, B. Bousquet, H. N. Brown, *et al.* [Muon g-2 Collaboration], *Phys. Rev. D* **73**, 072003 (2006).

[54] T. Teubner, K. Hagiwara, R. Liao, A. D. Martin, and D. Nomura, *Chin. Phys. C* **34**, 728 (2010).

[55] J. P. Miller, E. de Rafael, and B. L. Roberts, *Rep. Prog. Phys.* **70**, 795 (2007).

[56] T. C. Yuan, R. L. Arnowitt, A. H. Chamseddine, and P. Nath, *Z. Phys. C* **26**, 407 (1984).

[57] J. Lopez, D.V. Nanopoulos, and X. Wang, *Phys. Rev. D* **49**, 366(1994).

[58] U. Chattopadhyay and P. Nath, *Phys. Rev. D* **53**, 1648 (1996).

[59] T. Blazek, R. Dermisek, and S. Raby, *Phys. Rev. Lett.* **88**, 111804 (2002).

[60] H. Baer and J. Ferrandis, *Phys. Rev. Lett.* **87**, 211803 (2001).

[61] U. Chattopadhyay and P. Nath, *Phys. Rev. D* **65**, 075009 (2002).

[62] L. J. Hall, R. Rattazzi, and U. Sarid, *Phys. Rev. D* **50**, 7048 (1994); R. Hempfling, *Phys. Rev. D* **49**, 6168 (1994); M. Carena, M. Olechowski, S. Pokorski, and C. Wagner, *Nucl. Phys. B* **426**, 269 (1994); D. Pierce, J. Bagger, K. Matchev, and R. Zhang, *Nucl. Phys. B* **491**, 3 (1997).

[63] T. Ibrahim and P. Nath, *Phys. Rev. D* **67**, 095003 (2003) [*Phys. Rev. D* **68**, 019901 (2003)].

[64] B. Ananthanarayan, G. Lazarides, and Q. Shafi, *Phys. Rev. D* **44**, 1613 (1991).

[65] K. S. Babu, I. Gogoladze, P. Nath, and R. M. Syed, *Phys. Rev. D* **74**, 075004 (2006).

[66] R. L. Arnowitt, A. H. Chamseddine, and P. Nath, *Phys. Rev. Lett.* **50**, 232 (1983); A. H. Chamseddine, P. Nath, and R. L. Arnowitt, *Phys. Lett. B* **129**, 445 (1983); P. Nath, R. L. Arnowitt, and A. H. Chamseddine, HUTP-83/A077, NUB-2588a (unpublished 1983).

[67] S. Weinberg, *Phys. Rev. Lett.* **50**, 387 (1983).

[68] D. A. Dicus, S. Nandi, W. W. Repko, and X. Tata, *Phys. Rev. Lett.* **51**, 1030 (1983) [erratum, *ibid.* **51**, 1813 (1983)]; D. A. Dicus, S. Nandi, and X. Tata, *Phys. Lett. B* **129**, 451 (1983); J. M. Frere and G. L. Kane, *Nucl. Phys. B* **223**, 331 (1983); J. R. Ellis, J. M. Frere, J. S. Hagelin, G. L. Kane, and S. T. Petcov, *Phys. Lett. B* **132**, 436 (1983); J. S. Hagelin, G. L. Kane, and S. Raby, *Nucl. Phys. B* **241**, 638 (1984).

[69] S. Abel *et al.* [SUGRA Working Group Collaboration], hep-ph/0003154 (2000).

[70] P. Nath, B. D. Nelson, H. Davoudiasl, *et al.*, *Nucl. Phys. Proc. Suppl.* **200–202**, 185 (2010).

[71] P. Nath, R. L. Arnowitt, and A. H. Chamseddine, *Applied N=1 Supergravity*, World Scientific (1984); *Trieste Particle Phys.* **1** (1983) (QCD161:W626:1983).

[72] H. P. Nilles, *Phys. Rep.* **110**, 1 (1984).

[73] H. E. Haber and G. L. Kane, *Phys. Rep.* **117**, 75 (1985).

[74] D. A. Dicus, S. Nandi, and X. Tata, *Phys. Lett. B* **129**, 451 (1983) [erratum, *ibid.* **145**, 448 (1984)].

[75] H. Baer, K. Hagiwara, and X. Tata, *Phys. Rev. Lett.* **57**, 294 (1986).

[76] P. Nath and R. L. Arnowitt, *Mod. Phys. Lett. A* **2**, 331 (1987); R. L. Arnowitt, R. M. Barnett, P. Nath, and F. Paige, *Int. J. Mod. Phys. A* **2**, 1113 (1987); H. Baer, C. H. Chen, F. Paige and X. Tata, *Phys. Rev. D* **53**, 6241 (1996); V. D. Barger, C. Kao, and T. j. Li, *Phys. Lett. B* **433**, 328 (1998); E. Accomando, R. L. Arnowitt, and B. Dutta, *Phys. Lett. B* **475**, 176 (2000).

[77] C. S. Aulakh and R. N. Mohapatra, *Phys. Lett. B* **119**, 136 (1982); L. J. Hall and M. Suzuki, *Nucl. Phys. B* **231**, 419 (1984); A. Masiero and J. W. F. Valle, *Phys. Lett. B* **251**, 273 (1990).

[78] P. Fileviez Perez and S. Spinner, *Phys. Rev. D* **80**, 015004 (2009); V. Barger, P. Fileviez Perez, and S. Spinner, *Phys. Rev. Lett.* **102**, 181802 (2009); D. Feldman, P. Fileviez Perez, and P. Nath, *JHEP* **1201**, 038 (2012).

[79] N. Arkani-Hamed, G. L. Kane, J. Thaler, and L. T. Wang, *JHEP* **0608**, 070 (2006).

[80] D. Feldman, Z. Liu, and P. Nath, *Phys. Rev. Lett.* **99**, 251802 (2007) [*Phys. Rev. Lett.* **100**, 069902 (2008)]; D. Francescone, S. Akula, B. Altunkaynak, and P. Nath, *JHEP* **1501**, 158 (2015).

[81] B. Altunkaynak, B. D. Nelson, L. L. Everett, Y. Rao, and I. W. Kim, *Eur. Phys. J. Plus* **127**, 2 (2012).

[82] C. F. Berger, J. S. Gainer, J. L. Hewett, and T. G. Rizzo, *JHEP* **0902**, 023 (2009).

[83] J. S. Gainer, K. T. Matchev, and M. Park, *JHEP* **1506**, 014 (2015).

Further Reading

R. L. Arnowitt and P. Nath, in *Sao Paulo 1993, Proceedings, Particles and Fields*, pp. 3-63, [hep-ph/9309277].

H. Baer and X. Tata, Weak Scale Supersymmetry: From Superfields to Scattering Events, Cambridge University Press (2006).

M. Drees, R. Godbole, and P. Roy, *Theory and Phenomenology of Sparticles: An Account of Four-dimensional* N = 1 *Supersymmetry in High Energy Physics*, World Scientific (2004).

S. Weinberg, *The Quantum Theory of Fields*, vol. 3: *Modern Applications*, Cambridge University Press (2005).

15

CP Violation in Supergravity Unified Theories

15.1 Introduction

The history of CP symmetry and its violation dates back to the 1950s. It is intimately tied to the possibility of whether or not elementary particles can possess electric dipole moments (EDMs). Thus, till around 1950 it was generally accepted that elementary particles could not possess EDMs if there was a parity (P) symmetry. However, Purcell and Ramsay in 1950 pointed out for the first time that for nuclear forces there was no evidence for the parity symmetry, and that the parity assumption should be tested [1]. They also proposed that a search for the EDM would be a sensitive test of the parity assumption. Subsequently, Purcell and Ramsey along with their graduate student James Smith carried out such an experiment in 1951 which gave a null result but allowed them to put an upper limit on the neutron EDM of $d_n < 3 \times 10^{-20}$ cm, where e is the proton charge. This result was not published until much later [2], although it was cited in several publications at that time [3–5]. In 1957, after Wu et $al.$ [6] discovered the violation of parity following the work of Lee and Yang [5], the parity argument was no longer valid for the vanishing of the EDM. It was then argued by many theorists that the EDM could not exist because of time reversal symmetry. Soon thereafter, it was pointed out by Ramsey [7] and independently by Jackson and his collaborators [8] that there was no experimental evidence for such a symmetry and that a search for the EDM would provide a test for the symmetry. Since then, the search for violations of T-symmetry or of CP symmetry has been one of central interest in particle physics. In 1964, Cronin and Fitch and their associates [9] showed that CP symmetry was violated in the neutral kaon system. The importance of CP violation in a broader context was then shown by Sakharov [10], who pointed out that a violation of CP symmetry was one of the fundamental ingredients in the generation of a baryon excess in the universe. Currently, the study of CP violation is a topic of equal interest in both particle physics and cosmology.

15.2 Violation of CP in the Standard Model

The standard model of electroweak and strong interactions has two potential sources of CP violation, one of which appears in the electroweak sector and the other in the quantum chromodynamics (QCD) sector. We discuss the one in the electroweak sector first. In the charged current sector involving quarks, one has a CP phase appearing in the Cabibbo–Kobayashi–Maskawa (CKM) matrix V. It enters into the charged current interaction in the form

$$g_2 \bar{u}_i \gamma_\mu V_{ij} (1 - \gamma_5) d_j W^\mu + \text{h.c.}, \tag{15.1}$$

where u_i is the up quark ($u_i = u, c, t$) and d_i the down quark ($d_i = d, s, b$). In Eq. (15.1) V is the CKM matrix which is unitary, i.e.,

$$(VV^\dagger)_{ij} = \delta_{ij}. \tag{15.2}$$

The off-diagonal elements of the unitarity constraint (Eq. (15.2)) can be represented as unitarity triangles, and there are six such triangles, one of which reads

$$V_{ud} V_{ub}^* + V_{cd} V_{cb}^* + V_{td} V_{tb}^* = 0. \tag{15.3}$$

We define angles α, β, and γ of the unitarity triangle so that

$$\alpha = arg(-V_{td} V_{tb}^*/V_{ud} V_{ub}^*), \quad \beta = arg(-V_{cd} V_{cb}^*/V_{td} V_{tb}^*),$$
$$\gamma = arg(-V_{ud} V_{ub}^*/V_{cd} V_{cb}^*), \tag{15.4}$$

which then satisfy the relation

$$\alpha + \beta + \gamma = \pi. \tag{15.5}$$

CP violation in the quark sector can be described in a parametrization-independent way by the so-called Jarlskog invariant [11], which can be written in several forms, one of which is

$$J = Im(V_{us} V_{ub}^* V_{cb} V_{cs}^*). \tag{15.6}$$

An explicit form of the CKM matrix would involve four parameters which can be chosen to be three mixing angles and one phase. An alternative way is the Wolfenstein parametrization [12], which is given in terms of an expansion in λ which has the value $\lambda \simeq 0.226$ and the CKM matrix V can be written as a perturbation expansion in λ. Up to $O(\lambda^3)$, it has the form

$$\begin{pmatrix} 1 - \dfrac{\lambda^2}{2} & \lambda & A\lambda^3(\rho - i\eta) \\ -\lambda & 1 - \dfrac{\lambda^2}{2} & A\lambda^2 \\ A\lambda^3(1 - \rho - i\eta) & -A\lambda^2 & 1 \end{pmatrix}. \tag{15.7}$$

In this case, instead of three angles and a phase, one has λ, A, ρ, and η that parametrize the CKM matrix. In terms of these, the Jarlskog invariant takes the form $J \simeq A^2\lambda^6\eta$. Here, it is the dependence of J on η that brings in the CP violation.

In addition to the CP violation arising from the electroweak sector, the standard model also contains a source of CP violation arising from the QCD sector. This CP violation has the form

$$\delta L_{CP} = \theta\frac{\alpha_s}{8\pi}G\tilde{G}, \tag{15.8}$$

and is topological in origin. The strong CP can produce a very large contribution to the neutron EDM, and in order to suppress it below the current experimental limits one needs to constrain $\bar{\theta} = \theta + Arg\ Det(M_u M_d)$ so that $\bar{\theta} < 10^{-10}$. This fine tuning indicates that there is a more fundamental mechanism that suppresses strong CP violation. One possible mechanism is to have up quark mass vanish. However, lattice gauge analyses indicate that the up quark mass is non-vanishing. Another avenue to suppress strong CP is the Peccei–Quinn mechanism [13] and its refinements. One consequence of the Peccei–Quinn mechanism is the appearance of an axion [14,15]. Currently, there are severe limitations on the corridor in which the axion can exist. Many other attempts to suppress strong CP exist (e.g., [16,17]), and this area is still an active field of study.

15.3 Evidence for CP Violation

As mentioned earlier, historically CP violation was first observed in the kaon system, and it was via the decay

$$K_L \to \pi^- + \pi^+. \tag{15.9}$$

To explain why this implies CP violation we briefly review kaon physics. The kaon system consists of two states K^0 and $\bar{K}^0$ with strangeness $S = +1$ for K^0 and $S = -1$ for $\bar{K}^0$. From these, we can form eigenstates of CP. Thus, we choose $\bar{K}^0$ to be the CP conjugate of K^0 so that $CP|K^0\rangle = |\bar{K}^0\rangle$, and define

$$|K_{1,2}\rangle = \frac{1}{\sqrt{2}}(|K^0\rangle \pm |\bar{K}^0\rangle), \tag{15.10}$$

where $K_1(K_2)$ are CP-even (odd) states. Now, the decay of the neutral K contains a mode with a short decay life time called K_S so that

$$\tau_S = 0.89 \times 10^{-10}s, \tag{15.11}$$

where the decay modes of K_S are

$$K_S \to \pi^+\pi^-,\ \pi^0\pi^0. \tag{15.12}$$

The mode with the long lifetime is called K_L and has a lifetime

$$\tau_L = 5.2 \times 10^{-8}s, \tag{15.13}$$

and the decay modes of K_L are

$$K_L \to 3\pi, \; \pi l \nu. \tag{15.14}$$

The above decay modes would indicate that K_S is to be identified as the CP-even state K_1, and K_L is to be identified as the CP-odd stated K_2. If these were the only processes, CP would be conserved. However, it was observed by Cronin and Fitch and their collaborators [9] that K_L also has the decay

$$K_L \to \pi^+ \pi^-, \tag{15.15}$$

which is CP violating. The above observation was the first laboratory evidence for the existence of CP violation. The CP-violating decay can be understood by assuming that the K_S and K_L are not pure CP even and CP odd but, rather, mixtures of them so that

$$|K_S\rangle = p|K^0\rangle + q|\bar{K}^0\rangle, \quad |K_L\rangle = p|K^0\rangle - q|\bar{K}^0\rangle. \tag{15.16}$$

For $p/q = 1$ CP would be preserved in the $K^0 - \bar{K}^0$ system. But for $p/q \neq 1$ there would be CP violation.

Experimentally, a quantitative measure of CP violation is given by the parameters ϵ and ϵ', where ϵ is defined by

$$\epsilon = \frac{\langle (\pi\pi)_{I=0}|\mathcal{L}_W|K_L\rangle}{\langle (\pi\pi)_{I=0}|\mathcal{L}_W|K_S\rangle}, \tag{15.17}$$

where $\mathcal{L}_W$ is the Lagrangian for the weak $\Delta S = 1$ interactions and ϵ' is defined by

$$\epsilon' = \frac{\langle (\pi\pi)_{I=2}|\mathcal{L}_W|K_L\rangle}{\langle (\pi\pi)_{I=0}|\mathcal{L}_W|K_L\rangle} - \frac{\langle (\pi\pi)_{I=2}|\mathcal{L}_W|K_S\rangle}{\langle (\pi\pi)_{I=0}|\mathcal{L}_W|K_S\rangle}. \tag{15.18}$$

A determination of ϵ has existed for a long time, and is given by [18]

$$|\epsilon| = (2.288 \pm 0.011) \times 10^{-3}. \tag{15.19}$$

It is often referred to as a measure of indirect CP violation in the kaon system. The determination of ϵ' is more recent, and experimentally it is [18]

$$\epsilon'/\epsilon = (1.66 \pm 0.23) \times 10^{-3}. \tag{15.20}$$

ϵ' is often referred to as a measure of direct CP violation [19]. The experimental evidence for ϵ' rules out the so-called superweak theory of CP violation [20]. There are other channels where CP violation can be observed in the kaon system. One promising case is

$$K_L \to \pi^0 \nu \bar{\nu}. \tag{15.21}$$

A measurement of this decay will provide a determination of $V_{td}V_{ts}^*$. In the standard model the branching ratio for this process is predicted to be [21] $BR(K_L \to \pi^0 \nu \bar{\nu}) = (3.0 \pm 0.6) \times 10^{-11}$, which is about two orders of magnitude smaller than the current experimental limit [22], which is $BR(K_L \to \pi^0 \nu \bar{\nu}) < 1.7 \times 10^{-9}$. A test of the standard model prediction in this channel would require two orders of magnitude improvement in experiments. Much larger contributions to this branching ratio exist in models of physics beyond the standard model.

CP violation has also been observed in the $B^0 - \bar{B}^0$ system, where B^0 is a meson which is a bound state of a b-quark with either a d-quark (B_d^0) or with a s-quark (B_s^0). Analyses of various decay modes allow one to determine the angles α, β, and γ that enter into the unitarity triangle [18]:

$$\sin(2\beta) = 0.679 \pm 0.02 \tag{15.22}$$

$$\alpha = (89^{+4.4}_{-4.2})^0 \tag{15.23}$$

$$\gamma = (68^{+10}_{-11})^0, \tag{15.24}$$

while the Jarlskog invariant has the value

$$J = (2.96^{+0.2}_{-0.16}) \times 10^{-5}. \tag{15.25}$$

The third piece of evidence for CP violation in an indirect one. It relates to the excess of baryons over anti-baryons in the universe. The asymmetry can be expressed as the ratio [18]

$$\eta_B = (n_B - n_{\bar{B}})/n_\gamma \simeq 6 \times 10^{-10}, \tag{15.26}$$

where n_B is the baryon number density, $n_{\bar{B}}$ is the number density of anti-baryons, and n_γ is the number density of photons of the cosmic background radiation. Now, according to Sakharov [10] the generation of a baryon excess would require the following three conditions to be met: (1) the existence of a baryon-number-violating interaction, (2) the existence of a C- and CP-violating interaction, and (3) departure from thermal equilibrium. Condition (1) is easily met in grand unified models where baryon number violating interactions exist. The existence of a C- and CP-violating interaction is needed to create an imbalance between the production of particles versus antiparticles. Finally, baryon number violation and violation of C and CP are not sufficient, and to produce a net balance of baryons requires departure from thermal equilibrium. Since one basic element in the generation of a baryon excess is CP violation, the existence of a baryon excess provides indirect evidence that CP-violating interactions were at work in the era when the baryon excess was produced. However, it turns out that the CP violation that arises in the standard model from the CKM matrix is not large enough to generate the desired baryon excess, and additional sources of CP violation beyond those in the standard model are needed. Supersymmetric theories have a large number of such sources whose consequences will be explored in this chapter.

If there exist other sources of CP violation beyond those in the standard model, they are likely to show up in a variety of phenomena. One example of this is the EDM of leptons and quarks. In the standard model, the EDMs of the quarks and of the leptons are very small [23–27]. In the standard model, up to the three-loop level, the EDMs of the quarks have been calculated. For the down quark, one finds [28]

$$d_d/e = \frac{m_d m_c^2 \alpha_s G_F^2 \bar{\delta}}{108\pi^5} \left[\left(\gamma_{bc}^2 - 2\gamma_{bc} + \frac{\pi^2}{3} \right) \gamma_{Wb} + \frac{5}{8}\gamma_{bc}^2 - \left(\frac{335}{36} + \frac{2}{3}\pi^2 \right) \gamma_{bc} \right.$$
$$\left. - \frac{1231}{108} + \frac{7}{8}\pi^2 + 8\zeta_3 \right] + \mathcal{O}(m^2/M^2). \tag{15.27}$$

Here, $\gamma_{ab} \equiv \ln(m_a^2/m_b^2)$, ζ_3 is the zeta function, where $\zeta_3 \simeq 1.202$, $\bar{\delta}$ is the CP-violating invariant, and $\mathcal{O}(m^2/M^2)$ stands for terms that are suppressed by m_W or m_t. For the up quark, one has [28]

$$d_u/e = \frac{m_u m_s^2 \alpha_s G_F^2 \bar{\delta}}{216\pi^5} \left[\left(-\gamma_{bs}^2 + 2\gamma_{bs} + 2 - \frac{2\pi^2}{3} \right) \gamma_{Wb} - \gamma_{bc}\gamma_{cs}^2 + 2\gamma_{bc}\gamma_{cs} \right.$$
$$- \frac{5}{8}\gamma_{bs}^2 - \left(\frac{259}{36} + \frac{\pi^2}{3} \right) \gamma_{bs} \left(\frac{140}{9} + \pi^2 \right) \gamma_{cs}$$
$$\left. + -\frac{121}{108} + \frac{41}{36}\pi^2 - 4\zeta_3 \right] + \mathcal{O}(m^2/M^2). \tag{15.28}$$

The numerical value for the EDM of the quarks can be estimated from above so that [28]

$$d_d = -0.7 \times 10^{-34} \frac{m_d}{10 \text{ MeV}} \, e\,\text{cm}$$

$$d_u = -0.15 \times 10^{-34} \frac{m_u}{5 \text{ MeV}} \, e\,\text{cm}. \tag{15.29}$$

The neutron EDM which can be obtained by using an approximation which treats the neutron as a set of valence quarks and then uses the non-relativistic $SU(6)$ quark model, which gives

$$d_n = \frac{4}{3}d_d - \frac{1}{3}d_u. \tag{15.30}$$

Thus, the neutron EDM is also $10^{-34}e$ cm in size. For the electron, the EDM for the leptons arises only at the four-loop level. The analysis by Hoogeveen [26] (see also work by Pospelov and Khriplovich [29]) estimates d_e to be $\simeq 10^{-38}e$ cm. Experimentally, we have the current limits for the neutron EDM [30] as

$$|d_n| < 2.9 \times 10^{-26}e \text{ cm} \qquad (90\% \text{ confidence level}), \tag{15.31}$$

and for the electron EDM one has [31]

$$|d_e| < 8.7 \times 10^{-29}\text{ecm} \qquad (90\% \text{ confidence level}). \tag{15.32}$$

15.4 CP Violation in Supersymmetric Theories

In addition to the CP violation that can arise in non-supersymmetric theories such as via Yukawa couplings and strong CP violation, supersymmetric theories as well as theories based on strings and branes have additional sources of CP violation arising from soft terms as well as from the μ-parameter. Thus, in supersymmetric theories the μ-parameter, the trilinear couplings, and the gaugino masses are typically complex, and their phases govern CP violation. These CP-violating phases enter the slepton, squark, chargino, and neutralino masses and contribute to the EDM of the leptons and the quarks. The CP phases also affect a variety of other phenomena in supersymmetric theories such as the production of sparticles at colliders and their decays, and they can affect the magnetic moments of particles as well as flavor-changing processes such as $b \to s\gamma$ and $B_s \to \mu^+\mu^-$. We now discuss the specific way in which the sfermion masses and the masses of charginos and neutralinos are modified by the presence of CP phases.

In the presence of CP phases the selectron mass-squared matrix takes the form

$$M_{\tilde{e}}^2 = \begin{pmatrix} M_{\tilde{L}}^2 + m_e^2 - M_Z^2 \left(\frac{1}{2} - s_W^2\right)\cos 2\beta & m_e(A_e^* - \mu \tan\beta) \\ m_e(A_e - \mu^* \tan\beta) & m_{\tilde{E}}^2 + m_e^2 - M_Z^2 s_W^2 \cos 2\beta \end{pmatrix},$$

(15.33)

where A_e and μ are complex. We can diagonalize this mass-squared matrix so that

$$D_e^\dagger M_{\tilde{e}}^2 D_e = diag(M_{\tilde{e}1}^2, M_{\tilde{e}2}^2),$$

(15.34)

where the eigenvalues are given by

$$M_{\tilde{e}(1)}^2 = \frac{1}{2}(M_{\tilde{e}11}^2 + M_{\tilde{e}22}^2) - \frac{1}{2}[(M_{\tilde{e}11}^2 - M_{\tilde{e}22}^2)^2 + 4|M_{\tilde{e}21}^2|^2]^{\frac{1}{2}}$$

(15.35)

$$M_{\tilde{e}(2)}^2 = \frac{1}{2}(M_{\tilde{e}11}^2 + M_{\tilde{e}22}^2) + \frac{1}{2}[(M_{\tilde{e}11}^2 - M_{\tilde{e}22}^2)^2 + 4|M_{\tilde{e}21}^2|^2]^{\frac{1}{2}},$$

(15.36)

and where $M_{\tilde{e}ij}^2$ can be read from Eq. (15.33). The diagonalizing matrix D_e takes the form [32]

$$D_e = \begin{pmatrix} \cos\dfrac{\theta_e}{2} & -\sin\dfrac{\theta_e}{2} e^{-i\phi_e} \\ \sin\dfrac{\theta_e}{2} e^{i\phi_e} & \cos\dfrac{\theta_e}{2} \end{pmatrix}.$$

(15.37)

The phase ϕ_e is defined so that $M_{\tilde{e}21}^2 = |M_{\tilde{e}21}^2|e^{i\phi_e}$, and θ_e is defined so that $\tan\theta_e = 2|M_{\tilde{e}21}^2|/(M_{\tilde{e}11}^2 - M_{\tilde{e}22}^2)$, where we choose the range of θ_e so that $-\pi/2 \leq \theta_e \leq \pi/2$. θ_e and ϕ_e are given by

$$\tan\theta_e = \frac{2m_e|A_e m_0 - \mu^* \tan\beta|}{M_{\tilde{e}11}^2 - M_{\tilde{e}22}^2}$$

(15.38)

and

$$\sin \phi_e = \frac{m_0 |A_e| \sin \alpha_e + |\mu| \sin \theta_\mu \tan \beta}{|m_0 A_e - \mu^* \cot \beta|}, \tag{15.39}$$

where α_e is the phase of $A_e = |A_e| \exp(i\alpha_e)$. We discuss now the squark mass matrices with the inclusion of CP phases. Here, for the up squark mass-square matrix one has

$$M_{\tilde{u}}^2 = \begin{pmatrix} M_{\tilde{Q}}^2 + m_u^2 + M_Z^2 \left(\frac{1}{2} - Q_u s_W^2\right) \cos 2\beta & m_u(A_u^* - \mu \cot \beta) \\ m_u(A_u - \mu^* \cot \beta) & m_{\tilde{U}}^2 + m_u^2 + M_Z^2 Q_u s_W^2 \cos 2\beta \end{pmatrix}. \tag{15.40}$$

As for the slepton case, the squark mass-square matrix can be diagonalized by the unitary transformation

$$D_u^\dagger M_{\tilde{u}}^2 D_u = diag(M_{\tilde{u}1}^2, M_{\tilde{u}2}^2), \tag{15.41}$$

and D_u assumes a form similar to D_e so that

$$D_u = \begin{pmatrix} \cos \frac{\theta_u}{2} & -\sin \frac{\theta_u}{2} e^{-i\phi_u} \\ \sin \frac{\theta_u}{2} e^{i\phi_u} & \cos \frac{\theta_u}{2}. \end{pmatrix}. \tag{15.42}$$

Here, θ_u is given by

$$\tan \theta_u = \frac{2m_u |A_u m_0 - \mu^* \cot \beta|}{M_{\tilde{u}11}^2 - M_{\tilde{u}22}^2}, \tag{15.43}$$

where $(M_{\tilde{u}})_{ij}^2$ are elements of Eq. (15.40) and ϕ_μ is given by

$$\sin \phi_u = \frac{m_0 |A_u| \sin \alpha_u + |\mu| \sin \theta_\mu \cot \beta}{|m_0 A_u - \mu^* \cot \beta|}. \tag{15.44}$$

A very similar analysis holds for the down squark mass-squared matrix, where

$$M_{\tilde{d}}^2 = \begin{pmatrix} M_{\tilde{Q}}^2 + m_d^2 - M_Z^2 \left(\frac{1}{2} + Q_d s_W^2\right) \cos 2\beta & m_d(A_d^* - \mu \tan \beta) \\ m_d(A_d - \mu^* \tan \beta) & m_{\tilde{D}}^2 + m_d^2 + M_Z^2 Q_d s_W^2 \cos 2\beta \end{pmatrix}, \tag{15.45}$$

and one can carry out a diagonalization similar to the case for the up squark. Here, θ_d and the phase ϕ_d are given by

$$\tan \theta_d = \frac{2m_d |A_u m_0 - \mu^* \tan \beta|}{M_{\tilde{d}11}^2 - M_{\tilde{d}22}^2} \tag{15.46}$$

and

$$\sin \phi_d = \frac{m_0 |A_d| \sin \alpha_d + |\mu| \sin \theta_\mu \tan \beta}{|m_0 A_d - \mu^* \tan \beta|}. \tag{15.47}$$

Next, we discuss the chargino mass matrix. It has the form

$$M_C = \begin{pmatrix} \tilde{m}_2 & \sqrt{2} m_W \sin \beta \\ \sqrt{2} m_W \cos \beta & \mu \end{pmatrix}, \tag{15.48}$$

where $\tilde{m}_2$ and μ are, in general, both complex. This matrix is not symmetric and is not Hermitian. For the purpose of explicit diagonalization, we consider the case where $\tilde{m}_2$ is real and the only CP phase that enters the matrix is the phase of μ. The matrix M_C can be diagonalized by a biunitary transformation

$$U'^* M_C V^{-1} = M_D, \tag{15.49}$$

where U' and V are Hermitian matrices, which implies that

$$U'^* (M_C M_C^\dagger)(U'^*)^{-1} = diag\left(|m_{\tilde{\chi}_1^+}|^2, |m_{\tilde{\chi}_2^+}|^2\right) = V(M_C^\dagger M_C)V^{-1}. \tag{15.50}$$

The unitary matrix U' can be parameterized so that [32]

$$U' = \begin{pmatrix} \cos\dfrac{\theta_1}{2} & \sin\dfrac{\theta_1}{2} e^{i\phi_1} \\ -\sin\dfrac{\theta_1}{2} e^{-i\phi_1} & \cos\dfrac{\theta_1}{2} \end{pmatrix}, \tag{15.51}$$

where

$$\tan\theta_1 = 2\sqrt{2} m_W (\tilde{m}_2^2 - |\mu|^2 - 2m_W^2 \cos 2\beta)^{-1}$$
$$\times \left(\tilde{m}_2^2 \cos^2\beta + |\mu|^2 \sin^2\beta + |\mu|\tilde{m}_2 \sin 2\beta \cos\theta_\mu\right)^{\frac{1}{2}} \tag{15.52}$$

$$\tan\phi_1 = |\mu| \sin\theta_\mu \sin\beta \, (\tilde{m}_2 \cos\beta + |\mu| \cos\theta_\mu \sin\beta)^{-1}. \tag{15.53}$$

A very similar parametrization for V gives

$$V = \begin{pmatrix} \cos\dfrac{\theta_2}{2} & \sin\dfrac{\theta_2}{2} e^{-i\phi_2} \\ -\sin\dfrac{\theta_2}{2} e^{i\phi_2} & \cos\dfrac{\theta_2}{2} \end{pmatrix}, \tag{15.54}$$

where

$$\tan\theta_2 = 2\sqrt{2} m_W (\tilde{m}_2^2 - |\mu|^2 + 2m_W^2 \cos 2\beta)^{-1}$$
$$\times \left(\tilde{m}_2^2 \sin^2\beta + |\mu|^2 \cos^2\beta + |\mu|\tilde{m}_2 \sin 2\beta \cos\theta_\mu\right)^{\frac{1}{2}} \tag{15.55}$$

$$\tan\phi_2 = -|\mu| \sin\theta_\mu \cos\beta \, (\tilde{m}_2 \sin\beta + |\mu| \cos\theta_\mu \cos\beta)^{-1}. \tag{15.56}$$

We now note that while M_D is diagonal, it is not necessarily real and positive. We can accomplish this by defining $U = P \times U'$, where the matrix P is a phase matrix so that [32]

$$P = diag\left(e^{i\gamma_1}, e^{i\gamma_2}\right), \tag{15.57}$$

where γ_1 and γ_2 are chosen to exactly cancel the phases of the diagonal elements of M_D in Eq. (15.49). In this case we have

$$U^* M_C V^{-1} = diag(m_{\tilde{\chi}_1^+}, m_{\tilde{\chi}_2^+}), \tag{15.58}$$

where the diagonal elements are real and positive. The eigenvalues of the chargino mass matrix are now given by

$$M^2_{m_{\tilde{\chi}^+_1}} = \frac{1}{2}\left[\tilde{m}^2_2 + |\mu|^2 + 2m^2_W\right] - \frac{1}{2}\left[(\tilde{m}^2_2 - |\mu|^2)^2 + 4m^4_W\cos^2 2\beta + 4m^2_W\right.$$

$$\left. \times\, (\tilde{m}^2_2 + |\mu|^2 + 2\tilde{m}_2|\mu|\cos\theta_\mu\sin 2\beta)\right]^{\frac{1}{2}}, \tag{15.59}$$

$$M^2_{m_{\tilde{\chi}^+_2}} = \frac{1}{2}\left[\tilde{m}^2_2 + |\mu|^2 + 2m^2_W\right] + \frac{1}{2}\left[(\tilde{m}^2_2 - |\mu|^2)^2 + 4m^4_W\cos^2 2\beta + 4m^2_W\right.$$

$$\left. \times\, (\tilde{m}^2_2 + |\mu|^2 + 2\tilde{m}_2|\mu|\cos\theta_\mu\sin 2\beta)\right]^{\frac{1}{2}}. \tag{15.60}$$

We now turn to the neutralino mass matrix $M^0_{\tilde{\chi}}$, which is given by

$$M^0_{\tilde{\chi}} = \begin{pmatrix} \tilde{m}_1 & 0 & -M_Z\sin\theta_W\cos\beta & M_Z\sin\theta_W\sin\beta \\ 0 & \tilde{m}_2 & M_Z\cos\theta_W\cos\beta & -M_Z\cos\theta_W\sin\beta \\ -M_Z\sin\theta_W\cos\beta & M_Z\cos\theta_W\cos\beta & 0 & -\mu \\ M_Z\sin\theta_W\sin\beta & -M_Z\cos\theta_W\sin\beta & -\mu & 0 \end{pmatrix}. \tag{15.61}$$

Since $\tilde{m}_1$, $\tilde{m}_2$, and μ can be all complex, this matrix is complex, symmetric, and non-Hermitian. We can diagonalize it by a unitary transformation so that

$$X^T M^0_{\tilde{\chi}} X = diag(m_{\tilde{\chi}^0_1}, m_{\tilde{\chi}^0_2}, m_{\tilde{\chi}^0_3}, m_{\tilde{\chi}^0_4}). \tag{15.62}$$

15.5 The EDM of an Elementary Dirac Fermion

In the presence of CP-violating interactions, an elementary spin $1/2$ Dirac fermion ψ_f can have an EDM d_f which is defined by the relation

$$\mathcal{L}_{EDM} = -\frac{i}{2}d_f\bar{\psi}\sigma_{\mu\nu}\gamma_5\psi F^{\mu\nu}, \tag{15.63}$$

where $F_{\mu\nu}$ is the electric field strength. We note that in the non-relativistic limit keeping just the large component ψ_{Af} of the Dirac field ψ_f, the EDM formula takes the form

$$\mathcal{L}^{NR}_{EDM} = d_f\psi^\dagger_A\vec{\sigma}.\vec{E}\psi_A, \tag{15.64}$$

which corresponds to the common notion of what an EDM is. In a renormalizable field theory the EDM of the Dirac fermion cannot arise at the tree level. However, it can do so at the loop level. To show this, let us consider the case when the Dirac fermion has CP-violating couplings with a heavy Dirac field χ_i and a heavy scalar field S_a and that this coupling has the form

$$\mathcal{L} = L_{ia}\bar{\psi}_f P_L\chi_i S_a + R_{ia}\bar{\psi}_f P_R\chi_i S_a + \text{h.c.} \tag{15.65}$$

CP invariance of $\mathcal{L}$ is violated when

$$Im(L_{ia}R^*_{ia}) \neq 0. \tag{15.66}$$

Using the interaction of Eq. (15.65), an analysis of the EDM of the Dirac fermion ψ_f at the one-loop level gives [33]

$$d_f = \frac{m_i}{16\pi^2 m_a^2} Im(L_{ia}R_{ia}^*) \left[Q_i F_1\left(\frac{m_i^2}{m_a^2}\right) + Q_a F_2\left(\frac{m_i^2}{m_a^2}\right) \right], \qquad (15.67)$$

where $F_1(x)$ and $F_2(x)$ are the loop integrals and are given by

$$F_1(x) = \frac{1}{2(1-x)^2}\left(3 - x + \frac{2 \ln x}{1-x}\right)$$
$$F_2(x) = \frac{1}{2(1-x)^2}\left(1 + x + \frac{2x \ln x}{1-x}\right). \qquad (15.68)$$

15.6 The EDM of a Charged Lepton in Supersymmetry

We now discuss the EDM of a charged lepton in supersymmetry. Here, the contributions to the EDM arise from the exchange of a chargino and a sneutrino in the loop and from the exchange of a neutralino and a smuon in the loop (Fig. 15.1). For the case of exchange of charginos in the loop, the contribution to the EDM of a charged lepton is given by [32]

$$d_{e-\tilde{\chi}^+}^E = \frac{e\alpha_{EM}}{4\pi \sin^2\theta_W} m_{\tilde{\nu}e}^2 \sum_{i=1}^{2} m_{\tilde{\chi}_i^+} Im(\Gamma_{ei}) F_1\left(\frac{m_{\tilde{\chi}_i^+}^2}{m_{\tilde{\nu}e}^2}\right). \qquad (15.69)$$

In the above,

$$\Gamma_{ei} = (\kappa_e U_{i2}^* V_{i1}^*) = |\kappa_e| U_{R2i}^* U_{L1i}, \qquad (15.70)$$

where U and V are as defined by Eq. (15.58) and where $\kappa_e = m_e/(\sqrt{2}m_W \cos\beta)$. Here, Γ_{ei} can be easily seen to depend only on one combination of CP phases, i.e., the combination, $\xi_2 + \theta_\mu$, where ξ_2 is the phase of the gaugino mass $\tilde{m}_2$ and θ_μ is the phase of μ. To the above, we must add the neutralino exchange contribution, which is given by [32]

$$d_{f-\tilde{\chi}^0}^E = \frac{e\alpha_{EM}}{4\pi \sin^2\theta_W} \sum_{a=1}^{2}\sum_{i=1}^{4} Im(C_{fia}) \frac{m_{\tilde{\chi}_i^0}}{M_{\tilde{f}a}^2} Q_{\tilde{f}} F_2\left(\frac{m_{\tilde{\chi}_i^0}^2}{M_{\tilde{f}a}^2}\right), \qquad (15.71)$$

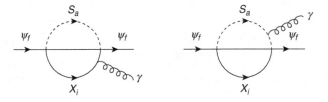

Figure 15.1. A generic diagram that shows the contribution to the EDM of a fermion from the exchange of a heavy fermion χ_i and a heavy scalar S_a.

where

$$C_{fia} = (c_1 X_{1i} D^*_{f1a} + c_2 X_{2i} D^*_{f1a} + \kappa_f X_{bi} D^*_{f2a})(c_3 X_{1i} D_{f2a} - \kappa_f X_{bi} D_{f1a}),$$
(15.72)

and where

$$c_1 = -\sqrt{2} \tan \theta_W (Q_f - T_{3f})$$
(15.73)

$$c_2 = -\sqrt{2} T_{3f}$$
(15.74)

$$c_3 = \sqrt{2} \tan \theta_W Q_f.$$
(15.75)

Further, in Eq. (15.72) $b = 3\,(4)$ for $T_{3q} = -\frac{1}{2}(\frac{1}{2})$. The lepton EDM depends on only the following combination of phases: $\xi_1 + \theta_\mu$ and $\alpha_f + \theta_\mu$, where α_f is the phase of the trilinear coupling A_f.

15.7 The EDM of Quarks

Next, we discuss the EDM of quarks. It consists of three separate types of contributions, which include contributions from the electric dipole operator (d_q^E), the chromoelectric dipole operator (d_q^C), and the purely gluonic dimension-six operator of Weinberg (d_q^G) [35]. The total contribution is then the sum of the three so that

$$d_q = d_q^E + d_q^C + d_q^G.$$
(15.76)

The diagrams that enter into each of these are different and so we discuss each of these separately below.

15.8 The EDM Operator Contribution to the EDM of Quarks

The Feynman diagram contributing to the EDM operator is similar to the one for leptons except that inside the loop one can have exchange contributions from the chargino, the gluino, and the neutralino (see Fig. 15.1). We first discuss the chargino exchange contribution, which involves exchange of the chargino and the squarks. The exchange involves two charginos and two squarks, so one has a double sum which gives [32]

$$d_{u\tilde{\chi}^+} = \frac{-e\alpha_{EM}}{4\pi \sin^2 \theta_W} \sum_{i,a=1}^{2} Im(\Gamma_{uia}) \frac{m_{\tilde{\chi}_i^+}}{M_{\tilde{d}a}^2} \left[(Q_u - Q_{\tilde{d}}) F_1 \left(\frac{m_{\tilde{\chi}_i^+}^2}{M_{\tilde{d}a}^2} \right) + Q_{\tilde{d}} F_2 \left(\frac{m_{\tilde{\chi}_i^+}^2}{M_{\tilde{d}a}^2} \right) \right].$$
(15.77)

Here, $F_1(x)$ and $F_2(x)$ are as defined by Eq. (15.68) and Γ_{uia} is given by

$$\Gamma_{uia} = \kappa_u V^*_{i2} D_{d1a}(U^*_{i1} D^*_{d1a} - \kappa_d U^*_{i2} D^*_{d2a}),$$
(15.78)

where

$$\kappa_d = \frac{m_d}{\sqrt{2} m_W \cos \beta}, \quad \kappa_u = \frac{m_u}{\sqrt{2} m_W \sin \beta}.$$
(15.79)

A more explicit form of Γ_{uia} is given by

$$\Gamma_{ui1(2)} = |\kappa_u|(\cos^2\theta_d/2)[U_{L2i}U^*_{R1i}] \mp \frac{1}{2}|\kappa_u\kappa_d|(\sin\theta_d)[U_{L2i}U^*_{R2i}]e^{i\{\xi_2-\phi_d\}}. \tag{15.80}$$

Here, $d_{u\tilde{\chi}^+}$ depends on two phase combinations which can be taken to be $\alpha_d + \theta_\mu$ and $\xi_2 + \theta_\mu$, where α_d is the phase of the trilinear coupling A_d. A very similar analysis holds for the down quarks, which again depend only on two phase combinations, which are similar to the case for the up quark but with α_d replaced by α_u.

Next, we discuss the gluino exchange contribution with the loop involving gluinos and squarks. We denote the squark eigenstates by $\tilde{q}_1$ and $\tilde{q}_2$. Then, we can write it in the form

$$d_{q\tilde{g}} = \frac{-2e\alpha_s}{3\pi}m_{\tilde{g}}Q_{\tilde{q}}\,Im(\Gamma^{11}_q)\left[\frac{1}{M^2_{\tilde{q}1}}F_2\left(\frac{m^2_{\tilde{g}}}{M^2_{\tilde{q}1}}\right) - \frac{1}{M^2_{\tilde{q}2}}F_2\left(\frac{m^2_{\tilde{g}}}{M^2_{\tilde{q}2}}\right)\right], \tag{15.81}$$

where $m_{\tilde{g}}$ is the gluino mass, $Q_{\tilde{q}}$ is the squark charge, $\alpha_s = g^2_s/4\pi$, and Γ^{1k}_q is given by

$$\Gamma^{1k}_q = e^{-i\xi_3}D_{q2k}D^*_{q1k}. \tag{15.82}$$

Further, an explicit analysis gives

$$Im(\Gamma^{11}_q) = -Im(\Gamma^{12}_q) \tag{15.83}$$
$$= \frac{m_q}{M^2_{\tilde{q}1} - M^2_{\tilde{q}2}}(m_0|A_q|\sin(\alpha_q - \xi_3) + |\mu|\sin(\theta_\mu + \xi_3)|R_q|).$$

Here, $d_{q\tilde{g}}$ depends on the combination $\alpha_q + \theta_\mu$ and $\xi_3 + \theta_\mu$ or some independent linear combination of the two, and $R_q = v_1/v^*_2(v_2/v^*_1)$ for $q = u(d)$.

15.9 Contribution of the Chromoelectric Dipole Moment to the EDM of Quarks

As mentioned above, the EDM of the quarks receives contributions not just from the electric dipole operator but also from the chromoelectric dipole operator and from the purely gluonic dimension-six operator. We discuss now the contribution from the chromoelectric dipole operator. It is given by the effective dimension-five operator

$$\mathcal{L}_I = -\frac{i}{2}\tilde{d}^C\bar{q}\sigma_{\mu\nu}\gamma_5 T^a q G^{\mu\nu a}, \tag{15.84}$$

where $G^{\mu\mu a}$ is the gluon field strength and T^a are the generators of $SU(3)_C$. There are three kinds of contributions to the chromoelectric dipole operators: these arise from the exchange of the gluino, the chargino, and the neutralino inside the loop (Fig. 15.2). Thus, the gluino contribution is given by [34]

$$\tilde{d}^C_{q\text{-}gluino} = \frac{g_s\alpha_s}{4\pi}\sum_{k=1}^{2}Im(\Gamma^{1k}_q)\frac{m_{\tilde{g}}}{M^2_{\tilde{q}k}}F_3\left(\frac{m^2_{\tilde{g}}}{M^2_{\tilde{q}k}}\right), \tag{15.85}$$

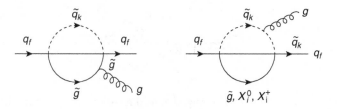

Figure 15.2. Diagrams that contribute to the chromoelectric dipole operator for a quark arising from the exchange of the chargino, neutralino, and gluino.

where the sum runs over the two squark states inside the loop and F_3 is defined by

$$F_3(x) = \frac{1}{6(x-1)^2}\left(10x - 26 + \frac{2x\ln x}{1-x} - \frac{18\ln x}{1-x}\right). \tag{15.86}$$

The chargino contribution is given by

$$\tilde{d}_{q-chargino}^{C} = \frac{-g^2 g_s}{16\pi^2}\sum_{a=1}^{2}\sum_{i=1}^{2} Im(\Gamma_{qia})\frac{m_{\tilde{\chi}_i^+}}{M_{\tilde{q}a}^2}F_2\left(\frac{m_{\tilde{\chi}_i^+}^2}{M_{\tilde{q}a}^2}\right), \tag{15.87}$$

where the sum runs over the two squark mass eigenstates and the two chargino mass eigenstates circulating inside the loop. Finally, the neutralino contribution is given by

$$\tilde{d}_{q-neutralino}^{C} = \frac{g_s g^2}{16\pi^2}\sum_{a=1}^{2}\sum_{i=1}^{4} Im(\eta_{qia})\frac{m_{\tilde{\chi}_i^0}}{M_{\tilde{q}a}^2}F_2\left(\frac{m_{\tilde{\chi}_i^0}^2}{M_{\tilde{q}a}^2}\right), \tag{15.88}$$

where the sum runs over the two squark mass eigenstates and four neutralino states circulating inside the loop. The CP violation in the above loop diagrams is encoded in the factors $Im(\Gamma_q^{1a})$, $Im(\Gamma_{qia})$, and $Im(\eta_{qia})$. Here, we have

$$\eta_{fik} = (a_0 X_{1i}D_{f1k}^* + b_0 X_{2i}D_{f1k}^* + \kappa_f X_{bi}D_{f2k}^*)(c_0 X_{1i}D_{f2k} - \kappa_f X_{bi}D_{f1k}), \tag{15.89}$$

where $b = 3(4)$ for $T_{3q} = -\frac{1}{2}$ ($\frac{1}{2}$), $a_0 = -\sqrt{2}\tan\theta_W(Q_f - T_{3f})$, $b_0 = -\sqrt{2}T_{3f}$, and $c_0 = \sqrt{2}\tan\theta_W Q_f$. In Eq. (15.89) D_f diagonalize the sfermion mass square matrix and X diagonalizes the neutralino mass matrix.

15.10 The EDM of the Quarks from the Purely Gluonic Dimension 6 Operator

Finally, we consider the contribution from the purely gluonic dimension-six operator [35], which is defined by

$$\mathcal{L}_{\text{dim6}} = -\frac{1}{6}\tilde{d}^G f_{\alpha\beta\gamma}G_{\alpha\mu\rho}G_{\beta\nu}^\rho G_{\gamma\lambda\sigma}\epsilon^{\mu\nu\lambda\sigma}, \tag{15.90}$$

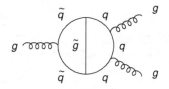

Figure 15.3. A CP-violating purely gluonic dimension-six operator with three external gluon lines arising from a two-loop diagram involving the exchange of quarks, squarks, and the gluino in the loops.

where $f_{\alpha\beta\gamma}$ are the structure constants of $SU(3)_C$ and $\epsilon^{\mu\nu\lambda\sigma}$ is the totally anti-symmetric tensor with $\epsilon^{0123} = +1$. Here, the phases that enter are the phase of the gluino mass ξ_3, the phase of μ, and the phase of A_t. The Feynman diagram contributing to the purely gluonic dimension-six operator is shown in Fig. 15.3. An analysis gives [34, 36]

$$
\tilde{d}^G = -3\alpha_s \left(\frac{g_s}{4\pi m_{\tilde{g}}}\right)^3 (m_t(z_1^t - z_2^t)Im(\Gamma_t^{12})F(z_1^t, z_2^t, z_t)
$$
$$
+ m_b(z_1^b - z_2^b)Im(\Gamma_b^{12})F(z_1^b, z_2^b, z_b)). \tag{15.91}
$$

Here

$$
z_\alpha^q = \left(\frac{M_{\tilde{q}\alpha}}{m_{\tilde{g}}}\right)^2, \quad z_q = \left(\frac{m_q}{m_{\tilde{g}}}\right)^2 \tag{15.92}
$$

and

$$
\Gamma_q^{1a} = e^{-i\xi_3} D_{q2a}D_{q1a}^*, \tag{15.93}
$$

while $F(z_1, z_2, z_3)$ is defined by the triple integral

$$
F(z_1, z_2, z_3) = \frac{1}{2}\int_0^1 dx \int_0^1 dv \int_0^1 dy\, x(1-x)v\frac{v_1 v_2}{w^4}, \tag{15.94}
$$

where

$$
v_1 = v(1-x) + z_3 x(1-x)(1-v) - 2vx[z_1 y + z_2(1-y)]
$$
$$
v_2 = (1-x)^2(1-v)^2 + v^2 - \frac{1}{9}x^2(1-v)^2
$$
$$
w = v(1-x) + z_3 x(1-x)(1-v) + vx[z_1 y + z_2(1-y)]. \tag{15.95}
$$

The function F simplifies considerably for the case when $m_{\tilde{q}}, m_{\tilde{g}} \gg m_q$. In this case F reduces to the following form

$$
F \simeq -\frac{m_{\tilde{g}}^2}{m_{\tilde{q}}^2} f(z_2^q), \tag{15.96}
$$

where $f(z)$ is defined by

$$
f(z) = \frac{1}{6(z-1)^2}[2(z-1)(11z-1) + (1 - 16z - 9z^2)\ln\, z]. \tag{15.97}
$$

To obtain the contribution from the chromoelectric dipole operator and from the purely gluonic dimension-six operator, one needs to use an approximation technique, since lattice QCD analyses of them do not yet exist. One technique used is the so-called naive dimensional analysis (NDA) [37], where one uses a "reduced coupling constant" to determine the contribution from a hadronic operator below the chiral symmetry breaking scale. Using these rules, the contributions of the electric dipole operator, the chromoelectric dipole operator, and the purely gluonic dimension-six operator are given by

$$d_q^E = d_q \eta^E$$
$$d_q^C = \frac{e}{4\pi} \tilde{d}_q^C \eta^C$$
$$d_q^G = \frac{eM}{4\pi} \tilde{d}_q^G \eta^G.$$
(15.98)

In the above, $M = 1.19$ GeV is the chiral symmetry breaking scale, and η^E, η^C, and η^G are renormalization group factors to evolve the EDMs from the electroweak scale to the hadronic scale. An estimate of their numerical values gives [38], $\eta^E \approx 0.61$ and $\eta^C \approx \eta^G \sim 3.4$. An alternative method to NDA is to estimate the contribution from the chromoelectric and the purely gluonic dimension-six operator is the use of QCD sum rules [39]. The EDM of the neutron can be obtained by using Eq. (15.30).

In general, the EDMs of the electron and of the neutron in supersymmetric theories with CP phases $\mathcal{O}(1)$ turn out to be far in excess of the current experimental limits. Several ways have been proposed to correct the situation. These include fine tuning the CP phases to be small [40], suppressing the EDMs by making the sparticle spectrum heavy [41], or by the cancellation mechanism [32, 34, 42–44], where various contributions cancel to reduce the EDM below the current experimental limits.

15.11 g-2 with Phases

A fermion of mass m and charge q has an intrinsic magnetic moment $\vec{\mu}$ given by

$$\vec{\mu} = g \left(\frac{q}{2m} \right) \vec{S}$$

where $\vec{S}$ is the spin, and g is Lande's g factor. One defines the anomalous magnetic moment by $a_f = (g - 2)/2$. Here, we consider the effect of CP phases on the anomalous magnetic moment of a fermion using Eq. (15.65). The one-loop contribution arises from Fig. 15.1 and is given by the sum

$$a_f = a_f^{\tilde{\chi}^0} + a_f^{\tilde{\chi}^-},$$
(15.99)

where $a_f^{\tilde{\chi}^0}$ arises from the neutralino and sfermion exchange while $a_f^{\tilde{\chi}^-}$ arises from the chargino and sneutrino exchange diagrams. In general, $a_f^{\tilde{\chi}^0} = a_{f1}^{\tilde{\chi}^0} + a_{f2}^{\tilde{\chi}^0}$

where [45–47]

$$a_{f1}^{\tilde{\chi}^0} = \sum_{ia} \frac{m_f}{8\pi^2 m_i} Re(R_{ia}L_{ia}^*)I_1\left(\frac{m_f^2}{m_i^2}, \frac{m_a^2}{m_i^2}\right), \tag{15.100}$$

and

$$I_1(\alpha, \beta) = -\int_0^1 dx \int_0^{1-x} dz \frac{z}{\alpha z^2 + (1 - \alpha - \beta)z + \beta}, \tag{15.101}$$

and where

$$a_{f2}^{\tilde{\chi}^0} = \sum_{ia} \frac{m_f^2}{16\pi^2 m_i^2}(|R_{ia}|^2 + |L_{ia}|^2)I_2\left(\frac{m_f^2}{m_i^2}, \frac{m_a^2}{m_i^2}\right) \tag{15.102}$$

and

$$I_2(\alpha, \beta) = \int_0^1 dx \int_0^{1-x} dz \frac{z^2 - z}{\alpha z^2 + (1 - \alpha - \beta)z + \beta}. \tag{15.103}$$

Similarly, $a_f^{\tilde{\chi}^-}$ consists of two terms, $a_f^{\tilde{\chi}^-} = a_{f1}^{\tilde{\chi}^-} + a_{f2}^{\tilde{\chi}^-}$, where

$$a_{f1}^{\tilde{\chi}^-} = \sum_{ia} \frac{m_f}{8\pi^2 m_i} Re(R_{ia}L_{ia}^*)I_3\left(\frac{m_f^2}{m_i^2}, \frac{m_a^2}{m_i^2}\right) \tag{15.104}$$

and

$$I_3(\alpha, \beta) = \int_0^1 dx \int_0^{1-x} dz \frac{1 - z}{\alpha z^2 + (\beta - \alpha - 1)z + 1}, \tag{15.105}$$

and where

$$a_{f2}^{\tilde{\chi}^-} = -\sum_{ia} \frac{m_f^2}{16\pi^2 m_i^2}(|R_{ia}|^2 + |L_{ia}|^2)I_4\left(\frac{m_f^2}{m_i^2}, \frac{m_a^2}{m_i^2}\right), \tag{15.106}$$

and

$$I_4(\alpha, \beta) = \int_0^1 dx \int_0^{1-x} dz \frac{z^2 - z}{\alpha z^2 + (\beta - \alpha - 1)z + 1}. \tag{15.107}$$

The above relations are exact without any approximations on the relative sizes of m_f, m_i, and m_a. The loop integrals simplify if we assume that the external fermion mass is much smaller than the masses circulating in the loop, i.e., $m_f \ll m_i, m_a$. However, later we will see that in order to consider the supersymmetric limit, we will need the full form for these integrals. Now, in the limit when $m_f \ll m_i, m_a$ we have the following:

$$I_1(0, x) = \frac{1}{2}G_1(x), I_2(0, x) = \frac{1}{6}G_2(x) \tag{15.108}$$

$$I_3(0, x) = -\frac{1}{2}G_3(x), \ I_4(0, x) = -\frac{1}{6}G_4(x), \tag{15.109}$$

where

$$G_1(x) = \frac{1}{(x-1)^3}(1 - x^2 + 2x \ln x) \tag{15.110}$$

$$G_2(x) = \frac{1}{(x-1)^4}(-x^3 + 6x^2 - 3x - 2 - 6x \ln x) \tag{15.111}$$

$$G_3(x) = \frac{1}{(x-1)^3}(3x^2 - 4x + 1 - 2x^2 \ln x) \tag{15.112}$$

$$G_4(x) = \frac{1}{(x-1)^4}(2x^3 + 3x^2 - 6x + 1 - 6x^2 \ln x). \tag{15.113}$$

Thus, under the above approximation for the case of the muon the chargino exchange gives

$$a_{f1}^{\tilde\chi^-} = \frac{m_\mu \alpha_{EM}}{4\pi \sin^2 \theta_W} \sum_{i=1}^{2} \frac{1}{M_{\tilde\chi_i^+}} Re(\kappa_\mu U_{i2}^* V_{i1}^*) G_3 \left(\frac{M_{\tilde\nu}^2}{M_{\tilde\chi_i^+}^2} \right) \tag{15.114}$$

and

$$a_{f2}^{\tilde\chi^-} = \frac{m_\mu^2 \alpha_{EM}}{24\pi \sin^2 \theta_W} \sum_{i=1}^{2} \frac{1}{M_{\tilde\chi_i^+}^2} (|\kappa_\mu U_{i2}^*|^2 + |V_{i1}|^2) G_4 \left(\frac{M_{\tilde\nu}^2}{M_{\tilde\chi_i^+}^2} \right), \tag{15.115}$$

where

$$\kappa_\mu = \frac{m_\mu}{\sqrt{2} M_W \cos \beta}. \tag{15.116}$$

The full form of the neutralino exchange without approximation gives

$$a_{f1}^{\tilde\chi^0} = \frac{m_\mu \alpha_{EM}}{2\pi \sin^2 \theta_W} \sum_{j=1}^{4} \sum_{k=1}^{2} \frac{1}{M_{\tilde\chi_j^0}} Re(\eta_{\mu j}^k) I_1 \left(\frac{m_\mu^2}{M_{\tilde\chi_j^0}^2}, \frac{M_{\tilde\mu_k}^2}{M_{\tilde\chi_j^0}^2} \right), \tag{15.117}$$

where

$$\eta_{\mu j}^k = -\left(\frac{1}{\sqrt{2}} [\tan \theta_W X_{1j} + X_{2j}] D_{1k}^* - \kappa_\mu X_{3j} D_{2k}^* \right)$$
$$\times (\sqrt{2} \tan \theta_W X_{1j} D_{2k} + \kappa_\mu X_{3j} D_{1k}) \tag{15.118}$$

and a_μ^{12} is given by

$$a_\mu^{12} = \frac{m_\mu^2 \alpha_{EM}}{4\pi \sin^2 \theta_W} \sum_{j=1}^{4} \sum_{k=1}^{2} \frac{1}{M_{\tilde\chi_j^0}^2} X_{\mu j}^k I_2 \left(\frac{m_\mu^2}{M_{\tilde\chi_j^0}^2}, \frac{M_{\tilde\mu_k}^2}{M_{\tilde\chi_j^0}^2} \right), \tag{15.119}$$

where

$$X_{\mu j}^k = \frac{m_\mu^2}{2M_W^2 \cos^2 \beta} |X_{3j}|^2$$

$$+ \frac{1}{2} \tan^2 \theta_W |X_{1j}|^2 (|D_{1k}|^2 + 4|D_{2k}|^2) + \frac{1}{2} |X_{2j}|^2 |D_{1k}|^2$$

$$+ \tan \theta_W |D_{1k}|^2 Re(X_{1j} X_{2j}^*)$$

$$+ \frac{m_\mu \tan \theta_W}{M_W \cos \beta} Re(X_{3j} X_{1j}^* D_{1k} D_{2k}^*)$$

$$- \frac{m_\mu}{M_W \cos \beta} Re(X_{3j} X_{2j}^* D_{1k} D_{2k}^*). \tag{15.120}$$

It is instructive to consider the supersymmetric limit of the chargino and neutralino exchange contributions. To achieve this limit, we set $M_{\tilde{\nu}} = 0$, $\tilde{m}_1 = 0 = \tilde{m}_2 = 0$, $\tan \beta = 1$, and $\mu = 0$. In this limit [45, 46]

$$U^* M_C V^{-1} = diag(M_W, M_W), \tag{15.121}$$

and the unitary matrices U and V take the form

$$U = \frac{1}{\sqrt{2}} \begin{pmatrix} 1 & 1 \\ -1 & 1 \end{pmatrix}, \; V = \frac{1}{\sqrt{2}} \begin{pmatrix} 1 & 1 \\ 1 & -1 \end{pmatrix}. \tag{15.122}$$

Further, in this limit

$$G_3(0) = -1, \; G_4(0) = 1, \tag{15.123}$$

and the chargino contributions $a_{f1}^{\tilde{\chi}^-}$ and $a_{f2}^{\tilde{\chi}^-}$ take on the following form:

$$a_{f1}^{\tilde{\chi}^-} = -\frac{\alpha_{EM}}{4\pi \sin^2 \theta_W} \frac{m_\mu^2}{M_W^2}$$

$$a_{f2}^{\tilde{\chi}^-} = \frac{\alpha_{EM}}{24\pi \sin^2 \theta_W} \frac{m_\mu^2}{M_W^2}. \tag{15.124}$$

Together, they give

$$a_\mu^{\tilde{\chi}^-} = -\frac{5\alpha_{EM}}{24\pi \sin^2 \theta_W} \frac{m_\mu^2}{M_W^2}. \tag{15.125}$$

For the muon the contribution from the W-exchange in the standard model is given by [48]

$$a_\mu^W = \frac{5m_\mu^2 G_F}{12\pi^2 \sqrt{2}}. \tag{15.126}$$

Using the equality

$$G_F = \pi \alpha_{em} / (M_W^2 \sqrt{2} \sin^2 \theta_W), \tag{15.127}$$

one finds

$$a_\mu^{\tilde{\chi}^-} + a_\mu^W = 0, \tag{15.128}$$

which shows that the sum of the W-exchange contribution and the chargino exchange contribution vanish in the supersymmetric limit.

Next, we discuss the contribution to the anomalous magnetic moment of the muon in the neutral sector. Here, the standard model gives

$$a_\mu^Z = \frac{m_\mu^2 G_F}{2\sqrt{2}\pi^2}\left(-\frac{5}{12} + \frac{4}{3}\left(\sin^2\theta_W - \frac{1}{4}\right)^2\right). \tag{15.129}$$

Regarding the contribution from the supersymmetric sector, in the supersymmetric limit two of the eigen values of the neutralino mass matrix vanish while the other two are $\pm M_Z$. To determine the loop contribution from this sector in the supersymmetric limit we need to choose the unitary transformation X which makes the non-vanishing eigenvalues positive definite so that

$$X^T M_{\tilde{\chi}^0} X = diag(0, 0, M_Z, M_Z). \tag{15.130}$$

The unitary matrix X that accomplishes this has the form

$$X = \begin{pmatrix} a & b & \dfrac{\sin\theta_W}{\sqrt{2}} & i\dfrac{\sin\theta_W}{\sqrt{2}} \\ a\tan\theta_W & b\tan\theta_W & -\dfrac{\cos\theta_W}{\sqrt{2}} & -i\dfrac{\cos\theta_W}{\sqrt{2}} \\ a & -\dfrac{1}{2}b\sec^2\theta_W & -\dfrac{1}{2} & \dfrac{i}{2} \\ a & -\dfrac{1}{2}b\sec^2\theta_W & \dfrac{1}{2} & -\dfrac{i}{2} \end{pmatrix}, \tag{15.131}$$

where

$$a = \frac{1}{\sqrt{3 + \tan^2\theta_W}}, \tag{15.132}$$

$$b = \frac{1}{\sqrt{1 + \tan^2\theta_W + \frac{1}{2}\sec^4\theta_W}}. \tag{15.133}$$

We notice that the last column of the matrix X has an overall $i\ (=\sqrt{-1})$ factor. It is there to make certain that the eigenvalues of $M_{\tilde{\chi}^0}$ are all positive definite. In the supersymmetric limit, one gets

$$a_{f1}^{\tilde{\chi}^0} = \sum_{j=3}^{4} \frac{m_\mu^2 \alpha_{EM}}{4\sqrt{2}\pi \sin^2\theta_W M_W M_Z}$$

$$[Re(X_{3j}X_{2j}) + \tan\theta_W Re(X_{3j}X_{1j}) - 2\tan\theta_W Re(X_{3j}X_{1j})] \tag{15.134}$$

and

$$a_{f2}^{\tilde{\chi}^0} = -2\sum_{j=3}^{4} \frac{m_\mu^2 \alpha_{EM}}{48\pi \sin^2\theta_W M_Z^2}$$

$$\times \left[5\tan\theta_W |X_{1j}|^2 + |X_{2j}|^2 + 2\tan\theta_W Re(X_{1j}X_{2j}^*)\right], \tag{15.135}$$

and substitution of the explicit form of X gives

$$a_{f1}^{\tilde{\chi}^0} = \frac{m_\mu^2 G_F}{2\sqrt{2}\pi^2}\left(\frac{1}{2}\right) \tag{15.136}$$

$$a_{f2}^{\tilde{\chi}^0} = -\frac{m_\mu^2 G_F}{2\sqrt{2}\pi^2}\left(\frac{4}{3}\sin^4\theta_W - \frac{2}{3}\sin^2\theta_W + \frac{1}{6}\right). \tag{15.137}$$

The sum of the above gives

$$a_\mu^{\tilde{\chi}^0} = -\frac{m_\mu^2 G_F}{2\sqrt{2}\pi^2}\left(\frac{4}{3}\sin^4\theta_W - \frac{2}{3}\sin^2\theta_W - \frac{1}{3}\right). \tag{15.138}$$

From Eqs. (15.129) and (15.138), one finds

$$a_\mu^Z + a_\mu^{\tilde{\chi}^0} = 0, \tag{15.139}$$

and so the sum of the Z-boson contribution and the neutralino contributions cancel. Thus, we find that as expected the loop contributions to $g_\mu - 2$ in the supersymmetric limit vanish.

15.12 Supersymmetry CP Phases and CP-Even–CP-Odd Mixing in the Neutral Higgs Boson Sector

Another important effect of CP-violating phases is their role in determining the mass spectrum and CP properties of the neutral Higgs fields arising due to mixings of the CP-even–CP-odd Higgs bosons [49–54]. Consider the low-energy effective potential consisting of a renormalization group improved tree potential and a loop correction so that

$$V(H_1, H_2) = V_0 + \Delta V, \tag{15.140}$$

where the tree potential V_0 is given by

$$V_0 = m_1^2|H_1|^2 + m_2^2|H_2|^2 + (m_3^2 H_1.H_2 + \text{h.c.}) + \frac{(g_2^2 + g_1^2)}{8}|H_1|^4$$
$$+ \frac{(g_2^2 + g_1^2)}{8}|H_2|^2 - \frac{g_2^2}{2}|H_1.H_2|^2 + \frac{(g_2^2 - g_1^2)}{4}|H_1|^2|H_2|^2. \tag{15.141}$$

Here, $m_1^2 = m_{H_1}^2 + |\mu|^2$, $m_2^2 = m_{H_2}^2 + |\mu|^2$, and $m_3^2 = |\mu B|$, and $m_{H_{1,2}}$ and B are the soft supersymmetry breaking parameters, and ΔV is the one-loop correction to the effective potential so that [55]

$$\Delta V = \frac{1}{64\pi^2}str\left(M^4(H_1, H_2)\left(\log\frac{M^2(H_1, H_2)}{Q^2} - \frac{3}{2}\right)\right). \tag{15.142}$$

Here, the sum runs over all particles including the standard model particles as well as extra particles that arise in the minimal supersymmetric standard model (MSSM), i.e., the sfermions, charginos, and neutralinos.

Mixings between CP-even and CP-odd Higgs bosons cannot occur at the tree level, but are possible when loop corrections to the effective potential are included. To calculate the mixings we use the one-loop effective potential as given by Eqs. (15.140), (15.141), and (15.142). We assume that the $SU(2)$ Higgs doublets $H_{1,2}$ have non-vanishing vacuum expectation values v_1 and v_2 so that we can write

$$(H_1) = \begin{pmatrix} (v_1 + \phi_1 + i\psi_1)\sqrt{2} \\ H_1^- \end{pmatrix}$$

$$(H_2) = e^{i\theta_H} \begin{pmatrix} H_2^+ \\ (v_2 + \phi_2 + i\psi_2)/\sqrt{2} \end{pmatrix}. \tag{15.143}$$

For the present case with the inclusion of CP-violating effects, the variations with respect to the fields ϕ_1 and ϕ_2 give the following:

$$-\frac{1}{v_1}\left(\frac{\partial \Delta V}{\partial \phi_1}\right)_0 = m_1^2 + \frac{g_2^2 + g_1^2}{8}(v_1^2 - v_2^2) + m_3^2 \tan\beta \cos\theta_H \tag{15.144}$$

$$-\frac{1}{v_2}\left(\frac{\partial \Delta V}{\partial \phi_2}\right)_0 = m_2^2 - \frac{g_2^2 + g_1^2}{8}(v_1^2 - v_2^2) + m_3^2 \cot\beta \cos\theta_H. \tag{15.145}$$

Similarly, the variations with respect to ψ_1 and ψ_2 give us the following:

$$\frac{1}{v_1}\left(\frac{\partial \Delta V}{\partial \psi_2}\right)_0 = m_3^2 \sin\theta_H = \frac{1}{v_2}\left(\frac{\partial \Delta V}{\partial \psi_1}\right)_0, \tag{15.146}$$

where the subscript 0 indicates that the quantities are evaluated at the point $\phi_1 = \phi_2 = \psi_1 = \psi_2 = 0$. Only one of the two equations in Eq. (15.146) is independent.

We now discuss the contributions arising from the top and the stop exchanges. The stop mass-squared matrix $M_{\tilde{t}}^2$ is given by

$$M_{\tilde{t}}^2 = \begin{pmatrix} M_Q^2 + h_t^2|H_2^0|^2 + (g_2^2 - g_1^2/3)/4(|H_1^0|^2 - |H_2^0|^2) & h_t(A_t^* H_2^{0*} - \mu H_1^0) \\ h_t(A_t H_2^0 - \mu^* H_1^{0*}) & M_U^2 + h_t^2|H_2^0|^2 + \frac{g_1^2}{3}(|H_1^0|^2 - |H_2^0|^2) \end{pmatrix},$$

$$\tag{15.147}$$

where $A_t = |A_t|e^{i\alpha_{A_t}}$. The contribution to the one-loop effective potential from the top and the stop exchanges is given by

$$\Delta V(\tilde{t}, t) = \frac{1}{64\pi^2}\left(\sum_{a=1,2} 6M_{\tilde{t}_a}^4\left(\log\frac{M_{\tilde{t}_a}^2}{Q^2} - \frac{3}{2}\right) - 12m_t^4\left(\log\frac{m_t^2}{Q^2} - \frac{3}{2}\right)\right). \tag{15.148}$$

The mass-squared matrix of the neutral Higgs bosons is defined by

$$M_{ab}^2 = \left(\frac{\partial^2 V}{\partial \Phi_a \Phi_b}\right)_0, \tag{15.149}$$

where $\Phi_a(a = 1-4)$ are given by

$$\{\Phi_a\} = \{\phi_1, \phi_2, \psi_1, \psi_2\}. \tag{15.150}$$

Here, one finds that θ_H that appears in Eq. (15.142) is determined by

$$m_3^2 \sin \theta_H = \frac{1}{2} \beta_{h_t} |\mu||A_t| \sin \gamma_t f_1(m_{\tilde{t}_1}^2, m_{\tilde{t}_2}^2), \tag{15.151}$$

where $\beta_{h_t} = 3h_t^2/16\pi^2$ and $\gamma_t = \alpha_{A_t} + \theta_\mu$, and $f_1(x, y)$ is defined by

$$f_1(x, y) = -2 + \log \frac{xy}{Q^4} + \frac{y+x}{y-x} \log \frac{y}{x}. \tag{15.152}$$

The tree and loop contributions to M_{ab}^2 are given by

$$M_{ab}^2 = M_{ab}^{2(0)} + \Delta M_{ab}^2, \tag{15.153}$$

where $M_{ab}^{2(0)}$ are the contributions at the tree level and ΔM_{ab}^2 are the loop contributions, where

$$\Delta M_{ab}^2 = \frac{1}{32\pi^2} Str \left(\frac{\partial M^2}{\partial \Phi_a} \frac{\partial M^2}{\partial \Phi_b} \log \frac{M^2}{Q^2} + M^2 \frac{\partial^2 M^2}{\partial \Phi_a \partial \Phi_b} \log \frac{M^2}{eQ^2} \right)_0, \tag{15.154}$$

and where $e = 2.718$. Computation of the 4×4 Higgs mass-square matrix in the basis of Eq. (15.150) gives

$$\begin{pmatrix} M_{11} + \Delta_{11} & -M_{12} + \Delta_{12} & \Delta_{13}s_\beta & \Delta_{13}c_\beta \\ -M_{12} + \Delta_{12} & M_{22} + \Delta_{22} & \Delta_{23}s_\beta & \Delta_{23}c_\beta \\ \Delta_{13}s_\beta & \Delta_{23}s_\beta & M_{33}s_\beta^2 & M_{33}s_\beta c_\beta \\ \Delta_{13}c_\beta & \Delta_{23}c_\beta & M_{33}s_\beta c_\beta & M_{33}c_\beta^2 \end{pmatrix}, \tag{15.155}$$

where $M_{11} = M_Z^2 c_\beta^2 + M_A^2 s_\beta^2$, $M_{12} = (M_Z^2 + M_A^2)s_\beta c_\beta$, $M_{22} = M_Z^2 s_\beta^2 + M_A^2 c_\beta^2$, $M_{33} = M_A^2 + \Delta_{33}$, and $(c_\beta, s_\beta) = (\cos \beta, \sin \beta)$. Here, the explicit Q-dependence has been absorbed in m_A^2, which is given by

$$m_A^2 = (\sin \beta \cos \beta)^{-1}(-m_3^2 \cos \theta + \frac{1}{2}\beta_{h_t}|A_t||\mu| \cos \gamma_t f_1(m_{\tilde{t}_1}^2, m_{\tilde{t}_2}^2)). \tag{15.156}$$

In the supersymmetric limit, $M_{\tilde{\chi}_i^0} = (0, 0, M_Z, M_Z)$, $(M_{h^0}, M_{H^0}) = (M_Z, 0)$, $M_{\tilde{\chi}_i^+} = M_{H^+} = M_W$, and $M_{\tilde{q}_i} = m_q$, and in this limit all the radiative corrections to the scalar potential vanish. By choice of an appropriate basis the 4×4 matrix of Eq. (15.155) can be put in a block diagonal form with the 4×4 matrix appearing with one block diagonal entry consisting of a 3×3 block, which is the block of interest. To show this, we consider the new basis consisting of ϕ_1, ϕ_2, and ψ_{1D} and ψ_{2D}, where

$$\psi_{1D} = \sin \beta \psi_1 + \cos \beta \psi_2$$
$$\psi_{2D} = -\cos \beta \psi_1 + \sin \beta \psi_2. \tag{15.157}$$

In this basis the field ψ_{2D} decouples from the other three fields and appears as a massless Goldstone field. The remaining 3×3 Higgs mass-squared matrix in the basis ϕ_1, ϕ_2, ψ_{1D}, and ψ_{2D} takes the form

$$M_{\text{Higgs}}^2 = \begin{pmatrix} M_{11} + \Delta_{11} & -M_{12} + \Delta_{12} & \Delta_{13} \\ -M_{12} + \Delta_{12} & M_{22} + \Delta_{22} & \Delta_{23} \\ \Delta_{13} & \Delta_{23} & M_A^2 + \Delta_{33} \end{pmatrix}. \tag{15.158}$$

The mixing between the CP-even and the CP-odd states occurs due to non-vanishing Δ_{13} and Δ_{23}. An experimental observation of such mixings would point to CP violation beyond the one in the standard model and, further, would imply the presence of large CP phases in the supersymmetric sector since Δ_{13} and Δ_{23} arise only in the presence of such phases.

Next, we consider the contributions of the chargino loops to the Higgs masses. Actually, the charginos, the W-boson, and the charged Higgs boson form a sub-sector, as it is the splittings among these particles that lead to a non-vanishing contribution to the one-loop effective potential. The one-loop correction from this sector is given by

$$\Delta V(\tilde{\chi}^+, W, H^+) = \frac{1}{64\pi^2} \Bigg[\sum_{a=1,2} (-4) M_{\tilde{\chi}_a^+}^4 \left(\log \frac{M_{\tilde{\chi}_a^+}^2}{Q^2} - \frac{3}{2} \right)$$
$$+ 6 M_W^4 \left(\log \frac{M_W^2}{Q^2} - \frac{3}{2} \right) + 2 M_{H^+}^4 \log \left(\frac{M_{H^+}^2}{Q^2} - \frac{3}{2} \right) \Bigg]. \tag{15.159}$$

In the presence of CP phases the chargino mass matrix has the form

$$M_C = \begin{pmatrix} |\tilde{m}_2| e^{i\xi_2} & g_2 H_2^0 \\ g_2 H_1^0 & |\mu| e^{i\theta_\mu} \end{pmatrix}. \tag{15.160}$$

For the purposes of the analysis it is more convenient to deal with the matrix $M_C M_C^\dagger$, where

$$M_C M_C^\dagger = \begin{pmatrix} |\tilde{m}_2|^2 + g_2^2 |H_2^0|^2 & g_2(\tilde{m}_2 H_1^{0*} + \mu^* H_2^0) \\ g_2(\tilde{m}_2^* H_1^0 + \mu H_2^{0*}) & |\mu|^2 + g_2^2 |H_1^0|^2 \end{pmatrix}. \tag{15.161}$$

The chargino eigenvalues are given by

$$M_{\tilde{\chi}_{1,2}^+}^2 = \frac{1}{2} [|\tilde{m}_2|^2 + |\mu|^2 + g_2^2(|H_2^0|^2 + |H_1^0|^2)]]$$
$$\pm \frac{1}{2} [(|\tilde{m}_2|^2 - |\mu|^2 + g_2^2(|H_2^0|^2 - |H_1^0|^2))^2 + 4 g_2^2 |\tilde{m}_2 H_1^{0*} + \mu^* H_2^0|^2]^{\frac{1}{2}}. \tag{15.162}$$

We note that in the supersymmetric limit $M_{\tilde{\chi}_{1,2}^+} = M_{H^+} = M_W$, and the loop correction (Eq. (15.159)) vanishes. Further, as already noted, the inclusion of the W and the H^+ exchange along with the chargino exchange is also needed to achieve an approximate Q-independence of the corrections to the Higgs masses and mixings from this sector. In this sense, $M_{\tilde{\chi}_a^+}$, H^+, and W form a sub-sector, and that is the reason for considering this set in Eq. (15.159). With the inclusion of the stop and the chargino contributions, θ_H is given by [54]

$$m_3^2 \sin \theta_H = \frac{1}{2}\beta_{h_t}|\mu||A_t|\sin \gamma_t f_1(m_{\tilde{t}_1}^2, m_{\tilde{t}_2}^2) - \frac{g_2^2}{16\pi^2}|\mu||\tilde{m}_2|\sin \gamma_2 f_1(m_{\tilde{\chi}_1}^2, m_{\tilde{\chi}_2}^2),$$
(15.163)

where $\gamma_2 = \xi_2 + \theta_\mu$, and there is a correction to the CP-odd Higgs boson mass from the chargino exchange contribution so that

$$\Delta m_A^2 = \frac{g_2^2}{16\pi^2}|\tilde{m}_2||\mu|\cos \gamma_2 f_1(m_{\tilde{\chi}_1^+}^2, m_{\tilde{\chi}_2^+}^2).$$
(15.164)

For Δ_{ij}, one has

$$\Delta_{ij} = \Delta_{ij\tilde{t}} + \Delta_{ij\tilde{\chi}^+} + \Delta_{ij\tilde{\chi}^0},$$
(15.165)

where $\Delta_{ij\tilde{t}}$ is the contribution from the stop exchange in the loops, $\Delta_{ij\tilde{\chi}^+}$ is the contribution from the chargino sector in the loops, and $\Delta_{ij\tilde{\chi}^0}$ is the contribution from the neutralino sector in the loops. We do not explicitly show the analysis of the chargino and the neutralino exchange contributions (details of which can be found in the literature [54]).

15.13 Effect of CP Phases on Neutralino Dark Matter

We now discuss the effect of CP phases on the direct detection of dark matter. The effective four-Fermi interaction that governs the scattering of neutralinos off quarks takes the form

$$\mathcal{L}_{eff} = \bar{\tilde{\chi}}\gamma_\mu\gamma_5\tilde{\chi}\bar{q}\gamma^\mu(AP_L + BP_R)q + C\bar{\tilde{\chi}}\tilde{\chi}m_q\bar{q}q$$
$$+ D\bar{\tilde{\chi}}\gamma_5\tilde{\chi}m_q\bar{q}\gamma_5 q + E\bar{\tilde{\chi}}i\gamma_5\tilde{\chi}m_q\bar{q}q + F\bar{\tilde{\chi}}\tilde{\chi}m_q\bar{q}i\gamma_5 q.$$
(15.166)

The fundamental cubic interaction which gives rise to the four-Fermi interaction above is given by

$$-\mathcal{L} = \bar{q}[C_{qL}P_L + C_{qR}P_R]\tilde{\chi}\tilde{q}_1 + \bar{q}[C'_{qL}P_L + C'_{qR}P_R]\tilde{\chi}\tilde{q}_2 + \text{h.c.}$$
(15.167)

where

$$C_{qL} = \sqrt{2}(\alpha_{q0}D_{q11} - \gamma_{q0}D_{q21})\ C_{qR} = \sqrt{2}(\beta_{q0}D_{q11} - \delta_{q0}D_{q21})$$
$$C'_{qL} = \sqrt{2}(\alpha_{q0}D_{q12} - \gamma_{q0}D_{q22}),\ C'_{qR} = \sqrt{2}(\beta_{q0}D_{q12} - \delta_{q0}D_{q22}),$$
(15.168)

and where α, β, γ, and δ are defined by

$$\alpha_{u(d)j} = \frac{gm_{u(d)}X_{4(3)j}}{2m_W \sin\beta(\cos\beta)}$$

$$\beta_{u(d)j} = eQ_{u(d)j}X_{1j}^{'*} + \frac{g}{\cos\theta_W}X_{2j}^{'*}(T_{3u(d)} - Q_{u(d)}\sin^2\theta_W)$$

$$\gamma_{u(d)j} = eQ_{u(d)j}X_{1j}' - \frac{gQ_{u(d)}\sin^2\theta_W}{\cos\theta_W}X_{2j}'$$

$$\delta_{u(d)j} = \frac{-gm_{u(d)}X_{4(3)j}^*}{2m_W \sin\beta(\cos\beta)}. \tag{15.169}$$

In the above, g is the $SU(2)_L$ gauge coupling and X_1' and X_2' are defined by

$$X_{1j}' = X_{1j}\cos\theta_W + X_{2j}\sin\theta_W$$
$$X_{2j}' = -X_{1j}\sin\theta_W + X_{2j}\cos\theta_W. \tag{15.170}$$

The effect of the CP-violating phases enters via the neutralino eigenvector components X_{ij} and via the matrix D_{qij} that diagonalizes the squark mass-squared matrix.

Integration over the squark fields leads to the effective Lagrangian for the scattering of neutralinos off quarks via the exchange of squarks. The effective interaction is given by [56,57].

$$\mathcal{L}_{eff} = \frac{1}{M_{\tilde{q}_1}^2 - M_{\tilde{\chi}}^2}\bar{\tilde{\chi}}[C_{qL}^*P_R + C_{qR}^*P_L]q\bar{q}[C_{qL}P_L + C_{qR}P_R]\tilde{\chi}$$

$$+ \frac{1}{M_{\tilde{q}_2}^2 - M_{\tilde{\chi}}^2}\bar{\tilde{\chi}}[C_{qL}^{*'}P_R + C_{qR}^{*'}P_L]q\bar{q}[C_{qL}'P_L + C_{qR}'P_R]\tilde{\chi} \tag{15.171}$$

Using the Fierz rearrangement, one can obtain the coefficients A, B, C, D, E, and F that appear in Eq. (15.166) in a straightforward fashion [56,57]. The first two terms (A, B) are spin-dependent interactions and arise from the Z-boson and the sfermion exchanges. For these, one has

$$A = \frac{g^2}{4M_W^2}[|X_{30}|^2 - |X_{40}|^2][T_{3q} - e_q\sin^2\theta_W] - \frac{|C_{qR}|^2}{4(M_{\tilde{q}_1}^2 - M_{\tilde{\chi}}^2)} - \frac{|C_{qR}'|^2}{4(M_{\tilde{q}_2}^2 - M_{\tilde{\chi}}^2)} \tag{15.172}$$

$$B = -\frac{g^2}{4M_W^2}[|X_{30}|^2 - |X_{40}|^2]e_q\sin^2\theta_W + \frac{|C_{qL}|^2}{4(M_{\tilde{q}_1}^2 - M_{\tilde{\chi}}^2)} + \frac{|C_{qL}'|^2}{4(M_{\tilde{q}_2}^2 - M_{\tilde{\chi}}^2)}. \tag{15.173}$$

Contributions to C, D, E, and F arise from sfermion exchange and from neutral Higgs exchange so that

$$C = C_{\tilde{f}} + C_{h^0} + C_{H^0}. \tag{15.174}$$

A derivation of these gives

$$C_{\tilde{f}}(u,d) = -\frac{1}{4m_q}\frac{1}{M_{\tilde{q}1}^2 - M_{\tilde{\chi}}^2}Re[C_{qL}C_{qR}^*] - \frac{1}{4m_q}\frac{1}{M_{\tilde{q}2}^2 - M_{\tilde{\chi}}^2}Re[C_{qL}'C_{qR}'^*] \tag{15.175}$$

$$C_{h^0}(u,d) = -(+)\frac{g^2}{4M_W M_{h^0}^2}\frac{\cos\alpha(\sin\alpha)}{\sin\beta(\cos\beta)}Re\,\sigma \tag{15.176}$$

$$C_{H^0}(u,d) = \frac{g^2}{4M_W M_{H^0}^2}\frac{\sin\alpha(\cos\alpha)}{\sin\beta(\cos\beta)}Re\,\rho, \tag{15.177}$$

where σ and ρ are defined by

$$\sigma = X_{40}^*(X_{20}^* - \tan\theta_W X_{10}^*)\cos\alpha + X_{30}^*(X_{20}^* - \tan\theta_W X_{10}^*)\sin\alpha \tag{15.178}$$

$$\rho = -X_{40}^*(X_{20}^* - \tan\theta_W X_{10}^*)\sin\alpha + X_{30}^*(X_{20}^* - \tan\theta_W X_{10}^*)\cos\alpha, \tag{15.179}$$

and where α, as usual, is the Higgs mixing angle. Finally, the terms D, E, and F are given by

$$D(u,d) = C_{\tilde{f}}(u,d) + \frac{g^2}{4M_W}\frac{\cot\beta(\tan\beta)}{m_{A_0}^2}Re\,\omega \tag{15.180}$$

$$E(u,d) = T_{\tilde{f}}(u,d) + \frac{g^2}{4M_W}\left[-(+)\frac{\cos\alpha(\sin\alpha)}{\sin\beta(\cos\beta)}\frac{Im\,\sigma}{m_{h^0}^2} + \frac{\sin\alpha(\cos\alpha)}{\sin\beta(\cos\beta)}\frac{Im\,\rho}{m_{H^0}^2}\right] \tag{15.181}$$

$$F(u,d) = T_{\tilde{f}}(u,d) + \frac{g^2}{4M_W}\frac{\cot\beta(\tan\beta)}{m_{A^0}^2}Im\,\omega. \tag{15.182}$$

Here, $T_{\tilde{f}}$ and ω are defined by

$$T_{\tilde{f}}(q) = \frac{1}{4m_q}\frac{1}{M_{\tilde{q}1}^2 - M_{\tilde{\chi}}^2}Im[C_{qL}C_{qR}^*] + \frac{1}{4m_q}\frac{1}{M_{\tilde{q}2}^2 - M_{\tilde{\chi}}^2}Im[C_{qL}'C_{qR}'^*] \tag{15.183}$$

and

$$\omega = -X_{40}^*(X_{20}^* - \tan\theta_W X_{10}^*)\cos\beta + X_{30}^*(X_{20}^* - \tan\theta_W X_{10}^*)\sin\beta. \tag{15.184}$$

In the limit when CP phases vanish, the results above limit to the standard neutralino–quark scattering in the absence of CP phases [58]. The effect of CP phases resides in the matrices D_q that diagonalize the squark mass-squared matrix and in the matrix X that diagonalizes the neutralino mass matrix.

15.14 Appendix: Renormalization Group Evolution of the EDM

The EDM is scale dependent, and we now discuss its evolution as we go down from the electroweak scale to scales below. The effective low-energy CP-violating

(CPV) Hamiltonian which generates the EDM is given by

$$\mathcal{H}_{CPV} = \sum_q C_E^q(\mu) O_E^q + \sum_q C_C^q(\mu) Q_C^q(\mu) + C_G(\mu) O_G(\mu), \qquad (15.185)$$

where we sum over the light quarks, and μ is the running scale. The μ-dependence must cancel between the operators O and the coefficients C. For our case, O_E^q, O_C^q, and O_G are given by

$$O_E^q = -\frac{i}{2} \bar{q} \sigma_{\mu\nu} \gamma_5 q F^{\mu\nu} \qquad (15.186)$$

$$O_C^q = -\frac{i}{2} \bar{q} \sigma_{\mu\nu} \gamma_5 T^\alpha q G_{\mu\nu\alpha} \qquad (15.187)$$

$$O_G = -\frac{1}{6} f_{\alpha\beta\gamma} G_{\alpha\mu\rho} G^\rho_{\beta\nu} G_{\gamma\lambda\sigma} \epsilon^{\mu\nu\lambda\sigma}. \qquad (15.188)$$

The renormalization group evolution at one loop for the electric dipole operator and for the color dipole operators is easily obtained using their anomalous dimension and using the fact that these operators are eigenstates under the renormalization group evolution. An evolution of these operators from $Q = M_Z$ to a lower scale μ gives

$$\mathcal{O}_i^q(\mu) = \Gamma^{-\gamma_i/\beta} \mathcal{O}_i^q(Q), \qquad (15.189)$$

where

$$\Gamma = \frac{g_s(\mu)}{g_s(Q)}$$

$$\gamma_E = 8/3$$

$$\gamma_C = (29 - 2N_f)/3$$

$$\beta = (33 - 2N_f)/3 \qquad (15.190)$$

and N_f is the number of light quarks at the scale μ.

The purely gluonic dimension-six operator obeys the following renormalization group equation [36, 59–62]

$$\mu \frac{\partial}{\partial \mu} \mathcal{O}_G = \frac{\alpha_s(\mu)}{4\pi} \left[\gamma_G \mathcal{O}_G + 6 \sum_q m_q(\mu) \mathcal{O}_C^q \right], \qquad (15.191)$$

where $\gamma_G = -3 - 2N_f$ while $\alpha_s(\mu)$ is the running gauge-couplings constant and $m_q(\mu)$ is the running quark mass which satisfy the renormalization group evolution equations

$$\mu \frac{\partial}{\partial \mu} g_s(\mu) = -\beta \frac{\alpha_s(\mu)}{4\pi} g_s(\mu) \qquad (15.192)$$

and

$$\mu \frac{\partial}{\partial \mu} m_q(\mu) = \gamma_m \frac{\alpha_s(\mu)}{4\pi} m_q(\mu), \qquad (15.193)$$

where $\gamma_m = -8$. A solution to the equations above shows that the operator $O_G(\mu)$ depends on $O_G(M)$ and $O_C^q(M)$. Since the renormalization group evolution must cancel between the operators and the coefficients, we have, for the case of the dipole operators, the following relation between the dipole moments at a high scale $Q = M_Z$ and a low scale μ:

$$d^{(E,C,G)}(\mu) \simeq \Gamma^{\gamma_{(E,C,G)}/\beta} d^{(E,C,G)}(M).\qquad(15.194)$$

One issue to keep in mind is that matching conditions must be used as we cross heavy quark thresholds. These are

$$d^G(m_Q^-) = d^G(m_Q^+) + d^C(m_Q)\frac{1}{8\pi}\frac{\alpha_s(m_Q)}{m_Q}.\qquad(15.195)$$

With the above renormalization group evolution we can compute the EDMs at low scales where EDM experiments are carried out. In general, there would be additional mixings among operators including between the electric and the chromoelectric dipole operators. A more detailed account of this is given by Degrassi *et al.* [38]. For further reading on topics related to CP violation and EDM see the literature [33, 63].

15.15 Problems

1. Show that the CP phases α, β, and γ defined by Eq. (15.4) satisfy the unitarity constraint

$$\alpha + \beta + \gamma = \pi.\qquad(15.196)$$

2. Show that the CP phase-dependent chargino mass matrix given by Eq. (15.48) can be diagonalized with the CP phase-dependent biunitary matrices given by Eqs. (15.49), (15.51), and (15.54)
3. Using the interaction Lagrangian of a heavy scalar with a Dirac fermion given by Eq. (15.65), determine the EDM of the Dirac fermion and verify that Eq. (15.67), which shows that the EDM is proportional to $Im(L_{ia}R_{ia}^*)$.
4. Show that the contribution of the gluino exchange contribution to the chromoelectric dipole operator is given by Eq. (15.85).
5. Verify that in the limit of no supersymmetry breaking the chargino exchange contribution is given by Eq. (15.125), and thus verify the cancellation of the chargino and the W exchange contributions as given in Eq. (15.128).

References

[1] The brief early history of CP recounted here is based on the published early works and in part on a private communication in 2007 with the late Norman Ramsay.
[2] J. H. Smith, E. M. Purcell, and N. F. Ramsey, *Phys. Rev.* **108**, 120 (1957).
[3] J. Smith, Ph.D. Thesis, Harvard University, 1951.
[4] N. F. Ramsey, Molecular Beams, Oxford University Press (1956).

[5] T. D. Lee and C. N. Yang, *Phys. Rev.* **104**, 254 (1956).

[6] C. S. Wu, E. Ambler, R. W. Hayward, D. D. Hoppes, and R. P. Hudson, *Phys. Rev.* **105**, 1413 (1957).

[7] N. F. Ramsey, *Phys. Rev.* **109**, 822 (1958).

[8] J. D. Jackson, S. B. Treiman, and H. W. Wyld, *Phys. Rev.* **106** 517 (1957).

[9] J. H. Christenson, J. W. Cronin, V. L. Fitch, and R. Turlay, *Phys. Rev. Lett.* **13**, 138 (1964).

[10] A. D. Sakharov, *Pisma Zh. Eksp. Teor. Fiz.* **5**, 32 (1967).

[11] C. Jarlskog, *Phys. Rev. Lett.* **55**, 1039 (1985).

[12] L. Wolfenstein, *Phys. Rev. Lett.* **51**, 1945 (1983).

[13] R. D. Peccei and H. R. Quinn, *Phys. Rev. D* **16**, 1791 (1977).

[14] S. Weinberg, *Phys. Rev. Lett.* **40**, 223 (1978).

[15] F. Wilczek, *Phys. Rev. Lett.* **40**, 279 (1978).

[16] S. M. Barr, *Phys. Rev. Lett.* **53**, 329 (1984).

[17] A. E. Nelson, *Phys. Lett. B* **136**, 387 (1984).

[18] J. Beringer *et al.* [Particle Data Group Collaboration], *Phys. Rev. D* **86**, 010001 (2012).

[19] S. Bertolini, M. Fabbrichesi, and J. O. Eeg, *Rev. Mod. Phys.* **72**, 65 (2000).

[20] L. Wolfenstein, *Phys. Rev. Lett.* **13**, 562 (1964).

[21] A. J. Buras, F. Schwab, and S. Uhlig, *Rev. Mod. Phys.* **80**, 965 (2008).

[22] V. V. Anisimovsky *et al.* [E949 Collaboration], *Phys. Rev. Lett.* **93**, 031801 (2004).

[23] K. Fujikawa and R. Shrock, *Phys. Rev. Lett.* **45**, 963 (1980).

[24] W. J. Marciano and A. I. Sanda, *Phys. Lett. B* **67**, 303 (1977).

[25] B. W. Lee and R. E. Shrock, *Phys. Rev. D* **16**, 1444 (1977).

[26] F. Hoogeveen, *Nucl. Phys. B* **341** (1990) 322;

[27] A. Soni and R. M. Xu, *Phys. Rev. Lett.* **69**, 33 (1992).

[28] A. Czarnecki and B. Krause, *Phys. Rev. Lett.* **78**, 4339 (1997).

[29] M. E. Pospelov and I. B. Khriplovich, *Sov. J. Nucl. Phys.* **53** (1991) 638.

[30] C. A. Baker, D. D. Doyle, P. GeHenbort, *et al.*, *Phys. Rev. Lett.* **97**, 131801 (2006).

[31] J. Baron *et al.* [ACME Collaboration], *Science* **343**(6168), 269 (2014).

[32] T. Ibrahim and P. Nath, *Phys. Rev. D* **57**, 478 (1998).

[33] T. Ibrahim and P. Nath, *Rev. Mod. Phys.* **80**, 577 (2008).

[34] T. Ibrahim and P. Nath, *Phys. Lett. B* **418**, 98 (1998).

[35] S. Weinberg, *Phys. Rev. Lett.* **63** (1989) 2333.

[36] J. Dai, H. Dykstra, R. G. Leigh, S. Paban, and D. Dicus, *Phys. Lett. B* **237**, 216 (1990) [erratum, *ibid.* **242**, 547 (1990)].

[37] A. Manohar and H. Georgi, *Nucl. Phys. B* **234**, 189 (1984).

[38] G. Degrassi, E. Franco, S. Marchetti, and L. Silvestrini, *JHEP* **0511**, 044 (2005).

[39] I. B. Khriplovich and K. N. Zyablyuk, *Phys. Lett. B* **383**, 429 (1996).

[40] J. Ellis, S. Ferrara, and D. V. Nanopoulos, *Phys. Lett. B* **114**, 231 (1982); W. Buchmuller and D. Wyler, *Phys. Lett. B* **121**, 321 (1983); F. del'Aguila, M. B. Gavela, J. A. Grifols, and A. Mendez, *Phys. Lett. B* **126**, 71 (1983); J. Polchinski and M. B. Wise, *Phys. Lett. B* **125**, 393 (1983); E. Franco and M. Mangano, *Phys. Lett. B* **135**, 445 (1984).

[41] P. Nath, Phys. Rev. Lett. **66**, 2565 (1991); Y. Kizukuri and N. Oshimo, *Phys. Rev. D* **46**, 3025 (1992).

[42] T. Ibrahim and P. Nath, *Phys. Rev. D* **58**, 111301 (1998).

[43] T. Falk and K Olive, *Phys. Lett. B* **439**, 71 (1998).

[44] M. Brhlik, G. J. Good, and G. L. Kane, *Phys. Rev. D* **59**, 115004 (1999).

[45] T. Ibrahim and P. Nath, *Phys. Rev. D* **61**, 095008 (2000).

[46] T. Ibrahim and P. Nath, *Phys. Rev. D* **62**, 015004 (2000).

[47] T. Ibrahim, U. Chattopadhyay, and P. Nath, *Phys. Rev. D* **64**, 016010 (2001).

[48] K. Fujikawa, B. W. Lee, and A. I. Sanda, *Phys. Rev. D* **6**, 2923 (1972); R. Jackiw and S. Weinberg, *Phys. Rev. D* **5**, 2473 (1972); G. Altarelli, N. Cabibbo, and L. Maiani, *Phys. Lett. B* **40**, 415 (1972); I. Bars and M. Yoshimura, *Phys. Rev. D* **6**, 374 (1972); W. A. Bardeen, R. Gastmans, and B. E. Lautrup, *Nucl. Phys. B* **46**, 315 (1972).

[49] A. Pilaftsis, *Phys. Lett. B* **435**, 88 (1998).

[50] A. Pilaftsis, *Phys. Rev. D* **58**, 096010 (1998).

[51] A. Pilaftsis and C. E. M. Wagner, *Nucl. Phys. B* **553**, 3 (1999).

[52] D. A. Demir, *Phys. Rev. D* **60**, 055006 (1999).

[53] S. Y. Choi, M. Drees, and J. S. Lee, *Phys. Lett. B* **481**, 57 (2000).

[54] T. Ibrahim and P. Nath, *Phys. Rev. D* **63**, 035009 (2001); *Phys. Rev. D* **66**, 015005 (2002); S. P. Martin, *Phys. Rev. D* **66**, 096001 (2002); M. S. Carena, J. R. Ellis, A. Pilaftsis, and C. E. M. Wagner, *Nucl. Phys. B* **586**, 92 (2000); M. Frank, T. Hahn, S. Heinemeyer, W. Hollik, H. Rzehak, and G. Weiglein, *JHEP* **0702**, 047 (2007).

[55] S. R. Coleman and E. J. Weinberg, *Phys. Rev. D* **7**, 1888 (1973).

[56] U. Chattopadhyay, T. Ibrahim, and P. Nath, *Phys. Rev. D* **60**, 063505 (1999).

[57] T. Falk, A. Ferstl, and K. A. Olive, *Phys. Rev. D* **59**, 055009 (1999).

[58] G. Jungman, M. Kamionkowski, and K. Griest, *Phys. Rep.* **267**, 195 (1996).

[59] E. Braaten, C. S. Li, and T. C. Yuan, *Phys. Rev. Lett.* **64**, 1709 (1990).

[60] E. Braaten, C. S. Li, and T. C. Yuan, *Phys. Rev. D* **42**, 276 (1990).

[61] R. L. Arnowitt, J. L. Lopez, and D. V. Nanopoulos, *Phys. Rev. D* **42**, 2423 (1990).

[62] G. Boyd, A. K. Gupta, S. P. Trivedi, and M. B. Wise, *Phys. Lett. B* **241**, 584 (1990).

[63] M. Pospelov and A. Ritz, *Ann. Phys.* **318**, 119 (2005).

16

Proton Stability in Supergravity Unified Theories

Baryon number in particle theory has been of interest for several decades. In 1929 Herman Weyl [1] conjectured such a conservation, and later in 1938 Stueckelberg [2] made it more concrete. This was done by postulating different transformation laws for light particles and heavy particles where the leptons were the light and the nucleons the heavy particles. The conservation of baryons was put on an equal footing to the conservation of the electric charge in further work by Wigner [3]. On the experimental side, the earliest work on putting a lower limit on the proton lifetime was by Goldhaber [4]. In 1954 Goldhaber proposed that spontaneous fission of Th^{232} after excitation by nucleon decay would be followed by a rearrangement energy due to the loss of a nucleon, and this would cause fission of the residual nucleus. This allowed Goldhaber to put a lower limit of 1.4×10^{14} years on the nucleon lifetime. This limit was soon improved to a lower limit of 1×10^{22} years for bound nucleons in a direct experiment using a liquid scintillation counter 30 m below the Earth's surface by Reines, Cowan, and Goldhaber [5] (for a review of the early history of experiment on proton lifetime limits see Gurr *et al.* [6] and Perkins [7], and for a broad overview of proton stability in grand unified theories, strings, and branes see Nath and Fileviez Perez [8]). On the theoretical side, an issue arose regarding the absolute conservation of baryon number. It was pointed out by Lee and Yang [9] that if heavy particle number is the result of a gauge invariance, there would be a long-range force associated with such a conservation, but no such long range force is observed. On the other hand, if baryon number is a global symmetry, there is no reason that a global symmetry would guarantee an absolute conservation of baryon number. Indeed, such a symmetry would be violated by anomalies [10] and by gravitational interactions [11–14], specifically worm hole effects [15]. The first concrete model for proton decay via a possible new superweak interaction was discussed by Yamaguchi in 1959 [16]. This work stimulated the first deep underground proton decay experiment in 1960 [7,17].

A further argument in favor of baryon number violation was set forth by Sakharov [18] to explain the excess of baryons over anti-baryons in the universe. As discussed in Chapters 1 within Big Bang cosmology, Sakharov postulated three criteria for the generation of a baryon excess. One of these criteria requires a fundamental interaction which violates baryon number, which along with violations of C and CP and non-equilibrium processes can lead to a baryon asymmetry. In the standard model, the baryon number is conserved at the classical level because the standard model has a $U(1)_B$ global symmetry at the classical level. However, in the standard model $U(1)_B$ is anomalous, and the global symmetry is broken at the quantum level [10], which leads to a non-conservation of the baryonic current J_B^μ. Thus, for n_f number of generations the divergence of the baryonic current is given by

$$\partial_\mu J_B^\mu = \frac{n_f g^2}{16\pi^2} tr\, F_{\mu\nu} \tilde{F}^{\mu\nu}, \tag{16.1}$$

where $F_{\mu\nu}$ is the gauge field strength and $\tilde{F}_{\mu\nu}$ is its dual, i.e.,

$$\tilde{F}_{\mu\nu} = \frac{1}{2}\epsilon_{\mu\nu\lambda\rho}\, F^{\lambda\rho}. \tag{16.2}$$

Here $\mu, \nu, \ldots$ refer to the four dimensions of Minkowskian space. Because of the anomaly in the baryonic current, a violation of baryon number can arise due to instanton transitions between degenerate $SU(2)_L$ gauge vacua. The instanton processes thus induce a B-violating effective Lagrangian which has the form (for details see, e.g., Espinosa [19]):

$$O_{eff} = c\left(\frac{1}{M_W}\right)^{14} e^{-\frac{2\pi}{\alpha_2}} \prod_{\acute{a}=1}^{3}(\epsilon_{abc}\, Q_{aL}^{\acute{a}}\, Q_{bL}^{\acute{a}}\, Q_{cL}^{\acute{a}}\, L_L^{\acute{a}}), \tag{16.3}$$

where $\acute{a}$ is the generation index and a, b, and c are color indices which run over $1, 2$, and 3. The operator of Eq. (16.3) preserves $B - L$ but violates baryon number B and lepton number L by three units, i.e., $\Delta B = 3 = \Delta L$. It is to be noted that the interaction carries a front factor $\exp(-2\pi/\alpha_2) \sim 10^{-173}$, which is very small and implies that the rate of any baryon- and lepton-number-violating process from this interaction regardless of other particulars will be very small.

However, baryon and lepton number violation can arise from grand unified theories [20–22] because of the feature that quarks and leptons belong to a common multiplet and they can convert from one to the other by exchange of heavy fields. Thus, one of the predictions of grand unified theories is the decay of the proton. After the emergence of the grand unified theories there was renewed interest for the search for proton decay, and several deep underground experiments were initiated, which include the Kolar Gold Field experiment [23], Nusex [24], Frejus [25], Soudan [26], IMB [27], Kamiokande [28], and Super-Kamiokande [29]. Experimentally, the proton decay experiment is still on going, and future experiments such as the Hyper-Kamiokande are at the proposal stage.

In $SU(5)$, the interactions preserve $B-L$, and thus irrespective of the details of other specifics of grand unified models one can write down the allowed dimension-six operators which violate lepton and baryon number but preserve $B-L$ using just the invariance under the standard model gauge group [30, 31]

$$O_1 = C_1\, \epsilon_{abc}\, \epsilon_{\alpha\beta}\, \overline{u^C_{a\acute{a}L}}\, \gamma^\mu\, Q_{b\alpha\acute{a}L}\, \overline{e^C_{\acute{b}L}}\, \gamma_\mu\, Q_{c\beta\acute{b}L} \tag{16.4}$$

$$O_2 = C_2\, \epsilon_{abc}\, \epsilon_{\alpha\beta}\, \overline{u^C_{a\acute{a}L}}\, \gamma^\mu\, Q_{b\alpha\acute{a}L}\, \overline{d^C_{\acute{b}L}}\, \gamma_\mu\, L_{c\beta\acute{b}L} \tag{16.5}$$

$$O_3 = C_2\, \epsilon_{abc}\, \epsilon_{\alpha\beta}\, \overline{d^C_{a\acute{a}L}}\, \gamma^\mu\, Q_{b\alpha\acute{a}L}\, \overline{u^C_{\acute{b}L}}\, \gamma_\mu\, L_{c\beta\acute{b}L} \tag{16.6}$$

$$O_4 = C_2\, \epsilon_{abc}\, \epsilon_{\alpha\beta}\, \overline{d^C_{a\acute{a}L}}\, \gamma^\mu\, Q_{b\alpha\acute{a}L}\, \overline{\nu^C_{\acute{b}L}}\, \gamma_\mu\, Q_{c\beta\acute{b}L}, \tag{16.7}$$

where a, b, and c are color indices, α and β are $SU(2)$ indices, $\acute{a}$ and $\acute{b}$ are generation indices, and C_i $(i = 1, \cdots, 4)$ are the dimensioned front factors with dimensions of inverse of the square mass where the size of the mass typically is the grand unification scale. For example, the operator of Eqs. (16.4) and (16.5) arise via the exchange of (X, Y) gauge bosons in $SU(5)$ while the operator of Eqs. (16.6) and (16.7) arise from the exchange of the (X', Y') gauge boson for the case of flipped $SU(5)$. All of the above operators are present in $SO(10)$. The operators of Eqs. (16.4)–(16.7) satisfy the selection rule $\Delta S/\Delta B = 0$ and -1 and, further, the fact that the operator conserves $B - L$ implies that a proton will decay into an anti-lepton. It is also of interest that proton decay, e.g., $p \to \pi^0 e^+$, would also test the charge equality of the proton charge and of the electron [32]. Since the electric charge conservation is the consequence of a $U(1)_{em}$ gauge invariance, the equality of the charges would have to be exact – which is another prediction of grand unified theories.

16.1 Proton Decay from Lepto-quarks

We now discuss proton decay arising from the exchange of the X and Y gauge bosons within $SU(5)$. Here, we have interactions of the gauge bosons with the $\bar{5}+10$ of matter fields. Let M'_i represent the $\bar{5}$-plets and let $M_{ij} = -M_{ji}$ represent the 10-plets of fields where $(i, j = 1$–$5)$ are $SU(5)$ indices. The Lagrangian interaction is given by

$$\mathcal{L}_{int} = g_5 \bar{M}'_i A_{ij} M'_j + g_5 (\overline{M}_{ij} A_{ik} M_{kj} + \overline{M}_{ij} A_{jk} M_{ik}), \tag{16.8}$$

where $A \equiv \gamma^\mu \sum_{x=1}^{24} A^x_\mu T_x$, and T_x are the generators of $SU(5)$ and A^x_μ are the $SU(5)$ gauge bosons. Using the explicit form of the Lie-valued quantity A, we can extract the baryon- and lepton-number-violating interactions arising from $X^c Y^c$ couplings of the 5-plet of matter. They are given by

$$\mathcal{L}_{BLV-5} = \frac{g_5}{\sqrt{2}} \left(\bar{d}_{R\alpha} \gamma^\mu X^c_{\mu\alpha} e^+_R - \bar{d}_{R\alpha} \gamma^\mu Y^c_{\mu\alpha} \nu^c_{eR} + \text{h.c.} \right). \tag{16.9}$$

Similarly, we can compute the baryon- and lepton-number-violating interactions arising from the gauge couplings of the 10-plet of matter fields in $SU(5)$. Here, we get

$$\mathcal{L}_{BLV\text{-}10} = \frac{g_5}{\sqrt{2}} \left(\bar{d}_{L\alpha} \gamma^\mu X^c_{\mu a} e^+_L + \epsilon_{abc} \bar{u}^c_{Lc} \gamma^\mu X^c_{\mu a} u_{Lb} \right.$$

$$\left. - \bar{u}_{L\alpha} \gamma^\mu Y^c_{\mu\alpha} \nu^c e^+_L + \epsilon_{abc} \bar{u}^c_{Lc} \gamma^\mu Y^c_{\mu a} d_{Lb} + \text{h.c.} \right). \qquad (16.10)$$

Combining the terms in Eqs. (16.9) and (16.10), we can write the total interaction so that

$$\mathcal{L}_{BLV} = \mathcal{L}_{BLV\text{-}5} + \mathcal{L}_{BLV\text{-}10}$$

$$= \frac{g_5}{\sqrt{2}} \left[X_{\mu a} J^\mu_{Xa} + Y_{\mu a} J^\mu_{Ya} + \text{h.c.} \right], \qquad (16.11)$$

where J^μ_{Xa} is

$$J^\mu_{Xa} = \bar{d}_{Ra} \gamma^\mu e^+_R + \bar{d}_{La} \gamma^\mu e^+_L + \epsilon_{abc} \bar{u}^c_{Lc} \gamma^\mu u_{Lb} \qquad (16.12)$$

and J^μ_{Ya} is

$$J^\mu_{Ya} = -\bar{u}_{La} \gamma^\mu e^+_L - \bar{d}_{Ra} \gamma^\mu \nu^c_R + \epsilon_{abc} \bar{u}^c_{Lc} \gamma^\mu d_{Lb}. \qquad (16.13)$$

Using these interactions we can determine the decay of the proton.

The interactions given above are, of course, at the grand unification scale, and we need to carry out a renormalization group evolution from the grand unification scale down to the low mass scale of ~ 1 GeV. This involves two renormalization group evolutions, one from the grand unification scale to the electroweak scale M_Z and the other from the electroweak scale down to around the hadronic scale. The first gives the so-called short distance renormalization factor A_S while the second gives the long-distance renormalization factor A_L. Estimates of A_S and A_L give, at the one-loop level [33–35],

$$A_S = \left(\frac{\alpha_1(M_Z)}{\alpha_5} \right)^{-23/198} \left(\frac{\alpha_2(m_Z)}{\alpha_5} \right)^{-\frac{3}{2}} \left(\frac{\alpha_3(M_Z)}{\alpha_5} \right)^{\frac{4}{9}} \qquad (16.14)$$

and

$$A_L = \left(\frac{\alpha_3(\mu_{\text{had}})}{\alpha_3(m_c)} \right)^{\frac{2}{9}} \left(\frac{\alpha_3(m_c)}{\alpha_3(m_b)} \right)^{\frac{6}{25}} \left(\frac{\alpha_3(m_b)}{\alpha_3(M_Z)} \right)^{\frac{6}{23}}. \qquad (16.15)$$

where $\mu_{\text{had}} = 1$ GeV. Numerically, estimates of A_S and A_L vary [33–36] so that A_S lies in the range 2.1–2.4 and A_L in the range 1.2–1.5. Thus, the estimates of $A_R = A_S A_L$ lie in the range 2.5–3.5. Two-loop long-range effects on the proton decay effective Lagrangian have also been determined, and are found to be significant [37]. Proton decay from the lepto-quark exchange has a number of possible modes such as $p \to e^+ \pi^0$, $\bar{\nu} \pi^+$, $\mu^+ K^0$, ... We show here the $p \to e^+ \pi^0$

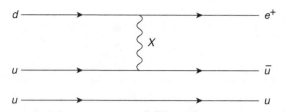

Figure 16.1. Proton decay diagram for the mode $p \to e^+ + \pi^0$.

decay diagram (Fig. 16.1), which is a prime mode for the discovery of proton decay. Using the interactions of X and Y gauge bosons, setting $M_X = M_Y$, and including the renormalization group factor, one finds

$$\Gamma(p \to e^+ \pi^0) = \frac{\beta_p^2 m_p}{32\pi f_\pi^2} \left(\frac{g_5^2 A_R}{M_X^2} \right)^2 (1 + D + F)^2, \qquad (16.16)$$

where β_p is the three-quark matrix element of the nucleon. More generally, α_p and β_p are matrix elements of the three-quark states between the nucleon and the vacuum state (e.g., [38]) which are defined by

$$\langle 0 | \epsilon_{abc} \epsilon_{\alpha\beta} u_{aR}^\alpha d_{bR}^\beta u_{Lc}^\gamma | p \rangle = \alpha_p u_L^\gamma \qquad (16.17)$$

$$\langle 0 | \epsilon_{abc} \epsilon_{\alpha\beta} u_{aL}^\alpha d_{bL}^\beta u_{Lc}^\gamma | p \rangle = \beta_p u_L^\gamma, \qquad (16.18)$$

where α_p and β_p have opposite sign but are expected to satisfy the constraint [39, 40] $|\alpha_p| = |\beta_p|$. Their numerical value has been the subject of various investigations over the years [38, 41]. The most reliable estimates of them come from lattice quantum chromodynamics analyses. For example, the JLQCD Collaboration gives [42]

$$|\alpha_p| = 0.0090(09)(^{+5}_{-19}) \text{ GeV}^3 \qquad (16.19)$$

$$|\beta_p| = 0.0096(09)(^{+6}_{-20}) \text{ GeV}^3, \qquad (16.20)$$

where the first error is statistical and the second error is systematic. This evaluation holds at the scale $Q \simeq 1$ GeV. The parameters F and D that enter into Eq. (16.16) can be obtained from an analysis of hyperon decays, which gives [43]

$$F + D = 1.2670 \pm 0.0030, \quad F - D = -0.341 \pm 0.016. \qquad (16.21)$$

The lifetime of the proton for the partial decay mode $p \to e^+ \pi^0$ is then given by [35]

$$\tau(p \to e^+ \pi^0) \simeq 7.2 \times 10^{34} \text{years} \left(\frac{A_R}{3.0} \right)^{-2} \left(\frac{\alpha_5}{1/24} \right)^2 \left(\frac{M_X}{10^{16} \text{ GeV}} \right)^4. \qquad (16.22)$$

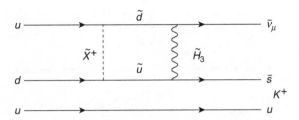

Figure 16.2. Supersymmetric decay mode of the proton $p \to \bar{\nu} + K^+$ via Higgsino triplet and chargino exchange. The Higgsino triplet $\tilde{H}_3$ vertices violate baryon and lepton number, and the chargino "dressing" converts quarks to squarks. There are additional diagrams with $\bar{\nu}_e$ and $\bar{\nu}_\tau$ final states. Also, the Cabibbo–Kobayashi–Maskawa (CKM) matrix elements appear at the $\tilde{\chi}^\pm$ vertices, allowing all three generations to enter into the loop.

A significant ambiguity can result theoretically since estimates of A_R, α_5 and M_X could have sizable errors. The current experiment gives us a lower limit to this decay of [44]

$$\tau(p \to e^+ \pi^0) > 1.4 \times 10^{34} \text{ years.} \tag{16.23}$$

For supersymmetry, the most dominant mode is the decay $p \to \bar{\nu} K^+$ (Fig. 16.2), for which the current experimental lower limit is [44]

$$\tau(p \to \bar{\nu} K^+) > 5.9 \times 10^{33} \text{ years.} \tag{16.24}$$

This mode is discussed in the next section.

16.2 Proton Decay in Supersymmetric Theories

Without R-parity conservation the superpotential will contain terms which preserve $SU(3)_C \times SU(2)_L \times U(1)_Y$ but violate B and L such as

$$W_{RV} = \lambda'_B u^c d^c d^c + \lambda'_L Q d^c L + \lambda''_L L L e^c + \lambda'''_L H_2 L. \tag{16.25}$$

The above interactions will lead to fast proton decay via exchange of sfermions, and would require an extreme constraint on the Yukawa couplings to be consistent with experiment, e.g.,

$$\lambda'_B \lambda'_L < O(10^{-27}). \tag{16.26}$$

Similar constraints also apply to the other couplings. Such dangerous decays can be eliminated if we impose R-parity invariance. However, if R-parity is a global symmetry, it is not necessarily preserved by gravitational interactions. Thus, for example, wormholes can generate baryon-number-violating dimension-four operators which would catalyze proton decay. However, if R-parity is a discrete symmetry which is a remnant of a gauge symmetry, then it will be preserved even in the presence of wormhole effects [45]. We will assume this to be the case. Now, even if baryon- and lepton-number-violating dimension-four operators are eliminated by R-parity invariance, one can generate R-parity violation from spontaneous breaking. Consider operators of the form below which might arise from Planck scale physics:

$$\frac{1}{M_{\text{Pl}}}\, u^c d^c d^c \nu^c, \qquad \frac{1}{M_{\text{Pl}}}\, QLd^c \nu^c. \qquad (16.27)$$

A spontaneous breaking which gives a vacuum expectation value to ν^c will once again violate R-parity and generate dangerous baryon- and lepton-number-violating dimension-four operators. If a vacuum expectation value of ν^c does occur, one would need to constrain such a vacuum expectation value to be very small, i.e., $\langle \nu^c \rangle / M_{\text{Pl}} < O(10^{-13})$.

Even with R-parity, B- and L-violating dimension-five operators will lead to proton decay. Thus, in supersymmetric grand unified models, on integration of the superheavy color triplet Higg fields, one finds operators of the type [46, 47]

$$\frac{1}{M_T} QQQL, \qquad \frac{1}{M_T} u^c u^c d^c e^c \qquad (16.28)$$

in the superpotential where the operators are suppressed by the color triplet Higgsino mass M_T. Because of this, supersymmetric grand unified models as well as string-based models produce instability for the proton above and beyond the instability generated by the lepto-quark exchange. Without proper constraints, such operators can produce a proton lifetime often in excess of the current experimental bounds. It is not difficult to find the constraint under which the color triplet Higgsino mediated proton decay can be suppressed. Assume there is an n-tuplet of color triplet Higgs multiplets. We work in the basis where the combinations of the Higgs multiplets that couple with the matter fields are labeled H_1 and $\bar{H}_1$. In this case, the interaction governing the color triplet Higgs fields with matter is given by

$$\bar{H}_1 J + \bar{K} H_1 + \bar{H}_i M_{ij} H_j, \qquad (16.29)$$

where J and K are quadratic in the matter fields which violate baryon and lepton number. From Eq. (16.29) one can easily deduce that the baryon- and lepton-number-violating dimension-five operators can be suppressed provided [48]

$$(M^{-1})_{11} = 0. \qquad (16.30)$$

One needs, in general, either discrete symmetries or non-standard embeddings of the standard model in a grand unified group to achieve Eq. (16.30). Now, very few models have automatic suppressions of the type in Eq. (16.30), and, in general, one needs a heavy color triplet Higgsino to suppress proton decay from dimension-five operators.

We now discuss proton decay from dimension-five operators in further detail (a more complete discussion of supersymmetric proton decay can be found in the literature [33, 49–58]). For specificity, we will consider the minimal $SU(5)$ grand unification theory, although all of the calculational apparatus can easily be adopted for another grand unified group. The interactions of the minimal $SU(5)$ are given by

$$W_Y = -\frac{1}{8} f_{1\acute{a}\acute{b}} \epsilon_{ijklm} H_1^i M_{\acute{a}}^{jk} M_{\acute{b}}^{lm} + f_{2\acute{a}\acute{b}} \bar{H}_{2i} \bar{M}_{\acute{a}j} M_{\acute{b}}^{ij}, \qquad (16.31)$$

where f_1 is a Yukawa coupling which gives mass to the up quarks and f_2 is the Yukawa coupling which gives mass to the down quarks and the leptons. More specifically, after spontaneous breaking of the grand unified symmetry and of the electro weak gauge group, one has

$$m_i^u = f_i^u \sin 2\theta_W / e) M_Z \sin \beta \qquad (16.32)$$

$$m_i^d = f_i^d (\sin 2\theta_W / e) M_Z \cos \beta. \qquad (16.33)$$

Further, after breakdown of the grand unified symmetry and integration over the color triplet Higgsino fields, one obtains the baryon- and lepton-number-violating dimension-five operator below the grand unification scale. Thus, one has

$$\mathcal{L}_5^L = \frac{1}{M_T} \epsilon_{abc} (P f_1^u V)_{\acute{a}\acute{b}} (f_2^d)_{\acute{c}\acute{d}} (\tilde{u}_{Lb\acute{a}} \tilde{d}_{Lc\acute{b}} (\bar{e}_{L\acute{c}}^c (V u_L)_{ad} + \cdots) + \text{h.c.}, \qquad (16.34)$$

where P_i are intergenerational phases so that $P_i = (e^{i\gamma_i})$, and $\sum_i \gamma_i = 0$ ($i = 1, 2, 3$). This is the chirality-left dimension-five operator with the structure LLLL. There is also another dimension-five operator of chirality type $RRRR$, which is given by

$$\mathcal{L}_5^R = -\frac{1}{M_T} \epsilon_{abc} (V^\dagger f^u)_{\acute{a}\acute{b}} (PV f^d)_{\acute{c}\acute{d}} (\bar{e}_{R\acute{a}}^c u_{Rab} \tilde{u}_{Rc\acute{c}} \tilde{d}_{Rb\acute{d}} + \ldots) + \text{h.c.} \qquad (16.35)$$

The LLLL and $RRRR$ dimension-five operators must be dressed by chargino, neutralino, and gluino exchange diagrams to produce baryon- and lepton-number-violating dimension-six operators. Thus, the dressing loops convert two sfermion fields into two fermion fields, i.e., convert squark/slepton fields into quark/lepton fields, which results in baryon- and lepton-number-violating four-fermion interaction. These operators allow for many possible decay modes of the proton, which include the following:

$$\bar{\nu}_i K^+, \ \bar{\nu}_i \pi^+, \ e^+ K^0, \ \mu^+ K^0, \ e^+ \pi^0, \ \mu^+ \pi^0, \ e^+ \eta, \ \mu^+ \eta. \qquad (16.36)$$

The decay widths depend on many factors, which include the Higgsino triplet mass, quark masses, and CKM matrix elements, as well as sfermion masses which enter into the dressing loop diagrams. The relative strength of the modes is largely determined by the relevant quark masses and the CKM factors that enter into the loop. A display of the dependence of various modes on the quark and CKM factors is given in Table 16.1. Numerically, one finds that the largest

Table 16.1 *Decay modes of the proton in*
supersymmetry unification

Mode	Quark factors	CKM factors
$\bar{\nu}_e K$	$m_d m_c$	$V_{11}^{\dagger} V_{21} V_{22}$
$\bar{\nu}_\mu K$	$m_s m_c$	$V_{21}^{\dagger} V_{21} V_{22}$
$\bar{\nu}_\tau K$	$m_b m_c$	$V_{31}^{\dagger} V_{21} V_{22}$
$\bar{\nu}_e \pi,\ \bar{\nu}_e\,\eta$	$m_d m_c$	$V_{11}^{\dagger} V_{21}^2$
$\bar{\nu}_\mu \pi,\ \bar{\nu}_\mu\,\eta$	$m_s m_c$	$V_{21}^{\dagger} V_{21}^2$
$\bar{\nu}_\tau \pi,\ \bar{\nu}_\tau\,\eta$	$m_b m_c$	$V_{31}^{\dagger} V_{21}^2$
eK	$m_d m_u$	$V_{11}^{\dagger} V_{12}$

decay widths arise with $\bar{\nu}K$ when the product of the quark and CKM factors are taken into account.

In general, the proton decay analysis involves many dressing loop diagrams, where the dressings occur via the exchange of the charginos, the neutralinos, and the gluino. Further, all three sfermion generations enter into the dressing loop diagrams. Among the dressings, the dominant contribution arises from the exchange of the charginos, and thus for simplicity we will focus on the chargino dressing loop diagrams here. Further, we will retain just the first two generations in the dressing loop diagrams. With this simplification, the proton decay width for the process $p \to \bar{\nu}_i K^+$ is given by the following:

$$\Gamma(p \to \bar{\nu}_i K^+) = \frac{\beta_p^2 m_N}{M_T^2 32\pi f_\pi^2} \left(1 - \frac{m_K^2}{m_N^2}\right)^2 |\mathcal{A}_{\nu_i K}|^2 A_L^2 A_S^2 \left|\left(1 + \frac{m_N(D+F)}{m_B}\right)\right|^2.$$

(16.37)

where $A_{\nu_i K}$ is given by

$$\mathcal{A}_{\nu_i K} = (\sin 2\beta M_W^2)^{-1} \alpha_2^2 P_2 m_c m_i^d V_{i1}^{\dagger} V_{21} V_{22} [F(\tilde{c}; \tilde{d}_i; \tilde{W}) + F(\tilde{c}; \tilde{e}_i; \tilde{W})]. \quad (16.38)$$

Here, P_2 is the relative phase of the second generation, and the functions $F(\tilde{c}; \tilde{d}_i; \tilde{W})$, etc., are loop functions that describe the dressings [50]. In general, both the LLLL and the $RRRR$ dimension-five operators contribute, with their relative contribution dependent on various factors. In general, the theoretical derivation of the proton decay widths into $\bar{\nu}K^+$ is very model dependent. Specifically, it is strongly dependent on the sparticle spectrum. Thus, for example, for large squark masses the proton decay lifetime for $\bar{\nu}K^+$ is very approximately given by

$$\tau(p \to \bar{\nu}K^+) \simeq C \frac{m_{\tilde{q}}^4}{m_{\tilde{\chi}^\pm}^2 \tan^2 \beta}. \tag{16.39}$$

This shows that large squark masses, small chargino masses, and small $\tan \beta$ tend to enhance the proton lifetime. The recent discovery of a Higgs boson at the Large Hadron Collider with a mass in the range 125–126 GeV [59,60] has implications for proton stability. Thus, a mass of the Higgs boson in the 125–126 GeV range implies a large loop correction to its tree value (which has an upper limit of M_Z). One way to achieve a large loop correction is to have large scalar masses which enter the loop. In turn, large scalar masses imply a relative enhancement of the proton lifetime, as seen from Eq. (16.39) [61]. In unified gauge theories there are constraints on proton lifetime, arising from the coupling constant unification. Thus, in $SU(5)$ it is straightforward to derive a sum rule using renormalization group evolution on the gauge-coupling constants. Assuming that the components $\Sigma_3 \equiv (3,1)(0)$ and $\Sigma_8 \equiv (1,8)(0)$ (where the components are labeled in the $SU(2) \times SU(3) \times U(1)$ decomposition) of the 24-plet Higgs field which break the $SU(5)$ grand unified symmetry have the same mass, one derives the sum rule of Eq. (9.92) [62]. It has been argued [63] that the constraint of Eq. (9.92) puts an upper limit on the Higgsino triplet mass so that $M_T \leq 3.6 \times 10^{15}$ GeV for the minimal $SU(5)$ case. However, the bound on M_T is relaxed if the Σ_3 and Σ_8 components have unequal masses [64]. In this case, Eq. (9.92) is modified so that

$$(3\alpha_2^{-1} - 2\alpha_3^{-1} - \alpha_1^{-1})(M_Z) = \frac{1}{2\pi}\left(\frac{12}{5} \ln\left(\frac{M_T}{M_Z}\left(\frac{M_{\Sigma_3}}{M_{\Sigma_8}}\right)^{5/2}\right) - 2 \ln\frac{M_{SUSY}}{M_Z}\right). \tag{16.40}$$

The splitting of Σ_3 and Σ_8 masses allows one to relax the constraint on M_T. Further, in $SO(10)$ the heavy thresholds are different, and the relation Eq. (9.92) would be significantly modified. In the analysis above, we discussed the decays of the proton into anti-leptons and pseudo-scalars. However, there are additional decay modes that arise from the same dimension-five baryon- and lepton-number-violating interactions. These are the modes containing anti-leptons and vector bosons such as $\bar{\nu}_i K^*$, $\bar{\nu}_i \rho$, $\bar{\nu}_i \omega$, eK^*, μK^*, $e\rho$, $\mu\rho$, $e\omega$, $\mu\omega$ ($i = e, \mu, \tau$). The analysis of these can be carried out along similar lines, and is discussed by Yuan [65].

As mentioned earlier, $SU(5)$ has baryon- and lepton-number-violating operators with $B - L = 0$. However, $SO(10)$ models have, in addition, baryon- and lepton-number-violating operators with $B - L = -2$ (e.g., [66,67]). These operators have interesting cosmological implications in terms of the generation of baryon asymmetry in the early universe (e.g., [66]). These operators have the property that they can also generate $B - L = -2$ proton decay processes such as $p \to \nu\pi^+$, $n \to e^-\pi^+$, e^-K^+ etc. Proton decay in such models, however, is typically suppressed unless special choices for the parameters are made [66].

16.3 Effective Lagrangian for Nucleon Decay

We now discuss the effective Lagrangian technique for converting the baryon- and lepton-number-violating interactions involving quarks into interactions involving mesons and baryons. After the dressing of the dimension-five operators which have chiral structure of the type LLLL and $RRRR$, one ends up with dimension-six operators in the Lagrangian with the chiral structure LLLL, $LLRR$, $RRLL$, and $RRRR$, which we can abbreviate to LL, LR, RL, and RR. Our main task is now to convert these interactions into an effective Lagrangian in terms of mesons and baryons which can be used directly for the derivation of proton decay matrix elements. Effective Lagrangian techniques use the transformation properties under $SU(3)_L \times SU(3)_R$ of the product of quark fields which enter into the dimension-six operators to derive the corresponding effective Lagrangian in terms of mesons and baryons. Now, the octet of mesons $\pi^\pm$, π^0, $K^\pm$, K^0, $\bar{K}^0$, and η can be associated with the pseudo-Goldstone bosons that arise from the breaking of the chiral $SU(3)_L \times SU(3)_R$ symmetry. The pseudo-Goldstone bosons can be represented by a 3×3 matrix $\mathcal{M}$ such that

$$\mathcal{M} = \sum_{a=1}^{8} \lambda_a \phi_a = \begin{pmatrix} \dfrac{\pi^0}{\sqrt{2}} + \dfrac{\eta}{\sqrt{6}} & \pi^+ & K^+ \\ \pi^- & -\dfrac{\pi^0}{\sqrt{2}} + \dfrac{\eta}{\sqrt{6}} & K^0 \\ K^- & \bar{K}^0 & -\sqrt{\dfrac{2}{3}}\eta \end{pmatrix}. \tag{16.41}$$

In addition to the mesons, one also has an octet of baryons which we can represent by

$$B = \sum_{a=1}^{8} \lambda_a B_a = \begin{pmatrix} \dfrac{\Sigma^0}{\sqrt{2}} + \dfrac{\Lambda}{\sqrt{6}} & \Sigma^+ & p \\ \Sigma^- & -\dfrac{\Sigma^0}{\sqrt{2}} + \dfrac{\Lambda}{\sqrt{6}} & n \\ \Xi^- & \Xi^0 & -\sqrt{\dfrac{2}{3}}\Lambda \end{pmatrix}, \tag{16.42}$$

Under $SU(3)_L \times SU(3)_R$, the matrix B transforms as follows:

$$B' = UBU^\dagger, \tag{16.43}$$

where U is a unitary matrix. To discuss the transformation properties of the pseudo-Goldstone bosons under $SU(3)_L \times SU(3)_R$ transformations it is convenient to introduce the object ξ defined by

$$\xi = \exp(i\mathcal{M}/f), \tag{16.44}$$

where ξ is assumed to have the transformation properties

$$\xi' = U_L \xi U^\dagger = U \xi U_R^\dagger. \tag{16.45}$$

Proton decay interactions contain $\Delta S = 0$ and $\Delta S = 1$ operators. The baryon- and lepton-number-violating $\Delta S = 0$ interaction is given by

$$\mathcal{L}_{\mathcal{BL}} = \; C_{RL}^{\nu_i} O_{RL}^{\nu_i} + C_{LL}^{\nu_i} O_{LL}^{\nu_i} + C_{RL}^{e_i} O_{RL}^{e_i} + C_{LR}^{e_i} O_{LR}^{e_i} + C_{LL}^{e_i} O_{LL}^{e_i} + C_{RR}^{e_i} O_{RR}^{e_i}, \tag{16.46}$$

where

$$O_{RL}^{\nu_i} = \epsilon_{abc} \overline{(d_{Ra})^C} u_{Rb} \overline{(d_{Lc})^C} \nu_{iL} \tag{16.47}$$

$$O_{LL}^{\nu_i} = \epsilon_{abc} \overline{(d_{La})^C} u_{Lb} \overline{(d_{Lc})^C} \nu_{iL} \tag{16.48}$$

$$O_{RL}^{e_i} = \epsilon_{abc} \overline{(d_{Ra})^C} u_{Rb} \overline{(u_{Lc})^C} e_{iL} \tag{16.49}$$

$$O_{LR}^{e_i} = \epsilon_{abc} \overline{(d_{La})^C} u_{Lb} \overline{(u_{Rc})^C} e_{iR} \tag{16.50}$$

$$O_{LL}^{e_i} = \epsilon_{abc} \overline{(d_{La})^C} u_{Lb} \overline{(u_{Lc})^C} e_{iL} \tag{16.51}$$

$$O_{RR}^{e_i} = \epsilon_{abc} \overline{(d_{Ra})^C} u_{Rb} \overline{(u_{Rc})^C} e_{iR}. \tag{16.52}$$

In the above, $a, b, c = 1, 2, 3$ are the color indices, and i is the generation index. We note that the product of the quarks fields in operators with the subscripts LL, LR, RL, and RR have the following types of transformations under $SU(3)_L \times SU(3)_R$:

$$LL : (8,1), \quad RR : (1,8)$$
$$LR : (3,3^*), \quad RL : (3^*,3). \tag{16.53}$$

We now construct combinations of meson and baryon fields which have the same transformation properties as the operators in Eq. (16.53). These combinations are as follows:

$$\begin{aligned}
\xi B \xi &\to U_L (\xi B \xi) U_R^\dagger & (3,3^*) \\
\xi^\dagger B \xi^\dagger &\to U_R (\xi^\dagger B \xi^\dagger) U_L^\dagger & (3^*,3) \\
\xi B \xi^\dagger &\to U_L (\xi B \xi^\dagger) U_L^\dagger & (8,1) \\
\xi^\dagger B \xi &\to U_R (\xi^\dagger B \xi) U_R^\dagger & (1,8).
\end{aligned} \tag{16.54}$$

Using the transformation properties of Eq. (16.54) the operators Eqs. (16.47)–(16.52) can be written in the following form:

$$O^{\nu_i}_{RL} = \alpha_p \overline{(\nu_{iL})^C} tr(P'\xi B_L \xi) \tag{16.55}$$

$$O^{\nu_i}_{LL} = \beta_p \overline{(\nu_{iL})^C} tr(P'\xi B_L \xi^\dagger) \tag{16.56}$$

$$O^{e_i}_{RL} = \alpha_p \overline{(e_{iL})^C} tr(P\xi B_L \xi) \tag{16.57}$$

$$O^{e_i}_{LR} = \alpha_p \overline{(e_{iR})^C} tr(P\xi^\dagger B_R \xi^\dagger) \tag{16.58}$$

$$O^{e_i}_{LL} = \beta_p \overline{(e_{iL})^C} tr(P\xi B_L \xi^\dagger) \tag{16.59}$$

$$O^{e_i}_{RR} = \beta_p \overline{(e_{iR})^C} tr(P\xi^\dagger B_R \xi), \tag{16.60}$$

where P and P' are projection operators [68]:

$$P = \begin{pmatrix} 0 & 0 & 0 \\ 0 & 0 & 0 \\ 1 & 0 & 0 \end{pmatrix}, \quad P' = \begin{pmatrix} 0 & 0 & 0 \\ 0 & 0 & 0 \\ 0 & 1 & 0 \end{pmatrix}. \tag{16.61}$$

In extracting the terms that contribute to $\Delta S = 0$ proton decay processes, one must take into account quadratic as well as cubic parts of the Lagrangian. Thus, the quadratic part of the B- and L-violating Lagrangian $\mathcal{L}_{B \mathcal{L} V}$ is given by

$$\mathcal{L}_{B \mathcal{L} V}^{(2)} = (\alpha_p C^{\nu_i}_{RL} + \beta_p C^{\nu_i}_{LL})\overline{(\nu_{iL})^C} n_L + (\alpha_p C^{e_i}_{RL} + \beta_p C^{e_i}_{LL})\overline{(e_{iL})^C} p_L$$
$$+ (\alpha_p C^{e_i}_{LR} + \beta_p C^{e_i}_{RR})\overline{(e_{iR})^C} p_R, \tag{16.62}$$

while the cubic part of Eqs. (16.55)–(16.60) is given by

$$\mathcal{L}_{B \mathcal{L} V}^{(3)} = \frac{i}{f}\{\alpha_p C^{\nu_i}_{RL} \left(\overline{(\nu_{iL})^C} p_L \pi^- - \overline{(\nu_{iL})^C} n_L \left(\frac{\pi^0}{\sqrt{2}} + \frac{\eta}{\sqrt{6}} \right) \right)$$

$$+ \beta_p C^{\nu_i}_{LL} \left(\overline{(\nu_{iL})^C} p_L \pi^- + \overline{(\nu_{iL})^C} n_L \left(-\frac{\pi^0}{\sqrt{2}} + \frac{3}{\sqrt{6}}\eta \right) \right)$$

$$+ \alpha_p C^{e_i}_{RL} \left(\overline{(e_{iL})^C} n_L \pi^+ + \overline{(e_{iL})^C} p_L \left(\frac{\pi^0}{\sqrt{2}} - \frac{1}{\sqrt{6}}\eta \right) \right)$$

$$- \alpha_p C^{e_i}_{LR} \left(\overline{(e_{iR})^C} n_R \pi^+ + \overline{(e_{iR})^C} p_R \left(\frac{\pi^0}{\sqrt{2}} - \frac{1}{\sqrt{6}}\eta \right) \right)$$

$$+ \beta_p C^{e_i}_{LL} \left(\overline{(e_{iL})^C} n_L \pi^+ + \overline{(e_{iL})^C} p_L \left(\frac{\pi^0}{\sqrt{2}} + \frac{3}{\sqrt{6}}\eta \right) \right)$$

$$- \beta_p C^{e_i}_{RR} \left(\overline{(e_{iR})^C} n_R \pi^+ + \overline{(e_{iR})^C} p_R \left(\frac{\pi^0}{\sqrt{2}} + \frac{3}{\sqrt{6}}\eta \right) \right). \tag{16.63}$$

In addition to the above, one also has terms in the Lagrangian that conserve baryon and lepton number. They are described by the following set of terms [69]:

$$\mathcal{L}_{BC} = \frac{1}{2i}(D - F)tr[\bar{B}\gamma^\mu\gamma_5 B\{\partial_\mu \xi \xi^\dagger - \partial_\mu \xi^\dagger \xi\}]$$

$$- \frac{1}{2i}(D + F)tr[\bar{B}\gamma^\mu\gamma_5\{\xi\partial_\mu\xi^\dagger - \xi^\dagger\partial_\mu\xi\}B] + \mathcal{L}_m. \tag{16.64}$$

Aside from the $\mathcal{L}_m$, these interactions are invariant under $SU(3)_L \times SU(3)_R$ transformations. The $\mathcal{L}_m$ is described by the following set of terms [68, 69]:

$$\mathcal{L}_m = a_1\, tr[\bar{B}\gamma_5(\xi^\dagger M_Q \xi^\dagger - \xi M_Q \xi)B] + a_2\, tr[\bar{B}\gamma_5 B(\xi^\dagger M_Q \xi^\dagger - \xi M_Q \xi)]. \quad (16.65)$$

Here, M_Q is the diagonal quark mass matrix, i.e., $M_Q = diag(m_u, m_d, m_s)$. These terms are not $SU(3)_L \times SU(3)_R$-invariant; rather, the coefficients of a_1 and a_2 transform like $(3, 3^*) + (3^*, 3)$. The $SU(3)_L \times SU(3)_R$ symmetry is then broken down to the residual $SU(3)_V$ symmetry. In the above, the matrix $\bar{B}$ is defined by

$$\bar{B} = \begin{pmatrix} \dfrac{\bar{\Sigma}^0}{\sqrt{2}} + \dfrac{\bar{\Lambda}}{\sqrt{6}} & \bar{\Sigma}^- & -\bar{\Xi}^- \\[2mm] \bar{\Sigma}^+ & -\dfrac{\bar{\Sigma}^0}{\sqrt{2}} + \dfrac{\bar{\Lambda}}{\sqrt{6}} & \bar{\Xi}^0 \\[2mm] \bar{p} & \bar{n} & -\sqrt{\dfrac{2}{3}}\bar{\Lambda} \end{pmatrix}. \quad (16.66)$$

Neglecting the quark masses, which make only a relatively small contribution, the relevant part of $\mathcal{L}_{BC}$ is

$$\mathcal{L}_{BC} = \left(\frac{D-F}{\sqrt{2}f}\bar{\Sigma}^0\gamma^\mu\gamma_5 p - \frac{D+3F}{\sqrt{6}f}\bar{\Lambda}^0\gamma^\mu\gamma_5 p + \frac{D-F}{\sqrt{2}f}\bar{\Sigma}^-\gamma^\mu\gamma_5 n \right)\partial_\mu K^-$$
$$+ \left(\frac{D-F}{f}\bar{\Sigma}^+\gamma^\mu\gamma_5 p - \frac{D-F}{\sqrt{2}f}\bar{\Sigma}^0\gamma^\mu\gamma_5 n - \frac{D+3F}{\sqrt{6}f}\bar{\Lambda}^0\gamma^\mu\gamma_5 n \right)\partial_\mu \bar{K}^0.$$
$$(16.67)$$

For simplicity, we introduce the notation

$$C'^{\nu_i, e_i}_{RL} = \alpha C^{\nu_i, e_i}_{RL}, \quad C'^{\nu_i, e_i}_{LL} = \beta C^{\nu_i, e_i}_{LL}, \quad C'^{e_i}_{LR} = \alpha C^{e_i}_{LR}, \quad C'^{e_i}_{RR} = \beta C^{e_i}_{RR}. \quad (16.68)$$

As an illustration, we give a couple of $\Delta S = 0$ nucleon decay modes. Specifically, we consider the neutrino–pion decay modes $p \to \bar{\nu}_i \pi^+$ and $n \to \bar{\nu}_i \pi^0$. The decay width for $p \to \bar{\nu}_i \pi^+$ is

$$\Gamma(p \to \bar{\nu}_i \pi^+) = (32\pi f^2 m_N^3)^{-1}(m_p^2 - m_{\pi^+}^2)^2|C'^{\nu_i}_{RL} + C'^{\nu_i}_{LL}|^2(1 + D + F)^2, \quad (16.69)$$

where f is the pion decay constant, which has the value $f \simeq 139$ MeV, while the decay width of the neutron decay mode $n \to \bar{\nu}_i \pi^0$ is

$$\Gamma(n \to \bar{\nu}_i \pi^0) = (32\pi f^2 m_N^3)^{-1}(m_n^2 - m_{\pi^0}^2)^2 \frac{1}{2}|C'^{\nu_i}_{RL} + C'^{\nu_i}_{LL}|^2(1 + D + F)^2. \quad (16.70)$$

The two decay widths can be related if we neglect the difference between the proton and the neutron mass and also neglect the mass difference between π^+ and π^0. In this case, one finds

$$\Gamma(n \to \bar{\nu}_i \pi^0) \simeq 0.5\Gamma(p \to \bar{\nu}_i \pi^+). \quad (16.71)$$

The effective Lagrangian method discussed above can be used to compute a variety of other $\Delta S = 0$ processes, i.e., $p \to e^+\pi^0$, $e^+\eta^0$, $n \to e^+\pi^-$, and

$n \to \bar{\nu}\eta^0$, and can be extended to include $\Delta S = 1$ processes, i.e., $p \to \bar{\nu}K^+$, $p \to e^+K^0$, and $n \to e^+\pi^-$ (e.g., [8]).

In supersymmetric theories the sparticle mass spectrum and the sparticle vertices are sensitive to CP phases. Since these enter into the analysis of proton lifetime, one finds that proton lifetime is also sensitive to CP phases [70]. In principle, proton decay can be used to discriminate between string and grand unified models. In grand unified models, interactions of both types, i.e., of the type arising from $10^2\overline{10}^2$ as well as from $10\overline{10}55$, contribute to proton decay, and consequently one has $p \to e_L^+\pi^0$ as well as $p \to e_R^+\pi^0$ decays. It has been pointed out that generic $SU(5)$ D-brane models allow only dimension-six operators of the type $10^2\overline{10}^2$, which lead to the proton decay mode $p \to \pi^0 e_L^+$, while an $SU(5)$ dimension-six operator of the type $10\overline{10}55$ is not allowed, which inhibits a proton decay mode of the type $p \to \pi^0 e_R^+$ [71]. The same reasoning leads one to the conclusion that the $SU(5)$ decay mode $p \to \pi^+\nu$ would not be allowed in generic D-brane models. There are, however, exceptions to this rule since certain special regions of intersecting D-brane models permit the dimension-six operator $10\overline{10}55$ [72]. We note that not all high-scale models necessarily lead to proton decay. Thus, in the flipped $SU(5) \times U(1)$ model [73, 74], baryon- and lepton-number-violating operators can be rotated away, and one does not have proton decay in such models (e.g., [8]). It is also possible to find minimal supersymmetric standard model (MSSM)-like models within D-branes where the baryon- and lepton-number-violating dimension-five operators are absent at the perturbative level [75].

16.4 Problems

1. Show that the baryon- and lepton-number-violating interactions arising from the gauge couplings of the $\bar{5}$-plet of matter fields in $SU(5)$ are given by

$$\mathcal{L}_{BLV-5} = \frac{g_5}{\sqrt{2}} \left(\bar{d}_{R\alpha}\gamma^\mu X_{\mu\alpha}^c e_R^+ - \bar{d}_{R\alpha}\gamma^\mu Y_{\mu\alpha}^c \nu_{eR}^c + \text{h.c.} \right) \tag{16.72}$$

and that of the 10-plet of matter fields in $SU(5)$ are given by

$$\mathcal{L}_{BLV-10} = \frac{g_5}{\sqrt{2}} \Big(\bar{d}_{L\alpha}\gamma^\mu X_{\mu\alpha}^c e_L^+ + \epsilon_{abc}\bar{u}_{Lc}^c\gamma^\mu X_{\mu a}^c u_{Lb}$$
$$- \bar{u}_{L\alpha}\gamma^\mu Y_{\mu\alpha}^c \nu^c e_L^+ + \epsilon_{abc}\bar{u}_{Lc}^c\gamma^\mu Y_{\mu a}^c d_{Lb} + \text{h.c.} \Big). \tag{16.73}$$

2. Show that on integration of the lepto-quark fields X and Y and using $M_X = M_Y$ the $B - L$-violating but the $B\&L$ conserving effective Lagrangian with

dimension-six operators in $SU(5)$ is given by

$$\mathcal{L}_{\text{BLV-eff}} = \frac{g_5^2}{2M_X^2} [2\epsilon_{abc} \bar{u}_{Lc}^c \gamma^\mu u_{Lb} (\bar{e}_L^+ \gamma_\mu d_{La})$$

$$+ \epsilon_{abc} \bar{u}_{Lc}^c \gamma^\mu u_{Lb} (\bar{e}_R^+ \gamma_\mu d_{Ra}) - \epsilon_{abc} \bar{u}_{Lc}^c \gamma_\mu d_{Lb} (\bar{\nu}_R^c \gamma_\mu d_{Ra}) + \text{h.c.}] \quad (16.74)$$

3. Show that for the supersymmetric case the constraint of Eq. (16.30) eliminates proton decay mediated by Higgsino color triplets.

4. Show that, analogous to Eq. (16.70), the decay width for $n \to \bar{\nu}\eta^0$ is given by

$$\Gamma(n \to \bar{\nu}_i \eta^0) = (32\pi f^2 m_N^3)^{-1} (m_n^2 - m_{\eta^0}^2)^2$$

$$\times \frac{3}{2} \left| C_{RL}'^{\nu_i} \left(-\frac{1}{3} - \frac{D}{3} + F \right) + C_{LL}'^{\nu_i} \left(1 - \frac{D}{3} + F \right) \right|^2. \quad (16.75)$$

5. Show that for $SU(5)$ for the case when the Σ_3 and Σ_8 masses are split, one has the relation

$$5\alpha_1^{-1}(M_Z) - 3\alpha_2^{-1}(M_Z) - 2\alpha_3^{-1}(M_Z)$$

$$= \frac{1}{2\pi} \left[12 \ln \frac{M_V^2 \sqrt{M_{\Sigma_3} M_{\Sigma_8}}}{M_Z^3} + 8 \ln \frac{M_S}{M_Z} \right]. \quad (16.76)$$

6. Show that in $SO(10)$ the couplings of the 16-plet of matter fields with the $\overline{126}$-plet of the Higgs field, i.e., the coupling $16_a 16_b \overline{126}_H$ in the superpotential, does not contribute to proton decay.

References

[1] H. Weyl, *Z. Phys.* **56**, 330 (1929) [*Surveys High Energ. Phys.* **5**, 261 (1986)].
[2] E. C. G. Stueckelberg, *Helv. Phys. Acta.* **11**, 299 (1939).
[3] E. P. Wigner, *Proc. Am. Phil. Soc.* **93**, 521 (1949).
[4] M. Goldhaber, in *Proceedings, Neutrino Physics and Astrophysics*, Boston (1988), pp. 486–489.
[5] F. Reines, C. L. Cowan, and M. Goldhaber, *Phys. Rev.* **96**, 1157 (1954).
[6] H. S. Gurr, W. R. Kropp, F. Reines, and B. Meyer, *Phys. Rev.* **158**, 1321 (1967).
[7] D. H. Perkins, *Ann. Rev. Nucl. Part. Sci.* **34**, 1 (1984).
[8] P. Nath and P. Fileviez Perez, *Phys. Rep.* **441**, 191 (2007).
[9] T. D. Lee and C. N. Yang, *Phys. Rev.* **98**, 1501 (1955).
[10] G. 't Hooft, *Phys. Rev. Lett.* **37**, 8 (1976).
[11] Y. B. Zeldovich, *Phys. Lett. A* **59**, 254 (1976).
[12] S. W. Hawking, D. N. Page, and C. N. Pope, *Phys. Lett. B* **86**, 175 (1979).
[13] D. N. Page, *Phys. Lett. B* **95**, 244 (1980).
[14] J. R. Ellis, J. S. Hagelin, D. V. Nanopoulos, and K. Tamvakis, *Phys. Lett. B* **124**, 484 (1983).
[15] G. Gilbert, *Nucl. Phys. B* **328**, 159 (1989).
[16] Y. Yamaguchi, *Proc. Theor. Phys.* **22**, 373 (1959).

[17] G. K. Backenstoss, H. Frauenfielder, B. D. Hyams, L. J. Koester, P. C. and Marin, *Nuovo Cim.* **16**, 749 (1960).

[18] A. D. Sakharov, *Pisma Zh. Eksp. Teor. Fiz.* **5**, 32 (1967) [*JETP Lett.* **5**, 24 (1967)] [*Sov. Phys. Usp.* **34**, 392 (1991)] [*Usp. Fiz. Nauk* **161**, 61 (1991)].

[19] O. Espinosa, *Nucl. Phys. B* **343**, 310 (1990).

[20] J. C. Pati and A. Salam, *Phys. Rev. Lett.* **31**, 661 (1973).

[21] J. C. Pati and A. Salam, *Phys. Rev. D* **8**, 1240 (1973).

[22] H. Georgi and S. L. Glashow, *Phys. Rev. Lett.* **32**, 438 (1974).

[23] M. R. Krishnaswamy *et al.*, in *Proceedings, Grand Unified Theories and Cosmology*, Tsukuba (1983), pp. 32–37.

[24] G. Battistoni, E. Bellotti, C. Bloise, *et al.*, *Nucl. Instrum. Meth. A* **245**, 277 (1986).

[25] C. Berger *et al.* [FREJUS Collaboration], *Nucl. Instrum. Meth. A* **262**, 463 (1987).

[26] J. L. Thron, *Nucl. Instrum. Meth. A* **283**, 642 (1989).

[27] R. Becker-Szendy *et al.*, *Nucl. Instrum. Meth. A* **324** (1993) 363.

[28] K. S. Hirata *et al.* [Kamiokande-II Collaboration], *Phys. Lett. B* **220**, 308 (1989).

[29] B. Viren [Super-Kamiokande Collaboration], hep-ex/9903029 (1999).

[30] S. Weinberg, *Phys. Rev. Lett.* **43**, 1566 (1979).

[31] F. Wilczek and A. Zee, *Phys. Rev. Lett.* **43**, 1571 (1979).

[32] M. Goldhaber and L. R. Sulak, *Comments Nucl. Part. Phys.* **10**(5), 215 (1981).

[33] J. R. Ellis, D. V. Nanopoulos, and S. Rudaz, *Nucl. Phys. B* **202**, 43 (1982).

[34] L. E. Ibanez and C. Munoz, *Nucl. Phys. B* **245**, 425 (1984).

[35] J. Hisano, hep-ph/0004266 (2000).

[36] R. Dermisek, A. Mafi, and S. Raby, *Phys. Rev. D* **63**, 035001 (2001).

[37] T. Nihei and J. Arafune, *Prog. Theor. Phys.* **93**, 665 (1995).

[38] S. Aoki *et al.* [JLQCD Collaboration], *Phys. Rev. D* **62**, 014506 (2000).

[39] S. J. Brodsky, J. R. Ellis, J. S. Hagelin, and C. T. Sachrajda, *Nucl. Phys. B* **238**, 561 (1984).

[40] M. B. Gavela, S. F. King, C. T. Sachrajda, G. Martinelli, M. L. Paciello, and B. Taglienti, *Nucl. Phys. B* **312**, 269 (1989).

[41] J. F. Donoghue and E. Golowich, *Phys. Rev. D* **26**, 3092 (1982).

[42] N. Tsutsui *et al.* [CP-PACS and JLQCD Collaborations], *Phys. Rev. D* **70**, 111501 (2004).

[43] N. Cabibbo, E. C. Swallow, and R. Winston, *Ann. Rev. Nucl. Part. Sci.* **53**, 39 (2003).

[44] K. S. Babu *et al.*, arXiv:1311.5285 [hep-ph] (2013).

[45] L. M. Krauss and F. Wilczek, *Phys. Rev. Lett.* **62**, 1221 (1989).

[46] S. Weinberg, *Phys. Rev. D* **26**, 287 (1982).

[47] N. Sakai and T. Yanagida, *Nucl. Phys. B* **197**, 533 (1982).

[48] R. L. Arnowitt and P. Nath, *Phys. Rev. D* **49**, 1479 (1994).

[49] S. Dimopoulos, S. Raby, and F. Wilczek, *Phys. Lett. B* **112**, 133 (1982).

[50] P. Nath, A. H. Chamseddine, and R. L. Arnowitt, *Phys. Rev. D* **32**, 2348 (1985).

[51] J. Hisano, H. Murayama, and T. Yanagida, *Nucl. Phys. B* **402**, 46 (1993).

[52] T. Goto and T. Nihei, *Phys. Rev. D* **59**, 115009 (1999).

[53] B. Bajc, P. Fileviez Perez, and G. Senjanovic, *Phys. Rev. D* **66**, 075005 (2002).

[54] D. Emmanuel-Costa, and S. Wiesenfeldt, *Nucl. Phys. B* **661**, 62 (2003).

[55] V. Lucas and S. Raby, *Phys. Rev. D* **55**, 6986 (1997).

[56] K. S. Babu, J. C. Pati, and F. Wilczek, *Nucl. Phys. B* **566**, 33 (2000).

[57] T. Fukuyama, A. Ilakovac, T. Kikuchi, S. Meljanac, and N. Okada, *Eur. Phys. J. C* **42**, 191 (2005).

[58] B. Dutta, Y. Mimura, and R. N. Mohapatra, *Phys. Rev. Lett.* **94**, 091804 (2005).

[59] G. Aad *et al.* [ATLAS Collaboration], *Phys. Lett. B* **716**, 1 (2012).

[60] S. Chatrchyan *et al.* [CMS Collaboration], *Phys. Lett. B* **716**, 30 (2012).

[61] M. Liu and P. Nath, *Phys. Rev. D* **87**(9), 095012 (2013) [arXiv:1303.7472 [hep-ph]].

[62] J. Hisano, H. Murayama, and T. Yanagida, *Phys. Rev. Lett.* **69**, 1014 (1992).

[63] H. Murayama and A. Pierce, *Phys. Rev. D* **65**, 055009 (2002).

[64] B. Bajc, P. Fileviez Pérez, and G. Senjanović, arXiv:hep-ph/0210374 (2002); I. Dorsner, P. Fileviez Perez, and G. Rodrigo, *Phys. Lett. B* **649**, 197 (2007).

[65] T. C. Yuan, *Phys. Rev. D* **33**, 1894 (1986).

[66] K. S. Babu and R. N. Mohapatra, *Phys. Rev. D* **86**, 035018 (2012)

[67] P. Nath and R. M. Syed, *Phys. Rev. D* **93**(5), 055005 (2016).

[68] M. Claudson, M. B. Wise, and L. J. Hall, *Nucl. Phys. B* **195**, 297 (1982).

[69] S. Chadha and M. Daniel, *Nucl. Phys. B* **229**, 105 (1983).

[70] T. Ibrahim and P. Nath, *Phys. Rev. D* **62**, 095001 (2000).

[71] I. R. Klebanov and E. Witten, *Nucl. Phys. B* **664**, 3 (2003).

[72] M. Cvetic and R. Richter, *Nucl. Phys. B* **762**, 112 (2007).

[73] S. M. Barr, *Phys. Lett. B* **112**, 219 (1982).

[74] J. P. Derendinger, J. E. Kim, and D. V. Nanopoulos, *Phys. Lett. B* **139**, 170 (1984).

[75] M. Cvetic, J. Halverson, and R. Richter, arXiv:0910.2239 [hep-th] (2009).

17

Cosmology, Astroparticle Physics, and Supergravity Unification

Observationally, the universe is homogeneous and isotropic at very large scales. A simple description of a homogeneous and isotropic universe is given by the Friedmann–Robertson–Walker (FRW) [1–3] line element ds^2, where

$$ds^2 = -(dt)^2 + R(t)^2 \left[\frac{dr^2}{1 - kr^2} + r^2 d\theta^2 + r^2 \sin^2 \theta \; d\phi^2 \right]. \qquad (17.1)$$

In the above, $R(t)$ is the so-called cosmic scale factor which describes the expansion of the universe, and the parameter k determines the curvature of the spatial hypersurface and describes the allowed geometries. Specifically, by a scale transformation, k takes on just three distinct values, i.e., $k = -1, 0$, and $+1$. In this case, r is unitless and $R(t)$ carries the unit of length. The three cases correspond to an open (hyperboloid), flat, and closed (elliptical) universe. In building a model of the universe, a form of the stress tensor is necessary, and we assume the following form:

$$T_{\mu\nu} = pg_{\mu\nu} + (p + \rho)u_\mu u_\nu, \qquad (17.2)$$

where p is the pressure, ρ is the energy density, and $u^\mu = dx^\mu/ds$ is taken to be

$$u^t = 1, u^i = 0, i = 1, 2, 3. \qquad (17.3)$$

From chapter 2 we recall the gravitational field equations:

$$\mathcal{R}_{\mu\nu} - \frac{1}{2} g_{\mu\nu} \mathcal{R} = -8\pi G_N T_{\mu\nu}. \qquad (17.4)$$

Alternatively, we can write Eq. (17.4) as

$$\mathcal{R}_{\mu\nu} = -8\pi G_N \mathcal{T}_{\mu\nu}, \qquad (17.5)$$

where

$$\mathcal{T}_{\mu\nu} \equiv T_{\mu\nu} - \frac{1}{2}g_{\mu\nu}T_\lambda^\lambda$$

$$= \frac{1}{2}(\rho - p)g_{\mu\nu} + (p + \rho)u_\mu u_\nu. \tag{17.6}$$

The components of the metric are given by Eq. (17.1),

$$g_{tt} = -1, g_{ij} = R^2(t)h_{ij}, \tag{17.7}$$

where $i, j = 1, 2,$ and 3 are indices for spatial co-ordinates and h_{ij} are given by

$$h_{rr} = \frac{1}{1 - kr^2}, h_{\theta\theta} = r^2, h_{\phi\phi} = r^2 \sin^2\theta$$

$$h_{ij} = 0, i \neq j. \tag{17.8}$$

We define the inverse h^{ij} so that

$$h^{ik}h_{kj} = \delta_j^i. \tag{17.9}$$

The affinities are given by

$$\Gamma_{jk}^i = \gamma_{jk}^i, \Gamma_{ij}^t = R\dot{R}h_{ij}, \Gamma_{tj}^i = \frac{\dot{R}}{R}\delta_j^i$$

$$\Gamma_{ti}^t = \Gamma_{tt}^t = \Gamma_{tt}^i = 0, \tag{17.10}$$

where γ_{jk}^i is the affinity corresponding to the metric h_{ij} so that

$$\gamma_{jk}^i = \frac{1}{2}h^{il}\left[h_{jl,k} + h_{kl,j} - h_{jk,l}\right], \tag{17.11}$$

where h^{ij} is the inverse defined by

$$h^{ik}h_{kj} = \delta_j^i. \tag{17.12}$$

The curvature tensor is given by

$$\mathcal{R}_{ij} = \mathcal{R}_{ij}^{(3)} - (2\dot{R}^2 + R\ddot{R})h_{ij}$$

$$\mathcal{R}_{tt} = \frac{3\ddot{R}}{R}$$

$$\mathcal{R}_{ti} = 0, \tag{17.13}$$

where $\mathcal{R}_{ij}^{(3)}$ is the curvature tensor corresponding to the metric h_{ij}. Now, we use the result (see Eq. (17.314))

$$\mathcal{R}_{ij}^{(3)} = -2kh_{ij}. \tag{17.14}$$

This leads to

$$\mathcal{R}_{ij} = -(R\ddot{R} + 2\dot{R}^2 + 2k)h_{ij}. \tag{17.15}$$

We also list the components of $\mathcal{T}_{\mu\nu}$, which are

$$\mathcal{T}_{tt} = \frac{1}{2}(\rho + 3p)$$
$$\mathcal{T}_{ti} = 0$$
$$\mathcal{T}_{ij} = \frac{1}{2}R^2 h_{ij}(\rho - p). \tag{17.16}$$

We look now at the components of the Einstein equations. The $\mu = 0 = \nu$ component gives

$$\mathcal{R}_{tt} = -8\pi G_N \mathcal{T}_{tt}, \tag{17.17}$$

which leads to the result

$$3\ddot{R} = -(4\pi G_N)(\rho + 3p)R. \tag{17.18}$$

Next, we look at the $\mu = i$ and $\nu = j$ component of the Einstein equations. Here, one finds

$$\mathcal{R}_{ij} = -8\pi G_N \mathcal{T}_{ij}, \tag{17.19}$$

which leads to

$$R\ddot{R} + 2\dot{R}^2 + 2k = (4\pi G_N)(\rho - p)R^2. \tag{17.20}$$

Using Eq. (17.18) in Eq. (17.20), we get

$$\dot{R}^2 + k = \frac{8\pi G_N}{3}\rho R^2. \tag{17.21}$$

Equations (17.18) and (17.21) are the two Friedman equations that describe a homogeneous and isotropic universe. We express them in their conventional form:

$$\left(\frac{\dot{R}}{R}\right)^2 = \frac{8\pi G}{3}\rho - \frac{k}{R^2} \tag{17.22}$$

$$\frac{\ddot{R}}{R} = -\frac{4\pi G}{3}(\rho + 3p). \tag{17.23}$$

A useful quantity in cosmology is the Hubble parameter, defined by

$$H(t) = \frac{\dot{R}(t)}{R(t)}. \tag{17.24}$$

Using Eq. (17.24), one finds that Eq. (17.21) takes the form

$$H^2(t) + kR^{-2}(t) = \frac{8\pi G_N}{3}\rho. \tag{17.25}$$

Next, we consider Eq. (17.21) and differentiate it with respect to time. This gives

$$2\dot{R}\ddot{R} = \frac{8\pi G_N}{3}\left[\frac{d\rho}{dt}R^2 + 2\rho R\dot{R}\right]. \tag{17.26}$$

Substitution of $\ddot{R}$ from Eq. (17.18) leads to the relation

$$\frac{d\rho}{dt} + 3H(\rho + p) = 0. \tag{17.27}$$

17.0.1 The Equation for Energy Conservation

The energy conservation equation has the form

$$0 = T^{\nu\mu}{}_{;\nu} = \frac{1}{\sqrt{g}}\partial_\nu(\sqrt{g}T^{\nu\mu}) + \Gamma^\mu_{\nu\lambda}T^{\nu\lambda}. \tag{17.28}$$

The $\mu = 0$ component gives

$$\frac{1}{\sqrt{g}}\partial_0(\sqrt{g}T^{00}) + \Gamma^0_{\nu\lambda}T^{\nu\lambda} = 0. \tag{17.29}$$

Now, it is easily seen that

$$\Gamma^0_{\nu\lambda}T^{\nu\lambda} = 3p\frac{\dot{R}}{R}$$

$$\frac{1}{\sqrt{g}}\partial_0(\sqrt{g}T^{00}) = \frac{1}{R^3}\frac{d}{dt}(R^3\rho), \tag{17.30}$$

which leads to the equation

$$\frac{d}{dt}(R^3\rho) + 3pR^2\dot{R} = 0. \tag{17.31}$$

or, alternatively, to the equation

$$\frac{d}{dR}(R^3\rho) = -3pR^2. \tag{17.32}$$

Assume we look at the universe when the energy density is dominated by non-relativistic matter which has negligible pressure. Setting $p = 0$ in Eq. (17.32), one finds

$$\rho \propto R^{-3}. \tag{17.33}$$

Next, assume we look at the universe when there is relativistic matter. Here, we have $p = \rho/3$ (see Eq. (17.53)), and in this case Eq. (17.32) gives

$$4\rho + R\frac{d\rho}{dR} = 0, \tag{17.34}$$

which implies that

$$\rho \propto R^{-4}. \tag{17.35}$$

For the case $p + \rho = 0$, Eq. (17.32) gives $dp/dR = 0$, which implies

$$p_{\text{vac}} = \text{constant.} \tag{17.36}$$

A quantity of interest is the critical matter density ρ_c, which is the matter density needed to close the universe. We can obtain an expression for it from Eq. (17.25), and for the case of a flat universe with $k = 0$ one has

$$\rho_c = \frac{3H_0^2}{8\pi G_N}$$
$$= 1.88 \times 10^{-29} \text{ g cm}^3 \, h_0^2, \tag{17.37}$$

where H_0 is the current value of the Hubble parameter and where h_0 is in units of 100 (km/s)/Mpc.

17.1 Particle Properties in the Early Universe

17.1.1 Chemical Potential

The statistics of a particle variety in the early universe is determined completely in terms of the particle mass, temperature, and chemical potential and whether the particle is a fermion or a boson. The chemical potential μ is a thermodynamic variable which is defined by the relation [4]

$$T \, dS = dU + p \, dV - \mu \, dN, \tag{17.38}$$

where S is the total entropy of the system, U is the total energy, V is the volume, and N is the number of particles. Thus, μ can be expressed as

$$\frac{\mu}{T} = - \left(\frac{\partial S}{\partial N} \right)_{U,V}. \tag{17.39}$$

Each of the particle species has a specific chemical potential, and for interactions involving different species of particles, one can define a chemical equilibrium. Thus, for a process involving particles a_i ($i = 1, \ldots, n$) in the initial state and particles b_j ($j = 1, \ldots, m$) in the final states,

$$\sum_i a_i \to \sum_j b_j, \tag{17.40}$$

chemical equilibrium implies the relation

$$\sum_i \mu_{a_i} \to \sum_j \mu_{b_j}. \tag{17.41}$$

Most often we are concerned with three quantities, i..e, the number density, the energy density, and the pressure related to the particle species. Let us consider first the number density n, which for a particle variety with g degrees of freedom, mass m, and chemical potential μ has a number density which is given by[*]

$$n = g \int \frac{d^3p}{(2\pi)^3} \frac{1}{e^{(E-\mu)/T} \pm 1}, \qquad (17.42)$$

where $E = (p^2 + m^2)^{1/2}$. Here, $+1$ corresponds to Fermi–Dirac statistics and -1 for Bose- Einstein statistics. Introducing variables $x = E/T$, $x_m = m/T$, and $x_\mu = \mu/T$, we can write Eq. (17.42) so that

$$n = \frac{gT^3}{2\pi^2} \int_{x_m}^{\infty} \frac{x^2 \, dx}{e^{(x-x_\mu)} \pm 1}. \qquad (17.43)$$

In the ultra-relativistic limit when $T \gg m, \mu$, and thus $x_m, x_\mu \to 0$, and Eq. (17.42) is replaced by

$$n = \frac{gT^3}{2\pi^2} \int_0^{\infty} \frac{x^2 \, dx}{e^x \pm 1}. \qquad (17.44)$$

The integrals are given in terms of the Riemann zeta function, which is defined by

$$\zeta(s) = \frac{1}{\Gamma(s)} \int_0^{\infty} \frac{x^{s-1} \, dx}{e^x - 1}. \qquad (17.45)$$

Using Eq. (17.45), it is easy to see that

$$\int_0^{\infty} \frac{x^{s-1} \, dx}{e^x + 1} = \Gamma(s)\zeta(s)[1 - \frac{1}{2^{s-1}}]. \qquad (17.46)$$

Thus, $s = 3$ gives the number densities in the ultra-relativistic region for the Bose–Einstein and Fermi–Dirac case as

$$n_{BE} = \frac{gT^3}{\pi^2} \zeta(3)$$

$$n_{FD} = \frac{3}{4} \frac{gT^3}{\pi^2} \zeta(3), \qquad (17.47)$$

where n_{BE} denotes the number density for the Bose–Einstein statistics, n_{FD} denotes the number density for the Fermi–Dirac statistics and $\zeta(3) \simeq 1.202$. The front factor of $(3/4)$ for the case n_{FD} arises from the factor $(1 - 1/2^{s-1})$ for $s = 3$ in Eq. (17.46).

We now look at the energy density ρ given by

$$\rho = g_f \int \frac{d^3p}{(2\pi)^3} \frac{E}{e^{(E-\mu)/T} \pm 1}$$

$$= \frac{gT^4}{2\pi^2} \int_{x_m}^{\infty} \frac{x^3 \, dx}{e^{(x-x_\mu)} \pm 1}, \qquad (17.48)$$

---

[*] We use the natural unit system where $c = k_B = 1$ (see Chapter 23).

and pressure p given by

$$p = g \int \frac{d^3p}{(2\pi)^3} \frac{p^2}{3E} \frac{1}{e^{(E-\mu)/T} \pm 1}$$
$$= \frac{gT^4}{6\pi^2} \int_{x_m}^{\infty} \frac{x^3(1 - x_m^2/x^2)dx}{e^{(x-x_\mu)} \pm 1}. \tag{17.49}$$

In the ultra-relativistic limit when $x_m = 0$ and $x_\mu = 0$, one gets

$$\rho_{BE} = \frac{g\pi^2}{30}T^4$$
$$\rho_{FD} = \left(\frac{7}{8}\right)\frac{g\pi^2}{30}T^4. \tag{17.50}$$

Here, the factor $7/8$ for ρ_{FD} arises from the factor $(1 - 1/2^{s-1})$ for $s = 4$ in Eq. (17.46). In the early universe there would be many species of particles i both bosonic and fermionic. Each of these particles i will have their own degrees of freedom g_i and their own temperature T_i. Thus, the total energy density will be the sum of them, and we can write

$$\rho = \frac{g_*\pi^2}{30}T^4$$
$$g_* = \sum_{i=\text{bose}} g_i \left(\frac{T_i}{T}\right)^4 + \sum_{i=\text{fermi}} \frac{7}{8}g_i \left(\frac{T_i}{T}\right)^4. \tag{17.51}$$

At thermal equilibrium, T_i will equal the photon temperature T, and Eq. (17.51) will reduce to

$$g_* = \sum_{i=\text{bose}} g_i + \sum_{i=\text{fermi}} \frac{7}{8}g_i. \tag{17.52}$$

From Eqs (17.48) and (17.49), the pressure in the relativistic limit can be obtained by taking the limit $x_m \to 0$, and is given by

$$p \simeq \frac{1}{3}\rho, \tag{17.53}$$

which is valid for both Bose–Einstein and Fermi–Dirac statistics.

17.1.2 Degrees of Freedom Count g_*

We count here the effective degrees of freedom g_* for the standard model and for the case of the minimal supersymmetric standard model (MSSM). For the standard model, the count of the bosonic and the fermionic degrees are given in Table 17.1. Here, we find that the bosonic degrees of freedom are $g_{\text{bose}}^{\text{SM}} = 28$ and the fermionic degrees of freedom are given by $g_{\text{fermi}}^{\text{SM}} = 90$. Thus, for temperatures $T > m_t$, g_* for the standard model is

$$g_*^{\text{SM}} = g_{\text{bose}}^{\text{SM}} + \frac{7}{8}g_{\text{fermi}}^{\text{SM}} = 106.75. \tag{17.54}$$

Table 17.1 *Bosonic and fermionic degrees of freedom count for the standard model particles.*

Particles							Total
Bosons	W^+	W^-	Z^0	γ	g	h^0	
g_i	3	3	3	2	16	1	$g_{\text{bose}}^{\text{SM}}$: 28
Leptons	e	μ	τ	ν_e	ν_μ	ν_τ	
g_i	4	4	4	2	2	2	18
Quarks	u	c	t	d	s	b	
g_i	12	12	12	12	12	12	72
							$g_{\text{fermi}}^{\text{SM}}$: 90

Table 17.2 *Fermionic and bosonic degrees of freedom count for the extra particles in the MSSM.*

Particles							Total
Sleptons	$\tilde{e}$	$\tilde{\mu}$	$\tilde{\tau}$	$\tilde{\nu}_e$	$\tilde{\nu}_\mu$	$\tilde{\nu}_\tau$	
g_i	4	4	4	2	2	2	18
Squarks	$\tilde{u}$	$\tilde{c}$	$\tilde{t}$	$\tilde{d}$	$\tilde{s}$	$\tilde{b}$	
g_i	12	12	12	12	12	12	72
Extra Higgs	H^+	H^-	H^0	A^0			
g_i	1	1	1	1			4
							Δg_{Bose} : 94
Higgsinos	$\tilde{H}_1^0$	$\tilde{H}_1^-$	$\tilde{H}_2^+$	$\tilde{H}_2^0$			
g_i	2	2	2	2			8
Gauginos	$\tilde{\lambda}_a (a = 1, 2, 3)$	λ_B	$\tilde{g}$				
g_i	$3 \times 2 = 6$	2	16	0	0	0	24
							Δg_{Fermi} : 32

Let us now extend the analysis to the MSSM. Here, the additional particles and their degrees of freedom are listed in Table 17.2, and we find that the MSSM brings in additional bosonic and fermionic degrees of freedom as follows: $\Delta g_{\text{bose}} = 94$ and $\Delta g_{\text{fermi}} = 32$. The total bosonic and fermionic degrees of freedom for the MSSM are then given by $g_{\text{bose}}^{\text{MSSM}} = 28 + 94 = 122$ and $g_{\text{fermi}}^{\text{MSSM}} = 90 + 32 = 122$. Thus, we see that there is an exact match between the bosonic and the fermionic degrees of freedom for the MSSM case, as it should be. We can now calculate g_* for the MSSM case. For temperatures above all standard model masses, and sparticle, masses we have

$$g_*^{\text{MSSM}} = 122 \times \left(1 + \frac{7}{8}\right) = 228.75. \tag{17.55}$$

17.1.3 Entropy in the Early Universe

An object of interest in the early universe is entropy. Thermodynamics governs the relation between total energy U, the temperature T, and pressure p, and is given by [4]

$$dU = T\,dS - p\,dV. \tag{17.56}$$

Writing $U = \rho V$ and $S = sV$, where s is the entropy density, we have

$$d\rho\,V + \rho\,dV = T\,ds V + Ts\,dV - p\,dV. \tag{17.57}$$

Equating the coefficients of dV we get

$$s = \frac{\rho + p}{T}. \tag{17.58}$$

Now, we consider the ultra-relativistic region where the relation $p = \rho/3$ holds (see Eq. (17.53)). Using this relation we have

$$\rho + p = \frac{g_* \pi^2 T^4}{30} + \frac{g_* \pi^2 T^4}{90}$$

$$= \frac{2}{45} g_* \pi^2 T^4. \tag{17.59}$$

Substitution of Eq. (17.59) into Eq. (17.58) gives

$$s = \frac{2}{45} g_s \pi^2 T^3, \tag{17.60}$$

where g_s is defined by

$$g_s = \sum_{\text{bose}} g_i \left(\frac{T_i}{T}\right)^3 + \frac{7}{8} \sum_{\text{fermi}} g_i \left(\frac{T_i}{T}\right)^3. \tag{17.61}$$

For the case when the particles are in thermal equilibrium with photons, one has $T_i(t) = T(t)$, and g_s (and g_*) becomes independent of temperature.

We now show that the total entropy in a volume $V = R^3$ is constant:

$$S = sV = \frac{\rho + p}{T} V, \tag{17.62}$$

where we have used Eq. (17.58). Next, we consider the time variation of S. We have

$$\frac{dS}{dt} = \frac{d}{dt}\left(\frac{\rho + p}{T} V\right). \tag{17.63}$$

From the above, we derive the result

$$\frac{T^2}{V}\frac{dS}{dt} = \frac{dT}{dt}\left(T\frac{\partial p}{\partial T} - \rho - p\right) + \left(\frac{d\rho}{dt} + 3H(\rho + p)\right).$$ (17.64)

Now, the second term on the right-hand side of Eq. (17.64) vanishes on using Eq. (17.27) while the first term on the right-hand side of Eq. (17.64) also vanishes (see Eq. (17.318)). Thus, we have

$$S = sR^3 = \text{constant}.$$ (17.65)

17.1.4 Particles versus Anti-particles in the Early Universe

We now look at the early universe where the temperature is high enough that the particles are relativistic and we can neglect the particle mass, and thus the particle statistics is determined by temperature and the chemical potential. For the photon the chemical potential is zero, and thus for photons the number density depends only on temperature. More generally, the distributions will also depend on the chemical potential, as discussed in Section 17.1. Thus, consider the process

$$1 + 2 \to 3 + 4$$ (17.66)

in thermal equilibrium. Using Eq. (17.41) we have

$$\mu_1 + \mu_2 = \mu_3 + \mu_4.$$ (17.67)

Since particles and anti-particles can annihilate into photons, we have that particles and their anti-particles have opposite chemical potentials. Thus, for example,

$$\mu_{\bar{q}_i} + \mu_{q_i} = 0, \mu_{e_i} + \mu_{\bar{e}_i} = 0.$$ (17.68)

Since the chemical potentials of particles and anti-particles are opposite to each other, their number densities at thermal equilibrium are not the same, and at temperature T the asymmetry between particles and anti-particles is given by

$$n_i - \bar{n}_i = \frac{g_i}{2\pi^2}\int_0^\infty dp\, p^2\left[(e^{(E_i - \mu_i)/T} \pm 1)^{-1} - (e^{(E_i + \mu_i)/T} \pm 1)^{-1}\right].$$ (17.69)

Here, g_i counts the degrees of freedom for a particle species i, n_i is the number density of particles at equilibrium at temperature T, and $\bar{n}_i$ is the number density of anti-particles at equilibrium at temperature T. Further, E_i, as usual, is given by $E_i = \sqrt{p^2 + m_i^2}$. Let us now determine $n_i - \bar{n}_i$ in the limit when $T \gg m_i$ and thus m_i can effectively be set to zero. Also, we assume that $\mu_i/T \ll 1$. Of course, in the limit $\mu_i/T \to 0$, the difference $n_i - \bar{n}_i$ vanishes. However, we want to retain the leading-order term in this difference. Suppressing the index i, we write Eq. (17.69) as

$$n - \bar{n} = \frac{g}{2\pi^2} \int_0^\infty dE \ E^2 \left[(e^{(E-\mu)/T} \pm 1)^{-1} - (e^{(E+\mu)/T} \pm 1)^{-1} \right]$$
$$= \frac{gT^3}{2\pi^2} \int_0^\infty dx \ x^2 \left[(e^{(x-x_\mu)} \pm 1)^{-1} - (e^{(x+x_\mu)} \pm 1)^{-1} \right].$$

$$(17.70)$$

where in the second line of the equation above we introduced the variables $x = E/T$ and $x_\mu = \mu/T$. An evaluation of the integrals gives

$$n - \bar{n} = \frac{gT^3}{6} \times \begin{cases} 2\frac{\mu}{T} + \left(\frac{\mu}{T}\right)^3 & \text{bosons} \\ \frac{\mu}{T} + \left(\frac{\mu}{T}\right)^3 & \text{fermions.} \end{cases}$$

$$(17.71)$$

17.2 Time and Temperature

In the early universe the scale factor $R(t)$ has the time dependence t^p, where $p = 1/2$ for a radiation-dominated universe and $p = 2/3$ for a non-relativistic matter-dominated universe. This leads to the relation $H = \dot{R}/R = 1/2t$ (see Eq. (17.133)) in the radiation-dominated era. We now determine the relation between time and temperature. To this end, we first determine the dependence of the relic density on time. Let us consider the case $k = 0$. In this case, the relation between $H(t)$ and the relic density is given by $H^2(t) = (8\pi G_N/3)\rho$. For the case of a radiation-dominated universe we have then the relation

$$\rho = \frac{3}{8\pi G_N} \frac{1}{4t^2}.$$

$$(17.72)$$

We also know that a radiation-dominated universe gives

$$\rho = \frac{\pi^2}{30} g_* T^4.$$

$$(17.73)$$

This leads to the following relation between temperature and time in the radiation dominated era:

$$T(t) = \left(\frac{45}{16\pi^3 G_N g_*} \right)^{1/4} t^{-\frac{1}{2}}.$$

$$(17.74)$$

17.3 Supersymmetry and Dark Matter

The most recent astrophysical observations show that up to 95% or more of the universe is dark matter or dark energy while the rest is visible matter. Of the 95% dark universe, approximately 25% is dark matter while the rest is dark energy. Among the dark matter candidates, axions [5–7] and weakly interacting massive particles (WIMPs) are the leading candidates. Relic density analyses show that WIMPs can produce the right amount of dark matter in the universe [8]. Within supergravity there are several possible WIMP candidates such as the gravitino [9, 10], the sneutrino, and the lightest neutralino. Indeed, soon after the formulation of supergravity grand unification [11], the neutralino was proposed as a possible candidate for dark matter [12]. Quite remarkably in supergravity grand unification, it is found that over most of the parameter space of supergravity models the lightest neutralino is indeed the lightest supersymmetric particle (LSP) [13]. Further, under the constraints of R-parity invariance the LSP becomes completely stable and thus truly a candidate for dark matter, validating its proposal [12]. In general, the neutralino is a linear combination of the gauginos and Higgsinos so that [14]

$$\chi^0 = \alpha \lambda_B + \beta \lambda_W + \gamma \tilde{H}_1^0 + \delta \tilde{H}_2^0. \tag{17.75}$$

Next, we discuss the derivation of the relic density of dark matter particles, assuming that the dark matter candidate is the neutralino.

17.3.1 Relic Density Analysis

In the early universe, various species of particles including the neutralino are produced and are in thermal equilibrium. The number density of these particles depends on temperature, and in order to compute their relic density at the current time we need to determine the equation that governs the number density. The equation that controls the number density is the Boltzmann equation, and we give below a brief outline of how it comes about. Let us assume that we have a gas of non-interacting particles in volume $V = R^3(t)$. As the universe expands, the total number remains constant. Thus, if n is the number density, we have that $n(t)R^3(t)$ is a constant as the universe expands. This leads to

$$\frac{d}{dt}(n(t)R^3(t)) = 0, \tag{17.76}$$

or, alternatively,

$$\frac{dn}{dt} + 3nH = 0.$$

$$\tag{17.77}$$

Let us assume we are dealing with dark mater particles and let us consider now the case with interactions. In this case, we will have processes where the dark matter particles scatter into the standard model particles. Further, there would be back-scattering. Thus, we can write

$$\frac{1}{R^3}\frac{d(nR^3)}{dt} = -an^2 + bn_0^2. \tag{17.78}$$

Here, n_0 is the number of particles at thermal equilibrium, and a can be identified with $\langle \sigma v \rangle$ where v is the relative velocity, σ is the annihilation cross-section, and

$\langle \sigma v \rangle$ denotes thermally averaged (σv), i.e.,

$$\langle \sigma v \rangle = \int_0^\infty dv v^2 (\sigma v) e^{-mv^2/4T} \Big/ \int_0^\infty dv v^2 e^{-mv^2/4T}. \qquad (17.79)$$

At thermal equilibrium the right-hand side of Eq. (17.78) vanishes, which gives $b = \langle v\sigma \rangle$, and Eq. (17.78) takes on the form [8]

$$\frac{dn}{dt} + 3Hn + \langle \sigma v \rangle (n^2 - n_0^2) = 0. \qquad (17.80)$$

For a dark matter particle such as a neutralino which is non-relativistic at the time of the freeze-out, n_0 is given by (see Eq. (17.317))

$$n_0 = g \left(\frac{mT}{2\pi} \right)^{3/2} e^{-\frac{m}{T}}, \qquad (17.81)$$

where g is the number of degrees of freedom of the dark matter particle, which for the case of the neutralino is 2, and m is the mass of the particle.

We have seen in the preceding discussion that, as the universe expands, one has conservation of entropy, i.e., $s(t)R^3(t)$ is a constant or, alternatively, the quantity $g_s T^3 R^3$ is a constant so that

$$g_s T^3 R^3(t) = C. \qquad (17.82)$$

It is useful to introduce a new function $f(t)$ so that

$$f(t) = \frac{n(t)}{g_s T^3}. \qquad (17.83)$$

We can then write

$$\frac{df(t)}{dt} = \frac{d}{dt} \left(\frac{n(t)R^3}{g_s T^3 R^3} \right). \qquad (17.84)$$

Using the fact that $g_s T^3 R^3$ is a constant independent of time (see Eq. (17.82)), we can write Eq. (17.84) as follows:

$$\frac{df(t)}{dt} = \frac{1}{g_s T^3} \left[\frac{dn}{dt} + 3Hn \right]. \qquad (17.85)$$

The Boltzmann equation (Eq. (17.80)) now takes the form

$$\frac{df(t)}{dt} = -g_s T^3 \langle \sigma v \rangle [f^2 - f_0^2], \qquad (17.86)$$

where we have defined

$$f_0 = \frac{n_0(t)}{g_s(t)T^3(t)} \qquad (17.87)$$

and n_0 is given by Eq. (17.81). It is convenient to introduce the parameter x so that

$$x = \frac{T(t)}{m}. \qquad (17.88)$$

Using x and the relation between $T(t)$ and time t given by Eq. (17.74), the Boltzmann equation takes the form

$$\frac{df}{dx} = m \left(\frac{45}{4\pi^3 G_N g_*} \right)^{1/2} g_s \langle \sigma v \rangle [f^2(x) - f_0^2(x)], \tag{17.89}$$

where f_0 is given by

$$f_0(x) = \frac{2}{g_s} \left(\frac{1}{2\pi x} \right)^{3/2} e^{-1/x}. \tag{17.90}$$

In writing Eq. (17.89) we have ignored the derivative term proportional to dg_*/dt. In the very early universe the annihilation rate Γ of the dark particles is much faster than the Hubble expansion (i.e., $\Gamma \gg H$). In this case, the dark particles are in thermal equilibrium with the background, and thus the function $f(x)$ assumes its equilibrium value $f_0(x)$. However, as the universe cools, the equilibrium is lost when the back-reactions become less important and can be neglected. This particular transition temperature, below which one can neglect the back-reactions, is the freeze-out temperature. Below the freeze-out temperature the quantity $f(x)$ will be determined by Eq. (17.89) but with the f_0 term neglected so that we have

$$\frac{df}{dx} = m \left(\frac{45}{4\pi^3 G_N g_*} \right)^{1/2} g_s \langle \sigma v \rangle f^2(x). \tag{17.91}$$

As stated, this relation holds when the back-reaction is neglected, which happens in the domain $\Gamma \ll H$. Now, Eq. (17.91) holds below the freeze-out temperature but it also holds approximately at the time of the freeze-out. However, at the time of the freeze-out we also had an approximate thermal equilibrium, and so we can obtain an approximate expression for the freeze-out temperature by replacing $f(x)$ by $f_0(x)$. Thus, an approximate expression for the freeze-out temperature can be obtained from the equation

$$\left(\frac{df_0}{dx} \right)_{x_f} = m \left(\frac{45}{4\pi^3 G_N g_*} \right)^{1/2} g_s \langle \sigma v \rangle_{x_f} f_0^2(x_f). \tag{17.92}$$

We can obtain the following relation for x_f from the above:

$$x_f^{-1} \simeq \ln \left[\frac{1}{\pi^3} \left(\frac{45}{8 G_N g_*(x_f)} \right)^{1/2} m \langle \sigma v \rangle_f x_f^{1/2} \right]. \tag{17.93}$$

This equation needs to be solved numerically but typically the value of x_f turns out to be of size $1/20$ for a significant set of parameter choices.

Integrating Eq. (17.91) from x_f to the current temperature $x_0 = T_\gamma/m$, where $T_\gamma = 2.73^0$ K, one has

$$\frac{1}{f(x_0)} = \frac{1}{f(x_f)} + I(x_f), \tag{17.94}$$

where

$$I(x_f) \equiv m \int_0^{x_f} dx \left(\frac{45}{4\pi^3 G_N}\right)^{1/2} \frac{g_s}{\sqrt{g_*}} \langle \sigma v \rangle. \tag{17.95}$$

Thus, we may write the relic density ρ_χ in the form

$$\rho_\chi = \left(\frac{4\pi^3 G_N}{45}\right)^{1/2} T_\gamma^3 \frac{g_s(x_0)}{\bar{J}(x_f)}, \tag{17.96}$$

where

$$\bar{J}(x_f) = \int_0^{x_f} dx \frac{g_s}{\sqrt{g_*}} \langle \sigma v \rangle(x). \tag{17.97}$$

A useful quantity is the ratio Ω_χ, defined so that

$$\Omega_\chi = \frac{\rho_\chi}{\rho_c}. \tag{17.98}$$

Here, ρ_c is the critical relic density needed to close the universe, as given by Eq. (17.37). We may write this ratio so that

$$\Omega_\chi h_0^2 = \frac{T_\gamma^3}{1.88 \times 10^{-29} \text{ g cm}^3} \left(\frac{4\pi^3 G_N}{45}\right)^{1/2} \frac{g_s(x_0)}{J(x_f)}. \tag{17.99}$$

An approximation often used is to move the factor $g_s/\sqrt{g_*}$ out of the integral in $J(x_f)$ using its freeze-out value, so that

$$\bar{J}(x_f) \simeq \frac{g_s(x_f)}{\sqrt{g_*}(x_f)} J(x_f). \tag{17.100}$$

We now come to an important fine point. Since x_f is small and we are in the non-relativistic regime, it had generally been thought that one could expand σv in a power series in v^2, $\sigma v = a + bv^2/6 + \cdots$ In this case, the thermal average becomes trivial to perform, i.e., $\langle \sigma v \rangle \cong a + bx$ (since $\langle v^2 \rangle = 6T/m_\chi$). However, it has been pointed out [15, 16] that this approximation can break down badly near a pole or threshold. The breakdown is particularly serious for the case at hand due to the narrowness of the h and Z poles [17]. To see what the problem is, consider the h-pole where σv has the Breit–Wigner form

$$\sigma v \cong \frac{A_h}{m_\chi^2} \frac{v^2}{[(v^2 - \varepsilon_h)^2 + \gamma_h^2]}. \tag{17.101}$$

Here $\varepsilon_h = (m_h^2 - 4m_\chi^2)/m_\chi^2$, $\gamma_h = m_h \Gamma_h/m_\chi^2$, and A_h is a constant. The width h is $\Gamma_h \simeq 2.5 \times 10^{-3}$ GeV, which implies that γ_h is very small. When we thermally average, we "smear" v^2, and if $\varepsilon_h > 0$ (i.e., $2m_\chi < m_h$) then for $v^2 \approx \varepsilon_h$ the integrand becomes very large, and a large enhancement in $\langle \sigma v \rangle$ can result. This will modify the estimate of $J(x_f)$ and produce significant changes in $\Omega_\chi h_0^2$.

We note that $J(x_f)$ involves a double integral over a pole, and this integration must be done with care because of the singular nature of the pole. A convenient procedure to circumvent this problem is to first do the x-integral analytically and obtain

$$J_h = \frac{A_h}{m_\chi^2 2\pi^{1/2}} \int_0^\infty d\xi \, e^{-\xi}\xi^{-\frac{1}{2}} \left\{ \frac{1}{2} \ln\left[\frac{(4\xi x_f - \varepsilon_h)^2 + \gamma_R^2}{\varepsilon_h^2 + \gamma_h^2} \right] \right.$$
$$\left. + \frac{\varepsilon_h}{\gamma_h} \left[\tan^{-1}\left(\frac{4\xi x_f - \varepsilon_h}{\gamma_h} \right) + \tan^{-1}\left(\frac{\varepsilon_h}{\gamma_h} \right) \right] \right\}. \qquad (17.102)$$

The remaining integral can then be done numerically without difficulty. A similar analysis can be carried out for the Z pole contributions. For the t-channel exchanges of sfermions, the expansion $\sigma v \cong a + bv^2/6$ is a good approximation, as we are not near a singularity. The above is the simplest picture, and several modifications can occur. For example, one can have co-annihilation with other sparticles near the freeze-out, which significantly modifies the relic density analyses. We note that data from colliders can be used to determine dark matter relic density in certain regions of the supergravity models if neutralinos are produced at colliders (e.g., [18]).

17.3.2 Dark Matter Detection

The direct detection of dark matter presents an interesting possibility. Since the halo of the Milky Way contains a population of WIMPs, they can scatter off nuclei, and such nuclear recoil can be detected. The detection of dark matter depends on the scattering cross-section, and an early estimation was made by Goodman and Witten [19]. The fundamental Lagrangian that governs the scattering is the interaction of the WIMP with the quarks inside the nucleon. Since the WIMP velocity is rather small (estimated at $\langle v \rangle \sim 300$ km/s), the WIMPs are highly non-relativistic, and the analysis of the scattering can be done by using the low-energy four-Fermi interaction. If the WIMP χ was a Dirac fermion, such a low-energy interaction would take the form

$$L_{eff}^D = \bar{\chi}(\lambda_1 + \lambda_2\gamma_5)\chi\bar{q}(\lambda_3 + \lambda_4\gamma_5)q + \bar{\chi}\gamma^\mu(\lambda_5 + \lambda_6\gamma_5)\chi\bar{q}\gamma_\mu(\lambda_7 + \lambda_8\gamma_5)q$$
$$+ \lambda_9\bar{\chi}\sigma^{\mu\nu}\chi\bar{q}\sigma_{\mu\nu}q. \qquad (17.103)$$

In MSSM-type models, Eq. (17.103) arises only after elimination of the heavy modes. For the case of the MSSM, this involves integration over the Z-boson, the Higgs boson, and the sfermion poles in the Feynman diagrams. Now, an integration over the Z-pole and the Higgs pole directly leads to terms of the type given on the right-hand side of Eq. (17.103). However, for the case of sfermion exchange, an integration on the sfermion pole gives an interaction of the type

$$\bar{\chi}(C_L P_L + C_R P_R)q\bar{q}(C_R^* P_L + C_L^* P_R)\chi. \qquad (17.104)$$

This interaction involves a four-Fermi interaction of the type $\bar{\chi}q\bar{q}\chi$, $\bar{\chi}\gamma_5 q\bar{q}\chi$, etc., which is not of the canonical four-Fermi interaction that appears in Eq. (17.103). In order to write these interactions in the form of Eq. (17.103), one needs a Fierz rearrangement, which is discussed in Section 22.6

For the case of a Majorana particle, one has

$$\bar{\chi}\gamma^{\mu}\chi = 0, \bar{\chi}\sigma^{\mu\nu}\chi = 0. \tag{17.105}$$

Further, the interactions where $\bar{q}\gamma_5 q$ and $\bar{\chi}\gamma_5\chi$ appear give relatively small contributions in the non-relativistic limit, and can be dropped. The effective interaction that governs the scattering of neutralinos from quarks in the non-relativistic limit is then given by

$$\mathcal{L}_{eff}^{M} = (\bar{\chi}\gamma^{\mu}\gamma^{5}\chi)\bar{q}\gamma^{\mu}(A_q P_L + B_q P_R)q + (\bar{\chi}\chi)C_q m_q \bar{q}q. \tag{17.106}$$

We now want to compute the neutralino–proton cross-section. Let us focus on the scalar interaction given by the last term of Eq. (17.106). For the light quarks (u, d, s) we define the matrix element

$$m_p f_i^p = \langle p | m_{q_i} \bar{q}_i q_i | p \rangle, \tag{17.107}$$

which gives the effective neutralino–proton interaction of the form

$$\left(\sum_{i=u,d,s} f_i^p C_i\right) m_p \bar{\chi}\chi \bar{p}p. \tag{17.108}$$

Shifman, Vainshtein, and Zakharov [20] have pointed out that the heavy quarks also contribute to the mass of the nucleon. This occurs via the anomaly. For each of the heavy quarks, one makes the following substitution:

$$m_Q \bar{Q}Q \rightarrow -\frac{\alpha_s}{12\pi}GG. \tag{17.109}$$

Thus, the trace of energy-momentum tensor for quarks and gluons is given by

$$m_u \bar{u}u + m_d \bar{d}d + m_s \bar{s}s + \sum_{Q=b,c,t} m_Q \bar{Q}Q - \frac{\beta(\alpha_s)}{4\alpha_s}GG, \tag{17.110}$$

where the last term arises from a triangle anomaly and $\beta(\alpha_s) = -7\alpha_s^2/(2\pi)$ for three heavy quarks (c, b, t).

Eliminating the heavy quarks, the trace reduces to

$$\theta_{\mu}^{\mu} = m_u \bar{u}u + m_d \bar{d}d + m_s \bar{s}s - \frac{9\alpha_s}{8\pi}GG. \tag{17.111}$$

The nucleon mass is the matrix element $m_n = \langle n | \theta_{\mu}^{\mu} | n \rangle$. Using the matrix element of Eq. (17.111) between nucleon states, one finds

$$\langle n | GG | n \rangle = -\frac{8\pi}{9\alpha_s} m_n \left(1 - \sum_{q=u,d,s} f_q^n\right)$$

$$\langle n | m_Q \bar{Q}Q | n \rangle = \frac{2}{27} m_n \left(1 - \sum_{q=u,d,s} f_q^n\right). \tag{17.112}$$

Including the contribution of the heavy quarks, the effective neutralino–proton interaction takes the form

$$\left(\sum_{i=u,d,s} f_i^p C_i + \frac{2}{27}\left(1 - \sum_{q=u,d,s} f_q^p\right) \sum_{a=c,b,t} C_a \right) m_p \bar{\chi}\chi \bar{p}p. \tag{17.113}$$

Next, we derive the contribution to the spin-independent cross-section for neutralino–proton scattering using the interaction of Eq. (17.113). We find that the neutralino–proton scattering cross-section for the interaction of Eq. (17.113) at threshold is given by

$$\sigma_{\chi-p}\,(\text{scalar}) = \frac{4}{\pi} m_r^2 \left(\sum_{i=u,d,s} f_i^p C_i + \frac{2}{27}\left(1 - \sum_{q=u,d,s} f_q^p\right) \sum_{a=c,b,t} C_a \right)^2, \tag{17.114}$$

where m_r is the reduced mass of the neutralino-proton system and the factor of 4 arises from the Majorana nature of the neutralinos. $C_{i,a}$ is the scattering amplitude for scattering off the quark i, a, and it contains contributions from the exchange of the Higgs bosons and the exchange of the sfermions. We can write a result similar to Eq. (17.114) for the χ–neutron scattering where f_i^n replaces f_i^p. Next, let us consider scattering of a neutralino off a nucleus of atomic number A consisting of Z number of protons. One can use the result of Eq. (17.114) and a similar result for χ – neutron scattering to get

$$\sigma_{\chi-N}(\text{scalar}) = \frac{4m_r^2}{\pi}\left(Z \sum_{i=u,d,s} f_i^p C_i + \frac{2}{27} Z(1 - \sum_{i=u,d,s} f_i^p) \sum_{a=c,b,t} C_a \right.$$
$$\left. + (A-Z) \sum_{i=u,d,s} f_i^n C_i + \frac{2}{27}(A-Z)(1 - \sum_{i=u,d,s} f_i^n) \sum_{a=c,b,t} C_a \right)^2. \tag{17.115}$$

We can rewrite the above result in the following form:

$$\sigma_{\chi-N}(\text{scalar}) = \frac{4m_r^2 A}{\pi}\left(\hat{f}\frac{m_u C_u + m_d C_d}{m_u + m_d} + \hat{f}\xi\Delta\frac{m_u C_u - m_d C_d}{m_u + m_d} + fC_s \right.$$
$$\left. + \frac{2}{27}(1 - f - \hat{f} - \hat{f}\xi\Delta\frac{m_u - m_d}{m_u + m_d}) \sum_{a=c,b,t} C_a \right)^2, \tag{17.116}$$

where we have introduced the notation $\Delta = (2Z - A)/A$, $f = f_s$, and $\hat{f} = \sigma_{\pi N}/m_p$ (see Eq. (17.117)). Numerically, $\hat{f} = 0.05 \pm 0.01$.

We now discuss the computation of f_i^p and f_i^n. These can be related to the parameters σ_0, ξ, and $\sigma_{\pi N}$, which can be determined by perturbation theory and by experiment. Thus define

$$\langle p|2^{-1}(m_u + m_d)(\bar{u}u + \bar{d}d|p\rangle = \sigma_{\pi N}$$

$$\xi = \frac{\langle p|\bar{u}u - \bar{d}d|p\rangle}{\langle p|\bar{u}u + \bar{d}d|p\rangle}$$

$$\sigma_0 = \sigma_{\pi N}(1 - y)$$

$$y \equiv \frac{\langle p|2\bar{s}s|p\rangle}{\langle p|\bar{u}u + \bar{d}d|p\rangle}. \tag{17.117}$$

It is also useful to define

$$x = \frac{\sigma_0}{\sigma_{\pi N}} = \frac{\langle p|\bar{u}u + \bar{d}d - 2\bar{s}s|p\rangle}{\langle p|\bar{u}u + \bar{d}d|p\rangle}. \tag{17.118}$$

From Eqs. (17.107), (17.117), and (17.118) it is straightforward to determine f_i^p and f_i^n in terms of x, ξ, and $\sigma_{\pi N}$. Thus, for f_i^p, one has

$$f_u^p = \frac{m_u}{m_u + m_d}(1 + \xi)\frac{\sigma_{\pi N}}{m_p}$$

$$f_d^p = \frac{m_d}{m_u + m_d}(1 - \xi)\frac{\sigma_{\pi N}}{m_p}$$

$$f_s^p = \frac{m_s}{m_u + m_d}(1 - x)\frac{\sigma_{\pi N}}{m_p}, \tag{17.119}$$

and f_i^n can be obtained from f_i^p by the switch $\xi \to -\xi$. It is also easily checked that $f_{u,d}^{p,n}$ satisfy the following sum rule:

$$f_u^p f_d^p = f_u^n f_d^n. \tag{17.120}$$

Currently, there are significant uncertainties in the determination of $\sigma_{\pi N}$, x, and ξ. Estimates give $\sigma_{\pi N} = 65 \pm 10$ MeV and $\sigma_0 = 35 \pm 6$ MeV while ξ can be related to x via baryon mass splittings [21], i.e.,

$$\xi = \frac{(\Xi^- + \Xi^0 - \Sigma^+ - \Sigma^-)x}{\Xi^- + \Xi^0 + \Sigma^+ + \Sigma^- - 2m_p - 2m_n}, \tag{17.121}$$

which leads to $\xi = 0.132 \pm 0.035x$ [22]. The quark mass ratios also contain uncertainties [23, 24], i.e.,

$$\frac{m_u}{m_d} = 0.553 \pm 0.043, \frac{m_s}{m_d} = 18.9 \pm 0.8. \tag{17.122}$$

These lead to the following numerical values for the f_i^p: $f_u^p = 0.021 \pm 0.004$, $f_d^p = 0.029 \pm 0.006$, $f_s^p = 0.21 \pm 0.12$, and for f_i^n one has $f_u^n = 0.016 \pm 0.003$, $f_d^n = 0.037 \pm 0.007$, and $f_s^n = 0.21 \pm 0.12$. Spin-dependent interactions provide an important tool for the identification of a dark matter candidate. For a recent analysis see the literature [25].

17.4 Inflation

The Big Bang cosmology is highly successful in explaining a vast amount of data. However, it is not a complete theory, specifically as it relates to the very early history of the universe. Within the standard Big Bang cosmology, one has at least three very significant puzzles. These relate to (1) the flatness problem, (2) the horizon problem, and (3) the monopole problem. Let us discuss these briefly. We begin with the flatness problem. The Friedmann equation (Eq. (17.22)) for the Robertson–Walker metric reads

$$\left(\frac{\dot{R}}{R}\right)^2 = \frac{8\pi G\rho}{3} - \frac{\kappa}{R^2}. \tag{17.123}$$

Since $H = \dot{R}/R$ and defining Ω so that

$$\Omega \equiv \frac{8\pi G\rho}{3H^2}, \tag{17.124}$$

one finds the relation

$$\Omega = 1 + k(R^2 H^2)^{-1}. \tag{17.125}$$

As shown below (see Eq. (17.133)) the time dependence of the scale factor is $R(t) \sim t^s$, where $s = 1/2$ for a radiation-dominated universe and $s = 2/3$ for a non-relativistic matter-dominated universe. This implies that $(R^2 H^2)^{-1} \sim t^{2-2s}$, which diverges with time as t increases. Thus, the observed flatness of the current universe , i.e., $\Omega \sim 1$, is possible only if $(\Omega - 1)$ is vanishingly small to a very high order, which requires a high degree of fine tuning and is unnatural.

Another problem of Big Bang cosmology is the horizon problem. The horizon is defined by the maximum distance travelled by a light signal. A light trajectory is defined by $ds^2 = 0 = -c^2\, dt^2 + R^2(t)dr^2$. The proper distance traveled by light in time dt is $c\, dt$. The comoving distance is defined to be $dr = c\, dt/R(t)$. Thus, the comoving distance traveled by the photon from the initial time t_i to the current time t is given by

$$\Delta r = \int_{t_i}^{t} \frac{c\, dt}{R(t)} \quad \text{(comoving distance)}. \tag{17.126}$$

From now on we set $c = 1$. As noted above, the scale factor is typically given by $R(t) = t^s$ with $s < 1$. Assuming $t_i = 0$, one finds the comoving horizon distance to be

$$\Delta r = \frac{t^{1-s}}{1-s}. \tag{17.127}$$

The physical horizon size (d_{Hor}) travelled by light is obtained by multiplication by $R(t)$ so that

$$d_{Hor} = \frac{t}{1-s} \sim \frac{s}{1-s} H^{-1}, \tag{17.128}$$

where we used $H = \dot{R}/R \sim st^{-1}$. The analysis implies that the horizon distance is increasing in time, and thus part of the universe in causal contact is increasing in time. However, the universe we observe is highly homogeneous and isotropic on large scales. For example, the cosmic microwave background temperature is about 3^0 K in all directions in the sky. Thus, it is difficult to understand the high degree of homogeneity of the universe in the framework of the standard Big Bang cosmology.

The third problem of Big Bang cosmology arises due to the possibility of producing stable superheavy particles, which can give large contributions to the energy density of the universe, making Ω at the current time much larger than unity. Thus, a more complete model needs to accommodate these observational constraints. A possible approach to resolving these issues is the idea of inflation, which requires the universe to go through a period of exponential expansion [26–31]. One may illustrate how inflation works by considering the line element of the Friedman–Robertson–Walker metric. Let us consider the general case where pressure p is parametrized as

$$p = \lambda_p \rho \tag{17.129}$$

so that

$$\frac{\ddot{R}}{R} = -\frac{4\pi G}{3} \rho(1 + 3\lambda_p). \tag{17.130}$$

We have seen that for relativistic matter $p = \frac{1}{3}\rho$, which gives $\lambda_p = 1/3$, and for non-relativistic matter one takes $p = 0$, which implies $\lambda_p = 0$. Further, we have the energy conservation equation

$$\dot{\rho} = -3H(\rho + p). \tag{17.131}$$

Let us now examine the time dependence of the scale factor for the case when matter is relativistic and when it is non-relativistic. Let us consider solutions of the type $R(t) = R_0 t^\alpha$ and $\rho(t) = \rho_0 t^\beta$. One finds that Eqs. (17.23) and (17.131) give the following determination for α and β: $\alpha = 2/(3(1 + \lambda_p))$ and $\beta = -2$. Thus, $\rho(t)$ and $R(t)$ have the following general behavior:

$$\rho(t) = \rho_0 t^{-2} \tag{17.132}$$
$$R(t) = R(t_0) t^{2/[3(1+\lambda_p)]}$$
$$= R(t_0) t^{1/2}, \qquad \text{relativistic}, \lambda_p = 1/3$$
$$= R(t_0) t^{2/3}, \qquad \text{non−relativistic}, \lambda_p = 0. \tag{17.133}$$

However, the time dependence of $R(t)$ is very different for the case if at some time in the past $H = \dot{R}/R$ was approximately constant. In that era, one finds

$$R(t) \sim R(t_0) e^{Ht}, \tag{17.134}$$

i.e., there is an exponential increase in the scale factor. A useful measure of the size of the exponential increase is the number of e-folds N defined so that

$$N = \ln \frac{R(t_0)}{R(t_i)}, \tag{17.135}$$

where t_0 is the current time and t_i is the initial time. Thus, we see that in this era one has an exponential increase in the scale factor of the universe. Using Eq. (17.134) in Eq. (17.135), one finds

$$N = \int_{t_i}^{t_0} dt \, H(t). \tag{17.136}$$

Typically, one needs ~50–60 e-foldings to get consistency with data.

Next, let us see how such a situation can be realized in an actual model. One approach is the so-called slow roll model involving scalar fields. Consider a scalar field $\phi(x)$ coupled to gravity given by the Lagrangian

$$\mathcal{L}_\phi = \sqrt{-g} \left(\frac{1}{2} g^{\mu\nu} \phi(x) \, \partial_\mu \partial_\nu \phi - V(\phi) \right). \tag{17.137}$$

Now, in a Robertson–Walker metric, the matter density and the pressure are given by (see Eqs. (17.315) and (17.316))

$$\rho = \frac{1}{2}\dot{\phi}^2 + \frac{1}{2}R^{-2}(\nabla\phi)^2 + V(\phi) \tag{17.138}$$

$$p = \frac{1}{2}\dot{\phi}^2 - \frac{1}{2}R^{-2}(\nabla\phi)^2 - V(\phi). \tag{17.139}$$

We will make the assumption that

$$R^{-2}(\nabla\phi)^2 \ll V(\phi), \tag{17.140}$$

so the gradient terms in Eqs. (17.138) and (17.139) can be dropped. Further, from Eqs. (17.138) and (17.139), one finds

$$\rho + p = \dot{\phi}^2, \tag{17.141}$$

From Eqs. (17.141) and (17.131) we have

$$\dot{\rho} = -3H\dot{\phi}^2. \tag{17.142}$$

while from Eq. (17.138) (ignoring the gradient term) we get

$$\dot{\rho} = \dot{\phi}\ddot{\phi} + V'(\phi)\dot{\phi}. \tag{17.143}$$

Thus, Eqs. (17.142) and (17.143) lead to the relation

$$\ddot{\phi} + 3H\dot{\phi} + V'(\phi) = 0, \tag{17.144}$$

where we have assumed $\dot{\phi} \neq 0$.

From Eq. (17.22) we get (when $k = 0$)

$$H^2 = \frac{8\pi G_N}{3} \left(\frac{1}{2}\dot{\phi}^2 + V(\phi) \right). \tag{17.145}$$

Taking the time derivative of Eq. (17.145), ignoring the $\ddot{\phi}$-term, and using Eq. (17.144) gives the relation

$$\dot{H} = -4\pi G_N \dot{\phi}^2. \tag{17.146}$$

For exponential expansion we need to have H almost constant during the period of the expansion. Specifically, we would like

$$\left| \frac{\dot{H}}{H} \right| \frac{1}{H} \ll 1. \tag{17.147}$$

Using Eqs. (17.145) and (17.146), the constraint of Eq. (17.147) can be met if

$$\dot{\phi}^2 \ll |V(\phi)|. \tag{17.148}$$

Under the constraint of Eq. (17.148), we find that Eqs. (17.138) and (17.139) give

$$\rho \simeq V(\phi), \; p \simeq -V(\phi)$$
$$\rho + p \simeq 0. \tag{17.149}$$

In addition to Eq. (17.147), we will make the assumption of slow roll, i.e., that

$$|\ddot{\phi}| \ll 3H|\dot{\phi}|. \tag{17.150}$$

Under the assumption of Eq. (17.150), Eq. (17.144) reduces to

$$3H\dot{\phi} + V'(\phi) \simeq 0. \tag{17.151}$$

Using Eqs. (17.145), (17.146), and (17.151), we find

$$\frac{|\dot{H}|}{H^2} = \frac{1}{16\pi G_N} \left(\frac{V'(\phi)}{V(\phi)} \right)^2. \tag{17.152}$$

Thus, the condition of Eq. (17.147) implies

$$\epsilon \equiv \frac{1}{16\pi G_N} \left(\frac{V'}{V} \right)^2$$
$$= \frac{1}{2} \left(\frac{M_{Pl}V'}{V} \right)^2 \ll 1. \tag{17.153}$$

Another slow-roll parameter involves the second derivative of the potential, and the constraint is expressed by

$$\eta \equiv \frac{1}{8\pi G_N}\left|\frac{V''(\phi)}{V(\phi)}\right|$$
$$= \left|\frac{M_{Pl}^2 V''(\phi)}{V(\phi)}\right| \ll 1. \tag{17.154}$$

The validity of Eq. (17.154) can be checked as follows. We take the time derivative of Eq. (17.151) and solve for $\ddot{\phi}$ so that

$$\ddot{\phi} = \frac{V''V'}{9H^2} - \left(\frac{V'^3}{48\pi G_N V^2}\right), \tag{17.155}$$

where we have used Eq. (17.151) to eliminate $\dot{\phi}$. The second term on the right-hand side of Eq. (17.155) is seen to be $\ll V'/3$ using Eq. (17.153). Further, on using Eq. (17.150) along with Eq. (17.155), we find

$$V''(\phi) \ll 9H^2, \tag{17.156}$$

and on using Eq. (17.145) under the constraint that $\dot{\phi}^2 \ll V$ along with Eq. (17.156), we find that the slow-roll constraint of Eq. (17.154) is satisfied. In addition, often we use a third slow-roll constraint which involves the third derivative of the potential such that $\xi^2 \ll 1$ where

$$\xi^2 \equiv \left(\frac{V'V''}{\kappa^4 V^2}\right). \tag{17.157}$$

Equations (17.153), (17.154), and (17.157) define the three standard slow-roll parameters of inflation.

We can also express the e-foldings as an integral over the scalar field that is involved in the slow roll. Thus, we can write

$$N_e = \int_{t_i}^{t_f} H \, dt = \int_{\phi_i}^{\phi_f} \frac{H \, d\phi}{\dot{\phi}}. \tag{17.158}$$

Next, using Eqs. (17.145) and (17.151) in Eq. (17.158), we find that e-foldings are given by

$$N_e = -\int_{\phi_i}^{\phi_f} \frac{8\pi G_N V(\phi)}{V'(\phi)} d\phi. \tag{17.159}$$

Equation (17.159) implies that the ratio V/V' must be negative, and since $V > 0$ for an acceptable potential, $V' < 0$. Now, from Eq. (17.153), we find that $|V/V'| \gg 1/\sqrt{16\pi G_N}$, which implies

$$\left|\frac{8\pi G_N V(\phi)}{V'(\phi)}\right| \gg \sqrt{4\pi G_N}. \tag{17.160}$$

The above also implies that

$$N_e \gg \sqrt{4\pi G_N} \, |\phi_f - \phi_i|. \tag{17.161}$$

The implication of Eq. (17.161) is that as long as $\Delta\phi = |\phi_f - \phi_i| \sim O(1/\sqrt{4\pi G_N})$ $\sim O(M_{\text{Planck}})$, a large number of e-foldings can occur. Thus, a scalar field coupled to gravity with a slow roll can generate conditions necessary for inflation.

We now consider a specific model for $V(\phi)$ so that

$$V(\phi) = \lambda\phi^p, \tag{17.162}$$

where $p = 2$ is the so-called chaotic inflation case. For the potential of Eq. (17.162) the two flatness conditions of Eqs. (17.153) and (17.154) give

$$|\phi| \gg \left|\frac{p}{\sqrt{16\pi G_N}}\right|, |\phi| \gg \left|\frac{\sqrt{p(p-1)}}{\sqrt{8\pi G_N}}\right|. \tag{17.163}$$

Both of these conditions are satisfied when

$$|\phi| \gg \frac{1}{\sqrt{4\pi G_N}}. \tag{17.164}$$

If we assume that $\phi_f \ll \phi_i$ and parametrize ϕ_i so that

$$\phi_i = \sqrt{\frac{n}{\pi G_N}}, \tag{17.165}$$

then the number of e-foldings are $N_e = 8n/p$, and for $n/p = 8$ we find $N_e = 64$. It is clear that the number of e-foldings depends on the choice of ϕ_i.

Since the scalar field is scaled by the Planck mass, it is relevant to ask if the energy density due to the scalar potential might exceed the Planck energy density. If this were to happen, quantum gravity effects would become relevant and the classical treatment would not be reliable. To avoid this possibility, we must impose on the parameters a constraint so that the quantum effects remain small. The condition that guarantees this is the constraint that

$$|V(\phi)| \ll (4\pi G_N)^{-2}. \tag{17.166}$$

As an illustration, let us consider the case $p = 4$. In this case, λ in Eq. (17.163) is a dimensionless parameter, and we can determine the constraint on λ so that the quantum gravity effects are negligible. Thus, using Eqs. (17.163) and (17.165) in Eq. (17.166), one finds that λ must obey the constraint $\lambda \ll (16n^2)^{-1}$. Let us assume that we want the number of e-foldings to be $N_e = 50$, which for $p = 4$ implies $n = 25$, and thus we find that $\lambda \ll 10^{-4}$, which is a significant constraint on the self-couplings of the scalar field ϕ.

17.4.1 The Higgs Field as an Inflaton Field

It has been proposed that the Higgs field of the standard model could itself be an inflaton field, and such a situation could be realized if one allows for non-minimal couplings of the Higgs field with the gravity field. Thus, we write the interaction of the Higgs field with gravity so that the action in the Jordan frame is given by [32][†]

$$A_J = \int d^4x \ \sqrt{-g} \left[-\frac{1}{2}(M_{Pl}^2 + \gamma h^2)R - \frac{1}{2}\partial_\mu h \partial^\mu h - \frac{\lambda}{4}(h^2 - v^2)^2 \right]. \quad (17.167)$$

As can be seen, we have here a non-minimal coupling to gravity. It is useful to express the action in the Einstein frame by making a conformal transformation. We define the metric in the Einstein frame with a tilde, i.e., $\tilde{g}_{\mu\nu}$, related to $g_{\mu\nu}$ in the Jordan frame by

$$\tilde{g}_{\mu\nu} = \Omega^2 g_{\mu\nu}, \quad (17.168)$$

where

$$\Omega^2 = 1 + \frac{\gamma h^2}{M_{Pl}^2}. \quad (17.169)$$

After this transformation, the kinetic energy of the Higgs field would no longer be minimal, and we need to introduce a transformation to get a canonical normalization for the Higgs field. Thus, we make the transformation from $h(x)$ to $\tilde{h}(x)$ so that

$$\frac{d\tilde{h}}{dh} = \left(\frac{\Omega^2 + 6\gamma^2 h^2/M_{Pl}^2}{\Omega^4} \right)^{1/2}. \quad (17.170)$$

With the above transformation one finds that the action in the Einstein frame takes the form

$$A_E = \int d^4x \sqrt{-\tilde{g}} \left(-\frac{M_{Pl}^2}{2}\tilde{R} - \frac{1}{2}\partial_\mu\tilde{h}\partial^\mu\tilde{h} - \tilde{V}(\tilde{h}) \right). \quad (17.171)$$

In the above, $\tilde{R}$ is the Einstein curvature scalar derived in the Einstein frame using the metric $\tilde{g}_{\mu\nu}$, and $\tilde{V}(x)$ is defined so that

$$\tilde{V}(\tilde{h}) = \frac{\lambda}{4}(h(\tilde{h})^2 - v^2)^2 \Omega(\tilde{h})^{-4}. \quad (17.172)$$

For the case when one has small values for the fields h, one has $\tilde{h} \sim h$, and here the deviations of Ω from unity are negligible. However, h and $\tilde{h}$ can differ by significant amounts for field configurations where h is large. Specifically, we will consider the case when

$$\tilde{h} \gg \sqrt{6}M_{Pl}. \quad (17.173)$$

In this case, the potential $\tilde{V}(\tilde{h})$ takes the form [32]

$$\tilde{V}(x) = \frac{\lambda M_{Pl}^4}{4\gamma^2} \left(1 + e^{-2\tilde{h}/\sqrt{6}M_{Pl}} \right)^{-2}. \quad (17.174)$$

[†] A Jordan frame is one where the Ricci scalar in the Lagrangian appears multiplied by a field-dependent factor while an Einstein frame is one where the Ricci scalar appears by itself without an additional factor.

It is now easily seen that the potential at large field configurations is very flat, and thus slow-roll inflation can occur. Under the constraint that $\Omega \gg 1$ and $h \gg v$, one has the following values for the slow-roll parameters ϵ, η, and ξ^2 [32]

$$\epsilon \equiv \frac{1}{2}\left(\frac{\tilde{V}'}{\kappa \tilde{V}}\right)^2 \simeq \frac{4}{3\kappa^4\gamma^2 h^4} \tag{17.175}$$

$$\eta \equiv \left(\frac{\tilde{V}''}{\kappa^2 \tilde{V}}\right) \simeq -\frac{4}{3\kappa^2\gamma h^2} \tag{17.176}$$

$$\xi^2 = \left(\frac{\tilde{V}'\tilde{V}'''}{\kappa^4 \tilde{V}^2}\right) \simeq \frac{16}{9\kappa^4\gamma^2 h^4}, \tag{17.177}$$

where the primes corresponds to differentiation with respect to $\tilde{h}$. Since $\Omega \gg 1$ implies $\kappa^2\gamma h^2 \gg 1$, the above equations satisfy the slow-roll conditions.

17.4.2 Inflation in Supergravity

We now discuss inflation in supergravity, specifically F-type inflation in which the F-part of the potential plays the central role in driving inflation. Recall that the supergravity scalar potential is given by

$$V = e^{\kappa^2 K}\left[(K^{-1})^j_i F_j F^i - 3\kappa^2|W|^2\right] + V_D, \tag{17.178}$$

where

$$F^i = W^i + \kappa^2 K^i W. \tag{17.179}$$

Let us assume that the D-terms do not play a role and their contribution in the potential can be ignored. In this case, differentiating Eq. (17.178) twice with respect to ϕ we have

$$V''(\phi) = \kappa^2 K'' V + \cdots, \tag{17.180}$$

where $\cdots$ refer to sub-leading terms. For a flat Kähler potential $K'' = 1$ and thus $\eta = 1$, which does not satisfy the slow-roll constraint, i.e., $\eta \ll 1$. This is often referred to as the η-problem of supergravity inflation with F-terms [33–35]. This problem does not apply to the supergravity inflation, which is driven by D-terms [36, 37] as we discuss in Section 17.4.4. However, even within F-term inflation, the η-problem can be overcome. We discuss a couple of such possibilities below.

One way to resolve the η-problem is to impose a Nambu–Goldstone shift symmetry. Thus, assume we have a Kähler potential which is invariant under the following shift symmetry for the superfield $\Phi(x, \theta)$ [38, 39]

$$\Phi(x, \theta) \to \Phi(x, \theta) + i\alpha, \tag{17.181}$$

where α is real. Let us assume that the Kähler potential is a function only of $\Phi + \Phi^*$. In this case we have $K(\Phi + \Phi^*)$ invariant under the shift symmetry.

Now, let $\phi(x)$ be the spin 0 component of $\Phi(x, \theta)$, which we write as

$$\phi(x) = \frac{1}{\sqrt{2}}(\zeta(x) + i\varphi(x)), \tag{17.182}$$

where we will identify φ as the inflaton. Let us now assume a superpotential of the form

$$W = \mu U \Phi, \tag{17.183}$$

where $U(x, \theta)$ is a different superfield and also assume a Kähler potential of the form

$$K = \frac{1}{2}(\Phi + \Phi^*)^2 + UU^*. \tag{17.184}$$

Here, one finds the kinetic energy for the scalars has the form

$$\mathcal{L}_{kin} = -\frac{1}{2}\partial_\mu\zeta\partial^\mu\zeta - \frac{1}{2}\partial_\mu\varphi\partial^\mu\varphi - \partial_\mu U\partial^\mu U^*, \tag{17.185}$$

while the scalar potential takes the form

$$V = \mu^2 e^{(\zeta^2 + |U|^2)}\left[\phi\phi^*(1 + |U|^4) + |U|^2\left\{1 - \phi\phi^* + 2\zeta^2(1 + \phi\phi^*)\right\}\right], \tag{17.186}$$

where $\phi\phi^* = (\zeta^2 + \varphi^2)/2$. Notice that the inflaton field φ is absent in the front e^K factor because of the shift symmetry. This resolves the η-problem of supergravity as long as we consider field configurations so that $|\zeta|, |U| \ll 1$ while $\varphi \gg 1$ in Planck units. In this case, we can approximate the inflaton scalar potential so that

$$V \simeq \mu^2\varphi^2, \tag{17.187}$$

which has the precise form of a chaotic potential as given by Eq. (17.163) with $p = 2$. At the end of inflation, the inflaton field decays. To accomplish the decay, one may introduce a coupling of the type

$$W_{UHH} = \lambda U H_1 H_2, \tag{17.188}$$

which leads to an interaction of the inflaton field with the Higgs field so that [38]

$$\varphi H_1 H_2 \tag{17.189}$$

The interaction of Eq. (17.189) will lead to inflaton decay and a reheating of the universe. Here, λ needs to be fine tuned to be rather small in order that the reheat temperature is not large enough to overproduce the gravitinos.

17.4.3 Hybrid Inflation

We now consider another approach to resolving the η-problem in F-type inflation. Here, we consider a superpotential of the form

$$W = \lambda\varphi(\phi_f\bar{\phi}_f - M^2), \tag{17.190}$$

where φ is the inflaton field and ϕ_f $(\bar{\phi}_f)$ are a conjugate pair of fields, often referred to as waterfall fields, while λ is a dimensionless coupling. We assume the Kähler potential to be flat so that

$$K = |\varphi|^2 + |\phi_f|^2 + |\bar{\phi}_f|^2. \tag{17.191}$$

For the model of Eqs. (17.190) and (17.191) the scalar potential takes the form

$$V_0 \simeq 2\lambda^2 |\varphi|^2 |\phi_f|^2 + \lambda^2 (|\phi_f|^2 - M^2)^2 \left(1 + \frac{\kappa^2}{2} |\varphi|^2 + 2\kappa^2 |\varphi|^2 + \kappa^4 |\phi_f|^4\right) + \cdots \tag{17.192}$$

It is now seen that in Eq. (17.192) the mass-squared term for the inflaton field has cancelled, which resolves the η-problem of supergravity in this model. Inflation occurs along the trajectory which is $\phi_f = \bar{\phi}_f = 0$.

Next, we take into account the one-loop correction to the potential to generate the curvature terms for the inflaton field. The one-loop correction to the potential is given by the Coleman–Weinberg term

$$\Delta V = \frac{1}{64\pi^2} \sum_i (-1)^{2J_i} (2J_i + 1) M_i^4 \left(\ln \frac{M_i^2}{Q^2} - \frac{3}{2}\right). \tag{17.193}$$

Along the inflation trajectory, one finds

$$V = V_0 + \Delta V = \lambda^2 M^4 + \frac{\lambda^4 M^4 \mathcal{N}}{32\pi^2} f(|\varphi|), \tag{17.194}$$

where $\mathcal{N}$ is the dimensionality of the representation of ϕ_f and $\bar{\phi}_f$, and $f(|\varphi|)$ is given by (see Eq. (17.321))

$$f(|\varphi|) = M^{-4} \left[(|\varphi|^2 + M^2)^2 \log \frac{\lambda^2 (|\varphi|^2 + M^2)}{Q^2} \right.$$
$$\left. + (|\varphi|^2 - M^2)^2 \log \frac{\lambda^2 (|\varphi|^2 - M^2)}{Q^2} - 2|\varphi|^4 \log \frac{\lambda^2 |\varphi|^2}{Q^2} - 3M^4 \right]. \tag{17.195}$$

During inflation, $\phi_f = 0 = \bar{\phi}_f$ and $\varphi > \varphi_c$, where $\varphi_c = M$. In this model, inflation continues until φ reaches ϕ_c. At this point the ϕ_f moves towards the true supersymmetric minimum which ends inflation.

Now, in the computation of ϵ and η, V' and V'' receive contributions only from ΔV while V is dominated by V_0. Thus, a derivation of ϵ and η gives the result

$$\epsilon \simeq \frac{1}{2} \left(\frac{\lambda^2}{32\pi^2}\right)^2 \mathcal{N}^2 (M_{Pl} f_\varphi(|\varphi|))^2 \tag{17.196}$$

$$\eta \simeq \left(\frac{\lambda^2}{32\pi^2}\right) \mathcal{N} (M_{Pl}^2 f_{\varphi\varphi}(|\varphi|)). \tag{17.197}$$

It is now seen that the ranges of parameters exist so that one may have $\epsilon \ll 1$ and $\eta \ll 1$, and slow-roll conditions are met.

17.4.4 D-term Inflation

In a D-term inflation [36, 37, 40] the F-term is suppressed, and since the D-term does not have an outside factor of $\exp(\kappa^2 K)$, one expects that the η-problem can be overcome. The main idea here is to extend the MSSM gauge group by one or more extra $U(1)$ factors, which brings additional D-terms into the potential. The extra $U(1)_X$ terms could be non-anomalous or anomalous where the anomaly is cancelled by additional terms not included. In the case of an anomalous $U(1)$, it is permissible to add a Fayet–Iliopoulos term to the Lagrangian such as $\xi_X D_X$. First, we consider the case of global supersymmetry, and we consider a set of fields $\phi_\pm$ which are charged under the $U(1)_X$ symmetry. We assume a superpotential for these of the form

$$W_\pm = m\phi_+ \phi_-. \tag{17.198}$$

The potential in this case consists of

$$V = m^2 \left(|\phi^+|^2 + |\phi^-|^2\right) + \frac{g_X^2}{2}\left(\sum_i Q_X^i|\tilde{f}_i|^2 + |\phi^+|^2 - |\phi^-|^2 + \xi_X\right)^2, \tag{17.199}$$

where $\tilde{f}_i$ are scalar matter fields with charges Q_X^i. Minimization of the potential leads to

$$\langle \phi^+ \rangle = 0$$

$$\langle \phi^- \rangle^2 = \xi_X - \frac{m^2}{g_X^2}$$

$$\langle D_X \rangle = \frac{m^2}{g_X^2}$$

$$\langle F_{\phi^+} \rangle = m\sqrt{\xi_X} + \cdots \tag{17.200}$$

The masses of the scalars here are given by

$$m_i^2 = g_X^2 Q_X^i \langle D_X \rangle, \tag{17.201}$$

and since $\langle D_X \rangle = m^2/g_X^2$, the scalar masses are $\mathcal{O}(m)$ and TeV size if $m \sim \mathcal{O}(\text{TeV})$. The gaugino masses would arise from higher-dimensional operators suppressed by the Planck mass, i.e.,

$$m_\lambda \sim \frac{1}{M_{Pl}^2}\langle F_{\phi^+}\phi^- + F_{\phi^-}\phi^+ \rangle \sim m\frac{\xi_X}{M_{Pl}^2}. \tag{17.202}$$

Thus, normal size sparticle masses will result from such an extension if $\xi_X \sim \mathcal{O}(M_{Pl}^2)$.

We now extend the above analysis to supergravity by promoting the $m\phi^+\phi^-$ term to $\phi\phi^+\phi^-$, where ϕ is neutral under $U(1)_X$ and will play the role of an inflaton. In this case we write

$$W = \phi\phi^+\phi^- + W_0, \qquad (17.203)$$

and assume a flat Kähler potential of the form

$$K = |\phi^+|^2 + |\phi^-|^2 + |\phi|^2 + K_0. \qquad (17.204)$$

The scalar potential in this case is given by

$$V = V_{hid} + e^{\kappa^2 K}\Big[|\phi\phi^+|^2 + |\phi\phi^-|^2 + |\phi^+\phi^-|^2 + \kappa^4\big(|\phi|^2 + |\phi^+|^2 + |\phi^-|^2\big)|W|^2$$

$$+ 3|\phi\phi^+\phi^-|^2 - 3|W_0|^2\Big] + \sum_a \frac{g_a^2}{2}\Big[|\phi^+|^2 - |\phi^-|^2 + \xi_a\Big]^2, \qquad (17.205)$$

and one finds that for ϕ greater than some critical value ϕ_c, the charged $\phi^\pm$ are stabilized at the origin $\phi^\pm = 0$. Their masses at $\phi^\pm = 0$ are positive for a large enough $\langle\phi\rangle$ but still significantly below the Planck scale. At the same time at $\phi^\pm = 0$ the potential in φ collapses to the form

$$V(\phi^\pm = 0) = V_{hid} + e^{\kappa^2 K_0}\kappa^2|W_0|^2\left[\kappa^2|\phi|^2 - 3\right] + \sum_a \frac{g_a^2}{2}\xi_a^2. \qquad (17.206)$$

Quite remarkably the spoiler F-term which was responsible for the η-problem has disappeared, and calculation of η now gives

$$|\eta| = \left|\frac{\partial_\phi^2 V}{\kappa^2 V}\right| \ll 1. \qquad (17.207)$$

Numerically, an η value so that $|\eta| \sim 0.01$ is acceptable. Taking $|\eta|$ to be 0.01, Eq. (17.207) implies

$$\frac{\sum_a g_a^2\xi^2}{\kappa^2 e^{\kappa^2 K}|W_0|^2} \gg 100. \qquad (17.208)$$

It is also straightforward to see that with $\phi \sim \sqrt{\xi}$ the computation of V' leads to $\epsilon \ll 1$. So, inflation can be successful. One finds that, in general, if the D-term dominates the F-term, i.e., if the D-term is larger by a factor of 100 or more than the F-term, then the D-term will drive inflation up to scales 10^{10-15} GeV. Inflation will terminate at the end of slow roll when ϕ falls below a certain threshold value. A more complete model is needed to describe the exit from inflation. Since the D-term inflation is driven by the Fayet–Iliopoulos term, we note that inclusion of a Fayet–Iliopoulos term in supergravity has some subtleties (e.g., [41–44]).

More recently, a number of works have appeared where one considers supergravity models with non-minimal scalar-curvature couplings of the type $\Phi(z, \bar{z})R$ [45–47]. Such models are often referred to as Jordan frame supergravity. They are in some sense supergravity analogs of the analyses of the Higgs inflation model [32] discussed earlier. Next, we discuss density fluctuations in a inflationary universe. For a broad discussion of this topic, see, e.g., Salopek *et al.* [48].

17.4.5 Inflation and Cosmological Perturbations

Inflation provides initial conditions for the computations of cosmological perturbations. Before we consider the perturbations of the Friedmann–Robertson–Walker geometry, let us consider as an exercise the case of a massless field in the gravitational background so that

$$A_\phi = \int d^4x \ \sqrt{-g} \ \left(-\frac{1}{2} \partial_\mu \phi \partial^\mu \phi \right). \tag{17.209}$$

We will consider the case of exponential expansion resulting from $H = \dot{R}/R$ being constant during the period of the expansion. It is found convenient to use a new variable, the conformal time ζ, instead of the cosmic time, t, where ζ is defined so that

$$\zeta = \int \frac{dt}{R(t)}. \tag{17.210}$$

In terms of the conformal time, the line element can be written in the form

$$ds^2 = R^2(\zeta)(-d\zeta^2 + \delta_{ij} \ dx^i \ dx^j). \tag{17.211}$$

Using the metric of Eq. (17.211), we can write Eq. (17.209) as follows:

$$A_\phi = \int d\zeta \ d^3x \frac{R^2(\zeta)}{2} (\phi'^2 - \vec{\nabla}\phi.\vec{\nabla}\phi), \tag{17.212}$$

where $\phi' \equiv \partial\phi/\partial\zeta$. Let us define a new variable $\chi(\zeta)$ so that

$$\chi = R\phi. \tag{17.213}$$

Differentiating with respect to ζ we get

$$R^2\phi'^2 = \chi'^2 + \frac{\chi^2 R'^2}{R^2} - \frac{2\chi\chi' R'}{R}. \tag{17.214}$$

Next, using the relation

$$\frac{\chi^2 R'^2}{R^2} - \frac{2\chi\chi' R'}{R} = \frac{\chi^2 R''}{R^2} - \left(\frac{\chi^2 R'}{R} \right)', \tag{17.215}$$

and discarding the total derivative term in the integral, we can write Eq. (17.212) in the following form:

$$A_\phi = \int d\zeta \ d^3x \frac{1}{2} \left(\chi'^2 - \vec{\nabla}\chi.\vec{\nabla}\chi + \frac{R''}{R}\chi^2 \right). \tag{17.216}$$

We quantize the system by imposing the canonical commutation relations. Thus, the canonical co-ordinates and momenta are $(\chi(\zeta, \vec{x}), \pi(\zeta, \vec{x}))$, where $\pi(\zeta, \vec{x}) = \chi'$ and the corresponding equal conformal time commutation relations are

$$[\chi(\zeta, \vec{x}), \pi(\zeta, \vec{x}')] = i\hbar\delta(\vec{x} - \vec{x}') \tag{17.217}$$

$$[\chi(\zeta, \vec{x}), \chi(\zeta, \vec{x}')] = 0 = [\pi(\zeta, \vec{x}), \pi(\zeta, \vec{x}')]. \tag{17.218}$$

Next, we carry out a Fourier expansion of $\chi(\zeta, \vec{x})$ so that

$$\chi(\zeta, \vec{x}) = \int \frac{d^3k}{(2\pi)^{3/2}} \left[a(\vec{k})\chi(\vec{k}, \zeta)e^{i\vec{k}.\vec{x}} + a^\dagger(\vec{k})\chi^*(\vec{k}, \zeta)e^{-i\vec{k}.\vec{x}} \right], \tag{17.219}$$

where $a(\vec{k})$ and $a(\vec{k})^\dagger$ are the creation and destruction operators which, using Eqs. (17.217) and (17.218), satisfy the commutation relations

$$[a(\vec{k}), a^\dagger(\vec{k}')] = \delta(\vec{k} - \vec{k}')$$

$$[a(\vec{k}), a(\vec{k}')] = 0 = [a^\dagger(\vec{k}), a^\dagger(\vec{k}')]. \tag{17.220}$$

The $\chi(\vec{k}, \zeta)$ satisfy the following equations of motion:

$$\chi''(\vec{k}, \zeta) + \left(k^2 - \frac{2}{\zeta^2} \right) \chi(\vec{k}, \zeta) = 0, \tag{17.221}$$

where we have used the result $R''/R = 2/\zeta^2$. This quantity acts like a mass-squared term which, however, is time dependent. Also, for the case when $H = \dot{R}/R$ is a constant, the scale factor is related to the conformal time so that

$$\zeta = \int \frac{dR}{R\dot{R}} = \frac{1}{H} \int \frac{dR}{R^2} \sim -\frac{1}{HR} \tag{17.222}$$

or

$$R(\zeta) = -(H\zeta)^{-1}. \tag{17.223}$$

We consider now a specific solution to Eq. (17.220) (see Eq. (17.323)):[‡]

$$\chi(\zeta, \vec{k}) = \sqrt{\frac{\hbar}{2k}} \left(1 - \frac{i}{k\zeta} \right) e^{-ik\zeta}. \tag{17.224}$$

Using this solution we can compute the correlation function $\langle 0|\phi(\vec{x})\phi(\vec{y})|0\rangle$. Thus, using the commutation relations of the creation and destruction operators, one finds

$$\langle 0|\phi(\vec{x})\phi(\vec{y})|0\rangle = \int d^3k \; e^{i\vec{k}.(\vec{x}-\vec{y})} \frac{|\chi(\zeta, \vec{k})|^2}{(2\pi)^3 R^2}. \tag{17.225}$$

One defines the power spectrum $\mathcal{P}_\phi(k)$ so that

$$\langle 0|\phi(\vec{x})\phi(\vec{y})|0\rangle = \int d^3k \; e^{i\vec{k}.(\vec{x}-\vec{y})} \frac{\mathcal{P}_\phi(k)}{4\pi k^3}. \tag{17.226}$$

[‡] This solution is referred to in the literature as the Bunch–Davies vacuum [49] on a curved space-time (see also work by Chernikov and Tegirov [50], and by Schomblond and Spindel [51]).

A comparison of Eqs. (17.225) and (17.226) gives

$$\mathcal{P}_\phi(k) = \frac{k^3}{2\pi^2} \frac{|\chi(\zeta, \vec{k})|^2}{R^2}.$$
(17.227)

Substitution of Eqs. (17.223) and (17.224) in Eq. (17.227) under the approximation $k|\zeta| \ll 1$ gives

$$\mathcal{P}_\phi(k) \simeq \hbar \left(\frac{H}{2\pi}\right)^2.$$
(17.228)

The limit $k|\zeta| \ll 1$ corresponds to the case when the wavelength is larger than the Hubble radius.

After the example above, we now consider the case of perturbations of the FLRW geometry. We can parametrize these perturbations as follows:

$$ds^2 = R^2 \left[-(1 + 2\alpha)d\zeta^2 + 2\beta_i \, dx^i \, d\zeta + (\delta_{ij} + h_{ij})dx^i \, dx^j \right].$$
(17.229)

Next, we want to decompose all the perturbations into scalar, vector, and tensor perturbations. To accomplish this, we carry out the following decompositions. For the fields β_i we write

$$\beta_i = \nabla_i \beta + \tilde{\beta}_i, \quad \nabla_i \tilde{\beta}^i = 0.$$
(17.230)

Here, β is the scalar component and $\tilde{\beta}_i$ is the vector component. In a similar fashion, we decompose the perturbations h_{ij} into scalar, vector, and tensor components as follows:

$$h_{ij} = 2\delta\delta_{ij} + 2\nabla_i\nabla_j\gamma + 2\Delta_{(i} \gamma_{j)} + \gamma_{ij},$$
(17.231)

where

$$\nabla_i\gamma^i = 0, \quad \nabla_i\gamma^{ij} = 0, \quad \gamma^{ij}\delta_{ij} = 0.$$
(17.232)

We will focus on the scalar perturbations of the background geometry. As seen above, these scalar components consist of α, β, γ, and δ.

Now, the metric perturbations are modified under the general co-ordinate transformations, i.e.,

$$x^\mu \to x^\mu + \xi^\mu, \quad \xi^\mu = (\xi^0, \xi^i).$$
(17.233)

The change in the components of the metric under the co-ordinate variations can be expressed in the form

$$\Delta(\delta g_{\mu\nu}) = -2\Delta_{(\mu}\xi_{\nu)},$$
(17.234)

where Δ denotes the co-ordinate variations. The variations of Eq. (17.234) can be used to obtain variations of the components α, β_i, and h_{ij} that appear in the metric perturbations. Thus, one has [52]

$$\Delta \alpha = -\xi^{0'} - H\xi^0 \tag{17.235}$$

$$\Delta \beta_i = \nabla_i \xi^0 - \xi_i' \tag{17.236}$$

$$\Delta h_{ij} = -2(\nabla_{(i}\xi_{j)} - H\xi^0 \delta_{ij}). \tag{17.237}$$

Next, we include the fluctuations of the scalar fields ϕ around the homogeneous background so that

$$\phi(\zeta, x_i) = \bar{\phi}(\zeta) + \delta\phi(\zeta, x_i). \tag{17.238}$$

Including the scalar fluctuations along with the perturbations of the background Friedmann–Robertson–Walker (FRW) metric, the action can be expanded around the perturbations as follows:

$$\mathcal{A}[\bar{\phi} + \delta\phi, g_{\mu\nu} + \delta g_{\mu\nu}] = \mathcal{A}^{(0)}[\bar{\phi}, g_{\mu\nu}] + \mathcal{A}^{(1)}[\bar{\phi}, g_{\mu\nu}; \delta\phi, \delta g_{\mu\nu}]$$
$$+ \mathcal{A}^{(2)}[\bar{\phi}, g_{\mu\nu}; \delta\phi, \delta g_{\mu\nu}] + \cdots, \tag{17.239}$$

where $\mathcal{A}^{(0)}$ contains only the homogenous part and $\mathcal{A}^{(1)}$ vanishes after use of the equations of motion. Thus, the relevant part of the action is $\mathcal{A}^{(2)}$. Now, it can be shown that the scalar perturbations enter in the combination

$$\sigma = R\left(\delta\phi - \frac{\phi'}{H}\delta\right). \tag{17.240}$$

This linear combination of the scalar field perturbation and the metric perturbation is gauge invariant and thus represents the real scalar dynamical variable. In terms of this variable, the quadratic part of the action takes the form

$$\mathcal{A}_\sigma = \frac{1}{2}\int d\zeta \, d^3x \left[\sigma'^2 + \vec{\nabla}\sigma \cdot \vec{\nabla}\sigma + \frac{b''}{b}\sigma^2\right], \tag{17.241}$$

where

$$b = R\frac{\phi'}{H}. \tag{17.242}$$

Now, in a slow roll, we expect the evolution of ϕ and H to be much slower than that of the cosmological scale factor $R(\zeta)$, and thus one has the relation

$$\frac{b''}{b} \simeq \frac{R''}{R}. \tag{17.243}$$

Inserting Eq. (17.243) into Eq. (17.241) one finds that the analysis from here on for the power spectrum is very similar to the case of Eq. (17.216). One also finds that the scalar power spectrum is given by

$$\mathcal{P}_s \simeq \frac{1}{4\pi^2}\left(\frac{H^4}{\dot{\phi}^2}\right)_{k_0}, \tag{17.244}$$

where $k_0 = RH$. Using Eqs. (17.144) and (17.145) and eliminating $\dot{\phi}$ using slow-roll conditions, one may write

$$\mathcal{P}_s = \frac{1}{12\pi^2}\left(\frac{V^3}{M_{Pl}^6 V'^2}\right)_{k_0} \tag{17.245}$$

Next, we consider tensor perturbations of the gravitational metric, which are given in Eq. (17.231) with the constraints on them discussed in Eq. (17.232). The contribution of the tensor perturbation to the action is given by

$$\mathcal{A}_\phi = \frac{M_{Pl}^2}{8} \int d\zeta \, d^3x \, R^2(\zeta) \partial^\mu \gamma_j^i \partial_\mu \gamma_i^j. \tag{17.246}$$

We can carry out a Fourier expansion of the tensor modes similar to that of the scalar mode, taking account of the tensor nature of the field so that

$$\gamma_j^i = \sum_\lambda \frac{1}{R} \int \frac{d^3k}{(2\pi)^{3/2}} e^{i\vec{k}\cdot\vec{x}} \chi_\lambda(\zeta, \vec{k}) e_j^i(\vec{k}, \lambda), \tag{17.247}$$

where $e_j^i(\vec{k}, \lambda)$ is the polarization vector and the summation is over the two polarizations λ of the tensor perturbation. Derivation of the tensor power spectrum gives

$$\mathcal{P}_t = \frac{2}{3\pi^2} \left(\frac{V}{M_{Pl}^4} \right)_{k_0}. \tag{17.248}$$

From Eqs. (17.245) and (17.248) the ratio of the tensor to the scalar power spectrum is given by

$$r = \frac{\mathcal{P}_t}{\mathcal{P}_s} = 8 \left(\frac{M_{Pl}^2 V'^2}{V^2} \right)_{k_0}^2 = 16\epsilon. \tag{17.249}$$

The scalar and the tensor power spectrum have a k-dependence which can be quantified by spectral indices. For the scalar power spectrum, one defines the spectral index n_s so that

$$n_s(k) - 1 = \frac{d \ln \mathcal{P}_s(k)}{d \ln k}. \tag{17.250}$$

Next, using the approximation that for slow roll we have $d \log k = d \log(RH) \simeq d \log(R)$, which gives $d\phi/d \ln k \simeq -M_{Pl}^2(V'/V)$, we find

$$\frac{dV}{d \ln k} = -M_{Pl}^2 \frac{V'^2}{V}, \quad \frac{dV'}{d \ln k} = -M_{Pl}^2 \left(\frac{V'V''}{V} \right). \tag{17.251}$$

Using these relations, it is then easily seen that

$$n_s - 1 = 2\eta - 6\epsilon. \tag{17.252}$$

Similarly, one can define the tensor spectral index

$$n_t(k) = \frac{d \ln \mathcal{P}_t(k)}{d \ln k}, \tag{17.253}$$

and an analysis similar to the case of the scalar spectral index gives

$$n_t(k) = -2\epsilon. \tag{17.254}$$

One can also compute the running of the spectral index with k. Thus, differentiating n_s with respect to $\ln k$ one finds

$$\frac{dn_s}{d\ln k} = 16\epsilon\eta - 2\xi - 24\epsilon^2, \tag{17.255}$$

where ξ is defined by Eq. (17.157). Recently, a joint analysis by BICEP2, the Keck Array, and Planck Data [53] has put stringent limits on the tensor/scalar ratio r, constraining many inflation models.

17.5 Baryogenesis and Leptogenesis

17.5.1 Baryon and Lepton number Violation in the Standard Model

In the standard model the leptonic current J_L^μ and the baryonic current J_B^μ are given by

$$J_L^\mu = \sum_i (\bar{\ell}_{L_i}\gamma^\mu \ell_{L_i} - \bar{e}^c_{L_i}\gamma^\mu e^c_{L_i})$$

$$J_B^\mu = \frac{1}{3}\sum_i (\bar{q}_{L_i}\gamma^\mu q_{L_i} - \bar{u}^c_{L_i}\gamma^\mu u^c_{L_i} - \bar{d}^c_{L_i}\gamma^\mu d^c_{L_i}). \tag{17.256}$$

Because of the presence of chiral fields in the standard standard model, the leptonic and the baryonic currents are not conserved. Specifically, one finds

$$\partial_\mu J_L^\mu = \partial_\mu J_B^\mu = \frac{N_f}{32\pi^2}(g^2 W_{\mu\nu}^a \tilde{W}^{a\mu\nu} - g'^2 B_{\mu\nu}\tilde{B}^{\mu\nu}), \tag{17.257}$$

where $W_{\mu\nu}^a$ and $B_{\mu\nu}$ are the field strengths for the $SU(2)_L$ gauge boson W_μ^a and the $U(1)_Y$ gauge boson B_μ; $\tilde{W}^{a\mu\nu}$ and $\tilde{B}^{\mu\nu}$ are their duals; and N_f is the number of generations. The above implies that $\partial_\mu(J_B^\mu - J_L^\mu) = 0$. The B and L charges Q_B and Q_L are defined by

$$Q_B = \int d^3x\, J_B^0, Q_L = \int d^3x\, J_L^0, \tag{17.258}$$

so that $Q_B - Q_L = 0$. Thus, the standard model interactions conserve $B - L$. On the other hand, the sum of the baryonic and the leptonic currents is not conserved, i.e., $\partial_\mu(J_B^\mu + J_L^\mu) \neq 0$, which implies that $B + L$ is violated by the standard model interactions. Specifically, we can write

$$\partial_\mu(J_B^\mu + J_L^\mu) = 2N_f\partial_\mu C^\mu, \tag{17.259}$$

where

$$C^\mu = -\frac{g^2}{16\pi^2}\epsilon^{\mu\nu\alpha\beta}W_{\nu a}\left(\partial_\alpha W_{\beta a} + \frac{g}{3}\epsilon_{abc}W_{\alpha b}W_{\beta c}\right) + \frac{g'^2}{32\pi^2}\epsilon^{\mu\nu\alpha\beta}B_\nu B_{\alpha\beta}, \tag{17.260}$$

which arises as a consequence of the anomaly. An integration of Eq. (17.257) from time t_i to t_f and over all space can be expressed as

$$\Delta Q_B = \int_{t_i}^{t_f} dt \int d^3x \; \partial_\mu J_B^\mu, \qquad (17.261)$$

where $\Delta Q_B = Q_B(t_f) - Q_B(t_i)$. We can write ΔQ_B as follows:

$$\Delta Q_B = N_f(n(t_f) - n(t_i)) \equiv N_f \Delta n, \qquad (17.262)$$

where $n(t)$ is defined by

$$n(t) = \frac{g^2}{96\pi^2} \int d^3x \; \epsilon_{ijk}\epsilon_{abc}W_a^i W_b^j W_c^k. \qquad (17.263)$$

Here, i, j, and k are spatial indices and a, b, and c are $SU(2)_L$ gauge group indices, and $n(t)$ is referred to in the literature as the Chern–Simons number. For non-abelian gauge theories, one has infinitely many vacuum states which are degenerate. These degenerate states are characterized by different values of n, and the difference Δn is an integer, i.e.,

$$\Delta n = \pm 1, \pm 2, \cdots \qquad (17.264)$$

Thus, Eq. (17.262) implies that $\Delta B = \Delta L = \pm 3, \pm 6, \cdots$ for $N_f = 3$. Now, $SU(2)$ instantons produce a $B + L$ violating effective operator involving 12 fermions of the standard model, and has the form

$$\prod_{i=1,2,3} (q_{L_i}q_{L_i}q_{L_i}\ell_{L_i}). \qquad (17.265)$$

This operator can produce a ΔB and ΔL in units of 3.

Let us first consider the transitions produced by this operator at zero temperature. Here, one has [54, 55]

$$\Gamma_{B+L} \sim e^{-S_{inst}} = e^{-\frac{4\pi}{\alpha}}. \qquad (17.266)$$

Clearly, this process is highly suppressed. The suppression is reduced if we consider the transition at a finite temperature. For finite temperatures but below the electroweak phase transition temperature, i.e., $T \leq T_{EWPT}$, one has the transition rate per unit volume given by

$$\frac{\Gamma_{B+L}}{V} = \left(\frac{M_W}{\alpha T}\right)^3 k M_W^4 e^{-\beta E_{sph}(T)}, \qquad (17.267)$$

where $\beta = 1/T$ and α is the fine structure constant, and E_{sph} is given by

$$E_{sph} \simeq \frac{8\pi}{g}\langle H(T)\rangle, \qquad (17.268)$$

where $\langle H(T) \rangle$ is the expectation value of the Higgs field at temperature T. This will produce the transition rate

$$\frac{\Gamma_{B+L}}{V} \sim e^{-\frac{M_W}{\alpha T}}. \tag{17.269}$$

Thus, one finds that below the electroweak phase transition temperature the transition rate is still very suppressed because of the $1/\alpha$ factor. However, the situation changes quite dramatically as we go to temperatures above the electroweak phase transition temperature. In this case, the transition no longer has an exponential suppression. Rather, for $T \geq T_{EWPT}$ one has

$$\frac{\Gamma_{B+L}}{V} \sim \alpha^5 \ln \alpha^{-1} T^4, \tag{17.270}$$

and the baryon-number-violating transitions are strong. These sphaleron processes play an important role in baryogenesis, as we later discuss.

As already mentioned in Chapter (1), achieving baryon asymmetry in the universe requires three conditions: violation of baryon number, violation of C and of CP, and departure from thermal equilibrium [56]. To see why the departure from thermal equilibrium is needed, we consider the thermal average of B at thermal equilibrium. We first note that baryon number B is odd under C and CP, and thus under CPT so that (CPT) B (CPT)$^{-1} = -B$. Assuming that CPT commutes with the Hamiltonian and using the oddness of B under (CPT), the thermal average of B is given by

$$\begin{aligned}
\langle B \rangle_T &= tr[B e^{-\beta H}] \\
&= -tr[(\text{CPT})B(\text{CPT})^{-1} e^{-\beta H}] \\
&= -tr[B e^{-\beta H}] = 0, \tag{17.271}
\end{aligned}$$

where in going from the second to the third line we have used the property that the CPT operator commutes with the Hamiltonian. Thus, $\langle B \rangle_T$ at thermal equilibrium vanishes, and one needs departure from thermal equilibrium to generate a baryon asymmetry. Now, the out-of-equilibrium processes to generate baryogenesis or leptogenesis can occur in different ways. One possible avenue to achieve it is via out-of-equilibrium decays of heavy particles. Another possibility is via electroweak phase transitions to achieve baryon asymmetry. Let us consider first the heavy particle decays. Assume we have a heavy particle U with mass m_U at the grand unified scale. At temperatures greater than M_U, the thermal abundance of this particle will be of size $\sim T^3$, i.e., $n_U \sim n_{\bar{U}} \sim T^3$. Let us now assume that the decay rate of U is much smaller than the expansion rate of the universe. Now, at temperatures $T_D \sim m_U$, the heavy particles would decouple from the thermal bath, and their number density below this temperature will continue to be $n_U^D \sim T_D^3$ even at temperatures below $T < T_D$. On the other hand, the number density of U in thermal equilibrium at temperature $T_0 < T_D$ when the particles are non-relativistic is $n_U^0 = n_{\bar{U}}^0 \sim (m_U T_0)^{3/2} e^{-m_U/T_0}$. Thus, $n_U^D > n_U^0$, which exhibits the out-of-equilibrium decays of the heavy particle U if its decay rate is slow compared with the expansion rate of the universe.

For the electroweak phase transition for the case of the standard model, one finds

$$\eta_B \simeq \frac{\alpha_w^4 T^3}{s} f_{CP}, \tag{17.272}$$

where η_B is defined in Section 15.3 and f_{CP} [57,58] is given by

$$f_{CP} = T_C^{-12}(m_t^2 - m_c^2)(m_t^2 - m_u^2)(m_c^2 - m_u^2)$$
$$\times (m_b^2 - m_s^2)(m_b^2 - m_d^2)(m_s^2 - m_d^2)s_{12}s_{23}s_{31}\sin\delta. \tag{17.273}$$

Here, $s_{ij} = \sin\theta_{ij}$ and θ_{ij} are the three mixing angles, and δ is the Cabibbo–Kobayashi–Maskawa (CKM) phase. T_C appearing in Eq. (17.273) is the temperature of the electroweak phase transition (typically around 100 GeV). With $T_C \sim 100$ GeV, one finds $f_{CP} \sim 10^{-20}$. An estimate for η_B gives $\eta_B \simeq 10^{-8} f_{CP}$, which using the estimate of f_{CP} for the standard model gives $\eta_B \simeq 10^{-28}$, whereas the desired value is $\eta_B \sim 10^{-10}$. Additionally, there are stringent constraints on the Higgs boson mass which are in violation of experiments. Analysis of baryogenesis in the MSSM relieves some of the tension, because in part there are new sources of CP violation in MSSM. However, achieving $\eta_B \sim 10^{-10}$ in the MSSM is still difficult.

17.5.2 Leptogenesis: The Non-supersymmetric Case

An attractive alternative to conventional baryogenesis (for reviews see the literature [59,60]) is baryogenesis via leptogenesis ([61–66]). The essential idea here is that if one can generate enough lepton asymmetry, then it can be converted into baryon asymmetry via sphaleron interactions which violate $B + L$ but preserve $B - L$. Leptogenesis is a natural consequence of the see-saw mechanism, which is a popular mechanism for the generation of small neutrino masses.

We start by considering leptogenesis in the standard model case with inclusion of Majorana neutrino fields N_i, and consider a Lagrangian of the form

$$\mathcal{L} = \lambda_{in} H \bar{N}_i P_L L_n + \lambda_{in}^* H^\dagger \bar{L}_n P_R N_i, \tag{17.274}$$

where N_i are the Majorana fields, and λ are, in general complex, and thus the λ-terms violate CP invariance. Further, $\mathcal{L}$ violates lepton number and $B - L$, and thus a Lagrangian of the above type has the general characteristics that might lead to the generation of baryon asymmetry via leptogenesis. The CP violation occurs in the decay of the Majoranas because of the overlap of the tree and the loop contribution.

One can define a CP asymmetry parameter so that

$$\epsilon = \frac{\sum_\alpha [\Gamma(N_i \to l_\alpha H) - \Gamma(N_i \to \bar{l}_\alpha H^\dagger)]}{\sum_\alpha [\Gamma(N_i \to l_\alpha H) + \Gamma(N_i \to \bar{l}_\alpha H^\dagger)]}. \tag{17.275}$$

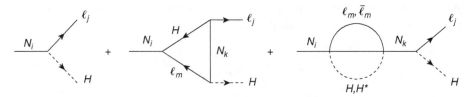

Figure 17.1. Leptogenesis producing lepton number asymmetry via decay of the right-handed neutrino (N) by interference between the tree, the vertex, and the self-energy or wave function diagrams.

Using the short-hand notation $\Gamma \equiv \Gamma(N_i \to l_\alpha H)$ and $\bar{\Gamma} \equiv \Gamma(N_i \to \bar{l}_\alpha H^\dagger)$ and suppressing the summation indices, we have $\epsilon \simeq (\Gamma - \bar{\Gamma})/(2\Gamma_{tree})$. Thus,

$$\epsilon \propto |g_{tree}T_{tree} + g_{loop}T_{loop}|^2 - |g^*_{tree}T_{tree} + g^*_{loop}T_{loop}|^2$$
$$= -4Im(g^*_{tree}g_{loop})Im(T^*_{tree}T_{loop}). \qquad (17.276)$$

We see from the above that a non-vanishing ϵ arises only as a consequence of the interference between the tree and the loop amplitudes. Further, the product of the tree and the loop couplings, i.e., $g^*_{tree}g_{loop}$, must be complex to achieve a non-vanishing ϵ, i.e., the couplings must be complex. But, as seen in Eq. (17.276), complex couplings are not sufficient, since the tree contribution alone cannot produce a leptonic asymmetry.

We now give a quantitative analysis of ϵ. Let us consider the process $N_i(k) \to L_n(p) + H(p')$. For the tree amplitude, we get

$$A_{tree}(N_i \to H + L_n) = i\lambda^*_{in}\bar{u}_L(p)P_R u_N(k). \qquad (17.277)$$

For the loop contribution arising from the vertex diagram in Fig. 17.1, we have

$$A_{loop} = -i\lambda_{im}\lambda^*_{km}\lambda^*_{kn} \int \frac{d^4q}{(2\pi)^4} \bar{u}_{Ln}(p)P_R \frac{\slashed{q}'}{q'^2} \frac{(-\slashed{s} + M_k)}{q^2(s^2 + M_k^2)} u_N(k), \qquad (17.278)$$

where

$$k = q + q' = p + p'$$
$$s = q - p = p' - q'. \qquad (17.279)$$

In order to obtain the imaginary part of the loop diagram we use Cutkosky rules [67] where we put the propagators for H and L on the mass shell, and more concretely we replace the propagator $1/(q^2 + m^2)$ by $-i\pi\delta(q^2 + m^2)$. Thus, the imaginary part of the loop amplitude becomes

$$-i\lambda_{im}\lambda^*_{km}\lambda^*_{kn} \int \frac{d^4q}{(2\pi)^4} \bar{u}_{Ln}(p)P_R(-i\pi)\delta(q^2)\slashed{q}'(-i\pi)\delta(q'^2)\frac{(-\slashed{s} + M_k)}{q^2(s^2 + M_k^2)} u_N(k). \qquad (17.280)$$

As discussed above for leptogenesis, we need interference between the tree and the loop term. This gives us the following:

$$g^*_{tree}g_{loop}Im(A^*_{tree}A_{loop}) = -(\lambda_{in}\lambda^*_{kn}\lambda_{im}\lambda^*_{km})$$

$$\times \int \frac{d^4q}{(2\pi)^4}\pi^2\delta(q^2)\delta(q'^2)tr(\not{p}P_R\not{q}'(-\not{s} + M_k)$$

$$(-\not{k} + M_i)P_L)\frac{1}{(s^2 + M_k^2)}. \tag{17.281}$$

An evaluation of the trace gives

$$tr(\not{p}P_R\not{q}'(-\not{s} + M_k)(-\not{k} + M_i)P_L) = -2M_iM_k(q' \cdot p). \tag{17.282}$$

Let us choose the momenta so that in the center-of-mass system we have

$$p = \frac{M_i}{2}(1, \hat{k}), p' = \frac{M_i}{2}(1, -\hat{k})$$

$$q = \frac{M_i}{2}(1, -(\sin\theta\hat{j} + \cos\theta\hat{k}))$$

$$q' = \frac{M_i}{2}(1, \sin\theta\hat{j} + \cos\theta\hat{k})$$

$$k = (M_i, \vec{0}). \tag{17.283}$$

Next, we write $d^4q = dq^0\, d^3q$ and carry out the integration using the delta functions. This gives

$$g^*_{tree}g_{loop}Im(A^*_{tree}A_{loop}) = \frac{(\lambda_{in}\lambda^*_{kn}\lambda_{im}\lambda^*_{km})}{32\pi}2M_iM_k\int_{-1}^{+1}d\cos\theta\frac{-q'.p}{s^2 - M_k^2}$$

$$= (\lambda_{in}\lambda^*_{kn}\lambda_{im}\lambda^*_{km})\frac{M_iM_k}{32\pi}\int_{-1}^{+1}d\cos\theta\frac{(1 - \cos\theta)}{1 + \cos\theta + 2x}$$

$$= -(\lambda_{in}\lambda^*_{kn}\lambda_{im}\lambda^*_{km})\frac{M_iM_k}{16\pi}\left[1 - (1+x)\ln\frac{1+x}{x}\right], \tag{17.284}$$

where $x \equiv M_k^2/M_i^2$. Now, the tree decay width is given by

$$\Gamma(N_i(k) \to L_n(p) + H(p')) \sim 2(\lambda_{in}\lambda^*_{in})p.k \sim (\lambda_{in}\lambda^*_{in})M_i^2. \tag{17.285}$$

This leads to ϵ_v arising from the vertex diagram being

$$\epsilon_v = \frac{1}{8\pi}\sum_k\frac{Im[(\lambda\lambda^\dagger)^2_{ik}]}{(\lambda\lambda^\dagger)_{ii}}\frac{M_k}{M_i}\left[1 - (1+x)\ln\frac{1+x}{x}\right]. \tag{17.286}$$

For the case $k = i$ we have

$$Im[(\lambda\lambda^\dagger)^2_{ii}] = 0. \tag{17.287}$$

Thus, only $k \neq i$ contributes, which means that we need at least *two* heavy neutrino states to have leptogenesis occur. Further, we have assumed that the state i is the lightest of all the neutrino states, i.e., $M_i < M_k$, and all the states should have masses which are greater than the sum of the intermediate states to which they connect. It is also clear that the couplings must be complex or else ϵ_v would vanish, as can be seen from Eq. (17.286). In the limit $x = M_k/M_i \gg 1$ we can expand the function in the bracket of Eq. (17.286), and we find that

$$\left[1 - (1+x)\ln\frac{1+x}{x}\right] \to -\frac{M_i^2}{2M_K^2}, \tag{17.288}$$

and Eq. (17.286) gives

$$\epsilon_v = -\frac{1}{16\pi} \sum_{k \neq i} \frac{Im(\lambda\lambda^\dagger)_{ik}^2)}{(\lambda\lambda^\dagger)_{ii}} \frac{M_i}{M_k}. \tag{17.289}$$

An additional contribution to the leptonic asymmetry arises from inclusion of the wave function diagram, and one can carry out a very similar analysis for the contribution arising from the wave function diagram. Here, one gets

$$\epsilon_w = -\frac{1}{8\pi} \sum_{k \neq i} \frac{Im[(\lambda\lambda^\dagger)_{ik}^2]}{(\lambda\lambda^\dagger)_{ii}} \frac{M_i(M_i + M_k)}{M_k^2 - M_i^2}. \tag{17.290}$$

Taking the limit $M_k \gg M_i$ in Eq. (17.290) and using Eq. (17.289), we get

$$\epsilon_L \equiv \epsilon_v + \epsilon_w \simeq -\frac{3}{16\pi} \sum_k \frac{Im[(\lambda\lambda^\dagger)_{ik}^2]}{(\lambda\lambda^\dagger)_{ii}} \frac{M_i}{M_k}. \tag{17.291}$$

Thus, inclusion of the wave function diagram increases the result of the asymmetry parameter arising from the vertex diagram by a factor of 3 [65].

Now, the number density of Majorana neutrinos in the early universe at temperature T is $n_N = \frac{3}{4}\zeta(3)\,(g_N T^3/\pi^2)$ (see Eq. (17.47)), where g_N is the neutrino degrees of freedom ($g_N = 2$), and thus the lepton asymmetry is given by

$$\eta_L \equiv \frac{\epsilon_L n_N}{s} = \epsilon_L \frac{3}{4}\zeta(3)\frac{g_N T^3}{\pi^2 s}, \tag{17.292}$$

where s is the entropy per unit volume and is given by Eq. (17.60). For the standard model case from Eq. (17.54), we have $g_s = 106.75$. Numerically, then, one finds

$$\eta_L \simeq 4.3 \times 10^{-3} \epsilon_L. \tag{17.293}$$

At the time of leptogenesis $B - L = -L$, and thus one may also analogously define η_{B-L} so that $\eta_{B-L} = -\eta_L$, and numerically

$$\eta_{B-L} \equiv -\frac{\epsilon_L n_N}{s} = -\epsilon_L \frac{3}{4}\zeta(3)\frac{g_N T^3}{\pi^2 s}. \tag{17.294}$$

We will see below that $B - L$ is conserved in sphaleron processes, and thus the above expression is useful in the derivation of baryon number asymmetry, which we discuss below.

To derive the baryon asymmetry we need to look at thermal equilibrium processes and the constraints on the chemical potentials. First, we consider the sphaleron processes ($\mathcal{O}_{\text{sph}} \sim \prod_{i=1,2,3} q_i q_i q_i L_i$) which give us one constraint on the chemical potential, i.e.,

$$\sum_i (3\mu_{q_i} + \mu_{L_i}) = 0, \tag{17.295}$$

where i runs over the generations. It is estimated that the sphaleron processes decouple at a temperature $T_{Sph} \sim [80 + 54(m_h/120 \text{ GeV})]$. We assume that the electroweak phase transition temperature T_{EWPT} lies above T_{Sph}, and thus the sphaleron processes are active at $T > T_{EWPT}$. We also assume that the total hypercharge of the universe is zero, which leads to the following condition on the chemical potentials:

$$Y_0 = \sum_i \left[2 \times 3 \times \tfrac{1}{6}\mu_{q_i} + 3 \times \tfrac{2}{3}\mu_{u_i} + 3 \times (-\tfrac{1}{3})\mu_{d_i} \right.$$
$$\left. + 2 \times \left(-\tfrac{1}{2}\right)\mu_{L_i} + (-1)\mu_{e_i} \right] + 2 \times 2 \times \left(\tfrac{1}{2}\right)\mu_H, \tag{17.296}$$

where we sum over the quark and lepton generations i. Inside the brace the factor of 3 for quarks indicates summing over colors, the factor of 2 for q, L and H counts two fields inside the doublets, and the additional factor of 2 for the Higgs is due to the bosonic nature of the Higgs. Thus, the hypercharge neutrality of the universe gives

$$\sum_i \left[\mu_{q_i} + 2\mu_{u_i} - \mu_{d_i} - \mu_{L_i} - \mu_{e_i} \right] + 2\mu_H = 0. \tag{17.297}$$

Finally, consider the Yukawa couplings

$$\mathcal{L}_{\text{Yukawa}} = g_{e_i} \bar{L}_i H e_i + g_{u_i} \bar{q}_i H^c u_i + g_{d_i} \bar{q}_i H d_i, \tag{17.298}$$

which yield the following relations among the chemical potentials:

$$\mu_H = \mu_{L_i} - \mu_{e_i} = \mu_{q_i} - \mu_{d_i} = \mu_{u_i} - \mu_{q_i}. \tag{17.299}$$

Assuming the chemical potentials are independent of i, one finds the following relations among the chemical potentials:

$$\mu_e = \frac{2N_f + 3}{6N_f + 3}\mu_L, \quad \mu_H = \frac{4N_f}{6N_f + 3}\mu_L$$
$$\mu_q = -\frac{1}{3}\mu_L, \mu_u = \frac{2N_f - 1}{6N_f + 3}\mu_L, \mu_d = -\frac{6N_f + 1}{6N_f + 3}\mu_L. \tag{17.300}$$

Now, define L so that $n_L \equiv LgT^2/6$ and $n_B \equiv BgT^2/6$, where we can express L and B as follows:

$$L = \sum_i (2\mu_{L_i} + \mu_{e_i})$$

$$B = \sum_i (2\mu_{q_i} + \mu_{u_i} + \mu_{d_i}). \tag{17.301}$$

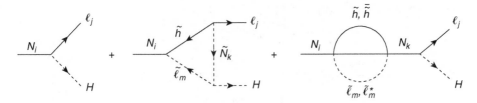

Figure 17.2. Generation of lepton number asymmetry via decay of the right-handed neutrino (N) by interference between the tree, the vertex, and the self-energy or wave function loop diagrams for the supersymmetric case.

Under the assumption of the chemical potentials being independent of the generation index, one finds

$$B = -\frac{4N_f}{3}\mu_L, L = \frac{14N_f^2 + 9N_f}{6N_f + 3}\mu_L$$

$$(B - L) = -\frac{N_f}{6N_f + 3}(22N_f + 13)\mu_L. \tag{17.302}$$

This gives

$$\frac{B}{B - L} = \frac{8N_f + 4}{22N_f + 13}$$

$$= \frac{28}{79}, \quad N_f = 3. \tag{17.303}$$

17.5.3 Leptogenesis: The Supersymmetric Case

Let us now consider leptogenesis for the supersymmetric case (see Fig. 17.2). Here, we have the superpotential

$$W = \frac{1}{2}M_i N_i N_i + \lambda_{ij}L_i^\alpha \epsilon_{\alpha\beta}H^\beta N_j. \tag{17.304}$$

In this case, we define the asymmetries so that

$$\tilde{\epsilon}_L^N \equiv \frac{\Gamma_{NL} - \Gamma_{N\bar{L}}}{\Gamma_{NL} + \Gamma_{N\bar{L}}}, \qquad \tilde{\epsilon}_{\tilde{L}}^N \equiv \frac{\Gamma_{N\tilde{L}} - \Gamma_{N\tilde{L}^*}}{\Gamma_{N\tilde{L}} + \Gamma_{N\tilde{L}^*}}$$

$$\tilde{\epsilon}_{\tilde{L}}^{\tilde{N}} \equiv \frac{\Gamma_{\tilde{N}\tilde{L}} - \Gamma_{\tilde{N}^*\tilde{L}^*}}{\Gamma_{\tilde{N}\tilde{L}} + \Gamma_{\tilde{N}^*\tilde{L}^*}}, \qquad \tilde{\epsilon}_{\tilde{L}}^{\tilde{N}^*} \equiv \frac{\Gamma_{\tilde{N}^*L} - \Gamma_{\tilde{N}\bar{L}}}{\Gamma_{\tilde{N}^*L} + \Gamma_{\tilde{N}\bar{L}}}. \tag{17.305}$$

Here for the vertex diagram (Fig. 17.2) one finds

$$\tilde{\epsilon}_L^N(v) = \tilde{\epsilon}_{\tilde{L}}^N(v) = \tilde{\epsilon}_{\tilde{L}}^{\tilde{N}}(v) = \tilde{\epsilon}_{\tilde{L}}^{\tilde{N}^*}(v), \tag{17.306}$$

and for the wave function (see Fig. 17.2) diagram one finds

$$\tilde{\epsilon}_L^N(w) = \tilde{\epsilon}_{\tilde{L}}^N(w) = \tilde{\epsilon}_{\tilde{L}}^{\tilde{N}}(w) = \tilde{\epsilon}_{\tilde{L}}^{\tilde{N}^*}(w). \tag{17.307}$$

Thus, defining $\tilde{\epsilon}_L^N = \tilde{\epsilon}_L^N(v) + \tilde{\epsilon}_L^N(w)$, the total $\tilde{\epsilon}_{N_i}$ is

$$\tilde{\epsilon}_{N_i} = \tilde{\epsilon}_L^{N_i} + \tilde{\epsilon}_{\tilde{L}}^{N_i} + \tilde{\epsilon}_{\tilde{L}}^{\tilde{N}_i} + \tilde{\epsilon}_{\tilde{L}}^{\tilde{N}_i^*} = 4\tilde{\epsilon}_L^{N_i}, \tag{17.308}$$

and its computation in the limit $M_k \gg M_i$ gives [65]

$$\epsilon_L \simeq -\frac{3}{2\pi}\sum_k \frac{Im[(\lambda\lambda^\dagger)_{ik}^2]}{(\lambda\lambda^\dagger)_{ii}}\frac{M_i}{M_k}. \tag{17.309}$$

There are in addition inverse processes which tend to wash out the baryon asymmetry discussed above. The dilution effect can lie in a wide range, i.e., $1-10^{-5}$, depending on the strength of the inverse processes [68]. A correct inclusion of these involves a solution of the Boltzmann equations. Leptogenesis can also occur from dark matter if dark matter carries a lepton number [69]. An interesting issue concerns cosmic coincidence, which relates to the fact that the ratio of dark matter relic density to the relic density of baryons in the universe is ∼5. One possibility is asymmetric dark matter, where dark matter and the lepton or baryon number excess have a common origin (e.g., [70–72] and for a review see Zurek [73]).

17.5.4 Grand Unified Scale Baryogenesis

One of the first attempts to understand baryogenesis was within in the context of grand unification (for a review see Kolb and Turner [74]). The reason for this is simple: grand unified theories automatically contain baryon and lepton number violation as well as sources of CP violation. Also, in the early universe an out-of-equilibrium processes can be easily achieved. However, after the discovery of the sphaleron processes grand unified baryogenesis induced by dimension-six operators went out of favor, since it was discovered that grand unified baryogenesis, for example, in $SU(5)$ while violating baryon number preserves the constraint $B - L = 0$. Thus, any baryon excess produced in grand unified baryogenesis is wiped out by sphaleron processes which violate $B + L$. Sphaleron processes are active in the energy region as low as 10^2 GeV, and thus any baryon asymmetry created above this scale will be wiped out by sphalerons. However, it turns out that $SO(10)$ grand unified models also have operators which are $B - L$ violating with $B - L = -2$ [75–77]. The baryon asymmetry generated by these operators will not be wiped out by the sphaleron processes, since sphaleron processes preserve $B - L$.

Grand unified theories based on $SO(10)$ offer the proper framework for the study of grand unified baryogenesis. Here, $B - L$ violation arises when the singlet of 126 or of $\overline{126}$ acquires a vacuum expectation value. To see this, we decompose 126 in terms of $SU(5) \otimes U(1)$ representations so that

$$126 = 1(-10) + \bar{5}(-2) + 10(-6) + \overline{15}(6) + 45(2) + \overline{50}(-2). \tag{17.310}$$

Here, the singlet field carries a $B - L$ quantum number, and thus formation of a vacuum expectation value for the singlet field violates $B - L$. We now consider the couplings of the 16-plet of matter with Higgs fields of the type $16 \cdot 16 \cdot 10$, $16 \cdot 16 \cdot 120$ and $16 \cdot 16 \cdot \overline{126}$. The $16 \cdot 16 \cdot \overline{126}$ interactions have the coupling of three singlet fields, $1_{16}.1_{16}.1_{\overline{126}}$. As discussed above, when $1_{\overline{126}}$ develops a vacuum expectation value there is a violation of $B - L$, and the mass term for the singlet in the 16-plet, i.e., $1_{16}.1_{16}\langle 1_{\overline{126}}\rangle$, violates $B - L$ by two units. Integration over the 1_{16} singlet leads to a number of higher-dimensional $B - L$ violating interactions. Specifically, in an $SO(10)$ model with a missing partner mechanism, one finds four-field, five-field, and six-field operators which carry $B - L = -2$. Thus, the $B - L = -2$ four-field operator in the superpotential is

$$L_\alpha L_\beta H_u^\alpha H_u^\beta, \tag{17.311}$$

suppressed by one power of a heavy mass. Equation (17.311) is the well-known Weinberg operator, which gives mass to the neutrinos. Next, there are several $B - L = -2$ five-field operators in the superpotential. Examples of these are

$$\epsilon^{abc} D_a^c D_b^c U_c^c L_\alpha H_u^\alpha, \epsilon^{\alpha\beta} L_\alpha L_\beta L_\gamma E^c H_u^\gamma, D_a^c Q^{a\alpha} L_\alpha L_\beta H_u^\beta. \tag{17.312}$$

The operators in Eq. (17.312) are suppressed by two powers of a heavy mass. As already stated, these $B - L = -2$ operators can be used to generate an excess of $B - L$ in the early universe which is not washed out by the sphaleron processes, and it appears possible to generate a sufficient amount of baryon asymmetry by this mechanism.

17.6 Problems

1. Consider a three-dimensional space with a metric in the spherical polar coordinates given by

$$h_{rr} = \frac{1}{1 - kr^2}, h_{\theta\theta} = r^2, h_{\phi\phi} = r^2 \sin^2\theta$$
$$h_{ij} = 0, \ i \neq j. \tag{17.313}$$

Show that the Ricci tensor $\mathcal{R}_{ij}^{(3)}$ is given by

$$\mathcal{R}_{ij}^{(3)} = -2kh_{ij}. \tag{17.314}$$

2. Show that for a scalar field in a Friedmann-Robertson–Walker universe with the Lagrangian given by Eq. (17.137) the matter density and the pressure are given by

$$\rho = \frac{1}{2}\dot{\phi}^2 + \frac{1}{2}R^{-2}(\nabla\phi)^2 + V(\phi) \tag{17.315}$$

$$p = \frac{1}{2}\dot{\phi}^2 - \frac{1}{2}R^{-2}(\nabla\phi)^2 - V(\phi). \tag{17.316}$$

3. Show that the number density n of a particle of mass m which is non-relativistic at temperature T in the early universe is given by

$$n = g \left(\frac{mT}{2\pi} \right)^{3/2} e^{-\frac{m}{kT}}. \tag{17.317}$$

4. Check the validity of the $(\mu/T)^3$ term given in Eq. (17.71).
5. Show that for the case of thermal equilibrium at high temperature, pressure and energy density satisfy the relation

$$T \frac{\partial p}{\partial T} - (\rho + p) \simeq 0. \tag{17.318}$$

6. Use Eq. (17.45) show that Eq. (17.46) holds.
7. A proper calculation of the relic density for the MSSM involves integration over the direct channel poles. Thus, one needs the integral $J(x_f) = \int_0^{x_f} \langle \sigma v \rangle dx$, where σ is the annihilation cross-section for two neutralinos and v is the relative velocity defined by $\sqrt{s} \simeq 2m_{\chi_1} + \frac{1}{4} m_{\chi_1} v^2$, where s is the square of the center-of-mass energy. x_f is defined so that $x_f = T_f/m_{\chi_1}$, where T_f is the freeze-out temperature and $\langle \sigma v \rangle$ is the thermally averaged cross-section. Show that, using $\langle \sigma v \rangle$ as given by Eq. (17.101), integration on dx leads to J_h as given by Eq. (17.102).
8. Using Eqs. (17.107) and (17.117) to determine f_i^p and f_i^n in terms of x, ξ, and $\sigma_{\pi N}$ show that Eqs. (17.119) and (17.120) hold.
9. Show that using the Lagrangian of Eq. (17.274) the wave function diagram in Fig. 17.1 gives a contribution to particle–anti-particle asymmetry which is

$$\epsilon_w = -\frac{1}{8\pi} \sum_{k \neq i} \frac{Im(\lambda \lambda^\dagger)_{ik}^2}{(\lambda \lambda^\dagger)_{ii}} \frac{M_i(M_i + M_k)}{M_k^2 - M_i^2}. \tag{17.319}$$

10. Consider the hybrid inflationary model with the potential for the inflaton field φ and the additional conjugate set of fields ϕ_f ($\bar{\phi}_f$) of the form

$$V = V_0 + \Delta V, \tag{17.320}$$

where V_0 is given by Eq. (17.192) and ΔV is given by Eq. (17.193). Show that, on the inflationary trajectory defined by $\phi_f = 0 = \bar{\phi}_f$, the potential has the form

$$V = V_0 + \Delta V = \lambda^2 M^4 + \frac{\lambda^4 \mathcal{N}}{32\pi^2} \left[(|\varphi|^2 + M^2)^2 \log \frac{\lambda^2(|\varphi|^2 + M^2)}{Q^2} \right.$$
$$\left. + (|\varphi|^2 - M^2)^2 \log \frac{\lambda^2(|\varphi|^2 - M^2)}{Q^2} - 2|\varphi|^4 \log \frac{\lambda^2 |\varphi|^2}{Q^2} - 3M^4) \right]. \tag{17.321}$$

11. Show that $\chi(\vec{k}, \zeta)$ defined by Eq. (17.219) along with the action of Eq. (17.216) gives the following equation of motion:

$$\chi''(\vec{k}, \zeta) + \left(k^2 - \frac{2}{\zeta^2} \right) \chi(\vec{k}, \zeta) = 0, \qquad (17.322)$$

where ζ is the conformal time.

12. Show that, on a curved space-time, $\chi(\vec{k}, \zeta)$ given by

$$\chi(\vec{k}, \zeta) = \sqrt{\frac{\hbar}{2k}} \left(1 - \frac{i}{k\zeta} \right) e^{-ik\zeta} \qquad (17.323)$$

satisfies the differential in Eq. (17.322) and the normalization $\chi\chi'^* - \chi^*\chi' = i\hbar$ holds.

References

[1] A. Friedman, *Z. Phys.* **10**, 377 (1922); *Z. Phys.* **21**, 326(1924).

[2] H. P. Robertson, *Astrophys. J.* **82**, 284 (1935); *Astrophys. J.* **83**, 187 (1935); *Astrophys. J.* **83**, 257(1936).

[3] A. G. Walker, *Proc. London Math. Soc.* **s2-42**, 90 (1937).

[4] See, e.g., H. B. Callen, *Thermodynamics*, John Wiley (1985).

[5] R. D. Peccei and H. R. Quinn, *Phys. Rev. Lett.* **38**, 1440 (1977).

[6] S. Weinberg, *Phys. Rev. Lett.* **40**, 223 (1978).

[7] F. Wilczek, *Phys. Rev. Lett.* **40**, 279 (1978).

[8] B. W. Lee and S. Weinberg, *Phys. Rev. Lett.* **39**, 165 (1977).

[9] H. Pagels and J. R. Primack, *Phys. Rev. Lett.* **48**, 223 (1982).

[10] S. Weinberg, *Phys. Rev. Lett.* **48**, 1303 (1982).

[11] A. H. Chamseddine, R. L. Arnowitt, and P. Nath, *Phys. Rev. Lett.* **49**, 970 (1982).

[12] H. Goldberg, *Phys. Rev. Lett.* **50**, 1419 (1983) [*Phys. Rev. Lett.* **103**, 099905 (2009)].

[13] R. L. Arnowitt and P. Nath, *Phys. Rev. Lett.* **69**, 725 (1992).

[14] J. R. Ellis, J. S. Hagelin, D. V. Nanopoulos, K. A. Olive, and M. Srednicki, *Nucl. Phys. B* **238**, 453 (1984).

[15] K. Griest and D. Seckel, *Phys. Rev. D* **43**, 3191 (1991).

[16] P. Gondolo and G. Gelmini, *Nucl. Phys. B* **360**, 145 (1991).

[17] P. Nath and R. L. Arnowitt, *Phys. Rev. Lett.* **70**, 3696 (1993).

[18] R. L. Arnowitt, B. Dutta, A. Gurrola, T. Kamon, A. Krislock, and D. Toback, *Phys. Rev. Lett.* **100**, 231802 (2008).

[19] M. W. Goodman and E. Witten, *Phys. Rev. D* **31**, 3059 (1985).

[20] M. Shifman, A. Vainshtein, and V. Zakharov, *Nucl. Phys. B* **147**, (1979) 385; 448. PLB 78, 443 (1978).

[21] H. Y. Cheng, *Phys. Lett. B* **219**, 347 (1989).

[22] A. Corsetti and P. Nath, *Phys. Rev. D* **64**, 125010 (2001); J. R. Ellis, A. Ferstl, and K. A. Olive, *Phys. Lett. B* **481**, 304 (2000).

[23] J. Gasser, H. Leutwyler, and M. E. Sainio, *Phys. Lett. B* **253**, 252 (1991).

[24] J. Gasser, H. Leutwyler, and M. E. Sainio, *Phys. Lett. B* **253**, 260 (1991).

[25] P. Agrawal, Z. Chacko, C. Kilic, and R. K. Mishra, arXiv:1003.1912 [hep-ph].

[26] A. H. Guth, *Phys. Rev. D* **23**, 347 (1981).

[27] A. A. Starobinsky, *Phys. Lett. B* **91**, 99 (1980).

[28] A. D. Linde, *Phys. Lett. B* **108**, 389 (1982).

[29] A. Albrecht and P. J. Steinhardt, *Phys. Rev. Lett.* **48**, 1220 (1982).

[30] A. D. Linde, *Phys. Lett. B* **129**, 177 (1983).

[31] A. D. Linde, *Phys. Rev. D* **49**, 748 (1994).

[32] F. L. Bezrukov and M. Shaposhnikov, *Phys. Lett. B* **659**, 703 (2008).

[33] E. J. Copeland, A. R. Liddle, D. H. Lyth, E. D. Stewart, and D. Wands, *Phys. Rev. D* **49**, 6410 (1994).

[34] E. D. Stewart, *Phys. Rev. D* **51**, 6847 (1995)

[35] H. Murayama, H. Suzuki, T. Yanagida, and J. Yokoyama, *Phys. Rev. D* **50**, 2356 (1994)

[36] P. Binetruy and G. R. Dvali, *Phys. Lett. B* **388**, 241 (1996)

[37] B. Kors and P. Nath, *Nucl. Phys. B* **711**, 112 (2005).

[38] M. Kawasaki, M. Yamaguchi, and T. Yanagida, *Phys. Rev. Lett.* **85**, 3572 (2000).

[39] M. Yamaguchi, *Class. Quant. Grav.* **28**, 103001 (2011).

[40] D. H. Lyth and A. Riotto, *Phys. Lett. B* **412**, 28 (1997)

[41] P. Binetruy, G. Dvali, R. Kallosh, and A. Van Proeyen, *Class. Quant. Grav.* **21**, 3137 (2004).

[42] H. Elvang, D. Z. Freedman, and B. Kors, *JHEP* **0611**, 068 (2006).

[43] K. R. Dienes and B. Thomas, *Phys. Rev. D* **81**, 065023 (2010).

[44] Z. Komargodski and N. Seiberg, *JHEP* **1007**, 017 (2010).

[45] M. B. Einhorn and D. R. T. Jones, *JHEP* **1003**, 026 (2010).

[46] S. Ferrara, R. Kallosh, A. Linde, A. Marrani, and A. Van Proeyen, *Phys. Rev. D* **82**, 045003 (2010).

[47] S. Ferrara, R. Kallosh, A. Linde, and M. Porrati, *Phys. Rev. D* **88**(8), 085038 (2013).

[48] D. S. Salopek, J. R. Bond, and J. M. Bardeen, *Phys. Rev. D* **40**, 1753 (1989).

[49] T. S. Bunch and P. C. W. Davies, *Proc. Roy. Soc. Lond. A* **360**, 117 (1978).

[50] N. A. Chernikov and E. A. Tagirov, *Ann. Poincaré Phys. Theor. A* **9**, 109 (1968).

[51] C. Schomblond and P. Spindel, *Ann. Poincaré Phys. Theor.* **25**, 67 (1976).

[52] D. Langlois, hep-th/0405053 (2004).

[53] P. A. R. Ade *et al.* [BICEP2 and Planck Collaborations], *Phys. Rev. Lett.* **114**, 101301 (2015).

[54] G. 't Hooft, *Phys. Rev. D* **14**, 3432 (1976) [erratum, *ibid.* **18**, 2199 (1978)].

[55] G. 't Hooft, *Phys. Rev. Lett.* **37**, 8 (1976).

[56] A. D. Sakharov, *Pisma Zh. Eksp. Teor. Fiz.* **5**, 32 (1967) [*JETP Lett.* **5**, 24 (1967)] [*Sov. Phys. Usp.* **34**, 392 (1991)] [*Usp. Fiz. Nauk* **161**, 61 (1991)].

[57] M. E. Shaposhnikov, *JETP Lett.* **44**, 465 (1986) [*Pisma Zh. Eksp. Teor. Fiz.* **44**, 364 (1986)].

[58] G. R. Farrar and M. E. Shaposhnikov, *Phys. Rev. Lett.* **70**, 2833 (1993) [*Phys. Rev. Lett.* **71**, 210 (1993)].

[59] A. G. Cohen, D. B. Kaplan, and A. E. Nelson, *Ann. Rev. Nucl. Part. Sci.* **43**, 27 (1993).

[60] A. Riotto and M. Trodden, *Ann. Rev. Nucl. Part. Sci.* **49**, 35 (1999).

[61] M. Fukugita and T. Yanagida, *Phys. Lett. B* **174**, 45 (1986).

[62] H. Murayama, H. Suzuki, T. Yanagida, and J. Yokoyama, *Phys. Rev. Lett.* **70**, 1912 (1993).

[63] M. Plumacher, *Z. Phys. C* **74**, 549 (1997).

[64] W. Buchmuller, P. Di Bari, and M. Plumacher, *Nucl. Phys. B* **643**, 367 (2002) [*Nucl. Phys. B* **793**, 362 (2008)].

[65] L. Covi, E. Roulet, and F. Vissani, *Phys. Lett. B* **384**, 169 (1996).

[66] A. Pilaftsis and T. E. J. Underwood, *Nucl. Phys. B* **692**, 303 (2004).

[67] R. E. Cutkosky, *J. Math. Phys.* **1**, 429 (1960).

[68] M. C. Chen, hep-ph/0703087 [HEP-PH] (2007).

[69] W. Z. Feng, A. Mazumdar, and P. Nath, *Phys. Rev. D* **88**(3), 036014 (2013).

[70] D. E. Kaplan, M. A. Luty, and K. M. Zurek, *Phys. Rev. D* **79**, 115016 (2009)

[71] W. Z. Feng, P. Nath, and G. Peim, *Phys. Rev. D* **85**, 115016 (2012)

[72] W. Z. Feng and P. Nath, *Phys. Lett. B* **731**, 43 (2014).

[73] K. M. Zurek, *Phys. Rep.* **537**, 91 (2014).

[74] E. W. Kolb and M. S. Turner, *Ann. Rev. Nucl. Part. Sci.* **33**, 645 (1983).

[75] S. Enomoto and N. Maekawa, *Phys. Rev. D* **84**, 096007 (2011)

[76] K. S. Babu and R. N. Mohapatra, *Phys. Rev. D* **86**, 035018 (2012)

[77] P. Nath and R. M. Syed, *Phys. Rev. D* **93**(5), 055005 (2016).

Further Reading

D. Bauman, "Cosmology," TASI-2009, e-Print: arXiv:0907.5424 [hep-th].

L. Bergström and A. Goobar, *Cosmology and Particle Astrophysics*, (2004).

W. Buchmuller, R. D. Peccei and T. Yanagida, *Ann. Rev. Nucl. Part. Sci.* **55**, 311 (2005).

G. Jungman, M. Kamionkowski, and K. Griest, *Phys. Rep.* **267**, 195 (1996).

E. W. Kolb and M. S. Turner, *The Early Universe*, Addison-Wesley, (1990).

A. R. Liddle and D. H. Lyth, *Cosmological Inflation and Large-scale Structure*, Cambridge University Press (2000).

A. D. Linde, *Particle Physics and Inflationary Cosmology*, Harwood (1990).

D.H. Lyth and A. Riotto, *Phys. Rep.* **314**, 1 (1998).

V. Mukhanov, *Physical Foundations of Cosmology*, Cambridge University Press (2005).

S. Weinberg, *Cosmology*, Oxford University Press (2007).

18

Extended Supergravities and Supergravities from Superstrings

So far we have discussed $N = 1$ supergravity theory in four dimensions. However, one may formulate supergravity for larger values of N in four dimensions, and some of the interesting cases are for $N = 2, 4$, and 8. The particle content of the supergravity multiplet for these cases is

- $N = 2$: graviton (1), gravitino (2), vector (1);
- $N = 4$: graviton (1), gravitino (4), vector (6), spin 1/2 (4), spin 0 (1);
- $N = 8$: graviton (1), gravitino (8), vector (28), spin 1/2 (56), spin 0 (70).

Thus, the supergravity multiplet for the case $N = 2$ consists of the graviton field, two spin 3/2 gravitini, and one vector field. One can think of the $N = 2$ as consisting of two $N = 1$ multiplets, one of which is the graviton multiplet consisting of spin 2 and spin 3/2 and the other, which is the gravitino multiplet, consisting of spin 3/2 and spin 1. Thus, one way to look at the $N = 2$ supergravity multiplet is to think of coupling the gravitino multiplet consisting of spin 3/2 and spin 1 to $N = 1$ supergravity, which results in $N = 2$ local supersymmetry. $N = 8$ is the supergravity with the largest value of N. For $N > 8$, one will get helicity states larger than two for which consistent interacting local theories cannot be formulated* At one time, the $N = 8$ supergravity theory was thought to be the ultimate unified theory until it was noted that the group $SO(8)$ does not accommodate the weak group $SU(2)_w$. One way of obtaining supergravity theories in four dimensions with higher N is to consider supergravity theories in higher dimensions and carry out a dimensional reduction or compactification to lower dimensions. Thus, the $N = 8$ theory can be shown to arise from compactification of 11-dimensional supergravity on a seven-dimensional torus T^7 [3]. We discuss supergravity theories formulated in higher dimensions below. For original

---

* Higher-spin theories have a long history [1]. For a recent discussion see Vasiliev [2].

works and for reviews on supergravities for $N > 1$ and for dimensions larger than four, see the literature [3–14].

18.1 Supergravity in Higher Dimensions

Before discussing the supergravity theories in higher dimensions, we discuss the spinors in higher space-time dimensions. To begin with, we look at the Clifford algebra in d dimensions, which is

$$\{\Gamma^a, \Gamma^b\} = 2\eta^{ab}\mathbb{1}. \tag{18.1}$$

We now discuss the case when the dimension of space-time is even and is of the form

$$d = 2n + 2. \tag{18.2}$$

In this case the minimum dimensionality of the spinor representation is 2^{n+1}. Thus, the minimum dimensionalities of the Dirac spinors in dimensions $d = 2, 4, 6, 8$, and 10 are $2, 4, 8, 16$, and 32, respectively. One can also define a Γ-matrix, which is the analog of the γ_5 matrix in four dimensions so that it involves the product of all the Γ-matrices, i.e.,

$$\Gamma = i^n \Gamma_0 \Gamma_1 \cdots \Gamma_{d-1}, \tag{18.3}$$

where Γ satisfies the condition

$$\Gamma^2 = \mathbb{1}. \tag{18.4}$$

The Γ-matrix anti-commutes with all the Γ-matrices so that

$$\{\Gamma, \Gamma^a\} = 0. \tag{18.5}$$

The Γ-matrix has eigenvalues ± 1.

We can define the generators of the Lorentz group in d dimensions in terms of the Γ-matrices so that

$$\Sigma^{ab} = -\frac{i}{4}[\Gamma^a, \Gamma^b]. \tag{18.6}$$

They satisfy the Lorentz algebra, i.e.,

$$\left[\Sigma^{ab}, \Sigma^{cd}\right] = -i\left[\eta^{ad}\Sigma^{bc} + \eta^{bc}\Sigma^{ad} - \eta^{bd}\Sigma^{ac} - \eta^{ac}\Sigma^{bd}\right]. \tag{18.7}$$

It is now easily checked that Γ commutes with Σ^{ab}, i.e.,

$$[\Gamma, \Sigma^{ab}] = 0. \tag{18.8}$$

In $d = 4$, one finds that the Dirac spinor of dimension 4 can be split into two Weyl spinors of dimension 2 each. On one Weyl spinor, Γ_5 has the eigenvalue $+1$ and on the other it has the eigenvalue -1. Thus, the two Weyl spinors are of opposite chirality. A very similar phenomenon is valid for other spinors in the above list. For instance, the 32-dimensional spinor in $d = 10$, can be split into two Weyl spinors, which we may label as $16_{\pm}$ since they are Weyl spinors with chirality $\pm$. Next, let us consider the case where we include Γ along with Γ^a ($a = 1, \ldots, 2n+2$). The $(2n+3)$ Γ-matrices form an irreducible representation of the Lorentz group. In this case, we have $d = 2n + 3$, and a spinor representation of dimension $2^{(d-1)/2}$.

18.1.1 $D = 11$ *Supergravity*

$D = 11$ is the largest space-time dimension where supergravity theories with states of spin $J \leq 2$ can exist. $D = 11$ supergravity is also thought to be the low-energy limit of M-theory, which is considered to be the mother of all string theories [15]. Further, supergravity theories in lower dimensions can be obtained by truncation of $D = 11$ supergravity. For all these reasons, $D = 11$, supergravity theory is of interest. $D = 11$ supergravity theory has a simple field content. It consists of the following:

- A vierbein e_μ^a: the vierbein field has $(D-1)(D-2)/2 - 1 = 44$ components.
- Majorana spin $3/2$ ψ_μ: in $D = 11$ the Majorana field has $\frac{1}{2} \cdot 32(D-3) = 128$ components.
- Anti-symmetric tensor $A_{\mu\nu\lambda}$: the totally anti-symmetric tensor with three indices in $D = 11$ has

$$\binom{9}{3} = 84$$

components.

The above count gives 128 bosonic and 128 fermionic components. The supergravity Lagrangian in $D = 11$ dimensions was constructed by Cremmer, Julia, and Scherk [4]. It is remarkably simple and compact. The tangent space indices are labeled by Latin indices, and the world indices are labeled by Greek indices. For matrices, Γ^a are chosen in the Majorana representation. Before giving the Lagrangian, we define some further notation. Corresponding to the totally anti-symmetric three-index tensor $A_{\mu\nu\rho}$ we define the field strength $F_{\mu\nu\lambda\rho}$ so that

$$F_{\mu\nu\lambda\rho} = 4\partial_{[\mu} A_{\nu\lambda\rho]}, \tag{18.9}$$

where the indices μ, ν, ... take on the values $0, 1, 2, \ldots, 10$. In the above, the anti-symmetrization within the bracket is defined over all permutations divided by their number.

The gamma matrices in $D = 11$ satisfy the Clifford algebra so that

$$\{\Gamma_\mu, \Gamma_\nu\} = 2\eta_{\mu\nu}, \tag{18.10}$$

where $\eta_{\mu\nu} = (-1, +1, \cdots, +1)$. There are two 32-dimensional irreducible representations of the algebra which are inequivalent. The difference arises depending on if

$$\Gamma_{12} \equiv \Gamma^0 \Gamma^1 \cdots \Gamma^{10} \tag{18.11}$$

is taken as $+1$ or -1. It is useful to introduce the notation on the anti-symmetrized product of gamma matrices:

$$\Gamma^{\mu_1 \cdots \mu_n} = \Gamma^{\mu_1} \cdots \Gamma^{\mu_n}. \tag{18.12}$$

Next, we introduce a covariant derivative of the spinor ψ_μ so that

$$D_\mu(\omega)\psi_\nu = \partial_\mu \psi_\nu + \frac{1}{4}\omega_{\mu ab}\Gamma^{ab}\psi_\nu, \tag{18.13}$$

where $\omega_{\mu ab}$ is the Lorentz connection and is given by

$$\omega_{\mu ab} = \omega^0_{\mu ab}(e) + K_{\mu ab}, \tag{18.14}$$

where $\omega^{(0)}_{\mu ab}$ is a function only of the vierbein, and $K_{\mu ab}$ contains the dependence on the Majorana fields and is given by

$$K_{\mu ab} = \frac{i}{4}\left[\bar{\psi}_\alpha \Gamma_{\mu ab}^{\ \ \alpha\beta}\psi_\beta - 2(\bar{\psi}_\mu \Gamma_b \psi_a - \bar{\psi}_\mu \Gamma_a \psi_b + \bar{\psi}_b \Gamma_\mu \psi_a)\right]. \tag{18.15}$$

The Lagrangian for 11D supergravity is given by [4]

$$2k_{11}^2 \mathcal{L}_{11} = eR(\omega) - \frac{e}{48}F_{\mu\nu\lambda\rho}F^{\mu\nu\lambda\rho}$$

$$+ \frac{1}{(144)^2}\epsilon^{\alpha_1\alpha_2\alpha_3\alpha_4\beta_1\beta_2\beta_3\beta_4\gamma_1\gamma_2\gamma_3}F_{\alpha_1\alpha_2\alpha_3\alpha_4}F_{\beta_1\beta_2\beta_3\beta_4}A_{\gamma_1\gamma_2\gamma_3}$$

$$- 2iV\bar{\psi}_\mu \Gamma^{\mu\nu\rho}D_\nu\left(\frac{\omega + \hat{\omega}}{2}\right)\psi_\rho$$

$$+ \frac{ie}{96}\left(\bar{\psi}_\mu \Gamma^{\mu\nu\alpha\beta\gamma\delta}\psi_\nu + 12\bar{\psi}^\alpha \Gamma^{\gamma\delta}\psi^\beta\right)(F_{\alpha\beta\gamma\delta} + \tilde{F}_{\alpha\beta\gamma\delta}). \tag{18.16}$$

In the above, $\tilde{F}$ is defined so that

$$\tilde{F}_{\mu\nu\lambda\rho} = F_{\mu\nu\lambda\rho} - 3\bar{\psi}_{[\mu}\Gamma_{\nu\lambda}\psi_{\rho]}, \tag{18.17}$$

and $\tilde{\omega}_{\mu ab}$ is defined by

$$\tilde{\omega}_{\mu ab} = \omega_{\mu ab} - \frac{i}{4}\bar{\psi}_\lambda \Gamma_{\mu ab}^{\ \ \lambda\rho}\psi_\rho. \tag{18.18}$$

The $D = 11$ Lagrangian is invariant under the general co-ordinate transformations, under supersymmetry transformations, and under local tangent space transformations.

One can obtain supergravity in $D = 10$ by reduction of the $D = 11$ theory. In $D = 10$, the maximal supersymmetry is $N = 2$, and here one has Majorana–Weyl spinors. For Majorana spinors, one has the constraint $\psi^C = \psi$, and for Weyl spinors one has the constraint $P_\pm \psi_\pm = \psi_\pm$, where $\psi_\pm = P_\pm \psi$ and where $P_+ = P_R$ and $P_- = P_L$. For Majorana–Weyl spinors, these constraints are simultaneously satisfied, i.e., one has

$$(\psi_\pm)^C = \psi_\pm. \tag{18.19}$$

The constraint of Eq. (18.19) can be satisfied in dimensions $D = 2$ mod 8 so that $D = 10$ supergravity theories have Majorana–Weyl spinors. For $N = 2$ in $D = 10$, one can construct a theory with spinors of opposite chirality or with the same chirality, i.e., one has $N = (1, 1)$ and $N = (2, 0)$, often labeled $N = IIA$ and $N = IIB$. The particle content for the two cases is as follows:

- $N = IIA$ supergravity: $g_{\mu\nu}$ (graviton), A_μ (graviphoton), ϕ (dilaton), $B_{\mu\nu}$ (rank-2 anti-symmetric tensor), $C^{(3)}_{\mu\nu\lambda}$ (rank-3 anti-symmetric tensor), $\chi^\pm$ (two Majorana–Weyl dilatinos, one left handed and one right handed), $\psi^\pm_\mu$ (two Majorana–Weyl gravitinos, one left handed and one right handed). Type IIA supergravity is non-chiral.
- $N = IIB$ supergravity: $g_{\mu\nu}$ (graviton), $B_{\mu\nu}$ and $C^{(2)}_{\mu\nu}$ (rank-2 anti-symmetric tensors), ϕ (dilaton), $C^{(0)}$ (axion), $C^{(4)}_{\mu\nu\alpha\beta}$ (self-dual rank-4 anti-symmetric tensor), χ (complex right-handed spinor), ψ_μ (complex left-handed vector spinor).

Each of the $N = 2$ supergravities above has the same number of bosonic and fermionic degrees of freedom, i.e., 128 (bosonic) + 128 (fermionic). In addition to the above, one can also generate an $N = 1$ supergravity in 10 dimensions.

Next, we discuss an explicit reduction of $D = 11$ supergravity to $D = 10$ for the $N = IIA$ case. Consider the 11th dimension, where $\sigma \equiv x^{10}$ is a circle of radius R so that $0 \leq \sigma \leq 2\pi R$, i.e.,

$$\tilde{x} = \{x, \sigma\} \tag{18.20}$$

$$x = \{x^0, x^1, \cdots, x^9\}. \tag{18.21}$$

In this case, we can carry out an expansion of $\phi(x, \sigma)$ of the form

$$\tilde{\phi}(x, \sigma) = \phi(x) + \Sigma_{-\infty}^{+\infty\prime} e^{ik\sigma R} \phi_k, \tag{18.22}$$

where the prime on Σ implies that the sum over k does not include $k = 0$.

If we are interested in the limit $R \to 0$, we may simply keep the $k = 0$ term. Next, we carry out splitting of the 11-dimensional quantities into $(10 + 1)$ dimensional quantities. Thus, the 11-dimensional vierbein $\tilde{e}^{\tilde{a}}_{\ \mu}$ can be written in the form

$$\tilde{e}^{\tilde{a}}_{\ \tilde{\mu}} = \begin{pmatrix} e^{c\phi} e^a_{\ \mu} & 0 \\ e^\phi A_\mu & e^\phi \end{pmatrix}, \tag{18.23}$$

which leads to the following form for the metric $\tilde{g}_{\tilde{\mu}\tilde{\nu}}$:

$$\tilde{g}_{\tilde{\mu}\tilde{\nu}} = \begin{pmatrix} e^{2c\phi} g_{\mu\nu} + e^{2\phi} A_\mu A_\nu & e^{2\phi} A_\nu \\ e^{2\phi} A_\mu & e^{2\phi} \end{pmatrix}. \tag{18.24}$$

Further, the 3-form in $D = 11$ $\tilde{A}_{\bar{\mu}\bar{\nu}\bar{\lambda}}$ can be decomposed into $10 + 1$ as follows:

$$\tilde{A}_{\mu\nu\lambda} = C^{(3)}_{\mu\nu\lambda} \tag{18.25}$$

$$\tilde{A}_{\mu\nu 10} = B_{\mu\nu}. \tag{18.26}$$

The field strength corresponding to $C^{(3)}_{\mu\nu\lambda}$ is $G_{\mu\nu\lambda\rho}$, and the field strength corresponding to $B_{\mu\nu}$ is $H_{\mu\nu\lambda}$, where

$$\tilde{F}_{\mu\nu\lambda\rho} = G_{\mu\nu\lambda\rho} \tag{18.27}$$

$$\tilde{F}_{\mu\nu\lambda 10} = H_{\mu\nu\lambda}. \tag{18.28}$$

The bosonic sector of the $10 + 1$ reduction of the $D = 11$ supergravity gives the following action, which is the bosonic action of type IIA supergravity [6, 8]

$$I^A_{10} = \int d^{10}x \ e(2k^2_{10})^{-1}\left[R(\omega) - \frac{1}{4}e^{\frac{9}{4}\phi}F_{\mu\nu}F^{\mu\nu} - \frac{9}{8}\partial_\mu\phi\partial^\mu\phi - \frac{1}{48}e^{\frac{3}{4}\phi}G_{\mu\nu\lambda\rho}G^{\mu\nu\lambda\rho}\right.$$
$$\left. - \frac{1}{12}e^{-\frac{3}{2}\phi}H_{\mu\nu\lambda}H^{\mu\nu\lambda} + \frac{e^{-1}}{(48)^2}\epsilon^{\mu_1\cdots\mu_{10}}B_{\mu_1\mu_2}G_{\mu_3\cdots\mu_6}G_{\mu_7\cdots\mu_{10}}\right], \tag{18.29}$$

where $F^{\mu\nu}$ is the field strength corresponding to A_μ, and $\kappa^2_{11} = 2\pi R\kappa^2_{10}$. Similarly, the fermionic sector can be derived. Type IIB supergravity cannot be obtained from a truncation of $D = 11$ supergravity [9–11, 16] but it can be realized by a duality transformation from type IIA supergravity. For related works, see the literature [12–15, 17].

18.2 Supergravities from Superstrings

Before we discuss how supergravity theories arise from strings, we give a brief historical introduction to strings (for a more extensive discussion, see the literature [18–24]). String theory arose as a consequence of the study of dual models in the 1960s which were formulated to understand the behavior of strong interactions. Thus, in a field theory a t-channel exchange of a spin J particle produces an amplitude $A(s,t)$ which behaves like s^J for large s while $J > 1$ violates the desirable behavior of amplitudes, and specifically the Froissart bound [25], which requires that the total cross-section not grow faster than $\sim \log^2(s/s_0)$, where s_0 is some constant. It also implies that the amplitudes not grow faster than $\sim s\log(s/s_0)$. One suggestion made to achieve conformity with the Froissart bound was to sum over an infinite number of poles in the direct and cross-channels. On the surface this appears to make things worse, but the sum of an infinite series may be very different from the sum of a few. A further suggestion made was that the sum over s-channel poles or t-channel poles may produce the full result. One may call this principle "duality," and Dolan, Horn, and Schmidt [26] demonstrated this for low values of s and t. An explicit analytic realization of the duality idea was achieved by Veneziano [27], where the scattering amplitude is given by

$$A(s,t) \sim \frac{\Gamma(-\alpha(s))\Gamma(-\alpha(t))}{\Gamma(-\alpha(s) - \alpha(t))}. \tag{18.30}$$

Here, $\alpha(s)$ is assumed to have, linear Regge trajectory, i.e., $\alpha(s) = \alpha(0) + \alpha's$, where $\alpha(0)$ is the intercept of the trajectory and α' is the Regge slope. The amplitude has interesting features, one of which is the absence of double poles – which is also a feature exhibited by quantum field theory. However, the Veneziano amplitude has problems as a model of strong interactions. First, it is found that the Veneziano scattering amplitudes have exponential fall-off at fixed angle at large energy while, experimentally, one finds only a polynomial fall-off consistent with parton models. Second, in general, the Veneziano model has negative norm states which violate unitarity. These states can be eliminated if one imposes the constraints

$$d = 26, \ \alpha(0) = 1. \tag{18.31}$$

Of course, the imposition of $d = 26$ makes the Veneziano model an unrealistic model of strong interactions even though the asymptotic behavior of the amplitude is acceptable because $\alpha(0) = 1$ for the leading trajectory.

The Veneziano model contained only bosons. In 1971, dual models were developed by Ramond [28] and by Neveu and Schwarz [29, 30] which contained fermions. However, in 1974 a new direction of investigation regarding the dual models opened up through the work of Scherk and Schwarz [31] and Yoneya [32], who proposed that the dual models could provide a framework for unified models including gravity. To see why this may make sense, consider the Virasoro–Shapiro [33, 34] model, which is given by the amplitude $A(s, t, u)$:

$$A(s,t,u) \sim \frac{\Gamma(-\alpha(s))\Gamma(-\alpha(t))\Gamma(-\alpha(u))}{\Gamma(-\alpha(s) - \alpha(t))\Gamma(-\alpha(t) - \alpha(u))\Gamma(-\alpha(u) - \alpha(s))}. \tag{18.32}$$

This amplitude corresponds to the scattering of a tachyon+tachyon $\rightarrow$ tachyon+tachyon. It has a massless mode, and at the pole the residue has the behavior $\sim (t - u)^2 \propto z^2$, which implies that the massless mode in the Virasoro–Shapiro amplitude has two components, one a spin 2 and the other a spin 0. The important observation of Scherk and Schwarz [31] and Yoneya [32] was to identify the spin 2 particle as the graviton. In fact, in the zero-slope limit (with $g_{VS}^2\alpha'$ held fixed) they obtained the Einstein theory of gravitation along with a massless scalar particle, which may be identified as a dilaton. They also conjectured that there could be $\sqrt{g}R^2$ and $\sqrt{g}R_{\mu\nu}R^{\mu\nu}$ type terms at higher orders. The above indicates that dual amplitudes could be the framework for unified models since gravity is an element of such a theory. Another benefit of such a viewpoint is that higher space-time dimensions are now more acceptable, keeping in mind that higher-dimensional space-time was being discussed as far back as the 1920s by Kaluza and Klein [35].

18.2.1 Bosonic Strings

However, to make progress towards finding the unified theory we need to go beyond the amplitudes and find the action that can lead to amplitudes that have the duality property. It turns out that if the duality property is indeed a basic property of amplitudes, then one is naturally lead to the idea of strings. This is because in the context of a two-dimensional string an s-channel scattering and a t-channel scattering are topologically equivalent. In fact, we can think of the tree level scattering of four particles as the scattering of four particles lying on the boundary of a two-dimensional world sheet which is topologically a disk or a sphere. This connection is made concrete by considering a string action on a world sheet described by the co-ordinates (σ, τ) where the space-time co-ordinates X^μ are treated as fields on the world sheet, i.e., one considers the quantity $X^\mu(\sigma, \tau)$. The Nambu–Goto [36] string is such an object, and the action of this two-dimensional string is given by

$$I = T \int d\sigma \; d\tau \left((\dot{X}^\mu)^2 (X^{\nu'})^2 - (\dot{X}^\mu X'_\mu)^2 \right)^{1/2}, \tag{18.33}$$

where

$$\dot{X}^\mu = \frac{\partial X^\mu}{\partial \tau}, \; X^{\mu'} = \frac{\partial X^\mu}{\partial \sigma}. \tag{18.34}$$

Here, X^μ has the dimension of length, and in order that the action be dimensionless, T must have units of $[\text{length}]^{-2}$ or units of $[\text{mass}]^2$. One may use Eq. (18.33) to derive scattering amplitudes and relate T to the Regge slope parameter α' so that

$$T = (2\pi\alpha')^{-1}. \tag{18.35}$$

We can simplify Eq. (18.33) by use of auxiliary fields ζ_{ij} $(i, j = 1, 2)$, which gives us the Polyakov action [37]

$$I = -\frac{T}{2} \int d^2\sigma \; \sqrt{\zeta} \zeta^{ij} \partial_i X^\mu \partial_j X^\nu \eta_{\mu\nu}, \tag{18.36}$$

where $d^2\sigma = d\sigma \; d\tau$, $\eta_{\mu\nu}$ is the metric in the Minkowski space, and

$$\zeta^{ik}\zeta_{kj} = \delta^i_j, \; \zeta = det(\zeta_{ij}). \tag{18.37}$$

Varying the action of Eq. (18.36) with respect to X^μ gives

$$\partial_i(\sqrt{\zeta}\zeta^{ij}\eta_{\mu\nu}\partial_j X^\nu) = 0 \tag{18.38}$$

and varying the action of Eq. (18.36) with respect to ζ^{ij} gives

$$\zeta_{ij}\zeta^{kl}\partial_k X^\mu \partial_l X_\mu - 2\partial_i X^\mu \partial_j X_\mu = 0. \tag{18.39}$$

The energy–momentum tensor on the world sheet is given by

$$T_{ij} = -\frac{2}{\sqrt{\zeta}}\frac{\delta I}{\delta \zeta^{ij}}$$
$$= T\left[\partial_i X^\mu \partial_j X_\mu - \frac{1}{2}\zeta_{ij}\zeta^{kl}\partial_k X^\mu \partial_l X_\mu\right], \tag{18.40}$$

and use of Eq. (18.39) in Eq. (18.40) shows that the stress energy tensor vanishes. From Eqs. (18.38) and (18.39), one may show that

$$\frac{1}{2}\sqrt{\zeta}\zeta^{ij}\partial_i X^\mu \partial_j X^\nu \eta_{\mu\nu} = \sqrt{det(\partial_i X^\mu \partial_j X^\nu \eta_{\mu\nu})}, \tag{18.41}$$

from which it follows that Eq. (18.36) reduces to Eq. (18.33).

The action of Eq. (18.36) possesses symmetries of the two-dimensional world sheet as well as symmetries of Minkowski space assumed to be d dimensional. These are:

- Two-dimensional general co-ordinate invariance on the world sheet. Here, one looks at the diffeomorphisms of the type

$$\sigma^i \to \sigma^{i'}(\sigma)$$
$$X^\mu(\sigma) \to X^{\mu'}(\sigma') = X^\mu(\sigma)$$
$$\zeta_{ij}(\sigma) \to \zeta'_{ij}(\sigma') = \frac{\partial \sigma^k}{\partial \sigma^{i'}}\frac{\partial \sigma^l}{\partial \sigma^{j'}}\zeta_{kl}(\sigma). \tag{18.42}$$

- Conformal invariance of the two-dimensional world sheet. Here, one has transformations on the two-dimensional metric so that

$$\zeta_{ij}(\sigma) \to C(\sigma)\zeta_{ij}(\sigma)$$
$$X^\mu(\sigma) \to X^\mu(\sigma). \tag{18.43}$$

The invariance of the action can be seen by noting that $\zeta^{ij} \to C^{-1}\zeta^{ij}$ and $\sqrt{\zeta} \to C\sqrt{\zeta}$ so that $\sqrt{\zeta}\zeta^{ij}$ is invariant under the conformal transformations. The conformal invariance is a consequence of the world sheet being a two-dimensional manifold. Thus, for an n-dimensional world sheet

$$\sqrt{\zeta}\zeta^{ij} \to C^{\frac{n}{2}-1}\sqrt{\zeta}\zeta^{ij}, \tag{18.44}$$

and thus $n = 2$ is required for invariance to hold.

- Poincaré symmetries on the D-dimensional Minkowski space. Here, one has invariance of the action under translations and rotations so that

$$X^\mu(\sigma) \to X^\mu(\sigma) + a^\mu$$
$$X^\mu(\sigma) \to \Lambda^\mu_{\ \nu}X^\nu(\sigma). \tag{18.45}$$

The transformation parameters a^μ and $\Lambda^\mu_{\ \nu}$ are independent of σ.

Let us now discuss the boundary conditions necessary to derive the string field equation from the action. We choose the gauge $\zeta^{ij} = \eta^{ij}$, where $\eta^{00} = -1$, $\eta^{11} = 1$, and $\eta^{01} = \eta^{10} = 0$ in the basis (τ, σ). In this case, Eq. (18.38) reduces to

$$\left(-\frac{\partial^2}{\partial\tau^2} + \frac{\partial^2}{\partial\sigma^2}\right)X^\mu = 0. \tag{18.46}$$

The deduction Eq. (18.46) requires that we impose the following boundary conditions:

$$\text{open} : X'_\mu(\tau, \sigma)|_{\sigma=\pi} = 0 = X'_\mu(\tau, \sigma)|_{\sigma=0} \tag{18.47}$$

$$\text{closed} : X'_\mu(\tau, \sigma) = X'_\mu(\tau, \sigma + \pi). \tag{18.48}$$

An open string is defined by Eqs. (18.46) and (18.47) while a closed string is defined by Eqs. (18.46) and (18.48). In addition, both for the open string as well as for the closed string, one must impose the vanishing of the world sheet stress tensor given by Eq. (18.40). The reason for this is that, since we have already fixed the gauge, we can no longer vary ζ_{ij} and obtain the constrain of $T_{ij} = 0$, which must be imposed from the outside. Now, Eq. (18.40) gives us the following T_{ij} components in the gauge $\zeta_{ij} = \eta_{ij}$:

$$T_{00} = T_{11} = \frac{T}{2}[\dot{X}_\mu \dot{X}^\mu + X^{\mu'} X'_\mu]$$

$$T_{01} = T_{10} = T\dot{X}^\mu X'_\mu. \tag{18.49}$$

Thus, only two independent constraints arise from the vanishing of the stress tensor, i.e.,

$$\dot{X}^\mu X'_\mu = 0 \tag{18.50}$$

$$\dot{X}^\mu \dot{X}_\mu + X^{\mu'} X'_\mu = 0. \tag{18.51}$$

In the above, we have written the string action using the Minkowskian metric $\eta_{\mu\nu}$. More generally, we can consider the string propagating in the background field $g_{\mu\nu}(X)$. Additionally, we can also introduce an anti-symmetric tensor $B_{\mu\nu}$ and a scalar field ϕ. Including these, we can write the string action so that

$$I = -\frac{T}{2}\int d^2\sigma\left[\sqrt{\zeta}\zeta^{ij}\partial_i X^\mu \partial_j X^\nu g_{\mu\nu}(X) + \epsilon^{ij}\partial_i X^\mu \partial_j X^\nu B_{\mu\nu}(X)\right.$$
$$\left. - \alpha'\sqrt{\zeta}R^{(2)}(\zeta)\phi(X)\right]. \tag{18.52}$$

Here, ϵ^{ij} is the two-index anti-symmetric tensor. Inclusion of the background fields destroys the conformal invariance of the action. So, one might ask what is needed to establish the conformal invariance of the theory with inclusion of $g_{\mu\nu}, B_{\mu\nu}$, and ϕ. One way to re-establish conformal invariance is to treat the background fields as coupling parameters and derive their renormalization group evolution, i.e., obtain the beta functions $\beta_{\mu\nu}(g)$, $\beta_{\mu\nu}(B)$, and $\beta_{\mu\nu}(\phi)$ and then set them to zero. A derivation to order $O(\alpha')$ gives the following set of constraints [38]

$$\beta_{\mu\nu}(g) = 0 : R_{\mu\nu} + \frac{1}{4}H_{\mu}{}^{\alpha\beta}H_{\nu\alpha\beta} - 2D_{\mu}D_{\nu}\phi = 0$$

$$\beta_{\mu\nu}(B) = 0 : D_{\alpha}H^{\alpha}{}_{\mu\nu} - 2D_{\alpha}\phi H^{\alpha}{}_{\mu\nu} = 0$$

$$\beta_{\mu\nu}(\phi) = 0 : \frac{d-26}{3\alpha'} + (D_{\mu}\phi)^2 - D_{\mu}D^{\mu}\phi + \frac{1}{4}R + \frac{1}{48}H_{\mu\nu\alpha}H^{\mu\nu\alpha} = 0, \quad (18.53)$$

where $H_{\mu\nu\alpha}$ is defined by

$$H_{\mu\nu\alpha} = \partial_{\mu}B_{\nu\alpha} + \partial_{\nu}B_{\alpha\mu} + \partial_{\alpha}B_{\mu\nu} \qquad (18.54)$$

and is the field strength for $B_{\mu\nu}$ and is totally anti-symmetric in the interchange of indices. From Eq. (18.53), we note that consistency requires $d = 26$ for bosonic strings. Quite remarkably, one can write an effective action where the consistency conditions of Eq. (18.53) with $d = 26$ arise as field equations. Such an effective action is

$$I = -\frac{1}{2\kappa^2}\int d^dx \sqrt{g}e^{-2\phi}\left(R - 4D_{\mu}\phi D^{\mu}\phi + \frac{1}{12}H_{\mu\nu\alpha}H^{\mu\nu\alpha}\right). \qquad (18.55)$$

The action of Eq. (18.55) holds only to $\mathcal{O}(\alpha')$, and there are additional corrections at higher orders in α'. These higher-order corrections, for instance, modify Einstein field equations by adding contributions that involve terms $\mathcal{O}(R^2)$ and higher-order terms. A closed-form solution that leads to Eq. (18.55) to $\mathcal{O}(\alpha')$ is not known. It should be noted that the fields $g_{\mu\nu}$, $B_{\mu\nu}$, and ϕ correspond to the zero modes of the bosonic string. The bosonic strings, however, suffer from the presence of a scalar tachyon.

18.2.2 Fermionic Strings

Let us now extend our analysis to include fermionic strings. To accomplish this, we include along with $X^{\mu}(\sigma, \tau)$ the fermionic co-ordinate $\psi^{\mu}(\sigma, \tau)$, and write

$$I = -\frac{T}{2}\int d^2\sigma \left(\partial_i X^{\mu}\partial^i X^{\mu} + \frac{1}{i}\bar{\psi}^{\mu}\rho^i\partial_i\psi^{\mu}\right), \qquad (18.56)$$

where ρ^i are Dirac matrices in two dimensions. In two dimensions the number of independent Dirac matrices is two, which we choose to be ρ^0 and ρ^1 defined by

$$\rho^0 = \begin{pmatrix} 0 & -i \\ i & 0 \end{pmatrix}, \quad \rho^1 = \begin{pmatrix} 0 & i \\ i & 0 \end{pmatrix}, \qquad (18.57)$$

where $\bar{\psi}^{\mu} = \psi^{\mu\dagger}\rho^0 = \psi^{\mu T}\rho^0$. Here, we expect a Dirac equation of the form

$$\rho^i\partial_i\psi^{\mu} = 0. \qquad (18.58)$$

Using the variables $\sigma_\pm$ where

$$\sigma_\pm = \frac{1}{2}(\sigma \pm \tau), \tag{18.59}$$

the Dirac equation takes the form

$$\begin{pmatrix} 0 & i\partial_- \\ i\partial_+ & 0 \end{pmatrix} \begin{pmatrix} \psi_- \\ \psi_+ \end{pmatrix} = 0, \tag{18.60}$$

which leads to the equations

$$\partial_-\psi_+ = 0 \tag{18.61}$$

$$\partial_+\psi_- = 0. \tag{18.62}$$

These imply that ψ_- is a function only of σ_-, and ψ_+ is a function only of σ_+.

Let us determine the precise boundary conditions we need to derive Eq. (18.58) from Eq. (18.56). Thus, variation of Eq. (18.56) gives

$$\begin{aligned}
\delta I &= \frac{iT}{2} \int d^2\sigma \left(\overline{\delta\psi^\mu} \rho^i \partial_i \psi^\mu + \bar{\psi}^\mu \rho^i \partial_i \delta\psi^\mu \right) \\
&= \frac{iT}{2} \int d^2\sigma \left(\overline{\delta\psi^\mu} \rho^i \partial_i \psi^\mu - \overline{\partial_i\psi^\mu} \rho^i \delta\psi^\mu + \partial_i J^i \right),
\end{aligned} \tag{18.63}$$

where

$$J^i = \bar{\psi}^\mu \rho^i \delta\psi^\mu. \tag{18.64}$$

For Eq. (18.58) to emerge, we need the boundary term in Eq. (18.63) to vanish. Regarding the boundary term, no constraints arise from integration on $d\tau$ as the variations vanish at $\tau \to \pm\infty$. The integration on $d\sigma$ gives the following constraint:

$$\bar{\psi}^\mu \rho^1 \delta\psi^\mu \Big|_{\sigma=0}^{\sigma=\pi} = (\psi_-^\mu \delta\psi_-^\mu - \psi_+^\mu \delta\psi_+^\mu)\Big|_{\sigma=0}^{\sigma=\pi} = 0. \tag{18.65}$$

Equation (18.65) can be satisfied for open and closed strings as follows. For open strings, choosing the phases so that at $\sigma = 0$ we have $\psi_-^\mu(\tau,0) = \psi_+^\mu(\tau,0)$, we have the following two possibilities at $\sigma = \pi$:

$$\psi_-^\mu(\tau,\pi) = \psi_+^\mu(\tau,\pi), \qquad \text{Ramond (R)} \tag{18.66}$$

$$\psi_-^\mu(\tau,\pi) = -\psi_+^\mu(\tau,\pi) \qquad \text{Neveu–Schwarz (NS).} \tag{18.67}$$

For closed strings we need to impose boundary conditions separately for ψ_+ and ψ_-. Thus, we have the following set of boundary conditions:

$$\psi_+^\mu(\tau,\sigma) = -\psi_+^\mu(\tau,\sigma+\pi) \qquad \text{(NS-NS)} \tag{18.68}$$

$$\psi_-^\mu(\tau,\sigma) = \psi_-^\mu(\tau,\sigma+\pi) \qquad \text{(RR)} \tag{18.69}$$

$$\psi_-^\mu(\tau,\sigma) = -\psi_-^\mu(\tau,\sigma+\pi) \qquad \text{(R-NS)} \tag{18.70}$$

$$\psi_+^\mu(\tau,\sigma) = \psi_+^\mu(\tau,\sigma+\pi) \qquad \text{(NS-R)} \tag{18.71}$$

The Lagrangian of Eq. (18.56) is invariant under the following infinitesimal transformations:

$$\delta X^\mu = \bar{\epsilon}\psi^\mu$$
$$\delta\psi^\mu = -i\rho^i\partial_i X^\mu\epsilon, \tag{18.72}$$

where ϵ is an infinitesimal anti-commuting Majorana spinor and has no dependence on σ and τ. Indeed, Eq. (18.72) exhibits $N = 1$ supersymmetry of the action of Eq. (18.56) on the world sheet. In the bosonic string, the constraints of conformal invariance require $D = 26$. For the fermionic string, the corresponding constraint is $D = 10$. Assume now that one wishes to extend the world sheet supersymmetry to larger values of N. This is possible, but the vanishing of the conformal anomaly determines the critical dimension so that

$$N = 2,\ D = 2\,;\ N = 4,\ D = -2. \tag{18.73}$$

As a consequence, the theories with world sheet supersymmetry $N > 1$ are not very useful for model building.

The model of Eq. (18.56) has a problem in that the theory contains a tachyon. This problem is overcome by a projection, i.e., the Gliozzi–Scherk–Olive (GSO) projection [39]. We now focus on string theories in 10 dimensions. The type of string theories one has in 10 dimensions is in part determined by the type of spinors one allows. Thus, in D dimensions a spinor has $2^{D/2}$ number of complex components. For example, for $D = 4$ one will have four complex components, which give us the Dirac spinor in four dimensions. For $D = 10$, one has $2^{D/2}$, which is a 32-dimensional spinor with complex components. The dimensionality of the spinor can be reduced by imposition of additional constraints. Thus, imposition of the Majorana constraint reduces the spinor to 32 real components. In 10 dimensions, one can impose Majorana and Weyl constraints simultaneously. The imposition of the Weyl constraint alone reduces the number of components in the spinor to 16 real components. These 16 real components obey the Dirac equation, which reduces the number to eight propagating modes. This number matches the number of transverse modes of X^μ. Thus, type I string theories involve a single Majorana–Weyl spinor and an $N = 1$ supersymmetry.

For type I string, it is straightforward to include a non-abelian symmetry. This can be done by use of Chan–Paton [40] factors, which arise as follows. Since type I is an open string, one can place charges at the two ends of an open string. For example, we could place a quark belonging to the n-plet representation of the gauge group $U(n)$ at one end and an anti-quark belonging to the $\bar{n}$ representation of $U(n)$ at the other end. In this case, a scattering involving k number of strings will have a group factor

$$Tr(\lambda_1\lambda_2\cdots\lambda_k), \tag{18.74}$$

where λ are the $U(n)$ generators in the n-dimensional representation of $U(n)$. However, not every group is allowed. It turns out that Chan–Paton [40] factors can be included consistently only for the classical groups, and exceptional groups are not allowed in weakly coupled theories.[*] Further, it turns out that the condition that there be no loop divergence and the theory be free of gauge and gravitational anomalies allows only the group $SO(32)$ for type I strings. As is obvious, Chan–Paton factors cannot be introduced in closed strings. In the heterotic string, a non-abelian gauge symmetry is introduced in a different way, as we will see below.

In type I string the spinor was subject to both the Majorana and the Weyl constraints. Consider now the case where we impose either the Majorana condition or the Weyl condition. Here, one will have two supercharges and $N = 2$ supersymmetry. Assume we impose the Majorana constraint only. In this case, we will have a theory with no chiral fermions, and this is the type IIA string. This theory is anomaly free since it is non-chiral and is derivable from reduction of an 11-dimensional supergravity, as discussed in Section 18.1. Next, assume we impose only the Weyl constraint. In this case, we again get an $N = 2$ supersymmetry, but one has Weyl fermions and so the theory is chiral. This is the type IIB string. Type IIB theory cannot be derived by reduction from a higher-dimensional theory.

Let us now discuss inclusion of non-abelian gauge symmetry in a closed string theory. In an open string one can put charges on the open end of strings, as discussed above. However, in a closed string one does not have that option. One possibility is to include internal symmetry indices on fermions. Assume we introduce Majorana spinors $\lambda^a (a = 1, \ldots, n)$. We could then take an action of the form

$$I = -\frac{1}{2\pi} \int d^2\sigma (\partial_i X_\mu \partial^i X^\mu - i\bar{\lambda}^a \rho^i \partial_i \lambda^a). \tag{18.75}$$

This action has an $SO(n)$ internal symmetry. In fact, a larger global symmetry exists since we can write the action in the form

$$I = -\frac{1}{2\pi} \int d^2\sigma (\partial_i X_\mu \partial^i X^\mu - \lambda^i_+ \partial_- \lambda^i_+ - \lambda^i_- \partial_+ \lambda^i_-). \tag{18.76}$$

The action of Eq. (18.76) has an $SO(n)_L \times SO(n)_R$ symmetry. For the model of Eq. (18.75), we can bosonize λ^a and get $n/2$ scalar fields, and thus we will have a total of $10 + n/2$ bosonic co-ordinates. Setting this equal to 26 to cancel the anomaly gives $n = 32$. Thus, the model of Eq. (18.75) has an $SO(32)_L \times SO(32)_R$ symmetry for $D = 10$. This model is related to the Shapiro–Virasoro model and suffers from the problem of tachyons.

---

[*] Away from weakly coupled limits, string junctions may also give rise to massless degrees of freedom. Since they have more end-points than the two of the open string, these junctions may give rise to a wider variety of gauge groups, and one may get exceptional groups from open strings [41–43].

However, there exists another way to include a non-abelian gauge symmetry in the closed string. The main idea behind this it that the left movers and the right movers can be treated differently. Assume we take the right movers to be of the type discussed earlier, which have world sheet supersymmetry and no tachyons. For the left-movers, we introduce gauge degrees of freedom. Thus, we consider the following model of the string:

$$I_{\rm h} = -\frac{1}{2\pi} \int d^2\sigma (\partial_i X_\mu \partial^i X^\mu - 2\psi_+^\mu \partial_- \psi_{\mu+} - 2\sum_{a=1}^n \lambda_-^a \partial_+ \lambda_-^a). \tag{18.77}$$

Here, X_R^μ, ψ_+^μ are the right movers and X_L^μ, λ_-^a are the left movers, where ψ_+^μ, λ_-^a are Majorana–Weyl fermions. The critical dimension here is $D = 10$, and X_R^μ and ψ_+^μ are related by supersymmetry, i.e.,

$$\delta X_R^\mu = \epsilon \psi_+^\mu, \quad \delta \psi_+^\mu = \epsilon \partial_+ X_R^\mu. \tag{18.78}$$

The left-moving sector is bosonic, and we need $n = 32$ to satisfy the consistency condition as discussed following Eq. (18.76), which implies that we have an $SO(32)$ symmetry. However, there is another possibility, which is that we have different boundary conditions for λ^a, which gives $E_8 \times E_8$.

18.2.3 The Field Content of Five Different String Theories

We discuss below the massless spectrum of the five different string theories, all of which reside in $D = 10$ space-time dimensions. These are type I, type IIA, type IIB, heterotic $SO(32)$, and heterotic $E_8 \times E_8$:

- Type I: this model has $N = 1$ supersymmetry in $D = 10$ dimensions. Here, the massless states consist of a vector field and chiral fermion. The vector field arises from the Neveu–Schwarz sector, which has eight independent components. The chiral spinor with eight real components arises from the Ramond sector. Further, we can increase the degrees of freedom by inclusion of Chan–Paton factors.
- Type IIA: this model has $N = 2$ supersymmetry and is non-chiral. The massless bosonic modes arise from the NS-NS and RR sectors, and the fermionic modes arise from the R-NS and NS-R sectors. Thus, the NS-NS sector gives a dilaton ϕ, an anti-symmetric tensor field $B_{\mu\nu}$ with dimension 28, and a traceless symmetric tensor $g_{\mu\nu}$, the graviton, with 35 independent components. In addition, the bosonic degrees of freedom arise from the RR sector, which consists of one vector $C_\mu^{(1)}$ with eight degrees of freedom, and an anti-symmetric 3-tensor $C_{\mu\nu\lambda}^{(3)}$ with 56 degrees of freedom. From the NS-R sector, one finds a chiral spinor χ of dimensionality 8 and a chiral spin 3/2 gravitino ψ^μ of dimensionality 56. In the R-NS sector, one finds the same set of massless

modes except that the chirality is opposite. Thus, the massless spectrum of type IIA consists of

(a) bosonic modes: $g_{\mu\nu}[35]$, $\phi[1]$, $B_{\mu\nu}[28]$; $C_\mu^{(1)}[8]$, $C_{\mu\nu\lambda}^{(3)}[56]$,

(b) fermionic modes: $\chi[8]$, $\psi^\mu[56]$; $\chi'[8]$, $\psi^{\mu'}[56]$,

where the fermionic modes χ, χ' and $\psi^\mu, \psi^{\mu'}$ are of opposite chirality.

- Type IIB: here, one has $N = 2$ supersymmetry, and the model is chiral. The massless modes in the NS-NS, NS-R, and R-NS sectors are the same as in the type IIA case. However, the RR sector is different. Here, we have the following set: a scalar field $C^{(0)}$ with one degree of freedom, a rank-2 anti-symmetric tensor $C_{\mu\nu}^{(2)}$ with 28 degrees of freedom, and a self-dual anti-symmetric 4-tensor $C_{\mu\nu\lambda\rho}^{(4)}$ with 35 degrees of freedom. We list them below:

(a) bosonic modes: $g_{\mu\nu}[35]$, $\phi[1]$, $B_{\mu\nu}[28]$; $C^{(0)}[1]$, $C_{\mu\nu}^{(2)}[28]$, $C_{\mu\nu\lambda\rho}^{(4)}[35]$,

(b) fermionic modes: $\chi[8]$, $\psi^\mu[56]$; $\chi'[8]$, $\psi^{\mu'}[56]$.

- $SO(32)$ or $E_8 \times E_8$ heterotic string: here, one has $N = 1$ supersymmetry in $D = 10$, and only the right movers have world sheet supersymmetry, so one has only NS and R sectors. Thus, from the 10 physical dimensions one has the following massless modes. From the NS sector we have a scalar $\phi[1]$, an anti-symmetric tensor $B_{\mu\nu}[28]$, and a graviton $g_{\mu\nu}[35]$. From the R sector one has a chiral fermion $\chi[8]$ and a gravitino $\psi^\mu[56]$. Thus, there are 64 bosonic and 64 fermionic states from the right movers. The massless states from the left movers arise from the compactified 16-torus. Here, one finds an equal number of degrees of freedom, which give a total of 496×8 degrees of freedom. In the fermionic sector, one finds an equal number of degrees of freedom, so the degrees of freedom in the bosonic and fermionic sectors are matched. This count holds for both $SO(32)$ and $E_8 \times E_8$ heterotic strings.

18.2.4 Supergravity Lagrangians for Type I and Heterotic Strings

- For type I string theory the low-energy theory produces a supergravity Lagrangian whose bosonic part is given by

$$I_{typeI} = \frac{1}{(2\pi)^7} \int d^{10}x \sqrt{g} \left[R - \frac{1}{2}(D_\mu\phi)^2 - \frac{1}{4}e^{\phi/2}F_{\mu\nu}^a F^{a\mu\nu} - \frac{1}{12}e^\phi H_{\mu\nu\alpha}H^{\mu\nu\alpha} \right].$$
(18.79)

- For heterotic string, the bosonic part of the low-energy theory gives

$$I_{het} = \frac{1}{(2\pi)^7} \int d^{10}x \sqrt{g} \left[R - \frac{1}{2}(D_\mu\phi)^2 - \frac{1}{4}e^{-\phi/2}F_{\mu\nu}^a F^{a\mu\nu} - \frac{1}{12}e^{-\phi}H_{\mu\nu\alpha}H^{\mu\nu\alpha} \right].$$
(18.80)

Here, the gauge group can be $SO(32) \times SO(32)$ or $E_8 \times E_8$. It may be noted that comparison of Eqs. (18.79) and (18.80) shows that while the dilation factors that appear in the last two terms on the right-hand side of Eq. (18.79) are $e^{\phi/2}$ and e^{ϕ}, the last two terms on the right-hand side of Eq. (18.80) have the factors $e^{-\phi/2}$ and $e^{-\phi}$. This indicates that the type I string and the heterotic string are related by an S-duality, where defining $g = e^{\phi}$ one has $g \to 1/g$.

For type IIA strings, the Lagrangian that arises from the zero-slope limit of strings is the same as given by Eq. (18.29). For type IIB strings, the supergravity Lagrangian for type IIB can be obtained by a duality transformation from type IIA.

18.2.5 Yang–Mills and Lorentz Group Chern–Simons 3-forms

The discovery that 10-dimensional superstring theories with gauge groups $SO(32)$ or $E_8 \times E_8$ have cancellation of gauge and gravitational anomalies [44] implies that they are mathematically consistent theories of quantum gravity. Thus, the hope expressed at the beginning of this subsection that superstring theories provide a possible framework for unified theories is further strengthened [18, 19, 44–46].

As discussed above, the zero-slope limit of type I and heterotic superstring theories gives $N = 1$ supergravity in 10 dimensions. In addition, one also finds coupling with $N = 1$ Yang–Mills theory [44]. This system is described by the following set of fields:

$$e_{\mu}{}^{a}, \psi_{\mu}, \chi, \phi, B_{\mu\nu}, \tag{18.81}$$

where $e_{\mu}{}^{a}$ is the vierbein, ψ_{μ} is the left-handed spinor, and χ is a right-handed spinor. ϕ is a scalar field, and $B_{\mu\nu}$ is a two-index anti-symmetric tensor. The field strength associated with the two-index anti-symmetric tensor will be denoted by $H_{\mu\nu\lambda}$. The Yang–Mills sector contains the fields

$$A_{\mu}^{i}, \lambda^{i}, \tag{18.82}$$

where A_{μ}^{i} is the gauge vector field (with i the group index) and λ^{i} is the left-handed spinor, both belonging to the adjoint representation of the gauge group, which is either $SO(32)$ or $E_8 \times E_8$. The coupling of $N = 1$ supergravity to Yang–Mills theory had previously been worked out in [47–49]. Here, it is found that the three-indexed quantity $H_{\mu\nu\lambda}$ appears in a specific combination, which is

$$H_{\mu\nu\lambda} + \omega_{\mu\nu\lambda}^{(Y)}, \tag{18.83}$$

where $\omega_{\mu\nu\lambda}^{(Y)}$ is a Chern–Simons gauge group 3-form defined by

$$\omega_{\mu\nu\lambda}^{(Y)} = Tr\left[A_{\mu}\left(F_{\nu\lambda} - \frac{2}{3}A_{\nu}A_{\lambda}\right)\right], \tag{18.84}$$

where $F_{\nu\lambda}$ is the field strength of the Yang–Mills gauge potential A_μ. However, it was found [43] that the low-energy limit of string theory gives the combination

$$H_{\mu\nu\lambda} + \omega^{(Y)}_{\mu\nu\lambda} - \omega^{(L)}_{\mu\nu\lambda}, \tag{18.85}$$

where $\omega^{(L)}_{\mu\nu\lambda}$ is the Lorentz Chern–Simons 3-form defined by

$$\omega^{(L)}_{\mu\nu\lambda} = tr\left[\omega_\mu R_{\nu\lambda} - \frac{2}{3}\omega_\mu \omega_\nu . \omega_\lambda\right]. \tag{18.86}$$

Here, ω_μ is the local Lorentz spin connection, $R_{\nu\lambda}$ is the Lorentz-curvature 2-form, and the trace is in the vector representation of $SO(1,9)$. Now, the Lorentz Chern–Simons 3-form is a higher-derivative term. Since the low-energy limit of the superstring theory is supersymmetric, such a term must be part of a supersymmetric multiplet. Thus, the Lorentz Chern–Simons 3-form must be supersymmetrized. This requires derivation of the Lagrangian coupling supergravity to Yang–Mills theory up to second order in κ. This has been done in a number of works [50–54].

18.3 Four-Dimensional String Models

Ten-dimensional superstring theories with $SO(32)$ and $E_8 \times E_8$ can compactify to four dimensions with $N = 1$ supersymmetry [55]. Let us focus on the heterotic string. Here, the compactification would produce models at scale M_c, which we expect to be the size of the string scale defined by [56]

$$M_{str} \approx 5 \times 10^{17} \text{ GeV}. \tag{18.87}$$

We then expect that at scales $Q \ll M_{str}$, one is in the regime of an effective field theory where physics can be described by an effective supergravity theory. This supergravity theory will have the Einstein gauge invariance as well as the unified symmetry of the parent string model. The resulting theory will be an $N = 1$ supergravity in four dimensions. Such an effective supergravity theory can be described by the techniques of applied supergravity, and different supergravity theories will be described by different Kähler potentials, superpotentials, and gauge kinetic energy functions. Thus, the formalism of applied supergravity and of supergravity grand unification as discussed in Chapters 12–14 is valid for the low-energy limit of most classes of string models.

We now discuss some generic features of an effective theory from strings. Since grand unification appears to have at least one success in that it explains the unification of gauge couplings, it appears desirable to have string grand unified theories whose low-energy limit will yield a supergravity grand unification. There have been many attempts to obtain grand unified models from strings (for a recent review, see Raby [57]). A promising new approach to grand unification is F-theory [58] which is essentially a generalization of type IIB strings.

Recent work regarding construction of models within F-theory can be found in the literature [59–63].

In the heterotic string framework, however, it is not essential that the unification of gauge couplings arises from grand unification. Specifically, one may have the standard model gauge group, which can emerge directly at the string scale so that [64]

$$g_i^2 k_i = g_{string}^2 = \left(\frac{8\pi G_N}{\alpha'} \right). \qquad (18.88)$$

Here, G_N is the Newtonian constant, α' is the Regge slope, and k_i is an integer if the subgroup is non-abelian but could be fractional if the subgroup is abelian (k_i are referred to as the Kac–Moody level of the subgroup i in string theory). In this case, one will have, in general, color neutral fractionally charged states unless the SM gauge group arises from an unbroken $SU(5)$ at the string scale, or unless $k > 1$ [65]. Since no color neutral fractionally charged particles are known, one would need to find a mechanism in this case which would make them supermassive, or alternatively produce bound states with integral charges by confining them.

One important issue concerns the unification of gauge couplings within heterotic strings consistent with LEP data. Thus, for heterotic string models, one has the gauge coupling relation:

$$\frac{16\pi^2}{g_i^2(M_Z)} = k_i \frac{16\pi^2}{g_{string}^2} + b_i \ln \left(\frac{M_{str}^2}{M_Z^2} \right) + \Delta_i. \qquad (18.89)$$

Here, b_i are for the subgroups using the particle content of the minimal supersymmetric standard model, and Δ_i contain both stringy and non-stringy effects. An immediate problem one encounters is that the use of the MSSM mass spectrum will produce a $\sin^2 \theta_W$, where θ_W is the weak angle, which will differ from experiment by many standard deviations. Various avenues have been explored to resolve this problem [66] such as by finding large-threshold corrections, using extra matter in vector-like representations, or using non-standard Kac–Moody levels. Another important difference in string models from supergravity grand unified models is in radiative breaking of the electroweak symmetry. We note that in ordinary supergravity models the parameter B_0 that appears in soft breaking in Eqs. (14.25) and (14.26) gets eliminated in favor of $\tan \beta$ when one considers the radiative breaking of the electroweak symmetry, as can be seen from Eq. (14.31). However, in string models both A_0 and B_0 are determined in terms of moduli vacuum expectation values. Specifically, B_0 is proportional to $e^{g_{st}^2/4}$ where $g_{st} = k_i g_i$ and k_i is the Kac–Moody level of the sub-group and g_i is the corresponding gauge-coupling constant. Consequently, the radiative symmetry breaking equation for B_0 (B at the electroweak scale) is actually an equation to determine $\tan \beta$ in terms of all other soft parameters and in terms of the moduli vacuum expectation values [67]. Alternatively, one may use the electroweak constraint for an empirical determination of α_{string} in terms of the soft parameters.

18.4 Gravitino Decay

In supergravity models where the gravitino is not the lightest supersymmetric particle it will be unstable and decay. However, the gravitino couples to matter only gravitationally or non-perturbatively, and thus its decay lifetime is expected to be long. If the decay of the gravitino occurs around or after the time of the Big Bang nucleosysthesis (BBN), it would upset the observed abundance of ^{4}He, and as a consequence BBN puts limits on the gravitino mass (for early work, see Coughlan *et al.* [68]). To illustrate this, let us consider the coupling of the gravitino with the spin $(1,1/2)$ multiplet. Here, one has the following interaction Lagrangian:

$$L_{int} = \frac{1}{4M_{Pl}} \bar{\psi}_\mu \sigma^{\nu\lambda} \gamma^\mu \lambda_a F^a_{\nu\lambda} + \text{h.c.}, \tag{18.90}$$

where a is an internal symmetry index. Specifically, consider the coupling with the photon, and for simplicity let us consider the case when the mixings of the gaugino $\tilde{\chi}$ with other spin $1/2$ neutral fields are ignored. In this case, the decay of the gravitino to $\gamma + \tilde{\chi}$ is given by [69]

$$\Gamma_{\tilde{G}\to\gamma\tilde{\chi}} = \frac{m^3_{\tilde{G}}}{4M^2_{Pl}} (1 - m^2_{\tilde{\chi}}/m^2_{\tilde{G}})^3. \tag{18.91}$$

The partial lifetime for this decay mode is

$$\tau_{\tilde{G}\to\gamma\tilde{\chi}} \sim 10^5 s \left(\frac{1 \text{ TeV}}{m_{\tilde{G}}}\right)^3, \tag{18.92}$$

where we neglected the phase-space factor, assuming $m_{\tilde{G}} \gg m_{\tilde{\chi}}$. The BBN time lies in the interval $10^{1\pm2}$ s. Thus, the BBN constraint $\tau_{\tilde{G}\to\text{all}} < 10^{-1}s$ implies $m_{\tilde{G}} \simeq 100$ TeV.

A situation similar to the gravitino decay exists regarding the decay of the super-Higgs field in supergravity models, where supersymmetry is broken by the super-Higgs mechanism. Here, one finds that the vacuum expectation value of the chiral scalar field that enters into the super-Higgs mechanism is $O(M_{Pl})$ while its mass is in the electroweak region. Like the gravitino, the super-Higgs field which resides in the hidden sector has no direct coupling with the fields of the visible sector since such couplings are forbidden, to avoid the problem of generating a new mass hierarchy problem in the visible sector. However, the super-Higgs field can couple with gravitational interactions with the visible sector fields. Analysis of its decay again indicates that the mass of the super-Higgs field should likely be in the 50–100 TeV range in order not to upset the BBN. Before the discovery of the Higgs boson mass at 126 GeV, such a large scalar mass was thought to be problematic. However, with the Higgs mass at 126 GeV, scalars in the desired mass range are possible. There is a significant amount of literature related to a discussion of these issues [70–72].

A similar situation occurs in Kaluza–Klein theories and in string theory where on compactification one has many massless scalars. Specifically, as noted earlier, supersymmetric strings are formulated in 10 dimensions, and one needs to compactify the 10-dimensional theory so that the compactified theory results in four-dimensional ordinary space-time ($\mathcal{R}_{1,3}$) and a compact six-dimensional manifold $\mathcal{M}_6$, i.e.,

$$\text{10-dimensional strings} \rightarrow \mathcal{R}_{1,3} \otimes \mathcal{M}_6. \tag{18.93}$$

The compact dimensions are small enough to have evaded observation thus far, but could become observable if the energies are large enough to probe the small scales of the compactified dimensions (situations of this type are discussed in Chapter 19). One of the commonly chosen compact space $\mathcal{M}_6$ is the so-called Calabi–Yau manifold, consisting of three complex dimensions. Deformations of the metric on this space consistent with the Calabi-Yau constraints have the form

$$\delta \, ds^2 = \delta g_{ab} \, dz^a \, d\bar{z}^b + \delta g'_{ab} \, dz^a \, dz^b + \text{h.c.} \tag{18.94}$$

The first set of fluctuations on the right-hand side of Eq. (18.94) keep intact the Kähler structure of the metric, and are referred to as the Kähler or volume moduli. The fluctuations arising from the second term on the right-hand side of Eq. (18.94) do not preserve the Kähler structure, and are referred to as the complex structure moduli. The number of Kähler (complex) moduli is given by the so-called Hodge numbers $h_{1,1}$ ($h_{2,1}$). In low-energy description of the compactified string, these fluctuations manifest as massless scalar fields. Typically, these scalars have no potential at leading order, so we have energy degenerate vacua and associated vacuum expectation values. The generation of potentials for these fields and fixing their vacuum expectation values is often referred to as moduli stabilization. Significant progress has been made in this direction since the work of Kachru, Kallosh, Linde, and Trivedi [73, 74]. In theories with large-radius compactification after moduli stabilization, one may have scalar fields with masses of electroweak size. These fields will couple to the standard model fields only gravitationally or non-perturbatively, and thus their decay could also violate the BBN constraint. This is the moduli problem of string theory. However, if the masses of the moduli fields are made large enough, one can avoid the BBN constraints, as discussed above.

18.5 Problems

1. Verify the dimensional reduction of 11-dimensional supergravity to $10 + 1$ given by Eq. (18.29).

2. Using Sterling's formula show that Eq. (18.30) with linear trajectories gives for large s and small t

$$A(s,t) \sim \Gamma(-\alpha(t))(-\alpha(s))^{\alpha(t)} e^{-\alpha(t)} \sim s^{\alpha(t)}. \qquad (18.95)$$

3. Show that an expansion of Eq. (18.30) alternatively over the sum of poles in the s-channel and in the t-channel leads to the following equality:

$$A(s,t) = -\sum_{n=0}^{\infty} \frac{(\alpha(s)+1)(\alpha(s)+2)\cdots(\alpha(s)+n)}{n!} \frac{1}{\alpha(t)-n}$$

$$= -\sum_{n=0}^{\infty} \frac{(\alpha(t)+1)(\alpha(t)+2)\cdots(\alpha(t)+n)}{n!} \frac{1}{\alpha(s)-n}. \qquad (18.96)$$

This exhibits the duality property of Eq. (18.30).

4. Show that for a conformally flat metric in two dimensions, i.e., for a metric of the form

$$g_{ij} = e^{\phi(\sigma)} \eta_{ij}, \quad i,j = 0,1, \qquad (18.97)$$

one has $\sqrt{g}R = \partial_i(\eta^{ij}\partial_j\phi)$ where $\eta_{ij} = \text{diag}(-1,1)$.

5. Show that the stress tensor for the fermionic string is given by

$$T_{ij} = T\left(\partial_i X^\mu \partial_j X_\mu - \frac{i}{4}\bar{\psi}^\mu(\rho_i\partial_j + \rho_j\partial_i)\psi_\mu - \frac{1}{2}\eta_{ij}(\text{trace})\right). \qquad (18.98)$$

6. Show that the fermionic string has a super-current which up to a multiplicative constant is given by

$$S_i = \rho^j \rho_i \psi^\mu \partial_j X_\mu \qquad (18.99)$$

and that it satisfies the constraints

$$\rho_i S^i = 0 = \partial_i S^i. \qquad (18.100)$$

References

[1] P. A. M. Dirac, *Proc. R. Soc. London, Ser. A* **155**, 447 (1936); M. Fierz and W. Pauli, *Proc. R. Soc. London, Ser. A* **173**, 211 (1939); V. L. Ginzburg, *JETPh*, **12**, 425 (1942); V. L. Ginzburg and I. E. Tamm, *J. Phys.* **11** (1947); E. S. Fradkin, *JETPh*, **20**, 27 (1950); 211; W. Rarita and J. Schwinger, *Phys. Rev.* **60**, 61 (1941); C. Fronsdal, *Phys. Rev. D* **18**, 3624 (1978); A. K. H. Bengtsson, I. Bengtsson, and L. Brink, *Nucl. Phys. B* **227**, 31 (1983); A. K. H. Bengtsson, I. Bengtsson, and L. Brink, *Nucl. Phys. B* **227**, 41 (1983); F. A. Berends, G. J. H. Burgers, and H. Van Dam, *Z. Phys. C* **24**, 247–254 (1984); F. A. Berends, G. J. H. Burgers, and H. van Dam, *Nucl. Phys. B* **260**, 295 (1985); E. S. Fradkin and M. A. Vasiliev, *Phys. Lett. B* **189**, 89–95 (1987); E. S. Fradkin and M. A. Vasiliev, *Nucl. Phys. B* **291**, 141 (1987).

[2] M. A. Vasiliev, arXiv:1603.01888 [hep-th] (2016).

[3] E. Cremmer and B. Julia, *Nucl. Phys. B* **159**, 141 (1979).

[4] E. Cremmer, B. Julia, and J. Scherk, *Phys. Lett. B* **76**, 409 (1978).

[5] P. Van Nieuwenhuizen, *Phys. Rep.* **68**, 189 (1981).

[6] M. Huq and M. A. Namazie, *Class. Quant. Grav.* **2**, 293 (1985) [erratum, *ibid.*, **2**, 597 (1985)].

[7] I. C. G. Campbell and P. C. West, *Nucl. Phys. B* **243**, 112 (1984).

[8] F. Giani and M. Pernici, *Phys. Rev. D* **30**, 325 (1984).

[9] J. H. Schwarz, *Nucl. Phys. B* **226**, 269 (1983).

[10] J. H. Schwarz and P. C. West, *Phys. Lett. B* **126**, 301 (1983).

[11] P. S. Howe and P. C. West, *Nucl. Phys. B* **238**, 181 (1984).

[12] B. de Wit, hep-th/0212245 (2002).

[13] H. Samtleben, *Class. Quant. Grav.* **25**, 214002 (2008).

[14] H. Nastase, arXiv:1112.3502 [hep-th] (2012).

[15] E. Witten, *Nucl. Phys. B* **443**, 85 (1995).

[16] J. Figueroa-O'Farrill and N. Hustler, *Class. Quant. Grav.* **30**, 045008 (2013).

[17] J. J. Fernandez-Melgarejo, arXiv:1311.7145 [hep-th] (2013).

[18] M. B. Green, J. H. Schwarz, and E. Witten, *Superstring Theory.* Vol. 1: *Introduction*, Cambridge University Press (2012).

[19] M. B. Green, J. H. Schwarz, and E. Witten, *Superstring Theory.* Vol. 2: *Loop Amplitudes, Anomalies and Phenomenology*, Cambridge University Press (1987).

[20] J. Polchinski, *String Theory.* Vol. 1: *An Introduction to the Bosonic String*, Cambridge University Press (2007).

[21] J. Polchinski, *String Theory.* Vol. 2: *Superstring Theory and Beyond*, Cambridge University Press (2007).

[22] E. Kiritsis, *String Theory in a Nutshell*, Princeton University Press (2007).

[23] B. Zwiebach, *A First Course in String Theory*, Cambridge University Press (2009).

[24] G.'t Hooft, www.staff.science.uu.nl/universitypresshooft101/lectures/stringnotes.pdf

[25] M. Froissart, *Phys. Rev.* **123**, 1053 (1961).

[26] R. Dolen, D. Horn, and C. Schmid, *Phys. Rev.* **166**, 1768 (1968).

[27] G. Veneziano, *Nuovo Cim. A* **57**, 190 (1968).

[28] P. Ramond, *Phys. Rev. D* **3**, 2415 (1971).

[29] A. Neveu and J. H. Schwarz, *Phys. Rev. D* **4**, 1109 (1971).

[30] A. Neveu and J. H. Schwarz, *Nucl. Phys. B* **31**, 86 (1971).

[31] J. Scherk and J. H. Schwarz, *Nucl. Phys. B* **81**, 118 (1974).

[32] T. Yoneya, *Lett. Nuovo Cim.* **8**, 951 (1973).

[33] M. A. Virasoro, *Phys. Rev.* **177**, 2309 (1969).

[34] J. A. Shapiro, *Phys. Lett. B* **33**, 361 (1970).

[35] Th. Kaluza, Sitzungober. Preuss. Akad. Wiss. Berlin, p.966 (1921); O. Klein, Z. Phys. **37**, 895(1926).

[36] T. Goto, *Prog. Theor. Phys.* **46**, 1560 (1971).

[37] A. M. Polyakov, *Phys. Lett. B* **103**, 207 (1981).

[38] C. G. Callan, Jr., E. J. Martinec, M. J. Perry, and D. Friedan, *Nucl. Phys. B* **262**, 593 (1985).

[39] F. Gliozzi, J. Scherk, and D. I. Olive, *Nucl. Phys. B* **122**, 253 (1977).

[40] J. E. Paton and H. M. Chan, *Nucl. Phys. B* **10**, 516 (1969).

[41] M. R. Gaberdiel and B. Zwiebach, *Nucl. Phys. B* **518**, 151 (1998); O. DeWolfe and B. Zwiebach, *Nucl. Phys. B* **541**, 509 (1999).

[42] A. Grassi, J. Halverson, and J. L. Shaneson, *JHEP* **1310**, 205 (2013); arXiv:1410.6817 [math.AG] (2014).

[43] J. Polchinski and E. Witten, *Nucl. Phys. B* **460**, 525 (1996).

[44] M. B. Green and J. H. Schwarz, *Phys. Lett. B* **149**, 117 (1984).

[45] M. B. Green and J. H. Schwarz, *Nucl. Phys. B* **181**, 502 (1981).

[46] D. J. Gross, J. A. Harvey, E. J. Martinec, and R. Rohm, *Phys. Rev. Lett.* **54**, 502 (1985).

[47] A. H. Chamseddine, *Phys. Rev. D* **24**, 3065 (1981).

[48] G. F. Chapline and N. S. Manton, *Phys. Lett. B* **120**, 105 (1983).

[49] E. Bergshoeff, M. de Roo, B. de Wit, and P. van Nieuwenhuizen, *Nucl. Phys. B* **195**, 97 (1982).

[50] A. H. Chamseddine and P. Nath, *Phys. Rev. D* **34**, 3769 (1986).

[51] L. J. Romans and N. P. Warner, *Nucl. Phys. B* **273**, 320 (1986).

[52] S. J. Gates, Jr. and H. Nishino, *Phys. Lett. B* **173**, 52 (1986).

[53] B. E. W. Nilsson, *Phys. Lett. B* **175**, 319 (1986).

[54] E. Bergshoeff, A. Salam, and E. Sezgin, *Nucl. Phys. B* **279**, 659 (1987).

[55] P. Candelas, G. T. Horowitz, A. Strominger, and E. Witten, *Nucl. Phys. B* **258**, 46 (1985).

[56] V. Kaplunovsky, *Nucl. Phys. B* **307**, 145 (1988) [erratum, *ibid.*, **382**, 436 (1992)].

[57] S. Raby, *Rep. Prog. Phys.* **74**, 036901 (2011).

[58] C. Vafa, *Nucl. Phys. B* **469**, 403 (1996).

[59] C. Beasley, J. J. Heckman, and C. Vafa, *JHEP* **0901**, 058 (2009).

[60] R. Donagi and M. Wijnholt, *Adv. Theor. Math. Phys.* **15**(6), 1523 (2011).

[61] J. J. Heckman, *Ann. Rev. Nucl. Part. Sci.* **60**, 237 (2010).

[62] M. J. Dolan, J. Marsano, and S. Schafer-Nameki, *JHEP* **1112**, 032 (2011).

[63] A. Grassi, J. Halverson, J. Shaneson, and W. Taylor, *JHEP* **1501**, 086 (2015).

[64] P. Ginsparg, *Phys. Lett. B* **197**, 139 (1987).

[65] A. Schellekens, *Phys. Lett.* **237**, 363 (1990).

[66] K. R. Dienes, *Phys. Rep.* **287**, 447 (1997).

[67] P. Nath and T. R. Taylor, *Phys. Lett. B* **548**, 77 (2002).

[68] G. D. Coughlan, W. Fischler, E. W. Kolb, S. Raby, and G. G. Ross, *Phys. Lett. B* **131**, 59 (1983).

[69] J. R. Ellis, J. E. Kim, and D. V. Nanopoulos, *Phys. Lett. B* **145**, 181 (1984).

[70] T. Moroi and L. Randall, *Nucl. Phys. B* **570**, 455 (2000).

[71] M. Kawasaki, K. Kohri, and N. Sugiyama, *Phys. Rev. Lett.* **82**, 4168 (1999).

[72] S. Hannestad, *Phys. Rev. D* **70**, 043506 (2004).

[73] S. Kachru, R. Kallosh, A. D. Linde, and S. P. Trivedi, *Phys. Rev. D* **68**, 046005 (2003).

[74] V. Balasubramanian, P. Berglund, J. P. Conlon, and F. Quevedo, *JHEP* **0503**, 007 (2005).

Further Reading

The reader is referred to the literature [18–24].

19

Specialized Topics

19.1 Models with Extra Dimensions

While we live in a four-dimensional space-time, it is possible that it is remnant of a higher-dimensional structure. This idea has a long history going back to the work on Kaluza and Klein [1–3]. The work in string theory over the past decades has given a new insight into this possibility. Thus, initially the string phenomenology was focussed mostly on the weakly coupled heterotic string where the string scale (M_{str}) and the Planck scale M_{Pl} have a fixed relation [4] in that $M_{str} \sim g_{srt} M_{Pl}$. However, the discovery of a variety of string dualities which connect the strong coupling regime of one string theory to the weak coupling limit of another has opened new possibilities for model building. One such avenue is the possibility that the strongly coupled $SO(32)$ heterotic string on compactification to four dimensions is related to a weakly coupled type I string compactified to four dimensions. In this case, the string scale is not necessarily close to the Planck scale but could be much lower [5, 6]. Although large extra dimensions have been investigated for quite some time [7], the possibility of TeV-scale strings provided further motivation for works on large extra dimensions [8–11]. Here, we give a brief discussion of such models, as they allow the exploration of a new realm of physics beyond the standard model [12–15]. To be more specific, we will discuss models where matter resides on a four-dimensional wall but gravity, gauge fields, and the Higgs fields propagate in higher dimensions. In this section, we will focus only on matter and gauge fields, and in the next section we will discuss gravity in higher dimensions. Thus, let us consider the case of an extra-dimensional theory with $D = 4 + d$ dimensions where matter resides on the four-dimensional wall and the gauge fields propagate in the bulk. To be specific, let us consider the case when $D = 5$. In this case, one can construct a minimal supersymmetric standard model (MSSM)-type Lagrangian of

the form

$$L_5 = -\frac{1}{4}\mathcal{F}_{AB}\mathcal{F}^{AB} - (D_A H_i)^\dagger (D^A H_i) - \bar{\psi}\frac{1}{i}\Gamma^A D_A \psi - V(H_i) + \Delta L_5, \quad (19.1)$$

where ΔL_5 denotes other terms in the extended MSSM Lagrangian which are not displayed. Here A, and B take on the values 0, 1, 2, 3, and 5, H_i ($i = 1, 2$) are the two Higgs doublets of MSSM Higgs fields, and $V(H_i)$ is the Higgs potential. The object D_A is the covariant derivative,

$$D_A = \partial_A - ig_5 \mathcal{A}_A. \quad (19.2)$$

Next, we compactify this theory on a half circle, i.e., on S^1/Z_2 with the radius of compatification R. After compactification, the spectrum of this theory will consist precisely of the MSSM spectrum in four dimensions, and in addition there will be Kaluza–Klein modes which form $N = 2$ multiplets in four dimensions. We may further carry out a breaking of the electroweak symmetry in five dimensions, which will generate electroweak masses for the W- and Z-bosons. Thus, after compactification we will have two types of masses, one of which arises from compactification from five to four dimensions and the other arising from electroweak symmetry breaking so that

$$M_n^2 = m_i^2 + n^2 M_R^2, \quad n = 1, 2, 3, \ldots, \infty, \quad (19.3)$$

where $n^2 M_R^2$ are the masses that arise from compactification of the fifth dimension (where $M_R = 1/R$) and m_i^2 are the masses that arise from electroweak symmetry breaking. Further, after compactification, we need to rescale the coupling constants. Thus, if one has a gauge coupling constant g_5^i in five-dimensional theory, the rescaled coupling constant g_i in four dimensions is given by

$$g_i = g_i^{(5)}/\sqrt{\pi R}. \quad (19.4)$$

The interactions of the zero modes and of the Kaluza–Klein excitations of the gauge bosons are given by

$$L_{int} = g_i j^\mu \left(A_{\mu i} + \sqrt{2} \sum_{n=1}^\infty A_{\mu i}^n \right), \quad (19.5)$$

where $A_{\mu i}$ are the zero modes and $A_{\mu i}^n$ are the Kaluza–Klein excitations of the gauge fields.

The Kaluza–Klein excitations of the W-boson modify the weak interactions in that they can be exchanged along with the W-boson. This affects the Fermi constant. Thus, an integration over the W-boson and its Kaluza–Klein excitations gives the effective Fermi constant to leading order in M_W/M_R to be [13]

$$G_F^{eff} \simeq G_F^{SM} \left(1 + \frac{\pi^2}{3} \frac{M_W^2}{M_R^2} \right). \quad (19.6)$$

One can extend the above analysis to the case when there are d number of extra dimensions. For simplicity, let us consider the case when one compactifies using Z_2 orbifolding for each of the extra dimensions to one common radius R and thus one common compactification mass M_R. In this case, we need to integrate on all the extra Kaluza–Klein excitations just as we did in the case for $d = 1$. Here, the effective Fermi constant is modified from the standard model case as follows:

$$G_F^{eff} = G_F^{SM} f_d \left(\frac{M_W^2}{M_R^2} \right), \tag{19.7}$$

where f_d is the modification factor arising from the contribution of all the Kaluza–Klein excitations, and is given by

$$f_d(x) = \int_0^\infty dt \, e^{-t} \left(\theta_3 \left(\frac{it}{x\pi} \right) \right)^d, \tag{19.8}$$

where for complex z with $Im \, z > 0$, $\theta_3(z)$ is defined by

$$\theta_3(z) = \sum_{k=-\infty}^{\infty} \exp(i\pi k^2 z). \tag{19.9}$$

For the case $d \geq 2$, the integral in Eq. (19.9) diverges. To achieve convergence we need to truncate the sum over the Kaluza–Klein states with a cutoff at the string scale. With inclusion of the cut-off, one can find approximations to Eq. (19.8) so that

$$f_d \left(\frac{M_W^2}{M_R^2} \right) \simeq \left(1 + \left(\frac{2\pi^2}{3} + 2\pi ln \left(\frac{M_{str}}{M_R} \right) \right) \left(\frac{M_W}{M_R} \right)^2 \right), \qquad d = 2 \tag{19.10}$$

$$\simeq \left(\frac{d}{d-2} \right) \frac{\pi^{d/2}}{\Gamma(1+d/2)} \left(\frac{M_{str}}{M_R} \right)^{d-2} \left(\frac{M_W}{M_R} \right)^2, \qquad d \geq 3. \tag{19.11}$$

There are two scales that enter into Eqs.(19.10) and (19.11). These are M_R and M_{str}, and they are constrained by gauge coupling unification, which must now occur in the TeV scale. The evolution of these couplings is given by [16, 17]

$$\alpha_i(M_Z)^{-1} = \alpha_U^{-1} + \frac{b_i}{2\pi} \ln \left(\frac{M_R}{M_Z} \right) - \frac{b_i^{KK}}{2\pi} \ln \left(\frac{M_{str}}{M_R} \right) + \Delta_i. \tag{19.12}$$

In Eq. (19.12), α_U is the effective unification scale, where $b_i = (-3, \, 1, \, 33/5)$ for $SU(3)_C \times SU(2)_L \times U(1)_Y$ in an $N = 1$ supersymmetric theory, which describes the gauge coupling evolution from scales Q to M_R. Above the scale M_R the evolution is described by $b_i^{KK} = (-6, \, -3, \, 3/5)$, which includes the contribution of the Kaluza–Klein states but of course without the contribution from the fermionic sector, which has no Kaluza–Klein excitations, and Δ_i denotes the threshold corrections. Satisfaction of the LEP data constrains M_R and M_{str}. Numerical estimates indicate that the lower limit on M_R consistent with the accurate measurement of the Fermi constant [18] and the LEP data is 1.6 TeV for $d = 1$, and 3.5, 5.7, and 7.8 TeV for $d = 2$, 3, and 4.

The Kaluza–Klein modes from extra dimensions also contribute to the anomalous magnetic moment of the muon. Thus, the W- and Z-boson excitations give the following contributions to a_μ [15]:

$$(\Delta a)_\mu^{WZ-KK} = \frac{G_F m_\mu^2}{\pi^2 2\sqrt{2}} \left[-\frac{5}{12} + \frac{4}{3} \left(\sin^2 \theta_W - \frac{1}{4} \right)^2 \right] \times \frac{f_d(x_Z)}{f_d(x_W) - 1}, \quad (19.13)$$

where $x_W = M_W/M_R$ and $x_Z = M_Z/M_R$. For $d = 1$ $f_1(x) \simeq 1 + (\pi^2/3)x$ while the $d \geq 2$ cases require a cut-off as given by Eqs. (19.10) and (19.11). Similarly, corrections arising from the Kaluza–Klein excitations of the photon are given by

$$(\Delta a)_\mu^{\gamma-KK} = \alpha \frac{\pi}{9} \frac{m_\mu^2}{M_R^2}, \qquad\qquad\qquad d = 1 \quad (19.14)$$

$$\simeq \frac{\alpha}{3\pi} \left(\frac{2\pi^2}{3} + 2\pi ln \frac{M_{str}}{M_R} \right) \frac{m_\mu^2}{M_R^2}, \qquad d = 2 \quad (19.15)$$

$$\simeq \frac{\alpha}{3\pi} \left(\frac{d}{d-2} \right) \frac{\pi^{d/2}}{\Gamma(1 + d/2)} \left(\frac{M_{str}}{M_R} \right)^{d-2} \frac{m_\mu^2}{M_R^2}, \qquad d \geq 3. \quad (19.16)$$

Including the constraint from the LEP data and the Fermi constant, the corrections to the muon anomalous magnetic moment from extra dimensions are too small to be detected.

Compactification of extra dimensions can be investigated at colliders if the Kaluza–Klein modes have masses within the reach of collider energies [19–23]. Thus, the Kaluza–Klein modes of the Z- and γ- bosons can be looked for in the Drell–Yan process $pp \to e^+ e^- + X$ where the Kaluza–Klein modes will appear as bumps in $d\sigma_{ll}/dm_{ll}$ versus m_{ll} plots where m_{ll} is the dilepton-invariant mass. In a similar fashion, the Kaluza–Klein modes of the $W^\pm$ bosons can be looked for in the process $pp \to \ell^\pm + \nu_\ell$, and the Kaluza–Klein modes of the gluon can be looked for in the process $pp \to jj + X$. Interestingly, colliders can discriminate among different types of compactifications if such resonances are observed. Thus, if we consider the compactification of d number of extra dimensions and assume that each of the d extra dimensions are compactified on S^1/Z_2 so that one has a compactification of the type $S^1/Z_2 \times S^1/Z_2 \times \cdots \times S^1/Z_2$. Further, let us assume that the radius of compactification is R in each case. So, here we have d number of Kaluza–Klein excitations which are degenerate and they overlap. In this case, the cross-section for the production of the Kaluza–Klein states will be enhanced by a factor of d^2. This means that the height of the resonance peak will be an indication of the number of extra dimensions relative to the $d = 1$ case. In addition to the above, collider signatures can allow one to discriminate among various types of compactifications.

19.2 Gravity in Extra Dimensions

As discussed in the previous subsection, our assumption is that while matter resides on the walls or branes, gravity propagates in the bulk. An interesting idea that has been proposed [9, 10] is that the relative weakness of gravity at long distances may be due to the presence of extra spatial dimensions which are large compared with the Planck length. Thus, assume that in $(4 + d)$ dimensions the Planck scale is given by M_{4+d} while the observed Planck scale in four dimensions is M_{Pl}. Let us also assume that we compactify the extra dimensions with a common compactification scale of radius R. In this case, the use of Gauss's theorem gives us the following relation between M_{Pl} and M_{4+d}:

$$M_{Pl} \sim R^{d/2} M_{4+d}^{1+d/2}. \tag{19.17}$$

Let us estimate the size of R when $M_{(4+d)} \sim 1$ TeV. For the case of one extra dimension, we find that $R \sim 10^{13}$ cm. Such a large R would modify Newtonian gravity for the solar system, and is therefore excluded. For $d = 2$, we find $R \sim 1$ mm, and this case is interesting since in the range below this scale Newtonian gravity has not been well tested, and thus deviations would represent new physics. Of course, there could be other sources of possible deviations (e.g., [24]) if such deviations from Newtonian gravity are observed. There are also collider and cosmological implications of such models [25]. We note that extensions of extra dimension models based on warped geometry have also been constructed, and the reader is directed to the original works [26–29] on this topic.

19.3 The Stueckelberg Mechanism in Supergravity and in Strings

Thus far we have discussed symmetry-breaking schemes where a gauge symmetry is broken by the Higgs mechanism, giving masses to the vector bosons. This mechanism works for breaking either an abelian or a non-abelian gauge group. However, for the case of an abelian gauge symmetry there is another mechanism for giving mass to the vector boson which predates the Higgs mechanism. This is the so-called Stueckelberg mechanism [30][*] There are a variety of models where extra $U(1)$ gauge symmetries other than those needed in the standard model are present. Such is the case for E_6 and $SO(10)$ grand unification, which contain extra $U(1)$ factors. Thus, $E_6 \supset SO(10) \times U(1)_\psi$ while $SO(10) \supset SU(5) \times U(1)_\chi$. Similarly, models based on D-branes typically have several $U(1)$ factors. Thus, a stack of n branes naturally leads to a $U(n)$ gauge group which factors to $SU(n) \times U(1)$, giving an extra $U(1)$. With several such stacks of D-branes, several $U(1)$ factors will emerge. The mass for an extra $U(1)$ gauge boson can arise from the Higgs mechanism if there exists a complex scalar field charged under the

---

[*] See the literature for a history of the Stueckelberg mechanism see [31] and related works [32,33].

$U(1)$ gauge group which undergoes spontaneous breaking, which gives a vacuum expectation value (VEV) to the scalar field and results in the gauge boson associated with the $U(1)$ gauge symmetry becoming massive. However, alternatively, the gauge boson can become massive by the Stueckelberg mechanism, which we now discuss.

In its simplest form the Stueckelberg mechanism works as follows. Consider the Lagrangian for a $U(1)$ gauge field A_μ interacting with a Dirac field ψ of the following form:

$$L_g = -\frac{1}{4}F_{\mu\nu}F^{\mu\nu} - \bar{\psi}\frac{1}{i}\gamma^\mu(\partial_\mu - igA_\mu)\psi - \frac{1}{2}M^2(A_\mu + \frac{1}{M}\partial_\mu\sigma)^2, \quad (19.18)$$

where σ is an axionic-like field. The Lagrangian of Eq. (19.18) is invariant under the local transformation

$$\psi \to e^{ig\lambda}\psi$$
$$\delta A_\mu \to \partial_\mu\lambda$$
$$\delta\sigma \to -M\lambda. \quad (19.19)$$

We can now add a gauge-fixing term so that

$$L_{gf} = -\frac{1}{2\xi}(\partial_\mu A^\mu + M\xi\sigma)^2, \quad (19.20)$$

and one then has

$$L_g + L_{gf} = -\frac{1}{4}F_{\mu\nu}F^{\mu\nu} - \bar{\psi}\frac{1}{i}\gamma^\mu D_\mu\psi - \frac{1}{2}M^2A_\mu A^\mu + \frac{1}{2\xi}(\partial_\mu A^\mu)^2$$
$$- \frac{1}{2}\partial_\mu\sigma\partial^\mu\sigma - \frac{1}{2}\xi M^2\sigma^2, \quad (19.21)$$

where $D_\mu = \partial_\mu - igA_\mu$ is the gauge-covariant derivative. The two fields A_μ and σ are now decoupled. In this way, the theory is both unitary and renormalizable.

The Stueckelberg mechanism can be viewed as a limit of the Higgs mechanism [33]. Consider a $U(1)$ gauge theory coupled to a Higgs field ϕ with a Higgs potential so that

$$L_0 = -\frac{1}{4}F_{\mu\nu}F^{\mu\nu} - (D^\mu\phi)^\dagger D_\mu\phi - V(\phi) + L_{gf}$$
$$V(\phi) = \mu^2(\phi^\dagger\phi) + \lambda(\phi^\dagger\phi)^2. \quad (19.22)$$

Here, a spontaneous breaking of the $U(1)$ gauge symmetry occurs when $\mu^2 < 0$, and $\lambda > 0$ and ϕ develops a VEV so that

$$\phi = \frac{1}{\sqrt{2}}(\rho + v)e^{-i\sigma/v}, \; v = \sqrt{-\mu^2/\lambda}. \quad (19.23)$$

Now, assume we take the limit $(-\mu^2, \lambda) \to \infty$ with $M = ev$ fixed. In this case, ρ becomes infinitely heavy and decouples from the rest of the system, and the residual Lagrangian is given by

$$L = -\frac{1}{4}F_{\mu\nu}F^{\mu\nu} - \frac{1}{2}M^2\left(A_\mu + \frac{1}{M}\partial_\mu\sigma\right)^2 + L_{gf}. \quad (19.24)$$

The limit considered above shows that the Higgs mechanism leads to the Stueckelberg mechanism in a very direct way.

Rather than being a curiosity, the Stueckelberg mechanism is in fact endemic in theories involving compactification of extra dimensions and in strings. Consider, for example, a $U(1)$ gauge theory in five dimensions which is compactified on a half circle of radius R. The kinetic energy of the $U(1)$ gauge field A_a ($a = 0, 1, 2, 3, 5$) in five dimensions with co-ordinates $z^a = (x^\mu, y)$ ($\mu = 0, 1, 2, 3$) is given by

$$L_5 = -\frac{1}{4} F_{ab}(z) F^{ab}(z) - \frac{1}{2\xi} (\partial_a A^a)^2, \quad a = 0, 1, 2, 3, 5. \tag{19.25}$$

One may decompose the gauge field A_a so that $A_a = (A_\mu(z), \phi(z))$ and expand it on the harmonics on the compact dimensions so that

$$A_\mu(z) = \sum_{n=0}^{\infty} A_\mu^{(n)}(x_\mu) f_n(y), \quad \sigma(z) = \sum_{n=0}^{\infty} \phi^{(n)}(x_\mu) g_n(y), \tag{19.26}$$

where $f_n(y)$ and $g_n(y)$ are harmonic functions on the interval $(0, 2\pi R)$ with appropriate periodicity conditions. The compactification results in one massless mode and an infinite number of massive (Kaluza–Klein) modes so that the compactified Lagrangian in four dimensions is given by

$$L_{4d} = \sum_{n=0}^{\infty} \left[-\frac{1}{4} F_{\mu\nu}^{(n)} F^{\mu\nu(n)} - \frac{1}{2} n^2 (M A_\mu^{(n)} + n \partial_\mu \phi^{(n)})^2 \right.$$
$$\left. - \frac{1}{2\xi} \left[(\partial_\mu A^{(n)\mu})^2 + 2nM \partial_\mu A^{(n)\mu} \phi^{(n)} + M^2 (\phi^{(n)})^2 \right] \right]. \tag{19.27}$$

Here, $M = 1/R$ is the inverse of the compactification radius. Expanding the above and choosing the gauge $\xi = 1$, one finds that terms bilinear in $A_\mu^{(n)}$ and $\phi^{(n)}$ form a total divergence and can be discarded, and thus the scalar fields $\phi^{(n)}$ decouple from the vector fields. The analysis above shows that it is the Stueckelberg mechanism which is operating to give masses to the vector bosons, as is explicit in the second term of Eq. (19.27), which results in an infinite tower of massive vector bosons. Thus, here it is clearly the Stueckelberg mechanism and not the Higgs mechanism which gives masses to the vector bosons.

A similar phenomenon arises when one couples a $U(1)$ gauge field A_μ to a 2-form field $B_{\mu\nu}$ so that [32–34]

$$L_0 = -\frac{1}{4} F_{\mu\nu} F^{\mu\nu} - \frac{1}{12} H^{\mu\nu\rho} H_{\mu\nu\rho} + \frac{M}{4} \epsilon^{\mu\nu\rho\sigma} F_{\mu\nu} B_{\rho\sigma}, \tag{19.28}$$

where $H_{\mu\nu\rho} = \partial_\mu B_{\nu\rho} + \partial_\nu B_{\rho\mu} + \partial_\rho B_{\mu\nu}$ is the field strength of the 2-form field $B_{\mu\nu}$. The last term may be written in the form [34] $-\frac{M}{6} \epsilon^{\mu\nu\rho\sigma} (H_{\mu\nu\rho} A_\sigma + \frac{1}{M} \sigma \partial_\mu H_{\nu\rho\sigma})$. It is then easily seen that an integration over σ gives L_0. Alternatively, however, one may solve for $H^{\mu\nu\rho}$, which gives

$$H^{\mu\nu\rho} = -M \epsilon^{\mu\nu\rho\sigma} \left(A_\sigma + \frac{1}{M} \partial_\sigma \sigma \right). \tag{19.29}$$

An integration on $H^{\mu\nu\rho}$ gives L_0 in the form [34]

$$L_1 = -\frac{1}{4}F_{\mu\nu}F^{\mu\nu} - \frac{M^2}{2}\left(A_\sigma + \frac{1}{M}\partial_\sigma\sigma\right)^2. \tag{19.30}$$

The simple exercise above shows that the presence of the Green–Schwarz-type term involving the Kalb–Ramond 2-form field $B_{\mu\nu}$ in Eq. (19.28) leads to a mass growth for the $U(1)$ gauge field. It is difficult to extend the Stueckelberg mechanism to the case of non-abelian gauge theories. The basic difficulty here is that the longitudinal component cannot be easily decoupled from the physical vector fields, which then leads to problems with the satisfaction of unitarity and renormalizability. Thus, the generation of masses for the non-abelian gauge fields requires the standard technique of the Higgs mechanism or a more esoteric extension of the simple Stueckelberg mechanism of the abelian case.

19.4 Stueckelberg Extension of the Standard Model

In spite the fact that the Stueckelberg mechanism has existed for a long time, its applications to particle physics are only recent. Thus, the first successful attempt to include the Stueckelberg mechanism in particle physics models was where the standard model gauge group was extended to include anxtra $U(1)_X$ gauge group [35]. This extension is done by including a linear combination of the $U(1)_Y$ gauge field B_μ and the $U(1)_X$ gauge field C_μ in the Stueckelberg combination and achieving gauge invariance by assigning shift transformations to the σ-field under both $U(1)_X$ and $U(1)_Y$. Thus, we write the extended standard model Lagrangian as

$$L = L_{SM} + L_{St}, \tag{19.31}$$

where L_{St} has the form

$$L_{St} = -\frac{1}{4}C_{\mu\nu}C^{\mu\nu} - \frac{1}{2}(\partial_\mu\sigma + M_1 C_\mu + M_2 B_\mu)^2 + g_X C_\mu J_X^\mu, \tag{19.32}$$

and where J_X^μ is a conserved current arising from the hidden sector. It is now seen that the combination involving the fields B_μ, C_μ, and σ is invariant under the $U(1)_X$ transformations

$$\delta_X B_\mu = 0, \quad \delta_X C_\mu = \partial_\mu\lambda_X, \quad \delta_X\sigma = -M_1\lambda_X, \tag{19.33}$$

and under the $U(1)_Y$ transformations

$$\delta_Y B_\mu = \partial_\mu\lambda_Y, \quad \delta_Y C_\mu = 0, \quad \delta_Y\sigma = -M_2\lambda_Y. \tag{19.34}$$

After spontaneous breaking of the electroweak symmetry, the above Lagrangian gives a mass-square matrix for the vector bosons which in the basis

$(V_\mu^{\mathrm{T}})_a = (C_\mu, B_\mu, A_\mu^3)_a$ takes the form

$$L_{mass} = -\frac{1}{2} V_{a\mu} M_{ab}^2 V_b^\mu \tag{19.35}$$

$$M_{ab}^2 = \begin{pmatrix} M_1^2 & M_1 M_2 & 0 \\ M_1 M_2 & M_2^2 + \frac{1}{4} g_Y^2 v^2 & -\frac{1}{4} g_Y g_2 v^2 \\ 0 & -\frac{1}{4} g_Y g_2 v^2 & \frac{1}{4} g_2^2 v^2 \end{pmatrix}. \tag{19.36}$$

In Eq. (19.36), g_2 and g_Y are the $SU(2)_L$ and $U(1)_Y$ gauge coupling constants, respectively, and $v = 2M_{\mathrm{W}}/g_2 = (\sqrt{2} G_F)^{-\frac{1}{2}}$, where M_{W} is the mass of the W-boson, and G_F is the Fermi constant.

The mass-square matrix of Eq. (19.36) contains three eigenmodes: one of these is massless, which is identified with the photon, while the other two are massive modes, and are the Z- and Z'-bosons, and their masses are given by

$$M_{\mp}^2 = \frac{1}{2}\left[M_1^2 + M_2^2 + \frac{1}{4} g_Y^2 v^2 + \frac{1}{4} g_2^2 v^2 \right]$$
$$\mp \frac{1}{2}\left[\left(M_1^2 + M_2^2 + \frac{1}{4} g_Y^2 v^2 + \frac{1}{4} g_2^2 v^2 \right)^2 - \left(M_1^2 (g_Y^2 + g_2^2) v^2 + g_2^2 M_2^2 v^2 \right) \right]^{\frac{1}{2}}. \tag{19.37}$$

We identify the Z- and Z'-masses so that $M_Z = M_-$ and $M_{Z'} = M_+$, and take the eigenvector basis to be $E^T = (Z'_\mu, Z_\mu, A_\mu^\gamma)^T$. $(V_\mu)_a$ and $(E_\mu)_a$ are then related by

$$V_{\mu a} = \sum_{b=1}^{3} \mathcal{O}_{ab} E_{\mu b} \tag{19.38}$$

$$\sum_{b,c=1}^{3} \mathcal{O}_{ba} M_{bc}^2 \mathcal{O}_{cd} = E_{ad}, \tag{19.39}$$

where $\mathcal{O}_{ab}$ is an orthogonal matrix parametrized by the three angles θ, ϕ, and ψ.[*] The angles θ, ϕ, and ψ are given by

$$\tan \phi = \frac{M_2}{M_1} \tag{19.41}$$

---

[*] The orthogonal matrix $\mathcal{O}_{ab}$ is parametrized by the rotation angles θ, ϕ, and ψ so that

$$\mathcal{O}_{ab} = \begin{bmatrix} \cos\psi\cos\phi - \sin\theta\sin\phi\sin\psi & -\sin\psi\cos\phi - \sin\theta\sin\phi\cos\psi & -\cos\theta\sin\phi \\ \cos\psi\sin\phi + \sin\theta\cos\phi\sin\psi & -\sin\psi\sin\phi + \sin\theta\cos\phi\cos\psi & \cos\theta\cos\phi \\ -\cos\theta\sin\psi & -\cos\theta\cos\psi & \sin\theta \end{bmatrix}_{ab}. \tag{19.40}$$

$$\tan\theta = \tan\theta_W \cos\phi \tag{19.42}$$

$$\tan\psi = \frac{\tan\theta \tan\phi M_W^2}{\cos\theta(M_{Z'}^2 - M_W^2(1 + \tan^2\theta))}, \tag{19.43}$$

where $\tan\theta_W = g_Y/g_2$.

The neutral current interaction is given by the sum $(g_2 A_\mu^3 J_2^{3\mu} + g_Y B_\mu J_Y^\mu + g_X C_\mu J_X^\mu)$. In terms of the eigenmodes one has

$$L_{NC} = Z'_\mu J_{Z'}^\mu + Z_\mu J_Z^\mu + A_\mu J_{em}^\mu, \tag{19.44}$$

where

$$J_{Z'}^\mu = \frac{\sin\psi}{\sqrt{g_2^2 + g_Y^2\cos^2\phi}}\left(\cos^2\phi g_Y^2 J_Y^\mu - g_2^2 J_2^{3\mu} - \frac{1}{2}\sin(2\phi)g_X g_Y J_X^\mu\right)$$
$$+ \cos\psi(\sin\phi g_Y J_Y^\mu + \cos\phi g_X J_X^\mu) \tag{19.45}$$

$$J_Z^\mu = \frac{\cos\psi}{\sqrt{g_2^2 + g_Y^2\cos^2\phi}}\left(\cos^2\phi g_Y^2 J_Y^\mu - g_2^2 J_2^{3\mu} - \frac{1}{2}\sin(2\phi)g_X g_Y J_X^\mu\right)$$
$$- \sin\psi(\sin(\phi)g_Y J_Y^\mu + \cos\phi g_X J_X^\mu) \tag{19.46}$$

$$J_{em}^\mu = \frac{g_2 g_Y \cos\phi}{\sqrt{g_2^2 + g_Y^2\cos^2\phi}}\left(J_Y^\mu + J_2^{3\mu} - \frac{g_X}{g_Y}\tan\phi J_X^\mu\right). \tag{19.47}$$

From the coupling of the photon to matter as given by Eq. (19.47), one finds that the relation of the electric charge to g_2 and g_Y is modified from what it is in the standard model, and it now reads

$$e = \frac{g_2 g_Y \cos\phi}{\sqrt{g_2^2 + g_Y^2\cos^2\phi}} \tag{19.48}$$

or, alternatively,

$$\frac{1}{e^2} = \frac{1}{g_2^2} + \frac{1 + \epsilon^2}{g_Y^2}, \tag{19.49}$$

where $\epsilon = M_2/M_1$. Fits to the precision electroweak data put a limit on ϵ, and one finds $\epsilon \le .05$, and with this constraint LEP I data can be satisfied with the same level of accuracy as fits to the standard model [36].

Assuming that the Z' boson does not decay into hidden sector particles, and making the approximation that $\sin\phi \ll 1$, and $\cos\psi \sim 1$, one finds that the total decay width of the Z' into standard model particles is given by

$$\Gamma\left(Z' \to \sum_i f_i \bar{f}_i\right) = M_{Z'}\, g_Y^2 \sin^2\phi \times \begin{cases} \frac{103}{288\pi}, & M_{Z'} \langle 2m_t \\ \frac{5}{12\pi}, & M_{Z'} \rangle 2m_t \end{cases}. \tag{19.50}$$

It is then easily checked that the Z' decay width is very narrow, lying in the MeV to GeV range for a Z' mass in the range of up to a few TeV for small ϕ. Thus, the observation of a sharp Z'-resonance such as the above will be a potential signal for a Stueckelberg gauge boson. Further, if such a resonance exists, it can be detected by the Drell–Yan process at colliders.

The reader may note that the Stueckelberg sector decouples from the standard model sector in the limit

$$\tan\phi, \ \tan\psi \longrightarrow 0 \tag{19.51}$$

$$\tan(\theta) \longrightarrow \tan\theta_{\mathrm{W}}. \tag{19.52}$$

In the above limit the Z-boson and the Z'-boson masses take the form

$$M_Z^2 \longrightarrow \frac{1}{4}v^2(g_2^2 + g_Y^2) \tag{19.53}$$

$$M_{Z'}^2 \longrightarrow M^2. \tag{19.54}$$

19.5 Supersymmetric Extension of the Stueckelberg Model

A supersymmetric extension of the above model can be given in a direct fashion [37, 38]. To accomplish this, we extend all vector fields to be the appropriate superfields, and introduce the chiral multiplets S and $\bar{S}$, which contain the axionic field. Thus, the extension of the MSSM Lagrangian to include the Stueckelberg mechanism is given by

$$L_{StMSSM} = L_{MSSM} + L_{St}, \tag{19.55}$$

where

$$L_{St} = L_{kin}(C) + \int d\theta^2 \, d\bar{\theta}^2 (M_1 C + M_2 B + S + \bar{S})^2. \tag{19.56}$$

Here, $L_{kin}(C)$ is the kinetic energy term for the vector superfield C (with components $(C_\mu, \lambda_C, \bar{\lambda}_C, D_C)$ in the two-component notation in the Wess–Zumino gauge), while the kinetic energy term for the vector supermultiplet B (with components $(B_\mu, \lambda_B, \bar{\lambda}_B, D_B)$ in the two-component notation in the Wess–Zumino gauge), is contained in the MSSM Lagrangian. The chiral multiplet S has the components $S = (\rho + i\sigma, \chi, F_S)$, and similarly for $\bar{S}$. The above Lagrangian is invariant under the following $U(1)_Y$ and $U(1)_X$ transformations:

$$\delta_Y B = \Lambda_Y + \bar{\Lambda}_Y$$
$$\delta_Y C = 0$$
$$\delta_Y S = -M_2 \Lambda_Y$$
$$\delta_X B = 0$$
$$\delta_X C = \Lambda_X + \bar{\Lambda}_X$$
$$\delta_X S = -M_1 \Lambda_X. \tag{19.57}$$

Inclusion of L_{St} brings in two additional Majorana spinors beyond those in the MSSM, which we construct as follows:

$$\psi_S = \begin{pmatrix} \chi_\alpha \\ \bar{\chi}^{\dot{\alpha}} \end{pmatrix}, \quad \lambda_X = \begin{pmatrix} \lambda_{C\alpha} \\ \bar{\lambda}_C^{\dot{\alpha}} \end{pmatrix}. \tag{19.58}$$

Including the four familiar gaugino and Higgsino states of the MSSM, which are $\lambda_Y, \lambda_3, \tilde{h}_1$, and $\tilde{h}_2$, we obtain the following six Majorana states in the Stueckelberg extension of the MSSM:

$$\psi_S, \lambda_X, \lambda_Y, \lambda_3, \tilde{h}_1, \tilde{h}_2. \tag{19.59}$$

The Majorana mass matrix in this basis is then given by

$$\begin{bmatrix} 0 & M_1 & M_2 & 0 & 0 & 0 \\ M_1 & \tilde{m}_X & 0 & 0 & 0 & 0 \\ M_2 & 0 & \tilde{m}_1 & 0 & -c_\beta s_W M_Z & s_\beta s_W M_Z \\ 0 & 0 & 0 & \tilde{m}_2 & c_\beta c_W M_Z & -s_\beta c_W M_Z \\ 0 & 0 & -c_\beta s_W M_Z & c_\beta c_W M_Z & 0 & -\mu \\ 0 & 0 & s_\beta s_W M_Z & -s_\beta c_W M_Z & -\mu & 0 \end{bmatrix}. \tag{19.60}$$

In the above, we have used the notation $(c_\beta, s_\beta) = (\cos\beta, \sin\beta)$, and $(c_W, s_W) = (\cos\theta_W, \sin\theta_W)$, where θ_W is the weak angle, μ is the Higgs mixing parameter and we have added a soft term m_X in the $U(1)_X$ gaugino sector. The mass matrix of Eq. (19.60) has six eigenvalues, and we denote the corresponding Majorana eigenstates by

$$(\xi_1^0, \xi_2^0; \chi_1^0, \chi_2^0, \chi_3^0, \chi_4^0). \tag{19.61}$$

In the above, the first two entries ξ_α^0 ($\alpha = 1, 2$) arise from the Stueckelberg sector, and the last four χ_a^0 ($a = 1, 2, 3, 4$) are essentially the four neutralino states of the MSSM. The enlargement of the neutralino states from four to six opens up the possibility that the lightest neutralino in the extended MSSM may lie in the hidden sector. In this case, the dark matter particle ξ_1^0 will lie mostly in the Stueckelberg sector and will be extra-weakly interacting [39].

19.6 Kinetic Mixing

It has been known for some time that the kinetic energy terms for two abelian gauge fields can mix, with such mixings arising from loops involving exchange of fields which are charged under both $U(1)$ [40–46]. Thus, in general, for abelian gauge groups with gauge fields A_1^μ and A_2^μ, the Lagrangian will assume the form

$$L_{km} = -\frac{1}{4}F_{1\mu\nu}F_1^{\mu\nu} - \frac{1}{4}F_{2\mu\nu}F_2^{\mu\nu} - \frac{\delta}{2}F_{1\mu\nu}F_2^{\mu\nu} + J'_\mu A_1^\mu + J_\mu A_2^\mu. \tag{19.62}$$

Here, δ parametrizes the mixing of the two $U(1)$, J_μ is the source from the visible sector, and J'_μ is the source from the hidden sector [35, 47]. In the limit of no mixing, A_1^μ couples to fields in the visible sector and A_1^μ couples to fields in the hidden sector. Returning to the case of the kinetic mixing of Eq. (19.62), we can make the following $GL(2)$ transformation, which takes us to the diagonal basis:

$$\begin{pmatrix} A_1^\mu \\ A_2^\mu \end{pmatrix} \to K_0 \begin{pmatrix} A'^\mu \\ A^\mu \end{pmatrix}, \tag{19.63}$$

where K_0 is defined by

$$K_0 = \begin{pmatrix} \dfrac{1}{\sqrt{1-\delta^2}} & 0 \\ \dfrac{-\delta}{\sqrt{1-\delta^2}} & 1 \end{pmatrix}. \tag{19.64}$$

There is, however, an arbitrariness in the choice of the transformation matrix. Thus, an equally good choice for the transformation matrix is $K = K_0 R$, where R is an orthogonal matrix so that

$$K = K_0 R, \quad R = \begin{pmatrix} \cos\theta & -\sin\theta \\ \sin\theta & \cos\theta \end{pmatrix}. \tag{19.65}$$

The angle θ is arbitrary but a convenient choice is $\theta = \arctan[\delta/\sqrt{1-\delta^2}]$. This leads to an asymmetric solution:

$$L_1^K = A^\mu \left[\frac{1}{\sqrt{1-\delta^2}} J_\mu - \frac{\delta}{\sqrt{1-\delta^2}} J'_\mu \right] + A^{\mu'} J'_\mu. \tag{19.66}$$

One may now identify A^μ as the photon field which couples with the source J_μ from the visible sector with normal strength; it also couples with the hidden sector source J'_μ but with a reduced strength which is proportional δ. On the other hand, the coupling of the field $A^{\mu'}$, usually referred to as the para photon, is only to the hidden sector current J'_μ and has no couplings to the visible sector.

We now consider the case when one has both a kinetic mixing and a Stueckelberg mechanism working simultaneously. Thus, we consider the Lagrangian interaction of Eq. (19.62) along with the Stueckelberg mass mixing terms below:

$$L_{St}^m = -\frac{1}{2} M_1^2 A_{1\mu} A_1^\mu - \frac{1}{2} M_2^2 A_{2\mu} A_2^\mu - M_1 M_2 A_{1\mu} A_2^\mu. \tag{19.67}$$

In this case, after diagonalizing the kinetic terms, we can also diagonalize the mass matrix by choosing θ so that

$$\theta = \arctan\left(\frac{\epsilon\sqrt{1-\delta^2}}{1-\epsilon\delta} \right). \tag{19.68}$$

With the choice of Eq. (19.68), both the kinetic energy and the mass terms are diagonal, and the interaction Lagrangian assumes the form

$$L_{St}^{int} = \frac{1}{\sqrt{1-2\epsilon\delta+\epsilon^2}} \left(J_\mu - \epsilon J'_\mu \right) A_\gamma^\mu + \frac{1}{\sqrt{1-2\epsilon\delta+\epsilon^2}}$$
$$\times \left(\frac{\epsilon-\delta}{\sqrt{1-\delta^2}} J_\mu + \frac{1-\epsilon\delta}{\sqrt{1-\delta^2}} J'_\mu \right) A_M^\mu, \tag{19.69}$$

where A_γ^μ is the photon field and A_M^μ is the massive vector boson field. Interestingly, here one finds that the photon field couples with the hidden sector with a strength proportional to ϵ rather than δ. Further, the massive vector field A_M^μ couples with both the visible sector and the hidden sector, in contrast to the para photon field, which coupled only to the hidden sector. One can extend the above analysis to include the kinetic mixing for the Stueckelberg extension of the standard model discussed in Section 19.4. Here, one finds that with inclusion of the kinetic mixing and mass mixing, the electric charge e is modified so that it depends on both ϵ and δ (see Eq. (19.72)). If matter in the hidden sector is absent, then the weak interactions of the model involve only a combination of ϵ and δ in the following combination $\bar{\epsilon}$ [48]:

$$\bar{\epsilon} = (\epsilon - \delta)/\sqrt{1 - \delta^2}. \tag{19.70}$$

19.7 Problems

1. Show by integration over the σ-field that the Lagrangian

$$L_2 = -\frac{1}{4}F_{\mu\nu}F^{\mu\nu} - \frac{1}{12}H^{\mu\nu\rho}H_{\mu\nu\rho} - \frac{M}{6}\epsilon^{\mu\nu\rho\sigma}\left(H_{\mu\nu\rho}A_\sigma + \frac{1}{M}\sigma\partial_\mu H_{\nu\rho\sigma}\right)$$

is equivalent to the Lagrangian of Eq. (19.28).

2. Show that for the quarks the Z'-couplings in the Stueckelberg model take on the form

$$\mathcal{L}_{Zq\bar{q}} = -\sqrt{g_2^2 + \cos^2\phi g_Y^2}$$
$$\times \left[Z'_\mu(J_3^\mu - \sin^2\theta J_{\rm em}^\mu)\sin\psi - Z'_\mu(J_{\rm em}^\mu - J_3^\mu)\sin\theta\tan\phi\cos\psi\right]. \tag{19.71}$$

3. Show that in the limit $|\psi| \ll |\phi|$, the ratio of the branching ratios for the Z' boson are given by $l\bar{l}/\nu\bar{\nu} = 5$, $b\bar{b}/\tau\bar{\tau} = 1/3$ and $u\bar{u}/d\bar{d} = 17/5$.

4. Show that in the $U(1)_X$ extension of the standard model electroweak sector the presence of both kinetic mixing and the Stueckelberg mass mixing of $U(1)_X$ and $U(1)_Y$, the unit of electric charge is modified so that

$$\frac{1}{e^2} = \frac{1}{g_2^2} + \frac{1 - 2\epsilon\delta + \epsilon^2}{g_Y^2}. \tag{19.72}$$

References

[1] Th. Kaluza, *Sitzungober. Preuss. Akad. Wiss. Berlin* 966 (1921); O. Klein, *Z. Phys.* **37**, 895 (1926).
[2] See also, T. Appelquist, A. Chodos, and P. G. O. Freund, *Modern Kaluza–Klein Theories, Frontiers in Physics*, Addison Wesley (1987).
[3] L. O'Raifeartaigh and N. Straumann, hep-ph/9810524 (1998).
[4] V. S. Kaplunovsky, *Nucl. Phys. B* **307**, 145 (1988) [*Nucl. Phys. B* **382**, 436 (1992)].
[5] E. Witten, *Nucl. Phys. B* **471**, 135 (1996).
[6] J. D. Lykken, *Phys. Rev. D* **54**, 3693 (1996).
[7] I. Antoniadis, *Phys. Lett. B* **246**, 377 (1990).

[8] I. Antoniadis and M. Quiros, *Phys. Lett. B* **392**, 61 (1997).

[9] N. Arkani-Hamed, S. Dimopoulos, and G. R. Dvali, *Phys. Lett. B* **429**, 263 (1998).

[10] I. Antoniadis, N. Arkani-Hamed, S. Dimopoulos, and G. R. Dvali, *Phys. Lett. B* **436**, 257 (1998).

[11] G. Shiu and S. H. H. Tye, *Phys. Rev. D* **58**, 106007 (1998).

[12] A. Delgado, A. Pomarol, and M. Quiros, *Phys. Rev. D* **60**, 095008 (1999).

[13] P. Nath and M. Yamaguchi, *Phys. Rev. D* **60**, 116004 (1999).

[14] M. Masip and A. Pomarol, *Phys. Rev. D* **60**, 096005 (1999).

[15] P. Nath and M. Yamaguchi, *Phys. Rev. D* **60**, 116006 (1999).

[16] K. R. Dienes, E. Dudas, and T. Gherghetta, *Phys. Lett. B* **436**, 55 (1998).

[17] D. Ghilencea and G. G. Ross, *Phys. Lett. B* **442**, 165 (1998).

[18] T. van Ritbergen and R. G. Stuart, *Phys. Rev. Lett.* **82**, 488 (1999).

[19] I. Antoniadis, K. Benakli, and M. Quiros, *Phys. Lett. B* **460**, 176 (1999).

[20] P. Nath, Y. Yamada, and M. Yamaguchi, *Phys. Lett. B* **466**, 100 (1999).

[21] E. A. Mirabelli, M. Perelstein, and M. E. Peskin, *Phys. Rev. Lett.* **82**, 2236 (1999).

[22] T. Han, J. D. Lykken, and R. J. Zhang, *Phys. Rev. D* **59**, 105006 (1999).

[23] G. F. Giudice, R. Rattazzi, and J. D. Wells, *Nucl. Phys. B* **544**, 3 (1999)

[24] H. Goldberg and P. Nath, *Phys. Rev. Lett.* **100**, 031803 (2008).

[25] N. Arkani-Hamed, S. Dimopoulos, and G. R. Dvali, *Phys. Rev. D* **59**, 086004 (1999).

[26] L. Randall and R. Sundrum, *Phys. Rev. Lett.* **83**, 3370 (1999).

[27] L. Randall and R. Sundrum, *Phys. Rev. Lett.* **83**, 4690 (1999).

[28] W. D. Goldberger and M. B. Wise, *Phys. Lett. B* **475**, 275 (2000).

[29] W. D. Goldberger and M. B. Wise, *Phys. Rev. Lett.* **83**, 4922 (1999).

[30] E. C. G. Stueckelberg, *Helv. Phys. Acta* **11**, 225 (1938).

[31] F. Cianfrani and O. M. Lecian, *Nuovo Cim. B* **122**, 123 (2007).

[32] M. Kalb and P. Ramond, *Phys. Rev. D* **9**, 2273 (1974).

[33] T. J. Allen, M. J. Bowick, and A. Lahiri, *Mod. Phys. Lett. A* **6**, 559 (1991).

[34] D. M. Ghilencea, L. E. Ibanez, N. Irges, and F. Quevedo, *JHEP* **0208**, 016 (2002).

[35] B. Kors and P. Nath, *Phys. Lett. B* **586**, 366 (2004).

[36] D. Feldman, Z. Liu, and P. Nath, *Phys. Rev. Lett.* **97**, 021801 (2006).

[37] B. Kors and P. Nath, *JHEP* **0412**, 005 (2004).

[38] B. Kors and P. Nath, *JHEP* **0507**, 069 (2005).

[39] D. Feldman, B. Kors, and P. Nath, *Phys. Rev. D* **75**, 023503 (2007).

[40] B. Holdom, *Phys. Lett. B* **166**, 196 (1986).

[41] B. Holdom, *Phys. Lett. B* **259**, 329 (1991).

[42] K. R. Dienes, C. F. Kolda, and J. March-Russell, *Nucl. Phys. B* **492**, 104 (1997).

[43] S. A. Abel and B. W. Schofield, *Nucl. Phys. B* **685**, 150 (2004).

[44] J. Kumar and J. D. Wells, *Phys. Rev. D* **74**, 115017 (2006).

[45] W. F. Chang, J. N. Ng, and J. M. S. Wu, *Phys. Rev. D* **74**, 095005 (2006).

[46] M. Ahlers, H. Gies, J. Jaeckel, J. Redondo, and A. Ringwald, *Phys. Rev. D* **77**, 095001 (2008).

[47] K. Cheung and T. C. Yuan, *JHEP* **0703**, 120 (2007).

[48] D. Feldman, Z. Liu, and P. Nath, *Phys. Rev. D* **75**, 115001 (2007).

20

The Future of Unification

Over the past several decades, significant progress has been made in our understanding of the fundamentals of particle interactions resulting in the formulation and testing of the standard model. This model describes the electroweak and the strong interactions to a remarkably good accuracy, and all of its predictions have been tested, including the recent discovery of the Higgs boson. The standard model can adequately describe the electroweak and the strong phenomena up to the TeV region. However, there are problems, some of which are empirical while others are theoretical, intrinsic to the model. On the empirical side, one finds that the gauge couplings fail to unify when extrapolated to high scales but come tantalizingly close to doing so. This phenomenon points to a missing piece in the model. On the theory side, there are also puzzles. The standard model contains a tachyonic mass term for the Higgs field which is needed to accomplish spontaneous breaking of the electroweak symmetry and give masses to the W and the Z gauge bosons and to the fermions. However, a tachyonic mass term requires an explanation, but in the standard model there is none. Another aspect of the standard model is even more serious, and relates to a quadratic divergence in the Higgs boson mass at the loop level requiring a cut-off. In a high-scale model the cut-off could be the Planck mass, requiring a fine tuning of 1 part in 10^{32}. Another potential problem for the standard model comes about from the measurement of the Higgs boson mass. As discussed in chapter 1, including the next-to-next leading-order corrections, one finds that in the standard model the vacuum is stable up to only $10^9 - 10^{10}$ GeV with a Higgs boson mass of ~ 126 GeV; above this scale the vacuum in unstable, which is not a desirable feature.

Quite remarkably, all four problems discussed above, which are intrinsic to the standard model, are resolved in supergravity grand unification. Thus, regarding the unification of gauge couplings, the inclusion of extra particles, i.e., sparticles, present in supersymmetry and supergravity grand unification modifies the renormalization group evolution and causes the electroweak and the strong gauge coupling constants to unify, which is consistent with data to a high degree of accuracy. Supergravity grand unification also resolves the problem of tachyons in the standard model. Thus, in supergravity grand unification, the Higgs boson mass gains a soft mass through spontaneous breaking of supersymmetry, and through the μ-term, which also has its origin in the spontaneous breaking of supersymmetry. At the grand unification scale, the Higgs boson has no tachyonic mass. However, renormalization group evolution leads to the mass square of the Higgs boson, which couples with the top quark to turn negative, which results in spontaneous breaking of the electroweak symmetry. Supersymmetry also provides a solution to the problem of quadratic divergence in the Higgs boson mass square in the standard model. In supersymmetry, the quadratic divergence is cancelled, and the dependence on the cut-off is only logarithmic. Finally, supergravity grand unification overcomes the problem of vacuum stability exhibited in the standard model, and a vacuum can be stable up to the Planck scale without any problem.

A key element that enters into resolving the problems alluded to above is the inclusion of sparticles, and thus discovery of sparticles is central to a test of supergravity grand unification. An explanation regarding the non-observation of sparticles thus far appears to lie in the measured mass of the Higgs boson at $\sim$126 GeV. Thus, at the tree level, the Higgs boson mass lies below M_Z, and a large loop correction is needed from the exchange of sparticle masses to generate the large correction needed. The largeness of the loop correction points to weak scale supersymmetry lying in the TeV region. The non-observation of sparticles thus far is consistent with the relatively large mass for the Higgs boson in the context of supersymmetry. However, sparticles must be found, to confirm the basic elements of supersymmetric theories and of supergravity grand unification in particular. Currently the Large Hadron Collider is operating at $\sqrt{s}$ = 13 TeV and collecting data. There is a high chance that we will see sparticle production. But a discovery of the full sparticle spectrum will require machines with larger center-of-mass energies such as a 100 TeV proton–proton collider. Further precision tests on the low-lying sparticles would require a linear $e^+ + e^-$ collider with a sufficiently large $\sqrt{s}$ so that the sparticles can be pair produced.

An added prediction of supergravity grand unification is that with R-parity it gives a candidate for dark matter which can be tested in direct and indirect dark matter experiments as well as via an analysis of sparticle decays at the Large Hadron Collider. If supergravity grand unification is experimentally confirmed, it would be the next step beyond the standard model, as it allows us to analyze phenomena from the electroweak scale up to the grand unification scale of $\sim 10^{16}$ GeV. This latter scale lies about two orders of magnitude below the Planck scale where gravity becomes strong. Thus, up to the grand unification scale the effects of quantum gravity tend to be small, and the framework of grand unification should work fairly well. However, unification above the grand unification scale would require inclusion of quantum gravity as an intrinsic element of the theory. This is what string theory does. However, string theory is a framework much like field theory, and is not a model—and it is a model that we seek to describe our universe.

A successful string model would be one where one goes from ten-dimensional string theory to a four-dimensional theory with just the standard model gauge group at low scales with three generations of quarks and leptons and possibly some extra matter consistent with gauge-coupling unification. The model should contain a mechanism that breaks supersymmetry and generates soft terms. In addition, we expect that a fully unified model would predict the precise value for the proton lifetime and give us a candidate for dark matter and predict its mass. Very significant progress has been made in string model building, and one hopes that we will narrow down the field of candidate models in the future. In the meantime, we should not overlook what string theory has already achieved. It has provided us with an answer to what is on the right-hand side of the Einstein equations: a question posed by Einstein himself in his paper "Physics and reality" in 1936 (see Chapter 1). Thus, a string model uniquely defines the right-hand side of the Einstein equations. There is a caveat, however, in that there are too many possibilities for doing so. In the future, one expects these possibilities to narrow down.

21

Appendices

21.1 Exterior Differential Forms

A scalar field $f(x)$ is a 0-form. A 1-form $f_1(x)$ is defined by

$$f_1(x) = \partial_\mu f \ dx^\mu. \tag{21.1}$$

Before defining a 2-form $f_2(x)$, we introduce a wedge product $dx^\mu \wedge dx^\nu$ so that

$$dx^\mu \wedge dx^\nu = -dx^\nu \wedge dx^\mu. \tag{21.2}$$

In addition to non-commutativity, the wedge product satisfies the following associative and distributive properties. Assume $\alpha, \beta,$ and γ are three 1-forms, then the associative property is

$$(\alpha \wedge \beta) \wedge \gamma = \alpha \wedge (\beta \wedge \gamma), \tag{21.3}$$

while the distributive property is

$$\alpha \wedge (\beta + \gamma) = \alpha \wedge \beta + \alpha \wedge \gamma. \tag{21.4}$$

With the above wedge product properties we define a 2-form $f_2(x)$ so that

$$f_2(x) = \frac{1}{2!} f_{\mu\nu}(x) dx^\mu \wedge dx^\nu. \tag{21.5}$$

More generally, we define a p-form f_p so that

$$f_p = \frac{1}{p!} f_{\mu_1 \cdots \mu_p} dx^{\mu_1} \wedge dx^{\mu_2} \wedge \cdots \wedge dx^{\mu_p}. \tag{21.6}$$

The product of a p-form f_p and a q-from f_q is defined so that

$$f_p \wedge f_q = \frac{1}{p!q!} f_{\mu_1 \cdots \mu_p} f_{\nu_1 \cdots \nu_q} dx^{\mu_1} \wedge \cdots \wedge dx^{\mu_p} \wedge dx^{\nu_1} \wedge \cdots \wedge dx^{\nu_q}. \tag{21.7}$$

Thus, the product $f_p \wedge f_q$ is a $p + q$ form and obeys the relation

$$f_p \wedge f_q = (-1)^{qp} f_q \wedge f_p. \tag{21.8}$$

Let us now look at the derivatives. As already seen in Eq. (21.1), the derivative of a 0-form $f(x)$,

$$df = \partial_\mu f dx^\mu,\tag{21.9}$$

gives a 1-form. The second exterior derivative in this case gives

$$d\, df = \partial_\nu \partial_\mu f\, dx^\nu \wedge dx^\mu = 0,\tag{21.10}$$

where we used the property that $\partial_\nu \partial_\mu$ is symmetric in the μ and ν interchange while the wedge product $dx^\nu \wedge dx^\mu$ is anti-symmetric. This is a general property of the exterior derivative d in that the operation of a d^2 on a p-form gives a null result, i.e.,

$$
\begin{aligned}
d^2 f_p(x) &= d\left[\frac{1}{p!}\partial_{[\nu_1} f_{\mu_1 \cdots \mu_p]}\, dx^{\nu_1} \wedge dx^{\mu_1} \wedge \cdots \wedge dx^{\mu_p}\right] \\
&= \frac{1}{p!}\partial_{[\nu_2}\partial_{\nu_1} f_{\mu_1 \cdots \mu_p]}\, dx^{\nu_2} \wedge dx^{\nu_1} \wedge dx^{\mu_1} \wedge \cdots \wedge dx^{\mu_p} \\
&= 0,
\end{aligned}\tag{21.11}
$$

where again we have used the fact that $\partial_{\nu_2}\partial_{\nu_1}$ are symmetric under the ν_1 and ν_2 interchange while the wedge product is anti-symmetric, which together give a null result. We note here the notation $f_{[\mu_1 \cdots \mu_p]}$ with a $1/p!$ factor, and sum over all permutations to make $f_{[\mu_1 \cdots \mu_p]}$ anti-symmetric in all indices. Thus, for $p = 2$ we have $f_{[\mu_1 \mu_2]} = (f_{\mu_1 \mu_2} - f_{\mu_2 \mu_1})/2$, and for $p = 3$ we have

$$f_{[123]} = \frac{1}{3!}(f_{123} + f_{231} + f_{312} - f_{132} - f_{213} - f_{321}).\tag{21.12}$$

A p-form f_p is called closed if

$$df_p = 0,\tag{21.13}$$

and a p-form is called exact if it can be written as

$$f_p = d\phi_{p-1}.\tag{21.14}$$

From the above, it follows that if f_p is an exact form, then $df_p = d^2\phi_{p-1} = 0$, and thus every exact form is closed. However, not every closed form is necessarily exact. There are situations where a closed form may be exact locally but not globally. The action of an exterior derivative on a product of a p-form and a q-form is given by

$$d(f_p \wedge f_q) = (df_p)f_q + (-1)^p f_p\, df_q.\tag{21.15}$$

Now, consider a curved d-dimensional space. In this case, a wedge product $dx^{\mu_1} \wedge \cdots \wedge dx^{\mu_d}$ can be expressed in terms of the epsilon symbol, i.e.,

$$dx^{\mu_1} \wedge \cdots dx^{\mu_d} \equiv \epsilon^{\mu_1 \cdots \mu_d}\, \sqrt{g}\, d^d x,\tag{21.16}$$

where $g = -det(g)$ and $\epsilon^{01 \cdots (p-1)} = +1$, and thus the d-form may be written as

$$f_d = \frac{1}{d!} f_{\mu_1 \cdots \mu_d} \epsilon^{\mu_1 \cdots \mu_d}\, \sqrt{g}\, d^d x.\tag{21.17}$$

In a d-dimensional space, a Hodge dual of $(dx^{\mu_1} \wedge dx^{\mu_2} \cdots dx^{\mu_p})$ is denoted by $^*(dx^{\mu_1} \wedge dx^{\mu_2} \cdots dx^{\mu_p})$, and is given by

$$^*(dx^{\mu_1} \wedge dx^{\mu_2} \cdots dx^{\mu_p}) = \frac{1}{(d-p)!} g^{\mu_1 \alpha_1} g^{\mu_2 \alpha_2} \cdots g^{\mu_p \alpha_p}$$

$$\times \epsilon_{\alpha_1 \cdots \alpha_p \alpha_{p+1} \cdots \alpha_d} dx^{\alpha_{p+1}} \wedge \cdots \wedge dx^{\alpha_d}. \qquad (21.18)$$

Thus, given a p-form f_p in a d-dimensional space, one can introduce a Hodge dual $^*f_p$ which is a $d - p$ form defined by

$$^*f_p = \frac{1}{p!(d-p)!} f_{\mu_1 \cdots \mu_p} g^{\mu_1 \alpha_1} g^{\mu_2 \alpha_2} \cdots g^{\mu_p \alpha_p}$$

$$\times \epsilon_{\alpha_1 \cdots \alpha_p \alpha_{p+1} \cdots \alpha_d} dx^{\alpha_{p+1}} \wedge \cdots \wedge dx^{\alpha_d}. \qquad (21.19)$$

Now, for the Yang–Mills field A_μ^α we introduce the Lie-valued object $A_\mu = A_\mu^\alpha T_\alpha$ and a 1-form A:

$$A = A_\mu \, dx^\mu. \qquad (21.20)$$

For the Yang–Mills field strength $F_{\mu\nu} = F_{\mu\nu}^\alpha T_\alpha$, which is given by

$$F_{\mu\nu} = \partial_\mu A_\nu - \partial_\nu A_\mu - i[A_\mu, A_\nu], \qquad (21.21)$$

we introduce a 2-form defined by

$$F = \frac{1}{2} F_{\mu\nu} \, dx^\mu \wedge dx^\nu. \qquad (21.22)$$

Next, we note that

$$A^2 = A_\mu A_\nu dx^\mu dx^\nu = \frac{1}{2}[A_\mu, A_\nu] dx^\mu \wedge dx^\nu. \qquad (21.23)$$

With the above, we can write the Yang–Mills field strength in the form

$$F = dA - iA^2. \qquad (21.24)$$

The Yang–Mills gauge transformations in the exterior differential notation take on the form

$$\delta\psi = i\theta\psi \qquad (21.25)$$

$$\delta A = d\theta - i[A, \theta], \qquad (21.26)$$

where $\theta = \theta^\alpha T_\alpha$ and θ^α are transformation parameters. Next, using Eq. (21.24) we deduce

$$[A, F] = [A, dA - iA^2] = A \, dA - dA \, A. \qquad (21.27)$$

Also, from Eq. (21.24) one finds

$$dF = d^2 A - -idA^2 \qquad (21.28)$$

$$= -i(dA \, A - A \, dA), \qquad (21.29)$$

where we used $d^2A = 0$. From the preceding two relations, one finds

$$dF - i[A, F] = 0. \tag{21.30}$$

This relation is the Bianchi identity for the Yang–Mills case.

Integration of a p-form can be defined as follows. Assume M is a p-dimensional surface and ∂M is its boundary. Further, assume we have

$$f_p = df_{p-1}. \tag{21.31}$$

Then, in this case

$$\int_M f_p = \int_{\partial M} f_{p-1}. \tag{21.32}$$

This is an extension of the familiar Stokes theorem in the exterior differential form.

Let us consider the case of a vector potential which is single valued and is pure gauge throughout the space, i.e., $A_\mu = \partial_\mu \phi$ so that $A = \partial_\mu \phi \, dx^\mu = d\phi$. Next, consider

$$dA = \frac{1}{2} F_{\mu\nu} dx^\mu \wedge dx^\nu. \tag{21.33}$$

In this case, $F = dA = d^2\phi = 0$. Thus, the entire space is free of electric and magnetic fields. This is the situation when the vector potential given by the one form is *globally exact*. Now, assume only part of the space is free of electric and magnetic fields. In this case, A is only locally exact but not globally. This fact has non-trivial consequences.

21.1.1 Problem

1. Show that the product of the p-form and its dual is given by

$$f_p \wedge^* f_p = \frac{\sqrt{g}}{p!} f^{\mu_1 \cdots \mu_p} f_{\mu_1 \cdots \mu_p} \, d^d x. \tag{21.34}$$

Thus show that the kinetic energy of the $U(1)$ gauge field

$$L_{kin} = -\frac{1}{4} \int d^d x \, \sqrt{g} F^{\mu\nu} F_{\mu\nu}. \tag{21.35}$$

may be written in the form

$$L_{kin} = -\frac{1}{2} \int F \wedge^* F. \tag{21.36}$$

Further Reading

L. Conlon, *Differentiable Manifolds*, Birkhuser (2001).

R. W. R. Darling, *Differential Forms and Connections*, Cambridge University Press (1994).

H. Flanders, *Differential Forms with Applications to the Physical Sciences*, Dover (1989).

M. T. Vaughn, *Introduction to Mathematical Physics*, Wiley-VCH (2006).

21.2 Two-component Notation

We summarize here the properties of the two-component notation used in Chapter 8. The component Weyl spinors are of two types: left-handed ξ_α and right-handed $\bar{\chi}^{\dot\alpha}$. In terms of these, a Dirac field is given by

$$\psi_D = \begin{pmatrix} \xi_\alpha \\ \bar{\chi}^{\dot\alpha} \end{pmatrix}, \tag{21.37}$$

and $\bar\psi_D$ is

$$\bar\psi_D = (\chi^\alpha, \bar\xi_{\dot\alpha}) \tag{21.38}$$

and

$$\bar\psi_D \psi_D = \chi^\alpha \xi_\alpha + \bar\xi_{\dot\alpha} \bar\chi^{\dot\alpha}. \tag{21.39}$$

We define invariant tensors $\epsilon^{\alpha\beta}$ so that $\epsilon^{12} = 1$ and $\epsilon^{\alpha\beta} = -\epsilon^{\beta\alpha}$, and define $\epsilon_{12} = -1$ and $\epsilon_{\alpha\beta} = -\epsilon_{\beta\alpha}$. $\epsilon_{\dot\alpha,\dot\beta}$ and $\epsilon^{\dot\alpha\dot\beta}$ are defined in an identical fashion. The epsilons then satisfy the relation

$$\epsilon^{\alpha\beta}\epsilon_{\beta\gamma} = \delta^\alpha_\gamma = \epsilon_{\gamma\beta}\epsilon^{\beta\alpha}, \tag{21.40}$$

with identical relations holding for the case with dotted indices. Each of the two parts on the right-hand side of Eq. (21.39) is separately Lorentz invariant so that

$$\xi^\alpha \chi_\alpha \to \xi^{\alpha'} \chi_{\alpha'} = \xi^\beta (L^{-1})_\beta{}^\alpha L_\alpha^\gamma \chi_\gamma = \xi^\beta (L^{-1}L)^\gamma_\beta \chi_\gamma = \xi^\beta \chi_\beta, \tag{21.41}$$

where L_α^β $(\alpha, \beta = 1, 2)$ is an element of an $SL(2, C)$ matrix which is complex and unimodular, i.e., $det\, L = 1$, and $L^\dagger L = 1$ and L^{-1} is the inverse of the matrix L. $\epsilon^{\alpha\beta}$ is an invariant tensor under the Lorentz transformation (see Problem 1 in Section 8.21). The following relations are then easily checked:

$$\xi^\alpha = \epsilon^{\alpha\beta}\xi_\beta, \xi_\alpha = \epsilon_{\alpha\beta}\xi^\beta, \tag{21.42}$$

and similarly

$$\xi^{\dagger\dot\alpha} = \epsilon^{\dot\alpha\dot\beta}\xi^\dagger_{\dot\beta}, \ \xi^\dagger_{\dot\alpha} = \epsilon_{\dot\alpha\dot\beta}\xi^{\dagger\dot\beta}. \tag{21.43}$$

One can define the scalar product of two-component spinors using the epsilon tensor. Thus, one has

$$\xi_1 \xi_2 \equiv \xi_1^\alpha \xi_{2\alpha}. \tag{21.44}$$

It is then easily checked that $\xi_2 \xi_1 = \xi_1 \xi_2$, where one uses the fact that ξ_1 and ξ_2 are Grassmann objects and anti-commute and one uses the anti-symmetry of the ϵ-tensor. Similarly, $\xi_1^\dagger \xi_2^\dagger = \xi_2^\dagger \xi_1^\dagger$. It is also easily checked that $(\xi_1 \xi_2)^* = \xi_1^\dagger \xi_2^\dagger$.

The transformations of a right-handed spinor $\bar\chi_{\dot\alpha}$ can be expressed as follows:

$$\bar\chi_{\dot\alpha} \to \bar\chi'_{\dot\alpha} = (L^*)_{\dot\alpha}{}^{\dot\beta} \bar\chi_{\dot\beta}, \tag{21.45}$$

while $\bar{\chi}^{\dot{\alpha}}$ can be expressed as

$$\bar{\chi}^{\dot{\alpha}} \to \bar{\chi}^{\dot{\alpha}'} = \bar{\chi}^{\dot{\beta}}(L^{-1})^*_{\dot{\beta}}{}^{\dot{\alpha}}. \tag{21.46}$$

Here, the indices are raised or lowered by the invariant tensors $\epsilon_{\dot{\alpha}\dot{\beta}}$ and $\epsilon^{\dot{\alpha}\dot{\beta}}$ so that

$$\bar{\chi}^{\dot{\alpha}} = \epsilon^{\dot{\alpha}\dot{\beta}}\bar{\chi}_{\dot{\beta}}, \bar{\chi}_{\dot{\alpha}} = \epsilon_{\dot{\alpha}\dot{\beta}}\bar{\chi}^{\dot{\beta}}. \tag{21.47}$$

As for the case of the undotted indices, one can form a Lorentz-invariant combination such as $\bar{\xi}_{\dot{\alpha}}\bar{\chi}^{\dot{\alpha}}$.

In order to write the Dirac equation in the two-component notation, we express the gamma matrices in the two-component form:

$$\gamma^{\mu} = \begin{pmatrix} 0 & \sigma^{\mu} \\ \bar{\sigma}^{\mu} & 0 \end{pmatrix}, \quad \gamma_5 = \begin{pmatrix} -\mathbb{1} & O \\ 0 & \mathbb{1} \end{pmatrix}, \tag{21.48}$$

where $\mathbb{1}$ is a 2×2 identity matrix and σ^{μ} and $\bar{\sigma}^{\mu}$ are given by

$$\sigma^0 = \bar{\sigma}^0 = \begin{pmatrix} -1 & 0 \\ 0 & -1 \end{pmatrix}, \quad \sigma^1 = -\bar{\sigma}^1 = \begin{pmatrix} 0 & 1 \\ 1 & 0 \end{pmatrix},$$

$$\sigma^2 = -\bar{\sigma}^2 = \begin{pmatrix} 0 & -i \\ i & 0 \end{pmatrix}, \quad \sigma^3 = -\bar{\sigma}^3 = \begin{pmatrix} 1 & 0 \\ 0 & -1 \end{pmatrix}. \tag{21.49}$$

The free Dirac Lagrangian in then expressed in the form

$$L_D = i\left[\xi_1^{\dagger}\bar{\sigma}^{\mu}\partial_{\mu}\xi_1 + \xi_2^{\dagger}\bar{\sigma}^{\mu}\partial_{\mu}\xi_2\right] - M\left(\xi_1\xi_2 + \xi_1^{\dagger}\xi_2^{\dagger}\right) - i\partial_{\mu}(\xi_2^{\dagger}\bar{\sigma}^{\mu}\xi_2). \tag{21.50}$$

In obtaining the above result we have used the following identity:

$$(\bar{\sigma}^{\mu})^{\dot{\alpha}\alpha} = \epsilon^{\dot{\alpha}\dot{\beta}}\epsilon^{\alpha\beta}(\sigma^{\mu})_{\beta\dot{\beta}}. \tag{21.51}$$

Using the above result, it is easy to see that

$$\xi_2^{\dagger}\bar{\sigma}_{\mu}\xi_2 = -\xi_2\sigma_{\mu}\xi_2^{\dagger}, \tag{21.52}$$

which is useful in writing the Dirac Lagrangian in the two-component notation.

21.3 Grassmann Co-ordinates

Grassmann co-ordinates enter into the path integral formulation involving fermionic fields and in the superspace formulation of supersymmetry and supergravity theories. We give below some elementary properties of Grassmann co-ordinates. We begin by considering a single fermionic co-ordinate θ which satisfies the property that

$$\theta^2 = 0. \tag{21.53}$$

A function $f(\theta)$ then has the expansion

$$f(\theta) = f_0 + \theta f_1, \tag{21.54}$$

since higher powers in θ vanish due to Eq. (21.53). Let us consider variation of $f(\theta)$ with respect to θ. We define left differentiation and right differentiation so that

$$\delta f = \delta\theta \frac{\delta_\ell f}{\delta\theta} = \frac{\delta_r f}{\delta\theta} \delta\theta, \tag{21.55}$$

where

$$\delta f = \delta\theta f_1 = f_1 \delta\theta, \qquad f_1 \text{ bosonic} \tag{21.56}$$
$$\delta f = \delta\theta f_1 = -f_1 \delta\theta, \qquad f_1 \text{ fermionic,} \tag{21.57}$$

which leads to

$$\frac{\delta_\ell f}{\delta\theta} = \frac{\delta_r f}{\delta\theta} = f_1, \qquad \text{if } f_1 \text{ bosonic} \tag{21.58}$$

$$\frac{\delta_\ell f}{\delta\theta} = -\frac{\delta_r f}{\delta\theta} = f_1, \qquad \text{if } f_1 \text{ fermionic.} \tag{21.59}$$

An integration over Grassmann co-ordinates is essentially the same as differentiation. Now, in the integration we require that the integration measure be invariant under translations, i.e.,

$$\theta = \theta' + \xi \tag{21.60}$$
$$d\theta = d\theta'. \tag{21.61}$$

Thus, we require

$$\int d\theta \, f(\theta) = \int d\theta' \, f(\theta'). \tag{21.62}$$

Using Eqs. (21.54) and (21.61) in (21.62), one finds that Eq. (21.62) can only hold if

$$\int d\theta = 0. \tag{21.63}$$

Further, the integration on θ is the same as differentiation, i.e,

$$\int d\theta = \frac{d}{d\theta}. \tag{21.64}$$

Thus, one has

$$\int d\theta\theta = 1. \tag{21.65}$$

Next, we consider a transformation on θ so that

$$\theta = A\theta'. \tag{21.66}$$

A priori we do not know the relationship between $d\theta$ and $d\theta'$, and so we assume

$$d\theta = J \, d\theta', \tag{21.67}$$

where J is a c-number. To determine J we require the validity of Eq. (21.62). Now, inserting Eq. (21.54) in Eq. (21.62) and using Eqs. (21.66) and (21.67), one finds

$$f_1 = AJf_1, \tag{21.68}$$

which leads to the result $J = 1/A$ and

$$d\theta = \frac{1}{A}d\theta'. \tag{21.69}$$

Let us now see how the Leibniz rule of differentiation works for the product of two functions of θ when one differentiates with respect to θ. Consider the product of two functions $f(\theta)g(\theta)$. Here,

$$\delta[f(\theta)g(\theta)] = f(\theta + \delta\theta)g(\theta + \delta\theta) - f(\theta)g(\theta)$$
$$= [f(\theta + \delta\theta) - f(\theta)]g(\theta + \delta\theta) + f(\theta)[g(\theta + \delta\theta) - g(\theta)]$$
$$= \delta\theta\frac{\partial_\ell f}{\partial\theta}g + f\delta\theta\frac{\partial_\ell g}{\partial\theta}. \tag{21.70}$$

The above leads to the following Leibz rule for the differentiation of two functions:

$$\frac{\partial_\ell(fg)}{\partial\theta} = \frac{\partial_\ell f}{\partial\theta}g \pm f\frac{\partial_\ell g}{\partial\theta}, \tag{21.71}$$

where $+$ denotes the case when f is bosonic and $-$ denotes the case when f is fermionic.

Let us now see how integration by parts works. Consider

$$\int d\theta \, f\frac{\partial_\ell g}{\partial\theta} = f_1g_1 = \int f_1 \, d\theta \, g = \int \frac{\partial_\ell f}{\partial\theta}d\theta \, g$$
$$= \mp \int d\theta\frac{\partial_\ell f}{\partial\theta}g, \tag{21.72}$$

where $-$ denotes the case when f is bosonic and $+$ denotes the case when f is fermionic. Next, we extend the above analysis to the case of many Grassmann variables. Consider a set of Grassmann variables θ_α ($\alpha = 1, \cdots, n$) which satisfy the anti-commutation relation

$$\{\theta_\alpha, \theta_\beta\} = 0, \quad \alpha, \beta = 1, \cdots, n. \tag{21.73}$$

A function $f(\theta)$ has the following expansion:

$$f(\theta) = f_0 + \sum_{\alpha=1}^{n} f_\alpha\theta_\alpha + \frac{1}{2!}\sum_{\alpha_1,\alpha_2=1}^{n} f_{\alpha_1\alpha_2}\theta_{\alpha_1}\theta_{\alpha_2} + \cdots + \frac{1}{n!}\sum_{\alpha_1\cdots\alpha_n} f_{\alpha_1\cdots\alpha_n}\theta_{\alpha_1}\cdots\theta_{\alpha_n}, \tag{21.74}$$

where

$$f_{\alpha_1\cdots\alpha_n} = \epsilon_{\alpha_1\cdots\alpha_n}f_{1\cdots n}, \tag{21.75}$$

and where $\epsilon_{\alpha_1\cdots\alpha_n} = +1$ for even permutations and -1 for odd permutations. The rules of differentiation and integration are direct extensions of the cases discussed above. Specifically, we have

$$\int d\theta_\alpha = 0 \tag{21.76}$$

$$\int d\theta_\alpha \, \theta_\beta = \delta_{\alpha\beta}. \tag{21.77}$$

Next, let us consider the transformations

$$\theta_\alpha = A_{\alpha\beta}\theta'_\beta. \tag{21.78}$$

To make the analysis explicit, let us consider the case of just two θ so that

$$f(\theta) = f_0 + f_1\theta_1 + f_2\theta_2 + f_{12}\theta_1\theta_2. \tag{21.79}$$

Here, one finds

$$\int d\theta_2 \, d\theta_1 \, f(\theta) = f_{12}. \tag{21.80}$$

Next, we assume that the measure changes so that

$$d\theta_2 \, d\theta_1 = J \, d\theta'_2 \, d\theta'_1. \tag{21.81}$$

In this case we have

$$\int d\theta_2 \, d\theta_1 \, f(\theta) = f_{12} \int d\theta_2 \, d\theta_1 \, \theta_1\theta_2$$

$$= Jf_{12} \int d\theta'_2 \, d\theta'_1 (A_{11}\theta'_1 + A_{12}\theta'_2)(A_{21}\theta'_1 + A_{22}\theta'_2)$$

$$= Jf_{12} \, det \, A. \tag{21.82}$$

Comparison of Eqs. (21.80) and (21.82) gives $J = 1/det \, A$. This result is in fact valid when one considers an arbitrary number of θ, and thus one has

$$\prod_\alpha d\theta_\alpha = \frac{1}{det \, A} \prod d\theta'_\alpha. \tag{21.83}$$

We note that the result of Eq. (21.83) is opposite to what one has in the bosonic case, where on the right-hand side one has $det \, A$ rather than $1/det \, A$.

Next, we examine what a Gaussian integral looks like with Grassmann coordinates. Let us consider the integral with integration over two θ:

$$I_2 = \int d\theta_2 \, d\theta_1 \, e^{-(i/2)\theta_\alpha M_{\alpha\beta}\theta_\beta}. \tag{21.84}$$

Using the anti-commuting property of the θ, we can expand the exponent so that

$$e^{-(i/2)\theta_\alpha M_{\alpha\beta}\theta_\beta} = 1 - \frac{i}{2}\theta_\alpha M_{\alpha\beta}\theta_\beta, \tag{21.85}$$

since all the higher terms in the expansion of the exponent vanish. The integration is now easily performed, and one gets

$$I_2 = -iM_{12} = -i(det \, M)^{1/2}. \tag{21.86}$$

The result above is opposite to what one has for a Gaussian integral with integration over bosonic co-ordinates, which gives the result $\sim 1/(det\, M)^{1/2}$. The result of Eq. (21.86) can be extended easily to the case with many Grassmann co-ordinates. For the case of $2n$ Grassmann co-ordinates, the integral corresponding to Eq. (21.84) is given by

$$I_{2n} = \int \prod_{\alpha=1}^{2n} d\theta_\alpha \; e^{-(i/2)\theta_\alpha M_{\alpha\beta}\theta_\beta}. \tag{21.87}$$

The matrix $M_{\alpha\beta}$ is anti-symmetric. It is possible to put this matrix in a block diagonal form by a real orthogonal transformation, where

$$\theta_\alpha = O_{\alpha\beta}\theta'_\beta \tag{21.88}$$

$$O^T O = I \tag{21.89}$$

$$\theta^T M \theta = \theta'^T M' \theta' \tag{21.90}$$

and the matrix M' has the form

$$M' = \begin{pmatrix} \begin{array}{cc|c|c} 0 & M'_{12} & & \\ -M'_{12} & 0 & & \\ \hline & & \begin{array}{cc} 0 & \cdot \\ \cdot & 0 \end{array} & \\ \hline & & & \begin{array}{cc} 0 & M'_{2n-1,2n} \\ -M'_{2n-1,2n} & 0 \end{array} \end{array} \end{pmatrix}. \tag{21.91}$$

Using the θ', we can write I_{2n} as follows:

$$I_{2n} = \int \prod_{\alpha=1}^{2n} d\theta'_\alpha \; e^{-i\left[\theta'_1 M'_{12}\theta'_2 + \cdots + \theta'_{2n-1} M'_{2n-1,2n}\theta'_{2n}\right]} \frac{1}{det(O)}. \tag{21.92}$$

Integration on the θ' co-ordinates can be easily performed, which gives

$$I_{2n} = P_f(M)$$

$$= \frac{1}{2^n n!} \sum_{p \subset S_{2n}} (-1)^p \prod_{i=1}^{n} M_{p(2i-1),p(2i)} \tag{21.93}$$

where $P_f(M)$ stands for the Pfaffian defined for a skew-symmetric matrix M, and where S_{2n} is the symmetric group and the sum is taken over all permutations p. We note that $det\, M = (P_f(M))^2$.

Next, we define the analog of the Dirac delta function for the case of Grassmann co-ordinates. For the case of a single Grassmann co-ordinate θ, we define $\delta(\theta)$ so that

$$\int d\theta \; f(\theta)\delta(\theta) = f(0). \tag{21.94}$$

It is now easily seen that in this case

$$\delta(\theta) = \theta, \tag{21.95}$$

since on using Eq. (21.54) one finds

$$\int d\theta f(\theta)\delta(\theta) = \int d\theta(f_0 + \theta f_1)\theta$$
$$= f_0 = f(0), \tag{21.96}$$

which is consistent with Eq. (21.94). For the case of n Grassmann co-ordinates, one has a direct extension, i.e.,

$$\delta(\theta) = \theta_1\theta_2\cdots\theta_n, \tag{21.97}$$

since in this case

$$\int d\theta_n \cdots d\theta_1\ \delta(\theta)f(\theta) = f(0). \tag{21.98}$$

Further Reading

F. A. Berezin, *Pure Appl. Phys.* **24**, 1 (1966).

R. L. Arnowitt, P. Nath, and B. Zumino, *Phys. Lett. B* **56**, 81 (1975).

S. J. Gates, M. T. Grisaru, M. Rocek, and W. Siegel, hep-th/0108200 (2001).

I. L. Buchbinder and S. M. Kuzenko, *Ideas and Methods of supersymmetry and Supergravity or a A Walk Through Superspace*, IOP (1995).

A. Salam and J. A. Strathdee, *Nucl. Phys. B* **76**, 477 (1974); *Nucl. Phys. B* **80**, 499 (1974); *Phys. Rev. D* **11**, 1521 (1975); *Nucl. Phys. B* **86**, 142 (1975).

21.4 Geometry on Supermanifolds

We give here a brief introduction to geometry on supermanifolds. The geometry on supermanifolds was initiated by Arnowitt, Nath, and Zumino [1, 2], and a detailed discussion is given by Dewitt [3]. First, we introduce a superspace with co-ordinates z^A consisting of Bose co-ordinates (x^μ) and Fermi co-ordinates (θ^α) so that

$$z^A = \{x^\mu, \theta^\alpha\}. \tag{21.99}$$

To index A, we associate an integer a which is either 0 or 1, depending on whether $z^A = x^\mu$ (Bose) or $z^A = \theta^\alpha$ (Fermi). To the quantity

$$Q = Q^{A_1\cdots A_i\cdots}_{B_1\cdots B_j\cdots} \tag{21.100}$$

we can associate a number $q = \sum_i a_i + \sum_j b_j$. The quantities Q_1 and Q_2 then obey the following multiplication rule:

$$Q_1 Q_2 = (-1)^{q_1 q_2} Q_2 Q_1. \tag{21.101}$$

Let us now consider the general co-ordinate transformation

$$z^{A'} = z^{A'}(z). \tag{21.102}$$

Under this transformation, a contravariant superfield vector V^A transforms as

$$V^{A'}(z') = R^{A'}_B V^B(z), \tag{21.103}$$

where $R^{A'}_B$ is the right derivative defined by $R^{A'}_B \equiv \partial_R z^{A'}/\partial z^B$. One may also define a contravariant superfield vector by the transformation

$$V^{A'}(z') = V^B(z) L^{A'}_B, \tag{21.104}$$

where $L^{A'}_B$ is the left derivative. The right and the left derivatives obey the contractions

$$R^B_{A'} R^{A'}_C = \delta^B_C = L^{A'}_C L^B_{A'}. \tag{21.105}$$

There exist two types of covariant vectors with transformations defined by

$$U'_A(z') = L^B_A U_B \tag{21.106}$$

$$\hat{U}'_A = \hat{U}_B R^B_A. \tag{21.107}$$

However, it is easily checked that, because of Eqs. (21.103)–(21.107), only the quantities $V^A U_A$ and $\hat{U}_A V^A$ form a scalar under contraction, while the contractions $U_A V^A$ and $V^A \hat{U}_A$ do not form a scalar. Thus, not every contraction is allowed if one wants to preserve the tensor properties.

We can follow essentially the same techniques as on the bosonic manifolds to define covariant derivatives by parallel transport. Thus, the right covariant differentiation is given by

$$V^A_{;B} = \partial_R V^A/\partial z^B + V^C \Gamma^A_{CB}. \tag{21.108}$$

For a scalar superfield ϕ, one has

$$\phi_{;A} = \partial \phi/\partial z^A. \tag{21.109}$$

We now assume that the generalized Leibniz rule for covariant differentiation of a product of two tensors Q_1 and Q_2 holds so that

$$(Q_1 Q_2)_{;A} = Q_1 Q_{2;A} + (-1)^{aq_2} Q_{1;A} Q_2. \tag{21.110}$$

Next, we consider the general co-ordinate transformations $z^{A'} = z^{A'}(z)$ which preserve the invariance of the line element:

$$ds^2 = dz^A g_{AB}\, dz^B, \tag{21.111}$$

where g_{AB} appears as the metric tensor of superspace and satisfies the symmetry property

$$g_{AB}(z) = \eta_{ab} g_{BA}(z), \quad \eta_{ab} = (-1)^{a+b+ab}. \tag{21.112}$$

The superfield $g_{AB}(z)$ has the following transformation properties:

$$g_{A'B'}(z') = L^C_{A'} g_{CD}(z) R^D_{B'}, \tag{21.113}$$

where $L_{A'}^C$ and $R_{B'}^D$ are the left and the right derivatives defined above and the transformations of Eq. (21.113) preserve the symmetry properties of Eq. (21.112). Thus, consider the infinitesimal transformations given by

$$z^{A'} = z^A + \xi^A(z), \tag{21.114}$$

which can be used in Eq. (21.113) to compute the gauge change $\delta g_{AB} \equiv g'_{AB}(z) - g_{AB}(z)$ so that

$$\delta g_{AB} = g_{AC}\xi^C_{,B} + \eta_{ab}g_{BC}\xi^C_{,A} + g_{AB,C}\xi^C. \tag{21.115}$$

Further, assuming $g_{AB;C} = 0$, defining the inverse metric g^{AC} so that $g^{AC}g_{CB} = \delta^A_B$, one finds that the affinity is uniquely determined in terms of the metric so that

$$\Gamma^C_{AB} = (-1)^{bc}\frac{1}{2}\left[(-1)^{bd}g_{AD,B} + (-1)^{ad}\eta_{ab}g_{BD,A} - g_{AB,D}\right]g^{DC}. \tag{21.116}$$

We note in passing that Γ^C_{AB} satisfies the symmetry property

$$\Gamma^C_{BA} = f_{abc}\Gamma^C_{AB}, \quad f_{abc} = (-1)^{a+b+ab+c(a+b)}. \tag{21.117}$$

As in the bosonic case, the Riemann–Christoffel tensor R^D_{ABC} can be obtained by a parallel transport around a closed infinitesimal curve, and one finds

$$R^D_{ABC} = -\Gamma^D_{AC,B} + (-1)^{bc}\Gamma^D_{AB,C} + (-1)^{b(c+d+e)}\Gamma^E_{AB}\Gamma^D_{EC}$$
$$- (-1)^{c(d+e)}\Gamma^E_{AC}\Gamma^D_{EB}. \tag{21.118}$$

There is only one independent tensor which can be obtained by contraction of the Riemann–Christoffel tensor, i.e.,

$$R_{AB} = (-1)^c R^C_{ABC}, \tag{21.119}$$

which leads to the explicit form

$$R_{AB} = -(-1)^c\Gamma^C_{AC,B} + (-1)^{c+bc}\Gamma^C_{AB,C} + (-1)^{c+bc}\Gamma^E_{AB}\Gamma^C_{EC} - (-1)^{ce}\Gamma^E_{AC}\Gamma^C_{EB}. \tag{21.120}$$

Similarly, one can define a curvature scalar R by the contraction

$$R = (-1)^b g^{BA} R_{AB}. \tag{21.121}$$

Next, we note the difference between a scalar $\phi(z)$ and a scalar density $D(z)$. Under general co-ordinate transformations, a scalar field is characterized by $\phi'(z') = \phi(z)$ so that under infinitesimal gauge transformations one has

$$\delta\phi(z) = \phi_{,A}\xi^A, \tag{21.122}$$

while a scalar density $D(z)$ is defined as an object that transforms according to the rule

$$\delta D = \sum\left[(-1)^a D\xi^A\right]_{,A}. \tag{21.123}$$

Because of this, the integral $I \equiv \int d^8z\, D(z)$ is an invariant, i.e., $\delta I = 0$ since δD is a total derivative. Further, it is easily shown that if $\phi(z)$ is a scalar and $D(z)$ is a scalar density, then $D(z)\phi$ is also a scalar density. In a similar fashion, one can define tensor densities by the multiplication of an ordinary tensor with a scalar density. Specifically, consider the gauge transformation of a contravariant vector, which is

$$\delta V^A(z) = V^A_{,B}\xi^B - \xi^A_{,B}V^B. \tag{21.124}$$

However, for the gauge transformations of a vector density defined by $\mathcal{V}^A = D(z)V^A(z)$, one has gauge transformations of the form

$$\delta\mathcal{V}^A(z) = \left[(-1)^{b+ab}\xi^B\mathcal{V}^A\right]_{,B} - \xi^A_{,B}\mathcal{V}^B. \tag{21.125}$$

From the above equation it follows that the quantity $(-1)^a\mathcal{V}^A_{,A}$ transforms like a scalar density. Thus, one has

$$(-1)^a\mathcal{V}^A_{;A} = (-1)^a\mathcal{V}^A_{,A}. \tag{21.126}$$

Using the above and the Leibniz rule for the covariant differentiation of a product of two superfields gives the relation for the covariant differentiation of the density, i.e.,

$$D_{;A} = D_{,A} - (-1)^{a+c}D\Gamma^{C}_{A\,C}, \tag{21.127}$$

where the affinity $\Gamma^{B}_{A\,C}$ is defined via the covariant differentiation as follows:

$$V^A_{;B} = V^A_{,B} + V^C\Gamma^{A}_{C\,B}. \tag{21.128}$$

The covariant differentiation of a vector density differs from the above due the presence of an additional term so that

$$\mathcal{V}^A_{;B} = \mathcal{V}^A_{,B} + \mathcal{V}^C\Gamma^{A}_{C\,B} - (-1)^{b+c}\mathcal{V}^A\Gamma^{C}_{B\,C}. \tag{21.129}$$

The above correctly gives the result $(-1)^a\mathcal{V}^A_{;A} = (-1)^a\mathcal{V}^A, A$. In Section 21.4.1 we have defined the superdeterminant g. Using that result we now establish that $\sqrt{-g}$ transforms like a scalar density. We can show this as follows. From Eq. (21.164) we find

$$\delta(det\ Q) = (det\ Q)tr\ Q^{-1}\delta Q. \tag{21.130}$$

Thus,

$$\delta\sqrt{-g} = \frac{1}{2}\sqrt{-g}\sum(-1)^a g^{AB}\delta g_{BA}. \tag{21.131}$$

Further, on using the result of Eq. (21.115), one finds

$$\delta\sqrt{-g} = \frac{1}{2}\sqrt{-g}\sum(-1)^a\left[\xi^A_{,A} + \frac{1}{2}g^{AB}g_{BA,C}\xi^C\right]. \tag{21.132}$$

The last term can be evaluated to be $\sqrt{-g}_{,C}\xi^C$, so that Eq. (21.132) takes the form

$$\delta\sqrt{-g} = \sum(-1)^a\left[\sqrt{-g}\xi^A\right]_{,A}. \tag{21.133}$$

With the above constructions, we can define an invariant action in curved super-space so that

$$I = \int d^8z \, \sqrt{-g}R. \tag{21.134}$$

Variation of the action of Eq. (21.134) can be written as follows:

$$\delta I = \int d^8z[(-1)^a\delta\zeta^{AB}R_{BA} + (-1)^a\zeta^{AB}\delta R_{BA}], \tag{21.135}$$

where

$$\zeta^{AB} \equiv \sqrt{-g}g^{AB}. \tag{21.136}$$

Let us discuss the difficult term first, which is the second term in the brackets on the right-hand side of Eq. (21.135) which involves the variation δR_{AB} under the gauge transformations. Following the same procedures as for the case of the bosonic manifolds (see Section 2.10), one can obtain a generalization of Palatini's formula in superspace, which reads

$$\delta R_{AB} = (-1)^{c+bc}\delta\Gamma^C_{AB;C} - (-1)^c\delta\Gamma^C_{AC;B}. \tag{21.137}$$

The second term on the right-hand side of Eq. (21.135) on using δR_{BA} given by Eq. (21.137) can be written as follows:

$$\int d^8z[-(-1)^{a+bc}\zeta^{AB}_{;C}\delta\Gamma^C_{BA} + (-1)^{c+d+bd}\zeta^{DB}_{;D}\delta\Gamma^C_{BC} + (-1)^aw^A_{;A}], \tag{21.138}$$

where w^A is a vector density and gives no contribution because of Eq. (21.126). Now, we recall from Section 2.10 that, in the Palatini method, the metric tensor and the connections are varied independently. Thus, the variation of the connection gives

$$\zeta^{AB}_{;C} = 0. \tag{21.139}$$

This equation can now be used to determine the affinity in terms of the metric. The variation of the action with respect to $\delta\zeta^{AB}$ gives the equations of motion in terms of the curvature tensor so that

$$R_{AB} = 0. \tag{21.140}$$

One can extend the action by the inclusion of the term $2\lambda \int d^8z \, \sqrt{-g}$, which leads to the field equation

$$R_{AB} = \lambda g_{AB}. \tag{21.141}$$

We can also derive the so-called Bianchi identities in superspace. We begin by writing the gauge transformations so that

$$\delta g_{AB}(z) = \xi_{A;B} + \eta_{ab}\xi_{B;A}. \tag{21.142}$$

A variation of the action using the above gives

$$\delta I = \int d^8z \, \sqrt{-g}(-1)^a\left[-R^{AB} + \frac{1}{2}g^{AB}R\right](\xi_{B;A} + \eta_{ab}\xi_{A;B}). \tag{21.143}$$

In the above, the indices on R^{AB} have been raised by use of the inverse metric, i.e., $R^{AB} = g^{AC} R_{CD} g^{DB}$. The use of Eq. (21.126) and the Leibniz rule allows one to derive the following identity:

$$(-1)^a \left(R^{AB} - \frac{1}{2} g^{AB} R \right)_{;A} = 0. \qquad (21.144)$$

Next, we consider sub-cases of the general gauge transformations in superspace. The sub-case of the Einstein gauge transformations can be obtained by setting

$$x^{\mu'} = x^{\mu'}(x), \ \theta' = \theta. \qquad (21.145)$$

The supergauge transformations of Eq. (21.115) also contain the Yang–Mills transformations as a sub-case. This can be seen by introducing Dirac spinor co-ordinates by adding a charge index q to get a doublet θ^α ($q = 1, 2$). The Dirac spinor is then described by the combination $\theta^{\alpha 1} - i\theta^{\alpha 2}$. We can also introduce an n-tuplet of spinor co-ordinates $\theta^{\alpha q a}$ ($a = 1, 2, \cdots, n$). Consider now the superspace transformations

$$\xi^A = (\xi^\mu = 0, \xi^\alpha = \lambda^a(x)(T_a \theta)^\alpha) \quad \text{(Yang–Mills)}, \qquad (21.146)$$

where $\lambda^a(x)$ are the gauge functions and T_a are the real anti-symmetric $SU(n)$ matrices in the Majorana spinor representation where $a = 1, \cdots, n^2 - 1$.

Expansion of the superspace metric $g_{AB}(z)$ when expanded in powers of θ contains in it the Einstein metric $g_{\mu\nu}(x)$ and the Rarita–Schwinger field ψ^μ as well as gauge fields. The above theory was the first extension of global supersymmetry to a local supersymmetry, and exhibited the important phenomenon that the extension of a global supersymmetry into a local supersymmetry necessarily brings in gravity. Thus, in this sense it was the first supergravity theory. The gauge group of this theory is $OSp(4|4)$. The particle content of this theory is rather complicated. The problem of achieving a simple particle content in a local supersymmetry was accomplished in subsequent work. Suggestions for further reading are listed at the end of this section.

21.4.1 The Superdeterminant

As discussed above, in order to write an invariant action in curved superspace one needs to define the superdeterminant, i.e., the determinant of a matrix Q_{AB} in superspace. Let us write the matrix Q_{AB} with $A, B = \mu, \nu, \cdots, \alpha\beta \cdots$ where μ, ν, etc., are bosonic indices and α, β, etc., are fermionic indices. We write Q in the form

$$Q = \begin{pmatrix} q_{\mu\nu} & r_{\mu\beta} \\ s_{\alpha\nu} & p_{\alpha\beta} \end{pmatrix}. \qquad (21.147)$$

Suppressing the indices, the inverse of this matrix is given by

$$Q^{-1} = \begin{pmatrix} q^{-1} + q^{-1}rt^{-1}sq^{-1} & -q^{-1}rt^{-1} \\ -t^{-1}sq^{-1} & t^{-1} \end{pmatrix}, \quad t \equiv p - sq^{-1}r, \qquad (21.148)$$

where q and p are assumed to have an inverse. One may recall that the determinant of an ordinary matrix obeys the following relation:

$$det\, Q = e^{tr(\ln Q)}. \qquad (21.149)$$

We adopt Eq. (21.149) as the basic definition of a determinant when extended to superspace except that we need to extend the concept of the trace for a matrix defined in superspace. Thus, we define

$$sdet\, Q = e^{s\,tr(\ln Q)}, \qquad (21.150)$$

where the supertrace is given by

$$str(Q) = \sum(-1)^a Q_{AA} = \underbrace{\sum Q_{\mu\mu}}_{\text{Bose}} - \underbrace{\sum Q_{\alpha\alpha}}_{\text{Fermi}}. \qquad (21.151)$$

Next, let us consider the trace of a product of two matrices Q_1 and Q_2 so that

$$str(Q_1 Q_2) = \sum(-1)^a (Q_1)_{AB}(Q_2)_{BA}. \qquad (21.152)$$

We use

$$(Q_1)_{AB}(Q_2)_{BA} = (-1)^{a+b}(Q_2)_{BA}(Q_1)_{AB}, \qquad (21.153)$$

and write Eq. (21.152) as

$$\begin{aligned} str(Q_1 Q_2) &= \sum(-1)^b (Q_2)_{BA}(Q_1)_{AB} \\ &= sTr(Q_2 Q_1), \end{aligned} \qquad (21.154)$$

which implies

$$str([Q_1, Q_2]) = 0. \qquad (21.155)$$

Let us now write $Q_i = e^{z_i}$ $(i = 1, 2)$ so that

$$sdet(Q_1 Q_2) = e^{z_1 + z_2 + C_{12}}, \qquad (21.156)$$

where C_{12} is the Baker–Hausdorf series of commutators. However, because of Eq. (21.155), C_{12} vanishes, and thus we have the result

$$\begin{aligned} sdet(Q_1 Q_2) &= e^{z_1 + z_2} \\ &= sdet(Q_1)\, sdet(Q_2). \end{aligned} \qquad (21.157)$$

Next, let us consider a matrix Q which is product of the two matrices Q_1 and Q_2:

$$Q = Q_1 Q_2. \qquad (21.158)$$

Using the matrix element of Eq. (21.147), we assume Q_1 and Q_2 have the following forms [2]:

$$Q_1 \equiv \begin{pmatrix} q & 0 \\ s & \mathbb{1} \end{pmatrix}, \quad Q_2 \equiv \begin{pmatrix} \mathbb{1} & q^{-1}r \\ 0 & p - sq^{-1}r \end{pmatrix}. \tag{21.159}$$

Using the result of Eq. (21.157), one finds

$$\begin{aligned}
sdet\ Q &= sdet(Q_1)sdet(Q_2) \\
&= det(q)det^{-1}(p - sq^{-1}r) \\
&= det(Q_{\mu\nu})det(Q^{-1})^{\alpha\beta}, \tag{21.160}
\end{aligned}$$

where in the last step we have used the result of Eq. (21.148). Specifically, the quantity $\sqrt{-g} \equiv [-det(g)]^{1/2}$ is defined so that

$$\sqrt{-g} = \left[-det(g_{\mu\nu})det((g^{-1})^{\alpha\beta})\right]^{1/2}. \tag{21.161}$$

21.4.2 Problems

1. Using parallel transport around a closed infinitesimal curve show that the Riemann–Christoffel tensor $R^D{}_{ABC}$ is given by

$$\begin{aligned}
R^D{}_{ABC} = &-\Gamma^D{}_{AC,B} + (-1)^{bc}\Gamma^D{}_{AB,C} \\
&+ (-1)^{b(c+d+e)}\Gamma^E{}_{AB}\Gamma^D{}_{EC} - (-1)^{c(d+e)}\Gamma^E{}_{AC}\Gamma^D{}_{EB}. \tag{21.162}
\end{aligned}$$

2. Show that in curved superspace the generalization of Palatini's formula is given by

$$\delta R_{AB} = (-1)^{c+bc}\delta\Gamma^C{}_{AB;C} - (-1)^c\delta\Gamma^C{}_{AC;B}. \tag{21.163}$$

3. Check the result of Eq. (21.148).

4. Show that in superspace, variation of the superdeterminant gives

$$\delta(det\ M) = (det\ M)Tr\ M^{-1}\ \delta M. \tag{21.164}$$

References

[1] P. Nath and R. L. Arnowitt, *Phys. Lett. B* **56**, 177 (1975).

[2] R. L. Arnowitt, P. Nath, and B. Zumino, *Phys. Lett. B* **56**, 81 (1975).

[3] Bryce S. Dewitt, *Supermanifiolds*, Cambridge University Press (1992).

Further Reading

R. L. Arnowitt and P. Nath, *Gen. Rel. Grav.* **7**, 89 (1976).

R. L. Arnowitt and P. Nath, *Phys. Lett. B* **78**, 581 (1978).

L. Brink, M. Gell-Mann, P. Ramond, and J. H. Schwarz, *Phys. Lett. B* **76**, 417 (1978).

P. Nath and R. L. Arnowitt, *Phys. Lett. B* **65**, 73 (1976).

P. Van Nieuwenhuizen, *Phys. Rep.* **68**, 189 (1981).

J. Wess and B. Zumino, *Phys. Lett. B* **66**, 361 (1977).

21.5 Group Theory

Group theory enters in a central way into theories of particle physics and gravity. The basic tenets of group theory are simple. A group G is defined to consist of a set of elements a, b, $\cdots$, e and a composition rule such as multiplication and to obey the following conditions: (1) the product of two elements produces another element which lies in the group, i.e., $a\ b = c$, where c belongs to the group; (2) the composition law is associative, i.e., $(ab)c = a(bc)$; (3) the group contains an identity element e such that $ae = ea = a$; (4) given any element of the set a there exists an inverse a^{-1} such that $aa^{-1} = a^{-1}a = e$. In general, the product is not commutative, i.e., ab does not equal ba. However, the condition that the product is commutative, i.e., $ab = ba$ leads to an abelian group. A group may have one or more subgroups H_a, each of which satisfies conditions 1–4. The elements of each of the subgroups will be a subset of elements the group G. A subgroup H is an invariant subgroup if given an element h that belongs to H, and an element g that belongs to G, the element ghg^{-1} is again an element of H. A group which has no non-trivial invariant subgroups with the possible exception of discrete ones is called simple, and a group which contains no non-trivial abelian subgroups with the possible exception of discrete ones is called semi-simple.

Lie groups are of special importance in particle physics. In Lie groups, the group elements instead of being a discrete set are functions of continuous parameters, $u = (u_1, u_2, \cdots, u_N)$ i.e., one has elements $a(u)$, with $a(0)$ being the identity. The product rule in this case reads

$$a(u)a(v) = a(w(u,v)), \tag{21.165}$$

where w is a function of u and v. We will assume that $a(u)$ is a continuous and differentiable function of u. Much of the information about Lie groups, i.e., aside from global properties, can be obtained by the study of the Lie group elements in the vicinity of the identity. For infinitesimal parameters u we can expand the element $a(u)$ around the identity as follows:

$$a(u) = a(0) + \sum_{i=1}^{N} u_i X_i + \sum_{i=1}^{N}\sum_{j=1}^{N} u_i u_j X_i X_j + \cdots, \tag{21.166}$$

where $X_i \equiv (\partial a/\partial u_i)_{u=0}$ and are the generators of infinitesimal group transformations. From an expansion of the quantity $a(v)^{-1}a(u)^{-1}a(v)a(u)$ and group properties, it is then straightforward to derive the following commutation relation on X_i $(i = 1, 2, \cdots, N)$

$$[X_i, X_j] = C_{ij}^{k} X_k, \quad i, j, k = 1, 2, \cdots N, \tag{21.167}$$

where C_{ij}^k are the structure constants that define the group. From Eq. (21.167), one deduces that C_{ij}^k must be anti-symmetric in the lower two indices, i.e.,

$$C_{ij}^k = -C_{ji}^k. \tag{21.168}$$

Additionally, the generators satisfy the Jacobi identity , i.e,

$$[[X_i, X_j], \ X_k] + [[X_j, X_k], \ X_i] + [[X_k, X_i], \ X_j] = 0, \tag{21.169}$$

which implies that the structure constants obey the constraint

$$C_{ij}^\ell \, C_{\ell k}^m \ + \ C_{jk}^\ell \, C_{\ell i}^m \ + \ C_{ki}^\ell \, C_{\ell j}^m = 0. \tag{21.170}$$

Let us introduce the matrices $(C_\ell)_k^m$ such that $(C_\ell)_k^m = C_{\ell k}^m$, etc., in which case we can rewrite Eq. (21.170) in the following form:

$$(C_i)_\ell^m (C_j)_k^\ell - (C_j)_\ell^m (C_i)_k^\ell = C_{ij}^\ell (C_\ell)_k^m. \tag{21.171}$$

Equation (21.171) shows that the set of N-matrices C_i, $(i = 1, 2, \cdots, N)$, satisfies the Lie algebra of Eq. (21.167). Here, the matrices are $N \times N$ dimensional, and so they belong to the adjoint representation of the group. The solutions to Eqs. (21.168) and (21.170) give the allowed Lie algebras that can exist. They have been fully classified, and we will return to this issue later in this discussion. Using the structure constants, one can define a metric tensor by

$$g_{ij} = C_{i\ell}^k C_{jk}^\ell. \tag{21.172}$$

For a semi-simple group, the determinant of the metric must be non-vanishing, i.e. $det(g_{ij}) \neq 0$, and in this case we define an inverse metric so that

$$g^{ik} g_{kj} = \delta_j^i. \tag{21.173}$$

In the study of angular momentum, one has an operator J^2 which commutes with all the generators of angular momentum, i.e., $[J^2, J_i] = 0$. One can define an analog of this for the more general case so that $C_2 = g_{ij} X^i X^j$, where C_2 commutes with all the X_i, i.e.,

$$[C_2, X_i] \ = \ 0. \tag{21.174}$$

More generally, one may define

$$C_n = C_{i_1 j_1}^{j_2} C_{i_2 j_2}^{j_3} \cdots C_{i_n j_n}^{j_1} X^{i_1} X^{i_2} \cdots X^{i_n}, \tag{21.175}$$

and it is straightforward to show that C_n commutes with all X_i. These are the Casimir operators.

Now, Eq. (21.167) looks different from the one we are used to in the study of angular momentum. There, we have three generators J_i $(i = 1, 2, 3)$, and these generators satisfy the commutation relation

$$[J_i, J_j] = i\epsilon_{ijk} \, J_k, \quad i, j, k = 1, 2, 3. \tag{21.176}$$

There are two main differences between Eq. (21.176) and (21.167). First, there is an $i \equiv \sqrt{-1}$ in front of the right-hand side of Eq. (21.176) while Eq. (21.167) does not have such a factor. Second, while C_{ij}^k are anti-symmetric in the lower two indices, ϵ_{ijk} is totally anti-symmetric in all the indices and, further, more it is real. It turns out that all the features of Eq. (21.176), i.e., the appearance of i, reality, and the total anti-symmetric nature of the structure constants when written in the form of Eq. (21.176), are generic. Thus, one can rewrite the commutation relations of the Lie algebra in a form where the structure constants are purely anti-symmetric and purely imaginary, and the Lie algebra then takes the form

$$[X_i, X_j] = i f_{ijk} X_k, \tag{21.177}$$

where f_{ijk} are real and totally anti-symmetric. It is the form in Eq. (21.177) which is most commonly used in particle physics. The analogy with the angular momentum relation goes even further. In angular momentum analysis it is often found convenient to introduce the raising and the lowering operator $J_\pm = J_1 \pm i J_2$, which along with J_3 gives us an alternative basis. They satisfy the commutation relations

$$[J_+, J_-] = J_3$$
$$[J_\pm, J_3] = \pm J_\pm. \tag{21.178}$$

It is useful to recast the general Lie algebra in this form. Let us first pick the largest set of generators H_i assumed ℓ in number which commute with each other, i.e., one has

$$[H_i, H_j] = 0, \; i, j = 1, 2, \cdots, \ell. \tag{21.179}$$

This allows us to write the remaining commutation relation in the following form:

$$[H_i, E_\alpha] = r_i(\alpha) E_\alpha$$
$$[E_\alpha, E_{-\alpha}] = r_i(\alpha) H_i$$
$$[E_\alpha, E_{\beta \neq -\alpha}] = N_{\alpha\beta} E_{\alpha+\beta}, \tag{21.180}$$

where E_α appear in conjugate pairs, i.e., one has $E_{\pm 1}, E_{\pm 2}, \cdots, E_{\pm(N-k)/2}$, and $r_i(\alpha)$ (called the roots) are real numbers so that the root $\vec{r}(\alpha)$ can be thought of as a vector in an ℓ-dimensional space. The generators (H_i, E_α) are in the so-called Cartan–Weyl basis. Thus, here H_i play the role of J_3 and $E_{\pm\alpha}$ the role of $J_\pm$ for the angular momentum case. The integer ℓ, which counts the number of commuting operators H_i, defines the rank of the group, and the commuting algebra of the group is called the Cartan subalgebra.

The study of roots plays an important role in the classification of Lie algebras. They satisfy a number of properties. First, $\vec{r}(\alpha) = -\vec{r}(-\alpha)$. Second, if $\vec{r}(\alpha)$ and $\vec{r}(\beta)$ are two roots, then the quantity $2(\vec{r}(\alpha), \vec{r}(\beta))/(\vec{r}(\alpha), \vec{r}(\alpha))$ is an integer. Also, if $\vec{r}(\alpha)$ and $\vec{r}(\beta)$ are two roots, then the quantity $\vec{r}(\beta) - 2\vec{r}(\alpha)(\vec{r}(\alpha), \vec{r}(\beta))/(\vec{r}(\alpha), \vec{r}(\alpha))$ is also a root. One can define an angle $\theta_{\alpha\beta}$ between the roots $\vec{r}(\alpha)$ and $\vec{r}(\beta)$ so that $\cos(\theta_{\alpha\beta}) = (\vec{r}(\alpha), \vec{r}(\beta))/((\vec{r}(\alpha), \vec{r}(\alpha)(\vec{r}(\beta), \vec{r}(\beta))^{1/2}$. The roots encode the structure constants, and, as mentioned already, the structure constants contain complete information regarding the nature and type of a Lie group. The study of the classification of Lie groups is simplified by use of the so-called simple roots, which we now define. The definition of simple roots includes what are called positive roots. A positive root is one whose first non-vanishing element is positive in an arbitrary basis. We will label the set of positive roots Σ^+. If a positive root cannot be decomposed into a sum of two positive roots, then such a root is called a simple root. Using the concept of simple roots, one can establish that a simple group of rank ℓ has ℓ number of simple roots called the Π-system. Further, if $\vec{r}(\alpha)$ and $\vec{r}(\beta)$ are two simple roots, the angle $\theta_{\alpha\beta}$ between them takes the following values:

$$\theta_{\alpha\beta} : 90°, 120°, 135°, 150°. \tag{21.181}$$

This leads to the result that $2\vec{r}(\alpha).\vec{r}(\beta)/\vec{r}(\alpha).\vec{r}(\alpha)$ and $2\vec{r}(\alpha).\vec{r}(\beta)/\vec{r}(\beta).\vec{r}(\beta)$ are integers. The above result has implications for the length ratio of simple roots. Specifically, one finds

$$\left(\frac{(\vec{r}(\alpha), \vec{r}(\alpha))}{(\vec{r}(\beta), \vec{r}(\beta))} \right) = \begin{array}{ll} 1, & \theta_{\alpha\beta} = 120° \\ 2, & \theta_{\alpha\beta} = 135° \\ 3, & \theta_{\alpha\beta} = 150°, \end{array} \tag{21.182}$$

while for the case $\theta_{\alpha\beta} = 90°$ the ratio is indeterminate. The set of simple roots in the Π-system allows one to express any positive root as a linear combination of simple roots with non-negative coefficients, i.e., if $\vec{r}(\alpha) \in \Sigma^+$ is a positive non-simple root, then we can express it in terms of the roots in the Π-system as follows:

$$\vec{r}(\alpha) = \sum_{\vec{r}(\beta) \in \Pi} n_\beta \vec{r}(\beta), \tag{21.183}$$

where n_β are non-negative integers. Using the simple roots, the full set of roots can be created. The total number of roots of a group is referred to as the dimensionality of the group. Further, simple roots can give a full classification of Lie groups. This can be done by using Dynkin diagrams. The procedure for constructing Dynkin diagrams is as follows. A simple root with the largest length is represented by an unfilled, open circle while a root with a smaller length is represented by a solid black circle. Thus, there are ℓ number of circles (black or white) for a group of rank ℓ. Two consecutive roots are connected by a single line if the angle between them is 120°, by a double line if the angle between them in 135°, or by a triple line if the angle between them is 150°. There is no line connecting two circles which have an angle between them of 90°.

The classification of Lie groups leads to two classes of Lie groups: (1) classical groups, and (2) exceptional groups. The classical groups of rank ℓ are classified as A_ℓ, B_ℓ, C_ℓ, and D_ℓ. Their familiar group names and dimensionalities are listed

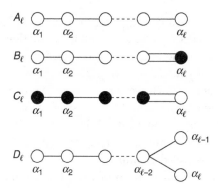

Figure 21.1. Classification of the irreducible root system of groups A_ℓ, B_ℓ, C_ℓ, and D_ℓ as Dynkin diagrams (from Wybourne in [5]).

below:

Cartan label	Group	No. of generators	
A_ℓ	$SU(\ell+1)$	$\ell^2 + 2\ell$	
$B_\ell(\ell \geq 2)$	$SO(2\ell+1)$	$2\ell^2 + \ell$	(21.184)
$C_\ell(\ell \geq 3)$	$Sp(2\ell)$	$2\ell^2 + \ell$	
$D_\ell(\ell \geq 4)$	$SO(2\ell)$	$2\ell^2 - \ell.$	

The Dynkin diagrams associated with the classical groups A_ℓ, B_ℓ, C_ℓ, and D_ℓ are shown in Fig. 21.1. As stated earlier, each circle represents a simple root. A single line connecting two simple roots means that the angle between the simple roots is $120°$ while two lines connecting two simple roots implies that the angle between the roots is $135°$. In any simple Lie algebra, at most two different length sizes for simple roots exist. For the groups A_ℓ and D_ℓ, all simple roots have the same length, and in this case each simple root is shown as an open circle. For groups B_ℓ and C_ℓ we have both open and solid circles. Here, the open circles are the simple roots with the larger length, and the solid circles are the roots with the shorter length.

We elaborate on these a bit more. The A_ℓ is the group $SU_{\ell+1}$, which is a group of unitary transformations in an $(\ell+1)$-dimensional complex space. The group B_ℓ is the group $O_{2\ell+1}$, and is the group of rotations in $(2\ell+1)$-dimensional space. C_ℓ is the symplectic group $Sp(2\ell)$, which is a group of unitary matrices U in a complex 2ℓ-dimensional space which satisfies the constraint $U^T K U = K$, with K an anti-symmetric and non-singular matrix. Finally, D_ℓ is the group $SO(2\ell)$ consisting of orthogonal rotations in a real 2ℓ-dimensional space. We now give some details of how the relations between the rank of a group and the dimensionality of the group arise.

- $A_\ell \sim SU(\ell+1)$, Here, we begin with $(\ell+1)$ number of unit vectors

$$\vec{u}_1, \vec{u}_2, \cdots, \vec{u}_{\ell+1} \tag{21.185}$$

which are all of *equal* length and satisfy the constraint

$$\vec{u}_1 + \vec{u}_2 + \cdots + \vec{u}_{\ell+1} = 0. \tag{21.186}$$

Next, define ℓ number of root vectors as differences such that

$$\vec{r}_\ell = \vec{u}_\ell - \vec{u}_{\ell+1}$$
$$\vec{r}_{(\ell-1)} = \vec{u}_{\ell-1} - \vec{u}_\ell$$
$$\cdot$$
$$\cdot$$
$$\cdot$$
$$\vec{r}_1 = \vec{u}_1 - \vec{u}_2. \tag{21.187}$$

We can represent the roots by open circles connected by single lines, since the angles between adjoining roots is $120°$. The Dynkin root diagram for this is shown in Fig. 21.1, labeled A_ℓ. While the number of simple roots is ℓ, the total number of roots is much larger. The full set of roots in this case is given by

$$\vec{u}_i - \vec{u}_j, \ i, j = 1, 2, \cdots, (\ell+1). \tag{21.188}$$

Taking into account $\pm$ signs for every root, which provides a factor of 2, the count gives $2 \times {}^{\ell+1}C_2 = \ell^2 + \ell$ for the total number of roots. Taking account of the rank of the group, the dimensionality of the group is given by $N = \ell^2 + 2\ell$.

- $B_\ell \sim SO(2\ell + 1), \ell \geq 2$. Here, we use ℓ number of unit vectors, all of equal length, i.e., we have

$$\vec{u}_1, \vec{u}_2, \cdots, \vec{u}_\ell, \tag{21.189}$$

which satisfy the constraint

$$\vec{u}_i \cdot \vec{u}_j = 0, i \neq j. \tag{21.190}$$

Next, define ℓ root vectors as follows:

$$\vec{r}_\ell = \vec{u}_\ell$$
$$\vec{r}_{(\ell-1)} = \vec{u}_{\ell-1} - \vec{u}_\ell$$
$$\cdot$$
$$\cdot$$
$$\cdot$$
$$\vec{r}_1 = \vec{u}_1 - \vec{u}_2. \tag{21.191}$$

In the above, $\vec{r}_\ell$ is taken to be the smallest root. We can represent the roots by open (solid) circles connected by single (double) lines. The Dynkin diagram for this is shown in Fig. 21.1, labeled B_ℓ. The total number of root vectors in this case consists of the following set:

$$\pm \vec{u}_i, \ \pm(\vec{u}_i - \vec{u}_j), \pm(\vec{u}_i + \vec{u}_j), \ i, j = 1, 2, \cdots, \ell. \tag{21.192}$$

Here, the set $\{\pm\vec{u}_i\}$ gives us 2ℓ number of roots, $\pm(\vec{u}_i - \vec{u}_j)$ and $\pm(\vec{u}_i + \vec{u}_j)$ each gives us $\ell(\ell-1)$, which together give $2\ell^2$ number of roots. Including the rank of the group, this gives us the dimensionality of the group as $N = 2\ell^2 + \ell$.

- $C_\ell \sim Sp(2\ell), \ell \geq 3$. As above, we begin with ℓ number of unit vectors all of *equal* length that satisfy the condition of Eq. (21.190). Next, define ℓ root vectors as follows:

$$\vec{r}_\ell = 2\vec{u}_\ell$$
$$\vec{r}_{(\ell-1)} = \vec{u}_{\ell-1} - \vec{u}_{\ell-2}$$
$$\cdot$$
$$\cdot$$
$$\cdot$$
$$\vec{r}_1 = \vec{u}_2 - \vec{u}_1. \tag{21.193}$$

In the above $\vec{r}_\ell$ is the largest root. Diagrammatically the simple root vectors are shown in Fig. 21.1 labeled C_ℓ. The count of the total number of root vectors in this case is similar to that for the B_ℓ case, and is $N = 2\ell^2 + \ell$.

- $D_\ell \sim SO(2\ell)$. Here also we consider ℓ number of unit vectors all of *equal* length that satisfy the condition of Eq. (21.190). In this case, the simple roots are given by

$$\vec{r}_\ell = \vec{u}_{\ell-1} + \vec{u}_\ell$$
$$\vec{r}_{(\ell-1)} = \vec{u}_{\ell-1} - \vec{u}_\ell$$
$$\cdot$$
$$\cdot$$
$$\cdot$$
$$\vec{r}_1 = \vec{u}_1 - \vec{u}_2. \tag{21.194}$$

In the above, $\underline{r}_\ell$ is the largest root. The Dynkin diagram for this is shown in Fig. 21.1 labeled D_ℓ. Here, the full set of root vectors is given by

$$\pm(\vec{u}_i - \vec{u}_j), \pm(\vec{u}_i + \vec{u}_j), \ i,j = 1,2,\cdots,(\ell). \tag{21.195}$$

The total number of root vectors is $2(\ell^2 - \ell)$, and including the rank of the group the dimensionality of the group in this case is given by $N = 2\ell^2 - \ell$.

In addition to the classical groups A_ℓ, B_ℓ, C_ℓ, and D_ℓ there are five more Lie groups called exceptional groups: G_2, F_4, E_6, E_7, and E_8 with ranks $2, 4, 6, 7$, and 8, respectively. Their dimensionalities are listed below:

Group	Dimensionality
G_2	14
F_4	52
E_6	78
E_7	133
E_8	248.

$$(21.196)$$

The Dynkin diagrams for these are shown in Fig. 21.2. From Fig. 21.2 we see that the groups E_6, E_7, and E_8 have simple roots all of the same length. The groups where all the simple roots have the same length are called *simply laced*. Thus, from Dynkin diagrams Figs. 21.1 and 21.2 we see that the groups A_ℓ, D_ℓ, E_6, E_7, and E_8 all have simple roots of equal length. These are sometimes referred to as the ADE system. From Fig. 21.2 we see that the exceptional group G_2 has two simple roots of unequal length with three lines connecting the two simple roots, and thus the angle between them is $150°$, with the root shown as an open circle being the simple root of larger length. Also from Fig. 21.2 we see that F_4 has two simple roots which are open circles and two simple roots which are solid circles, again with the open circles being of the larger length.

As discussed already, a specification of the simple roots allows one to compute the full set of roots for the group. To this end, it is useful to define the so-called Cartan matrix. Before doing so, the following results should be kept in mind, i.e., that if $\vec{\alpha}_1$ and $\vec{\alpha}_2$ are two simple roots, then

$$2\frac{\vec{\alpha}_i \cdot \vec{\alpha}_j}{\vec{\alpha}_i \cdot \vec{\alpha}_i} = -k, \quad i \neq j, \tag{21.197}$$

where k is a positive integer or zero and can take only the value $0, 1, 2$, or 3. The Cartan matrix is then defined as follows. Let $\vec{\alpha}_1, \vec{\alpha}_2, \cdots, \vec{\alpha}_\ell$ be the set of simple roots for a group of rank ℓ. The Cartan matrix in this case is given by

$$A_{ij} = 2\frac{\vec{\alpha}_i \cdot \vec{\alpha}_j}{\vec{\alpha}_i \cdot \vec{\alpha}_i}. \tag{21.198}$$

The diagonal elements of this matrix are all 2 while the off-diagonal elements can only be $0, 1, 2$, and 3. As an illustration, let us construct the Cartan matrix for $A_\ell \sim SU_{l+1}$. Let us label the simple roots as α_i $(i = 1, 2, \cdots, \ell)$. In this case, all the simple roots are of equal length, and the angle between two adjacent simple roots is $120°$. Thus, here we have

$$A_{(i+1)i} = A_{i(i+1)} = 2\frac{\vec{\alpha}_i \cdot \vec{\alpha}_{i+1}}{\vec{\alpha}_i \cdot \vec{\alpha}_i} = 2\cos(120°) = -1, \tag{21.199}$$

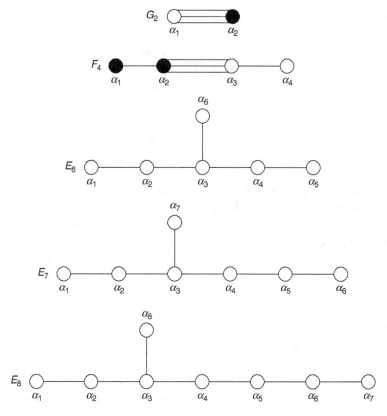

Figure 21.2. Dynkin diagrams for the five exceptional groups: G_2, F_4, E_6, E_7, and E_8 (From Wybourne in [5]).

while all other elements of the matrix vanish. The Cartan matrix in this case takes on the form

$$
(A_{ij}) = \begin{bmatrix}
2 & -1 & 0 & \cdots & 0 & 0 \\
-1 & 2 & -1 & \cdots & 0 & 0 \\
0 & -1 & 2 & \cdots & 0 & 0 \\
\cdot & \cdot & \cdot & \cdots & \cdot & \cdot \\
0 & 0 & 0 & \cdots & 2 & -1 \\
0 & 0 & 0 & \cdots & -1 & 2
\end{bmatrix}.
\tag{21.200}
$$

As another illustration, let us derive the Cartan matrix for the exceptional group F_4. The Dynkin diagram is shown in Fig. 21.2, and the simple roots are labeled left to right as $\vec{\alpha}_1, \vec{\alpha}_2, \vec{\alpha}_3$, and $\vec{\alpha}_4$. $\vec{\alpha}_1$ and $\vec{\alpha}_2$ are of equal length, and they are connected by a single line, which means that the angle between them is $120°$. Thus, $A_{12} = A_{21} = -1$. Similarly, $\vec{\alpha}_3$ and $\vec{\alpha}_4$ are of equal length and are connected by a single line, and so $A_{34} = A_{43} = -1$. The roots $\vec{\alpha}_2$ and $\vec{\alpha}_3$ are connected by two lines, so the angle between them is $135°$. Since $\vec{\alpha}_2$ has a shorter length than $\vec{\alpha}_3$, we have

$$\frac{|\vec{\alpha}_3|}{|\vec{\alpha}_2|} = \sqrt{2}. \tag{21.201}$$

It then follows that

$$A_{23} = 2\frac{\vec{\alpha}_2 . \vec{\alpha}_3}{\vec{\alpha}_2 \cdot \vec{\alpha}_2} = 2\frac{|\vec{\alpha}_3|}{|\vec{\alpha}_2|}\cos(135°) = -2, \tag{21.202}$$

and for the element A_{32} one gets $A_{32} = \frac{1}{2}A_{23} = -1$. Thus, the Cartan matrix in this case is given by

$$(A_{ij}) = \begin{bmatrix} 2 & -1 & 0 & 0 \\ -1 & 2 & -2 & 0 \\ 0 & -1 & 2 & -1 \\ 0 & 0 & -1 & 2 \end{bmatrix}. \tag{21.203}$$

We note that unlike the case of $A_\ell \sim SU(\ell+1)$ the Cartan matrix in this case in not symmetric. The simple roots can be obtained using the Cartan matrix.

We now turn to the question of how one may generate all the roots of a semi-simple group given just the simple roots. A geometrical method is given by the Weyl reflection method. An ℓ-dimensional root system can be thought of as spanning an ℓ-dimensional Euclidean space. For each root $\vec{\alpha}_i$, one can define an $(\ell-1)$-dimensional hyperplane which is orthogonal to $\vec{\alpha}_i$ and passes through the origin. A reflection of another root $\vec{\alpha}_j$ across this hyperplane produces a new root vector $\vec{\alpha}_k$ given by

$$\vec{\alpha}_k = \vec{\alpha}_j - 2\frac{(\vec{\alpha}_j \cdot \vec{\alpha}_i)}{(\vec{\alpha}_i \cdot \vec{\alpha}_i)}\vec{\alpha}_i. \tag{21.204}$$

This procedure when repeated for each of the simple roots and more than once if needed produces all the roots of the semi-simple group. We next discuss representation theory. A representation is a linear transformation that maps elements $a, b, c, \cdots, e$ into matrices $D(a), D(b), D(c), \cdots, D(e)$ ($D(e) = \mathbb{1}$ where $\mathbb{1}$ is a unit matrix) on a vector space V such that the relation $ab = c$ has the correspondence $D(a)D(b) = D(c)$ and $D(a^{-1}) = (D(a))^{-1}$. A representation is fully reducible if every map can be put in block diagonal form. Irreducible representations are those for which this is not possible.

Our focus is on representations of the Lie groups. So, to get started we look back at the case of the angular momentum again. Here, H is J_3. Let us consider an eigenstate of J_3, which we label as $|j\rangle$ so that $j_3|j\rangle = j|j\rangle$. We call $|j\rangle$ the highest-weight state. Next, by the action of J_- on $|j\rangle$ we can create a state $J_-|j\rangle$ with a lower weight. We continue this process until we reach the lowest-weight state $|-j>$. In this case, the dimensionality is $d(j) = (2j+1)$. A similar idea is useful for obtaining the dimensionality of irreducible representations for the more general case where instead of just one J_3 (rank 1) we are dealing with a rank-ℓ group with ℓ number of commuting operators. Here, we introduce a weight vector $\vec{m}$ and define the weight vector $\vec{m}$ to be higher than the weight vector $\vec{m}'$ provided the vector $(\vec{m} - \vec{m}')$ has a first non-vanishing component which is positive. One then defines a highest-weight state $\underline{\Lambda}$ to be the one which is higher than any other weight state.

Some results which are valid for the more general case of representation of groups of rank ℓ are now provided. Given a weight vector $\vec{m}$ and a simple root $\underline{r}(\alpha)$, the quantity $\vec{m} \cdot \vec{r}(\alpha) / \vec{r}(\alpha) \cdot \vec{r}(\alpha)$ is either an integer or a half integer. Second, as in the case of angular momentum, an irreducible representation is described uniquely by the highest weight $\vec{\Lambda}$. Defining $j_\alpha = \underline{\Lambda} \cdot \vec{r}(\alpha) / \vec{r}(\alpha) . \vec{r}(\alpha)$, the highest-weight $\vec{\Lambda}$ is then given by $(j_1, j_2, \cdots, j_\ell)$, where $j_1, j_2, \cdots$ can be considered as analogs of j for the angular momentum case. To correspond to the prevalent notation, we will label an irreducible representation as $\underline{\Lambda}(a_1, a_2, \cdots, a_\ell)$, where $a_1 = 2j_1$, etc. We are interested in the dimensionality $d(\vec{\Lambda})$. This can be derived by the use of the Weyl formula, which is given by

$$d(\Lambda) = \prod_{\alpha \in \Sigma^+} \frac{\sum_i q_\alpha^i n_{\alpha_i}(1 + a_i)}{\sum_i q_\alpha^i n_{\alpha_i}}, \tag{21.205}$$

where Σ^+ contains all the positive roots, and q_α^i is defined so that a positive root is decomposed in terms of simple roots as:

$$\vec{\alpha}(\vec{\alpha} \in \Sigma^+) = \sum_{\alpha_i \in \Pi} q_\alpha^i \vec{\alpha}_i, \tag{21.206}$$

and n_{α_i} is the ratio of the length square of a simple root in units of the length square of the smallest length simple root, i.e.,

$$n_{\alpha_i} = \frac{\langle \vec{\alpha}_i, \vec{\alpha}_i \rangle}{\langle \vec{\alpha}_s, \vec{\alpha}_s \rangle}, \tag{21.207}$$

where $\vec{\alpha}_i$ is a simple root and $\vec{\alpha}_s$ is the shortest-length simple root. Now, from the Dynkin diagrams, we infer the possible values of n_{α_i}, and we list them below. As mentioned after Eq. (21.196), the groups where in the Dynkin diagrams simple roots are connected by a single line are simply laced, and those with two lines are doubly laced, etc. From Figures 21.1 and 21.2 we find

$$A_\ell, D_\ell, E_6, E_7, E_8 \ [ADE] : \qquad n_{\alpha_i} = 1$$
$$B_\ell, C_\ell : \qquad n_{\alpha_i} = 1, 2. \tag{21.208}$$

The entire problem of the derivation of the dimensionality of an irreducible representation then reduces to a determination of the positive roots and expressing the positive roots in terms of the simple roots to determine the positive integers q_α^i for any positive root $\vec{\alpha}_i \in \Sigma^+$.

We will now apply Eq. (21.205) to some illustrative examples. Consider the case $A_2 \sim SU(3)$. Its positive roots consist of $\vec{\alpha}_1, \vec{\alpha}_2, \vec{\alpha}_3 = \vec{\alpha}_1 + \vec{\alpha}_2$ where $\vec{\alpha}_1, \vec{\alpha}_2$ are simple roots. All the roots in this case are of equal length, and so $\langle \vec{\alpha}_i, \vec{\alpha}_i \rangle$ factors out of the numerator and the denominator. Next, using Eq. (21.206) we have $q^i_{\alpha_1} = (1,0), q^i_{\alpha_2} = (1,0),$ and $q^i_{\alpha_3} = (1,1),$ which immediately gives the dimensionality

$$d_{A_2}(a_1, a_2) = (a_1 + 1)(a_2 + 1)\left(1 + \frac{a_1 + a_2}{2}\right). \qquad (21.209)$$

We now give an illustration of the Weyl formula for B_2. Here, the positive roots are $\vec{\alpha}_1, \vec{\alpha}_2, \vec{\alpha}_3 = \vec{\alpha}_1 + \vec{\alpha}_2,$ and $\vec{\alpha}_4 = \vec{\alpha}_1 + 2\vec{\alpha}_2,$ where $\vec{\alpha}_1$ and $\vec{\alpha}_2$ are the simple roots and the root $\vec{\alpha}_1$ is the root with the largest length so that $n_{\alpha_1} = \langle \vec{\alpha}_1, \vec{\alpha}_1 \rangle / \langle \vec{\alpha}_2, \vec{\alpha}_2 \rangle = 2$ while $n_{\alpha_2} = 1.$ The quantities that appear in the Weyl formula are $\bar{q}^i_\alpha \equiv q^i_\alpha \langle \alpha_i, \alpha_i \rangle$ $(i = 1, 2).$ For $B_2,$ one has q^i_α $(\alpha = 1\text{--}4)$ given by $q^i_1 = (1,0), q^i_2 = (0,1), q^i_3 = (1,1),$ and $q^i_4 = (1,2),$ which give $\bar{q}^i_1 = (2,0), \bar{q}^i_2 = (0,1), \bar{q}^i_3 = (2,1), \bar{q}^i_4 = (2,2),$ respectively, and $\bar{q}_\alpha \equiv \sum_i \bar{q}^i_\alpha$ given by $\bar{q}_1 = 2, \bar{q}_2 = 1, \bar{q}_3 = 3,$ and $\bar{q}_4 = 4.$ The contributions from the four positive roots give the following product:

$$\left(\frac{2(1 + a_1)}{2}\right)_1 (1 + a_2)_2 \left(\frac{(1 + a_1) + (1 + a_2)}{3}\right)_3 \left(\frac{2(1 + a_1) + 2 \times 2(1 + a_2)}{4}\right)_4,$$

where the subscript indicates the positive root which gives the term. Simplification gives the result for the irreducible representation with Dynkin indices (a_1, a_2) of the group B_2:

$$d_{B_2}(a_1, a_2) = (1 + a_1)(1 + a_2)\left(1 + \frac{a_1 + a_2}{2}\right)\left(1 + \frac{a_1 + 2a_2}{3}\right). \qquad (21.210)$$

We now give an illustration of the Weyl formula for the exceptional group G_2. Here, the positive roots are given by $\vec{\alpha}_1, \vec{\alpha}_2, \vec{\alpha}_3 = \vec{\alpha}_1 + \vec{\alpha}_2, \vec{\alpha}_4 = \vec{\alpha}_1 + 2\vec{\alpha}_2, \vec{\alpha}_5 = \vec{\alpha}_1 + 3\vec{\alpha}_2,$ and $\vec{\alpha}_6 = 2\vec{\alpha}_1 + 3\vec{\alpha}_2,$ where $\vec{\alpha}_1$ and $\vec{\alpha}_2$ are the simple roots and the root $\vec{\alpha}_1$ is longer in length that the root $\vec{\alpha}_2$ such that $\langle \vec{\alpha}_1, \vec{\alpha}_1 \rangle / \langle \vec{\alpha}_2, \vec{\alpha}_2 \rangle = 3.$ Since in the derivation of the dimensionality only the length squares appear, we can set $\langle \vec{\alpha}_2, \vec{\alpha}_2 \rangle = 1$ and $\langle \vec{\alpha}_1, \vec{\alpha}_1 \rangle = 3.$ Now, the elements of q^i_α $(\alpha = 1 - 6)$ are given by $q^i_1 = (1,0), q^i_2 = (0,1), q^i_3 = (1,1), q^i_4 = (1,2), q^i_5 = (1,3),$ and $q^i_6 = (2,3),$ which give $\bar{q}^i_1 = (3,0), \bar{q}^i_2 = (0,1), \bar{q}^i_3 = (3,1), \bar{q}^i_4 = (3,2), \bar{q}^i_5 = (3,3),$ and $\bar{q}^i_6 = (6,3),$ respectively, where, as noted before, $\bar{q}^i_\alpha = q^i_\alpha \langle \alpha_i, \alpha_i \rangle$ $(i = 1, 2).$ The quantities $\bar{q}_\alpha = \sum_i \bar{q}^i_\alpha$ are then $\bar{q}_1 = 3, \bar{q}_2 = 1, \bar{q}_3 = 4, \bar{q}_4 = 5, \bar{q}_5 = 6,$ and $\bar{q}_6 = 9.$ The contributions from these six positive roots give the following product:

$$\left(\frac{3(1 + a_1)}{3}\right)_1 (1 + a_2)_2 \left(\frac{3(1 + a_1) + (1 + a_2)}{4}\right)_3 \left(\frac{3(1 + a_1) + 2(1 + a_2)}{5}\right)_4$$
$$\times \left(\frac{3(1 + a_1) + 3(1 + a_2)}{6}\right)_5 \left(\frac{2 \times 3(1 + a_1) + 3(1 + a_2)}{9}\right)_6, \qquad (21.211)$$

where the subscript indicates the positive root which gives the term. Simplification gives the result for the irreducible representation with Dynkin indices

(a_1, a_2) of the group G_2:

$$d_{G_2}(a_1, a_2) = \frac{1}{5!}(1+a_1)(1+a_2)(a_1+a_2+2)(2a_1+a_2+3)(3a_1+a_2+4)(3a_1+2a_2+5).$$
(21.212)

We note that from the Cartan root diagrams that one can deduce local isomorphism between Lie algebras. Thus, for example,

$$B_2 \sim C_2$$
$$A_3 \sim D_3$$
$$D_2 \sim A_1 \times A_1.$$
(21.213)

At the beginning of this section we discussed simple and semi-simple groups. We expand further on them here. As discussed earlier, simple groups are those which do not have any invariant subgroup and thus cannot be written as products of groups. Thus, the Georgi–Glashow grand unification model is based on the simple group $SU(5)$. More generally, the groups $A_n(SU_{n+1}), B_n(SO_{2n+n}), C_n(Sp_{2n})$, and $D_n(SO_{2n})$ under the restrictions on n given by Eq. (21.184) are all simple groups. The group $U(n)$ can be written as $SU(n) \otimes U(1)$, and is not simple. Also as discussed at the beginning of this section, a semi-simple group is one which can be written as a product of groups, but this product should not contain a $U(1)$ factor. Thus, the Pati–Salam group $SU(4)_C \otimes SU(2)_L \otimes SU(2)_R$ is semi-simple.

21.5.1 Problems

1. From the $U(2)$ generators 1 and $\sigma_i/2, (i = 1, 2, 3)$, where σ_i are the Pauli matrices and the $U(3)$ generators 1 and $\lambda_\alpha/2, (\alpha = 1-8)$, where λ_α are the Gell–Mann matrices, one can compose the generators of a $U(6)$ algebra. Show that this set consists of $1, \sigma_i/2, \lambda_\alpha/2$, and $\sigma_i\lambda_\alpha/2$.

2. The N^2 generators of the unitary group $U(N)$ can be taken to be U_j^i $(i, j = 1, 2, \cdots, N)$ with elements given by

$$(U_j^i)_{k\ell} = \delta_\ell^i \delta_j^k.$$
(21.214)

Show that they satisfy the algebra

$$[U_j^i, U_\ell^k] = \delta_\ell^i U_j^k - \delta_j^k U_\ell^i.$$
(21.215)

For $SU(N)$, define traceless matrices T_j^i so that

$$T_j^i = U_j^i - \frac{1}{N}\delta_j^i \mathbb{1}.$$
(21.216)

Show that T_j^i satisfy the same algebra as the U_j^i. Note that T_j^i are not Hermitian. Using the set T_j^i obtain a new set which is a linear combination of T_j^i, which are Hermitian.

3. For the special orthogonal group $SO(N)$ one may take as generators the following set of $N(N-1)/2$ objects:

$$(M_{\alpha\beta})_{\mu\nu} = -i(\delta_{\mu\alpha}\delta_{\nu\beta} - \delta_{\nu\alpha}\delta_{\mu\beta}), \qquad (21.217)$$

where $1 \leq \mu, \nu, \alpha, \beta \leq N$. Show that for N even the $N/2$ number of generators $M_{12}, M_{34}, M_{56}, \cdots, M_{N-1,N}$ and for N odd the $(N-1)/2$ number of generators $M_{12}, M_{34}, M_{56}, \cdots, M_{N-2,N-1}$ are commuting and form a Cartan subalgebra.

Further Reading

R. Cahn, *Semi-Simple Lie Algebras and their Representatoins*, Benjamin-Cummings (2006).

H. Georgi, *Lie Groups in Particle Physics*, Benjamin-Cummings (1982).

A. Salam, The formalism of lie groups (unpublished).

R. Slansky, *Phys. Rep.* **79**, 1 (1981).

B. G. Wybourne, *Classical Groups for Physicists*, John Wiley (1974).

21.6 Couplings of $N = 1$ Supergravity with Matter and Gauge Fields

As discussed in Chapter 12, the construction of supergravity grand unification requires couplings of an arbitrary number of chiral fields with vector fields and then coupling them to supergravity. Some of the main results of this coupling were given in Chapter 12. Here, we give further details of the construction. As already mentioned in Chapter 12, a very general discussion of how to couple a single chiral field to gauge fields and to supergravity has been discussed by Cremmer *et al.* [1]. The same task for the coupling an arbitrary number of chiral fields to gauge fields and to supergravity has also been accomplished [2–4]. Further details of this construction are discussed by Cremmer *et al.* [3], Nath *et al.* [4], and in Bagger and Witten [5].

The full Lagrangian of the theory is given by

$$\mathcal{L} = \mathcal{L}_V + \mathcal{L}_F + \mathcal{L}_D, \qquad (21.218)$$

where $\mathcal{L}_V$ is the Lagrangian for the vector multiplet given by Eq. (11.45), $\mathcal{L}_F$ is given by Eq. (12.2), and $\mathcal{L}_D$ is given by Eq. (12.4). We note that $\mathcal{L}_D$ contains the supergravity Lagrangian, and so we do not need to include it in Eq. (21.218) as a separate term. In the analysis we assume the supergravity Lagrangian with the minimal set of auxiliary fields, i.e., we assume $\mathcal{L}_{SG}$ as given by Eq. (11.20).

Our main aim in this appendix is to couple an arbitrary number of chiral fields with the gauge multiplet and then couple them to supergravity. We will work in the Wess–Zumino gauge, where the vector multiplet has the components

$V = (V_\mu, \lambda, D)$. In this gauge we have

$$e^{g_\alpha V} = (1,\ 0,\ 0,\ g_\alpha V_\mu,\ g_\alpha \lambda,\ g_\alpha D - \frac{1}{2} g_\alpha^2 V_\mu V^\mu), \qquad (21.219)$$

and the gauge-invariant interaction of the chiral fields is given by

$$\frac{1}{4} \left(\Gamma_{1a} (e^{g_\alpha V})^a{}_b \Gamma_2^b + 1 \leftrightarrow 2 \right). \qquad (21.220)$$

This interaction can easily be generalized to a set of arbitrary number of chiral fields by replacing Γ_1 by Γ_I, which is a product of $\Gamma_{a_1} \cdots \Gamma_{a_m}$, and Γ_2 is replaced by Γ_{II}, which is a product of $\Gamma^{b_1} \cdots \Gamma^{b_m}$, where the product of the Γ is obtained by using Eq. (11.33). The general interaction can be written in a compact form as

$$\frac{1}{2} \left(\Gamma_I \left(A e^{(g_\alpha V)} \right) \Gamma_{II} + \text{h.c.} \right), \qquad (21.221)$$

where A has indices $A^{a_1 \cdots a_m}{}_{b_1 \cdots b_n}$.

The components of the vector multiplet arising from the product of Eq. (21.221) are [4]

$$C = \frac{1}{2} \phi$$

$$\xi = i \left(\phi_{,a} \chi^a - \phi^a_{,} \chi_a \right)$$

$$v = - \left(\phi_a h^a - \phi_{,ab} \bar{\chi}^a \chi^b \right)$$

$$V_\mu = \frac{i}{2} \left(\phi_{,a} \mathcal{D}_\mu z^a - \phi^a_{,} \mathcal{D}_\mu z_a - 2 \phi^a_{,b} \overline{\chi^a} \gamma^\mu \chi^b \right)$$

$$\lambda = -i \phi^a_{,b} h_a \chi^b + i \phi^{ab}_{,c} \left(\overline{\chi^a} \chi_b \right) \chi^c + i \phi^a_{,b} h^b \chi_a$$
$$\qquad - i \phi_{,ab}{}^c \left(\overline{\chi_a} \chi^b \right) \chi_c - i \phi^a_{,b} \gamma^\mu \hat{\mathcal{D}}_\mu z_a \chi^b + i \phi^a_{,ab} \gamma^\mu \hat{\mathcal{D}}_\mu z^b \chi_a$$
$$\qquad + \frac{g_\alpha}{2} \lambda^\alpha \phi_{,a} (T^\alpha)^a{}_b z^b$$

$$D = \phi^a_{,b} h_a h^b - \phi^{ab}_{,c} \overline{\chi^a} \chi_b h^c - \phi_{,ab}{}^c \overline{\chi_a} \chi^b h_c$$
$$\qquad + \phi_{,ab}{}^{cd} \overline{\chi_a} \chi^b \cdot \overline{\chi^c} \chi_d - \phi^a_{,b} \hat{\mathcal{D}}_\mu z_a \hat{\mathcal{D}}^\mu z^b$$
$$\qquad - \phi^a_{,b} \overline{\chi^a} \gamma^\mu \overleftrightarrow{\hat{\mathcal{D}}}_\mu \chi^b - \phi^a_{,bc} \overline{\chi^a} \gamma^\mu \chi^b \hat{\mathcal{D}}_\mu z^c$$
$$\qquad - \phi_{,a}{}^{bc} \overline{\chi_a} \gamma^\mu \chi_b \hat{\mathcal{D}}_\mu z_c + \frac{g_\alpha}{2} D^\alpha \phi_{,a} (T^\alpha)^a{}_b z^b$$
$$\qquad - i g_\alpha \phi^a_{,b} \left(\bar{\lambda}^\alpha \chi^b (T^\alpha z)_a - \overline{\chi^a} \lambda^\alpha (T^\alpha z)^b \right), \qquad (21.222)$$

where $\phi_{,a} = \partial \phi / \partial z^a$ and $\phi^a_{,} = \partial \phi / \partial z_a$. Further, the function ϕ that appears in Eq. (21.222) is defined in Eq. (12.7). In Eq. (21.222) the script derivatives are the gauge-covariant form of the Latin ones. Thus, for example,

$$\hat{\mathcal{D}}_\mu z^a = \hat{D}_\mu z^a - \frac{i}{2} g_\alpha V_\mu^\alpha (T^\alpha)^a{}_b z^b. \qquad (21.223)$$

Next we use Eqs. (12.4) and (21.222) to construct the D-term contribution to the Lagrangian which is locally supersymmetric and is also gauge invariant and with no more than two derivatives. This Lagrangian is given by [4]

$$
\begin{aligned}
e^{-1}\mathcal{L}_D = {}& \phi^a_{,\,b}h_a h^b - \phi^{ab}_{,\,c}\overline{\chi^a}\chi_b h^c - \phi_{,ab}\overline{\chi_a}\chi^b h_c \\
& + \phi^c_{,ab}\overline{\chi}_a\chi^b\overline{\chi^c}\chi_d - \phi^a_{,\,b}\hat{\mathcal{D}}_\mu z_a \hat{\mathcal{D}}^\mu z^b \\
& - \phi^a_{,\,b}\overline{\chi^a}\gamma^\mu\overleftrightarrow{\hat{\mathcal{D}}}_\mu\chi^b - \overline{\chi}^a\gamma^\mu\chi^b\left(\phi^a_{,\,bc}\hat{\mathcal{D}}_\mu z^c - \phi^{ac}_{,\,b}\hat{\mathcal{D}}_\mu z_c\right) \\
& - g_\alpha\phi^a_{,\,b}\left(\overline{\lambda^\alpha}\chi^b(T^\alpha z)_a - \overline{\chi^a}\lambda^\alpha(T^\alpha z)^b\right) \\
& + \frac{1}{2}g_\alpha D^\alpha\phi_{,a}(T^\alpha z)^a + \frac{1}{2}\kappa\phi^a_{,\,b}\overline{\psi}_\mu\gamma_\mu(h_a\chi^b + h^b\chi_a) \\
& - \frac{1}{2}\kappa\left(\overline{\psi}_\mu\gamma^\mu\chi^c\overline{\chi^a}\chi_b\phi^{ab}_{,\,c} - \overline{\chi^c}\gamma^\mu\psi_\mu\overline{\chi}_a\chi^b\phi^c_{,ab}\right) \\
& - \frac{\kappa}{2}\phi^a_{,\,b}\left(\overline{\psi}_\mu\gamma^\mu\gamma^\nu\chi^b\hat{\mathcal{D}}_\nu z_a - \overline{\chi}^a\gamma^\nu\gamma^\mu\psi_\mu\hat{\mathcal{D}}_\nu z^b\right) \\
& - i\kappa\frac{g_\alpha}{2}\overline{\psi}_\mu\gamma_5\gamma^\mu\lambda^\alpha\phi_{,a}(T^\alpha z)^a \\
& + \frac{\kappa}{3}\left[u^*(\phi_{,a}h^a - \phi_{,ab}\overline{\chi}_a\chi^b) + u(\phi^{,a}_{,}h_a - \phi^{ab}\overline{\chi^a}\chi_b)\right] \\
& + \left(i\frac{\kappa}{3}A^\mu + e^{-1}\frac{\kappa^2}{8}\epsilon^{\mu\nu\rho\sigma}\overline{\psi}_\nu\gamma_\rho\psi_\sigma\right)\left(\phi_{,a}\hat{\mathcal{D}}_\mu z^a - \phi^{,a}\hat{\mathcal{D}}_\mu z_a - 2\phi^a_{,\,b}\overline{\chi^a}\gamma_\mu\chi^b\right) \\
& + \frac{4}{3}\kappa\left(D_\mu\overline{\psi}_\nu s^{\mu\nu}\chi^a\phi_{,a} - \phi^{,a}\overline{\chi^a}s^{\mu\nu}D_\mu\psi_\nu\right) \\
& - \frac{\kappa^2}{3}\phi\left(-\frac{R}{2\kappa^2} - \frac{e^{-1}}{2}\epsilon^{\mu\nu\rho\sigma}\overline{\psi}_\mu\gamma_5\gamma_\nu D_\rho\psi_\sigma - \frac{1}{3}|u|^2 + \frac{1}{3}A_\mu A^\mu\right) \\
& - \frac{\kappa^3}{8}\epsilon^{\mu\nu\rho\sigma}e^{-1}\overline{\psi}_\mu\gamma_\nu\psi_\rho\left(\overline{\psi}_\sigma\chi^a\phi_{,a} - \overline{\chi^a}\psi_\sigma\phi^{,a}\right),
\end{aligned}
\tag{21.224}
$$

where $s^{\mu\nu}$ are as defined after Eq. (11.11). All the gauge covariant derivatives in Eq. (21.224) contain torsion parts. In addition to the above, we can add to the Lagrangian self-interactions of the chiral multiplets. Such self-interactions are described by the superpotential term $W(\Gamma)$, where the general form of the superpotential term is given by Eq. (12.6). The components of $W(\Gamma)$ are given by

$$
W(\Gamma^a) = \left(W(z^a),\ W_{,a}\chi^a,\ W_{,a}h^a - 2W_{,ab}\overline{\chi_a}\chi^b\right).
\tag{21.225}
$$

On using Eq. (12.2) and Eq. (21.225), the interaction which is locally supersymmetric which arises from the superpotential term is given by

$$
e^{-1}\mathcal{L}_F = \frac{1}{2}\left(W_{,a}h^a - 2W_{,ab}\overline{\chi_a}\chi^b + \kappa u W + \kappa W_{,a}\overline{\psi}_\mu\gamma^\mu\chi^a + \kappa^2 W\overline{\psi}_\mu s^{\mu\nu}\psi_{\nu R} + \text{h.c.}\right).
\tag{21.226}
$$

To proceed further, we note that Eq. (21.218), contains auxiliary fields, which are u, A_μ, h^a, and D^α. In order to obtain the effective Lagrangian containing only the physical fields, we must eliminate these auxiliary fields by using field equations. Some of the auxiliary field contributions are hidden inside covariant derivatives (e.g., see Eqs. (11.46), (11.48), and (11.28)). To extract their contributions, we expand the covariant derivatives retaining their torsionless parts, and we have [4]

$$
\begin{aligned}
e^{-1}\mathcal{L} =\ & \frac{\phi}{6}\left(R(\omega(e)) + e^{-1}\kappa^2\epsilon^{\mu\nu\rho\sigma}\overline{\psi_\mu}\gamma_5\gamma_\nu D_\rho(\omega(e))\psi_\sigma + \frac{2}{3}\kappa^2|u|^2 - \frac{2}{3}\kappa^2 A_\mu A^\mu\right) \\
& - \phi^a_{,\,b}\left(\mathcal{D}_\mu z_a \mathcal{D}_\mu z^b + \bar{\chi}^a\,\overleftrightarrow{\slashed{\mathcal{D}}(\omega(e))}\,\chi^b - h_a h^b\right) \\
& + \kappa\phi^a_{,\,b}\left(\mathcal{D}_\nu z^b\overline{\chi^a}\gamma^\mu\gamma^\nu\psi_\mu + \bar\psi_\mu\gamma^\nu\gamma^\mu\chi^b\mathcal{D}_\nu z_a\right) \\
& - \overline{\chi^a}\gamma^\mu\chi^b\left(\phi^a_{,\,bc}\mathcal{D}_\mu z^c - \phi^{\,ac}_{,\,b}\mathcal{D}_\mu z_c\right) \\
& - \phi^{ab}_{,\,c}\overline{\chi^a}\chi_b h^c - \phi^c_{,ab}\bar\chi_a\chi^b h_c + \frac{g\alpha}{2}D^\alpha\phi_{,a}(T^\alpha z)^a \\
& - ig_\alpha\phi^a_{,\,b}\left((T^\alpha z)_a\overline{\lambda^\alpha}\chi^b - \overline{\chi^a}\lambda^\alpha(T^\alpha z)^b\right) \\
& - i\kappa\frac{g_\alpha}{4}\bar\psi_\mu\gamma_5\gamma^\mu\lambda^\alpha\phi_{,a}(T^\alpha z)^a \\
& + \frac{\kappa}{3}\left(u^*(\phi_{,a}h^a - \phi_{,ab}\bar\chi_a\chi^b + \frac{3}{2}W^*) + u(\phi_{,}{}^a h_a - \phi_{,}{}^{ab}\overline{\chi^a}\chi_b + \frac{3}{2}W)\right) \\
& + i\frac{\kappa}{3}A^\mu\left(\phi_{,a}\mathcal{D}_\mu z^a - \phi^a_{,}\mathcal{D}_\mu z_a - \kappa\phi_{,a}\bar\psi_\mu\chi^a + \kappa\phi^a_{,}\overline{\chi^a}\psi_\mu\right. \\
& \qquad\qquad \left. + \phi^a_{,\,b}\overline{\chi^a}\gamma_\mu\chi^b - \frac{3}{4}\overline{\lambda^\alpha}\gamma_\mu\gamma_5\lambda^\alpha\right) \\
& - e^{-1}\frac{\kappa^2}{8}\epsilon^{\mu\nu\rho\sigma}\bar\psi_\mu\gamma_\nu\psi_\rho\left(\phi_{,a}\mathcal{D}_\sigma z^a - \phi^a_{,}\mathcal{D}_\sigma z_a\right) \\
& + \frac{4}{3}\kappa\left(\overline{D_\mu(\omega(e))\psi_\nu}s^{\mu\nu}\chi^a\phi_{,a} - \phi^a_{,}\overline{\chi^a}s^{\mu\nu}D_\mu(\omega(e))\psi_\nu\right) \\
& + \frac{1}{2}\left(W_{,a}h^a - W_{,ab}\overline{\chi_a}\chi^b + \kappa W_{,a}\bar\psi_\mu\gamma^\mu\chi^a + \kappa^2 W\bar\psi_\mu s^{\mu\nu}\psi_{\nu R}\right. \\
& \qquad\quad \left. W^{*a}_{,}h_a - W^{*ab}_{,}\bar\chi_a\chi_b + \kappa W^{*a}_{,}\overline{\chi^a}\gamma^\mu\psi_\mu + \kappa^2 W^*\bar\psi_\mu s^{\mu\nu}\psi_{\nu L}\right) \\
& + \frac{1}{6}\kappa^2 e^{-1}\phi\partial_\mu(e\bar\psi_\nu\gamma^\nu\psi^\mu) \\
& - \frac{1}{4}F_{\mu\nu}{}^\alpha F^{\mu\nu\alpha} - \frac{1}{2}\overline{\lambda^\alpha}\gamma^\mu D_\mu(w(e))\lambda^\alpha + \frac{1}{2}D^\alpha D^\alpha - \frac{1}{2}\kappa\bar\psi_\mu s^{\kappa\lambda}\gamma^\mu\lambda^\alpha F_{\kappa\lambda}{}^\alpha \\
& + e^{-1}\mathcal{L}_{4\text{-Fermi}},
\end{aligned}
\tag{21.227}
$$

where $\mathcal{L}_{4\text{-Fermi}}$ contains terms with four fermions, which include the gravitino, gaugino, and chiral fermion fields, and is given by [4]

$$
\begin{aligned}
e^{-1}\mathcal{L}_{4\text{-Fermi}} =\ & \phi^{\,cd}_{,ab}(\bar\chi_a\chi^b)(\bar\chi^c\chi_d) - \kappa^2\phi^a_{,\,b}(\overline{\chi^a}\psi_\mu)(\bar\psi_\mu\chi^b) \\
& - \frac{1}{8}e^{-1}\kappa^2\epsilon^{\mu\nu\rho\sigma}\bar\psi_\mu\gamma_\nu\psi_\rho\left(\phi^a_{,\,b}\overline{\chi^a}\gamma_\sigma\chi^b + \frac{1}{2}\overline{\lambda^\alpha}\gamma_5\gamma_\sigma\lambda^\alpha\right)
\end{aligned}
$$

$$+ \frac{1}{4}\kappa^2 \overline{\lambda^\alpha}\gamma^\mu s^{\nu\rho}\psi_\mu \bar\psi_\nu \gamma_\rho \lambda^\alpha$$

$$+ \frac{1}{96}\kappa^4 \phi \left((\bar\psi_\mu \gamma_\nu \psi_\sigma + 2\bar\psi_\nu \gamma_\mu \psi_\rho)\overline{\psi^\mu}\gamma^\nu \psi^\rho - 4(\bar\psi_\mu \gamma^\nu \psi_\nu)^2 \right)$$

$$- \frac{1}{6}\kappa^3 \left(\overline{\psi^\nu}\gamma^\rho \psi_\rho - \frac{1}{4}e^{-1}\epsilon^{\mu\nu\rho\sigma}\bar\psi_\mu \gamma_\rho \psi_\sigma \right)(\phi^a \overline{\chi^a}\psi_\nu)$$

$$- \frac{1}{6}\kappa^3 \left(\overline{\psi^\nu}\gamma^\rho \psi_\rho + \frac{1}{4}e^{-1}\epsilon^{\mu\nu\rho\sigma}\bar\psi_\mu \gamma_\rho \psi_\sigma \right)(\bar\psi_\nu \chi^a \phi_{,a}). \qquad (21.228)$$

The Fierz identities of Eqs. (21.239) and (21.240) have been used in obtaining Eqs. (21.227) and (21.228). The Lagrangian of Eq. (21.227) contains the auxiliary fields u, A_μ, h^a, and D^α. In the analysis of extracting the auxiliary fields it is found convenient to use the combination $(u - \frac{1}{2}\kappa K_{,a}h^a)$ as an independent variable, where K is defined by Eq. (12.8).

It is useful to segregate parts of the Lagrangian containing the auxiliary fields. Using the set of auxiliary fields $u - \frac{1}{2}\kappa K_{,a}h^a, h^a, A_\mu$, and D^α, terms in the Lagrangian of Eq. (21.227) with auxiliary parts are given by Eq. (21.235). On varying $\mathcal{L}^{aux}$ given by Eq. (21.235) with respect to the auxiliary fields and then eliminating them, one finds that Eq. (21.235) can be written in the form given by Eq. (21.237).

However, to get the proper interpretation of the Lagrangian we need a Weyl rescaling. This is so because in Eq. (21.227) one finds that instead of the scalar curvature R one has $(\phi/6)R$ appearing. Thus, a Weyl rescaling is needed to separate out the graviton from ϕ. A rescaling that accomplishes this is the following:

$$e_\mu{}^r \to e^{\kappa^2 K/12}e_\mu{}^r. \qquad (21.229)$$

Now, under the above Weyl rescaling $\omega_\mu{}^{rs}(e)$ and $R(\omega(e))$ transform as shown in Eq. (21.238). The Lagrangian can be written in a more compact form if we combine W and K in a specific combination defined by

$$\mathcal{G} \equiv -\frac{\kappa^2}{2}K - \log\left(\frac{\kappa^6}{4}|W|^2\right). \qquad (21.230)$$

The effective potential can now be written in terms of $\mathcal{G}$ alone. Thus, we have

$$e^{-1}V = -\frac{1}{\kappa^4}e^{-\mathcal{G}}\left((\mathcal{G}^{-1})^a{}_b \mathcal{G}_{,a}\mathcal{G}_,{}^b + 3\right) + \frac{g_\alpha^2}{8\kappa^4}\left(\mathcal{G}_{,a}(T^\alpha z)^a\right)^2. \qquad (21.231)$$

In deducing the above we have used the gauge invariance of W, i.e.,

$$\delta W(z) = \epsilon^\alpha \left(W_{,a}(T^\alpha z)^a\right) = 0. \qquad (21.232)$$

Next, we discuss the fermionic part of the Lagrangian, which turns out to be significantly more complicated than the bosonic part of the Lagrangian. We begin with the gravitino field. Here, we find that the gravitino field is mixed with spin 1/2 fields, and this mixing is of the form

$$\phi_{,}{}^{a}\overline{\chi^{a}}s^{\mu\nu}D_{\mu}\psi_{\nu} + \text{h.c.} \tag{21.233}$$

Because of this mixing, we need to redefine the gravitino field. Additionally, a rescaling of the fermionic fields $\psi_{\mu}, \lambda^{\alpha}$, and χ^{a} is needed in order that we obtain the correct normalizations for these fields. The necessary transformations to accomplish this are given by [4]

$$\begin{pmatrix} \psi_{\mu L} \\ \psi_{\mu R} \end{pmatrix} = \begin{pmatrix} \left(\dfrac{W}{W^{*}}\right)^{1/4}\left[e^{\kappa^{2}K/24}\psi_{\mu L}^{\text{n}} + \dfrac{\kappa}{6}\gamma_{r}(e_{\mu}^{r})^{\text{n}}K_{,}{}^{a}(\chi_{a})^{\text{n}}\right] \\ \left(\dfrac{W^{*}}{W}\right)^{1/4}\left[e^{\kappa^{2}K/24}\psi_{\mu R}^{\text{n}} + \dfrac{\kappa}{6}\gamma_{r}(e_{\mu}^{r})^{\text{n}}K_{,a}(\chi^{a})^{\text{n}}\right] \end{pmatrix}$$

$$\begin{pmatrix} \chi^{a} \\ \lambda_{L}^{\alpha} \\ \lambda_{R}^{\alpha} \end{pmatrix} = \begin{pmatrix} e^{-\kappa^{2}K/24}\left(\dfrac{W^{*}}{W}\right)^{1/4}(\chi^{a})^{\text{n}} \\ e^{-\kappa^{2}K/8}\left(\dfrac{W}{W^{*}}\right)^{1/4}(\lambda_{L}^{\alpha})^{\text{n}} \\ e^{-\kappa^{2}K/8}\left(\dfrac{W^{*}}{W}\right)^{1/4}(\lambda_{R}^{\alpha})^{\text{n}} \end{pmatrix}. \tag{21.234}$$

In the above, the variables labelled with a superscript "n" arise from the use of the new vierbein after the transformation given in Eq. (21.229). Next, we eliminate χ^{a} etc. in terms of $(\chi^{a})^{\text{n}}$, etc. by substitution of Eq. (21.234) into Eqs. (21.227) and (21.237). This substitution leads to the fermionic part of the Lagrangian, which can be found in the literature [3,4]. The couplings discussed above can also be obtained using superconformal techniques [6,7].

21.6.1 Problems

1. It is useful to segregate parts of the Lagrangian containing the auxiliary fields. Using the set of auxiliary fields $u - \frac{1}{2}\kappa K_{,a}h^{a}, h^{a}, A_{\mu}$, and D^{α} show that the auxiliary fields in Eq. (21.227) are given by

$$\begin{aligned} e^{-1}L^{aux} = {} & \frac{\kappa^{2}}{9}\phi\left|u - \frac{\kappa}{2}K_{,a}h^{a}\right|^{2} - \frac{\kappa^{2}}{6}\phi K_{b}^{a}h_{a}h^{b} \\ & + \frac{1}{2}\left[\kappa W\left(u - \frac{\kappa}{2}K_{,a}h^{a}\right) - \frac{\kappa^{2}}{9}\phi A_{\mu}A^{\mu}\right. \\ & \left. + W\left(\frac{\kappa^{2}}{2}W_{,a} + \frac{W_{,a}}{W}\right)h^{a} + \text{h.c.}\right] \\ & + \frac{\kappa^{2}}{6}\phi\left\{\left[\frac{\kappa}{3}\left(u - \frac{\kappa}{2}K_{,c}h^{c}\right)\left(K^{ab} - \frac{\kappa^{2}}{6}K_{,}{}^{a}K_{,}{}^{b}\right)\right.\right. \\ & \left.\left. + h^{c}\left(K_{,c}^{\ ab} - \frac{\kappa^{2}}{3}K_{,}{}^{a}K_{,}{}^{b}{}_{c}\right)\right](\overline{\chi^{a}}\chi_{b}) + \text{h.c.}\right\} \end{aligned}$$

$$- i\frac{\kappa^3}{18}\phi A^\mu \left[K_{,a}\hat{\mathcal{D}}_\mu z^a - K_{,}{}^a\hat{\mathcal{D}}_\mu z_a + \left(K_{,}{}^a{}_b - \frac{\kappa^2}{6}K_{,}{}^a K_{,b} \right) \right.$$

$$\left. \times \overline{\chi^a}\gamma_\mu \chi^b + \frac{9}{2\kappa^2\phi}\overline{\lambda^\alpha}\gamma_\mu\gamma_5\lambda^\alpha \right]$$

$$+ \frac{1}{2}D^\alpha D^\alpha + \frac{g_\alpha}{2}D^\alpha \phi_a (T^\alpha)^a{}_b z^b. \tag{21.235}$$

2. Varying $u - \frac{1}{2}\kappa K_{,a}h^a$, h^a, A_μ, and D^α independently in Eq. (21.235), obtain the equations for these auxiliary fields. Show that they are given by

$$h^a = \frac{3}{\kappa^2\phi}W^*(K^{-1})^a{}_b \left(\frac{\kappa^2}{2}K_{,}{}^b + \frac{W_{,}{}^{*b}}{W^*} \right)$$

$$+ \left((K^{-1})^a{}_b K_{,}{}^b{}_{ce} - \frac{\kappa^2}{3}\delta^a{}_c K_{,e} \right) \bar{\chi}_c \chi^e$$

$$u - \frac{\kappa}{2}K_{,a}h^a = -\frac{9}{2\kappa\phi}W^* - \frac{\kappa}{2} \left(K_{,ab} - \frac{\kappa^2}{6}K_{,a}K_{,b} \right) \overline{\chi_a}\chi^b$$

$$A_\mu = -i\frac{\kappa}{4} \left[K_{,a}\hat{\mathcal{D}}_\mu Z^a - K_{,}{}^a\hat{\mathcal{D}}_\mu Z_a + \left(K_{,}{}^a{}_b - \frac{\kappa^2}{6}K_{,}{}^a K_{,b} \right) \right.$$

$$\left. \times \overline{\chi^a}\gamma^\mu \chi^b + \frac{9}{2\kappa^2\phi}\overline{\lambda^\alpha}\gamma_\mu\gamma_5\lambda^\alpha \right]$$

$$D^\alpha = -\frac{1}{2}g_\alpha\phi_{,a}(T^\alpha Z)^a. \tag{21.236}$$

3. Next, eliminate the auxiliary fields given by Eq. (21.236) in Eq. (21.235), and show that Eq. (21.235) takes the following form:

$$e^{-1}\mathcal{L}^{aux} = -\frac{9}{4\phi} \left| W + \frac{\kappa^2\phi}{9} \left(K^{,ab} - \frac{\kappa^2}{6}K_{,}{}^a K_{,}{}^b \right) \overline{\chi^a}\chi_b \right|^2$$

$$+ \frac{6}{\kappa^2\phi}(K^{-1})^a{}_b \left[\frac{1}{2}W \left(\frac{W_{,a}}{W} + \frac{\kappa^2}{2}K_{,a} \right) \right.$$

$$+ \frac{\kappa^2}{6}\phi \left(K_{,}{}^{ce}{}_a - \frac{\kappa^2}{3}K_{,}{}^c{}_a K_{,}{}^e \right) \overline{\chi^c}\chi_e \right]$$

$$\times \left[\frac{W^*}{2} \left(\frac{W^*_{,}{}^b}{W^*} + \frac{\kappa^2}{2}K_{,}{}^b \right) + \frac{\kappa^2\phi}{6} \left(K_{,}{}^b{}_{c'e'} - \frac{\kappa^2}{3}K_{,}{}^b{}_{c'}K_{,e'} \right) \overline{\chi_{c'}}\chi^{e'} \right]$$

$$- \frac{\kappa^4\phi}{144} \left[K_{,a}\hat{\mathcal{D}}_\mu Z^a - K_{,}{}^a\hat{\mathcal{D}}_\mu Z_a + \left(K_{,}{}^a{}_b - \frac{\kappa^2}{6}K_{,}{}^a K_{,b} \right) \right.$$

$$\left. \overline{\chi^a}\gamma^\mu \chi^b + \frac{3}{2\kappa^2\phi}\overline{\lambda^\alpha}\gamma_\mu\gamma_5\lambda^\alpha \right]^2$$

$$- \frac{g_\alpha^2}{8} \left(\phi_{,}(T^\alpha Z)^a \right)^2. \tag{21.237}$$

4. Consider a Weyl transformation of Eq. (21.229) on the vierbein. Show that under this transformation the connection $\omega_\mu{}^{rs}(e)$ and the Ricci scalar $R(\omega(e))$ transform as follows:

$$\omega_\mu{}^{rs}(e) \to \omega_\mu{}^{rs}(e) + \frac{\kappa^2}{12}\left(e_\mu{}^r e^{\nu s} - e_\mu{}^s e^{\nu r}\right)\partial_\nu K$$

$$\frac{\phi}{6}R(\omega(e)) \to -\frac{1}{2\kappa^2}R(\omega(e)) - \frac{\kappa^2}{48}(\partial_\mu K)(\partial^\mu K). \qquad (21.238)$$

5. Show that the following Fierz identities hold:

$$\overline{\chi^a}\gamma^\mu\chi^b\psi_\mu\chi^c = \frac{1}{2}\bar{\psi}_\mu\chi_a\bar{\chi}_b\chi^c \qquad (21.239)$$

$$\bar{\chi}s^{\mu\nu}s^{\rho\sigma}\psi_\nu C_{\mu\rho\sigma} = \frac{1}{2}\bar{\chi}\psi_\nu\left(\overline{\psi^\nu}\gamma^\rho\psi_\rho - \frac{1}{4}e^{-1}\epsilon^{\mu\nu\rho\sigma}\bar{\psi}_\mu\gamma_\rho\psi_\sigma\right), \qquad (21.240)$$

where $C_{\mu\nu\sigma}$ is defined by

$$C_{\mu\rho\sigma} = \left(\bar{\psi}_\mu\gamma_\rho\psi_\sigma - \bar{\psi}_\mu\gamma_\sigma\psi_\rho + \bar{\psi}_\rho\gamma_\mu\psi_\sigma\right). \qquad (21.241)$$

References

[1] E. Cremmer, B. Julia, J. Scherk, P. van Nieuwenhuizen, S. Ferrara, and L. Girardello, *Phys. Lett. B* **79**, 231 (1978).

[2] A. H. Chamseddine, R. L. Arnowitt, and P. Nath, *Phys. Rev. Lett.* **49**, 970 (1982).

[3] E. Cremmer, S. Ferrara, L. Girardello, and A. Van Proeyen, *Nucl. Phys. B* **212**, 413 (1983).

[4] P. Nath, R. Arnowitt, and A. H. Chamseddine, *Applied N=1 Supergravity*, Trieste Lecture Series vol. I, Appendix A, World Scientific (1984).

[5] J. Bagger and E. Witten, *Phys. Lett. B* **118**, 103 (1982).

[6] M. Kaku, P. K. Townsend, and P. van Nieuwenhuizen, *Phys. Rev. D* **17**, 3179 (1978).

[7] T. Kugo and S. Uehara, *Nucl. Phys. B* **222**, 125 (1983).

21.7 Further Details on the Couplings of the MSSM/Supergravity Lagrangian

In this appendix we give further details on the MSSM/applied-supergravity Lagrangian discussed in Chapters 10 and 14.

21.7.1 Electromagnetic A_μ Couplings of the Sfermions and of the Charged Higgs Bosons

The electromagnetic couplings of $\tilde{f}$ and of $H^\pm$ are given by

$$\mathcal{L}_{A_\mu} = +ie\sum_i Q_{f_i}\left(\tilde{f}_{1i}^* \overset{\leftrightarrow}{\partial}_\mu \tilde{f}_{1i} + \tilde{f}_{2i}^* \overset{\leftrightarrow}{\partial}_\mu \tilde{f}_{2i}\right)A^\mu + ie(H^- \overset{\leftrightarrow}{\partial}_\mu H^+)A_\mu, \qquad (21.242)$$

where Q_{f_i} are the charges of the sfermions and $\tilde{f}_{1i}$ and $\tilde{f}_{i2}$ are mass eigenstates. The sfermion mass squared matrix is diagonalized so that mass diagonal states $\tilde{f}_1$ and $\tilde{f}_2$ are related to the left and right chiral fields so that $\tilde{f}_1 = \cos\theta_f \tilde{f}_L + \sin\theta_f \tilde{f}_R$ and $\tilde{f}_2 = -\sin\theta_f \tilde{f}_L + \cos\theta_f \tilde{f}_R$.

21.7.2 Z-boson Couplings with Sfermions and Higgs Bosons

The Z-boson couplings with $\tilde{f}$ and $H^\pm$ are given by

$$
\mathcal{L}_Z = +i\frac{g}{\cos\theta_W} \sum_i \left[(T_{3f_i}\cos^2\theta_{f_i} - \sin^2\theta_W Q_{f_i})(\tilde{f}_{1i}^* \overset{\leftrightarrow}{\partial}_\mu \tilde{f}_{1i}) \right.
$$
$$
+ (T_{3f_i}\sin^2\theta_{f_i} - \sin^2\theta_W Q_{f_i})(\tilde{f}_{2i}^* \overset{\leftrightarrow}{\partial}_\mu \tilde{f}_{2i})
$$
$$
\left. - T_{3f_i}\cos\theta_{f_i}\sin\theta_{f_i}(\tilde{f}_{1i}^* \overset{\leftrightarrow}{\partial}_\mu \tilde{f}_{2i} + \tilde{f}_{2i}^* \overset{\leftrightarrow}{\partial}_\mu \tilde{f}_{1i}) \right] Z^\mu
$$
$$
+ \frac{g}{2\cos\theta_W}[\cos(\alpha-\beta)(h^0 \overset{\leftrightarrow}{\partial}_\mu A^0) + \sin(\alpha-\beta)(H^0 \overset{\leftrightarrow}{\partial}_\mu A^0)]Z^\mu
$$
$$
+ i\frac{g}{\cos\theta_W}\left(\frac{1}{2} - s_W^2\right)(H^- \overset{\leftrightarrow}{\partial}_\mu H^+)Z^\mu. \tag{21.243}
$$

where α and β are as defined in Chapter 14.

21.7.3 W-boson Couplings with Sfermions and Higgs Bosons

The W-boson couplings with $\tilde{f}$ and $H^\pm$ are given by

$$
\mathcal{L}_W = +i\frac{g}{\sqrt{2}} \sum_i \left[\cos\theta_{f_{ui}}\cos\theta_{f_{di}}(\tilde{f}_{ui1}^* \overset{\leftrightarrow}{\partial}^\mu \tilde{f}_{di1}) - \cos\theta_{f_{ui}}\sin\theta_{f_{di}}(\tilde{f}_{ui1}^* \overset{\leftrightarrow}{\partial}^\mu \tilde{f}_{di2}) \right.
$$
$$
\left. - \sin\theta_{f_{ui}}\cos\theta_{f_{di}}(\tilde{f}_{ui2}^* \overset{\leftrightarrow}{\partial}^\mu \tilde{f}_{di1}) + \sin\theta_{f_{ui}}\sin\theta_{f_{di}}(\tilde{f}_{ui2}^* \overset{\leftrightarrow}{\partial}^\mu \tilde{f}_{di2}) \right] W_\mu^+ + \text{h.c.}
$$
$$
- \left\{ \frac{g}{2}i[\cos(\alpha-\beta)(h^0 \overset{\leftrightarrow}{\partial}^\mu H^-) + \sin(\alpha-\beta)(H^0 \overset{\leftrightarrow}{\partial}^\mu H^-)]W_\mu^+ \right.
$$
$$
\left. - \frac{g}{2}(A^0 \overset{\leftrightarrow}{\partial}^\mu H^-)W_\mu^+ + \text{h.c.} \right\}. \tag{21.244}
$$

21.7.4 Charged Higgs Couplings with Fermions and Sfermions

The charged Higgs couplings with fermions in the four component notation are given by

$$
\mathcal{L}_{H^\pm} = \frac{g}{\sqrt{2}M_W} \sum_i \left[m_{ei}\tan\beta(\bar{\nu}_i P_R e_i) + m_{di}\tan\beta(\bar{u}_i P_R d_i) \right.
$$
$$
\left. + m_{ui}\cot\beta(\bar{u}_i P_L d_i) \right] H^+ + \text{h.c.} \tag{21.245}
$$

The charged Higgs couplings with sfermions are given by

$$
\begin{aligned}
\mathcal{L}_{H^\pm} = -\frac{g}{\sqrt{2}M_W} \Big\{ &[(M_W^2 \sin 2\beta - m_e^2 \tan\beta)\tilde{\nu}^*\tilde{e}_L + m_e(A_e \tan\beta - \mu)\tilde{\nu}^*\tilde{e}_R] \\
&+ [M_W^2 \sin 2\beta - m_d^2 \tan\beta - m_u^2 \cot\beta]\tilde{u}_L^*\tilde{d}_L \\
&+ [m_u(A_u \cot\beta - \mu)\tilde{u}_R^*\tilde{d}_L + m_d(A_d \tan\beta - \mu)\tilde{u}_L^*\tilde{d}_R] \\
&+ \frac{\sqrt{2}m_u m_d}{\sin 2\beta}\tilde{u}_R^*\tilde{d}_R \Big\} H^+ + \text{h.c.}
\end{aligned}
\tag{21.246}
$$

21.7.5 Cubic Couplings of h^0 and H^0 with Fermions and Sfermions

We first look at the cubic couplings of the light Higgs boson h^0 with fermions and sfermions. The couplings with fermions are given by

$$
\mathcal{L}_{h^0 ff} = -\frac{gm_u \cos\alpha}{2M_W \sin\beta}\bar{u}uh^0 + \frac{gm_d \sin\alpha}{2M_W \cos\beta}\bar{d}dh^0 + \frac{gm_e \sin\alpha}{2M_W \cos\beta}\bar{e}eh^0.
\tag{21.247}
$$

The couplings of h^0 with sfermions are given by

$$
\begin{aligned}
\mathcal{L}_{\tilde{f}\tilde{f}h^0} = &-\left(\frac{gm_u^2 \cos\alpha}{M_W \sin\beta} - g_c M_Z \sin(\alpha+\beta)\left(\frac{1}{2} - \frac{2}{3}s_W^2\right)\right)\tilde{u}_L^*\tilde{u}_L h^0 \\
&+\left(\frac{gm_d^2 \sin\alpha}{M_W \cos\beta} - g_c M_Z \sin(\alpha+\beta)\left(\frac{1}{2} - \frac{1}{3}s_W^2\right)\right)\tilde{d}_L^*\tilde{d}_L h^0 \\
&-\left(\frac{gm_u^2 \cos\alpha}{M_W \sin\beta} - \frac{2}{3}g_c M_Z \sin(\alpha+\beta)s_W^2\right)\tilde{u}_R^*\tilde{u}_R h^0 \\
&+\left(\frac{gm_d^2 \sin\alpha}{M_W \cos\beta} - \frac{1}{3}g_c M_Z \sin(\alpha+\beta)s_W^2\right)\tilde{d}_R^*\tilde{d}_R h^0 \\
&+\frac{gm_u}{2M_W \sin\beta}(A_u \cos\alpha - \mu\sin\alpha)(\tilde{u}_L^*\tilde{u}_R + \tilde{u}_R^*\tilde{u}_L)h^0 \\
&-\frac{gm_d}{2M_W \cos\beta}(A_d \sin\alpha - \mu\cos\alpha)(\tilde{d}_L^*\tilde{d}_R + \tilde{d}_R^*\tilde{d}_L)h^0 \\
&+\frac{g_Z}{2}M_Z \sin(\alpha+\beta)\tilde{\nu}^*\tilde{\nu}h^0 \tag{21.248} \\
&+\left(\frac{gm_e^2 \sin\alpha}{M_W \cos\beta} - g_c M_Z \sin(\alpha+\beta)\left(\frac{1}{2} - s_W^2\right)\right)\tilde{e}_L^*\tilde{e}_L h^0 \\
&+\left(\frac{gm_e^2 \sin\alpha}{M_W \cos\beta} - g_c M_Z \sin(\alpha+\beta)s_W^2\right)\tilde{e}_R^*\tilde{e}_R h^0 \\
&-\frac{gm_e}{2M_W \cos\beta}(A_e \sin\alpha - \mu\cos\alpha)(\tilde{e}_L^*\tilde{e}_R + \tilde{e}_R^*\tilde{e}_L)h^0. \tag{21.249}
\end{aligned}
$$

where $g_c = g/cos\theta_w$.

The H^0 couplings can be obtained from the h^0 couplings. The transformations that accomplish this are as follows:

h^0 couplings		H^0 couplings
$\sin\alpha$	$\rightarrow$	$-\cos\alpha$
$\sin(\alpha+\beta)$	$\rightarrow$	$-\cos(\alpha+\beta)$
$\cos\alpha$	$\rightarrow$	$\sin\alpha$
$\cos(\alpha+\beta)$	$\rightarrow$	$\sin(\alpha+\beta).$

$$(21.250)$$

21.7.6 Cubic Couplings of A^0 with Fermions and Sfermions

The couplings of A^0 with fermions and sfermions are given by

$$
\begin{aligned}
\mathcal{L}_{A^0 ff + A^0 \tilde{f}\tilde{f}} = {}& i\left[\frac{gm_u}{2M_W}\cot\beta\,\bar{u}\gamma_5 u + \frac{gm_d}{2M_W}\tan\beta\,\bar{d}\gamma_5 d + \frac{gm_e}{2M_W}\tan\beta\,\bar{e}\gamma_5 e\right] A^0 \\
& - i\frac{g}{2M_W}\Big[m_e(A_e\tan\beta - \mu)(\tilde{e}_L^*\tilde{e}_R - \tilde{e}_R^*\tilde{e}_L) \\
& + m_d(A_d\tan\beta - \mu)(\tilde{d}_L^*\tilde{d}_R - \tilde{d}_R^*\tilde{d}_L) \\
& + m_u(A_u\cot\beta - \mu)(\tilde{u}_L^*\tilde{u}_R - \tilde{u}_R^*\tilde{u}_L)\Big] A^0.
\end{aligned}
$$

$$(21.251)$$

21.7.7 Sfermion–Sfermion–Vector–Vector Couplings

$$
\begin{aligned}
-\mathcal{L}_{\tilde{f}\tilde{f}VV} = {}& e^2\sum_i Q_{f_i}^2\left(\tilde{f}_{Li}^*\tilde{f}_{Li} + \tilde{f}_{Ri}^*\tilde{f}_{Ri}\right) A_\mu A^\mu \\
& + \frac{2eg}{\cos\theta_W}\sum_i\left[Q_{f_i}\left(T_{3f_i} - \sin^2\theta_W Q_{f_i}\right)\tilde{f}_{Li}^*\tilde{f}_{Li} - \sin^2\theta_W Q_f^2\tilde{f}_{Ri}^*\tilde{f}_{Ri}\right]A_\mu Z^\mu \\
& + \frac{g^2}{\cos^2\theta_W}\sum_i\left[\left(T_{3f_i} - \sin^2\theta_W Q_{f_i}\right)^2\tilde{f}_{Li}^*\tilde{f}_{Li} + \sin^4\theta_W Q_f^2\tilde{f}_{Ri}^*\tilde{f}_{Ri}\right]Z_\mu Z^\mu \\
& + \frac{g^2}{2}\left[\sum_i\tilde{f}_{Li}^*\tilde{f}_{Li}\right]W_\mu^+ W^{-\mu} \\
& + \frac{eg}{3\sqrt{2}}\sum_i\left(\tilde{u}_{Li}^*\tilde{d}_{Li}W_\mu^+ + \tilde{d}_{Li}^*\tilde{u}_{Li}W_\mu^-\right)A^\mu \\
& - \frac{g^2\sin\theta_W}{3\sqrt{2}}\tan\theta_W\sum_i\left(\tilde{u}_{Li}^*\tilde{d}_{Li}W_\mu^+ + \tilde{d}_{Li}^*\tilde{u}_{Li}W_\mu^-\right)Z^\mu \\
& + eg_3\sum_i Q_{qi}\left(\tilde{q}_{Li}^*\lambda_a\tilde{q}_{Li} + \tilde{q}_R^*\lambda_a\tilde{q}_R\right)G_{\mu a}A^\mu \\
& + \frac{g_3 g}{\cos\theta_W}\sum_i\left[\tilde{q}_{Li}^*(T_{3qi} - \sin^2\theta_W Q_{qi})\lambda_a\tilde{q}_{Li} - \sin^2\theta_W Q_{qi}\tilde{q}_R^*\lambda_a\tilde{q}_R\right]G_{\mu a}Z^\mu
\end{aligned}
$$

$$+ \frac{gg_3}{\sqrt{2}} \sum_i \left[\tilde{u}^*_{Li} \lambda_a \tilde{d}_{Li} W^{+\mu} + \tilde{d}^*_{Li} \lambda_a \tilde{u}_{Li} W^{-\mu} \right] G_{\mu a}$$

$$+ \frac{g_3^2}{4} \sum_i \left(\tilde{q}^*_{Li} \lambda_a \lambda_b \tilde{q}_{Li} + \tilde{q}^*_{Ri} \lambda_a \lambda_b \tilde{q}_{Ri} \right) G_{\mu a} G_b^\mu. \tag{21.252}$$

21.7.8 Gluon and Gluino Couplings

Gluinos couple to quarks and to squarks and to glucons. These couplings are given by

$$\mathcal{L} = i g_3 \sum_i \tilde{q}^*_i T_a \overset{\leftrightarrow}{\partial}_\mu \tilde{q}_i G^\mu_a - i \frac{g_3}{2} f_{abc} \bar{\tilde{G}}_a \gamma^\mu \tilde{G}_b G_{\mu c}$$

$$+ \sqrt{2} g_3 \sum_{q=u,d} \left(\bar{q} T_a P_L \tilde{G}_a \tilde{q}_R + \bar{\tilde{G}}_a P_R \tilde{T}_a \tilde{q}^*_R - \bar{q} T_a P_R \tilde{G}_a \tilde{q}_L - \bar{\tilde{G}}_a P_L q \tilde{T}_a \tilde{q}^*_L \right),$$

$$\tag{21.253}$$

where $\tilde{G}_a$ is the gluino field, $a, b, c = 1 - 8$ are the $SU(3)$ color indices and $\tilde{T}_a$ stands for the transpose of T_a. In addition, there are a variety of other couplings not shown here. Suggestions for further reading are listed below.

Further Reading

H. Baer and X. Tata, *Weak Scale Supersymmetry: From Superfields to Scattering Events*, Cambridge University Press (2006).

M. Drees, R. Godbole, and P. Roy, *Theory and Phenomenology of Sparticles: An Account of Four-dimensional N=1 Supersymmetry in High Energy Physics*, World Scientific (2004).

J. F. Gunion, H. E. Haber, G. L. Kane, and S. Dawson, *Front. Phys.* **80**, 1 (2000) [errata, hep-ph/9302272 (1993)].

J. F. Gunion and H. E. Haber, *Nucl. Phys. B* **272**, 1 (1986) [erratum, *ibid.* **402**, 567 (1993)].

H. E. Haber and G. L. Kane, *Phys. Rep.* **117**, 75 (1985).

M. Kuroda, arXiv: hep-ph/9902340 (1999).

P. Nath, R. L. Arnowitt, and A. H. Chamseddine, *Model Independent Analysis*, HUTP-83/A077, NUB-2588a (1983).

P. Nath, R. Arnowitt, and A. H. Chamseddine, *Applied N=1 Supergravity*, Trieste Lecture Series vol. I, World Scientific (1984).

22

Notation, Conventions, and Formulae

22.1 Lagrangians

Variation of a Lagrangian with no more than one deivative on the fields

$$\delta\mathcal{L}(\chi_i, \partial_\mu\chi_i) = \left(\frac{\partial\mathcal{L}}{\partial\chi_i} - \partial_\mu\left(\frac{\partial\mathcal{L}}{\partial_\mu\chi_i}\right)\right)\partial\chi_i + \partial_\mu\left(\frac{\partial\mathcal{L}}{\partial(\partial_\mu\chi_i)}\delta\chi_i\right)$$

$$\text{Spin 0}: \ \mathcal{L}_\phi = -\phi_a^\mu\partial_\mu\phi_a + \frac{1}{2}(\phi_a^\mu\phi_{\mu a} - m_\phi^2\phi_a^2)$$

$$\text{Spin 1 massive}: \ \mathcal{L}_\rho = -\frac{1}{2}G_a^{\mu\nu}(\partial_\mu v_{\nu a} - \partial_\nu v_{\mu a}) + \frac{1}{4}G_a^{\mu\nu}G_{\mu\nu a} - \frac{1}{2}m_\rho^2 v_a^\mu v_{\mu a}$$

$$\text{Maxwell}: \ \mathcal{L}_A = -\frac{1}{2}F^{\mu\nu}(\partial_\mu A_\nu - \partial_\nu A_\mu) + \frac{1}{4}F^{\mu\nu}F_{\mu\nu}$$

$$\text{Dirac}: \ \mathcal{L}_D = -\bar{\psi}_D\left(\frac{1}{i}\gamma^\mu\partial_\mu + m\right)\psi_D. \tag{22.1}$$

The Dirac gamma matrices are

$$\gamma^\mu\gamma^\nu + \gamma^\nu\gamma^\mu = -2\eta^{\mu\nu}\mathbb{1}$$
$$\eta^{\mu\nu} = diag(-1,1,1,1)$$
$$\gamma_5 = i\gamma^0\gamma^1\gamma^2\gamma^3$$
$$\gamma_5^\dagger = \gamma_5, \ \gamma_5^2 = \mathbb{1}$$
$$\gamma^0(\gamma^\mu)^\dagger\gamma^0 = \gamma^\mu$$
$$\sigma^{\mu\nu} \equiv \frac{i}{2}[\gamma^\mu, \gamma^\nu]. \tag{22.2}$$

The Dirac representation of the γ-matrices is

$$\gamma^0 = \begin{pmatrix} \mathbb{1} & \mathbb{0} \\ \mathbb{0} & -\mathbb{1} \end{pmatrix}, \quad \gamma^i = \begin{pmatrix} \mathbb{0} & \sigma^i \\ -\sigma^i & \mathbb{0} \end{pmatrix}$$

$$\gamma_5 = \begin{pmatrix} \mathbb{0} & \mathbb{1} \\ \mathbb{1} & \mathbb{0} \end{pmatrix}, \quad \sigma^{0i} = i\begin{pmatrix} \mathbb{0} & \sigma^i \\ \sigma^i & \mathbb{0} \end{pmatrix}. \tag{22.3}$$

The Pauli matrices σ^i are given by

$$\sigma^1 = \begin{pmatrix} 0 & 1 \\ 1 & 0 \end{pmatrix}, \quad \sigma^2 = \begin{pmatrix} 0 & -i \\ i & 0 \end{pmatrix}, \quad \sigma^3 = \begin{pmatrix} 1 & 0 \\ 0 & -1 \end{pmatrix}. \tag{22.4}$$

The γ-matrices have the following properties:

$$\gamma^\nu \gamma^\mu \gamma_\nu = 2\gamma^\mu, \quad \gamma^\lambda \gamma^\mu \gamma^\nu \gamma_\lambda = 4\eta^{\mu\nu} \mathbb{1}$$
$$tr(\gamma^\mu) = 0 = tr(\gamma_5)$$
$$tr(\gamma^\mu \gamma^\nu) = -4\eta^{\mu\nu}, \quad tr(\gamma_5 \gamma^\mu \gamma^\nu) = 0$$
$$tr(\gamma^\mu \gamma^\nu \gamma^\lambda \gamma^\rho) = 4(\eta^{\mu\nu}\eta^{\lambda\rho} - \eta^{\mu\lambda}\eta^{\nu\rho} + \eta^{\mu\rho}\eta^{\nu\lambda}). \tag{22.5}$$

The Levi–Civita tensor $\epsilon^{\mu\nu\lambda\rho}$ is defined so that it is totally anti-symmetric in all of its indices and cyclic permutations do not change its value so that

$$\epsilon^{0123} = +1$$
$$\epsilon_{\mu\nu\lambda\rho} = -\epsilon^{\mu\nu\lambda\rho}. \tag{22.6}$$

Using the Levi–Civita tensor we have the following identities:

$$\gamma_5 = \frac{-i}{4!} \epsilon_{\mu\nu\lambda\rho} \gamma^\mu \gamma^\nu \gamma^\lambda \gamma^\rho$$
$$tr(\gamma_5 \gamma^\mu \gamma^\nu \gamma^\lambda \gamma^\rho) = -4i\epsilon^{\mu\nu\lambda\rho}. \tag{22.7}$$

In addition to the Dirac representation of the γ-matrices there are two other representations which are also commonly used. One of these is the Weyl or chiral representation and the other is the Majorana representation. For the Weyl representation in the basis where we write

$$\psi = \begin{pmatrix} \psi_L \\ \psi_R \end{pmatrix}, \quad \psi_L = \frac{1}{2}(1 - \gamma_5)\psi, \quad \psi_R = \frac{1}{2}(1 + \gamma_5)\psi, \tag{22.8}$$

the gamma matrices take the form

$$\gamma^0 = \begin{pmatrix} 0 & 1 \\ 1 & 0 \end{pmatrix}, \quad \gamma^i = \begin{pmatrix} 0 & \sigma^i \\ -\sigma^i & 0 \end{pmatrix}, \quad \gamma_5 = \begin{pmatrix} -1 & 0 \\ 0 & 1 \end{pmatrix}. \tag{22.9}$$

22.2 U(3) Matrices

The $U(3)$ matrices λ_a are defined as follows:

$$\lambda_1 = \begin{pmatrix} 0 & 1 & 0 \\ 1 & 0 & 0 \\ 0 & 0 & 0 \end{pmatrix}, \ \lambda_2 = \begin{pmatrix} 0 & -i & 0 \\ i & 0 & 0 \\ 0 & 0 & 0 \end{pmatrix}, \ \lambda_3 = \begin{pmatrix} 1 & 0 & 0 \\ 0 & -1 & 0 \\ 0 & 0 & 0 \end{pmatrix}$$

$$\lambda_4 = \begin{pmatrix} 0 & 0 & 1 \\ 0 & 0 & 0 \\ 1 & 0 & 0 \end{pmatrix}, \ \lambda_5 = \begin{pmatrix} 0 & 0 & -i \\ 0 & 0 & 0 \\ i & 0 & 0 \end{pmatrix}, \ \lambda_6 = \begin{pmatrix} 0 & 0 & 0 \\ 0 & 0 & 1 \\ 0 & 1 & 0 \end{pmatrix}$$

$$\lambda_7 = \begin{pmatrix} 0 & 0 & 0 \\ 0 & 0 & -i \\ 0 & i & 0 \end{pmatrix}, \ \lambda_8 = \sqrt{\frac{1}{3}} \begin{pmatrix} 1 & 0 & 0 \\ 0 & 1 & 0 \\ 0 & 0 & -2 \end{pmatrix}, \ \lambda_9 = \sqrt{\frac{2}{3}} \begin{pmatrix} 1 & 0 & 0 \\ 0 & 1 & 0 \\ 0 & 0 & 1 \end{pmatrix}, \quad (22.10)$$

where

$$Tr(\lambda_a \lambda_b) = 2\delta_{ab}, \quad a, b = 1, \ldots, 9, \quad (22.11)$$

and where λ_1 to λ_8 are the Gell–Mann matrices. The non-vanishing components of f_{abc} and d_{abc} for $SU(3)$ are as follows:

abc	f_{abc}	abc	d_{abc}	abc	d_{abc}
123	1	118	$\frac{1}{\sqrt{3}}$	366	$-\frac{1}{2}$
147	$\frac{1}{2}$	228	$\frac{1}{\sqrt{3}}$	377	$-\frac{1}{2}$
157	$-\frac{1}{2}$	338	$\frac{1}{\sqrt{3}}$	448	$-1/2\sqrt{3}$
246	$\frac{1}{2}$	146	$\frac{1}{2}$	558	$-1/2\sqrt{3}$
257	$\frac{1}{2}$	157	$\frac{1}{2}$	668	$-1/2\sqrt{3}$
345	$\frac{1}{2}$	247	$-\frac{1}{2}$	778	$-1/2\sqrt{3}$
367	$-\frac{1}{2}$	256	$\frac{1}{2}$	888	$-1/\sqrt{3}$
458	$\frac{\sqrt{3}}{2}$	344	$\frac{1}{2}$	$ij9$	$\delta_{ij}\sqrt{2/3}$
678	$\frac{\sqrt{3}}{2}$	355	$\frac{1}{2}$		$i, j = 1, \cdots, 8,$

$$(22.12)$$

while all other components of f_{abc} and d_{abc} vanish.

22.3 The Standard Model Relations

Fields	$(SU(3), SU(2), U_Y(1))$
q_L	$\left(3, 2, \dfrac{1}{6}\right)$
$u_R^\dagger$	$\left(3, 1, -\dfrac{2}{3}\right)$
$d_R^\dagger$	$\left(3, 1, \dfrac{1}{3}\right)$
ℓ_L	$\left(3, 2, -\dfrac{1}{2}\right)$
$\ell_R^\dagger$	$(1, 1, 1)$
H	$\left(1, 2, \dfrac{1}{2}\right)$
B_μ	$(1, 1, 0)$
$A_{\mu a}, a = 1 - 3$	$(1, 3, 0)$
$g_{\mu A}, A = 1 - 8$	$(8, 1, 0).$

$$(22.13)$$

The brackets in the second column give the $SU(3)$ and $SU(2)$ representations and the hypercharge quantum number Y where $Q = T_3 + Y$.

$$\langle H^0 \rangle = \frac{v}{\sqrt{2}}$$
$$v \simeq 246 \text{ GeV}$$
$$M_W = \frac{1}{2} g v$$
$$M_Z = \frac{v}{2} \sqrt{g^2 + g'^2}$$
$$G_F^{-\frac{1}{2}} = 2^{\frac{1}{4}} v. \qquad (22.14)$$

The weak angle θ_W is defined by

$$\sin^2 \theta_W = \frac{g'^2}{g^2 + g'^2}. \qquad (22.15)$$

The $SU(2)$ coupling constant g is related to the Fermi constant G_F by

$$\frac{G_F}{\sqrt{2}} = \frac{g^2}{8 M_W}. \qquad (22.16)$$

In the standard model, the gauge coupling constants at one loop have the following evolution:

$$\mu\frac{d\alpha_i}{d\mu} = -\frac{1}{2\pi}b_i + \cdots$$

$$(b_1, b_2, b_3) = \left(\frac{41}{10}, -\frac{19}{6}, -7\right). \tag{22.17}$$

22.4 MSSM Fields and Quantum Numbers

For the minimal supersymmetric standard model (MSSM) the superfields labeled with a hat that enter are

Superfields	$(SU(3), SU(2), U_Y(1))$
$\hat{q}_L$	$\left(3, 2, \frac{1}{6}\right)$
$\hat{u}_L^c$	$\left(3, 1, -\frac{2}{3}\right)$
$\hat{d}_L^c$	$\left(3, 1, \frac{1}{3}\right)$
$\hat{\ell}_L$	$\left(3, 2, -\frac{1}{2}\right)$
$\hat{\ell}_L^c$	$(1, 1, 1)$
$\hat{H}_1$	$\left(1, 2, -\frac{1}{2}\right)$
$\hat{H}_2$	$\left(1, 2, \frac{1}{2}\right)$
$\hat{B}$	$(1, 1, 0)$
$\hat{A}_a, a = 1-3$	$(1, 3, 0)$
$\hat{g}_A, A = 1-8$	$(8, 1, 0).$

$$\tag{22.18}$$

In the MSSM, b_i are

$$(b_1, b_2, b_3) = \left(\frac{33}{5}, 1, -3\right). \tag{22.19}$$

22.5 SU(5)

The coupling constants g_1, g_2, and g_3 unify at the grand unification scale where

$$g_1 = \sqrt{\frac{5}{3}}g' \tag{22.20}$$

$$\sin^2 \theta_W = \frac{g_1^2}{g_1^2 + \frac{5}{3}g_2}$$

$$= \frac{3}{8} \quad \text{(GUT scale)}. \tag{22.21}$$

For the fundamental representation F and for the adjoint representation A of $SU(N)$ the quadratic Casimir is

$$C_2(F) = \frac{N^2 - 1}{2N} \tag{22.22}$$

$$C_2(A) = N. \tag{22.23}$$

22.6 The Fierz Rearrangement

For any u_1, u_2, u_3, and u_4 which are either Dirac or Majorana spinors, one has in general

$$\bar{u}_1 \Gamma^A u_2 \bar{u}_3 \Gamma^B u_4 = \sum_{CD} C_{CD}^{AB} \bar{u}_1 \Gamma^C u_4 \bar{u}_3 \Gamma^D u_2, \tag{22.24}$$

where Γ^A are a set of 16 independent gamma matrices and where C_{CD}^{AB} is given by

$$C_{CD}^{AB} = \frac{-(+)1}{16} Tr[\Gamma^C \Gamma^A \Gamma^D \Gamma^B]. \tag{22.25}$$

Here, the plus sign is for the case of commuting u spinors and the minus sign is for the case when they anti-commute. A convenient set to choose for Γ^A is the following:

$$\Gamma^A = \{1, \gamma^0, i\gamma^i, i\gamma^0 \gamma_5, \gamma^i \gamma_5, \gamma_5, i\sigma^{0i}, \sigma^{ij}\}. \tag{22.26}$$

The Γ^A are normalized so that

$$tr(\Gamma^A \Gamma^B) = 4\delta^{AB}. \tag{22.27}$$

Using the above, one finds that for q a Dirac and χ a Majorana spinor, one has the following Fierz rearrangement relations:

$$\bar{q}\chi\bar{\chi}q = -\frac{1}{4}\left(\bar{q}q\bar{\chi}\chi + \bar{q}\gamma_5 q\bar{\chi}\gamma_5\chi - \bar{q}\gamma^\mu\gamma_5 q\bar{\chi}\gamma_\mu\gamma_5\chi\right)$$

$$\bar{q}\gamma_5\chi\bar{\chi}\gamma_5 q = -\frac{1}{4}\left(\bar{q}q\bar{\chi}\chi + \bar{q}\gamma_5 q\bar{\chi}\gamma_5\chi + \bar{q}\gamma^\mu\gamma_5 q\bar{\chi}\gamma_\mu\gamma_5\chi\right)$$

$$\bar{q}\gamma_5\chi\bar{\chi}q = -\frac{1}{4}\left(\bar{q}q\bar{\chi}\gamma_5\chi + \bar{q}\gamma^\mu q\bar{\chi}\gamma_\mu\gamma_5\chi + \bar{q}\gamma_5 q\bar{\chi}\chi\right)$$

$$\bar{\chi}\gamma_5 q\bar{q}\chi = \frac{1}{4}(\bar{\chi}\gamma^\mu\gamma_5\chi\bar{q}\gamma_\mu q - \bar{\chi}\chi\bar{q}\gamma_5 q - \bar{\chi}\gamma_5\chi\bar{q}q), \tag{22.28}$$

where we have used the fact that, for χ a Majorana spinor,

$$\bar{\chi}\gamma^\mu\chi = 0$$
$$\bar{\chi}\sigma^{\mu\nu}\chi = 0. \tag{22.29}$$

22.7 Dimensional Regularization

Dimensional regularization is a convenient way to regularize a Feynman integral, and it often allows a quicker way to extract the finite part of the integral. Essentially, the technique is to first compute the integral not in four dimensions but in dimensions $d < 4$ where it is finite, and then continue on to $d = 4$. Thus, an integral $\int d^4p/(p^2 + L^2)^2$ is logarithmically divergent while $\int d^dp/(p^2 + L^2)^2$ $(d < 4)$ is not. In order to make the above transition, it is convenient to go to Euclidean space by defining $p^0 = iq^0$ and $\vec{p} = \vec{q}$ so that we can replace the integral

$$I_4 = \int d^4p \frac{1}{(p^2 + L^2)^2} \tag{22.30}$$

by the integral

$$I_d = i \int d^dq \frac{1}{(q^2 + L^2)^2}. \tag{22.31}$$

Let us first consider d to be an integer. It is useful to go to spherical co-ordinates so that

$$(q_1, \cdots, q_d) \to (Q, \theta_1, \cdots, \theta_{d-1}), \tag{22.32}$$

where Q is the length of the n-dimensional vector in Euclidean space and the ranges of the angles are $0 \le \theta_1 \le 2\pi$ and $0 \le \theta_{2,3,\cdots,d-1} \le \pi$. In this case, the volume elements for the two are connected by

$$d^dq = J \, dQ. \prod_{i=1}^{d-1} d\theta_i, \tag{22.33}$$

where

$$J = Q^{d-1} \sin\theta_2 (\sin\theta_3)^2 \cdots (\sin\theta_{d-1})^{d-2}. \tag{22.34}$$

To derive the solid angle we integrate on the angles using the relation

$$\int_0^\pi \sin^m\theta \, d\theta = \sqrt{\pi}\frac{\Gamma(m/2 + \frac{1}{2})}{\Gamma(m/2 + 1)}, \tag{22.35}$$

which leads to the solid angle

$$\int d\Omega_d = \frac{2\pi^{d/2}}{\Gamma(d/2)}. \tag{22.36}$$

The quantity Ω_d can be continued analytically to non-integer values of d and also when d is complex. Using the above result, I_d can be written in the form

$$I_d = i\frac{2\pi^{d/2}}{\Gamma(d/2)}\int_0^\infty dQ\frac{Q^{d-1}}{(Q^2 + L^2)^2}. \tag{22.37}$$

Using the general formula

$$\int_0^\infty du\frac{u^a}{(u^2 + \alpha^2)^b} = \frac{1}{2}\frac{\Gamma((a+1)/2)\Gamma(b-(a+1)/2)}{\Gamma(b)(\alpha^2)^{b-(a+1)/2}}, \tag{22.38}$$

one finds

$$I_d = \frac{i\pi^{d/2}\Gamma(2-\frac{d}{2})}{(L^2)^{2-d/2}}. \tag{22.39}$$

It is now seen that I_d has simple poles for even values of d larger than or equal to 4. Thus, one cannot simply set $d = 4$ to get to four dimensions. The pole part implies that one has a divergence which was anticipated. In a quantum field theory we must remove this divergence by addition of a counter-term to the Lagrangian. We can derive the necessary counter-terms needed by deriving the pole part of I_d, and then adding terms which would cancel it. After such a cancellation, one will be left with the finite part. So, defining $\epsilon = d/2 - 2$ and the expansion $a^\epsilon = e^{\epsilon \ln a} = 1 + \epsilon \ln a + O(\epsilon^2)$ one can find the pole part and the finite part to be derived. Thus, the pole part I_d^P is given by

$$(I_d^P)_{d\to 4} = \frac{2i\pi^2}{4 - d}, \tag{22.40}$$

and the finite part I_4^F can be obtained by subtracting the pole part from I_d and taking the limit $d \to 4$, which gives

$$I_4^F = (I_d - I_d^P)_{d\to 4} = -i\pi^2 \ln(L^2) + C, \tag{22.41}$$

where C is a constant. We note that the constant reflects the ambiguity of continuation from four to d dimensions since a function I_d' defined by

$$I_d' = i\int d^dq\, a^{d-4}\frac{1}{(q^2 + L^2)^2} \tag{22.42}$$

is an equally valid continuation of I_4. Such an ambiguity is not limited to dimensional regularization but is symptomatic of regularization schemes in general.

Next, we consider the inclusion of spinors in the analysis, which requires inclusion of γ-matrices in Feynman diagrams. Now, in four dimensions the gamma matrices γ^μ satisfy the relation Eq. (22.2). We will continue to use this relation as we move away from four dimensions. However, the trace will be affected so that

$$\gamma^\mu\gamma_\mu = -\delta^\mu_\mu \mathbb{1} = -d\mathbb{1}, \tag{22.43}$$

while the trace of the unit matrix remains unchanged, i.e., $tr(\mathbb{1}) = 4$. It is also easily checked that

$$\gamma^\mu\gamma^\nu\gamma_\mu = (d - 2)\gamma^\nu. \tag{22.44}$$

Also, one may verify that the following identities hold:

$$\gamma^\lambda \gamma^\mu \gamma^\nu \gamma_\lambda = 4\eta^{\mu\nu} \mathbb{1} + (4-d)\gamma^\mu \gamma^\nu$$
$$\gamma^\lambda \gamma^\mu \gamma^\nu \gamma^\alpha \gamma_\lambda = 2\gamma^\alpha \gamma^\nu \gamma^\mu - (4-d)\gamma^\mu \gamma^\nu \gamma^\alpha.$$

$$(22.45)$$

It is easy to extend the above analysis to the derivation of other integrals:

$$\int \frac{d^d p}{[p^2 + 2q \cdot p + L^2]^\alpha} = i \frac{\pi^{d/2} \Gamma(\alpha - d/2)}{\Gamma(\alpha)} \frac{1}{(L^2 - q^2)^{\alpha - d/2}}. \qquad (22.46)$$

Differentiation with respect to q_μ gives

$$\int \frac{d^d p\, p^\mu}{[p^2 + 2q \cdot p + L^2]^\alpha} = -i \frac{\pi^{d/2} \Gamma(\alpha - d/2)}{\Gamma(\alpha)} \frac{q^\mu}{(L^2 - q^2)^{\alpha - d/2}}, \qquad (22.47)$$

while two differentiations lead to

$$\int \frac{d^d p\; p^\mu p^\nu}{[p^2 + 2q \cdot p + L^2]^\alpha} = i \frac{\pi^{d/2}}{\Gamma(\alpha)(L^2 - q^2)^{\alpha - d/2}}$$
$$\times \left[\Gamma\left(\alpha - \frac{d}{2}\right) q^\mu q^\nu + \frac{1}{2}\Gamma\left(\alpha - 1 - \frac{d}{2}\right)(L^2 - q^2)\eta^{\mu\nu} \right].$$

$$(22.48)$$

Contraction with $\eta_{\mu\nu}$ gives

$$\int \frac{d^d p\; p^2}{[p^2 + 2q \cdot p + L^2]^\alpha} = i \frac{\pi^{d/2}}{\Gamma(\alpha)(L^2 - q^2)^{\alpha - d/2}}$$
$$\times \left[\Gamma\left(\alpha - \frac{d}{2}\right) q^2 + \frac{d}{2}\Gamma\left(\alpha - 1 - \frac{d}{2}\right)(L^2 - q^2) \right]. \qquad (22.49)$$

The following expansions are useful. For small x, $\Gamma(x)$ has the expansion

$$\Gamma(x) = \frac{1}{x} - \gamma + \frac{x}{2}\left[\gamma^2 + \frac{\pi^2}{3} - \psi'(1)\right] + O(x^2), \qquad (22.50)$$

where ψ' is the derivative of the digamma function $\psi(x)$ defined by

$$\psi(x) = \frac{d}{dx}\ln \Gamma(x) = \frac{\Gamma'(x)}{\Gamma(x)}, \qquad (22.51)$$

and γ is the Euler–Mascheroni constant ($\gamma \simeq 0.5772$) related to ψ by $\psi(1) = -\gamma$. A generalization of the above near other poles of the gamma function is given by

$$\Gamma(-n+x) = \frac{(-1)^n}{n!}\left[\frac{1}{x} + \psi(n+1) + \frac{x}{2}\left(-\psi'(n+1) + \psi(n+1)^2 + \frac{\pi^2}{3}\right)\right] + O(\epsilon^2). \qquad (22.52)$$

Further Reading

C. G. Bollini and J. J. Giambiagi, *Nuovo Cim. B* **12**, 20 (1972).

G. 't Hooft and M. Veltman, *Nucl. Phys. B* **44**(1), 189–213 (1972).

G. Leibbrandt, *Rev. Mod. Phys.* **47**(4), 849 (1975).

23

Constants and Units

23.1 Physical Constants

Speed of light in vacuum: $c = 299\ 792\ 458$ ms^{-1}

Gravitational constant: $G_N = 6.67428(67) \times 10^{-11}$ m^3 kg^{-1}s^{-2}

$$= 6.70881(67) \times 10^{-39}\ \hbar c\ (\text{GeV}c^{-2})^{-2}$$

Planck's constant: $h = 6.62606896(33) \times 10^{-34}$ J s

(Planck's constant)$/2\pi$: $\hbar = 1.054571628(53) \times 10^{-34}$ J s

$$= 6.58211899(16) \times 10^{-22}\ \text{MeV s}$$

Reduced Planck mass: $M_{Pl} = (8\pi G_N/\hbar c)^{-1/2}$

$$= 2.43 \times 10^{18} \text{GeV}c^2$$

Fine structure constant: $\alpha = e^2/\hbar c = 1/137.035$

Fermi constant: $G_F/(\hbar c)^3 = 1.16637(1) \times 10^{-5}$ GeV^{-2}

Boltzmann constant: $k_B = 1.3806504(24) \times 10^{-23}$ JK^{-1}

$$= 8.617343(15) \times 10^{-5}\ \text{eVK}^{-1}$$

Electron mass: $m_e = 0.51099$ MeV

Proton mass: $m_p = 938.2720$ MeV

W $-$ boson mass: $M_W = (80.399 \pm 0.023)$ GeV

Z $-$ boson mass: $M_Z = (91.1876 \pm 0.0021)$ GeV

Strong coupling constant: $\alpha_s\ (M_Z) = 0.1184(7)$

Electron electric dipole moment: $d_e < (7 \pm 7) \times 10^{-28}$ ecm

Neutron electric dipole moment: $d_n < 0.29 \times 10^{-25}$ ecm,

confidence limit (CL) $= 90\%$

Some useful conversions are

$$1 \text{ eV} = 1.602\ 176\ 487(40) \times 10^{-19} \text{ J}$$
$$1 \text{ eV}/c^2 = 1.782\ 661\ 758(44) \times 10^{-36} \text{ kg}$$
$$1 \text{ dyne} \equiv 10^{-5} \text{ N}$$
$$1 \text{ erg} \equiv 10^{-7} \text{ J}.$$

The values of the physical constants listed are from K. A. Olive *et al.* (Particle Data Group), "Review of Particle Physics," *Chin. Phys. C* **38**, 090001 (2014).

23.2 Natural Units

Often, it is convenient to use units where the natural physical constants assume unit values. For instance, if we assume units so that

$$c = \hbar = 1, \tag{23.1}$$

then in these units energy can be expressed as inverse length since $L = \hbar c/E = 1/E$, which gives

$$\frac{1}{\text{GeV}} = 1.97327 \times 10^{-14} \text{ cm}.$$

One may also assume as fundamental units the Boltzmann constant and Newton's constant so that

$$k_B = 1 = G_N. \tag{23.2}$$

Equations (23.1) and (23.2) taken together are often referred to as Planck natural units. Here, the units of mass, length, temperature, and time using just the four physical constants are given by

$$M_P = \left(\frac{c\hbar}{G_N}\right)^{1/2}, \ L_P = \left(\frac{\hbar G_N}{c^3}\right)^{1/2}, \ T_P = \left(\frac{\hbar c^5}{G_N k_B^2}\right)^{1/2}, \ t_P = \left(\frac{\hbar G_N}{c^5}\right)^{1/2}.$$

The numerical values of Planck units in terms of metric units are as follows:

$$M_P = 2.18 \times 10^{-8} \text{ kg}, \ L_P = 1.616 \times 10^{-35} \text{ m}$$
$$T_P = 1.42 \times 10^{32} \text{ K}, \ t_P = 5.4 \times 10^{-44} \text{ s}.$$

These are often referred to as the Planck mass, Planck length, Planck temperature, and Planck time, respectively. There are several other systems of natural units, and thus the choice of natural units is not unique.

23.3 Astrophysical Parameters

Planck length:	$\sqrt{\dfrac{\hbar G_N}{c^3}}$	$1.61620(10) \times 10^{-35}$ m
Baryon-to-photon ratio :	$\eta = n_b/n_\gamma$	$5.7\text{--}6.7 \times 10^{-10}$ (95% CL)
Hubble parameter:	H_0	$100\, h$ km s^{-1} Mpc^{-1}
		$= h \times (9.777752 \text{ Gyr})^{-1}$
Scale factor for Hubble expansion:	h	$0.673(12)$
Critical density of the universe:	$\rho_c = \dfrac{3H_0^2}{8\pi G_N}$	$1.87847(23) \times 10^{-29}\, h^2$ g cm^{-3}
		$= 1.05375(13)\times$ $10^{-5}\, h^2 (\text{GeV}/c^2)$ cm^{-3}
Number density of baryons:	n_b	$2.1\text{--}2.7 \times 10^{-7}$ cm^{-3}
Baryon density of the universe:	$\Omega_b = \rho_b/\rho_c$	$0.0499(22)$
Cold dark matter density of universe:	$\Omega_{\text{cdm}} = \rho_{\text{cdm}}/\rho_c$	$0.265(11)$
Sum of neutrino masses:	$\sum m_\nu$	< 0.23 eV
Neutrino density of the universe:	Ω_ν	< 0.0055 (95% CL)
Age of the universe:	t_{univ}	13.81 ± 0.05 Gyr

23.4 Cross-section Units

$$1 \text{ barn} \equiv 10^{-24} \text{ cm}^2$$
$$1 \text{ nanobarn (nb)} \equiv 10^{-33} \text{ cm}^2$$
$$1 \text{ picobarn (pb)} \equiv 10^{-36} \text{ cm}^2$$
$$1 \text{ femtobarn (fb)} \equiv 10^{-39} \text{ cm}^2$$

23.5 Numerical Constants

$$\pi = 3.14159$$

Euler's constant: $e = 2.71828$

Zeta function: $\zeta(t) = \sum_{n=1}^{\infty} \dfrac{1}{n^t}$, $\quad \zeta(2) = \dfrac{\pi^2}{6}$, $\zeta(3) \simeq 1.202$

Gamma function: $\Gamma(x) = \displaystyle\int_0^{\infty} t^{x-1} e^{-t} dt$, $\quad \Gamma(\tfrac{1}{2}) = \sqrt{\pi}$

24

Further Reading

The following is a list of books and review articles which the reader may find useful for cross-referencing the material discussed in this book and for further reading.

24.1 Field Theory and Gauge Theories

1. E. S. Abers and B. W. Lee, "Gauge theories," *Phys. Rep.* **9**, 1 (1973).
2. T.-P. Cheng and L.-F. Li, *Gauge Theory of Elementary Particle Physics*, Oxford University Press (1984).
3. P. H. Frampton, *Gauge Field Theories*, Benjamin-Cummings (1987).
4. M. E. Peskin and D. V. Schroeder, *An Introduction to Quantum Field Theory*, Addison-Wesley (1995).
5. C. Quigg, *Gauge Theories of the Strong, Weak, and Electromagnetic Interactions*, 2nd edn, Princeton University Press (2013).
6. M. D. Schwartz, *Quantum Field Theory and the Standard Model*, Cambridge University Press (2013).
7. M. Srednicki, *Quantum Field Theory*, Cambridge University Press (2007).
8. S. Weinberg, *The Quantum Theory of Fields*, vols I and II, Cambridge University Press (2000).
9. A. Zee, *Quantum Field Theory in a Nutshell*, Princeton University Press (2010).

24.2 Supersymmetry and Supergravity

1. R. Arnowitt, A. Chamseddine, and P. Nath, *Applied N = 1 Supergravity*, World Scientific (1984).
2. H. Baer and X. Tata, *Weak Scale Supersymmetry*, Cambridge University Press (2006).

3. D. Bailin and A. Love, *Supersymmetric Gauge Field Theory and String Theory*, Institute of Physics (1994).

4. P. Binetruy, *Supersymmetry: Theory, Experiment and Cosmology*, Oxford University Press (2006).

5. I. L. Buchbinder and S. M. Kuzenko, *Ideas and Methods of Supersymmetry and Supergravity: A Walk Through Superspace*, Institute of Physics (1995).

6. D. J. H. Chung, L. L. Everett, G. L. Kane, S. F. King, J. D. Lykken, and L. T. Wang, *Phys. Rep.* **407**, 1 (2005).

7. B. S. Dewitt, *Supermanifiolds*, Cambridge University Press (1992).

8. M. Dine, *Supersymmetry and String Theory: Beyond the Standard Model*, Cambridge University Press (2007).

9. M. Drees, R. Godbole, and P. Roy, *Theory and Phenomenology of Sparticles*, World Scientific (2004).

10. D. Z. Freedman and A. Van Proeyen, *Supergravity*, Cambridge University Press (2012).

11. S. J. Gates, M. T. Grisaru, M. Rocek, and W. Siegel, "Superspace or one thousand and one lessons in supersymmetry," *Front. Phys.* **58**, 1 (1983).

12. H. E. Haber and G. L. Kane, "The search for supersymmetry: probing physics beyond the standard model," *Phys. Rep.* **117**, 75 (1985).

13. T. Ibrahim and P. Nath, "CP violation from standard model to strings," *Rev. Mod. Phys.* **80**, 577 (2008).

14. S. P. Martin, "A Supersymmetry primer," *Adv. Ser. Direct. High Energy Phys.* **21**, 1 (2010).

15. R. N. Mohapatra, *Unification and Supersymmetry: The Frontiers of Quark–Lepton Physics*, Springer (1992).

16. P. Nath and P. Fileviez Perez, "Proton stability in grand unified theories, in strings and in branes," *Phys. Rep.* **441**, 191 (2007).

17. P. Van Nieuwenhuizen, "Supergravity," *Phys. Rep.* **68**, 189 (1981).

18. H. P. Nilles, *Phys. Rep.* **110**, 1 (1984).

19. P. Ramond, *Journeys Beyond the Standard Model*, Perseus (1999).

20. M. A. Shifman, "Nonperturbative dynamics in supersymmetric gauge theories," *Prog. Part. Nucl. Phys.* **39**, 1 (1997) [hep-th/9704114].

21. M. F. Sohnius, "Introducing supersymmetry," *Phys. Rep.* **128**, 39 (1985).

22. P. P. Srivastava, *Supersymmetry and Superfields and Supergravity: An Introduction*, Adam-Hilger (1986).

23. J. Terning, *Modern Supersymmetry: Dynamics and Duality*, Oxford University Press (2006).

24. S. Weinberg, *The Quantum Theory of Fields*, Vol. 3: *Supersymmetry*, Cambridge University Press (2000).

25. J. Wess and J. Bagger, *Supersymmetry and Supergravity*, Princeton University Press (1992).

26. P. C. West, *Introduction to Supersymmetry and Supergravity*, World Scientific (1990).

24.3 Relativity

1. J. L. Anderson, *Principles of Relativity Physics*, Academic Press (1967).
2. S. Carrol, *Spacetime and Geometry: An Introduction to General Relativity* Addison-Wesley (2003).
3. R. D'Inverno, "Introducing Einstein's Relativity", Oxford University Press (1992).
4. M. Gasperini and Venzo De Sabbata, "Introduction to Gravitation", World Scientific Publishing Company (1986).
5. C. W. Misner, K. S. Thorne, and J. A. Wheeler, *An Introduction to General Relativity and Cosmology* Freeman (1973).
6. R. M. Wald, *General Relativity*, Chicago University Press (1984).
7. S. Weinberg, *Gravitation and Cosmology: Principles and Applications of the General Theory of Relativity*, John Wiley (1972).

24.4 Cosmology

1. D. Bauman, "Cosmology," www.damtp.cam.ac.uk/user/db275/Cosmology/ Lectures.pdf.
2. L. Bergström and A. Goobar, *Cosmology and Particle Astrophysics*, 2nd edn., Springer (2004).
3. W. Buchmuller, R. D. Peccei, and T. Yanagida, "Leptogenesis as the origin of matter," *Annu. Rev. Nucl. Part. Sci.* **55**, 311 (2005).
4. G. Jungman, M. Kamionkowski, and K. Griest, "Supersymmetric dark matter," *Phys. Rep.* **267**, 195 (1996).
5. E. W. Kolb and M. S. Turner, *The Early Universe*, Addison-Wesley (1990).
6. A. R. Liddle and D. H. Lyth, *Cosmological Inflation and Large-Scale Structure*, Cambridge University Press (2000).
7. A. D. Linde, *Particle Physics and Inflationary Cosmology*, Harwood (1990).
8. D. H. Lyth and A. Riotto, "Particle physics models of inflation and the cosmological density perturbation," *Phys. Rep.* **314**, 1 (1999).
9. V. Mukhanov, *Physical Foundations of Cosmology*, Cambridge University Press (2005).
10. S. Weinberg, *Cosmology*, Oxford University Press (2007).

24.5 Other Recommended Publications

1. F. A. Berezin, "The method of second quantization," *Pure Appl. Phys.* **24**, 1 (1966).
2. R. Cahn, *Semi-Simple Lie Algebras and Their Representations*, Benjamin Cummings (1984).
3. L. Conlon, *Differentiable Manifolds*, Birkhuser (2001).

4. R. W. R. Darling, *Differential Forms and Connections*, Cambridge University Press (1994).
5. K. R. Dienes, "String theory and the path to unification: a review of recent developments," *Phys. Rep.* **287**, 447 (1997).
6. H. Flanders, *Differential Forms with Applications to the Physical Sciences*, Dover (1989).
7. H. Georgi, *Lie Groups in Particle Physics*, Benjamin Cummings (1982).
8. P. Langacker, *The Standard Model and Beyond*, CRC Press (2010).
9. G. G. Ross, *Grand Unified Theories*, Benjamin Cummings (1984).
10. R. Slansky, "Lie Groups for model building," *Phys. Rep.* **79**, 1 (1981).
11. M. T. Vaughn, *Introduction to Mathematical Physics*, Wiley-VCH (2006).
12. B. G. Wybourne, *Classical Groups for Physicists*, John Wiley (1974).

Author Index

Subject Index